Designing Audio Power Amplifiers

This comprehensive book on audio power amplifier design will appeal to members of the professional audio engineering community as well as the student and enthusiast. *Designing Audio Power Amplifiers* begins with power amplifier design basics that a novice can understand and moves all the way through to in-depth design techniques for very sophisticated audiophile and professional audio power amplifiers. This book is the single best source of knowledge for anyone who wishes to design audio power amplifiers. It also provides a detailed introduction to nearly all aspects of analog circuit design, making it an effective educational text.

Develop and hone your audio amplifier design skills with in-depth coverage of these and other topics:

- Basic and advanced audio power amplifier design
- Low-noise amplifier design
- Static and dynamic crossover distortion demystified
- Understanding negative feedback and the controversy surrounding it
- Advanced NFB compensation techniques, including TPC and TMC
- Sophisticated DC servo design
- MOSFET power amplifiers and error correction
- Audio measurements and instrumentation
- Overlooked sources of distortion
- SPICE simulation for audio amplifiers, including a tutorial on LTspice®
- SPICE transistor modeling, including the VDMOS model for power MOSFETs
- Thermal design and the use of ThermalTrak™ transistors
- Four chapters on class D amplifiers, including measurement techniques
- Professional power amplifiers
- Switch-mode power supplies (SMPS).

Bob Cordell is an electrical engineer who has been deeply involved in audio since his adventures with vacuum tube designs in his teen years. He is an equal-opportunity designer to this day, having built amplifiers with vacuum tubes, bipolar transistors and MOSFETs. Bob is also a prolific designer of audio test equipment, including a high-performance THD analyzer and many purpose-built pieces of audio gear. He has published numerous articles and papers on power amplifier design and distortion measurement in the popular press and in the *Journal of the Audio Engineering Society*. In 1983 he published a power amplifier design combining vertical power MOSFETs with error correction, achieving unprecedented distortion levels of less than 0.001% at 20 kHz. He also consults in the audio and semiconductor industries. Bob is also an avid DIY loudspeaker builder, and has combined this endeavor with his electronic interests in the design of powered audiophile loudspeaker systems. Bob and his colleagues have presented audiophile listening and measurement workshops at the Rocky Mountain Audio Fest and the Home Entertainment Show. As an electrical engineer, Bob has worked at Bell Laboratories and other related telecommunications companies, where his work has included design of integrated circuits and fiber optic communications systems. Bob maintains an audiophile website at www.cordellaudio.com where diverse material on audio electronics, loudspeakers and instrumentation can be found.

Designing Audio Power Amplifiers

Second Edition

Bob Cordell

Routledge
Taylor & Francis Group

NEW YORK AND LONDON

Second edition published 2019
by Routledge
52 Vanderbilt Avenue, New York, NY 10017

and by Routledge
2 Park Square, Milton Park, Abingdon, Oxon, OX14 4RN

Routledge is an imprint of the Taylor & Francis Group, an informa business

First edition published by McGraw-Hill 2011

Library of Congress Cataloging-in-Publication Data
Names: Cordell, Bob, author.
Title: Designing audio power amplifiers / Bob Cordell.
Description: Second edition. | New York, NY : Routledge, [2019] |
    Includes bibliographical references and index.
Identifiers: LCCN 2018058879 | ISBN 9781138555457 (hardback : alk.
    paper) | ISBN 9781138555440 (pbk. : alk. paper)
Subjects: LCSH: Audio amplifiers—Design and construction. | Power
    amplifiers—Design and construction.
Classification: LCC TK7871.58.A9 C675 2019 | DDC 621.389/33—dc23
LC record available at https://lccn.loc.gov/2018058879

ISBN: 978-1-138-55545-7 (hbk)
ISBN: 978-1-138-55544-0 (pbk)

Typeset in Times New Roman
by Apex CoVantage, LLC

This book is dedicated to my son Jon and his family, of whom I am so proud.

Contents

Preface

This Second Edition of *Designing Audio Power Amplifiers* expands and updates the First Edition, with five new chapters and significant expansion of many of the original chapters. There have been many developments in audio power amplifier design since the release of the First Edition and there are some important topics that deserve more depth of coverage.

Designing Audio Power Amplifiers is written to address many advanced topics and important design subtleties. At the same time, however, it has enough introductory and tutorial coverage to allow designers relatively new to the field to absorb the material of the book without being overwhelmed. It is targeted to professionals in the audio field as well as enthusiasts and students learning electronics. To this end, the book starts off at a relaxing pace that helps the reader develop an intuitive feel and understanding for amplifier design. Although this book covers advanced subjects, highly involved mathematics is kept to a minimum—much of that is left to the academics. Design choices and decisions are explained and analyzed. Practical amplifier circuits of numerous different topologies and circuit features are described in depth and many are accompanied by performance measurements.

This is not just a cookbook; it is intended to teach the reader how to think about power amplifier design and understand the many concepts and nuances, then analyze and synthesize the many possible variations of amplifier design. Nor is it focused on numerous variations of just one simple topology or design philosophy.

The design of modern high-performance audio power amplifiers touches on most aspects of electronic design, including solid state devices, feedback theory, low noise design, thermal analysis, switching power supplies, laboratory measurement and circuit simulation, to name a few. As such, skills acquired in power amplifier design can provide a sound educational basis for the study of more specialized areas in electronics. Analog circuit design is covered broadly and in depth.

I have divided the book into six parts. Part 1 introduces audio power amplifier design and includes the basics. This part is designed to be readable and friendly to those with less technical background while still providing a very sound footing for the more detailed design discussions that follow. In this part I show how a simple power amplifier design evolves in several steps to a modern architecture, describing how performance deficiencies are mitigated with circuit improvements at each step in the evolution. Even experienced designers may gain valuable insights here. A new chapter covers the design and construction of a complete amplifier based on the evolved designs. A large chapter is devoted to transistor characteristics and circuit design, including many key circuit building blocks. Part 1 concludes with a chapter on negative feedback principles and a new chapter on low-noise amplifier design.

Part 2 delves into the design of advanced power amplifiers with state-of-the-art performance. Crossover distortion, one of the most problematic distortions in power amplifiers, is covered in depth. Special attention is paid to dynamic crossover distortion, which is less well understood. A new second chapter covering the all-important output stage is included, where additional circuits like the DoubleCross™ output stage are discussed. This section also includes a detailed treatment of lateral and vertical MOSFET power amplifiers, error correction techniques, advanced feedback compensation, ultra-low distortion drive circuits and DC servos.

Part 3 covers those real-world design considerations that influence sound quality and reliability, including power supplies and grounding, short circuit and safe area protection, and amplifier behavior when driving difficult loads. A new chapter is dedicated to in-depth coverage of switch-mode power supplies (SMPS), including power factor correction. Electromagnetic interference (EMI) ingress and egress via the input, output and mains ports of the amplifier are also treated here. In like manner, conductive and radiated emissions suppression and regulatory limits are discussed. Amplifier thermal design, thermal stability and temperature compensation are covered, including simple thermal models and simulations.

SPICE simulation can be very important to power amplifier design, and its use is described in detail in Part 4. Even those with no SPICE experience will learn how to use this valuable tool, helped along by a tutorial chapter with in-depth coverage of the free LTspice® simulator, with ready-to-run amplifier simulations and transistor models available at www.cordellaudio.com. Feedback loop gain simulations employing the Middlebrook and Tian probes are explained. A full chapter describes how you can create your own accurate SPICE models for BJT, JFET and MOSFET transistors, many of which are poorly modeled by manufacturers. Numerous approaches to distortion measurement are also explained in Part 4. I've also described some techniques for achieving the high sensitivity required to measure the low-distortion designs discussed in the book. Less well-known distortion measurements, such as TIM, PIM and IIM, are also covered here. In the quest for meaningful correspondence between listening and measurement results, other non-traditional amplifier tests are also described.

Part 5, "Topics in Amplifier Design," covers all of those other important matters that do not fit neatly into the other parts. Advanced designers as well as audiophiles will find many interesting topics in this part. Some of the controversies in audio, such as the use of negative feedback, are addressed here. For balance, the design of amplifiers without negative feedback is covered as well. Integrated circuit power amplifiers and drivers are also discussed. A new chapter devoted to modern professional power amplifiers is included. This chapter covers all of the special design considerations and features necessary in professional audio amplifiers, including microcomputer control, DSP, integrated functionality, networked control and *Audio over Internet Protocol* (AoIP).

Class D amplifiers are playing a more important role in audio amplification as every day passes. They have enjoyed vast improvements in performance over the last several years and can be expected to improve much further in the future. Four chapters in Part 6 cover this exciting technology. PWM and Sigma Delta modulators, negative feedback, noise shaping, class D shortcomings and special measurement techniques for class D amplifiers are discussed.

In summary, many of the following topics covered in *Designing Audio Power Amplifiers* should prove especially interesting to readers familiar with earlier texts:

- Ultra-low distortion amplifier topologies
- Low-noise amplifier design
- High-performance feedback compensation techniques
- Lateral and vertical MOSFET power amplifiers
- Output stage error correction circuits
- Switch-mode power supplies (SMPS)
- Thermal analysis of BJT and MOSFET output stages
- Output transistors with temperature tracking diodes
- Integrated circuit amplifiers and drivers

- SPICE simulation and modeling for amplifier design
- Amplifier measurement instrumentation and techniques
- PC-based instrumentation for amplifier measurements
- How amplifiers misbehave and why they sound different
- Sources of distortion in class D amplifiers
- PWM, sigma delta and direct digital class D amplifiers
- Measurement techniques for class D amplifiers

No single text can cover all aspects of audio power amplifier design. It is my hope that an experienced designer, a student or an enthusiast who seeks to learn more about audio amplifier design and circuit design in general will find this book most helpful. I also hope that this text will provide a sound basis for those wishing to learn analog circuit design.

Acknowledgments

My Lord and Savior Jesus Christ has given me the peace and guidance that allowed me to complete this undertaking. My mother inspired me by writing her autobiography. My father supported me in my audio and electronics activities beginning before my teen years, helping me to purchase electronic kits and showing me how to put them together.

Those many authors of other texts on audio that have gone before me have truly been an inspiration and have shown how good engineering can be applied to audio while making their writings understandable to those without a formal engineering background.

I owe a special debt of gratitude to Andy Connors and many others for providing insight and encouragement. I am also grateful to Jan Didden and Peter Smith for their generous support. Gene Pitts, past editor of *Audio* magazine, was pivotal in enabling me to get my start in writing about audio. There are too many other friends and colleagues to list, but I wish especially to thank the members of the Audio Engineering Society and the DIYaudio Forum (www.diyaudio.com) for all that I have learned from them. The audiophile fraternity is alive and well.

Finally I wish to thank the professional group at Taylor & Francis for turning my dream into reality.

Part 1

Audio Power Amplifier Basics

Part 1 introduces audio power amplifier design and covers the basics. This part is written to be readable and friendly to those with less technical background while still providing a very sound footing for the more detailed design issues to follow. That footing includes discussions of transistor operation, important circuit building blocks, noise, negative feedback and the different amplifier classes. In Chapter 3 we show how a simple power amplifier design can be evolved in several steps to become a modern architecture with very good performance. At each step we describe how performance deficiencies are mitigated with circuit improvements. In Chapter 4, building a particular amplifier design is covered in detail. Part 1 closes with a chapter that summarizes in a succinct way the many issues that should be addressed in the power amplifier design process. This serves as a preamble for the more detailed chapters that follow in the later parts. Even experienced designers will gain valuable insights in Part 1.

Chapter 1

Introduction

1. INTRODUCTION

Audio power amplifier design is both an art and a science, in more ways than one. Solid-state power amplifiers have been around since the late 1960s, and yet new designs still proliferate. Questions about relating sonic performance to measured performance still abound. This is not just limited to the high end where audio mystique has a strong influence. While there is a tremendous amount of science to the design of audio power amplifiers, there are also many nuances that demand attention to detail. At times, it is difficult to separate the influence of experience from just plain art. There are also things we still do not understand fully, and this is where the art aspect of amplifier design flourishes.

This book is not meant to be a survey. Topologies of historical or narrow interest are ignored in favor of deeper coverage of important nuances in relevant contemporary designs. I have sought to touch on virtually every amplifier design subject, but some are treated in less depth when there is a better treatment elsewhere.

There are thousands of variations on amplifier architectures out there, and it would be impossible to study all of them. For that reason, there is a strong focus on deeply understanding the more popular architectures in a way that conveys enough understanding so that the reader can analyze and even conceive many variants, some of which may be very different from the ones covered here.

1.1 Organization of the Book

There is no right or wrong way to organize the enormous amount of material on power amplifier design. The approach taken here is to ramp up your confidence first, with emphasis on how to think about amplifier designs and analyze them. For this reason, some details and nuances are postponed to later parts of the book.

I begin with a very basic amplifier and show how it works and how to analyze it. I also discuss its shortcomings. The approach includes emphasis on examples and plugging in the numbers to evaluate design approaches and see how well they actually perform.

A strong attempt has been made in the early chapters to avoid distracting you with side trips along the way to understanding amplifier design. As you progress through the book, an adequate amount of just-in-time tutorial material is presented to aid the less experienced reader. This includes material on transistors and building block circuits that is written to be easily digestible. Chapters on noise and negative feedback provide further foundation material for amplifier design. Even the experienced designer will find some welcome nuggets of detail here.

Power amplifier design is introduced by describing a very basic design. That design is analyzed and then followed through many stages of improvements. The explanation of the reasons for and results of the improvements illustrates how to think about amplifier design and understand many of the trade-offs. Actual simulated distortion results are presented at each stage of the evolution of the design.

Once you are made comfortable with amplifier design and analysis, Part 2 delivers the meat of high-performance design and a much deeper understanding of performance-limiting factors and sources of distortion.

By the end of Part 2, you have gone deeply through the design of amplifiers that perform superbly on the lab bench. Alas, the real world is introduced in Part 3. There is indeed a large gap between a superlative laboratory amplifier and one that will perform that way and survive in the real world. Protection circuits, EMI filtering, power supplies, grounding and many other things need to be taken into account, and that is the priority of Part 3.

Tools are very important to the successful design of power amplifiers, and these are covered in Part 4. Some tools are physical, like distortion analyzers, while others take the form of software, like the SPICE simulator. SPICE is most valuable in the design process before the amplifier is built, helping to evaluate and sort out the many possible combinations of circuit architectures from which to choose. SPICE can then be used in optimizing the selected design.

After the amplifier is built, it must be thoroughly evaluated for distortion and other behavior, and this is where different kinds of instrumentation and measurement techniques come into play. Here is where the rubber meets the road in terms of measuring an amplifier in ways that may have some correlation with how well they sound. To some extent, amplifiers sound different because they misbehave differently. The ability to expose and sort out amplifier misbehavior is a key to making a fine-sounding amplifier.

Part 5 of the book is titled "Topics in Amplifier Design." Here a wide variety of subjects of keen interest are covered. For example, a book on audio power amplifier design would not be complete without discussion of the pros and cons of negative feedback; its pervasive use and sometimes-controversial reputation demand it. Amplifier design myths and common misunderstandings are also discussed. Why amplifiers sound different and how that may correlate to measurable differences are also considered. This part of the book also covers other types of amplifiers, such as balanced amplifiers and amplifiers without negative feedback. The special needs and design features of professional power amplifiers are also covered in Part 5.

Part 6 covers class D amplifiers. Traditional amplifiers have served us well for many decades, and will continue to do so. However, class D amplifiers are the wave of the future. They are smaller and far more efficient. As a result, they generate less heat, making them ideal for multi-channel Home Theater receivers. Sound quality has historically been a problem for class D, but that has improved greatly in the last decade. Their design requires a somewhat different skill set. Part 6 will introduce you to these new design challenges and discuss measurements.

1.2 The Role of the Power Amplifier

The power amplifier in an audio system converts the line-level signal to a large signal that can drive the loudspeaker. The line-level signal required to produce maximum power is typically on the order of 1 to 3 V RMS and it is not expected to supply much current to the power amplifier. A typical power amplifier will have an input impedance of greater than 10 kΩ. A 100-W power amplifier driving 8 Ω will need to produce about 28 V RMS at about 3.5 A RMS at full power with a sine wave. Thus, it is the job of the power amplifier to produce both relatively high voltage and high current. A very common range of power amplifier voltage gains is on the order of 20 to 30. One-volt RMS into a power amplifier with a gain of 20 will produce 50 W into an 8-Ω load. Ideally, the power amplifier has very low output impedance so that it essentially acts like a voltage source driving the load. The power amplifier's role in the system is illustrated in Figure 1.1.

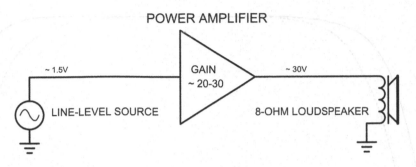

Figure 1.1 Power Amplifier Driving a Load

1.3 Basic Performance Specifications

The performance specifications listed by the manufacturer of an audio power amplifier range from a very sparse set to a fairly detailed list. The primary specifications include maximum power, frequency response, noise and distortion.

Rated Output Power

Maximum output power is almost always quoted for a load of 8 Ω and is often quoted for a load of 4 Ω as well. A given voltage applied to a 4-Ω load will cause twice the amount of current to flow, and hence twice the amount of power to be delivered. Ideally, the output voltage of the power amplifier is independent of the load, both for small signals and large signals. This implies that the maximum power into a 4 Ω load would be twice that into an 8 Ω load. In practice, this is seldom the case, due to power supply sag and limitations on maximum available output current.

The correct terminology for power rating is *continuous average sine wave power*, as in *100-W continuous average sine wave power*. However, many often take the liberty of using the term *W RMS*. Although technically incorrect, this wording simply is referring to the fact that the power would have been measured by employing a sine wave whose RMS AC voltage was measured on a long-term basis. There are other ways of rating power that are sometimes used because they provide larger numbers for the marketing folks, but we will ignore them here. When you hear terms like *peak power* just realize that these are not the same as the more rigorous *continuous average power rating*.

Frequency Response

The frequency response of a power amplifier must extend over the full audio band from 20 Hz to 20 kHz within a reasonable tolerance. Modern amplifiers usually far exceed this range, with frequency response from 5 Hz to 200 kHz not the least bit uncommon. The frequency response for such an amplifier is illustrated with the solid curve in Figure 1.2. While the tolerance assigned to the frequency response of loudspeakers is often ±3 dB, the tolerance associated with power amplifiers is usually +0 dB, −3 dB or tighter. Specifying where an amplifier is down by 3 dB from the nominal 0 dB reference is the conventional way of specifying the bandwidth of a system. This is often referred to as the *3-dB bandwidth.*

The frequency response for a less capable amplifier is shown with the dashed curve in Figure 1.2. This amplifier has a 3-dB bandwidth from 10 Hz to 80 kHz. Its response is down 1 dB at 20 Hz and 0.5 dB at 20 kHz.

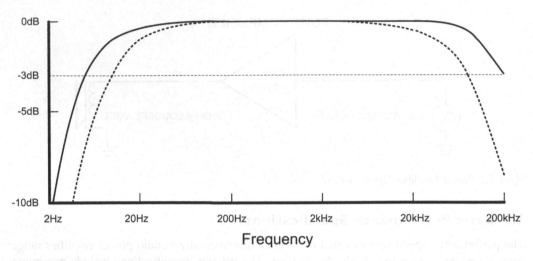

Figure 1.2 Amplifier Frequency Response

Noise

It is important that power amplifiers produce low noise, since the noise they make is always there, independent of the volume control setting and the listening level. This is particularly so when the amplifiers are used with high-efficiency loudspeakers. The noise is usually specified as being so many decibels down from either the maximum output power or with respect to 1 W. The former number will be larger by 20 dB for a 100-W amplifier, so it is often the one that manufacturers like to cite. The noise referenced to 1 W into 8 Ω (or, equivalently, 2.83 V RMS) is the one more often measured by reviewers.

The noise specification may be unweighted or weighted. Unweighted noise for an audio power amplifier will typically be specified over a full 20 kHz bandwidth (or more). Weighted noise specifications take into account the ear's sensitivity to noise in different parts of the frequency spectrum. The most common one used is *A weighting*, illustrated in Figure 1.3. Notice that the weighting curve is up about +1.2 dB at 2 kHz and down 3 dB at approximately 500 Hz and 10 kHz.

The A-weighted noise specification for an amplifier will usually be quite a bit better than the unweighted noise because the weighted measurement tends to attenuate noise contributions at higher frequencies and hum contributions at lower frequencies. A very good amplifier might have an unweighted signal-to-noise ratio (S/N) of 90 dB with respect to a 1-W output into 8 Ω, while that same amplifier might have an A-weighted S/N of 105 dB with respect to 1 W. A fair amplifier might sport 65 dB and 80 dB S/N figures, respectively. The A-weighted number will usually be 10–20 dB better than the unweighted number.

Distortion

The most common distortion specification is *total harmonic distortion (THD)*. It will usually be specified at one or two frequencies or over a range of frequencies. It will be typically specified at a given power level with the amplifier driving a specified load impedance. A good 100-W amplifier might have a 1-kHz THD (referred to as *THD-1*) of 0.005% at 100 W into 8 Ω. That same amplifier might have a 20 kHz THD (*THD-20*) of 0.02% up to 100 W into 8 Ω. Although 1-kHz THD is at a frequency in the middle of the audible frequency range where hearing sensitivity is

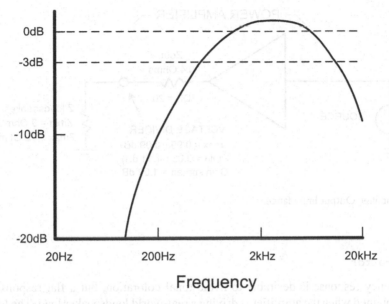

Figure 1.3 A-Weighting Frequency Response

high, it is not very difficult to achieve low THD figures at 1 kHz. Good THD-20 performance is much more difficult to achieve and is generally a better indicator of amplifier performance.

In practice, the harmonic distortion specification will be described as *THD+N*, where the *N* refers to noise. This reflects the way in which THD is most often measured. When measuring THD-1, a 1-kHz *fundamental* sine wave is applied to the amplifier input. The 1-kHz fundamental in the output signal is then removed by a very sharp notch filter. Everything else, both distortion harmonics and noise, is measured, giving rise to the THD+N specification. At higher power testing levels, the true THD will often dominate the noise, but at lower power levels the measurement may often reflect the noise rather than the actual THD being measured. Graphs that show rising THD+ N at lower power levels can be misleading. The rising level may actually be noise rather than distortion. This is because a fixed noise voltage becomes a larger percentage of the level of the fundamental as the fundamental decreases in amplitude at lower power levels. There are many other power amplifier distortion specifications, and these will be covered in detail in later chapters in this book.

The Federal Trade Commission (FTC) long ago tried to wrap things up in a single statement that would largely capture power, distortion and bandwidth together [1]. It would read something like "100-W continuous average power from 20 Hz to 20 kHz with less than 0.02% total harmonic distortion." This was a reasonably comprehensive and honest way to describe the most basic capability of an amplifier. It is unfortunate that it has fallen into disuse by many manufacturers. Part of the reason was that it also required that the amplifier could be run at 1/3 rated power into 8 Ω for an extended period of time without overheating. Operating at 1/3 rated power is close to the point where most amplifiers dissipate the most heat, and it was expensive for many amplifier manufacturers to provide enough heat sinking to meet this requirement.

1.4 Additional Performance Specifications

There is almost no limit to the number of useful performance specifications for an audio power amplifier, but the following are a few that are a bit less basic and yet quite useful.

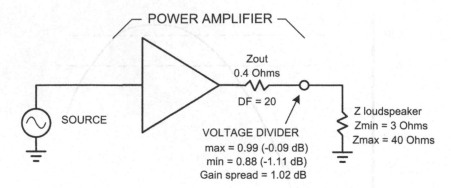

Figure 1.4 Amplifier Output Impedance

Damping Factor

A flat frequency response is desirable to avoid tonal coloration, but a flat response may not always be obtained when the amplifier is driving a real-world loudspeaker load. The load imped-ance of real loudspeakers can vary dramatically as a function of frequency, while the output impedance of the power amplifier is non-zero. A voltage divider is thus formed by the ampli-fier's output impedance and the loudspeaker input impedance, as illustrated in Figure 1.4. Here the amplifier is modeled with an ideal amplifier with zero output impedance in series with impedance Z_{out} that describes its actual output impedance. This is referred to as a Thévenin equivalent circuit.

This is where the *damping factor* (DF) comes into play. In spite of its important-sounding name, this is just a different way of expressing the output impedance of the amplifier. While amplifiers ideally act like voltage sources with zero output impedance, they all have finite output impedance. The term *damping factor* came from the fact that a loudspeaker is a mechanically resonant system; the low output impedance of an amplifier damps that resonance via the resis-tance of the loudspeaker's voice coil and *electromotive force*. An amplifier with higher output impedance will provide less damping of the loudspeaker cone motion because it adds to the total amount of resistance in the circuit.

Damping factor is defined as the ratio of 8 Ω to the actual output impedance of the amplifier. Thus, an amplifier with an output impedance of 0.2 Ω will have a DF of 40. Most vacuum tube amplifiers have a DF of less than 20, while many solid-state amplifiers have a DF in excess of 100. It is important to bear in mind that the DF is usually a function of frequency, often being larger at low frequencies. This is consistent with the need to dampen the cone motion of woofers, but ignores the influence of the DF on frequency response at higher frequencies. Many loud-speakers have a substantial peak or dip in their impedance at or near their crossover frequencies. This could result in coloration if the amplifier DF is low.

The effect of damping factor and output impedance on frequency response must not be under-estimated in light of the large impedance variations seen in many contemporary loudspeakers. It is not unusual for a loudspeaker's impedance to dip as low as 3 Ω and rise as high as 40 Ω across the audio band. Consider this wildly varying load against the 0.4-Ω output impedance of a vacuum tube amplifier with a DF of 20. This will cause an audible peak-to-peak frequency response variation of ±0.5 dB across the audio band.

Dynamic Headroom

Unlike a sine wave, music is impulsive and dynamic. Its power peaks are often many times its average power. This ratio is often referred to as the *crest factor*. *Dynamic headroom* refers to the fact that an amplifier can usually put out a greater short-term burst of power than it can on a continuous basis. The primary cause of this is power supply sag, which is a reflection of power supply regulation. The power supply voltages will initially remain high and near their no-load values for a brief period of time during heavy loading due to the energy storage of the large reservoir capacitors. Under long-term conditions, the voltage will sag and less maximum power will be available.

Consider an amplifier that clips at 100 W into 8 Ω on a continuous test basis. If this amplifier has a power supply with 10% regulation from no-load to full load (which is fairly good), the available power supply voltage will be about 10% higher during a short-term signal burst. This will result in a short-term power capability on the order of 120 W, since power goes as the square of voltage.

Dynamic headroom is a two-edged sword. It is good to have it because music tends to have an average power level much lower than the brief peak power levels it can demand (referring again to the crest factor). It is nice to have 20% to 40% more power available when it is needed for those brief peaks. On the other hand, a large amount of dynamic headroom is often symptomatic of an amplifier with a sloppy power supply.

Slew Rate

Slew rate is a measure of how fast the output voltage of the amplifier can change under large-signal conditions. It is specified in volts per microsecond. Slew rate is an indicator of how well an amplifier can respond to high-level transient program content. A less capable amplifier might have a slew rate of 5 V/μs, whereas a really high-performance amplifier might have a slew rate on the order of 50–300 V/μs. For a given type of program material, a higher-power amplifier needs to have a higher slew rate to do as well as a lower-power amplifier, since its voltage swings will be larger. A 100-W amplifier driving a loudspeaker whose efficiency is 85 dB will need to have 3.16 times the amount of slew rate capability as a 10-W amplifier driving a 95-dB speaker to the same sound pressure level.

As a point of reference, the maximum voltage rate of change of a 20-kHz sine wave is 0.125 V/μs per volt peak. This means that a 100-W amplifier that produces a level of 40 V peak at 20 kHz must have a slew rate of at least 5 V/μs. In practice a much larger value is desirable for low-distortion performance on high-frequency program content. Although technically imprecise, the rate of change of a signal is often referred to as its slew rate for convenience.

The slew rate capability of audio power amplifiers received a lot more attention after the term *transient intermodulation distortion* (TIM) was coined and studied intensely during the 1970s and early 1980s [2, 3, 4]. This was largely another way of describing high-frequency distortion that resulted from slew rate deficiency. The TIM controversy will be discussed in greater detail in Chapter 28.

Output Current

Output current is another lesser-known amplifier specification that can have a strong influence on sonic quality. As we will see later, the complex reactive loudspeaker load presented to an amplifier can demand larger currents than the rated resistive load with which an amplifier is often tested. Add to this the fact that many loudspeakers have impedances that dip well below their rated impedance, and we have a recipe for high current demands.

Minimum Load Impedance

Related to output current capability is the specification of minimum stable or safe output impedance that an amplifier can drive. Although there are many 4-Ω-rated loudspeakers out there
(whose impedance often dips below 3 Ω), there are many amplifiers in AV receivers that are not
able to properly drive a 4-Ω load. This is partly because cramming five or more amplifiers into
one enclosure that can properly drive 4-Ω loads while being able to deliver over 100 W each
into 8 Ω is quite difficult and surely more expensive. This is much less of a problem with stereo
amplifiers, where heat removal for only two channels is necessary. However, even amplifiers that
are rated to drive 4-Ω loads may at times find themselves with too little current drive capability
to drive some contemporary loudspeakers. High-end loudspeakers are often designed with little
regard for what it takes to drive them.

1.5 Output Voltage and Current

Here we will briefly touch on the reality of output voltage and current swing that an amplifier
may have to deliver in practice. Table 1.1 shows the RMS value of the sine wave voltage, the
peak voltage, the peak current and the reserve current required for the popular 8-Ω resistive load
as a function of power.

The reserve current listed below is simply a factor of 3 greater than the peak current required
of a resistive load and represents the reality of driving difficult reactive loudspeaker loads with
non-sinusoidal waveforms. In Section 1 of Chapter 22 we will see where this somewhat arbitrary
factor of 3 comes from. The reserve current can be assumed to occur only in a brief time interval
under fairly rare circumstances.

This data gives a glimpse of what is necessary for the amplifier to provide. Notice the very
substantial voltage swings, and implied power supply voltages, required for a 400-W amplifier.
The peak and reserve currents are also into the tens of amperes at 400 W. This is just the beginning of the story, however. Table 1.2 shows what the same amplifier would encounter when
driving a 4-Ω load. Here we have assumed that the drive signal has remained the same and only
the load impedance has dropped. We have also implicitly assumed that the amplifier has ideal
power supply regulation, so all of the power numbers are doubled.

Given the nature of some of today's high-end loudspeakers, some have argued that really high-
performance amplifiers should be rated for power delivery into 2 Ω (at least for short intervals).

Table 1.1 Voltage and Current into an 8-Ω Load

Power, W	V_{RMS}	V_{peak}	I_{peak}	$I_{reserve}$
50	20	28	3.5	10.5
100	28	40	5.0	15.0
200	40	56	7.0	21.0
400	57	80	10.0	30.0

Table 1.2 Voltage and Current into a 4-Ω Load

Power, W	V_{RMS}	V_{peak}	I_{peak}	$I_{reserve}$
100	20	28	7	21
200	28	40	10	30
400	40	56	14	42
800	57	80	20	60

Table 1.3 Voltage and Current into a 2-Ω Load

Power, W	V_{RMS}	V_{peak}	I_{peak}	$I_{reserve}$
200	20	28	14	42
400	28	40	20	60
800	40	56	28	84
1600	57	80	40	120

Indeed, the testing done in some amplifier technical reviews regularly subjects power amplifiers to a 2-Ω resistive load test. The figures for output current become almost bewildering under these conditions (see Table 1.3).

An important point here is that there are amplifiers sold every day that are rated at up to 400 W per channel into 8 Ω, and designers implement such amplifiers every day. The sobering point is that if at the same time the designer thinks in terms of his or her amplifier being 2-Ω compatible, the potential demanded burst current could on occasion be quite enormous.

1.6 Basic Amplifier Topology

Figure 1.5 shows a simplified three-stage audio power amplifier design. This is a descendant of the Lin topology introduced in the 1950s by Dr. H. C. Lin of RCA Laboratories. Even the simple amplifier of Figure 1.5 includes several evolutionary improvements over the original Lin amplifier, such as the use of an input differential pair, a current source load for the VAS and a complementary output stage. Some examples of evolved versions of the Lin topology published in the late 1960s and early 1970s can be found in these references [5–8]. Although other arrangements

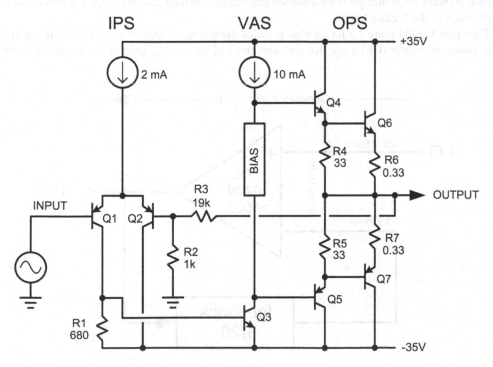

Figure 1.5 Simple Three-Stage Power Amplifier

have appeared through the years, the Lin topology and its many derivatives account for the vast majority of power amplifier designs, and it will be the focus of most of this book.

Transistors Q1 and Q2 form the input differential pair. This arrangement is often called a *long-tailed pair* (LTP) because it is supplied with a so-called tail current from a very high-impedance circuit like the current source shown. We will often take the liberty of referring to the amplifier's *input stage* as the *IPS*. The input differential amplifier usually has a fairly low voltage gain, typically ranging from 1 to 15.

The IPS compares the applied input signal to a fraction of the output of the amplifier and provides the amount of signal necessary for the remainder of the amplifier to create the required output. This operation forms the essence of the negative feedback loop. The fraction of the output to which the input is compared is determined by the voltage divider consisting of R3 and R2. If the fraction is 1/20 and the forward gain of the amplifier is large, then very little difference need exist between the input and the signal fed back to the IPS in order to produce the required output voltage. The gain of the amplifier will then be very nearly 20. This is referred to as the *closed-loop gain* of the *amplifier* (CLG or A_{cl}).

This simplified explanation of how negative feedback works is illustrated in Figure 1.6. The core of the amplifier that provides all of the *open-loop gain* (OLG or A_{ol}) is shown as a gain block symbol just like an operational amplifier. For purposes of illustration, it is shown with a gain of 1000. The feedback network is shown as a block that attenuates the signal being fed back by a factor of 20. Suppose the output of the amplifier is 20 V. The amount fed back will then be 1 V. The input across the differential inputs of the gain block will be 20 mV if the forward gain is 1000. The required input from the input terminal will then be 1.02 V. This simplified approach to looking at a feedback circuit is sometimes referred to as *input-referred* feedback analysis because we start at the output and work our way back to the input to see what input would have been required to produce the assumed output. The closed-loop gain is thus 20/1.02 = 19.6. This is just 2% shy of what we would get if we assumed that the closed-loop gain were just the inverse of the attenuation in the feedback path.

Transistor Q3 in Figure 1.5 forms what is called the *voltage amplifier stage* (VAS). It is a high-gain *common-emitter* (CE) stage that provides most of the voltage gain of the amplifier. Notice

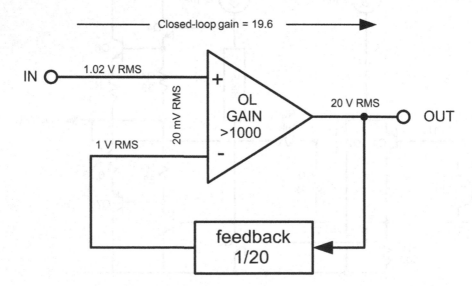

Figure 1.6 Negative Feedback Operation

that it is loaded with a current source rather than a resistor so as to provide the highest possible gain. It is not unusual for the VAS to provide a voltage gain of 100 to 10,000. This means that the difference signal needed to drive the input stage does not need to be very large to drive the output to its required level. If the difference signal is close to zero, and 1/20 of the output is compared to the input, it follows that the output would be almost exactly 20 times the input.

The *output stage* (OPS) is composed of transistors Q4 through Q7. Its main job is to provide buffering in the form of current gain between the output of the VAS and the loudspeaker load. Most output stages have a voltage gain of approximately unity. The output stage here consists essentially of two pairs of *emitter followers* (EF), one for each polarity of the output swing. This is called a complementary push-pull output stage. Transistors Q4 and Q5 are referred to as the *drivers*, while Q6 and Q7 are the output devices.

The two-stage OPS, like this one, will typically provide a current gain between 500 and 10,000. This means that an 8-Ω load resistance will look like a load resistance between 4000 Ω and 80,000 Ω to the output of the VAS. Other output stages, like so-called *Triples*, can provide current gain of 100,000 to 1 million, greatly reducing the load on the VAS.

This OPS is the classic *class B* output stage used in most audio power amplifiers. The upper output transistor conducts on positive half-cycles of the signal when it is necessary to source current to the load. The bottom output transistor conducts on the negative half-cycle when it is necessary to sink current from the load. The signal thus follows a different path through the amplifier on different halves of the signal. This of course can lead to distortion.

The box labeled *bias* provides a DC bias voltage that overcomes the turn-on base-emitter voltage drops (V_{be}) of the driver and output transistors. It also keeps them active with a small quiescent bias current even when no current is being delivered to the load. This bias circuit is usually referred to as the *bias spreader*. The output stage bias current creates a small region of overlapping conduction between the positive and negative output transistors. This smooths the transition from the upper transistors to the lower transistors (and vice versa) when the output signal goes from positive to negative and the output stage goes from sourcing current to the load to sinking current from the load. We'll have much more to say about this crossover region and the distortion that it can create in Chapters 7 and 12. Because there is a small region of overlap where both transistors are conducting, this type of output stage is often referred to as a *class AB* output stage.

If the bias spreader is set to provide a very large output stage idle bias current, both the top and bottom output transistors will conduct on both half-cycles of the signal. One will be increasing its current as the other decreases its current, with the difference flowing into the load. In this case we have a so-called *class A* output stage. The fact that the signal is then always taking the same path to the output (consisting of two parallel paths) tends to result in less distortion because there is no crossover from one half of the output stage to the other as the signal swings from positive to negative. The price paid is very high power dissipation as a result of the high output stage bias current.

Actual operation of the amplifier of Figure 1.5 is quite simple. The input differential amplifier compares the input voltage to a scaled-down version of the output voltage and acts to make them essentially the same. This action applies to both stabilization of the DC operating points and the processing of AC signals. In the quiescent state transistors Q1 and Q2 are conducting the same amount of current, in this case 1 mA each. The resulting voltage drop across R1 is just enough to turn on Q3 to conduct 10 mA, balancing the current supplied to its collector by the 10 mA current source.

Now suppose the output is more positive than it should be. The voltage at the base of PNP transistor Q1 will then be negative with respect to the scaled version of the output voltage at the base of Q2. A more negative voltage at the base of a PNP transistor causes it to conduct more current. Transistor Q1 will thus conduct more current and increase the voltage drop across R1. This will increase the voltage at the base of NPN transistor Q3. A more positive voltage at the base of an NPN transistor causes it to conduct more current. Transistor Q3 will thus turn on harder. This will cause an imbalance between Q3's collector current and the 10-mA current source. Q3's increased collector current will thus pull the voltage at its collector node more negative. This will drive the bases of the driver and output transistors more negative. Their emitters will follow this negative voltage change, causing the output of the amplifier to go more negative. The result will be that the initially assumed positive error in the output voltage will be corrected.

1.7 Summary

We've seen what an audio power amplifier has to do, and we've seen the basic design described in qualitative terms. In the next chapter we'll learn a bit about bipolar transistors and the simple circuit building blocks that make up a power amplifier. Equipped with this knowledge, we will then analyze in some detail the workings of the basic power amplifier.

References

1. Federal Trade Commission (FTC), "Power Output Claims for Amplifiers Utilized in Home Entertainment Products," CFR 16, Part 432, 1974.
2. Matti Otala, "Transient Distortion in Transistorized Audio Power Amplifiers," *IEEE Transactions on Audio and Electroacoustics*, vol. AU-18, pp. 234–239, 1970.
3. Walter G. Jung, Mark L. Stephens and Craig C. Todd, "Slewing Induced Distortion and Its Effect on Audio Amplifier Performance—With Correlated Measurement Listening Results," AES preprint no. 1252 presented at the 57th AES Convention, Los Angeles, 1977.
4. Robert R. Cordell, "Another View of TIM," *Audio*, February–March 1980.
5. Dan Meyer, "Tigers that Roar," *Popular Electronics*, pp. 51–63, 99, July 1969.
6. Dan Meyer, "Assembling a Universal Tiger," *Popular Electronics*, pp. 31–45, October 1970.
7. Dan Meyer, "The Plastic Tiger Audio Power Amplifier," *Popular Electronics*, pp. 27–34, 100, October 1971.
8. "Build a 4-Channel Power Amplifier," *Radio Electronics*, pp. 39–42, March 1973, and pp. 62–68, April 1973.

Chapter 2

Power Amplifier Basics

2. INTRODUCTION

In this chapter we'll look at the design of a basic power amplifier in detail. Some information about transistors will first be discussed, followed by a simple analysis of the basic building block circuits that are inevitably used to build a complete amplifier circuit. This will provide a good foundation for the detailed analysis of the basic amplifier that follows. Chapter 3 will then take us on a tour of amplifier design, evolving and assessing a design as its performance is improved to a high level.

2.1 Bipolar Transistors

The *bipolar junction transistor* (BJT) is the primary building block of most audio power amplifiers. This section is not meant to be an exhaustive review of transistors, but rather presents enough knowledge for you to understand and analyze transistor amplifier circuits. More importantly, transistor behavior is discussed in the context of power amplifier design, with many relevant tips along the way.

Current Gain

If a small current is sourced into the base of an NPN transistor, a much larger current flows in the collector. The ratio of these two currents is the current gain, commonly called *beta* (β) or h_{fe}. Similarly, if one sinks a small current from the base of a PNP transistor, a much larger current flows in its collector.

The current gain for a typical small-signal transistor often lies between 50 and 200. For an output transistor, β typically lies between 20 and 100. Beta can vary quite a bit from transistor to transistor and is also a mild function of the transistor current and collector voltage.

Because transistor beta can vary quite a bit, circuits are usually designed so that their operation does not depend heavily on the particular value of β for its transistors. Rather, the circuit is designed so that it operates well for a minimum value of β and better for very high β. Because β can sometimes be very high, it is usually bad practice to design a circuit that would misbehave if β became very high. The *transconductance* (gm) of the transistor is actually the more predictable and important design parameter (as long as β is high enough not to matter much). For those unfamiliar with the term, *transconductance* of a transistor is the change in collector current in response to a given change in base-emitter voltage, in units of Siemens, S (amps per volt).

$$gm = \Delta I_c / \Delta V_{be} \tag{2.1}$$

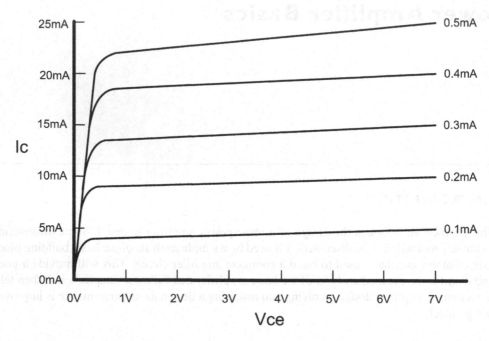

Figure 2.1 Transistor Collector Current Characteristics

The familiar collector current characteristics shown in Figure 2.1 illustrate the behavior of transistor current gain. This family of curves shows how the collector current increases as collector-emitter voltage (V_{ce}) increases, with base current as a parameter. The upward slope of each curve with increasing V_{ce} reveals the mild dependence of β on collector-emitter voltage. The spacing of the curves for different values of base current reveals the current gain. Notice that this spacing tends to increase as V_{ce} increases, once again revealing the dependence of current gain on V_{ce}. The spacing of the curves may be larger or smaller between different pairs of curves. This illustrates the dependence of current gain on collector current. The transistor shown has β of about 50.

Beta can be a strong function of current when current is high; it can decrease quickly with increases in current. This is referred to as *beta droop* and can be a source of distortion in power amplifiers. A typical power transistor may start with a beta of 70 at a collector current of 1 A, but have its β fall to 20 or less by the time I_c reaches 10 A. This is especially important when the amplifier is called on to drive low load impedances. This is sobering in light of the current requirements illustrated in Table 1.3.

Base-Emitter Voltage

The bipolar junction transistor requires a certain forward bias voltage at its base-emitter junction to begin to conduct collector current. This turn-on voltage is usually referred to as V_{be}. For silicon transistors, V_{be} is usually between 0.5 and 0.7 V. The actual value of V_{be} depends on the transistor device design and the amount of collector current (I_c). It also depends on the size of the device and the temperature of the device [1–4].

The base-emitter voltage increases by about 60 mV for each decade of increase in collector current. This reflects the logarithmic relationship of V_{be} to collector current. For the popular

2N5551, for example, V_{be} = 600 mV at 100 µA and rises to 720 mV at 10 mA. This corresponds to a 120 mV increase for a two-decade (100:1) increase in collector current.

Tiny amounts of collector current actually begin to flow at quite low values of forward bias (V_{be}). Indeed, the collector current increases exponentially with V_{be}. That is why it looks like there is a fairly well-defined turn-on voltage when collector current is plotted against V_{be} on linear coordinates. It becomes a remarkably straight line when the log of collector current is plotted against V_{be}. Some circuits, like multipliers, make great use of this logarithmic dependence of V_{be} on collector current.

Put another way, the collector current increases exponentially with base-emitter voltage, and we have the approximation

$$I_c = I_S e^{(V_{be}/V_T)} \tag{2.2}$$

where the voltage V_T is called the *thermal voltage*. Here V_T is about 26 mV at room temperature and is proportional to absolute temperature. This plays a role in the temperature dependence of V_{be}. However, the major cause of the temperature dependence of V_{be} is the strong increase with temperature of the *saturation current* I_S. This ultimately results in a negative temperature coefficient of V_{be} of about −2.2 mV/°C [1,4].

Expressing base-emitter voltage as a function of collector current, we have the analogous approximation

$$V_{be} = V_T \ln (I_c/I_S) \tag{2.3}$$

where ln (I_c/I_S) is the natural logarithm of the ratio I_c/I_S. The value of V_{be} here is the *intrinsic* base-emitter voltage, where any voltage drops across physical base resistance and emitter resistance are not included.

The base-emitter voltage for a given collector current typically decreases by about 2.2 mV for each degree Centigrade increase in temperature. This means that when a transistor is biased with a fixed value of V_{be}, the collector current will increase as temperature increases. As collector current increases, so will the power dissipation and heating of the transistor; this will lead to further temperature increases and sometimes a vicious cycle called *thermal runaway*. This is essentially positive feedback in a local feedback system.

The V_{be} of power transistors will start out at a smaller voltage at a low collector current of about 100 mA, but may increase substantially to 1 V or more at current in the 1 to 10 A range. At currents below about 1 A, V_{be} typically follows the logarithmic rule, increasing by about 60 mV per decade of increase in collector current. As an example, V_{be} might increase from 550 mV at 150 mA to 630 mV at 1 A. Even this is more than 60 mV per decade.

For collector current above about 1 A, the V_{be} versus I_c curve for a power transistor often begins to behave linearly, like a resistance. In the same example, V_{be} might increase to about 1.6 V at 1 A. This would correspond to effectively having a resistance of about 0.1 Ω in series with the emitter. The actual emitter resistance is not necessarily the physical origin of the increase in V_{be}. The voltage drop across the base resistance RB due to base current is often more significant. This voltage drop will be equal to $RB(I_c/\beta)$. The effective contribution to resistance as seen at the emitter by RB is thus RB/β. The base resistance divided by β is often the dominant source of this behavior.

Consider a power transistor operating at I_c = 10 A with a base resistance of 4 Ω, an operating β of 50 and an emitter resistance of 20 mΩ. Base current will be 200 mA and voltage drop across the base resistance will be 0.8 V. Voltage drop across the emitter resistance will be 0.2 V. Adding

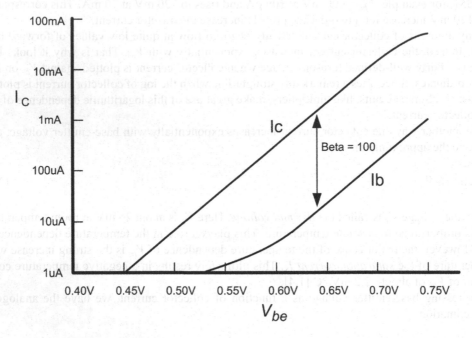

Figure 2.2 Transistor Gummel Plot

the intrinsic V_{be} of perhaps 660 mV, the base-emitter voltage becomes 1.66 V. It is thus easy to see how rather high V_{be} can develop for power transistors at high operating currents.

The Gummel Plot

If the log of collector current is plotted as a function of V_{be}, the resulting diagram is very revealing. As mentioned above, it is ideally a straight line. The diagram becomes even more useful and insightful if base current is plotted on the same axes. This is now called a *Gummel plot* [1–2]. It sounds fancy, but that is all it is. The magic lies in what it reveals about the transistor. A Gummel plot is shown in Figure 2.2.

In practice, neither the collector current nor the base current plots are straight lines over the full range of V_{be}, and the bending illustrates various non-idealities in the transistor behavior. The vertical distance between the lines corresponds to the β of the transistor, and the change in distance shows how β changes as a function of V_{be} and, by extension, I_c. The curves in Figure 2.2 illustrate the typical loss in transistor current gain at both low and high current extremes.

Transconductance

While transistor current gain is an important parameter and largely the source of its amplifying ability, the transconductance of the transistor is perhaps the most important characteristic used by engineers when doing actual design. Transconductance, denoted as *gm*, is the ratio of the change in collector current to the change in base voltage.

The unit of measure of transconductance is the Siemens (S), which corresponds to a current change of 1 A for a voltage change of 1 V. This is the inverse of the measure of resistance, the ohm (it was once called the *mho, ohm* spelled backward). If the base-emitter voltage of a transistor is increased by 1 mV, and as a result the collector current increases by 40 μA, the transconductance of the transistor is 40 millisiemens (mS).

The transconductance of a bipolar transistor is governed by its collector current. This is a direct result of the exponential relationship of collector current to base-emitter voltage. The slope of that curve increases as I_c increases; this means that transconductance also increases. Transconductance is given simply as

$$gm = I_c/V_T \tag{2.4}$$

where V_T is the thermal voltage, typically 26 mV at room temperature. At a current of 1 mA, transconductance is 1 mA/26 mV = 0.038 S.

The inverse of gm is a resistance. Sometimes it is easier to visualize the behavior of a circuit by treating the transconductance of the transistor as if it were a built-in dynamic emitter resistance re'. This resistance is just the inverse of gm, so we have

$$re' = V_T/I_c = 0.026/I_c \quad \text{(at room temperature)} \tag{2.5}$$

In the above case $re' = 26\ \Omega$ at a collector current of 1 mA.

An important approximation that will be used frequently is that $re' = 26\ \Omega/I_c$ where I_c is expressed in milliamperes. If a transistor is biased at 10 mA, re' will be about 2.6 Ω. The transistor will act as if a change in its base-emitter voltage is directly impressed across 2.6 Ω; this causes a corresponding change in its emitter current and very nearly the same change in its collector current. This forms the basis of the common-emitter (CE) amplifier.

It is important to recognize that $gm = 1/re'$ is the *intrinsic* transconductance, ignoring the effects of base and emitter resistance. Actual transconductance will be reduced by emitter resistance RE and RB/β being added to re' to arrive at net transconductance. This is especially important in the case of power transistors.

Input Resistance

If a small change is made in the base-emitter voltage, how much change in base current will occur? This defines the effective input resistance of the transistor. The transconductance dictates that if the base-emitter voltage is changed by 1 mV, the collector current will change by about 40 μA if the transistor is biased at 1 mA. If the transistor has a beta of 100, the base current will change by 0.38 μA. Note that the β here is the effective current gain of the transistor for small changes, which is more appropriately referred to as the *AC current gain* or *AC beta* (β_{AC}). It is also known as the small-signal gain, h_{fe}. The effective input resistance in this case is therefore about 1 mV/0.38 μA = 2.6 kΩ. The effective input resistance is just β_{AC} times re'.

Early Effect

The Early effect manifests itself as finite output resistance at the collector of a transistor and is the result of the current gain of the transistor being a function of the collector-base voltage. The collector characteristic curves of Figure 2.1 show that the collector current at a given base current increases with increased collector voltage. This means that the current gain of the transistor is increasing with collector voltage. This also means that there is an equivalent output resistance in the collector circuit of the transistor.

The increase of collector current with increase in collector voltage is called the *Early effect*. If the straight portions of the collector current curves in Figure 2.1 are extrapolated to the left, back to the X axis, they will intersect the X axis at a negative voltage. The value of this voltage is called the *Early voltage, VA*. The slope of these curves represents the output resistance *ro* of the device.

Typical values of *VA* for small-signal transistors lie between 20 and 200 V. A very common value of *VA* is 100 V, as for the 2N5551. The output resistance due to the Early effect decreases with increases in collector current. A typical value of this resistance for a small-signal transistor operating at 1 mA is on the order of 100 kΩ.

The Early effect is especially important because it acts as a resistance in parallel with the collector circuit of a transistor. This effectively makes the net load resistance on the collector smaller than the external load resistance in the circuit. As a result, the gain of a common-emitter stage decreases. Because the extra load resistance is a function of collector voltage and current, it is a function of the signal and is therefore nonlinear and so causes distortion.

The Early effect can be modeled as a resistor *ro* connected from the collector to the emitter of an otherwise "perfect" transistor [1]. The value of *ro* is

$$ro = \frac{(VA + V_{ce})}{I_C} \tag{2.6}$$

I should point out that this equation is a very simplified description of the Early effect and that even many standard versions of SPICE don't support the advanced models needed for accurate modeling of the effect.

For the 2N5551, with a *VA* of 100 and operating at V_{ce} = 10 V and I_c = 10 mA, *ro* comes out to be 11 kΩ. The value of *ro* is doubled as the collector voltage swings from very small voltages to a voltage equal to the Early voltage.

It is important to understand that this resistance is not, by itself, necessarily the output resistance of a transistor stage, since it is not connected from collector to ground. It is connected from collector to emitter. Any resistance or impedance in the emitter circuit will significantly increase the effective output resistance caused by *ro*.

The Early effect is especially important in the VAS of an audio power amplifier. In that location the device is subjected to very large collector voltage swings and the impedance at the collector node is quite high due to the usual current source loading and good buffering of the output load from this node.

A 2N5551 VAS transistor biased at 10 mA and having no emitter degeneration will have an output resistance on the order of 14 kΩ at a collector-emitter voltage of 35 V. This would correspond to a signal output voltage of 0 V in an arrangement with ±35 V power supplies. The same transistor with 10:1 emitter degeneration will have an output resistance of about 135 kΩ.

At a collector-emitter voltage of only 5 V (corresponding to a −30 V output swing) that transistor will have a reduced output resistance of 105 kΩ. At a collector-emitter voltage of 65 V (corresponding to a +30 V output swing), that transistor will have an output resistance of about 165 kΩ. These changes in output resistance as a result of signal voltage imply a change in gain and thus second harmonic distortion.

Because the Early effect manifests itself as a change in the β of the transistor as a function of collector voltage, and because a higher-β transistor will require less base current, it can be argued that a given amount of Early effect has less influence in some circuits if the β of the transistor is high. A transistor whose β varies from 50 to 100 due to the Early effect and collector voltage swing will have more effect on circuit performance in many cases than a transistor whose β varies from 100 to 200 over the same collector voltage swing. The variation in base current will be less in the latter than in the former. For this reason, the product of β and *VA* is an important *figure of merit* (FOM) for transistors. In the case of the 2N5551, with a current gain of 100 and an Early voltage VA of 100 V, this FOM is 10,000 V. The FOM for bipolar transistors often lies in the range of 5000 V to 50,000 V.

Early effect FOM $= \beta * VA$ (2.7)

The value of $\beta * VA$ tends to be constant for a given transistor type; high-β parts have lower VA, and vice versa.

Junction Capacitance

All BJTs have base-emitter capacitance (C_{be}) and collector-base capacitance (C_{cb}). This limits the high-frequency response, but also can introduce distortion because these junction capacitances are a function of voltage.

The base, emitter and collector regions of a transistor can be thought of as plates of a capacitor separated by non-conducting regions. The base is separated from the emitter by the base-emitter junction, and it is separated from the collector by the base-collector junction. Each of these junctions has capacitance, whether it is forward-biased or reverse-biased. Indeed, these junctions store charge, and that is a characteristic of capacitance.

A reverse-biased junction has a so-called *depletion region*. The depletion region can be thought of roughly as the spacing of the plates of the capacitor. With greater reverse bias of the junction, the depletion region becomes larger. The spacing of the capacitor plates is then larger, and the capacitance decreases. The junction capacitance is thus a function of the voltage across the junction, decreasing as the reverse bias increases.

This behavior is mainly of interest for the collector-base capacitance C_{cb}, since in normal operation the collector-base junction is reverse-biased while the base-emitter junction is forward-biased. It will be shown that the effective capacitance of the forward-biased base-emitter junction is quite high.

The variance of semiconductor junction capacitance with reverse voltage is taken to good use in *varactor diodes*, where circuits are electronically tuned by varying the reverse bias on the varactor diode. In audio amplifiers, the effect is an undesired one, since capacitance varying with signal voltage represents nonlinearity. It is obviously undesirable for the bandwidth or high-frequency gain of an amplifier stage to be varying as a function of the signal voltage.

The collector-base capacitance of the popular 2N5551 small-signal NPN transistor ranges from a typical value of 5 pF at 0 V reverse bias (V_{cb}) down to 1 pF at 100 V. For what it's worth, its base-emitter capacitance ranges from 17 pF at 0.1 V reverse bias to 10 pF at 5 V reverse bias. Remember, however, that this junction is usually forward-biased in normal operation. The junction capacitances of a typical power transistor are often about two orders of magnitude larger than those of a small-signal transistor.

Speed and f_T

The AC current gain of a transistor falls off at higher frequencies in part due to the need for the input current to charge and discharge the relatively large capacitance of the forward-biased base-emitter junction.

The most important speed characteristic for a BJT is its f_T, or *transition frequency* [1–2]. This is the frequency where the AC current gain β_{AC} falls to approximately unity. For small-signal transistors used in audio amplifiers, f_T will usually be on the order of 50 to 300 MHz. A transistor with a low-frequency β_{AC} of 100 and an f_T of 100 MHz will have its β_{AC} begin to fall off (be down 3 dB) at about 1 MHz. This frequency is referred to as f_β.

The effective value of the base-emitter capacitance of a conducting BJT can be shown to be approximately

$$C_{be} = gm/\omega_T \qquad (2.8)$$

where ω_T is the radian frequency equal to $2\pi f_T$ and gm is the transconductance.
Because $gm = I_c/V_T$, one can also state that

$$C_{be} = I_c/(V_T * \omega_T) \qquad (2.9)$$

This capacitance is often referred to as C_π for its use in the so-called hybrid pi model. Because transconductance increases with collector current, so does C_{be}. For a transistor with a 100 MHz f_T and operating at 1 mA, the effective base-emitter capacitance will be about 61 pF.

Power transistors usually have a much lower value of f_T, often in the range of 1 to 8 MHz for conventional power devices. The effective base-emitter capacitance for a power transistor can be surprisingly large. Consider a power transistor whose f_T is 2 MHz. Assume it is operating at $I_c = 1$ A. Its transconductance will be $I_c/V_T = 1.0/0.026 = 38.5$ S. Its ω_T will be 12.6 Mradians/s. Its C_{be} will be

$$C_{be} = gm/\omega_T = 3.1 \ \mu F$$

Needless to say, this is a real eye-opener!

This explains why it can be difficult to turn off a power transistor quickly. Suppose the current gain of the power transistor is 50, making the base current 20 mA. If that base-current drive is removed and the transistor is allowed to turn off, the V_{be} will change at a rate of

$$I_b/C_\pi = 0.02/3.1 \times 10^{-6} = 6.4 \ mV/\mu s$$

Recall that a 60 mV change in V_{be} will change the collector current by a factor of about 10. This means that it will take about 9 μs for the collector current to fall from 1A to 0.1 A. This illustrates why it is important to actively pull current out of the base to turn off a transistor quickly. This estimate was only an approximation because it was assumed that C_π was constant during the discharge period. It was not, since I_c was decreasing. However, the base current, which was the discharge current in this case, was also decreasing during the discharge period. The decreasing C_π and the decreasing base current largely cancel each other's effects, so the original approximation was not too bad.

In a real circuit there will usually be some means of pulling current out of the base, even if it is just a resistor from base to emitter. This will help turn off the transistor more quickly.

In order to decrease the collector current of the transistor from 1A to 0.1 A, C_π must be discharged by 60 mV. Recognizing that the capacitance will decrease as the collector current is brought down, the capacitance can be approximated by using an average value of one-half, or about 1.5 μF. Assume high transistor β so that the base current that normally must flow to keep the transistor turned on can be ignored. If a constant base discharge current of 30 mA is employed, the time it takes to ramp down the collector current by a decade can be estimated as follows:

$$T = C * V/I = 3.0 \ \mu s$$

This is still quite a long time if the amplifier is trying to rapidly change the output current as a result of a large high-frequency signal transient. Here the average rate of change of collector current is about 0.3 A/μs. To put this in perspective, assume an amplifier is driving 40 V peak into a 4 Ω load at 20 kHz. The voltage rate of change is 5 V/μs and the current rate of change must be 1.25 A/μs.

Unfortunately, just as BJTs experience beta droop at higher currents, so they also suffer from f_T *droop* at higher currents. A good conventional power transistor might start off with f_T of 6 MHz at 1 A, be down in f_T by 20% at 3 A, and be all the way down to 2 MHz at 10 A. At the same time, BJTs also suffer f_T droop at lower collector-emitter voltages while operating at high currents. This compounds the problem when an output stage is at a high-amplitude portion of a high-frequency waveform and delivering high current into the load. Under these conditions, V_{ce} might be as little as 5 V or less and device current might be several amperes.

So-called *ring emitter transistors* (RETs) and similar advanced BJT power transistor designs can have f_T in the 20-MHz to 80-MHz range. However, they also suffer from f_T droop at high currents. A typical RET might start out with an f_T of 40 MHz at 1 A, and maintain it quite well to 3 A, then have it crash to 4 MHz or less at 10 A. The RET devices also lose f_T at low current. At 100 mA, where they may be biased in a class AB output stage, the same RET may have f_T of only 20 MHz.

The Hybrid Pi Model

Those more familiar with transistors will recognize that much of what has been discussed above is the makeup of the hybrid pi small-signal model of the transistor, shown in Figure 2.3. The fundamental active element of the transistor is a voltage-controlled current source, namely a *transconductance*. Everything else in the model is essentially a passive *parasitic* component. AC current gain is taken into account by the base-emitter resistance r_π. The Early effect is taken into account by ro. Collector-base capacitance is shown as C_{cb}. Current gain roll-off with frequency (as defined by f_T) is modeled by C_π. The values of these elements are as described above. This is a small-signal model; element values will change with the operating point of the transistor.

The Ideal Transistor

Operational amplifier circuits are often designed by assuming an ideal op amp, at least initially. In the same way a transistor circuit can be designed by assuming an "ideal" transistor. This is

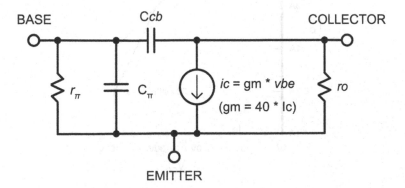

Figure 2.3 Hybrid Pi Model

like starting with the hybrid pi model stripped of all of its passive parasitic elements. The ideal transistor is just a lump of transconductance. As needed, relevant impairments, such as finite β, can be added to the ideal transistor. This usually depends on what aspect of performance is important at the time.

The ideal transistor has infinite current gain, infinite input impedance and infinite output resistance. It acts as if it applies all of the small-signal base voltage to the emitter through an internal intrinsic emitter resistance re'.

Safe Operating Area

The *safe operating area* (SOA) for a transistor describes the safe combinations of voltage and current for the device. This area will be bounded on the X axis by the maximum operating voltage and on the Y axis by the maximum operating current. The SOA is also bounded by a line that defines the maximum power dissipation of the device. Such a plot is shown for a power transistor in Figure 2.4, where voltage and current are plotted on log scales and the power dissipation limiting line becomes the outermost straight line.

Unfortunately, power transistors are not just limited in their safe current-handling capability by their power dissipation. At higher voltages they are more seriously limited by a phenomenon called *secondary breakdown*. This is illustrated by the more steeply sloped inner line in Figure 2.4.

Although there are many different ways to specify SOA, perhaps the single most indicative number for audio power amplifier design is the amount of current the transistor can safely sustain for at least 1 second at some high collector-emitter voltage such as 100 V. In the absence of secondary breakdown, a 150-W power transistor could sustain a current of 1.5 A at 100 V. In reality, this number may only be 0.5 A, corresponding to only 50 W of dissipation. Secondary breakdown causes the sustainable power dissipation at high voltages to be less than that at low voltages.

Secondary breakdown results from localized hot spots in the transistor. At higher voltages the depletion region of the collector-base junction has become larger and the effective base region

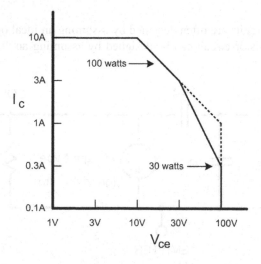

Figure 2.4 Safe Operating Area

has become thinner. Recall that the collector current of a transistor increases as the junction temperature increases if the base-emitter voltage is held constant. A localized increase in the power transistor's base-emitter junction temperature will cause that area to hog more of the total collector current. This causes that local area to become hotter, conduct even more current and become hotter still; this leads to a localized thermal runaway.

SOA is very important in the design of audio amplifier output stages because the SOA can be exceeded, especially when the amplifier is driving a reactive load. This can lead to the destruction of the output transistors unless there are safe area protection circuits in place. There will be a much deeper examination of power transistor SOA in Chapter 19, including discussion of the higher value of SOA that a transistor can withstand for shorter periods of time.

2.2 JFETs

JFETs operate on a different principle than BJTs. Picture a bar of n-type doped silicon connected from source to drain [5–8]. This bar will act like a resistor. Now add a *p-n* junction somewhere along the length of this bar by adding a region with p-type doping. This is the gate. As the p-type gate is reverse-biased, a *depletion region* will be formed, and this will begin to pinch off the region of conductivity in the n-type bar. This reduces current flow. This is called a *depletion-mode* device. The JFET is nominally on, and its degree of conductance will decrease as reverse bias on its gate is increased until the channel is completely pinched off.

The reverse gate voltage where pinch-off occurs is referred to as V_p or as the *threshold voltage* V_t (not to be confused with the thermal voltage V_T). The threshold voltage is often on the order of $-.5V$ to $-4V$ for most small-signal N-channel JFETs. Note that control of a JFET is opposite to the way a BJT is controlled. The BJT is normally off and the JFET is normally on. The BJT is turned on by application of a forward bias to the base-emitter junction, while the JFET is turned off by application of a reverse bias to its gate-source junction.

The reverse voltage that exists between the drain and the gate can also act to pinch off the channel. At V_{dg} greater than the threshold voltage, the channel will be pinched in such a way that the drain current becomes self-limiting. In this region the JFET no longer acts like a resistor, but rather like a voltage-controlled current source. These two operating regions are referred to as the *linear region* and the *saturation region*, respectively. JFET amplifier stages usually operate in the saturation region.

JFET I_d Versus V_{gs} Behavior

Figure 2.5a shows how drain current changes as a function of gate voltage in the saturation region; Figure 2.5b illustrates how transconductance changes as a function of drain current in the same region. The device is one-half of an LSK489 dual JFET [6–7]. Threshold voltage for this device is nominally about -1.8 V.

The JFET *I-V* characteristic (I_d versus V_{gs}) obeys a square law, rather than the exponential law applicable to BJTs. The simple relationship below is valid for $V_{ds} > V_t$ and does not take into account the influence of V_{ds} that is responsible for output resistance of the device.

$$I_d = \beta(V_{gs} - V_t)^2 \tag{2.10}$$

The equation is valid only for positive values of $V_{gs} - V_t$. The factor β governs the transconductance of the device. When $V_{gs} = V_t$, the $V_{gs} - V_t$ term is zero and no current flows. When $V_{gs} = 0V$, the term is equal to V_t^2 and maximum current flows.

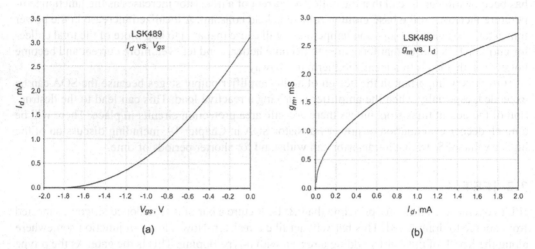

Figure 2.5 (a) JFET Drain Current as a Function of Gate Voltage (b) Transconductance as a Function of Drain Current

The maximum current that flows when $V_{gs} = 0$ V and $V_{ds} \gg V_t$ is referred to as I_{DSS}, a key JFET parameter usually specified on datasheets. Under these conditions the channel is at the edge of pinch-off and the current is largely self-limiting. In this case it is the reverse bias of the gate junction with respect to the drain that is pinching off the channel. The parameter β is the transconductance coefficient and is related to I_{DSS} and V_t. The value of I_{DSS} for the LSK489 is about 3.1 mA, and the value of β is about 0.9 mA/V^2.

$$\beta = I_{DSS}/V_t^2 \tag{2.11}$$

The parameter β can also be expressed in mS/V; this means that if gm is plotted as a function of V_{gs}, a straight line will result. With some manipulation of Equation 2.10, it can be seen that the transconductance of the JFET is proportional to the square root of the drain current.

$$gm = 2\sqrt{\beta I_d} \tag{2.12}$$

This is different from the behavior of a BJT, where gm is proportional to collector current. Transconductance for a JFET at a given operating current is smaller than that of a BJT by a factor of 10 or more in many cases. The transconductance for the LSK489 at $I_d = 1$ mA is about 2 mS. The gm of a BJT at $I_c = 1$ mA is about 40 mS, greater by a factor of 20. The larger LSK389 has $V_t = -0.54$ V, $I_{DSS} = 8.4$ mA, and gm of 11.3 mS at 1 mA.

The JFET turn-on characteristic is much less abrupt than that of a BJT. Absent degeneration of a BJT, the input voltage range over which the JFET is reasonably linear is much greater than that of a BJT. The collector current of the BJT increases by a factor of about 2 for every increase of 18 mV in V_{be}. Between 0.75 mA and 1.5 mA, V_{gs} of an LSK489 changes by about 370 mV.

2.3 Power MOSFETs

MOSFETs have an insulated gate rather than a *p-n* junction gate. The gate is insulated from the underlying silicon channel by a silicon dioxide layer across which the controlling electric field is developed. Unlike the JFET, the most common MOSFETs are *enhancement-mode* devices, meaning that the gate must have a positive voltage (for N-channel devices) applied to it to cause or increase conduction. The devices are normally in the *off* non-conducting state. Like the JFET, MOSFETs are square-law devices [9–11]. Power MOSFETs used in audio power amplifier output stages are either of the lateral or vertical types.

Lateral MOSFET Structure

The modern power MOSFET is made possible by many of the same advanced techniques that are employed in MOS large-scale integrated circuits, including fine-line photolithography, self-aligned polysilicon gates and ion implantation. Two planar structures, one lateral MOSFET and one vertical DMOS, are currently the most suitable devices for audio applications. Both are available in complementary pairs, offer suitable current and voltage ratings and are realized with a cellular structure that provides the equivalent of thousands of small-geometry MOSFETs connected in parallel.

The structure of the lateral power MOSFET is illustrated in Figure 2.6a. The N-channel device shown is similar to small-signal MOSFETs found in integrated circuits, except that a lightly doped n-type drift region is placed between the gate and the n+ drain contact to increase the drain-to-source breakdown voltage by decreasing the gradient of the electric field. Current flows laterally from drain to source when a positive bias on the silicon gate inverts the p-type body region to form a conducting n-type channel. Note that the arrows in Figure 2.6 illustrate the direction of carrier flow rather than conventional current flow. The device is fabricated by a self-aligned process where the source and drain diffusions are made using the previously formed gate as part of the mask. Alignment of the gate with the source and drain diffusions thus occurs naturally, and the channel length is equal to the gate length less the sum of the out-diffusion distances of the source and drain regions under the gate. Small gate structures are required to realize the short channels needed to achieve high transconductance and low *on* resistance.

While providing high breakdown voltage, the lightly doped drift region tends to increase *on* resistance. This partly explains why higher-voltage power MOSFETs tend to have higher *on* resistance. A further disadvantage of this structure is that all of the source, gate and drain interconnect lies on the surface, resulting in fairly large chip area for a given amount of active channel area, which in turn limits transconductance per unit area. Series gate resistance also tends to be fairly high (about 40 Ω) as a result, limiting maximum device speed. Lateral power MOSFETs

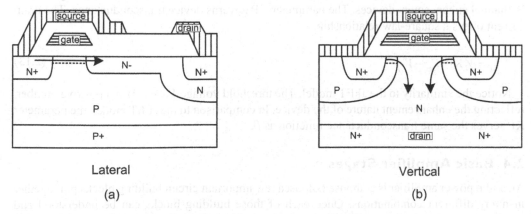

Lateral   Vertical

(a)   (b)

Figure 2.6 Structures of N-Channel Lateral and Vertical Power MOSFETs

have been widely used in audio amplifiers. Examples of this structure are the Hitachi 2SK-134 (N-channel) and 2SJ-49 (P-channel). Desirable features of these devices include a threshold voltage of only a few tenths of a volt and a zero temperature coefficient of drain current versus gate voltage at a drain current of about 100 mA, providing good bias stability.

Vertical MOSFET Structure

A more advanced power MOSFET design is the vertical DMOS structure illustrated in Figure 2.6b. When a positive gate bias inverts the p-type body region into a conducting N-channel, current initially flows vertically from the drain contact on the back of the chip through the lightly doped n-type drift region to the channel, where it then flows laterally through the channel to the source contact. The double-diffused structure begins with an n-type wafer that includes a lightly doped epitaxial layer. The p-type body region and the n+ source contact are then diffused into the wafer in that order. Because both diffusions use the same mask edge on either side of the gate, channel length is the difference of the out-diffusion distances of the body and source regions. As a result, short channels are easily realized without heavy dependence on photolithographic resolution. Short channels permit high transconductance and low *on* resistance. The geometry and dimensions of the n-type drift region are such that its effective resistance can be much smaller than that of the drift region for the lateral devices. This also aids in achieving low *on* resistance while retaining high voltage capability.

The vertical DMOS structure is much more compact and area efficient than the lateral structure because the source metalization covers the entire surface; the polysilicon gate interconnect is buried under the source metalization. Also, each gate provides two channels, one on each side. The amount of active channel area for a given chip area is thus higher than for the lateral geometry. The fact that source metalization areas can occupy virtually an entire side of the chip leads to high current capability. Finally, the length of the gate can be greater in this structure because it does not directly control channel length. This feature, combined with the compact structure, results in lower series gate resistance (about 6 Ω) and higher speed. Because of its many advantages, the planar vertical DMOS structure is the main-line power MOSFET technology. Examples of this cellular structure are the International Rectifier IRFP240 (N-channel) and IRFP9240 (P-channel).

MOSFET I_d Versus V_{gs}

The threshold voltage V_t is positive for enhancement mode N-channel devices and negative for P-channel enhancement devices. The parameter KP governs device transconductance. The drain current obeys a square-law relationship.

$$I_d = \frac{1}{2}\,\mathrm{KP}\left(V_{gs} - V_t\right)^2 \tag{2.13}$$

Notice the similarity to the JFET model. The threshold voltage V_t (VTO) is a positive number, reflecting the enhancement nature of the device. In comparison to the JFET model, the parameter KP serves the same transconductance function as β.

2.4 Basic Amplifier Stages

An audio power amplifier is composed of just a few important circuit building blocks put together in many different combinations. Once each of those building blocks can be understood and analyzed, it is not difficult to do an approximate analysis by inspection of a complete power amplifier. Knowledge of how these building blocks perform and bring performance value to the table permits the designer to analyze and synthesize circuits.

Common-Emitter Stage

The common-emitter (CE) amplifier is possibly the most important circuit building block, as it provides basic voltage gain. Assume that the transistor's emitter is at ground and that a bias current has been established in the transistor. If a small voltage signal is applied to the base of the transistor, the collector current will vary in accordance with the base voltage. If a load resistance R_L is provided in the collector circuit, that resistance will convert the varying collector current to a voltage. A voltage-in, voltage-out amplifier is the result, and it likely has quite a bit of voltage gain. A simple common emitter amplifier is shown in Figure 2.7a.

The voltage gain will be approximately equal to the collector load resistance times the transconductance gm. Recall that the intrinsic emitter resistance $re' = 1/gm$. Thus, more conveniently, assuming the ideal transistor with intrinsic emitter resistance re', the gain is simply R_L/re'.

Consider a transistor biased at 1 mA with a load resistance of 5000 Ω and a supply voltage of 10 V, as shown in Figure 2.7a. The intrinsic emitter resistance re' will be about 26 Ω. The gain will be approximately 5000/26 = 192.

This is quite a large value. However, any loading by other circuits that are driven by the output has been ignored. Such loading will reduce the gain.

The Early effect has also been ignored. It effectively places another resistance ro in parallel with the 5 kΩ load resistance. This is illustrated with the dashed resistor drawn in the figure. As mentioned earlier, ro for a 2N5551 operating at 1 mA and relatively low collector-emitter voltages will be on the order of 100 kΩ, so the error introduced by ignoring the Early effect here will be about 5%.

Because re' is a function of collector current, the gain will vary with signal swing and the gain stage will suffer from some distortion. The gain will be smaller as the current swings low and the output voltage swings high. The gain will be larger as the current swings high and the output voltage swings low. This results in second harmonic distortion.

If the input signal swings positive so that the collector current increases to 1.5 mA and the collector voltage falls to 2.5 V, re' will be about 17.3 Ω and the incremental gain will be 5000/17.3 = 289.

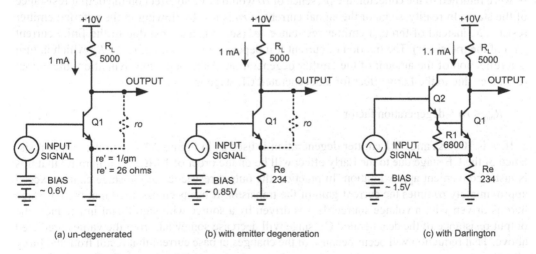

(a) un-degenerated   (b) with emitter degeneration   (c) with Darlington

Figure 2.7 Common-Emitter Amplifiers

If the input signal swings negative so that the collector current falls to 0.5 mA and the collector voltage rises to 7.5 V, then re' will rise to about 52 Ω and incremental gain will fall to 5000/52 = 96. The incremental gain of this stage has thus changed by over a factor of 3 when the output signal has swung 5 V peak-to-peak. This represents a high level of distortion.

If external emitter resistance is added as shown in Figure 2.7b, then the gain will simply be the ratio of R_L to the total emitter circuit resistance consisting of re' and the external emitter resistance R_e. Since the external emitter resistance does not change with the signal, the overall gain is stabilized and is more linear. This is called *emitter degeneration*. It is a form of local negative feedback.

The CE stage in Figure 2.7b is essentially the same as that in 2.7a but with a 234-Ω emitter resistor added. This corresponds to 10:1 emitter degeneration because the total effective resistance in the emitter circuit has been increased by a factor of 10 from 26 Ω to 260 Ω. The nominal gain has also been reduced by a factor of 10 to a value of approximately 5000/260 = 19.2.

Consider once again what happens to the gain when the input signal swings positive and negative to cause a 5-V peak-to-peak output swing. If the input signal swings positive so that the collector current increases to 1.5 mA and the collector voltage falls to 2.5 V, total emitter circuit resistance R_e will become 17 + 234 = 251 Ω, and the incremental gain will rise to 5000/251 = 19.9.

If the input signal swings negative so that the collector current falls to 0.5 mA and the collector voltage rises to 7.5 V, then R_e will rise to about 234 + 52 = 286 Ω and incremental gain will fall to 5000/286 = 17.4. The incremental gain of this stage has now swung over a factor of 1.14:1, or only 14%, when the output signal has swung 5-V peak-to-peak. This is indeed a much lower level of distortion than occurred in the un-degenerated circuit of Figure 2.7a. This illustrates the effect of local negative feedback without resort to any negative feedback theory.

We thus have, for the CE stage, the approximation

$$\text{Gain} = R_L/(re' + R_e) \tag{2.14}$$

where R_L is the net collector load resistance and R_e is the external emitter resistance. The emitter degeneration factor is defined as $(re' + R_e)/re'$. In this case that factor is 10.

Emitter degeneration also mitigates nonlinearity caused by the Early effect in the CE stage. As shown by the dotted resistance ro in Figure 2.7b, most of the signal current flowing in ro is returned to the collector by way of being injected into the emitter. If 100% of the signal current in ro were returned to the collector, the presence of ro would have no effect on the output resistance of the stage. In reality, some of the signal current in ro is lost by flowing in the external emitter resistor R_e instead of through emitter resistance re' (some is also lost due to the finite current gain of the transistor). The fraction of current lost depends on the ratio of re' to R_e, which in turn is a reflection of the amount of the emitter degeneration. As a rough approximation, the output resistance due to the Early effect for a degenerated CE stage is

$$R_{out} \sim ro * \text{degeneration factor} \tag{2.15}$$

If ro is 100 kΩ and 10:1 emitter degeneration is used as in Figure 2.7b, then the output resistance of the CE stage due to the Early effect will be on the order of 1 MΩ. Bear in mind that this is just a convenient approximation. In practice, the output resistance of the stage cannot exceed approximately ro times the current gain of the transistor. It has been assumed that the CE stage here is driven with a voltage source. If it is driven by a source with significant impedance, the output resistance of the degenerated CE stage will decrease somewhat from the values predicted above. That reduction will occur because of the changes in base current that result from the Early effect.

Darlington Common-Emitter Stage

Figure 2.7c shows a CE stage that uses a Darlington transistor pair. The Darlington connection adds Q2 and R1 in front of Q1. Q2 acts like an emitter follower in front of Q1 with the exception that its collector is connected to that of Q1 and is thus powered by the load resistor. Pull-down resistor R1 establishes a minimum current in Q2, here about 100 μA, preserving its speed. The Darlington connection essentially forms a super-transistor whose β is on the order of the product of the βs of Q1 and Q2, often in the range of 10,000. If the impedance seen looking into the base of Q1 in Figure 2.7b is 23.4 kΩ with β of 100 for Q1, then the impedance seen looking into the base of Q2 in Figure 2.7c may be over 2 MΩ. Note that the *on* voltage for the Darlington connection is 2 V_{be}, or about 1.3 V.

Bandwidth of the Common-Emitter Stage and Miller Effect

The high-frequency response of a CE stage will be limited if it must drive any load capacitance. This is no different than when a source resistance drives a shunt capacitance, forming a first-order low-pass filter. A pole is formed at the frequency where the source resistance and reactance of the shunt capacitance are the same; this causes the frequency response to be down 3 dB at that frequency. The reactance of a capacitor is equal to $1/(2\pi fC) = 0.159/(fC)$. The −3 dB frequency f_3 will then be $0.159/(RC)$.

In Figure 2.7a the output impedance of the CE stage is approximately that of the 5 kΩ collector load resistance. Suppose the stage is driving a load capacitance of 100 pF. The bandwidth will be dictated by the low-pass filter created by the output impedance of the stage and the load capacitance. A pole will be formed at

$$f_3 = 1/(2\pi R_L C_L) = 0.159/(5 \text{ k}\Omega * 100 \text{ pF}) = 318 \text{ kHz}$$

As an approximation, the collector-base capacitance should also be considered part of C_L. The bandwidth of a CE stage is often further limited by the collector-base capacitance of the transistor when the CE stage is fed from a source with significant impedance. The source must supply current to charge and discharge the collector-base capacitance through the large voltage excursion that exists between the collector and the base. This phenomenon is called the *Miller effect*.

Suppose the collector-base capacitance is 5 pF and assume that the CE stage is being fed from a 5-kΩ source impedance R_S. Recall that the voltage gain G of the circuit in Figure 2.7a was approximately 192. This means that the voltage across C_{cb} is 193 times as large as the input signal (bearing in mind that the input signal is out of phase with the output signal, adding to the difference). This means that the current flowing through C_{cb} is 193 times as large as the current that would flow through it if it were connected from the base to ground instead of base to collector. The input circuit thus sees an effective input capacitance C_{in} that is $1 + G$ times that of the collector-base capacitance. This phenomenon is referred to as *Miller multiplication* of the capacitance. In this case the effective value of C_{in} would be 965 pF.

The base-collector capacitance effectively forms a shunt feedback circuit that ultimately controls the gain of the stage at higher frequencies where the reactance of the capacitor becomes small. As frequency increases, a higher proportion of the input signal current must flow to the collector-base capacitance as opposed to the small fixed amount of signal current required to flow into the base of the transistor. If essentially all input signal current flowed through the collector-base capacitance, the gain of the stage would simply be the ratio of the capacitive reactance of C_{cb} to the source resistance

$$G = X_{Ccb}/R_S = 1/(2\pi fC_{cb})R_S = 0.159/(fC_{cb}R_S)$$

This represents a value of gain that declines at 6 dB per octave as frequency increases. This decline will begin at a frequency where the gain calculated here is equal to the low-frequency gain of the stage. Collector-base capacitance C_{cb} is nonlinear; it decreases as V_{ce} increases. This can cause distortion, since it causes the incremental gain of the stage to vary with signal, especially at high frequencies. The Miller effect is often used to advantage in providing the high-frequency roll-off needed to stabilize a negative feedback loop. This is referred to as *Miller compensation*. Miller effect compensation is usually implemented with a Miller capacitor C_M that is significantly larger than C_{cb}.

Differential Amplifier

The differential amplifier is illustrated in Figure 2.8. It is much like a pair of common emitter amplifiers tied together at the emitters and biased with a common current. This current is called the *tail current*. The arrangement is often referred to as a *long-tailed pair (LTP)*.

The differential amplifier routes its tail current to the two collectors of Q1 and Q2 in accordance with the voltage differential across the bases of Q1 and Q2. If the base voltages are equal, then equal currents will flow in the collectors of Q1 and Q2. If the base of Q1 is more positive than that of Q2, more of the tail current will flow in the collector of Q1 and less will flow in the collector of Q2. This will result in a larger voltage drop across the collector load resistor R_{L1} and a smaller voltage drop across load resistor R_{L2}. Output A is thus inverted with respect to Input A, while Output B is non-inverted with respect to Input A.

Visualize the intrinsic emitter resistance *re'* present in each emitter leg of Q1 and *Q2*. Recall that the value of *re'* is approximately 26 Ω divided by the transistor operating current in milliamperes. With 1 mA flowing nominally through each of Q1 and Q2, each can be seen as having an emitter resistance *re'* of 26 Ω. Note that since *gm* = 1/*re'* is dependent on the instantaneous transistor current, the values of *gm* and *re'* are somewhat signal dependent, and indeed this represents a nonlinearity that gives rise to distortion.

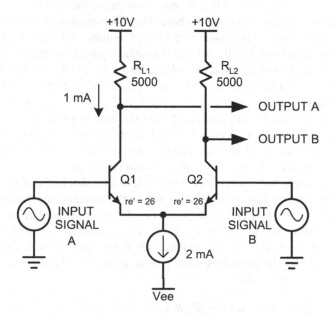

Figure 2.8 Differential Amplifier

Having visualized the ideal transistor with emitter resistance *re'*, one can now assume that the idealized internal emitter of each device moves exactly with the base of the transistor, but with a fixed DC voltage offset equal to V_{be}. Now look what happens if the base of Q1 is 5.2 mV more positive than the base of Q2. The total emitter resistance separating these two voltage points is 52 Ω, so a current of 5.2 mV/52 Ω = 0.1 mA will flow from the emitter of Q1 to the emitter of Q2. This means that the collector current of Q1 will be 100 μA more than nominal, and the collector current of Q2 will be 100 μA less than nominal. The collector currents of Q1 and Q2 are thus 1.1 mA and 0.9 mA, respectively, since they must sum to the tail current of 2.0 mA (assuming very high β for the transistors).

This 100 μA increase in the collector current of Q1 will result in a change of 500 mV at Output A, due to the collector load resistance of 5000 Ω. A 5.2 mV input change at the base of Q1 has thus caused a 500-mV change at the collector of Q1, so the stage gain to Output A in this case is approximately 500/5.2 = 96.2. More significantly, the stage gain defined this way is just equal numerically to the load resistance of 5000 Ω divided by the total emitter resistance *re'* = 52 Ω across the emitters.

Had additional external emitter degeneration resistors been included in series with each emitter, their value would have been added into this calculation. For example, if 48-Ω emitter degeneration resistors were employed, the gain would then become 5000/(26 + 48 + 48 + 26) = 5000/200 = 34. This approach to estimating stage gain is a very important back-of-the-envelope concept in amplifier design. In a typical amplifier design, one will often start with these approximations and then knowingly account for some of the deviations from the ideal. This will be evident in the numerous design analyses to follow.

It was pointed out earlier that the change in transconductance of the transistor as a function of signal current can cause distortion. Consider the situation where a negative input signal at the base of Q1 causes Q1 to conduct 0.5 mA and Q2 to conduct 1.5 mA. The emitter resistance *re'* of Q1 is now 26/0.5 = 52 Ω. The emitter resistance *re'* of Q2 is now 26/1.5 = 17.3 Ω. The total emitter resistance from emitter to emitter has now risen from 52 Ω in the case above to 69.3 Ω. This results in a reduced gain of 5000/69.3 = 72.15. This represents a reduction in gain by a factor of 0.75, or about 25%. This is an important origin of distortion in the LTP. The presumed signal swing that caused the imbalance of collector currents between Q1 and Q2 resulted in a substantial decrease in the incremental gain of the stage. More often than not, distortion is indeed the result of a change in incremental gain as a function of instantaneous signal amplitude.

The gain of an LTP is typically highest in its balanced state and decreases as the signal goes positive or negative away from the balance point. This symmetrical behavior is in contrast to the asymmetrical behavior of the common-emitter stage, where the gain increases with signal swing in one direction and decreases with signal swing in the other direction. To first order, the symmetrical distortion here is third harmonic distortion, while that of the CE stage is predominantly second harmonic distortion.

Notice that the differential input voltage needed to cause the above imbalance in the LTP is only on the order of 25 mV. This means that it is fairly easy to overload an LTP that does not incorporate emitter degeneration. This is of great importance in the design of most power amplifiers that employ an LTP input stage.

Suppose the LTP is pushed to 90% of its output capability. In this case Q1 would be conducting 0.1 mA and Q2 would be conducting 1.9 mA. The two values of *re'* will be 260 Ω and 14 Ω, for a total of 274 Ω. The gain of the stage is now reduced to 5000/274 = 18.25. The nominal gain of this un-degenerated LTP was about 96.2. The incremental gain under these large signal conditions is down by about 80%, implying gross distortion.

As in the case of the CE stage, adding emitter degeneration to the LTP will substantially reduce its distortion while also reducing its gain. In summary we have the approximation

$$\text{Gain} = R_{L1}/2 * (re' + R_e) \tag{2.16}$$

where R_{L1} is a single-ended collector load resistance and R_e is the value of external emitter degeneration resistance in each emitter of the differential pair. This gain is for the case where only a single-ended output is taken from the collector of $Q1$. If a differential output is taken from across the collectors of $Q1$ and $Q2$, the gain will be doubled. For convenience, the total emitter-to-emitter resistance in an LTP, including the intrinsic re' resistances, will be called R_{LTP}. In the example above,

$$R_{LTP} = 2(re' + R_e) \qquad (2.17)$$

Emitter Follower

The emitter follower (EF) is essentially a unity voltage gain amplifier that provides current gain. It is most often used as a buffer stage, permitting the high impedance output of a CE or LTP stage to drive a heavier load.

The emitter follower is illustrated in Figure 2.9a. It is also called a common collector (CC) stage because the collector is connected to an AC ground. The output pull-down resistor R1 establishes a fairly constant operating collector current in Q1. For illustration, a load resistor R2 is being driven through a coupling capacitor. For AC signals, the net load resistance R_L at the emitter of Q1 is the parallel combination of R1 and R2. If re' of Q1 is small compared to R_L, virtually all of the signal voltage applied to the base of Q1 will appear at the emitter, and the voltage gain of the emitter follower will be nearly unity.

The signal current in the emitter will be equal to V_{out}/R_L, while the signal current in the base of Q1 will be this amount divided by the β of the transistor. It is immediately apparent that the input impedance seen looking into the base of Q1 is equal to the impedance of the load multiplied by the current gain of Q1. This is the most important function of the emitter follower.

As mentioned above, the voltage gain of the emitter follower is nearly unity. Suppose R1 is 9.4 kΩ and the transistor bias current is 1 mA. The intrinsic emitter resistance re' will then be about 26 Ω. Suppose R2 is 1 kΩ, making net R_L equal to 904 Ω. The voltage gain of the emitter follower is then approximately

$$G = R_L/(R_L + re') = 0.97$$

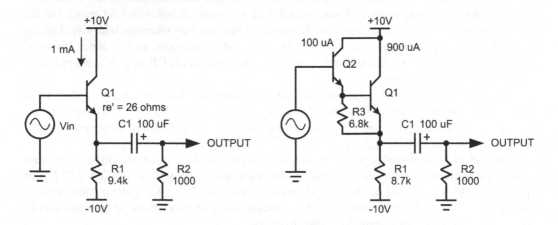

(a) conventional emitter follower   (b) Darlington emitter follower

Figure 2.9 Emitter Followers

At larger voltage swings the instantaneous collector current of Q1 will change with signal, causing a change in re'. This will result in a change in incremental gain that corresponds to distortion. Suppose the signal current in the emitter is 0.9 mA peak in each direction, resulting in an output voltage of about 814 mV peak. At the negative peak swing, emitter current is only 0.1 mA and re' has risen to 260 Ω. Incremental gain is down to about 0.78. At the positive peak swing the emitter current is 1.9 mA and re' has fallen to 13.7 Ω; this results in a voltage gain of 0.985.

Voltage gain has thus changed by about 21% over the voltage swing excursion; this causes considerable second harmonic distortion. One solution to this is to reduce R1 so that a greater amount of bias current flows, making re' a smaller part of the gain equation. This of course also reduces net R_L somewhat. A better solution is to replace R1 with a constant current source.

The transformation of low-value load impedance to much higher input impedance by the emitter follower is a function of the current gain of the transistor. The β is a function of frequency, as dictated by the f_T of the transistor. This means, for example, that a resistive load will be transformed to impedance at the input of the emitter follower that eventually begins to decrease with frequency as β_{AC} decreases with frequency. A transistor with a nominal β of 100 and f_T of 100 MHz will have an f_β of 1 MHz. The AC β of the transistor will begin to drop at 1 MHz. The decreasing input impedance of the emitter follower thus looks capacitive in nature, and the phase of the input current will lead the phase of the voltage by an amount approaching 90 degrees.

The impedance transformation works both ways. Suppose we have an emitter follower that is driven by a source impedance of 1 kΩ. The low-frequency output impedance of the EF will then be approximately 1 kΩ divided by β, or about 10 Ω. However, the output impedance will begin to rise above 1 MHz where β begins to fall. Impedance that increases with frequency is inductive. Thus, Z_{out} of an emitter follower tends to be inductive at high frequencies.

Now consider an emitter follower that is loaded by a capacitance. This can lead to instability, as we will see. The load impedance presented by the capacitance falls with increasing frequency. The amount by which this load impedance is multiplied by β_{AC} also falls with frequencies above 1 MHz. This means that the input impedance of the emitter follower is ultimately falling with the square of frequency. It also means that the current in the load, already leading the voltage by 90 degrees, will be further transformed by another 90 degrees by the falling transistor current gain with frequency. This means that the input current of the emitter follower will lead the voltage by an amount approaching 180 degrees. When current is 180 degrees out of phase with voltage, this corresponds to a *negative resistance*. This can lead to instability, since the input impedance of this emitter follower is a frequency-dependent negative resistance under these conditions. This explains why placing a resistance in series with the base of an emitter follower will sometimes stabilize it; the positive resistance adds to the negative resistance by an amount that is sufficient to make the net resistance positive.

There is one more aspect of emitter follower behavior that pertains largely to its use in the output stage of a power amplifier. It was implied above that if an emitter follower was driven from a very low impedance source that its output impedance would simply be re' of its transistor. This is not quite the whole story. Transistors have finite base resistance. The output impedance of an emitter follower will actually be the value of re' plus the value of the base resistance divided by β of the transistor. This can be significant in an output stage. Consider a power transistor operating at 100 mA. Its re' will be about 0.26 Ω. Suppose that transistor has a base resistance of 5 Ω and a current gain of 50. The value of the transformed base resistance will be 0.1 Ω. This is not insignificant compared to the value of re' and must be taken into account in some aspects of design. This can also be said for the emitter resistance of the power transistor, which may range from 0.01 Ω to 0.1 Ω.

Figure 2.9b shows an emitter follower employing a Darlington transistor pair. The input impedance of that emitter follower pair, with the load shown, could be as high as 10 MΩ or more. The simplicity of the emitter follower, combined with its great ability to buffer a load, accounts for it being the most common type of circuit used for the output stage of power amplifiers. An emitter follower will often be used to drive a second emitter follower in a Darlington arrangement to achieve even larger amounts of current gain and buffering. Such a pair of transistors, each with a current gain of 50, can increase the impedance seen driving a load by a factor of 2500. Such an output stage driving an 8-Ω load would present an input impedance of 20,000 Ω.

Cascode

A cascode stage is implemented by Q2 in Figure 2.10a. The cascode stage is also called a common base stage because the base of its transistor is connected to AC ground. Here the cascode is being driven at its emitter by a CE stage comprising Q1. The most important function of a cascode stage is to provide isolation. It provides near-unity current gain, but can provide very substantial voltage gain. In some ways it is like the dual of an emitter follower.

A key benefit of the cascode stage is that it largely keeps the collector of the driving CE stage at a constant potential. It thus isolates the collector of the CE stage from the large swing of the output signal. This eliminates most of the effect of the collector-base capacitance of Q2, resulting in wider bandwidth due to suppression of the Miller effect. Similarly, it mitigates distortion caused by the nonlinear collector-base junction capacitance of the CE stage, since very little voltage swing now appears across the collector-base junction to modulate its capacitance.

The cascode connection also avoids most of the Early effect in the CE stage by nearly eliminating signal swing at its collector. A small amount of Early effect remains, however, because the signal swing at the base of the CE stage modulates the collector-base voltage slightly.

If the current gain of the cascode transistor is 100, then 99% of the signal current entering the emitter will flow in the collector. The input-output current gain is thus 0.99. This current transfer factor from emitter to collector is sometimes referred to as the *alpha* of the transistor.

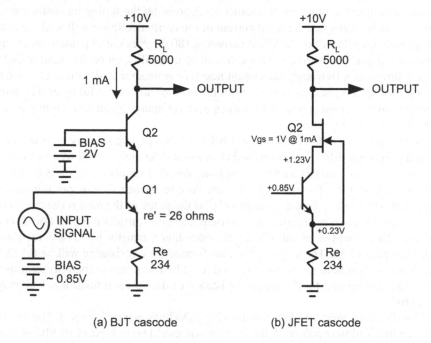

(a) BJT cascode   (b) JFET cascode

Figure 2.10 Cascodes

The Early effect resistance *ro* still exists in the cascode transistor. It is represented as a resistance *ro* connected from collector to emitter. Suppose *ro* is only 10 kΩ. Is the output impedance of the collector of the cascode 10 kΩ? No, it is not.

Recall that 99% of the signal current entering the emitter of the cascode re-appears at the collector. This means that 99% of the current flowing in *ro* also returns to the collector. Only the lost 1% of the current in *ro* results in a change in the net collector current at the collector terminal. This means that the net effect of *ro* on the collector output impedance in the cascode is roughly like that of a 1 MΩ resistor to ground. This is why the output impedance of cascode stages is so high even though Early effect still is present in the cascode transistor.

$$R_{out} = \beta * ro \tag{2.18}$$

$$ro = (VA + V_{ce})/I_c \tag{2.19}$$

$$ro \approx VA/I_c \quad \text{at low } V_{ce} \tag{2.20}$$

$$R_{out} = \beta * VA/I_c \tag{2.21}$$

Notice that the product of β and *VA* is the Early effect figure of merit mentioned previously. The output resistance of a cascode is thus the FOM divided by the collector current.

$$R_{out} = FOM/I_c \tag{2.22}$$

Figure 2.10b illustrates a JFET cascode. Because the N-channel JFET will have its source positive with respect to its gate, V_{ce} for Q1 will equal V_{gs} for the JFET, here on the order of 1 V. Indeed, V_{ce} for Q1 is nearly constant if the signal current does not change V_{gs} very much. With Q2 operating in pinch-off, the output impedance will be extremely high. This is especially advantageous in a high-gain stage with a current-source load. Unfortunately, the JFET cascode requires careful JFET selection and suffers from the wide variability of threshold voltage among JFETs of the same part number. A JFET with high threshold voltage will eat into negative swing headroom, while a JFET with low threshold voltage will deprive Q1 of adequate collector voltage.

2.5 Current Mirrors

Figure 2.11a depicts a very useful circuit called a current mirror [12]. If a given amount of current is sourced into Q1, that same amount of current will be sunk by Q2, assuming that the emitter degeneration resistors R1 and R2 are equal, that the transistor V_{be} drops are the same and that losses through base currents can be ignored. The values of R1 and R2 will often be selected to drop about 100 mV to ensure decent matching in the face of unmatched transistor V_{be} drops, but this is not critical.

If R1 and R2 are made different, a larger or smaller multiple of the input current can be made to flow in the collector of Q2. In practice, the base currents of Q1 and Q2 cause a small error in the output current with respect to the input current. In the example above, if transistor β is 100, the base current I_b of each transistor will be 50 μA, causing a total error of 100 μA, or 2% in the output current.

Figure 2.11b shows a variation of the current mirror that minimizes errors due to the finite current gain of the transistors. Here emitter follower Q3, often called a *helper* transistor, provides current gain to minimize that error. Optional resistor R3 assures that a small minimum amount of current flows in Q3 even if the current gains of Q1 and Q2 are very high. Sometimes a small

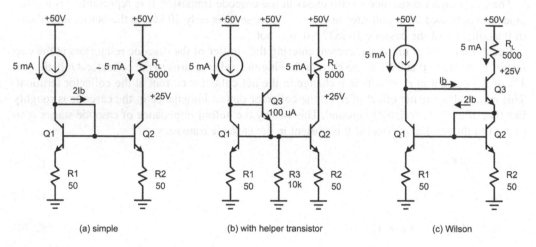

(a) simple   (b) with helper transistor   (c) Wilson

Figure 2.11 Current Mirrors

capacitor, on the order of 470 pF in the example shown, is connected from base to emitter of helper transistor Q3 to reduce high-frequency peaking by reducing the effective f_T of Q3. Note that the input node of the current mirror now sits one V_{be} higher above the supply rail than in Figure 2.11a.

Many other variations of current mirrors exist, such as the *Wilson* current mirror shown in Figure 2.11c [13]. The Wilson current mirror includes transistors Q1, Q2 and Q3. Input current is applied to the base of Q3 and is largely balanced by current flowing in the collector of Q1. Input current that flows into the base of output transistor Q3 will turn Q3 on, with its emitter current flowing through Q2 and R2. Q1 and Q2 form a conventional current mirror. The emitter current of Q3 is mirrored and pulled from the source of input current by Q1.

Any difference between the current of Q1 and the input current is available to drive the base of Q3. If the input current exceeds the mirrored emitter current of Q3, the base voltage of Q3 will increase, causing the emitter current of Q3 to increase and self-correct the situation with feedback action. The equilibrium condition can be seen to be when the input current and the output current are the same, providing an overall 1:1 current mirror function.

Notice that in normal operation all three of the transistors operate at essentially the same current, namely the supplied input current. Ignoring the Early effect, all of the base currents will be the same if the betas are matched. Assume that each base current is I_b and that the collector current in Q1 is equal to I. It can be quickly seen that the input current must then be $I + I_b$ and that the emitter current of Q3 must be $I + 2I_b$. It is then evident that the collector current of Q3, which is the output current, will be $I + I_b$, which is the same as the input current. This illustrates the precision of the input-output relationship when the transistors are matched.

Transistor Q3 acts much like a cascode, and this helps the Wilson current mirror to achieve high output impedance. Transistors Q1 and Q2 operate at a low collector voltage, while output transistor Q3 will normally operate at a higher collector voltage. Thus, the Early effect will cause the base current of Q3 to be smaller, and this will result in a slightly higher voltage-dependent output current. This is reflected in the output resistance of the Wilson current mirror. Operation of the Wilson current mirror presumes that the β of all three transistors is the same, as it largely would be on an integrated circuit chip. This may not be the case in discrete circuits, and the Wilson current mirror might not be as effective in discrete applications.

2.6 Current Sources and Voltage References

Current sources are used in many different ways in a power amplifier, and there are many different ways to make a current source. The distinguishing feature of a current source is that it is an element through which a current flows wherein that current is independent of the voltage across that element. The current source in the tail of the differential pair is a good example of its use.

Most current sources are based on the observation that if a known voltage is impressed across a resistor, a known current will flow. A simple current source is shown in Figure 2.12a. The voltage divider composed of R2 and R3 places 2.7 V at the base of Q1. After a V_{be} drop of 0.7 V, about 2 V is impressed across emitter resistor R1. If R1 is a 400-Ω resistor, 5 mA will flow in R1 and very nearly 5 mA will flow in the collector of Q1. The collector current of Q1 will be largely independent of the voltage at the collector of Q1, so the circuit behaves as a decent current source. The load resistance R_L is just shown for purposes of illustration. The output impedance of the current source itself (not including the shunting effect of R_L) will be determined largely by the Early effect in the same way as for the CE stage. Early effect in Q1 causes changes in Q1's base current with Q1's collector voltage, which in turn changes the voltage at the base because of the impedance of the R2–R3 voltage divider. Output impedance will be increased if smaller values for R2 and R3 are used. The output impedance for this current source is found by SPICE simulation to be about 290 kΩ.

In Figure 2.12b, R3 is replaced with a pair of silicon diodes. Here roughly one diode drop is impressed across R1 to generate the desired current. The circuit employs 1N4148 diodes biased with the same 0.5 mA used in the voltage divider in the first example. Together they drop only about 1.1 V, and about 0.38 V is impressed across the 75-Ω resistor R1. The output impedance of this current source is approximately 300 kΩ, about the same as the one above.

Turning to Figure 2.12c, R3 is replaced instead with a Green LED, providing a convenient voltage reference of about 1.8 V, putting about 1.1 V across R1 [14]. Once again, 0.5 mA is used to bias the LED. The output impedance of this current source is about 750 kΩ. It is higher than in the design of Figure 2.12b because there is effectively more emitter degeneration for Q1 with the larger value of R1. The temperature coefficient of the forward voltage drop of the LED is about −2.0 mV/°C. This almost completely negates the negative TC of the V_{be} of Q1, making

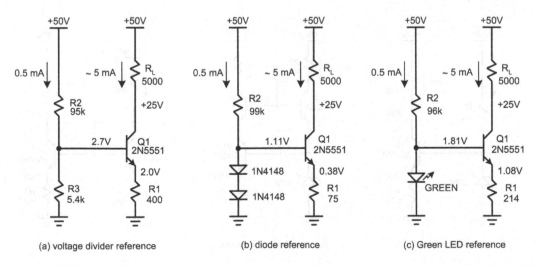

(a) voltage divider reference   (b) diode reference   (c) Green LED reference

Figure 2.12 Current Sources

the current source quite temperature-stable [15]. The LED can act as a photo-sensor, so in some circuits and situations hum can be introduced if the LED is exposed to room lighting.

R3 is replaced with a 6.2-V Zener diode in Figure 2.13a. This puts about 5.5 V across R1. The output impedance of this current source is about 2 MΩ, quite a bit higher than the earlier arrangements due to the larger emitter degeneration for Q1. The price paid here is that the base of the transistor is fully 6.2 V above the supply rail, reducing headroom in some applications. The positive temperature coefficient of the 6.2V-Zener diode and the negative TC of Q1's V_{be} make the current source have an overall positive TC. However, Zener diodes are noisy, so the Zener should be bypassed with a capacitor or filtered with an R-C network before application to the base of Q1. The latter approach is much more effective because the impedance of the Zener is quite low, and would therefore require a larger bypass capacitor.

It should be noted that Zener diodes that operate above about 5.6 V are not true Zeners, but rather are operating as avalanche diodes. This explains their positive TC and their noisy attributes. The positive TC will tend to cancel the TC of a BJT that is in series with it. The predominant breakdown mechanism below about 5.6 V is the Zener breakdown, and it has a negative TC [16].

The Zener-based current source is conveniently cascoded by adding two transistors, as shown in Figure 2.13b. This results in extremely high output impedance.

In Figure 2.13c, a current mirror fed from a known supply voltage is used to implement a current source. Here a 1:1 current mirror is used and 5 mA is supplied from the known power rail. The output impedance of this current source is about 230 kΩ. Only 0.25 V is dropped across R1 (corresponding to 10:1 emitter degeneration), and the base is only 1 V above the rail. Output impedance will be increased if larger values are used for R1 and R3, but this will come at the expense of larger voltage drop in the mirror emitter circuits and less voltage headroom at the output of the mirror.

Figure 2.14a illustrates a clever two-transistor feedback circuit that is used to force one V_{be} of voltage drop across R1. It does so by using transistor Q2 to effectively regulate the current of Q1. If the current of Q1 is too large, Q2 will be turned on harder and pull down on the base of Q1, adjusting its current downward appropriately. A current of 0.5 mA is supplied to bias the current source. This current flows through Q2. The output impedance of this current source is an impressive 3 MΩ, similar to that which a cascode would yield. This circuit achieves higher output impedance than the Zener-based version and yet only requires the base of Q1 to be 1.4 V above the power rail.

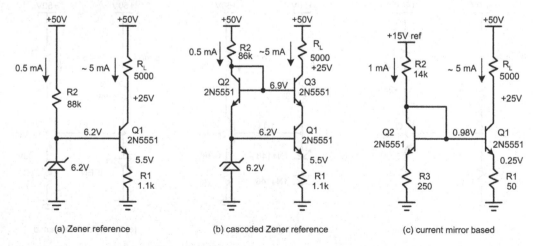

(a) Zener reference   (b) cascoded Zener reference   (c) current mirror based

Figure 2.13 Current Sources (Continued)

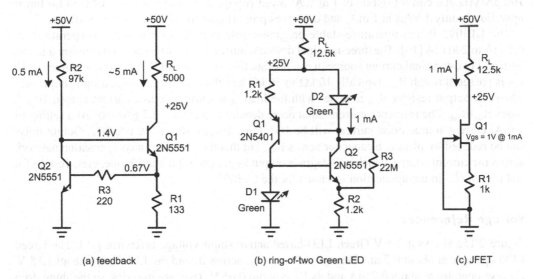

Figure 2.14 Current Sources (Continued)

This circuit will work satisfactorily even if less than 0.5 mA (one-tenth the output current) is supplied as bias for Q2, but then the output impedance will fall to a lower value and the "quality" of the current source will suffer somewhat. This happens because at lower collector current, Q2 has less transconductance and its feedback control of the current variations in Q1 as a result of the Early effect is less strong. If the bias current is reduced to 0.1 mA, for example, the output impedance falls to about 1 MΩ.

This circuit can also be used to place an over-current limit on a CE transistor stage implemented with Q1. This is sometimes done in the VAS of a power amplifier. However, if the nominal operating voltage at the emitter of the CE stage is enough to cause a bit of conduction in Q2, this can make Q2 appear to have a rather nonlinear effective current gain, leading to distortion. Because it is a feedback circuit, albeit with a tight loop of only two transistors, stability must be considered. Sometimes a small resistor, R3, is included in series with the base of Q2 to increase the phase margin of the loop.

Figure 2.14b illustrates what is called a ring-of-two current source [17]. It uses two Green LEDs in a positive feedback arrangement while implementing a two-terminal current source that can float. The current that biases D1 is itself regulated and made quite temperature-independent by D2 and Q1, as in Figure 2.12c. Notice that the circuit requires a start-up resistor, R3. Interestingly, the current flowing in R3 flows back into the emitter of Q2, greatly mitigating the effect of R3 on current source impedance.

A JFET current source is shown in Figure 2.14c. At a given current within its operating range, an N-channel JFET will have its source positive by V_{gs} with respect to its gate. If V_{gs} is 1 V at $I_d = 1$ mA, 1 V will appear across R1, here 1 kΩ. Since Q1 will be operating in its pinch-off region, output impedance will be very high. Unfortunately, the wide variation in threshold voltage and I_{DSS} among JFETs of the same type results in wide variation of the current.

Numerous three-terminal and two-terminal IC current sources are available as well, but most seem to have some shortcomings. The LM134 is a three-terminal programmable reference that applies $V_{set} = 64$ mV across a current-setting resistor to establish the current value [18]. V_{set} is proportional to absolute temperature (PTAT), so temperature stability is not great. It can operate with as little as 1 V across it. Because of the way that it operates, the LM134 is fairly noisy, producing

100 pA/√Hz at a current setting of 1 mA. Allowed voltage slew rate across the LM134 for linear operation is only 1 V/μs at 1 mA, and decreases proportionately at lower current settings.

The LT3092 is a temperature-stable programmable current source that can operate from 0.5 mA to 200 mA [19]. The three-terminal device requires two resistors to set its current and can float as a two-terminal current source. It can operate from 1.2 V to 40 V. A reference current of 10 μA is passed through R_{set}, typically 10 kΩ to 50 kΩ and the resulting voltage drop is impressed across an output resistor R_{out} in series with the load. The current source current is then 10 μA times R_{set}/R_{out}. The reference current input noise density is typically 2.7 pA/√Hz. At a setting of 1 mA, current source noise current will be 100 times this, or about 270 pA/√Hz. Output noise can be reduced by placing a capacitor across R_{set}, but this then requires a compensation network across the current source, which will degrade output impedance at higher frequencies. Testing for stability *in situ* in the application is a must for the LT3092.

Voltage References

Figure 2.15a shows a 2.5 V Green LED-based active shunt voltage reference [15]. The Green LED operates at about 0.7 mA because R1 has 1 V_{be} across it, and the LED drops about 1.85 V. Q1 also operates at about 0.7 mA and its V_{be} is about 0.65 V. They are in series, so the shunt drop is about 2.5 V. R3 is an optional trimming resistor. Both the LED and V_{be} have similar negative temperature coefficients, so the TC of the reference is about −4 mV/°C [16, 20]. The action of Q1 and Q2 greatly reduces the impedance of the shunt reference. A small increase in applied voltage will cause the collector current of Q1 to increase and that of Q2 to increase greatly, implying that the output impedance is very low. Put another way, a small increase in applied current will flow mainly through Q1 to its collector, where it will flow into the base of Q2 and be amplified. Q2 will increase its current as much as necessary to keep the voltage drop across the shunt reference from increasing by all but the smallest amount.

Figure 2.15b shows a Zener-based 6.9-V shunt regulator that works on the same principles as in Figure 2.15a in order to provide very low impedance. Here the 6.2-V Zener diode is in series with the 0.7-V V_{be} of Q1 to establish the 6.9 V drop. D1 operates in avalanche mode, and it has a positive temperature coefficient of about +2.0 mV/°C, which largely cancels the V_{be} TC of Q1, resulting in a temperature-stable voltage reference.

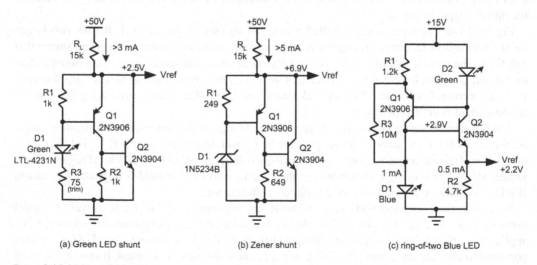

(a) Green LED shunt   (b) Zener shunt   (c) ring-of-two Blue LED

Figure 2.15 Voltage References

A so-called *ring-of-two* voltage reference is shown in Figure 2.15c [17, 21]. Its operation is similar to that of the ring-of-two current source depicted in Figure 2.14b. Here a Blue LED with voltage drop of about 2.9 V at 1 mA is used. When the V_{be} of Q2 is subtracted, the resulting reference voltage across R2 is about 2.2 V. The Blue LED has a negative TC of about −4 mV/°C, so the reference TC is about −1.8 mV/°C. If an additional V_{be} is placed in series with the emitter of Q2, the reference voltage across R2 would drop to about 1.5 V and the TC would go positive to about +0.4 mV/°C.

Several IC voltage references are available as well. The LM329 is a fixed 6.9-V precision shunt reference that acts like a precision Zener diode with very low dynamic resistance of only 1 Ω up to 8 kHz [22]. It can operate from 0.6 mA to 15 mA. The LM329 has a moderate noise level of 75 nV/√Hz.

The TL431 is a programmable shunt reference for voltage values ranging from 2.5 V to 36 V [23]. A simple resistive voltage divider that connects to its V_{ref} terminal sets its voltage. The device operates by shunting current as necessary to keep its V_{ref} terminal at its reference voltage of 2.5 V. For a 2.5-V reference, the V_{ref} terminal is simply connected to its positive terminal. The device can operate from 1 mA to 100 mA. Its dynamic impedance is typically less than 0.4 Ω up to 50 kHz. The temperature coefficient of its reference voltage is about −0.25 mV/°C. The TL431 is fairly noisy, with input-referred noise at its V_{ref} terminal as high as 125 nV/√Hz.

2.7 Complementary Feedback Pair (CFP)

Another important two-transistor connection that achieves current gain on the order of the product of the β of two transistors is shown in Figure 2.16a. This circuit is called a *complementary feedback pair* (CFP), also known as the *Sziklai* pair after George Sziklai of RCA. It acts like a single transistor and can usually be used as such. Unlike the Darlington, its turn-on voltage is only 1 V_{be}. Like the Darlington, its saturation voltage is on the order of 0.8 V. It is easy to see that collector current in Q1 is multiplied by β of Q2. The CFP is a feedback circuit; Q2 will conduct as much current as necessary to keep Q1's current equal to approximately V_{be} of Q2 divided by

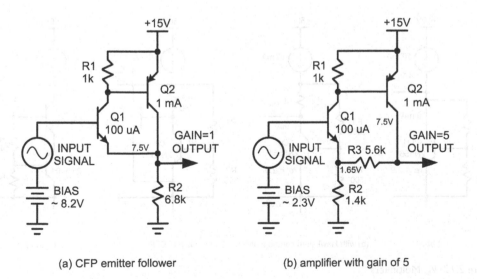

(a) CFP emitter follower   (b) amplifier with gain of 5

Figure 2.16 Complementary Feedback Pair

R1 (here about 680 μA) plus the collector current of Q2 divided by β of Q2. The feedback process greatly increases the effective transconductance of Q1 by a factor on the order of β of Q2. The CFP implements an emitter follower in Figure 2.16a. The very high gm of the CFP greatly reduces the distortion of the emitter follower that can result from gm variations as a function of signal. Because the CFP is a feedback circuit, oscillation is possible under some conditions. As with a simple Darlington, Q1 can operate at very small collector current, on the order of 10 μA if Q2 is operating at 1 mA with β of 100. This can cause Q1 to be slow and can make turn-off of Q2 slow. Resistor R1 serves to set a minimum current for Q1, just as in the Darlington.

The CFP can be connected to provide gain, as shown in Figure 2.16b, where the circuit is set for a gain of 5. The combination of R3 and R2 implements a simple 5:1 voltage divider in the feedback path to the emitter of Q1. Because Q1 forces the input signal to appear at its emitter, if the output is not 5 times the input, an error current will flow in Q1 that will correct the output to the proper 5X replica of the input signal. Because R3 creates an error *current*, this arrangement implements a very simplified version of a so-called *current feedback amplifier* (CFA).

2.8 V_{be} Multiplier

Figure 2.17a shows what is called a V_{be} *multiplier*. This circuit is used when a voltage drop equal to some multiple of V_{be} drops is needed. This circuit is most often used as the bias spreader for power amplifier output stages, partly because its voltage is conveniently adjustable.

In the circuit shown, the V_{be} of Q1 is multiplied by a factor of approximately 4 because the voltage divider formed by R1 and R2 places about one-fourth of the collector voltage at the base of Q1. When the voltage at the collector is at four V_{be}, one V_{be} will be at the base, just enough to turn on Q1 by the amount necessary to carry the current supplied. Here about 1 mA flows through the resistive divider while about 9 mA flows through Q1. When the V_{be} multiplier is used as a bias spreader, R2 will be made adjustable with a trim pot. As R2 is made smaller, the amount of bias voltage is increased. Notice that if for some reason R2 fails open, the voltage across the V_{be} multiplier falls to about one V_{be}, failing in the safe direction. In practice the V_{be} multiplication ratio will be a bit greater than 4 due to the base current required by Q1.

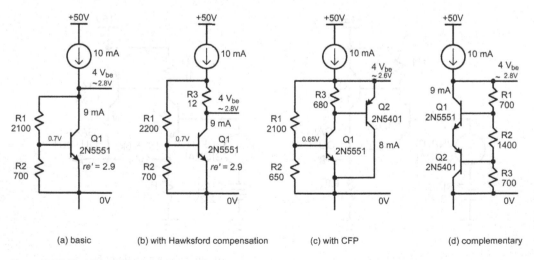

(a) basic   (b) with Hawksford compensation   (c) with CFP   (d) complementary

Figure 2.17 V_{be} Multipliers

The impedance of the V_{be} multiplier depends on the transconductance of Q1. It is about 4 times re' of Q1. At 9 mA, re' is 2.9 Ω, so ideally the impedance of the multiplier would be about 11.6 Ω. In practice, SPICE simulation shows it to be about 25 Ω. This larger value is mainly a result of the finite current gain of Q1.

The impedance of the V_{be} multiplier rises at high frequencies. This is a result of the fact that the impedance depends on a negative feedback process. The amount of feedback decreases at high frequencies and the impedance-reducing effect is lessened. The impedance of the V_{be} multiplier in Figure 2.17a is up by 3 dB at about 2.3 MHz and doubles for every octave increase in frequency thereafter. The V_{be} multiplier is usually shunted by a capacitor of 0.1 to 10 μF. A shunt capacitance of as little as 0.1 μF eliminates the increase in impedance at high frequencies.

The modified V_{be} multiplier shown in Figure 2.17b includes R3 to reduce the effective impedance of the circuit [24, 25, 26, 27]. As the current applied to the V_{be} multiplier increases, the total drop across the V_{be} multiplier increases, but so does the drop across R3. The voltage drop across R3 acts to reduce the "output" voltage of the circuit. If the value of R3 is the same as the DC impedance of the V_{be} multiplier, the change in output voltage with applied current will be nearly compensated.

A CFP V_{be} multiplier is shown in Figure 2.17c. The very high effective transconductance of the CFP compound transistor greatly reduces the impedance of the V_{be} multiplier.

A complementary V_{be} multiplier is shown in Figure 2.17d. Here an NPN and a PNP transistor are connected in series with a single three-resistor voltage-dividing network. In some amplifier designs, one of the transistors will be mounted on the heat sink to provide temperature compensation of a fraction of the total voltage drop, while the other transistor responds only to the ambient temperature.

2.9 Operational Amplifiers

The very simple discrete operational amplifier of Figure 2.18 can be implemented by a simple combination of some of the circuit building blocks that have been discussed thus far. Q1 and Q2 form a differential input stage (IPS) loaded by a current mirror built with Q4 and Q5. The 200 μA tail current for the IPS is provided by Q3, which is part of a feedback current source formed with Q4. The voltage amplifier stage (VAS) is formed by a Darlington common-emitter transistor loaded by a 500-μA feedback current source consisting of Q7 and Q8. The output stage (OPS) is a complementary emitter follower biased at 2 mA by the voltage drops of diode-connected transistors Q11 and Q12 and R8. Compensation capacitor C1 (100 pF) sets the gain-bandwidth to about 2.5 MHz and allows slew rate of about 1.8 V/μs to be achieved. Open-loop gain is simulated to be over 100 dB at DC.

2.10 Amplifier Design Analysis

Here we apply the understanding of transistors and circuit building blocks to analyze the basic power amplifier. Having accomplished this, we will be well armed to explore, evolve and analyze the amplifier design steps that will be taken to achieve high performance in the next chapter.

Figure 2.19 is a schematic of a basic 50-W power amplifier that includes the three stages that appear in most solid-state power amplifiers.

- Differential input stage (IPS) comprising Q1–Q3
- Voltage amplification stage (VAS) comprising Q4, Q6 and Q7
- Output stage (OPS) comprising Q8–Q11

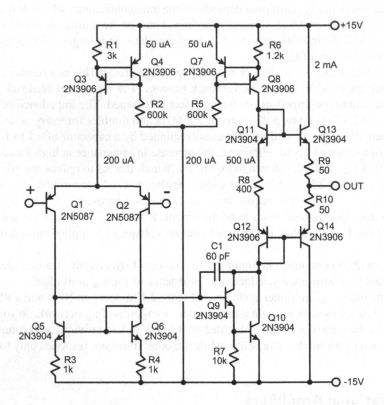

Figure 2.18 Simple Operational Amplifier

The design also includes a bias spreader implemented with Q5 connected as a V_{be} multiplier. Some details like coupling capacitors, input networks and output networks have been left out for simplicity.

The amplifier of Figure 2.19 will be described in simple terms, so that those who are less familiar with circuit design will quickly come to understand its behavior. Those who are already familiar with these concepts can relax and skim through this section.

Basic Operation

This simple design is a more detailed version of the basic amplifier design illustrated in Figure 1.5. As shown, it is a DC-coupled design, so that even DC changes at the input will be amplified and presented at the output.

Input Stage

The input signal is applied to the input differential pair at the base of Q1. A fraction of the output signal is coupled via the negative feedback path to the other differential input at the base of Q2. Transistor Q3 implements a 2-mA current source that provides tail current to the differential pair. This input arrangement is often called a *long-tailed pair* (LTP).

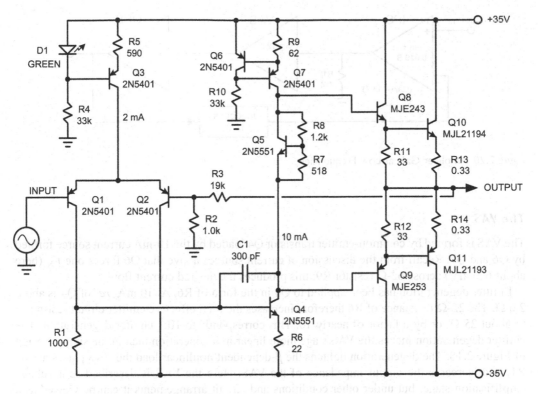

Figure 2.19 A Basic 50-W Power Amplifier

Feedback resistors R2 and R3 implement a voltage divider that feeds back 1/20 of the output signal to the input stage. The forward path gain of the amplifier in the absence of negative feedback is called the *open-loop gain* (A_{ol}). If the open-loop gain is large, then the error signal across the bases of Q1 and Q2 need only be very small to produce the desired output. If the signals at the bases of Q1 and Q2 are nearly equal, then the output of the amplifier must be 20 times that of the input, resulting in a *closed-loop gain* (A_{cl}) of 20. This is just a very simplified explanation of the negative feedback process.

The approximate low-frequency gain of the input stage is the ratio of the net collector load resistance divided by the total emitter-to-emitter resistance R_{LTP} (which includes the intrinsic emitter resistance *re'* of Q1 and Q2). With each transistor biased at 1 mA, the intrinsic emitter resistance *re'* is about 26 Ω each, so the total emitter-to-emitter resistance is 52 Ω. If we assume that β of the following VAS transistor Q4 is infinity, the net IPS collector load resistance is just that of R1. The DC gain of the IPS is then 1000/52 = 19. In practice the finite β of Q4 reduces this to about 13.7 if we assume that β of Q4 is 100.

IPS gain ≈ 13.7

The amplifier at low frequencies is illustrated in Figure 2.20 where the input stage is shown as a block of transconductance with *gm* = 1/52 Ω = 0.019 S. The R1' load of 714 Ω on the IPS is just the parallel combination of R1 and the estimated input impedance of the VAS.

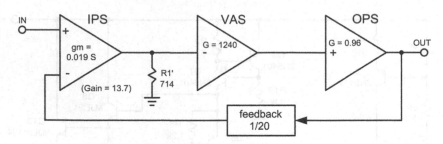

Figure 2.20 Amplifier Gain at Low Frequencies

The VAS

The VAS is formed by common-emitter transistor Q4 loaded by the 10-mA current source formed by Q6 and Q7. Recall from the discussion of current sources above that Q6 forces one V_{be} (here about 620 mV) across 62-Ω resistor R9; this produces the desired current flow.

Emitter degeneration has been applied to Q4 in the form of R6. At 10 mA, re' of Q4 is about 2.6 Ω. The 22-Ω resistance of R6 therefore increases the total effective emitter circuit resistance to about 25 Ω, or by a factor of nearly 10. This corresponds to 10:1 emitter degeneration. The emitter degeneration makes the VAS stage more linear in its operation than in the simple design of Figure 2.18. The degeneration lightens the β-dependent nonlinear load the VAS places across R1, and increases the output impedance of the VAS. Here the VAS is described as a voltage amplification stage, but under other conditions and circuit arrangements it can be viewed as a current gain stage at low frequencies or as a transimpedance stage (current in, voltage out) at high frequencies when the Miller compensation provided by C1 is in effect.

If the β of Q4 is assumed to be 100, the input impedance of the VAS will be about 100 * 25 Ω = 2500 Ω. This impedance is in parallel with R1, making the actual load on the first stage approximately 714 Ω. The voltage gain of the input stage is therefore close to 13.7. The loading of Q4 thus plays a substantial role in determining the first-stage gain. It would play a far greater role if Q4 were not degenerated by R6. In that case the impedance seen looking into the base of Q4 would be only about 260 Ω (β = 100, times re' = 2.6 Ω).

The low-frequency gain of the VAS will be set by the ratio of its collector load impedance to its total effective emitter resistance of about 25 Ω. The VAS collector load impedance in this simple design is dominated by the loading of the output stage, since the output impedance of the current source is quite high.

The output stage is a complementary Darlington arrangement, buffering the loudspeaker load impedance as seen by the VAS collector circuit. The amount of buffering depends on the current gain of the driver and output transistors. Throughout these discussions it will be assumed that small-signal transistors have current gain of 100 and power transistors have current gain of 50. If the driver and output transistor current gains are assumed to be 100 and 50 respectively, the buffering factor will be the product of these, or about 5000. If the load impedance at the output is assumed to be 8 Ω, this will appear as a load impedance of 40,000 Ω to the VAS collector circuit. The gain of the VAS is thus on the order of 40,000/25 = 1600. This simple analysis based on transistor current gain ignores the loading effect due to signal current flowing in R11 and R12. The gain of the output stage is less than unity, so these transistors are not fully bootstrapped. Their effect in Figure 2.19 could reduce the load impedance on the VAS by about a factor of 2. In some driver stages, R11 and R12 are replaced with a single resistor connecting the emitters of Q8 and Q9 to establish driver bias current. In this case, the additional loading effect described above is largely eliminated.

In the discussion on the Early effect in Section 2.1 it was shown that a common-emitter stage like the VAS used here (10 mA bias, 10:1 emitter degeneration) has an intrinsic output impedance of about 135 kΩ. Recall that the output resistance ro for a transistor is

$$ro = (VA + V_{ce})/I_c$$

and that R_{out} for a common-emitter stage with degeneration is approximated by

$$R_{out} = ro * \text{degeneration factor}$$

The Early voltage for the 2N5551 is about 100 V. At a collector current of 10 mA and a V_{ce} of 35 V, ro for the 2N5551 will be about 13.5 kΩ. With 10:1 degeneration, R_{out} of the CE stage will be about 135 kΩ, or about 3.4 times that of the load imposed by the output stage.

The net collector load impedance is the parallel combination of the 135-kΩ Early effect resistance and the 40-kΩ effective external load resistance, or 31 kΩ. The output resistance of the current source has been ignored because it is much higher. The estimated voltage gain of the VAS is equal to 31 kΩ/25 = 1240.

VAS gain ≈ 1240

This VAS voltage gain of 1240 is shown in Figure 2.20.

Open-Loop Gain

If the voltage gain of the output stage is approximately unity, the forward gain of the amplifier (without considering negative feedback) is the product of the input stage gain and the VAS gain, or about 13.7 * 1240 ≈ 17,000. Recall that we refer to this as the open-loop gain A_{ol}. In reality the gain of the output stage is about 0.96 when driving an 8-Ω load, so the open-loop gain is a bit less.

Open-loop gain ≈ 16,300

Because the feedback network attenuates the signal by a factor of 20, the gain around the feedback loop, or loop gain, is about 816. This corresponds to about 58 dB of negative feedback (NFB) at low frequencies.

Loop gain ≈ 816

NFB ≈ 20 log(816) ≈ 58 dB

If the amplifier is producing 20 V at its output, the error signal across the bases of Q1 and Q2 needs only to be 20/16,300 = 1.2 mV. Earlier it was asserted that the closed-loop gain would be approximately 20 on the basis of the feedback network attenuating the output by a factor of 20 and the required differential input to the amplifier being small. Because the input only needs to be 1.2 mV (compared to the input signal level of 1 V), it is apparent that this is a very good approximation.

Miller Feedback Compensation

Capacitor C1 is the so-called Miller compensation capacitor C_M. It plays a critical role in stabilizing the global negative feedback loop around the amplifier. It does this by rolling off the high-frequency gain of the amplifier so that the gain around the feedback loop falls below unity before enough phase lag builds up to cause instability. This will be discussed in much more detail in Chapter 6.

The effect of Miller compensation capacitor C1 on the open-loop gain is illustrated in Figure 2.21. Here the topology of the amplifier has effectively transitioned from that of Figure 2.20 to that of Figure 2.21 as the analysis has gone from low frequencies to high frequencies. In Figure 2.21 the combined gain of the input stage and the VAS is equal to the product of the transconductance of the IPS and the impedance of C1. At high frequencies the gain is dominated by C1 rather than by R1'.

The capacitor controls the high-frequency AC gain of the VAS by forming a shunt feedback loop around the VAS transistor. At higher frequencies, virtually all signal current from the LTP input stage flows through C1. This creates a voltage drop across C1 that becomes the output voltage of the VAS. At this point the VAS is acting like a so-called Miller integrator, where the output voltage is the integral of the input current.

While at low frequencies the gain of the combined input stage and VAS is set by the product of the individual voltage gains of those two stages, at higher frequencies that combined gain is set by the ratio of the impedance of C1 to the total emitter resistance ($1/gm$) in the LTP. That is because the signal current from the input stage is inversely proportional to the total LTP emitter resistance R_{LTP}. Since the impedance of C1 is inversely related to frequency, the gain set by it will decrease at 6 dB per octave as frequency increases. The frequency at which the AC gain based on this calculation becomes smaller than the DC gain is where the roll-off of the amplifier's open-loop frequency response begins.

Assume for the moment an operating frequency of 20 kHz. At this frequency the reactance of C_1 is $1/(2\pi * 20\ kHz * 300\ pF) = 26{,}500\ \Omega$. If all of the signal current provided by the LTP passes through C1, then the gain of the combined input and VAS stage at this frequency is

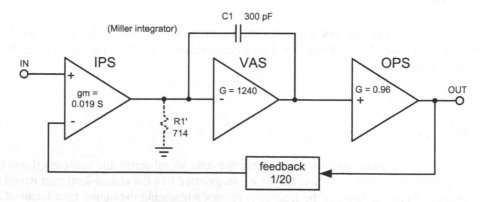

Figure 2.21 Amplifier Gain at High Frequencies

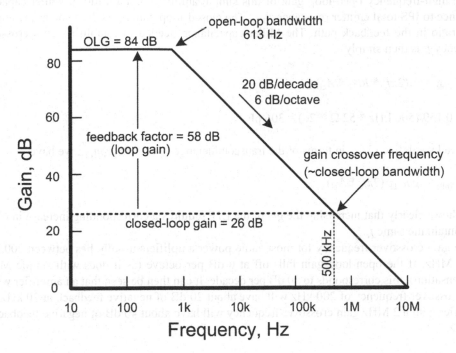

Figure 2.22 Bode Plot of the Amplifier

26,500/52 = 510. This is considerably less than the low-frequency open-loop gain of 17,000. This means that the capacitor is dominating the gain at this frequency. This further supports the validity of the assumption that essentially all signal current from the LTP flows through the capacitor at this frequency.

The frequency where the gain around the negative feedback loop becomes unity is called the *gain crossover frequency* f_c, or simply the *unity loop-gain frequency (ULGF)*. This is illustrated by Figure 2.22, which is called a *Bode plot*. It shows the various gains of the amplifier in an idealized form with straight lines.

The open-loop gain starts out at 84 dB at low frequencies and begins to fall off at 20 dB per decade starting at about 613 Hz. This frequency corresponds to the open-loop bandwidth and is the frequency where the behavior of the amplifier transitions from that of Figure 2.20 to that of Figure 2.21.

The closed-loop gain of 20, corresponding to 26 dB, is shown as the horizontal dotted line. Where it intersects the falling open-loop gain line is the *gain crossover frequency*. This also corresponds approximately to the actual closed-loop bandwidth of the amplifier. Here, that crossing occurs at 500 kHz. The distance between the closed-loop gain line and the open-loop gain curve represents the amount of negative feedback, often called *loop gain* because it is the gain around the feedback loop.

The gain crossover frequency f_c is chosen to be low enough to assure adequate feedback loop stability. There is a trade-off between stability and distortion here. Making the gain crossover frequency higher results in more negative feedback at high frequencies and less high-frequency distortion. Making the gain crossover frequency too high jeopardizes loop stability. For this amplifier f_c has been chosen to be 500 kHz as a reasonable compromise between distortion reduction and stability.

The high-frequency open-loop gain of this simple amplifier is the ratio of Miller capacitor reactance to IPS total emitter resistance R_{LTP}. The closed-loop gain A_{cl} is the same as the attenuation ratio in the feedback path. The Miller capacitance needed to establish a gain crossover frequency f_c is then simply

$$C_{Miller} = 1/(2\pi f_c * R_{LTP} * A_{cl})$$

$$= 0.159/(500 \text{ kHz} * 52 \ \Omega * 20) = 306 \text{ pF}$$

If instead we define C_{Miller} in terms of the transconductance of the IPS, gm_{LTP}, we have

$$C_{Miller} = gm_{LTP}/(2\pi f_c * A_{cl})$$

This shows clearly that increasing the gm of the LTP requires a corresponding increase in C_{Miller} to maintain the same f_c.

The gain crossover frequency for most audio power amplifiers usually lies between 200 kHz and 2 MHz. If the open-loop gain falls off at 6 dB per octave (as it does with simple Miller compensation), this corresponds to 20 dB per decade. It can then be seen that an amplifier with a gain crossover frequency of 200 kHz will have about 20 dB of negative feedback at 20 kHz. An amplifier with a 2 MHz gain crossover frequency will have about 40 dB of negative feedback at 20 kHz.

The Output Stage

The output stage is a class AB complementary Darlington arrangement consisting of emitter follower drivers Q8 and Q9 followed by output devices Q10 and Q11. Emitter resistors R11 and R12 set the idle current of the drivers at about 20 mA. The output stage emitter resistors R13 and R14 provide thermal bias stability and also play a role in controlling crossover distortion. These resistors will also be referred to as R_E. The output stage provides a voltage gain of slightly less than unity. Its main role is to buffer the output of the VAS with a large current gain. If driver transistor betas are assumed to be 100 and output transistor betas are assumed to be 50, the combined current gain of the output stage is 5000. When driving the 8-Ω load as shown, the load impedance seen by the VAS looking into the output stage will be about 40,000 Ω.

On positive half-cycles of the signal, Q8 and Q10 conduct current and transport the signal to the output node by sourcing current into the load. On negative half-cycles, Q9 and Q11 conduct current and transport the signal to the output node by sinking current from the load. When there is no signal, a small idle bias current of approximately 100 mA flows from the top NPN output transistor through the bottom PNP output transistor. As will be seen shortly, the value of 100 mA discussed here is on the high side of the theoretical optimum, making the output stage a bit over-biased. A key observation is that the signal takes a different path through the output stage on positive and negative half-cycles. If the voltage or current gains of the top and bottom parts of the output stage are different, distortion will result. Moreover, the "splice point" where the signal current passes through zero and crosses from one path to the other can be tricky, and this can lead to so-called crossover distortion.

The voltage gain of the output stage is determined by the voltage divider formed by the output stage emitter follower output impedance and the loudspeaker load impedance. The output

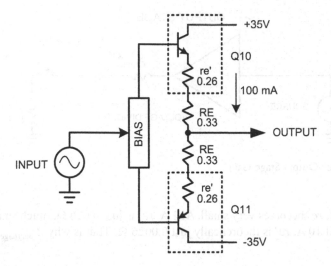

Figure 2.23 Push-Pull Output Stage

impedance of each half of the output stage is approximately equal to *re'* plus R_E. To be exact, any ohmic emitter resistance in the transistor should be considered to be a part of R_E. This is illustrated in Figure 2.23.

Since the two halves of the output stage act in parallel when they are both active at idle and under small-signal conditions, the net output impedance will be about half that of each side.

$$Z_{\text{out(small signal)}} \approx (re'_{\text{idle}} + R_E)/2$$

If the output stage is biased at 100 mA, then *re'* of each output transistor will be about 0.26 Ω. The summed resistance for each side will then be 0.26 + 0.33 = 0.59 Ω. Both output halves being in parallel will then result in an output impedance of about 0.3 Ω. Because voltage gain is being calculated, these figures assume that the output stage is being driven by a voltage source. If the load impedance is 8 Ω, then the voltage gain of the output stage will be 8/(8 + 0.3) = 0.96, as shown in Figures 2.20 and 2.21. If instead the load impedance is 4 Ω, the gain of the output stage will fall to 0.93. The voltage divider action governing the output stage gain is illustrated in Figure 2.24.

Bear in mind that the small-signal gain of the output stage has been calculated at its quiescent bias current. The value of *re'* for each of the output transistors will change as transistor currents increase or decrease, giving rise to complex changes in the output stage gain. Moreover, at larger signal swings only one half of the output stage is active. This means that the output impedance under those conditions will be approximately *re'* + R_E rather than half that amount. These changes in incremental output stage gain as a function of output signal current cause what is called *static crossover distortion*.

$$Z_{\text{out(large signal)}} \approx re'_{\text{high current}} + R_E \approx R_E$$

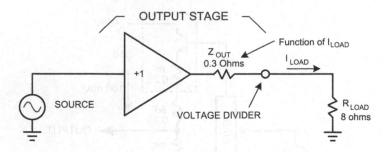

Figure 2.24 Amplifier Output Stage Gain

At high current, re' becomes very small. At 1 A, re' is just 0.026 Ω, much smaller than a typical value of R_E. At 10 A, re' is theoretically just 0.0026 Ω. That is why $Z_{out\,(large\ signal)} \approx R_E$. If R_E is chosen so that

$$R_E = re'_{idle}$$

then

$$Z_{out(large\ signal)} \approx Z_{out(small\ signal)} \approx R_E$$

and crossover distortion is minimized by making the large-signal and small-signal output stage gains approximately equal [28]. With R_E equal to 0.33 Ω in Figure 2.23, the required bias current to satisfy this condition would actually be approximately 79 mA. This is only a compromise solution and does not eliminate static crossover distortion because the equality does not hold at intermediate values of output current as the signal passes through the crossover region. This variation in output stage gain as a function of output current is illustrated in Figure 2.25.

Output Stage Bias Current

The quiescent bias current of the output stage plays a critical role in controlling crossover distortion. It is important that the right amount of bias current flows through the output stage, from top to bottom, when the output is not delivering any current to the load. Notice that together the two driver and two output transistors in Figure 2.19 require at least four V_{be} voltage drops from the base of Q8 to the base of Q9 to begin to turn on. Any additional drop across the output emitter resistors will increase the needed bias spreading voltage.

The optimum class AB quiescent bias for a conventional output stage like this is that amount of current that produces a voltage drop of approximately 26 mV across each of the output emitter resistors [28]. That is referred to as the Oliver condition. Recall that

$$re' = V_T/I_c = 26 \text{ mV}/I_c$$

Then

$$I_c = 26 \text{ mV}/re' = 26 \text{ mV}/R_E$$

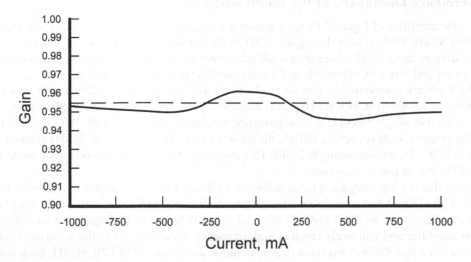

Figure 2.25 Output Stage Gain Versus Output Current

and

$$V_{RE} = I_c * R_E = 26\,mV$$

This amount of bias current makes *re'* of the output transistor equal to the resistance of its associated emitter resistor. With 0.33-Ω emitter resistors, this corresponds to about 79 mA. In the example design here I have chosen to over-bias the output stage slightly to a current of 100 mA. This means that small-signal gain will be slightly larger than large-signal gain in this case. This is evident in Figure 2.25.

There is a caveat to the assertion that the optimum bias point occurs when 26 mV is dropped across each output emitter resistor. Recall from Section 2.1 that the actual output impedance of an emitter follower is slightly greater than *re'*. The additional resistance results from physical ohmic base and emitter resistances inside the transistor. This additional resistive component acts as an extension of the external emitter resistor R_E. This means that the optimum voltage drop across the external emitter resistor will be somewhat less than 26 mV.

The required bias voltage for the output stage is developed across the bias spreader comprising a V_{be} multiplier built around Q5. In practice R7 is adjusted to set the output stage bias current to the desired value.

The objective of the bias spreader design is temperature stability of the bias point of the output stage. The temperature coefficient of the voltage produced by the V_{be} multiplier should match approximately that of the base-emitter junction voltages of the driver and output transistors. Since the V_{be} of a transistor decreases by about 2.2 mV/°C, it is important for thermal bias stability that these junction drops track one another reasonably. The output transistors will usually heat up the most. Because they are mounted on a heat sink, Q5 should also be mounted on the heat sink so that it is exposed to the same approximate temperature. This approach is only an approximation, because the drivers are often not mounted on the heat sink and because the temperature of the heat sink changes more slowly in time than that of the power transistor junctions. Q5 is often mounted on the case of one of the power transistors so that it can react more quickly to changes in the temperature of the power transistor.

Performance Limitations of the Simple Amplifier

The basic amplifier of Figure 2.19 has a decent low-frequency open-loop gain of about 16,300, or about 84 dB. With a closed-loop gain of 20 (26 dB) it has a feedback factor of about 816, corresponding to about 58 dB of negative feedback. However, its open-loop gain has not been made very linear and it is very vulnerable to β variations with signal in the output stage.

The feedback compensation was set to obtain a moderate negative feedback gain crossover frequency f_c of 500 kHz. This will typically result in a closed-loop 3-dB bandwidth of about 500 kHz. Selection of f_c = 500 kHz is what governed the choice of C1 at 300 pF. With f_c = 500 kHz and the assumed 6 dB per octave roll-off, the amount of negative feedback at 20 kHz is about 500 kHz/20 kHz = 25, corresponding to 28 dB. This means that there will be less distortion correction at 20 kHz than at low frequencies.

Notice that the input stage can never deliver more than ± 1 mA with respect to its nominal bias point. If all of this 1-mA swing goes into charging or discharging C1, the maximum voltage rate of change across C1 will be 1 mA/300 pF = 3.3 V/μs. This is very inadequate for virtually any power amplifier and will likely result in high frequency distortion often referred to as *slewing-induced distortion (SID)* or *transient intermodulation distortion (TIM)* [29, 30, 31]. Even with a demanded voltage rate of change well below the slew rate limit of 3.3 V/μs, the un-degenerated input differential stage will become nonlinear and produce high-frequency distortion.

These limitations will all be addressed in the next chapter as the design is evolved to a high-performance architecture.

References

1. Paul Horowitz and Winfield Hill, *The Art of Electronics*, 3rd ed., Cambridge University Press, Cambridge, 2015.
2. Adel Sedra and Kenneth Smith, *Microelectronic Circuits*, 6th ed., Oxford University Press, Oxford, 2010.
3. Giuseppe Massobrio and Paolo Antognetti, *Semiconductor Device Modeling with SPICE*, 2nd ed., McGraw-Hill, New York, 1993.
4. Bob Pease, "What's All This V_{BE} Stuff Anyhow?" in *Analog Circuits: World Class Designs*, Newnes, Oxford, 2008.
5. Application Note, "Field Effect Transistors in Theory and Practice," Freescale Semiconductor, AN211A.
6. LSK389, LSK489 and LSK689 JFET Datasheets, Linear Integrated Systems. Available at www.linear systems.com.
7. Bob Cordell, "LSK489 Ultra Low Noise JFET," Application Note, Linear Integrated Systems. Available at www.linearsystems.com.
8. Bob Cordell, "LSK689 Ultra Low Noise P-Channel Dual JFET," Application Note, Linear Integrated Systems. Available at www.linearsystems.com.
9. Vrej Barkhordarian, "Power MOSFET Basics," International Rectifier. Available at www.infineon.com.
10. Alpha and Omega Semiconductor, "Power MOSFET Basics." Available at www.aosmd.com.
11. Application Note, "MOSFET Basics," AN-9010, ON Semiconductor/Fairchild. Available at www.onsemi.com.
12. Robert Widlar, "Some Circuit Design Techniques for Linear Integrated Circuits," *IEEE Transactions on Circuit Theory*, vol. 12, no. 4, December 1965.
13. G. R. Wilson, "A Monolithic Junction FET-n-p-n Operational Amplifier," *IEEE Journal of Solid-State Circuits*, vol. SC-3, pp. 341–348, 1968.
14. Walt Jung, "Sources 101: Audio Current Regulator Tests for High Performance," Part 1: Basics of Operation, AudioXpress, 2007. Available at audioXpress.com.
15. Walt Jung, "GLED431: An Ultra Low Noise LED Reference Cell," December 2015. Available at www.waltjung.org.
16. Ken Walters and Mel Clark, "Zener Voltage Regulation with Temperature," MicroNotes Series No. 203, Microsemi Corp. Available at www.microsemi.com.
17. P. Williams, "Ring-of-two Reference," *Wireless World*, pp. 318–322, July 1967.

18. Datasheet, "LM134 3-Terminal Adjustable Current Source," Texas Instruments/National Semiconductor.
19. Datasheet, "LT3092 200mA 2-Terminal Programmable Current Source," Linear Technology.
20. E. Fred Schubert, *Light Emitting Diodes*, 2nd ed., Cambridge University Press, New York, 2006.
21. Peter A. Lefferts, "LED Used as Voltage Reference Provides Self-Compensating Temp Coefficient," *Electronic Design*, p. 92, February 15, 1975.
22. Datasheet, "LM329 Precision Reference," Texas Instruments/National Semiconductor.
23. Datasheet, "TL431 Precision Programmable Reference," ON Semiconductor. Available at www.onsemi.com.
24. Malcolm J. Hawksford, "Optimization of the Amplified Diode Bias Circuit for Audio Amplifiers," *Journal of the Audio Engineering Society*, vol. 32, no. 1–2, pp. 31–33, January–February 1984.
25. Walter G. Jung, *Op Amp Applications Handbook*, Analog Devices, Elsevier/Newnes, Burlington, MA, 2005.
26. Bruce Carter, *Op Amps for Everyone*, 4th ed., Newnes, Oxford, 13.
27. Glen Ballou, *Handbook for Sound Engineers*, 5th ed., Focal Press, Burlington, MA, 2015.
28. Barney M. Oliver, "Distortion in Complementary Pair Class B Amplifiers," *Hewlett Packard Journal*, pp. 11–16, February 1971.
29. Matti Otala, "Transient Distortion in Transistorized Audio Power Amplifiers," *IEEE Transactions on Audio and Electroacoustics*, vol. AU-18, pp. 234–239, September 1970.
30. Walter G. Jung, Mark L. Stephens and Craig C. Todd, "Slewing Induced Distortion and Its Effect on Audio Amplifier Performance—With Correlated Measurement Listening Results," AES preprint No. 1252 presented at the 57th AES Convention, Los Angeles, May 1977.
31. Robert R. Cordell, "Another View of TIM," *Audio*, February–March 1980.

Chapter 3

Power Amplifier Design Evolution

3. INTRODUCTION

In this chapter I begin by describing and analyzing a simple power amplifier and then progress through various improvements to it. The rationale for the performance-improving changes is described at each step of the way. The amplifier does not include all of the usual circuit details needed in a real amplifier, such as input AC coupling and filtering, output stage protection circuits and output stage networks, although some of those will be touched on briefly in the last section of the chapter. Nevertheless, it serves to illustrate the design choices and process in evolving a design to a high performance level. At each step of the evolution, the total harmonic distortion (THD) of the amplifier is evaluated by SPICE simulation.[1, 2] The amplifiers described throughout the evolution are not meant to be built, but rather to serve as illustrations of how the circuits and performance change throughout the evolution. However, some details not shown on the schematic are mentioned in the text. Section 3.12 covers many of the details not shown in the schematics, such as required input and output networks. Later, in Chapter 4, a complete version of one of the amplifiers in the evolution is built and measured performance is shown.

3.1 About Simulation

The use of SPICE simulation can save hours in reaching the point where you can build a working amplifier. Intuition is not always right when it comes to circuit design, and SPICE helps here. A great benefit of SPICE simulation is that it allows one to look inside an amplifier design and see all of the node voltages and path currents, both for DC and AC signals. All of the amplifier examples in this chapter were simulated using the LTspice® simulator (discussed extensively in Chapters 23 and 24) and the performance characteristics are derived from those simulations [2]. Caution must be exercised in relying on the results of simulation, since device models are not perfect and a simulation does not take into account real-world matters like parasitic capacitances and inductances and capacitive and magnetic coupling effects, just to name a few.

3.2 The Basic Power Amplifier

Figure 3.1 is the simple 50-W amplifier analyzed in the last chapter. It has many performance limitations as a result of its simple design.

While not shown in the interest of simplicity, the bias spreader should be bypassed with a 10-μF electrolytic capacitor from collector to emitter of Q5 in a real amplifier. Similarly, Q5 should be mounted on the heat sink or on one of the output transistors so that it can track the temperature of the output stage for bias temperature compensation. Q5 is chosen to be a transistor in an insulated TO–126 package for easy mounting to the heat sink or an output transistor. R7 sets

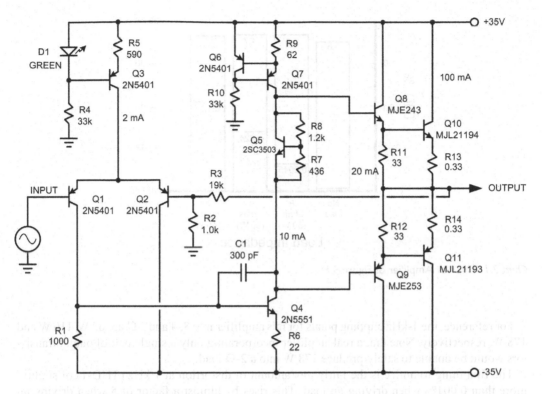

Figure 3.1 A Basic 50 W Power Amplifier

the value of the output stage bias current, and its value shown is the value used in simulation to obtain the chosen 100-mA output stage bias. In practice, R7 would comprise a fixed resistor and a 10-turn trim pot in series. The bias spreader shown is very rudimentary, and more discussion of bias spreader design can be found in the chapter on thermal design. LED reference diode D1 should be bypassed to minimize its noise contribution. R10 should be split and the junction of the two resistors should be bypassed to the positive rail to keep power supply noise from being injected into the Q6–Q7 VAS current source. These comments apply to all of the example designs shown in this chapter.

The performance of the amplifier was evaluated by simulating it with a version of the SPICE program called *LTspice*® [1, 2]. This free program is very useful for audio electronics design and its use is described in detail in Chapter 23. Of special interest is its ability to perform distortion analysis. Because simulations do not always capture or model all of the distortion mechanisms that can exist in the real world, results can sometimes be a bit optimistic. Nevertheless, they do provide valuable comparisons among different circuit design arrangements.

The 50-W amplifier was simulated under various conditions to evaluate its total harmonic distortion (THD). Performance of the first version of the amplifier is shown in Chart 3.1. THD results are presented at 1 kHz and 20 kHz at 50 W when driving an 8-Ω load. This corresponds to a peak output voltage of 28 V. For comparison, results are also shown for the same voltage output with a no-load condition. Distortion at slightly reduced output voltages when the amplifier drives a 4-Ω load and a 2-Ω load are also shown, exposing the effects of heavier loading. Those voltages correspond to power levels of 90 W at 4 Ω and 145 W at 2 Ω.

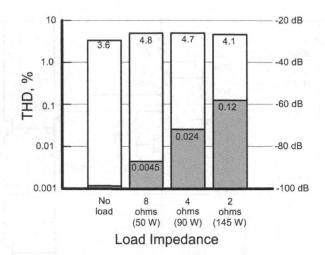

Chart 3.1 THD for Amplifier of Figure 3.1

For reference, the 1-kHz clipping points for this amplifier into 8, 4 and 2 Ω are 62 W, 110 W and 178 W, respectively. Note that a real amplifier incorporating only a single pair of output transistors would be unable to safely produce 178 W into a 2-Ω load.

The first thing to notice is the fairly low amount of distortion at 1 kHz (THD-1) of slightly more than 0.001% when driving no load. This rises by almost a factor of 5 when driving an 8-W load, reflecting the loading on the output stage. The higher peak currents required to drive the 4-Ω and 2-Ω loads are reflected in even higher amounts of distortion. At a drive level of 17 V RMS into a 2-Ω load, the peak current is 12 A. At this current level the output transistors are suffering considerable beta droop, contributing significantly to the distortion. It is very revealing to evaluate distortion under conditions ranging from no-load to a heavy load. This helps isolate the distortion contributions of the output stage from those of the remainder of the amplifier.

At 20 kHz distortion is very high, even under a no-load condition. This is a direct reflection of the inadequate slew rate of the amplifier. Recall from Chapter 2 that the estimated slew rate of this amplifier is 3.3 V/μs. A 28 V peak sine wave at 20 kHz has a maximum rate of change of 3.5 V/μs. This means that at 20 kHz this amplifier is just into slew rate limiting when it is attempting to deliver 50 W. The THD-20 results are in the range of 4%.

3.3 Adding Input Stage Degeneration

There are several different ways in which to evolve this amplifier and improve it. There is nothing sacred about the order in which the improvements are made. Here I have chosen to make the first improvement by adding emitter degeneration to the input stage LTP to improve the slew rate and reduce high-frequency distortion. The IPS tail current source has been changed to a Zener-based design to illustrate a different implementation that provides higher output impedance. The Zener should be bypassed to suppress its noise contribution. Notice that in every schematic going forward the parts have been re-numbered for purposes of clarity in the schematics. This means that a given resistor number, for example, does not refer to the resistor with the same function in a previous schematic.

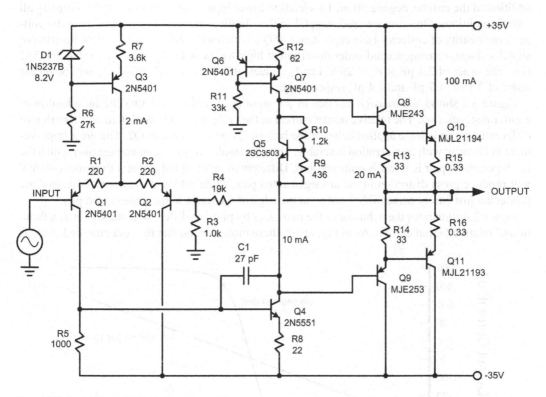

Figure 3.2 Input Stage Emitter Degeneration

The design of Figure 3.2 differs from that of Figure 3.1 by the addition of emitter degeneration resistors R1 and R2 and by reducing C1 from 300 pF to 27 pF. The pair of 220-Ω emitter degeneration resistors implements 10:1 degeneration of the input differential pair by increasing the total emitter-to-emitter resistance R_{LTP} from 52 Ω to about 500 Ω. This reduces its transconductance by a factor of 10. In order to keep the negative feedback gain crossover frequency f_c at the same 500 kHz for equivalent stability, C1 must be reduced by approximately that same 10:1 factor.

Recall the relationship described in Chapter 2 for Miller compensation.

$$C_{Miller} = 1/(2\pi f_c * R_{LTP} * A_{cl})$$

A_{cl} is the closed-loop gain, R_{LTP} is the total emitter-to-emitter LTP resistance including re', and f_c is the desired gain crossover frequency for the negative feedback loop.

$$C1 = C_{Miller} = 0.159/(500 \text{ kHz} * 520 \ \Omega * 20) = 30.6 \text{ pF}$$

By this calculation C1 must be reduced to about 30 pF (here 27 pF). Notice that the same maximum ±1 mA is still available from the input stage to charge and discharge C1. This now results in an achievable theoretical slew rate of about 1 mA/27 pF = 37 V/μs. This is a much more respectable value of slew rate for an audio power amplifier. Here it is assumed that all the available signal current from the input stage flows through C1. In reality, some flows into R1 and some flows into the base of Q4, reducing slew rate somewhat from the value calculated above to a simulated value of 32 V/us. The input stage has been made significantly more linear by the addition of the emitter degeneration. This leads to lower input stage distortion under virtually all signal conditions. One caveat to reducing C1 in this circuit is increased susceptibility to the voltage nonlinearity of collector-base capacitance of Q4, which will cause f_c to be a slight function of signal voltage, causing second-order distortion at high frequencies. Typical C_{cb} for the 2N5551 is on the order of 2.1 pF at V_{ce} = 20 V. For V_{ce} values of 5 V, 50 V and 100 V, C_{cb} will be on the order of 3.1 pF, 1.7 pF and 1.4 pF, respectively.

Figure 3.3 shows the transfer function of the input stage before and after the introduction of emitter degeneration. The relative output current at the collector of Q1 is plotted as a function of differential input voltage applied between the base of Q1 and the base of Q2. The great improvement in linearity with degeneration is notable. The very small range of input voltage over which the un-degenerated pair is linear is quite obvious. Differential input signals greater than about 40 mV will produce gross distortion in the un-degenerated pair, while differential inputs of 400 mV are producing just barely perceptible bending at the edge of the range in the degenerated pair.

Figure 3.4 illustrates the behavior of the two cases by plotting differential stage gain as a function of relative output current. As in Figure 3.3, these plots assume that the collector load at Q1 is

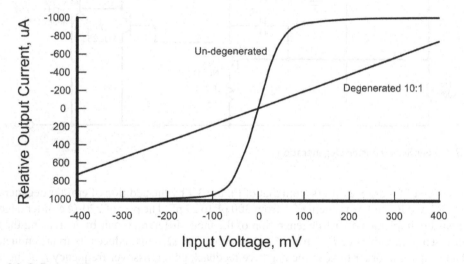

Figure 3.3 Differential Stage Current Versus Input Voltage

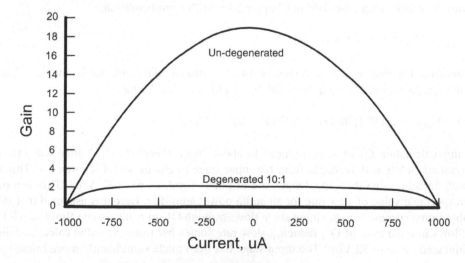

Figure 3.4 Differential Stage Gain Versus Output Current

only the 1-kΩ resistor R5. Two things are immediately apparent. First, the gain at the quiescent point has been reduced by the expected factor of 10 when degeneration is introduced. Second, the gain in the un-degenerated case is a strong function of the output current. This was also apparent in Figure 3.3 when one recognizes that the gain of the stage corresponds to the slope of the transfer function curve. At an output current of only half its maximum value, the gain in the un-degenerated case has fallen from 19 to 14.

The Bode plot in Figure 3.5 illustrates how the gain variation with signal in the un-degenerated input stage also affects dynamically the frequency responses in the amplifier. The shaded region shows the variation in open-loop gain as the output current from the input stage varies from its quiescent point to half its maximum signal value. Because the closed-loop bandwidth is changing with signal, the in-band phase response of the amplifier will also change with the signal. This phenomenon has been termed *phase intermodulation* (PIM) *distortion*, [3, 4] and will be discussed in more detail in Chapter 28.

There is a price to be paid for employing emitter degeneration in the input stage. The DC gain of the input stage was previously 13.7; it is now only about 1.3. This combined with the VAS-OPS gain of about 1200 yields an open-loop gain of about 1600 and a loop gain of about 82.

IPS gain ≈ 1.3

NFB ≈ 38 dB

The reduced IPS gain means that the amplifier's low-frequency open-loop gain and negative feedback factor have been reduced by a factor of 10 (20 dB). The amplifier now has only 38 dB of negative feedback available at low frequencies. This reduces the ability of the negative feedback to reduce distortion caused by the VAS and the OPS. Feedback at 20 kHz is still about the same, approximately 28 dB. This change in the open-loop frequency response of the amplifier is illustrated in Figure 3.6. The important thing to notice is that the high-frequency gain behavior

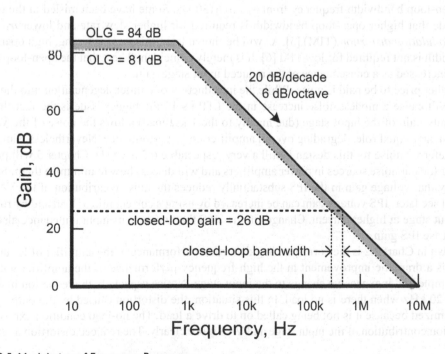

Figure 3.5 Modulation of Frequency Response

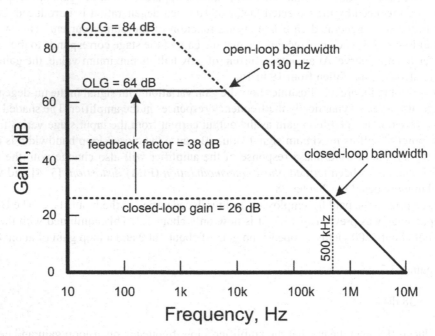

Figure 3.6 Reduced Open-Loop Gain with Degeneration

above the open-loop bandwidth frequency remains unchanged because the value of the Miller compensation capacitor was changed to account for the reduced input stage gain.

Notice that a side effect of the reduced open-loop gain at low frequencies is an increase in the open-loop bandwidth frequency from 613 to 6130 Hz. Some have been misled in the past to conclude that higher open-loop bandwidth is required for higher slew rate and lower *transient intermodulation distortion* (TIM) [5]. As will be shown later, this view is wrong; high open-loop bandwidth is not required for low TIM [6]. It is merely coincidental here that the open-loop bandwidth increased as a consequence of the reduced input stage gain.

Another price to be paid is noise. While the introduction of emitter degeneration into the input stage will cause a modest noise increase in the LTP itself, the bigger issue is the fact that the near-unity gain of the input stage (due mainly to the 1-kΩ load) allows the noise of the VAS to play a nearly equal role, degrading overall amplifier noise performance. Nevertheless, simulated input-referred noise for this design is still a very respectable 6.5 nV/√Hz. Chapter 5 will present a closer look at noise sources in power amplifiers and will discuss how to minimize them. Suffice it to say that voltage gain in the IPS substantially reduces the noise contribution of the VAS. As we will see later, IPS voltage gain can be increased by using a current mirror load and by running the input stage at higher current. Changes in the VAS that increase its input impedance also help to increase IPS gain.

Shown in Chart 3.2 is the corresponding distortion performance of the amplifier of Figure 3.2. There is a dramatic improvement in the high-frequency performance of the amplifier with this very simple and inexpensive change to the input stage. Notice especially the reduction in distortion at 20 kHz when there is no load. In this situation, the distortion caused by the output stage is minimized because it is not being called on to drive a load. The no-load condition exposes the distortion contribution of the input stage and VAS more clearly. The reduced distortion at 20 kHz

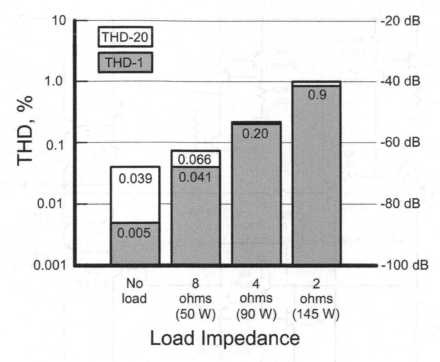

Chart 3.2 THD for Amplifier of Figure 3.2

is a direct result of the greatly increased slew rate capability of the amplifier and the improved linearity of the input stage as a result of the introduction of the emitter degeneration.

Unfortunately, the distortion at 1 kHz has worsened under all loading conditions. This is a direct result of the reduction in negative feedback that has occurred because of the emitter degeneration introduced into the input stage.

It is interesting to observe that the distortion at 20 kHz is not much different from the distortion at 1 kHz. This is not always what we would expect. As can be seen in Figure 3.6, there is still more global negative feedback present at 1 kHz than at 20 kHz. This would normally lead one to expect lower distortion at 1 kHz.

The reason for this seemingly anomalous behavior is that the dominant source of distortion in this version of the amplifier is the VAS. At low frequencies the VAS does indeed have more negative feedback surrounding it via the normal global negative feedback loop. However, at higher frequencies the loss of global negative feedback around the VAS is replaced by greater local negative feedback introduced by the Miller compensation capacitor. The two effects largely cancel, leaving the distortion caused by the VAS about the same at low and high frequencies.

3.4 Adding a Darlington VAS

The next logical evolution is to try and get back some of that open-loop gain that was lost by the introduction of the input stage emitter degeneration. This can be accomplished by adding an emitter follower in front of the VAS transistor to provide some current gain. This change is shown in the design of Figure 3.7, where Q6 has been added to buffer the input of the VAS. This is often referred to as a Darlington VAS, although in the strict sense the transistor connection would only be a Darlington if the collector of Q6 were connected to the collector of Q7. There is a good

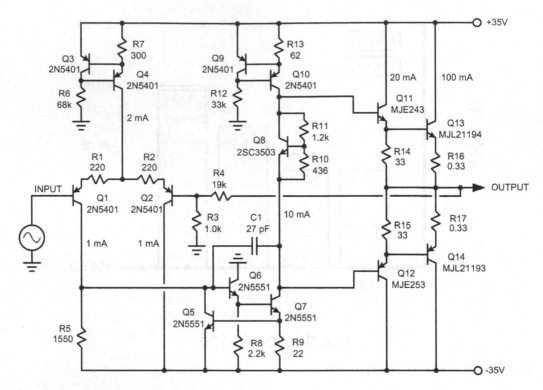

Figure 3.7 Amplifier with Darlington VAS Added

reason to connect Q6's collector to ground. This connection eliminates the large signal voltage that would otherwise appear at the collector of Q6. This signal voltage would cause nonlinear current flow into the base of Q6 as a result of the nonlinear nature of Q6's collector-base capacitance.

The inclusion of Q6 increases the input impedance of the new VAS to about 100 kΩ, assuming β of 100 for Q6 and Q7. The input impedance is partly determined by the choice of the 2.2-kΩ emitter resistor R8, which provides a healthy 0.5 mA of turn-off current to the base of Q7. This is needed to sustain a high slew rate in the face of the collector-base capacitance of Q7. If Q7 has a C_{cb} of 2.5 pF, this turn-off current will sustain a slew rate of about 0.5 mA/2.5 pF = 200 V/μs, far more than what is needed. If R8 is connected to the emitter of Q7 instead of the negative rail, it will be bootstrapped and the input impedance of the VAS will be further increased. In this case, R8 should be reduced to about 1200 Ω to maintain Q6's bias current at 0.5 mA.

The inclusion of Q6 also raises the node voltage at the output of the LTP to about 1.7 V above the negative rail. This calls for the use of a larger LTP load resistor R5. With R5 now at 1550 Ω, the input stage voltage gain is back up to about 3. Combined with the VAS-OPS gain of 1200, the open-loop gain becomes 3600 and the loop gain becomes 180, corresponding to about 45 dB. This approximate 7-dB increase in negative feedback as compared to the design of Figure 3.2 will tend to reduce low-frequency distortion by nearly 7 dB.

IPS gain ≈ 3.0

NFB ≈ 45 dB

Transistor Q5 has also been added to limit the maximum current of Q7 if the amplifier clips on negative-going signals. Under such clipping conditions, Q1 will conduct the full 2 mA of LTP tail current and will attempt to raise the node voltage at the base of Q12 to almost 3.1 V above the negative rail, severely overloading the VAS. Current limiter transistor Q5 turns on only if the emitter current of Q7 exceeds about 27 mA. This protection is desirable because the addition of the Darlington transistor Q6 made substantial overdrive of Q7 possible. In many cases the current source loading the VAS will limit the current of Q7 to about 10 mA; however, this will not be the case in the event of an output short circuit. It will also not be the case with some safe area protection circuits that shunt the VAS output node to the output node when they protect. Even in a normal clipping scenario, Q6 will attempt to deliver high current to the emitter of Q7 through Q7's base-emitter junction, and Q5 will limit this current by limiting the total emitter current of Q7 to about 27 mA. At the nominal VAS bias current of 10 mA, the base-emitter junction of Q5 sees only 0.22 V, so Q5's very small conduction of less than 100 nA will introduce very little nonlinearity due to it stealing signal current.

Another important improvement has taken place with the introduction of the Darlington VAS. Prior to the use of the Q6 buffer, the collector-base capacitance of Q7 was effectively in parallel with the Miller compensation capacitor C1 in the earlier design. Unfortunately, the collector-base junction capacitance of a transistor is nonlinear, going from a high value when collector-base voltage is low to a smaller value when V_{cb} is high. The resulting change in net compensation capacitance with signal causes the high-frequency gain of the amplifier to change with signal, corresponding to a source of high-frequency nonlinear distortion. The isolation afforded by buffer Q6 suppresses this effect. It is important to note that this suppression would not occur if the collector of Q6 were connected to the collector of Q7, as in a true Darlington.

Chart 3.3 lists the corresponding distortion performance of the amplifier of Figure 3.7. Note the significant improvement in performance at both low and high frequencies provided by this very inexpensive addition of two small-signal transistors Q5 and Q6. Both low- and high-frequency

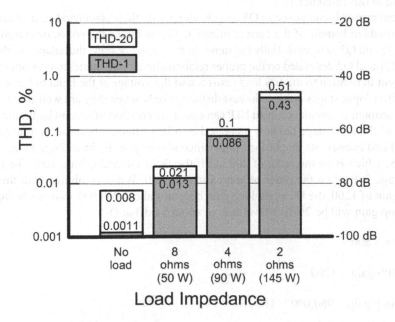

Chart 3.3 THD for the Amplifier of Figure 3.7

distortion when driving an 8-Ω load have decreased by a factor of about 3. This is largely due to the increase in input stage gain. The no-load distortion at 20 kHz has been significantly reduced from 0.039% to 0.008%. The 1-kHz distortion has also been improved by a factor of almost 5. These improvements are a result of the increase in low-frequency feedback factor. Distortion is higher with 4-Ω and 2-Ω loads, however. This is a reflection of the output stage struggling to drive the low-value load impedances.

It is still the case that THD-1 and THD-20 are still about the same at all load impedances. This is a reflection of the fact that VAS and output stage distortion is still dominating performance.

The tail current source in Figure 3.7 has also been changed to a feedback current source for purposes of illustration. Note that some have proposed using the base voltage of the VAS feedback current source to bias the tail current source transistor as well, saving a transistor and a resistor (Q3 and R6). This is a bad idea. Biasing Q4 with what it sees as a fixed potential destroys the high output impedance it would enjoy if it were connected as an independent feedback current source. That will degrade IPS PSRR. Such a connection also invites a feedback interaction from the VAS to the IPS, since the voltage at the base of Q10 will change slightly with signal due to Early effect in Q10.

3.5 Input Stage Current Mirror Load

The voltage gain of the input stage has been limited by the load impedance at its output collector in the designs that have been discussed to this point. This impedance is governed by load resistor R5 in parallel with the input impedance of the VAS in Figure 3.7. The introduction of the Darlington VAS greatly reduced the loading due to the latter, but only reduced the influence of the collector load resistance by a factor of about 1.5. The single-ended use of just one of the output collector signals from the LTP is also wasteful of gain.

Figure 3.8 shows the addition of a current mirror to serve as the load for the input stage. The amplifier circuit evolved to this point, where an IPS current mirror and Darlington VAS are included, has been referred to as the *Thompson* topology. That circuit was employed in monolithic operational amplifiers beginning in the late 1960s. A description of its operating principles can be found in this reference [7].

The current mirror is composed of Q5 and Q6 along with their associated emitter degeneration resistors. An added benefit of the current mirror is that it forces the collector currents of input transistors Q1 and Q2 to be essentially the same. In the earlier designs the balance of the collector currents of Q1 and Q2 depended on the proper relationship among numerous parameters, such as the tail current in relation to the IPS load resistor and the voltage at the input to the VAS. Differential amplifier input stages produce lowest distortion only when they are well balanced. Even a fairly small amount of imbalance in an LTP can cause the creation of second harmonic distortion.

The gain of the input stage has now increased by a large amount because its transconductance has doubled and because its output load impedance is now just the input impedance of the Darlington VAS, which is on the order of 200 kΩ if the βs of Q8 and Q9 are 100. The DC gain of the input stage is now on the order of approximately 800. When combined with the estimated VAS-OPS gain of 1200, the DC open-loop gain becomes about 960,000, corresponding to nearly 120 dB. Loop gain will be 26 dB below that, or about 94 dB at DC.

IPS gain $\approx$ 800

VAS-OPS gain $\approx$ 1200

Open-loop gain $\approx$ 960,000 (~112 dB)

Feedback factor $\approx$ 48,000 (~94 dB)

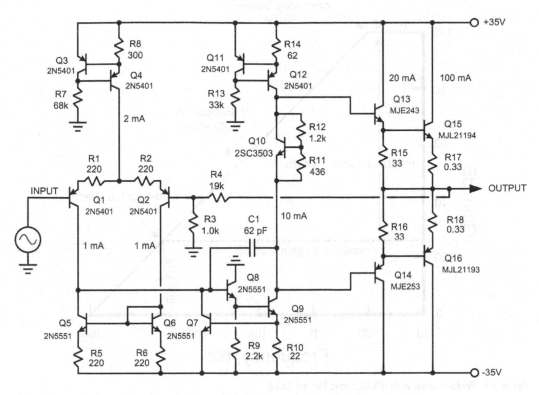

Figure 3.8 Input Stage with Current Mirror Load (Thompson Topology)

The open-loop gain and amount of negative feedback at lower frequencies are much greater as compared to the case in Figure 3.7. The total open-loop gain of this version of the amplifier is illustrated in Figure 3.9.

The use of both signal currents from the collectors of the LTP effectively doubles the transconductance of the LTP. This means that C1 must be doubled to about 60 pF in order to maintain the negative feedback gain crossover frequency at about 500 kHz.

As an aside, it should be mentioned that the current mirror chosen here is a very simple one. A slightly more complex current mirror arrangement could be employed to maintain the collector voltages of both Q14 and Q15 at more nearly the same value, further improving symmetry and performance. One of the other current mirrors described in Chapter 2 would help accomplish this. The current mirror of Figure 2.9b would be a good choice. Chart 3.4 lists the corresponding distortion performance of the amplifier of Figure 3.8.

Note the very significant improvement in performance at both low and high frequencies provided by this very inexpensive addition of the current mirror load. No-load THD-1 has been reduced by a factor of 16, and no-load THD-20 has been reduced by a factor of over 6. These reductions in distortion are attributable to two things. First, the input stage does not have to work as hard to drive the VAS to produce a given output level. Second, there is now a greater amount of negative feedback.

Especially notable is the very low amount of THD-20 under no-load conditions. This suggests that the IPS-VAS combination is capable of very good performance even out to high frequencies. It now becomes much clearer how the output stage is limiting both low-frequency and high-frequency performance as the load becomes heavier.

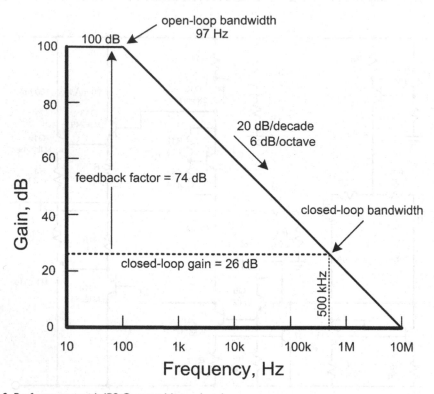

Figure 3.9 Performance with IPS Current Mirror Load

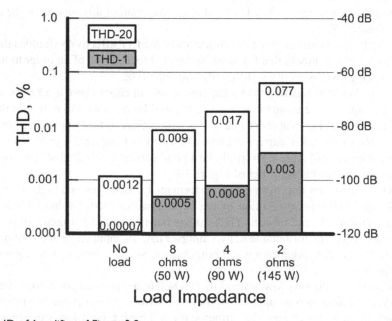

Chart 3.4 THD of Amplifier of Figure 3.8

3.6 The Output Triple

With this section the design evolution begins in earnest to improve the output stage. The performance comparisons thus far have shown that performance degrades as the amplifier goes from a no-load condition to a heavy-load condition. This is almost always a sign of distortion that originates in the output stage or in the way the output stage loads the VAS.

The use of two stages of emitter follower current gain in the output stage is simply not sufficient for driving lower-impedance loads with the highest quality. To the extent possible, it is very desirable to isolate the high impedance output of the VAS from the loudspeaker load.

Figure 3.10 shows the use of an output stage that is a triple emitter follower, often just called a *Triple*. This output stage was popularized by Bart Locanthi, and is also known as the *Locanthi T circuit* [8, 9]. The extra emitter follower stage provides an additional amount of buffering for the VAS stage in the form of higher current gain by a factor of about 100.

Transistors Q13 and Q14 act as pre-driver emitter followers that provide the extra current gain and buffering. This increases the input impedance of the output stage to a typical value of about 4 MΩ when an 8-Ω load is being driven. With the Triple, the total current gain in the output stage is on the order of 500,000. The extra buffering is especially effective in mitigating the effects of beta droop at high currents in the output transistors.

The VAS gain was previously estimated to be governed by an output stage load resistance of 40 kΩ in parallel with a 135-kΩ Early effect output resistance of the VAS transistor, for a net load of 31 kΩ. The ratio of 31 kΩ to the effective VAS emitter resistance provided the VAS voltage gain figure of 1240. When combined with an output stage gain of about 0.96 when driving an 8-Ω load, the VAS-OPS gain became about 1200.

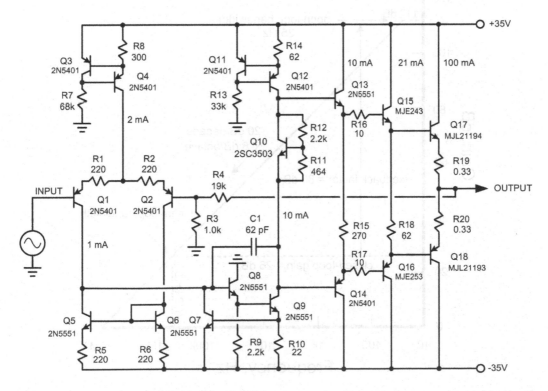

Figure 3.10 Output Stage with Triple Emitter Follower

The 40-kΩ output stage load resistance has now become 4 MΩ with the introduction of the Triple, for a net VAS load resistance on the order of 130 kΩ. With the effective emitter resistance in the VAS at about 25 Ω, the VAS gain now becomes 130 kΩ/25 = 5200. This is an increase of about 12 dB in VAS gain.

IPS gain ≈ 800

VAS gain ≈ 5200

OPS gain ≈ 0.96

Open-loop gain ≈ 4,160,000 (~132 dB)

Feedback factor ≈ 208,000 (~106 dB)

The frequency responses of the amplifier employing the Triple output stage are shown in the Bode plot of Figure 3.11. It is important to notice that although the amount of open-loop gain and negative feedback have increased substantially, the high-frequency portion of the frequency responses remains the same, including the gain crossover frequency and closed-loop bandwidth. This is again due to the fact that the Miller compensation capacitor is controlling the gain at high frequencies. In this context, that means all frequencies above 25 Hz. The gain at low frequencies is actually poorly controlled because it depends heavily on β of many transistors and on the Early

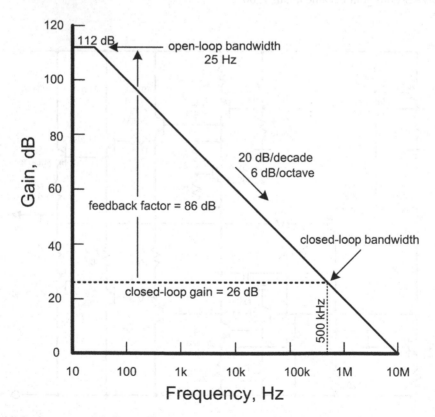

Figure 3.11 Performance with Output Triple

effect. For this reason, the estimated value of low-frequency gain of 132 dB is indeed a very rough estimate. What is important is that the number is large, even if it is uncertain.

A second benefit of the Locanthi T output stage architecture is the way in which the bias current in the pre-driver and driver stages is established. In each case a single resistor is connected from emitter to emitter of the respective stage. This can be seen in the way that R15 and R18 are connected. This gives the output stage its T-like appearance. More importantly, this connection of the bias resistors causes the pre-driver and driver emitter follower stages to stay turned on throughout the signal swing. They are operating in class A. In the earlier designs, where the bias resistors were connected to the output rail, the driver transistors turned off for one-half of the signal cycle, just like the output transistors.

As with many of the other improvements made in the evolution of this amplifier, this improvement has been made with the addition of only two inexpensive small-signal transistors, Q13 and Q14. Yet another advantage of the Triple is that the VAS no longer needs to drive the larger driver transistors that are usually required to supply the fairly high base currents of the output transistors. This often means that the smaller pre-driver transistors will load the high-impedance VAS output node with less nonlinear collector-base junction capacitance.

Chart 3.5 illustrates the distortion performance of the amplifier of Figure 3.10. The distortion scale has been shifted down by a factor of 10 compared to the previous chart. The minimum distortion on the vertical scale is now 0.00001% (–140 dB).

Note the very significant improvement in performance at low frequencies provided by this very inexpensive addition of two small-signal transistors. As expected, this improvement is especially notable when driving the 2-Ω load, where THD-1 has been reduced by a factor of 14 as compared to the same design using only a Darlington output stage.

The improvement at 20 kHz is less dramatic, ranging from a reduction factor of 2.5 when driving an 8-Ω load down to a factor of 1.5 when driving a 2-Ω load. This suggests that at high

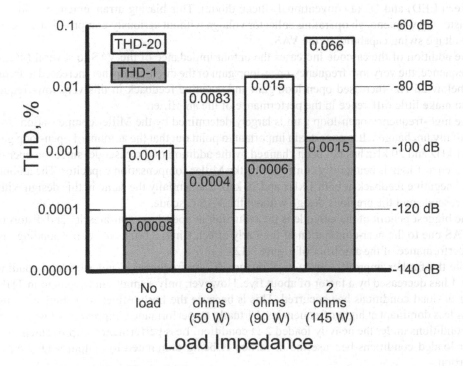

Chart 3.5 THD for Amplifier of Figure 3.10

frequencies the performance is being limited largely by output stage crossover distortion. In Chapter 12 it will be shown how such crossover distortion can be reduced.

Note that while the introduction of the Triple increased the amount of negative feedback at very low frequencies by about 12 dB, a comparison of Figures 3.9 and 3.11 reveals that it did not increase the amount of global negative feedback at either 1 kHz or 20 kHz. The reductions in distortion occurred as a result of the Triple making the VAS-OPS combination more linear.

Base stopper resistors R17 and R18 have been added between the pre-drivers and drivers to improve stability of the output stage by assuring a resistive load for the pre-driver emitter followers at high frequencies. This will be of greater importance in designs where larger driver transistors with higher C_{cb} are employed.

3.7 Cascoded VAS

It was shown in the earlier discussion on the Early effect that the transistor has a very finite output resistance characteristic as a result of the current gain of the transistor increasing as the collector voltage increases. Since the current gain of the device is changing with signal, this is a nonlinear effect.

The load impedance presented by the output stage is now much larger due to the use of the Triple, so the intrinsic output impedance of the VAS matters much more. Recall that the estimated Early effect output resistance of the VAS is about 135 kΩ, while the Triple output stage presents a load of about 4 MΩ when it is driving an 8-Ω load. Indeed, the addition of the VAS cascode would not have made much sense until the Triple output stage was introduced into the design.

Figure 3.12 illustrates the addition of a cascode transistor Q10 to the VAS. The base of Q10 is held at a fixed potential of about 2.5 V above the negative supply rail by the combination of D1 (a Green LED) and D2 (a conventional silicon diode). This biasing arrangement provides VAS transistor Q9 with enough operating collector voltage without seriously compromising the negative voltage swing capability of the VAS.

The addition of the cascode increases the output impedance of the VAS to several MΩ. As a consequence, the very low-frequency open-loop gain of the circuit is further increased at frequencies below 25 Hz. Increased open-loop gain and negative feedback in this very low-frequency region make little difference in the performance of the amplifier.

The high-frequency open-loop gain is largely determined by the Miller compensation, so it is essentially unchanged. It is once again important to point out that the amount of open-loop gain at both 1 kHz and 20 kHz has not been changed by the addition of the cascode, since the open-loop gain even at 1 kHz is being fully controlled by the Miller compensation capacitor. The amount of global negative feedback at both 1 kHz and 20 kHz is essentially the same in this design with the VAS cascode and the previous design without the VAS cascode.

The biggest benefit of the cascode is the reduction in open-loop nonlinearity and distortion in the VAS due to the near-elimination of the Early effect. Chart 3.6 lists the corresponding distortion performance of the amplifier of Figure 3.12.

Note the further improvement in performance at low frequencies under no-load conditions. THD-1 has decreased by a factor of about five. However, only a small improvement in THD-20 under no-load conditions has occurred. This is because the Early effect contribution to distortion is less dominant at higher frequencies. Virtually no performance improvement is seen under any conditions under the heavily loaded 2-Ω condition. Less performance improvement is seen under loaded conditions because the distortion is being dominated by output stage crossover distortion.

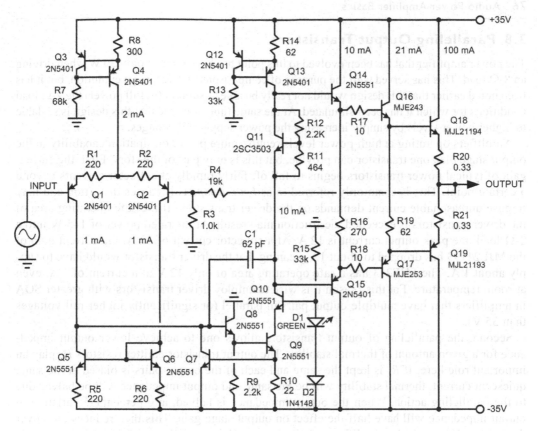

Figure 3.12 Amplifier with Cascoded VAS

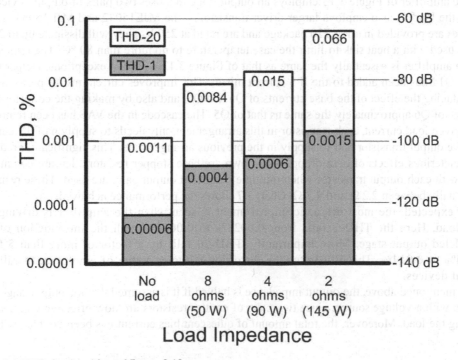

Chart 3.6 THD for Amplifier of Figure 3.12

3.8 Paralleling Output Transistors

The power amplifier that has been evolved to this point has been rated at only 50 W when driving an 8-Ω load. This has served well the purpose of comparisons during design evolution, but it has been noted earlier that this design would not really be able to support (at full power) the 2-Ω load conditions for which it has been simulated. At the same time, it is true that this design is scalable to higher power levels by simply increasing the power supply rail voltages.

Amplifiers operating at high power levels require more power dissipation capability in the output stage than one transistor can provide, but this is only part of the story. First, the current gain of typical power transistors begins to fall off fairly rapidly at collector currents beyond a certain point. This is commonly referred to as *beta droop*. This causes distortion and may impose unreasonable current demands on the driver transistors. If the safe operating area of the driver transistors is exceeded, destruction may result. At the rated power of 145 W into a 2-Ω load, the peak output current is 12 A. At a collector current of 12 A, the current gain of the MJL21193 has drooped to about 12, meaning that the driver transistor would have to supply about 1 A. The MJE243 has a safe operating area of only 12 V at a current of 1 A, even at room temperature. For this reason it is wise to employ driver transistors with greater SOA in amplifiers that have multiple output pairs in parallel (or significantly higher rail voltages than 35 V).

Second, the paralleling of output transistors allows one to achieve lower output impedance for a given amount of thermal stability. The output transistor emitter resistors R_E play an important role here. If R_E is kept the same and each of the output pairs is biased at the same quiescent current, thermal stability will be the same and output impedance will be halved due to the paralleling action. When the output impedance is halved, any percentage variation in output impedance will have half the effect on output stage gain. This then reduces crossover distortion. Of course, the amplifier now dissipates twice as much power under the quiescent condition.

The amplifier of Figure 3.13 employs an output stage that uses two pairs of output devices to drive the load. It also employs larger driver transistors—the MJE15032 and MJE15033. These devices are provided in a TO-220 package and are rated at 250 V. They will dissipate up to 28 W when used with a heat sink to limit the case temperature to no more than 80 °C. The remainder of the amplifier is essentially the same as that of Figure 3.12 with two exceptions. Helper transistor Q14 has been added to the IPS current mirror. This improves current mirror performance by reducing the effect of the base currents of Q5 and Q6, and also by making the collector-base voltage of Q6 approximately the same as that of Q5. The cascode in the VAS has been removed. At a given load current, each transistor in this arrangement only needs to supply half the current that the output transistor had to supply in the previous arrangements. This significantly reduces the deleterious effects of beta droop. Not shown are base stopper resistors that are commonly used with each output transistor when multiple paralleled output pairs are used. These resistors are usually between 2.2 Ω and 4.7 Ω. Chart 3.7 shows the performance achieved.

As expected, the most dramatic improvement is seen when the amplifier is driving the 2-Ω load. Here the THD-1 falls from 0.0023% to 0.0006% with the introduction of the paralleled output stage. More importantly, THD-20 falls by a factor of more than 5 from 0.074% to 0.014%. This illustrates the distortion-reducing value of employing paralleled output devices.

As mentioned above, the output impedance is halved if it is assumed that the output stage was driven with a voltage source, since two pairs of emitter resistors are now effectively in parallel driving the load. Moreover, the total amount of quiescent bias current has been doubled without

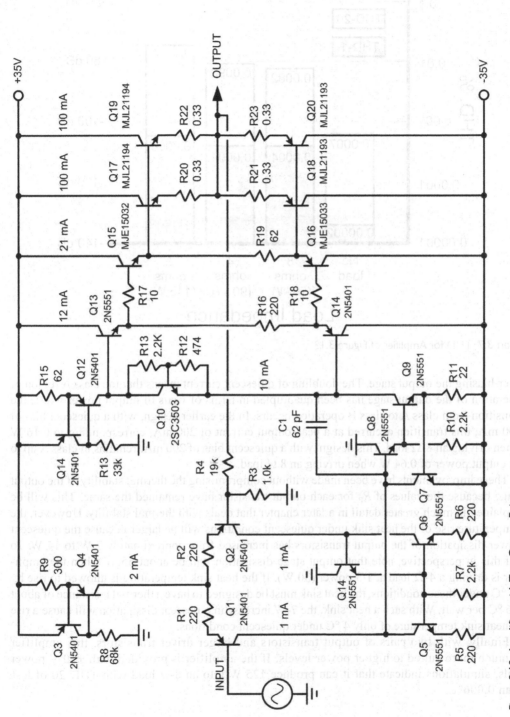

Figure 3.13 Amplifier with Two Output Pairs

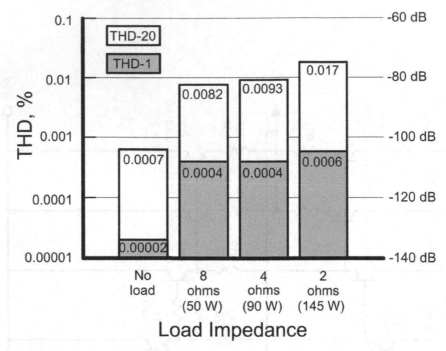

Chart 3.7 THD for Amplifier of Figure 3.13

over-biasing the output stage. The doubling of quiescent current means that the class A region of operation of the output stage has been quadrupled in terms of watts of output power where the transition from class A to class B operation occurs. In the earlier design, with a quiescent bias of 100 mA, this transition occurred at a peak output current of 200 mA, corresponding to 0.16 W when driving an 8-Ω load. This design, with a quiescent bias of 200 mA, remains in class A up to an output power of 0.64 W when driving an 8 Ω load.

These improvements have been made without compromising the thermal stability of the output stage because the values of R_E for each output transistor have remained the same. This will be explained in much greater detail in a later chapter that deals with thermal stability. However, the temperature rise of the heat sink under quiescent conditions will be larger because the quiescent power dissipation of the output transistors has increased from approximately 7 W to 14 W. To put this in perspective, note that output stage dissipation will be about 55 W when the amplifier is driving a 4-Ω load at 1/3 power (30 W). If the heat sink temperature is allowed to rise by 35 °C under these conditions, the heat sink must be designed to have a thermal resistance of about 0.6 °C per watt. With such a heat sink, the 7-W increase in quiescent dissipation will cause a rise in heat sink temperature of only 4 °C under quiescent conditions.

Finally, with two pairs of output transistors and larger driver transistors, this amplifier is much more suited to higher power levels. If the amplifier is provided with ±50 V power rails, simulations indicate that it can produce 125 W into an 8-Ω load with THD-20 of less than 0.006%.

3.9 Higher-Power Amplifiers

Thus far the evolution of a 50-W power amplifier has been illustrated. Keeping the discussion to a fairly low-power design has simplified some things and allowed a level playing field to be maintained throughout the evolution of the design. We now take a look at what happens as the power capability of the design is increased.

Here the amplifier design is evolved to one rated at 300 W when driving an 8-Ω load, with correspondingly higher powers when driving loads of 4 Ω and 2 Ω. Such an amplifier requires higher power supply rail voltages to accommodate the 69.3-V peak output swing required for 300 W with an 8-Ω load. Here ±75-V rails (at full power) have been specified. The closed-loop gain of the amplifier has been increased to 31.6 (30 dB) to allow it to be driven to full power by a 1.55-V RMS input signal.

The amplifier of Figure 3.14 employs an output stage that uses four pairs of output devices to drive the load. This is done in order to support the higher peak current that will occur. This also provides the necessary higher output stage safe operating area (SOA) and power dissipation capability. When driving its achievable 1000 W into a 2-Ω load, peak output current will be almost 35 A, or nearly 9 A for each of the four output devices.

Some other changes have been made to accommodate the higher supply voltage and the higher power dissipation that will result. The VAS and pre-driver transistors have been replaced with the 2SC3503 and 2SA1381 transistors. These devices are rated at 300 V and are provided in a TO-126 package that can dissipate more power than the TO-92 package of the 2N5551/2N5401 pair used in the earlier examples. These devices also feature a fairly high f_T of 150 MHz and a very high Early voltage of over 500 V. The VAS transistors in this example each dissipate 750 mW, while the pre-driver transistors each dissipate 1.4 W.

The MJE15032 and MJE15033 drivers have also been replaced by the MJL21193 and MJL21194 transistors (same as the output transistors). This change provides the higher driver SOA and power dissipation capability needed for the higher-power amplifier. These devices are provided in a TO-264 package and are rated at 250 V. The bias current in the drivers has been nearly quadrupled from 21 mA to 81 mA to provide more turn-off current for the larger output stage. As a result, the driver transistors now dissipate about 6 W each. Output stage emitter resistors have been reduced to 0.22 Ω and each output transistor is now biased at 120 mA to provide 26 mV across each emitter resistor. This further reduces crossover distortion. Each output transistor dissipates about 9 W in the quiescent state. The complete output stage thus dissipates about 87 W at idle.

Sometimes the kind of quality low-noise transistors that would be preferred for use in the input stage cannot withstand high rail voltages. While this is not a serious problem for 50-W to 100-W amplifiers, it does become an issue for larger power amplifiers with higher rail voltages, sometimes in the 60-V to 100-V range. This problem is a special concern when JFET devices are used in the input stage. Although not used in this example, a common solution to this problem is to cascode the input stage so that the collector or drain voltages of the LTP may reside at a lower voltage like 15 V. The input stage has also been improved by adding emitter followers in front of the LTP, increasing open-loop input impedance of the amplifier and greatly reducing input bias current. They also act to lighten the load on the all-important feedback node. The impedance of the feedback network has been cut in half and higher-dissipation resistors have been used in the network to accommodate the higher power levels. Placing 2-W feedback resistors R6 and R7 in series not only increases power-handling capability, but also reduces any voltage modulation effects in the resistors.

Chart 3.8 lists the corresponding distortion performance of the amplifier of Figure 3.14. Although it is operating at 6 times the 50-W power level of the amplifier of Figure 3.13, distortion levels are quite comparable and in some cases even smaller. Notably, THD-20 has fallen from 0.014% to 0.0084%. This is a further reflection of reduced crossover distortion resulting from the doubled-size and halved-output impedance of the output stage.

3.10 Crossover Distortion

Crossover distortion has been mentioned many times during this evolution of the amplifier design, but amplifier performance has always been compared on the basis of full-rated power into the

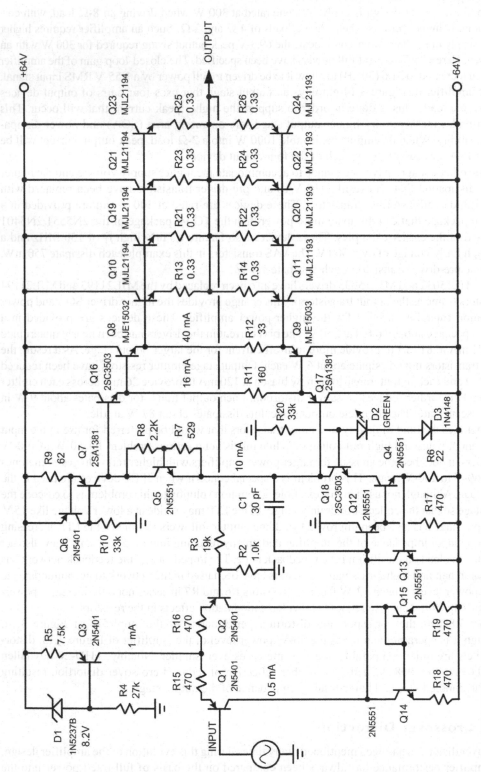

Figure 3.14 High-Power Version of the Amplifier

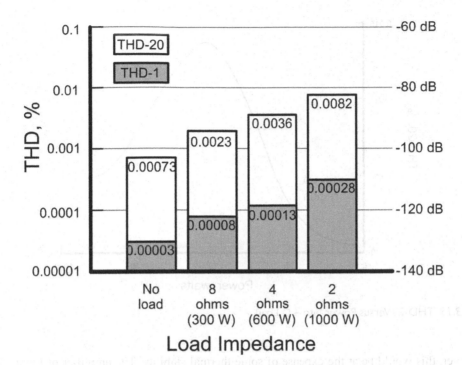

Chart 3.8 THD for Amplifier of Figure 3.14

different load impedances. Before leaving the discussion on amplifier evolution, it is instructive to show the effect of crossover distortion.

It is generally understood that crossover distortion becomes worse at high frequencies and when lower-impedance loads are being driven. THD-20 is a better indicator of crossover distortion because there is less negative feedback available at 20 kHz and its harmonic frequencies to reduce distortion. THD-20 also captures both static and dynamic (switching) crossover distortion.

Crossover distortion characteristically becomes worse at lower power levels. Figure 3.15 shows THD-20 as a function of power when the amplifier of Figure 3.14 is driving a 4-Ω load. The graph illustrates how crossover distortion peaks at a relatively low power. Here the crossover distortion manifests itself as a peak in THD-20 at a power level of 11 W for an amplifier that is rated at 360 W into this 4-Ω load.

Full power distortion thus does not tell the whole story. Here THD-20 peaks at 0.014% when THD-20 at maximum power is only 0.0084%. Crossover distortion in this example results in THD-20 that is greater by a factor of 1.7.

Figure 3.15 also illustrates where the transition occurs from class A operation to class B operation for this amplifier when driving the 4-Ω load. With an idle bias current of 400 mA in the output stage, this amplifier can drive about 800 mA into the load before either the top or bottom power transistors go into cutoff. This 800 mA of peak current into a 4-Ω load corresponds to 1.28 W.

It was mentioned earlier in this chapter that the output stage of the amplifier designs up through Figure 3.13 were not fully optimized for low crossover distortion. The per-transistor operating current of 100 mA represents an approximate 26% over-biasing compared to the theoretical optimum that places a voltage drop of 26 mV across the emitter resistors [10]. The choice of 0.33-Ω emitter resistors is also not optimal. Smaller emitter resistors and correspondingly higher bias currents would reduce crossover distortion by further reducing output stage output impedance.

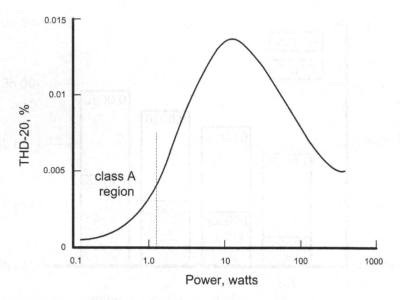

Figure 3.15 THD-20 Versus Power into 4 Ω Load

However, this would be at the expense of some thermal stability. The amplifier of Figure 3.14 employs smaller 0.22-Ω emitter resistors and is biased to place 26 mV across them in order to satisfy the Oliver condition [10]. Much more will be said about crossover distortion in Chapter 12.

3.11 Performance Summary

Chart 3.9 shows a summary of distortion performance for all the example amplifiers discussed in this chapter. In all cases the load being driven is 4 Ω and each amplifier is being driven at the maximum power level that was used in the individual discussions. This power level was 90 W for all cases except the high-power amplifier of Figure 3.14, where the power level was 360 W. THD-1 decreased by a factor of 600 from beginning to end, while THD-20 decreased by a factor of 900 from beginning to end.

3.12 Completing an Amplifier

The amplifier versions described have been deliberately simplified in some ways. Before leaving this chapter, I will discuss a couple of necessary additions to these amplifier cores to make them at least somewhat complete.

Much of this added circuitry is illustrated in Figure 3.16, where the core amplifier is merely shown as a block of open-loop gain. Feedback network resistors R2 and R3 that were included in the core amplifier designs above are shown here for clarity.

Input Network

Resistor R_{in} and capacitor C_{in} form a first-order input low-pass filter that keeps out unwanted radio frequencies. This filter is often designed to limit the bandwidth of the complete amplifier to a frequency that is somewhat less than the actual closed-loop bandwidth of the amplifier proper. A

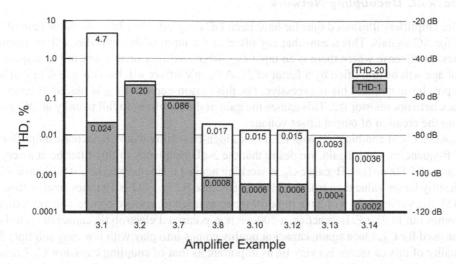

Chart 3.9 Performance Comparison of the Amplifiers

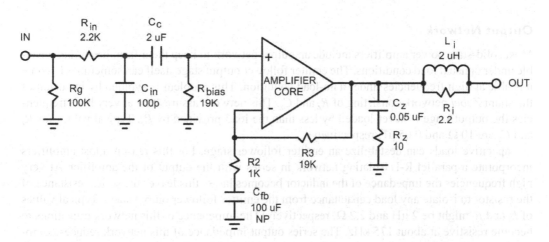

Figure 3.16 A More Complete Amplifier

typical 3-dB frequency for the input filter might be 270 kHz, as shown in Figure 3.16 with R_{in} = 2.2 kΩ and C_{in} = 270 pF. Resistor R_g keeps the input terminal from floating at DC.

Coupling capacitor C_c blocks any DC that might be present from the source, while R_{bias} provides a return path for the input bias current of the amplifier's input stage. This keeps the non-inverting input node of the amplifier near 0 V. Notice that if Q1 in the amplifier is biased at 0.5 mA and it has a beta of 100, its base current will be approximately 0.005 mA. This will cause a voltage drop of 95 mV across R_{bias}. The value of R_{bias} will often be set to equal that of R3 so that voltage drops across these two resistors balance out any DC offset. In this case, R_{bias} would be set to 19 kΩ. The capacitance of C_c against R_{bias} forms a high-pass filter whose 3-dB frequency we wish to keep below 5 Hz so as not to introduce significant frequency response roll-off at 20 Hz. Here a 2-μF capacitor provides a lower 3-dB frequency of 4 Hz. This capacitor should be of high quality (e.g., polypropylene), and performance would be even better if it were 5 μF. However, there is a cost and size issue here. Later we will see that some other trade-offs can be made.

Feedback AC Decoupling Network

The core amplifiers illustrated thus far have been DC coupled. They have the same gain of 20 at DC as for AC signals. This means that any offset at the input of the amplifier will be amplified 20 times. In the case where there is an input R_{bias} whose voltage drop is not fully compensated, that voltage will be amplified by a factor of 20. A 10-mV offset will become a 200-mV offset at the output, for example. This is excessive. For this reason capacitor C_{fb} is placed in series with feedback network resistor R2. This causes the gain of the amplifier to fall to unity at DC, greatly reducing the creation of output offset voltage.

Of course, C_{fb} in combination with R2 forms a high-pass filter that will decrease amplifier gain at low frequencies. Once again, we desire that the 3-dB frequency of this filter be at a very low frequency of 5 Hz or less. Because C_{fb} is working against a smaller resistor, it will have to be a significantly larger value than was used for C_c. Because R2 at 1 kΩ is 19 times smaller than R_{bias} at 19 kΩ, the value of C_{fb} will have to be 19 times as large to achieve comparable performance. This comes out to 38 μF. In practice, a 100-μF non-polarized electrolytic capacitor would typically be used for C_{fb}. Once again, capacitor quality comes into play with few easy solutions here. The quality of this capacitor is every bit as important as that of coupling capacitor C_c. Capacitor distortion is discussed in Chapter 16. We will later see some discussion of alternatives to the use of this capacitor in Chapter 10 where DC servos are discussed.

Output Network

Most solid-state power amplifiers include an output network to keep them from becoming unstable under unusual load conditions. The emitter follower output stage itself can sometimes become unstable at high frequencies under a no-load condition. This problem is avoided by inclusion of the shunt *Zobel* network consisting of R_z and C_z. This network assures that at very high frequencies the output stage is never loaded by less than the load provided by R_z. Typical values for R_z and C_z are 10 Ω and 0.05 μF, respectively.

Capacitive loads can destabilize an emitter follower stage. For this reason, most amplifiers incorporate a parallel R-L isolating network in series with the output of the amplifier. At very high frequencies the impedance of the inductor becomes large; this leaves the series resistance of the resistor to isolate any load capacitance from the emitter follower output stage. Typical values of L_i and R_i might be 2 μH and 2.2 Ω, respectively. The impedance of this network transitions to become resistive at about 175 kHz. The series output impedance of this network reduces damping factor at high frequencies and can cause a reduction in high-frequency response. The impedance of a 2-μH inductor at 20 kHz is about 0.25 Ω. This corresponds to a damping factor of 32 even if the output impedance of the amplifier proper is zero. When a 4 Ω load is being driven, this will cause a frequency response droop at 20 kHz of about 0.06 dB. Much more will be said about these output networks in a later chapter.

Base Stopper Resistors

Small resistors are often placed in series with the individual bases of the output transistors when multiple output pairs are connected in parallel. These resistors, often on the order of 2.2 Ω to 4.7 Ω, improve the stability of the output stage when multiple devices are connected in parallel. They have a minor effect on circuit performance that should be taken into consideration. For example, they create a small voltage drop that is dependent on the beta of the output transistors. If $\beta = 50$ and the device is biased at 120 mA, base current will be 2.4 mA and 11 mV will be dropped across a 4.7-Ω resistor. The ohmic impedance seen looking back into the emitter of the output transistor will also be increased approximately by the value of the resistor divided by β, in this case by 0.1 Ω. This can influence the value of optimum bias for minimization of crossover distortion, since this 0.1 Ω will tend to look like part of R_E.

Power Supply Decoupling

A single pair of positive and negative power supply rails has been used to power all of the circuits in the amplifiers that have been described. This was done for simplicity. In a practical amplifier the IPS and VAS circuits will usually have an R-C filter inserted in the power supply rail to filter their supply. This provides them with a much cleaner power supply rail than the high-current portion of the rail that supplies the output stage. The high-current portion of the rail usually has considerable ripple and noise on it. It is also very important that power supply decoupling is used to avoid sneak feedback paths through the power supply rails that might cause instability or even oscillation.

3.13 Summary

We've seen how the basic power amplifier topology can be evolved with straightforward steps into a design with fairly high performance. The main purpose of this chapter has been to show the thinking that goes into such an evolution. Although the results achieved here are quite good, this is just the beginning of high-performance design. There are many more nuances and topologies to be considered in the following chapters. Other approaches include JFET input stages, complementary push-pull VAS designs, MOSFET output stages and many more variations and improvements. Moreover, the designs evolved here did not address many real-world issues like protection circuits, power supplies and many other such matters.

References

1. Larry W. Nagel and Donald O. Pederson, "Simulation Program with Integrated Circuit Emphasis," Proceedings of Sixteenth Midwest Symposium on Circuit Theory, Waterloo, Canada, April 12, 1973; available as Memorandum No. ERL-M832, Electronics Research Laboratory, University of California, Berkeley.
2. LTspice®, developed and distributed by Linear Technology Corporation/Analog Devices, Inc. Available at www.analog.com.
3. Matti Otala, "Conversion of Amplitude Nonlinearities to Phase Nonlinearities in Feedback Audio Amplifiers," Proceedings of IEEE International Conference on Acoustics, Speech and Signal Processing, Denver, CO, 1980, pp. 498–499.
4. Robert R. Cordell, "Phase Intermodulation Distortion—Instrumentation and Measurements," *Journal of the Audio Engineering Society*, vol. 31, March 1983.
5. Matti Otala, "Transient Distortion in Transistorized Audio Power Amplifiers," *IEEE Transactions on Audio and Electroacoustics*, vol. AU-18, pp. 234–239, September 1970.
6. Robert R. Cordell, "Another View of TIM," *Audio*, February–March 1980.
7. R. W. Russell and James E. Solmon, "A High-Voltage Monolithic Operational Amplifier," *IEEE Journal of Solid State Circuits*, vol. SC-6, no. 6, p. 352, December 1971.
8. Bart Locanthi, "Operational Amplifier Circuit for Hi-Fi," *Electronics World*, January 1967, pp. 39–41.
9. Bart Locanthi, "An Ultra-low Distortion Direct-current Amplifier," *Journal of the Audio Engineering Society*, vol. 15, no. 3, pp. 290–294, July 1967.
10. Barney M. Oliver, "Distortion in Complementary Pair Class B Amplifiers," *Hewlett Packard Journal*, pp. 11–16, February 1971.

Chapter 4

Building an Amplifier

4. INTRODUCTION

Here we describe what it takes to build a complete and working amplifier by covering a working design that can be built. The amplifiers in Chapter 3 were incomplete and for purposes of illustration only. This design will gel those concepts and satisfy the appetite of those who may not want to read the rest of the book until later. Real-world design, build and test issues are discussed in one place on a simple amplifier. This amplifier is not sophisticated, but yields very respectable performance and it is designed to be robust. It includes necessary items to make a complete amplifier and improvements to elevate its quality at minimum increase in complexity and cost. Various aspects of the amplifier will be referred to in some of the later chapters, and it will be referred to as the BC-1 amplifier.

The amplifier employs two output pairs and produces about 125 watts into an 8-Ω load with ±52 V rails (under load). It is possible to design for lower or higher power by changing only the rail voltages and adding additional output pairs and heat sinking for rated power levels above 150 watts. Key measured performance characteristics include:

- 125 W into 8-Ω load with ±52 V power rails
- THD+N 0.0008% at 1kHz, 0.004% at 20 kHz
- 9 nV/$\sqrt{\text{Hz}}$ input-referred noise
- Un-weighted S/N of 97 dB in 80-kHz noise bandwidth
- A-weighted S/N of 100 dB

4.1 The Basic Design

The amplifier is much like that of Figure 3.13. Some component value changes and other improvements and additions make it a practical amplifier. The schematic of the front-end is shown in Figure 4.1. These revisions and additions include:

- Gain set to 28 (29 dB) so that 1V RMS generates 100 W into 8 Ω
- Complete input, feedback and output networks
- Quasi-differential input/feedback network
- Diode clamps across the feedback network capacitor for fault protection
- Diode clamps across input differential pair for transistor protection
- Diode clamps across the IPS differential outputs for cleaner clipping
- ULGF increased to 1 MHz
- VAS output Zobel network for increased stability
- Dual V_{be} multiplier bias spreader with bypass/reservoir capacitor

- Current limiting diodes from bias spreader to output rail
- Output transistor base stopper resistors
- Faster output transistors (MJL3281/1302)
- Reduced output emitter resistor values to 0.22 Ω
- Rail clamp diodes added to amplifier output
- Rail reversal protection diodes
- Progressive rail decoupling

Functional Additions

- Amplifier protection circuits
- Loudspeaker protection circuits
- Power on/off muting

4.2 The Front-End: IPS, VAS and Pre-Drivers

The amplifier front-end (AFE) is shown in Figure 4.1. It includes all of the amplifier circuitry up to and including the pre-drivers. It is implemented on its own board. The output stage, including drivers, is implemented on a second board that is mounted on the heat sink. The protection circuits are implemented on a third board.

The front-end is much like that of Figure 3.13, but with the addition of an input network, clamping diodes, a dual bias spreader and current limiting diodes.

Input Network

The main part of the power amplifier can be viewed as a power operational amplifier. Looking closely at the input and feedback networks, one can see that the amplifier is configured somewhat like a single op-amp differential amplifier. This provides some common-mode rejection to signals that exist at both the shield and center conductor of the input cable.

The first input low-pass filter formed by R2 and C1 has two purposes. First, it provides a roll-off at 23 MHz. This is the first line of defense against high-frequency EMI at the input. Second, it terminates the input interconnect at approximately its characteristic impedance at high frequencies (68 Ω is a compromise between the typical values of 50 Ω and 75 Ω for shielded interconnect). The second, and more conventional, input low-pass filter is formed by R4 and C3 with a corner frequency at 720 kHz. The 10-μF NP coupling capacitor and 27-kΩ shunt return resistor create a roll-off at 0.6 Hz.

Resistor R1 (4.7 Ω) isolates the RCA connector ground and circuit ground. This reduces the effects of ground loops. The 1-kΩ negative feedback shunt resistor R6 is returned to the RCA connector ground through C4. This arrangement, in combination with the small attenuation of R4 against R5, gives the amplifier some of the properties of a differential input. The feedback resistor comprises two 1-W metal film resistors in series (R7 and R8). This minimizes voltage and temperature distortions in this critical location. Shunt resistor R6 is also a 1-W metal film type. Together, these resistances establish the closed-loop gain to be 27.3, not including the 0.5-dB loss created by R4 and R5. The reasonably low impedance of the feedback network provides a good compromise between low noise and the amplifier input resistance of 28 kΩ.

The feedback resistors consist of one 13.3-kΩ resistor in series with one 14.0-kΩ resistor, both available in E48 or E96 series values. An approximate center tap is made available by the use of two feedback resistors in series. It can be used for phase margin and gain margin testing to be described later. Reduced feedback network impedance can be used to reduce noise a bit, but this will require a smaller input return resistor unless DC offset is to be incurred.

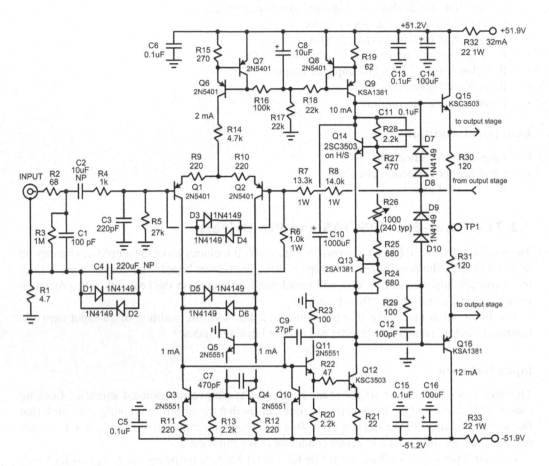

Figure 4.1 Amplifier Front-End

The IPS input return resistor is made equal to the feedback resistor value. This balances out the effect of input transistor base currents if they are equal. At 1 mA collector current, base current of the IPS is 10 μA for transistor beta of 100. This 10 μA flowing through 27 kΩ creates a 270-mV drop. If beta of the two input transistors is matched to 10%, the offset voltage will be only 27 mV. This underscores the need for reasonable beta matching of the input pair.

Feedback Network

A non-polarized electrolytic decoupling capacitor C4 has been added to the feedback network to bring the gain down to unity at DC. This minimizes DC offset at the output. The 220-μF, 50-V NP capacitor creates a 0.7-Hz low-frequency corner against R6. It contributes virtually no electrolytic capacitor distortion because of its relatively high voltage rating for this application. This capacitor is made by Nichicon (UES1H221MHM). Nevertheless, it is often recommended that such electrolytic capacitors in critical locations be bypassed with a film capacitor. Although not shown on the schematic, 1-μF, 63-V polypropylene capacitors have been placed across both C2 and C4. Bypassing C4 also minimizes any effects of inductance in C4 on the feedback network at high frequencies. Clamping diodes across C4 prevent high voltages from appearing in the event of a fault where the output is stuck at a rail. Although C4 can withstand 50 V, such a voltage applied to the LTP could damage it.

Input Stage and VAS

The design includes a degenerated LTP with a current mirror load and a Darlington VAS, as shown in Figure 4.1. The tail current is 2 mA and the degeneration resistor values are 220 Ω. Base-to-base protection clamp diodes prevent the input differential pair transistors from ever seeing a large base-emitter reverse bias that could cause Zener breakdown of the base-emitter junctions under fault conditions. Base-emitter breakdown is only 6 V for the 2N5551. Breakdown conduction can damage the input transistors and cause permanent beta degradation. The diodes chosen are 1N4149, which are similar to the venerable 1N4148, but with lower capacitance. The IPS and VAS current sources share a common intermediate filtered voltage (25 V) from which they are energized.

The current mirror includes an emitter follower "helper" transistor that supplies the base current for the mirror transistors. This improves mirror accuracy and causes the collectors of both IPS transistors to be at the same DC potential as a result of the $2\text{-}V_{be}$ input voltage of the Darlington VAS. Emitter degeneration voltage drops in the current mirror and VAS are made the same. The equal IPS collector voltages enable the use of anti-parallel 1N4149 clamp diodes across the IPS collectors to limit and equalize the voltage swings under clipping conditions. This mitigates some clipping artifacts and limits over-current in the VAS Darlington transistor under negative clipping conditions. A 470-pF compensation capacitor has been placed from base to emitter of the helper transistor to reduce its role at very high frequencies, improving local stability without compromising the bandwidth of the current mirror. It also reduces HF interactions between the current mirror and VAS Miller compensation at high frequencies.

The two-transistor VAS enjoys good performance due to the speed and minimal Early effect of the KSC3503 main transistor. The VAS current limiting transistor is a 2N5551 with $V_{be} = 0.66$ V @ 2 mA. In combination with R21 = 22 Ω, this creates a 30 mA current limit.

The 2T VAS emitter follower transistor can sometimes be subjected to excessive current when negative clipping occurs. In a normal arrangement, when negative clipping occurs, the IPS output will try to turn on the VAS transistor harder. This is especially the case when a current mirror load is used. Nearly all of the IPS tail current will try to flow into the base of the first transistor of the 2T VAS. That transistor will source significant current from its emitter into the base of the saturated VAS transistor, ultimately flowing into its emitter and emitter degeneration resistor. Collector current spikes of at least 20 mA can occur in the first VAS transistor. Since this transistor has nearly the full rail voltage across it, there may be safe area concerns. A limiting resistor that operates in conjunction with the IPS output clamps is placed in series with the base of the VAS transistor.

The limited voltage swing from the IPS as a result of the clamp diodes limits the amount of current that can flow in Q11 because of the drop across the VAS emitter resistor. When negative clipping occurs, the input to Q11 increases by only one diode drop. However, this alone does not sufficiently limit collector current in Q11. To further limit the current, a 47-Ω resistor is placed in series between the emitter of Q11 and the base of Q12. Excess current flowing from Q11 into the base of now-saturated Q12 will create a voltage drop across this resistor that will create a self-limiting effect on the current in Q11. Current will be limited to a much safer 6 mA. With 52-V rails, Q11's dissipation will briefly be about 300 mW.

Compensation capacitor C9 has been decreased to 27 pF, setting the gain crossover frequency (ULGF) to just below 1 MHz. C9 should be a ceramic COG capacitor rated at 100 V or more for best linearity. A small Zobel network has been added in shunt with the VAS output node to improve stability at frequencies above the ULGF. It ensures that the impedance at this node remains positive resistive at very high frequencies. This is especially helpful when output Triples are employed. The Zobel has little effect on the global compensation scheme and detracts minimally from slew rate. Note that on positive clipping the Miller shunt feedback around the VAS becomes inactive and the VAS output impedance may go high, potentially causing instability in the output stage. The VAS output node Zobel network prevents this.

The pre-driver emitter resistor is actually made up of two equal resistors (R30, R31) in series. This provides a center tap for testing purposes. It allows the feedback loop to be closed from the pre-driver output. This is especially handy for testing the front-end board by itself.

Bias Spreader

In contrast to the simple V_{be} multiplier bias spreader illustrated in the earlier amplifier examples, a two-transistor dual V_{be} multiplier is used. Most amplifiers employ a single V_{be} multiplier. If all of the pre-driver, driver and output transistors are mounted together on the main heat sink, then a single-transistor bias spreader with its transistor also mounted on the heat sink works well, since all of the transistors will be at about the same case temperature.

Here a different approach to mounting the pre-driver and driver transistors has been chosen. The pre-driver transistors are not mounted on the main heat sink. They are thus at a different temperature from the driver and output transistors. The pre-driver transistors are best compensated with a separate V_{be} multiplier. One of the V_{be} multiplier transistors is mounted on the heat sink and the other is mounted on the circuit board.

The dual bias spreader design employs a complementary pair of V_{be} multiplier transistors connected in series. One transistor is used to temperature compensate the driver and output transistors while the other compensates the pre-driver transistors. A ten-turn bias trim pot is included in one of the V_{be} multipliers. A TO-126 NPN V_{be} multiplier transistor (2SC3503) is mounted on top of one of the power transistors with the same screw that holds the power transistor to the heat sink. Thermal grease is used at the interface. The insulated TO-126 case of the KSC3503 can be easily mounted to the heat sink or onto the case of one of the output power transistors. It serves to temperature-compensate the output transistors and the drivers with a spreader voltage of ~ 4 V_{be}. A TO-126 PNP V_{be} multiplier transistor (KSA1381) is sandwiched between the pre-driver transistors on the front-end board, with thermal grease at the interfaces of the three transistors. A single screw with washers on either side holds the three TO-126 transistors together. Its spreader voltage is ~ 2 V_{be} to furnish the bias required by the pre-drivers. Total dissipation of the two pre-drivers is about 1.3 W. A small on-board heat sink is mounted to the group of three transistors.

The dual bias spreader arrangement provides good stability during warm-up and does not subject the pre-drivers to the potentially significant thermal variations of the main heat sink that are a function of power output. The pre-driver transistors operate at nearly constant power in class A, so there is no reason to subject them to such thermal variations. The dual V_{be} bias spreader approach is especially helpful to bias stability in Triple EF output stages because of the number of V_{be} drops that are stacked up in the Triple EF.

An unusually large 1000-μF bypass capacitor has been added to the bias spreader. The large-value capacitor acts like a reservoir capacitor for the bias spreader in the event that current through the VAS goes to zero during positive clipping. This minimizes the impact of bias spreader collapse under these conditions. Such a collapse can create a time-dependent discharge of bias voltage, resulting in biasing error hangover effects that may persist after the clipping interval. This occurs on only one-half clipped cycle with a single-ended VAS, since the VAS current source does not cut off on negative clipping. This can happen on both half-cycles with a push-pull VAS.

Bias spreader collapse can be significant on large-amplitude low-frequency signals, whose half-cycle could last 25 ms at 20 Hz. Some fraction of that half-cycle may clip the amplifier.

Assume a 15 ms clipping interval and a maximum allowable bias spreader sag of 26 mV (to half idle bias) for sizing the capacitor. The main cause of spreader sag is the discharge of the spreader bypass capacitor by the spreader resistors. This assumes that under positive clipping conditions no current is flowing in the bias spreader transistor. If 1mA is being passed through the V_{be} multiplier resistors and a 1000-μF spreader bypass capacitor is being employed, the collapse will be limited to 15 mV.

Current Limiting

Two series-connected pairs of 1N4149 natural current limiting diodes are connected from the ends of the bias spreader to the output rail. I also refer to these as "flying catch diodes." These diodes clamp the VAS output voltage relative to the output rail if the voltage drop across the output emitter resistors exceeds a certain value. This prevents the output stage from delivering more than a set value of maximum current. Consider a total spreader voltage of about 3.8 V and a diode turn-on voltage of 0.6 V. The voltage at one end of the spreader needs to move by about half the spreader voltage plus a diode drop before the limiting diodes conduct and limit further current increase. In this case, that would be about 2.5 V. With 0.22-Ω emitter resistors, each output transistor will be limited to peak current of about 11 A. The short-circuit protection circuit limits the duration of this peak to a very brief interval. If this interval is exceeded, the loudspeaker relay will be opened. In reality, the peak current limit will be a bit smaller than 11 A, since this simple analysis did not take into account the increased V_{be} drops in the output stage under these conditions. For this reason, two diodes in series are used.

In the case of a Triple, where the bias spread is about 6 V_{be}, the clamp diode pair is reverse-biased by 3 V_{be} when no output current is flowing. If enough current flows to cause a 3-V_{be} plus two-diode drop change at either end of the bias spreader, the diodes will become forward-biased and prevent any further increase in output current. Five junction drops here is about 3 V. Much of this change will end up across R_E. If impressed across R_E = 0.22 Ω, this corresponds to a current limit of about 13.6 A per output transistor. At these high currents there will be some voltage drop in the base stopper resistors and increased V_{be} of the output transistors. As a result, the actual current limit is smaller, at about 11 A. With two output pairs, total output current will thus be 22 A. This high current-limiting threshold allows the amplifier to briefly deliver 44 V into a 2-Ω load.

4.3 Output Stage: Drivers and Outputs

The output stage and output network are shown in Figure 4.2. The output stage is a Locanthi Triple with two output pair [1, 2]. The emitter resistor values have been reduced to 0.22 W, making the quiescent current for each pair 118 mA when the Oliver condition is satisfied (26 mV across R_E) [3]. This reduces crossover distortion and increases the class A region of operation. Peak class A current is 4 * 118 mA = 472 mA, corresponding to about 1 W into 8 Ω.

The driver emitter resistor is made up of two equal resistors (R36, R37) in series in order to provide a center tap for testing purposes. Capacitor C27 in combination with R36 and R37 forms a Zobel network to enhance output stage stability by forcing the net load on the driver emitter followers to be largely resistive at all frequencies. The healthy 29-mA driver bias current helps turn off the output transistors quickly by depleting stored charge in the base. Base stopper resistors have been added in series with each output transistor base to suppress HF oscillations that can sometimes occur when multiple output pairs are connected in parallel. These resistors are 2.2 Ω, 1-W metal oxide film (MOF) types.

The output stage emitter resistors must pass significant current and be capable of dissipating several watts. It is also desirable that they be non-inductive. It has often been common practice to use 5-W non-inductive wire-wound power resistors for R_E. In this amplifier 0.22-Ω, 3-W MOF

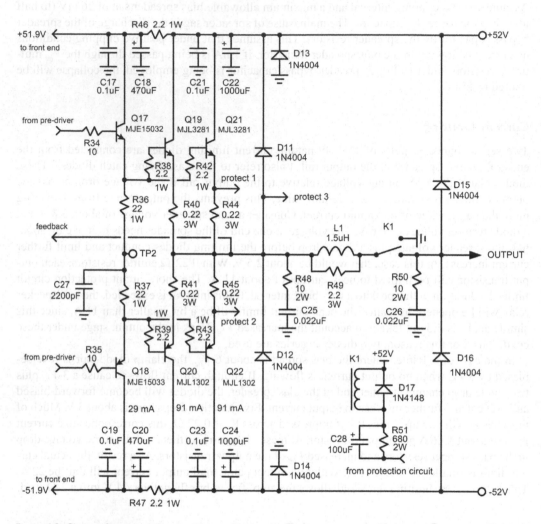

Figure 4.2 Output Stage and Output Network

resistors are used. They have almost no inductance and provide adequate power dissipation capability. The use of wire-wound resistors has been avoided in this amplifier.

Total power dissipation in the R_E resistors can be calculated by recognizing that the net output resistance due to R_E will dissipate a fraction of the power being delivered to the load in proportion to the series resistance they create in the output stage compared to the load resistance. In this amplifier the net output resistance contributed by the R_E resistors is 0.11 Ω under large-signal conditions. This is 2.8% of a 4-W load. If the amplifier is delivering 250 W into 4 Ω, the total dissipation in the 4 R_E resistors will be 7 W, and dissipation in each resistor will be less than 2 W. This amount of dissipation will occur only in continuous full-power testing. For best power dissipation, the emitter resistors should be mounted vertically (one lead coming off the top and down the side) or axially 1/4-inch off the board. MOF resistors are capable of passing a large amount of current for brief portions of a cycle without damage or degradation. In the above example, each resistor is called upon to conduct brief peak current of 5.6 A, and drop 1.2 V, for peak dissipation of about 7 W.

Protection diodes D11 and D12 are connected from the output node to each power rail to prevent the output transistors from being reverse-biased if an inductive load should ever cause the output node to snap to a voltage beyond the rail voltage. Reverse diodes (D13, D14) to ground from each rail prevent the rail voltage from going to a reverse polarity in the event of a fault condition, such as the failure of one rail.

Proper power supply rail decoupling is paramount in all amplifier implementations. Here the power rails are bypassed with 1000-μF right at the power transistors to minimize the amount of high-frequency current that will flow through the wiring back to the power supply. Instead, the signal current tends to circulate locally through the bypass capacitors. The ground ends of the capacitors are connected together and then their junction node is routed to the rest of the ground line to encourage local circulation of the currents and minimize the net amount of this circulating current that passes through the main ground line. All electrolytic decoupling capacitors are bypassed with 0.1-μF metalized polypropylene film capacitors.

The power rails have been progressively decoupled from the output stage back to the pre-drivers and IPS-VAS. This further improves stability of the output Triple by suppressing inadvertent coupling at high frequencies from the output transistors back to the drivers and pre-drivers. The low R-C impedances of these decoupling networks also act like Zobel networks to damp high-frequency resonance in the rails due to wiring inductance. Finally, the added rail filtering provides a cleaner power supply to the more sensitive IPS-VAS circuits. The DC voltage drop across these R-C networks is kept below 1 V in order to preserve voltage headroom for the earlier stages.

Output Network

The R-C Zobel network provides a minimum resistive load to the output stage at high frequencies, ensuring stability. Its value is relatively uncritical, but a combination of 0.022 μF and 10 Ω is used here. It should be located physically close to the output transistors, with minimal wiring inductance. At 31.6 V RMS, corresponding to about 125 W into 8 Ω, power dissipation of the 10-Ω resistor will be 0.6 W at 40 kHz. A 2-W metal oxide film (MOF) resistor is used, avoiding the need for a wire-wound resistor that might be inductive. This is adequate for most normal operating conditions and full-power testing below 40 kHz. The resistor should be mounted axially 1/4 inch off the board. A serious high-level parasitic oscillation that persists may fry the resistor, but this will be the least of one's problems.

The L-R series network consisting of L1 and R49 preserves stability with capacitive loads by isolating them from the output stage at high frequencies. Relatively small values of 2.2 Ω and 1.5 μH provide adequate isolation for this amplifier. The network becomes resistive above about 230 kHz. The impedance of the coil is about 0.2 Ω at 20 kHz. This will cause a loss of less than 0.03 dB at 20 kHz when driving an 8-Ω load. Because the impedance of the coil rises to 0.2 Ω at 20 kHz, DF at 20 kHz will be limited to 40.

Dissipation of R49 is a concern when continuous high-frequency testing is being done at high power levels. For 250 W into 4 Ω at 20 kHz, 1.6 V RMS will appear across the 1.5-μH coil, resulting in 1.2 W of dissipation in the resistor. This power dissipation rises as the square of frequency. Dissipation will be about 4.8 W for continuous full-power testing at 40 kHz with a 4-Ω load. A 3-W MOF resistor is used here in recognition that high dissipation will be a rare occurrence in normal operation. This avoids the use of a wire-wound resistor. Full-power testing at higher frequencies like 40 kHz should be kept to less than 20 seconds. Full-power testing into a 2-Ω load should be limited to 30 kHz. These resistors should be mounted axially 1/4 inch off the board. Resistor dissipation is also of concern if the amplifier breaks into a continuous oscillation at very high frequencies, but in this case overheating of the resistor might be the least of one's problems.

The 1.5-μH output coil consists of 12 turns of 18-AWG magnet wire with a diameter of 0.5 inch. The coil length will also be about 0.5 inch. Such a coil can be free-standing, without a former. Do not wind the coil on R49. Coil doping or the equivalent should be used to hold the windings in place and damp any tendency to vibration. Self-bonding magnet wire is also available, where an adhesive layer on the surface can be activated with a heat gun. Coil resistance is about 10 mΩ, and will result in some power dissipation. Coil dissipation will be in proportion to the coil resistance as a fraction of the load impedance. If a 10-mΩ coil is loaded with 4 Ω, then the coil will dissipate 0.25% of the power in the load. If the power output is 250 W, the coil will dissipate 0.6 W. This is reasonably small and will only occur in continuous sine wave testing. The coil should be kept physically away from any ferrous materials to the extent possible, such as the enclosure if it is steel.

Speaker relay K1 provides muting, short-circuit protection and loudspeaker protection. The circuits that control it are described later. When K1 is open its NC contact shunts the loudspeaker to ground. An optional second 0.022-μF/10-W Zobel network is mounted across the loudspeaker terminals. It further damps the loudspeaker side at high frequencies and helps to absorb EMI energy right at the chassis output terminals. It also provides some snubbing of the speaker relay contacts when the relay opens, reducing the degree of arcing if current is flowing when the contacts open. Rail clamp diodes D15 and D16 prevent any inductive flyback from the loudspeaker from exceeding the rail voltage when the relay contacts open. The Panasonic SPDT ALZN1B24 is a good choice for the relay. It is rated at 16 A, and has a 24-V, 1440-Ω coil.

Output Stage Layout

The high collector currents in a class AB output stage are very nonlinear, with a half-wave-rectified shape when the signal is a sinusoid. These currents can create nonlinear magnetic fields that may cause significant distortion if they are coupled into other parts of the amplifier signal path, especially the feedback path. The sum of the collector currents of an NPN-PNP output pair must be nearly a linear representation of the signal, since this sum is nearly equal to the actual amplifier output current. This means that it is desirable to have these currents sum together with as little intervening wiring as possible, and for the loop formed by this wiring to be as small as possible.

It is also important to have a minimum of these nonlinear currents flowing in the power rails or the ground. Ideally, both transistors in an output pair should be next to each other on the heat sink. Their collectors should be strongly AC-coupled together with tight wiring. In designs with multiple output pairs, having all of the NPN output transistors grouped together on one side, and having the PNP transistors grouped together on the other side, is less optimal. In such a case, the path lengths to merge the resulting positive and negative rail currents are longer. Always keep in mind that current will follow the path of least impedance.

4.4 Heat Sink and Thermal Management

The heat sink must be sized to prevent the power transistors from becoming too hot under anticipated worst-case operating conditions. A useful criteria for selecting the heat sink thermal resistance rating is that the heat sink not get hotter than 60 °C when the amplifier is delivering 1/3 rated power into an 8-Ω load. This is fairly conservative, and in compliance with the original FTC requirement [4]. It is less conservative when driving a 4-Ω load. A heat sink at 60 °C is not dangerous to touch; you can keep touching it for about 5 seconds. Limiting the heat sink temperature to 60 °C also keeps the junctions of the output transistors reasonably cool, preserving their safe operating area. A TO-220 IC temperature sensor mounted on the heat sink opens the speaker relay if the heat sink temperature reaches 70 °C. A mains thermal cutout switch can be used instead, if desired.

A typical class AB output stage dissipates about 46% of its rated power when operating at 1/3 continuous power. A 150-W version of the amplifier will thus dissipate about 70 W when driving 50 W into an 8-Ω load. With an ambient temperature of 25 °C, the temperature rise for the heat sink must be less than 35 °C, so the thermal resistance of the heat sink should be about 0.5 °C/W. With this size heat sink, the amplifier should not be operated at high continuous average power for extended periods of time into 4-Ω or 2-Ω loads.

Optional Driver Circuit Heat Spreader

For alternative physical designs where the drivers are not mounted on the main heat sink, the pre-driver and driver transistors can be mounted together on a common heat spreader. The heat spreader can be implemented with a 1 inch × 3 inch × 1/8 inch piece of aluminum oriented vertically. A copper strip can also be used. With pre-drivers operating at 12 mA and drivers operating at 30 mA, total power dissipation is about 5 W if rail voltages are ±58 V under no-signal conditions. Here the V_{be} multiplier transistor for the pre-drivers and drivers would be mounted on the heat spreader and would provide about 4 V_{be} of bias voltage.

4.5 Protection Circuits

Current limiting and short-circuit protection act to prevent destruction of the output transistors in the event of fault conditions. Loudspeaker protection safeguards the expensive loudspeakers in the event of an amplifier fault condition, such as one where a large DC voltage would appear at the output. Turn-on/off muting prevents thumps from getting through to the loudspeakers when the amplifier is power cycled. An over-temperature sensor opens the speaker relay in the event that the heat sink becomes too hot.

The protection circuits are implemented on a third board that can be mounted on top of the output board. The speaker relay and its series resistor are mounted on the output board. A suitcase jumper can connect the series resistor to ground, so that the relay can be closed for testing in the absence of the protection board. Relatively few connections are required to feed information to the protection board. The schematics of the protection circuit board are shown in Figures 4.3, 4.4 and 4.5.

Features of the protection circuit include:

*   Current limiting of 22 A, allowing 480-W bursts into 2 Ω
*   Speaker relay
*   Short-circuit protection
*   DC offset protection
*   Rail voltage monitoring
*   On/off muting
*   Over-temperature detection

Current Limiting and Short-Circuit Protection

The difference between current limiting and short-circuit protection is that the former limits output current, while the latter opens the speaker relay in the event of a short circuit (and here keeps it open for 3 seconds). The natural current limiting implemented at the pre-driver bases is instantaneous, and results in current clipping, while the short-circuit protection circuit incorporates a delay time constant so that it is not activated for brief current spikes that are part of program material.

It is important that the short-circuit protection circuit long-term (e.g., 1 second) trigger threshold be set to a lower value of current than that of the current limiting circuit. Otherwise the short-circuit protection circuit will never trigger and currents at the level of the current limiting value will be allowed to persist indefinitely.

Output Transistor Safe Area

The MJL3281/MJL1302 output transistors have a safe operating area (SOA) rating of 3 A at 55 V for 1 second. For this amplifier, two pairs will safely source 6 A for 1 second into a short circuit. Of course, it is likely that the power supply will sag to less than 55 V in much less than 1 second when 6 A is being sourced into a short circuit unless the power supply is extremely stiff. Note that 6 A at 55 V corresponds to 330 W peak into a 9-Ω load. For short intervals, like 50 ms, the devices can withstand significantly more current and/or voltage. They can withstand 6 A at 55 V and 3.5 A at 80 V, for 50 ms, for example.

In the early stages of a short circuit event, the transistors will conduct much more current, as limited by the current limiting circuit. This will trip the short-circuit protection and open the speaker relay within less than 50 ms. The 50-ms SOA is thus the most important parameter for short-circuit protection in this design.

Current Limiting

As described earlier, flying catch diodes have been added between each end of the bias spreader and the output node to implement natural current limiting. If sufficient current flows through the top output emitter resistors, the VAS output node will rise high enough to turn on the clamping diodes at the bottom end of the bias spreader, shunting VAS current to the output node and limiting peak output current. The opposite happens if sufficient current flows through the bottom emitter resistors. In this design, these diodes limit the peak current to 11 A per output transistor, so the amplifier can produce 22 A peak.

The limit of 11 A per output device is almost 4 times the 1-second SOA of 3 A at 55 V. This current limiting alone will not provide protection against a short circuit fault, unless other means are used to keep the duration of this fault to much less than 1 second. That is the role of the output short-circuit protection arrangement, wherein the speaker relay is opened in a fraction of a second.

A peak current of 22 A into a 2-Ω load will result in output voltage of 44 V, corresponding to 968 W peak, or 484 W average. This places a reasonable limit on maximum output power into 2 Ω that is about 1.6 times the maximum (voltage-limited) output for a 4-Ω load. These numbers apply only to brief bursts that do not cause significant rail sag.

Loudspeaker Relay

The amplifier employs a speaker relay that opens during muting, short-circuit, DC offset faults and over-temperature conditions. The 16-A speaker relay is mounted on the output circuit board and controlled from the protection circuit board. The loudspeaker is connected to the swinger of the relay. When the relay is open, the swinger will be connected to ground, shunting the loudspeaker to ground and absorbing any EMF energy. This also prevents a contact arc from delivering current into the loudspeaker, directing it instead to ground. The optional second Zobel network acts as a damper for the relay contacts, reducing the possibility of arcing when the relay opens. It also damps the loudspeaker during the brief interval when the swinger is open while transitioning from the NO contact to the NC contact.

A 24-V relay is used. Most relays of this type require about 0.5 W for the coil, corresponding to about 24 mA and a coil resistance of about 1000 Ω. The relay used here has a 1440-Ω coil that requires 17 mA. The relay driver circuit pulls the relay coil and series resistor combination down to about +8 V when the relay is energized. The series resistor is thus required to supply 24 mA (with an average relay) with about 20 V across it when the rails are at 52 V under nominal 8-Ω full-load conditions. This corresponds to about 830 Ω. A 680-Ω resistor is used to provide adequate relay current under slightly worse conditions when the power supply sags at high power into a 4-Ω load.

If the rail voltages rise to 60 V under no-signal conditions, about 31 mA will flow with a 1000-Ω coil, and dissipation in R51 will be about 650 mW, allowing the use of a 2-W resistor for good margin. If a less-sensitive relay is used (e.g., lower coil resistance), a smaller value of series resistance may have to be used. A 100-μF electrolytic capacitor is placed across the series resistor to create a brief high turn-on voltage to strongly activate the contact closure to aid cleaning of the contacts. K1 is closed by Q5, Q6 and Zener diode D6 when the voltage across muting delay capacitor C5 is charged to about 8.2 V.

If the signal path passes through the soft iron material of the relay frame, distortion can result. Many high-current relays are constructed this way. Try to avoid this type of relay. A suitable relay is the Panasonic SPDT ALZN1B24. It features 16-A SPDT contacts and a 1440-Ω coil that requires 17 mA. It is available from Mouser Electronics.

Short-Circuit Protection

The protection and speaker relay control circuits are shown in Figures 4.3, 4.4 and 4.5. The short-circuit detection circuit is shown in Figure 4.3, with a single output transistor pair depicted. Output current is sensed as corresponding to the voltage drop across the 0.22-Ω R_E resistors. In a class AB amplifier with a single-pair output stage, the output current will all flow through one of the resistors, while the drop across the other resistor is essentially zero. This observation applies to situations where the output current is substantially greater than the bias current.

The protection circuit monitors the emitter-to-emitter voltage, V_{E-E}, so that a positive value of this voltage indicates output current flowing in either direction (i.e., its absolute value). Indeed, with a sine wave current, V_{E-E} has the appearance of a full-wave-rectified signal. For a square wave output current, V_{E-E} would look like a constant DC voltage with the exception of narrow dips when polarity changes. Here the short-circuit protection is triggered when the average value of V_{E-E} rises to about one V_{be} after a given amount of time.

Q1 straddles the emitter nodes and will begin to conduct current when a filtered version of V_{E-E} reaches one V_{be}, which is about 0.55 V at 100 μA and 25 °C for the 2N5551. This voltage across 0.22 Ω corresponds to 2.5 A. R1 and R2 (470 Ω) limit base current and, in combination with C1 (22 μF), provide delay with a time constant of about 22 ms. The collector current of Q1 is mirrored by Q3 and Q4 to the input of the comparator IC2-A (Figure 4.4) with a load resistance of 6.8 kΩ. When Q1 conducts more than 100 μA, the comparator quickly discharges C4 and opens the speaker relay. Q2 limits the current in Q1 to about 2 mA.

The long-term short circuit trigger level, I_{sc}, is about 2.7 A (100 μA, through R2 and R5, drops an additional 80 mV). The 22-ms delay time constant will allow an over-current of 4 A to persist for about 22 ms. The 50-ms SOA of the MJL3281/1302 transistors is 6 A at 55 V and 3.5 A at 80 V. An amplifier with rails that rise to 80 V under no-signal conditions may need to have a lower 50-ms trigger threshold current. One way to accomplish this is to use a transistor for Q2 that has a smaller V_{be} at 100 μA.

Also note that V_{be} conveniently decreases with temperature. One could choose a TO-126 device for Q1 and mount it on the heat sink so that its ambient temperature is closer to 40 °C, decreasing

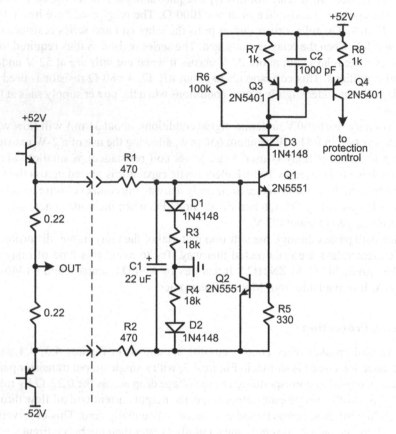

Figure 4.3 Short Circuit Detection

nominal V_{be} by about 33 mV. In this case, the over-current threshold will decrease further when the heat sink is hot and the SOA of the power transistors has decreased. Other circuit revisions can also be used to decrease the trigger threshold for a given V_{be}.

A larger over-current of 11 A will open the relay after only about 12 ms. These times do not include the inherent release time of the relay, which is specified as a maximum of 10 ms for the recommended Panasonic relay. The current limiting circuit prevents the current of one output transistor from exceeding 11 A at any time. Bipolar power transistors can tolerate relatively high current for very short periods of time, even when V_{ce} is equal to the rail voltage. The output current capability is doubled for the complete amplifier here, as it employs two output pairs (current in only one pair is monitored).

It is often desirable that I_{sc} be a function of the output voltage. There will be little or no voltage at the output under a short circuit condition. This can play a key role in distinguishing a short circuit from a legitimate high-power, high-current situation of significant duration. A high-power square wave during testing could also masquerade as a short circuit if I_{sc} is not increased for high output voltage swings. R3 and R4, in combination with D1 and D2, act to increase I_{sc} by pulling some current through R1 or R2 when the output swing is large. Note that R3 and R4 conduct no current at all until the output is at least one diode drop away from zero. In Figure 4.3, this arrangement increases I_{sc} to over 7 A for a peak output swing of 34 V.

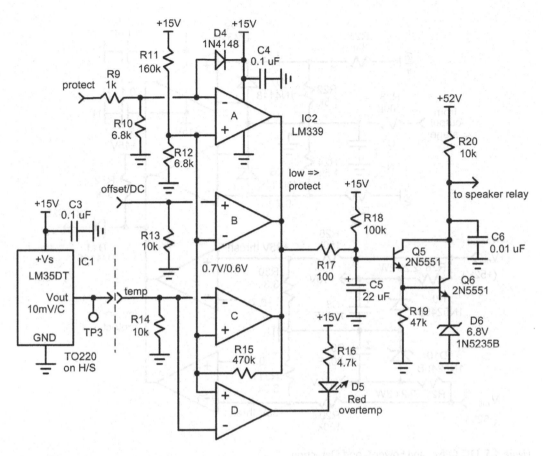

Figure 4.4 Protection Control Circuit

Protection Control Circuit and Turn-On Muting

The protection control circuit is shown in Figure 4.4. It includes four sections of an LM339 comparator to merge all of the protection functions for control of the speaker relay. If any of the open-collector outputs of sections A, B or C are low, C5 will be discharged and the Darlington relay driver will release the relay. When all three outputs are high, C5 will charge through R18. When its voltage reaches about 8.2 V, after about 3 seconds, the relay driver will turn on and close the speaker relay. This provides the muting delay at turn-on to prevent audible thumps.

Each of the A, B and C sections has a threshold of 0.7 V established by R11 and R12. Section A implements short-circuit protection. Section B implements DC offset and power-good protection based on a signal from Figure 4.5.

Over-Temperature Protection

Section C implements over-temperature protection. If the heat sink temperature reaches 70 °C, the loudspeaker relay will open. A TO-220 LM35DT Centigrade temperature sensor, IC1, is mounted to the heat sink. It produces an output voltage of +10 mV/°C across its 10-kΩ load

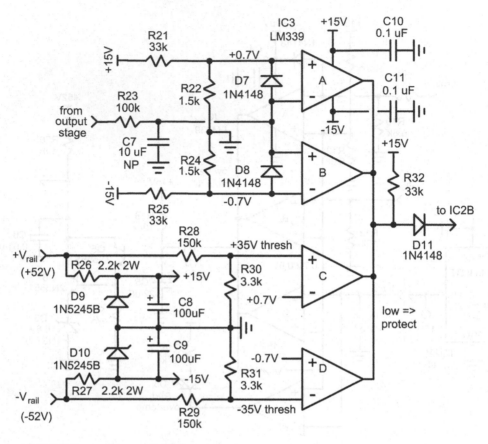

Figure 4.5 DC Offset and Power-Good Detection

resistor. If this voltage reaches 700 mV, comparator IC2-C discharges C5 and opens the speaker relay. Hysteresis is provided by IC2-D to prevent cycling. The relay will not be closed again until the heat sink temperature has fallen below 60 °C. Section D is used to illuminate a Red LED when over-temperature is sensed.

DC Offset Protection

A large DC offset can appear at the output under certain fault conditions, such as when an output transistor has failed with a collector-emitter short. This can destroy an expensive loudspeaker. DC offset protection is implemented with two sections of an LM339 comparator (IC3-A and B) configured as a window comparator, as shown in Figure 4.5. The comparator is fed from the output of the amplifier through series resistor R23 and delaying capacitor C7, with a time constant of 1 second. If DC offset larger than 700 mV persists for more than the time permitted by the delay, the loudspeaker relay will be opened. A DC offset of about 1.1 V in either direction will trigger protection in 1 second. The values of R23 and C7 are chosen to be large enough to prevent a 10-Hz sine wave at full power from triggering offset protection. With the values shown, if the output is stuck at a 55-V rail, the DC offset protection will be triggered in about 13 ms.

Rails OK Protection and Turn-Off Muting

This protection circuit prevents the speaker relay from closing unless and until the rails are above ±35 V. This will come into play at turn-on. More importantly, if a rail fuse blows on one polarity of rail, the speaker will be immediately disconnected. This circuit also opens the speaker relay at turn-off before the amplifier becomes dysfunctional and might cause a turn-off thump. This circuit employs sections C and D of comparator IC3.

Powering the Protection Circuits

The protection circuits are powered by a ±15-V supply implemented with Zener diode shunt regulators D9 and D10. The 2.2-kΩ, 2-W series resistors provide 17 mA to the ±15-V rails.

Fuses

The speaker relay obviates the need for a speaker fuse. Power supply rail fusing can be used as an optional last line of defense in the event of a fault condition. If only one rail fuse blows, the rail-reversal protection diodes prevent the amplifier circuitry from being subjected to potentially damaging reverse voltages. In this amplifier, rail fusing is not used. Instead, the mains fuse is relied upon in the event of a serious over-current fault that is not handled by the protection circuits.

4.6 Power Supply

The power supply is very simple, incorporating a 250-VA toroidal transformer with an 84-V center-tapped secondary and a single bridge rectifier. It is shown in Figure 4.6. The mains side includes an IEC power connector, an X capacitor across the line, a DPST power switch and a 5-A fuse. IEC power connectors that incorporate EMI filters and sometimes a fuse and/or power switch are also available and convenient. No inrush control is employed in this design. As shown, the supply is designed to provide ±52-V rails under load at 125 W into an 8-Ω load.

The transformer can be of toroidal or conventional construction. The secondary voltage is uncritical, but will affect achievable power output. For an amplifier with only two output pairs, the power supply rail voltages should not be greater than about ±55 V at full power into an 8-Ω load. This corresponds to an amplifier with an 8-Ω power rating of about 150 W (clipping at about 163 W). Reservoir capacitors should have a voltage rating at least 10 V greater than the anticipated no-load rail voltage. This amplifier draws about 2 A from each rail when delivering

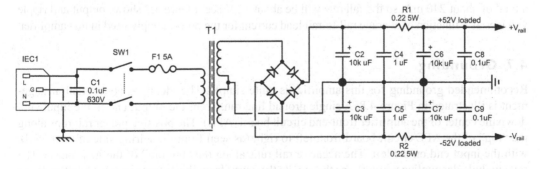

Figure 4.6 Power Supply

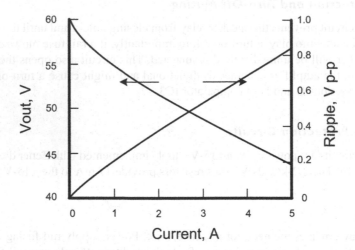

Figure 4.7 Power Supply Voltage and Ripple

its rated 125 W into 8 Ω. A typical power supply with an effective output resistance of 2 Ω/rail will have its rail voltages increase by about 4 V to about ±56 V under no-signal conditions. The transformer should have a rating of at least 500 VA for a 2-channel, 125-W per channel amplifier.

The bridge rectifier is a conventional 25-A, 400-V device. While not shown in the figure, snubber networks placed across each diode in the bridge are recommended. These networks can comprise 0.1-μF, 400-V capacitors in series with an optional 1-Ω resistor. Alternatively, four 25-A ultra-fast soft recovery diodes in TO-220 or TO-247 packages can be used at additional expense. The TO-247 Vishay HFA25PB60PBF HEXFRED® is a good choice. Snubbers are not required if these rectifiers are employed.

Two pairs of 10,000-μF reservoir capacitors connected in a π filtering arrangement provide a total of 20,000 μF for each rail. Each 0.22-Ω series resistor creates a low-pass filter that is down 3 dB at 72 Hz and down 9 dB at the 120 Hz ripple frequency (with respect to the ripple voltage at the first reservoir capacitor). Attenuation is much greater at the higher rectification harmonics. The total amount of reservoir capacitance is uncritical, but more is usually better. Very large reservoir capacitors may dictate the use of an inrush control arrangement to avoid blowing the mains fuse.

Bleeder resistors are not used. Output stage bias current will pull the rails down until the output stage can no longer conduct bias current. This will be below about ±20 V. Two output pairs will draw a total of about 240 mA, so the fall rate will be about 12 V/sec. Figure 4.7 shows output and ripple voltages as a function of average rail-to-rail load current for the power supply used in this amplifier.

4.7 Grounding

Recommended grounding for this amplifier is quite simple. The idealized conceptual arrangement is illustrated in Figure 4.8. A single ground line runs from the output end to the input end, down the center of the amplifier front-end circuit board (AFE). The positive power rail runs along the "top" of the AFE circuit board from left to right (as seen from the wiring side of the board), with the input end on the left. The negative rail runs along the "bottom" of the AFE board. The rails include decoupling networks as they make their way from the output stage back to the driver stage and then pre-driver and input stages. The ground ends of each pair of decoupling capacitors are connected directly together at the center of the board, and then that node is connected

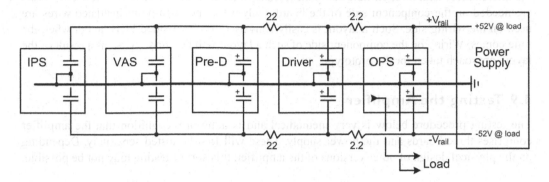

Figure 4.8 Grounding Architecture

by a single wire or trace to the ground bus that runs down the center of the board. Electrolytic decoupling capacitors are locally bypassed by 0.1-μF film or ceramic capacitors. The inevitable realities of physical layout will of course force some deviation from this idealized topology, especially on the output board.

The positive and negative rails, and ground, are connected from the power supply to the amplifier with three wires twisted together. Star-quad microphone cable can be used for this short distance. Perhaps non-intuitively, the smaller-AWG wire over this small distance is actually a good thing, as a small amount of resistance adds to the filtering effect of the capacitance on the output board and also encourages circulating class-AB currents to resolve locally at the output board (current takes the path of least impedance). The loudspeaker output and return terminals are both connected directly to the output circuit board, with the return connected directly to the power ground at the circuit board. Output signal currents are resolved at the circuit board.

This arrangement is not a traditional "star ground" wherein the speaker and other major grounds are returned separately to a single point ground at the power supply. In essence, the star ground is located on the amplifier circuit board. This is sometimes called a star-on-star approach.

Within the amplifier circuit board, the grounding arrangement can be described as a "tree" ground as illustrated in Figure 4.8. The grounding progresses from the output end of the board back to the input end. Most of the branches are vertical conductors to the power supply decoupling capacitors located along the "top" and "bottom" of the circuit board.

4.8 Building the Amplifier

Simulate the amplifier first. Go to Chapter 23 and read about using SPICE if necessary. Start by simulating some of the circuits in Chapter 2 or on the author's website. Finally, several simulation files for this amplifier can be found at *cordellaudio.com*. The beauty of simulation is that you get to know how the circuit behaves before you build it, and what to expect when you build it. You will know what voltages to expect at every node in the circuit.

Prototyping

I always build the prototype of one channel on perf board instead of just going straight to a printed wiring board. Of course, as many do, one can go right to a printed wiring board and hope that major changes are not going to be necessary. I also find that a neat printed-wiring-board-like perf board layout can lead to a quicker, better PWB layout down the line. I am usually able to put about 90% of the interconnect on the wiring side of the board without crossovers, implementing most of the interconnect with through-hole component leads soldered to each other. Very few crossovers are needed on the component side of the board. Only a few point-to-point insulated wires are used on the wiring side. Such a layout is easily converted to a two-sided PCB layout when the time comes. Wiring on the component side of such a layout will be fairly sparse as a result of the layout approach taken for the prototype.

4.9 Testing the Amplifier

The testing procedure below is very methodical and is done in recognition that the amplifier comprises three boards and the power supply. These will first be tested separately. Depending on the physical design of other versions of the amplifier, this sort of testing may not be possible.

Initial Inspection and Passive Tests

Once the amplifier is built, it is very important to test it with patience and diligence in an orderly fashion. Don't just turn the whole thing on and hope there will be no magic smoke. Inspect and poke around a lot before powering up a prototype. Take your time and inspect every component and its connection and associated wiring for things like polarity, shorts, open joints, values, etc. Do this with an illuminated magnifier. Are any components missing? Are any transistors or diodes or polarized electrolytic capacitors installed in the wrong orientation?

Inspect and verify every path connection on the breadboard or PWB to make sure that it physically corresponds to the schematic. This is painstaking and boring, but it is well worth it for a first-time build. I cannot tell you how many times this has saved me. Not all missing or shorted connections will have fatal consequences, but finding them now will save you hours of troubleshooting later.

It is also wise to poke around with an ohmmeter to see that it registers about what you expect it to. This will often reveal shorts and opens, and sometimes wrong resistor values. This, of course, is in-circuit testing, so some knowledge of the circuit is necessary for you to know what to expect when you probe any two points. If you don't get what you expect, find out why. It may be reflective of a circuit fault or it may be that a junction has been forward-biased by the ohmmeter current. In any case, satisfy yourself that you are getting what you should, even if it requires some further study of how the circuit should work. Sometimes an old-fashioned battery-operated ohmmeter works better for this than a DVM; see how the meter you are using reacts to reading a diode by itself in both directions.

Preparation for Testing

Initial testing of the prototype amplifier is done in several stages in a methodical approach. The power supply is tested first, by itself. The front-end board is tested next, in isolation. The feedback loop is closed by a jumper, which connects the center-tap of the pre-driver emitter resistors to the feedback resistor. For this test, the board is powered directly from the amplifier power supply. Some rudimentary tests are next performed on the output board by itself. The protection board is then tested. Finally, the three modules and power supply are connected together for testing of the complete amplifier.

Power Supply Test

Unlike some, I do not power up the amplifier with a Variac. Test the power supply by itself. This is a simple step, but worth the small amount of time it takes. Connect temporary 1-kΩ, 5-W bleeder resistors to the output rails. Apply power and make sure that all voltage readings are as expected. The rail voltages should come up quickly at turn-on. Remove the power. The rail voltages should fall slowly at turn-off. A total of 20,000 µF on each rail and the 1-kΩ, 5-W bleeder resistors should result in a time constant of about 20 seconds, meaning that the rail voltage should fall to about 1/3 of its energized value within about 20 seconds after turn-off. Allow the rail voltages to fall below 3 V. This may take 2 minutes or more. The bleeder resistors are somewhat optional, but do add to safety in testing.

Front-End Test

Initial testing of the IPS-VAS and pre-drivers should be done without the output stage module. This reduces the chances that the output transistors (and possibly others) will be destroyed in the event of a serious problem in the IPS-VAS/pre-driver circuits. This testing is done with the feedback loop closed by connecting feedback resistor R8 to TP1, the junction of the pre-driver emitter resistors (R30, R31), instead of to the output rail. Short the input to ground. Connect a temporary substitute KSC3503 V_{be} multiplier transistor to the terminals for Q14, which normally resides on the main heat sink. Adjust the bias pot full counter-clockwise to the point corresponding to lowest bias current setting.

Apply power and check the voltage at the junction of the pre-driver emitter resistors, looking for nearly zero volts within less than ±50 mV. Probe the emitters of Q15 and Q16. They should be roughly plus and minus 0.9 V, respectively. If the amplifier passes this test, this is a very good indication that the circuit is operating properly. Probe this point with a scope to see if there is any evidence of an oscillation. Use a 10:1 scope probe or a 1:1 probe with a 100-Ω resistor in series at the tip to prevent introducing instability from probe capacitance.

Now probe numerous nodes in the front-end to see that the voltages are as expected, based on voltages that were observed in simulation. Look especially at the voltage drop across resistors that are connected to the voltage rails, especially the emitter resistors of current sources and VAS transistors to verify that the proper currents are flowing. Probe the voltage at every emitter and every collector. Probe the voltages at the two bases of the input stage; they should be between +50 and +300 mV, and within ±20 mV of each other. I always make such measurements with a 1-kΩ resistor in series at the end of the DVM probe to make sure that probe capacitance does not inadvertently cause an oscillation when touching a sensitive node.

Check the voltages at the top and bottom of the bias spreader. The magnitudes of these bias voltages should be about the same, at about ±1.5 V with respect to ground with the pot set for minimum bias. Adjust the 10-turn bias pot clockwise by a couple of turns and observe increased voltages. Adjust the pot full clockwise and observe voltages on the order of ±2.4 V. Check the voltages at the emitters of the pre-driver transistors and observe that they are on the order of ±1.7 V. Reduce the bias setting to the minimum and observe that these voltages are on the order of ±0.9 V. As a final DC check, touch all of the transistors and verify that none is hot.

Initial Small-Signal Front-End Test

With the front-end working properly, perform small-signal and large-signal testing of the circuit with an audio generator, scope and AC voltmeter. The audio generator and AC voltmeter should operate to at least 1 MHz. The scope should go up to at least 20 MHz. If you don't have this equipment, obtain it. Ebay can be your friend here, especially for a good used, working oscilloscope. Alternatively, shop around for new equipment. Some good PC-based oscilloscopes are available at reasonable prices. Web searches with terms like *PC-based oscilloscope* or other terms like *audio oscillator* or *function generator* can bring up many good choices.

Connect the scope and AC voltmeter to the output (junction of pre-driver emitter resistors). With the input grounded, there should be a clean, noiseless line on the scope with its sensitivity set to 1 V/division. There should be no evidence of any parasitic oscillation or line ripple. The AC voltmeter should read less than 10 mV RMS. Remove the input short and see the same performance with the open input.

Apply 100 mV RMS at 1 kHz to the amplifier input and observe a clean sine wave at the output that measures about 2.7 V RMS. Sweep the frequency from 20 Hz to 20 kHz and observe a flat frequency response within +0, −0.2 dB with respect to the 1-kHz amplitude. Sweep the frequency from 20 Hz to 1 MHz and observe frequency response within +1, −10 dB with respect to the 1-kHz level. Observe that the output amplitude is down 3 dB at about 550 kHz. Lift C3. HF response should be down 3 dB at about 1.1 MHz. Replace C3.

Apply a 200-mV peak-to-peak 10-kHz square wave to the front-end and observe a well-behaved 5.5-V peak-to-peak square wave on the scope, with minimal ringing and overshoot. A multi-cycle burst of decaying ringing following the edges of the square wave is an indication that there is a problem with the feedback compensation. Ringing is usually accompanied with a peak at some frequency in the small-signal frequency response. Rise time from 10–90% should be about 0.5 μs.

Initial Large-Signal Front-End Test

The amplifier IPS-VAS/pre-driver circuits should now be tested with large signals. Set the scope to 10 V or 20 V per division, using a 10:1 probe if necessary. Set the generator frequency to 1 kHz and gradually increase the level up to 1 V RMS. The signal should remain clean and rise to 27.3 V RMS. Increase the level gradually until the output exhibits minor clipping. The signal should remain clean, with no sign of oscillation, even before and after the clipping points. The output level at this point should be on the order of 30 to 40 V RMS, depending on rail voltages. The peak voltage at clipping should be no more than 3 V below the rail voltage. Positive and negative clipping should be reasonably symmetrical. There should be little evidence of power supply ripple modulation. Increase the signal level further until harder clipping is evident and observe that the circuit is still behaving well.

Protection Circuit Test

Prepare the protection board for testing. Connect the DC offset input end of R23 to ground. Connect a 2.2-kΩ, 5-W resistor from the speaker relay output to the +52-V supply. Connect the *protect 1* and *protect 2* input terminals both to ground. Connect a voltmeter from the relay output to ground. Connect the board to the power supply.

Apply power. The relay voltage at the collector of Q6 should drop to about +8 V after about 7 seconds. Bear in mind that the relay itself is not connected; this node is just being pulled up by R20. Verify that the voltages at C8 and C9 are +15 V and −15 V, respectively. Apply +1 V to the DC offset input. The relay voltage should go to the +52-V rail voltage within 1 second. Remove the voltage from the DC offset input and connect the input to ground. The relay voltage should drop down to +8 V after about 7 seconds. Apply −1 V to the DC offset input. The relay voltage should go to the rail within 1 second. Remove the voltage from the offset input and connect the input to ground. The relay voltage should drop down after about 7 seconds.

Apply +0.8 V to the temperature sensor input. The relay voltage should go high immediately. Remove the voltage from the sensor input. The relay voltage should drop down after about 7 seconds. Apply +1 V to the *protect 1* pin. The relay voltage should go high immediately. Remove the voltage from the *protect 1* pin and connect the pin to ground. The relay voltage should drop down after about 7 seconds. Apply −1 V to the *protect 2* pin. The relay voltage should go high immediately. Remove the voltage from the *protect 2* pin and connect the pin to ground. The relay voltage should drop down after about 7 seconds. Connect a voltmeter to the +52-V rail. Turn off the power. The relay voltage should go high at about the time when the rail voltage has fallen to +35 V. This completes testing of the protection board.

Output Module Test

Here some brief testing of the output module and the protection circuit board is done. Temporarily connect the driver inputs of the output module to ground through 100-Ω resistors. Connect the power supply to the module and apply power. All transistors should remain cold and the driver emitter resistor center tap and the output rail should be at ground potential. The output bus should be at ground potential. Connect the loudspeaker relay series resistor R51 to ground and observe that the relay closes.

Remove power and connect the protection circuit board to the output module. Apply power and verify that there is a 7-second delay before the speaker relay closes. Remove power and observe that the relay opens after several seconds.

Test the Complete Amplifier

With a known-good, thoroughly tested IPS-VAS/pre-driver, it is now time to test the amplifier as a whole, with the output stage connected to the front-end. Having spent the time in testing as described above, it is now much less likely that destruction will occur when the output stage is included. Remove the temporary feedback connection that was made to the center of the pre-driver emitter resistors and connect R8 to the output rail. Remove the temporary 100-Ω resistors at the driver base inputs of the output module. Connect the pre-driver outputs to the output module. Set the bias pot counter-clockwise for minimum bias. Short the amplifier input. Apply power and verify that there is a 7-second delay before the speaker relay closes. Observe 0 V ±50 mV at the amplifier output. Measure and record the rail voltages and observe that they are not significantly lower than they were in earlier tests. Verify that the voltages at the rail filter capacitors are approximately the same as those at the power supply, with voltage drops of less than 1.5 V as the measurement points move from the amplifier output stage toward the amplifier input stage. Kill the power and verify that the speaker relay opens when the power supply rails fall below about ±35 V.

Bias Adjustment

Apply power. While probing across one of the NPN output transistor emitter resistors with the DC voltmeter, gradually turn the bias pot clockwise, increasing the bias setting, until there is 10 mV across the emitter resistor. Observe that there is no oscillation and that the output of the amplifier is within ±50 mV of ground. Place the DVM across each of the other emitter resistors and observe that there is between 1 mV and 20 mV across each of them. Ideally they should all have 10 mV across them, but V_{be} mismatches among the output transistors may introduce differences.

Increase the bias until there is on average about 20 mV across all of the emitter resistors. Record the voltage seen across each emitter resistor. Alternatively, measure the emitter-to-emitter voltage (V_{E-E}) for each pair and adjust the bias pot for an average of 40 mV. Note that these voltages being less than the theoretical 26 mV is intentional. Check the rail voltages and observe that they have sagged just a bit due to the fact that the output stage is now drawing over 180 mA. Observe that the power transistors will begin to get warm. Leave the amplifier on for 30 minutes and then observe that the voltages across the emitter resistors have not changed by more than 10 mV from their previous readings.

Measure the output voltage of the IC Centigrade temperature sensor at TP3. It should read somewhere between 350 mV and 500 mV, corresponding to the heat sink temperature being between 35 °C and 50 °C, depending on the thermal resistance of your heat sink.

No-Load Test

Connect the scope and AC voltmeter to the output rail of the amplifier (ahead of the output coil). Repeat the small-signal testing that was done on the IPS-VAS/pre-driver circuits earlier, using a 200-mV RMS input signal. This should generate 5.45 V RMS at the output. Similar results to the earlier small-signal tests should be seen. There should be no evidence of oscillation or high-frequency gain peaking greater than 1 dB. Repeat the large-signal testing that was done earlier on the IPS-VAS/pre-driver, observing similar results. The peak voltage at which clipping occurs may be up to 1.4 V less than seen before.

Small-Signal Test Under Load

Reduce the input signal level to zero and connect an 8-Ω, 50-W dummy load to the amplifier output. It can be a heat-sink-mounted power resistor. It is not important that it be non-inductive. Repeat the tests that were conducted above for small signals, with an AC voltmeter and scope connected to the output terminal of the amplifier. Expect similar results. Amplifier gain should not be reduced by more than 0.2 dB from the no-load value. Frequency response should be down by 3 dB at about 360 kHz ±50 kHz. This includes the bandwidth-limiting contributions of the input filter and the output L-R network. Accuracy of the 1.5-μH inductor will not have a significant effect on amplifier performance, but it will have a significant effect on the frequency at which response is down 3 dB. If the output is probed ahead of the L-R network, the response should be down 3 dB at about 560 kHz ±100 kHz.

Large-Signal Test Under Load

Set the signal generator to zero volts at 1 kHz. Output power levels greater than 50 W should be limited to less than 30 seconds to avoid overheating of the load resistor. Repeat the large-signal testing that was done above under no-load conditions. Clipping signal voltages will be a few volts less than observed under no-load conditions due to power supply sag and signal voltage drop across the R_E resistors. Measure the rail voltages at just under clipping at 1 kHz and record them. Compare them to the no-load values and observe the amount of power supply rail sag under full load. Although not critical, it is desirable that the amount of sag from no-load to the loaded condition should be less than about 5 V. This is a measure of the stiffness of the power supply.

The amplifier should also be tested under large-signal conditions with a 4-Ω load. Running this test at near full power is optional and requires the use of a higher-wattage load resistor. The amplifier may output up to 250 W or more in this test. A second 8-Ω, 50-W resistor can be placed in parallel with the original 8-Ω load. This will suffice as long as the duration of the test is limited to about 30 seconds.

Temperature Test

Operate the amplifier at approximately 1/3 power with an 8-Ω load at 1 kHz for 15 minutes. This corresponds to about 50 W (20 V RMS), and is the approximate power level at which amplifier dissipation is maximum. Periodically touch the heat sink with your index finger. At no point should it get so hot that you cannot keep your finger on the heat sink for at least 3 seconds. This will verify that the heat sink temperature does not rise above about 60 °C under those conditions. Probe TP3 and read the heat sink temperature as the voltage reported by the LM35 multiplied by 100 °C. The voltage reading should be less than 0.65 V.

Capacitive Load Test

Amplifier stability under capacitive loading conditions is verified in this test. Apply a 200-mV peak-to-peak 20-kHz square wave to the input of the amplifier with an 8-Ω load. Sequentially apply 1000-pF, 0.01-μF, 0.1-μF and 1-μF capacitances across the load and observe on the scope that there are no parasitic oscillation bursts anywhere on the output signal. There should be no multi-cycle ringing bursts at the edges of the square wave that do not decay fully within 3 cycles. Some overshoot and minimal ringing may be seen on some of these tests. No overshoot or ringing should be seen with 1000 pF or 0.01 μF loading. With 0.1-μF loading, 6% overshoot with no ringing may be seen. With 1-μF loading, 74% overshoot with one cycle of ringing may be seen. If the output is probed before the L-R network, only 2% overshoot should be seen, confirming that virtually all of the overshoot at the output is due to the highly damped resonance of the L-R network and the load capacitance.

Repeat these tests with a 1.2-V peak-to-peak square wave at the input to verify large-signal stability with a 33-V peak-to-peak output. Results should appear similar. With 0.1-μF loading, the amplifier is called upon to deliver 6.5-A peaks into the load. With 1-μF loading, the peak current delivered is on the order of 20 A. This latter condition approaches or exceeds the current limiting value of the amplifier.

DC Offset Protection Test

Briefly connect a 1-MΩ resistor from the positive rail to the junction of R23 and C7 on the protection board. The speaker relay should open in less than 1 second. It should close about 7 seconds after the 1-MΩ resistor is removed. Repeat the test with the resistor connected to the negative rail.

Short-Circuit Test

This is the scary test. It is optional, and not for the faint of heart. Failing this test may result in magic smoke. Briefly connect a 47-kΩ resistor from the positive rail to the base node of Q1 on the protection board. The speaker relay should open. After removal of the 47-kΩ resistor, the speaker relay should close in about 7 seconds. This procedure tests most of the short-circuit protection circuitry.

With the amplifier input shorted, short the output terminals. You should hear the speaker relay open quickly. Immediately remove the short circuit when you hear the relay open. About 3 seconds later you should hear the speaker relay click again as it closes. Check the output for less than ±50 mV offset. Apply 100 mV RMS at 1 kHz to the input. Repeat the above test with a scope attached to the output. The 1-kHz sine wave should re-appear after the relay closes following the removal of the output short. Note that this protection circuit will "retest" for a short circuit about every 7 seconds.

Stability Evaluation

Adequate phase and gain margins can be verified by checking frequency response or square wave response with the feedback network temporarily modified. This is done with input filter capacitors C1 and C3 disconnected and with the amplifier output measured before the L-R output network so that the wideband response of the amplifier is measured. The connection at the junction of feedback resistors R7 and R8 is altered to introduce added phase lag or to decrease closed-loop gain. These tests do not measure phase and gain margin as such, but rather verify that there is enough.

Shunt R8 with a 10-kΩ resistor to increase closed-loop gain by 3 dB. Shunt the junction of R7 and R8 to ground with a 39-pF capacitor. This introduces a pole and 3 dB of loss at 1 MHz. Existing phase margin has now been reduced by 45 degrees. If the amplifier does not oscillate, it has at least 45 degrees of phase margin. Measure the amount of peaking in the frequency response. Peaking of 3 dB would suggest a phase margin of 45 degrees. Expected peaking in this test should be on the order of 8 dB at about 1 MHz. Response will be down 3 dB at about 2 MHz.

Remove the resistor and capacitor added above. Short R8. This will decrease closed-loop gain by about 6 dB. If the amplifier does not oscillate, it will have at least 6 dB of gain margin. Under this condition, closed-loop bandwidth will be on the order of 3.6 MHz with no peaking.

Further Testing

You now have a working amplifier and you can proceed with other performance tests, such as THD, as described later in this book.

4.10 Troubleshooting

If the amplifier does not work at any point in the process, troubleshooting is necessary. A great deal of the testing outlined above is also done for troubleshooting. This is particularly so for voltage measurements. If an amplifier is built for the first time and passes the initial testing in the order described above, it is very likely that it will work. Of course, if problems occur at one of those steps, troubleshooting is called for. The sequencing of the tests helps to localize where the problem is. If an already-built amplifier does not work, or has never worked, the troubleshooting described here is especially helpful.

Always start by carefully inspecting the amplifier, often with an illuminated magnifying glass. Look for parts that appear to be damaged in any way. Look for shorts or opens. Verify proper part values and part orientation.

If the amplifier is known to have experienced damage, as with burned parts or suspected blown transistors, it is especially important to make sure that any and all blown transistors are replaced before powering up the unit with the replaced components. Transistors can often be evaluated in-circuit for having been blown. The collector-base or base-emitter junction is usually shorted if the transistor is blown (unlike a fuse, which blows open). I always check with an old-fashioned battery-operated passive ohmmeter, often on the 100-Ω range. The meter will usually apply enough current to forward-bias a junction. A shorted junction will read zero ohms. A good junction will usually give a reading about 3/4 the way up the meter scale.

If the amplifier blows up when turned on, and if there are no bad transistors, be especially careful to make sure that the bias pot has been set to minimum bias and that the bias spreader is working properly. Check to see if the amplifier operates if the output stage is disconnected and with the negative feedback taken from the pre-driver emitter resistor center tap. This connection will often be tolerant of bias spreader voltage set too high, and allows verification that the bias spreader voltage has been set for low bias.

If the output of the amplifier is stuck at one rail, measure the voltages at each signal path node to see if they are nominal, stuck high of normal or stuck low of normal. Always note the correspondence between input and output voltages of a stage. For an inverting stage, the input and output voltages should be of opposite relative polarity. Find where in the complete feedback loop the voltage level or polarity is not what it is supposed to be given the state of the input voltage to that stage.

4.11 Performance

The performance of the amplifier is summarized here. With 52-V rails (under load) the amplifier clips at about 135 W into an 8-Ω load. Depending on power supply stiffness, the 52-V rail may rise to 60 V under a no-load condition, suggesting momentary clipping power of perhaps 190 W.

Frequency Response

The 3-dB bandwidth extends from 1 Hz to 360 kHz with an 8-Ω load. The high-frequency 3-dB point is dominated by the input filter and output network. Without the output network, the HF response extends to about 560 kHz. Without the output network or the input filter, the HF response extends to about 1.1 MHz with no peaking. Measured frequency response for no-load and loads of 8 Ω, 4 Ω and 2 Ω is shown in Figure 4.9. Square wave response at 10 kHz is shown in Figure 4.10 (1-V p-p, 8-Ω load). Bandwidth of the amplifier for 4-Ω and 2-Ω loads is about 285 kHz and 180 kHz, respectively.

Square Wave Response

Apply a 200-mV peak-to-peak 10-kHz square wave to the amplifier and observe a well-behaved 5.5-V peak-to-peak square wave at the output. Fairly sharp edges and no ringing should be observed. Repeat the test with a 100-kHz square wave. Rise time (10% to 90%) should be about 1.2 μs.

Slew Rate

Slew rate for the amplifier is +48 V/μs and −54 V/μs when measured without the input low-pass filter and before the output L-R network. Slew rate is measured with a 40-V peak-to-peak, 10-kHz square wave.

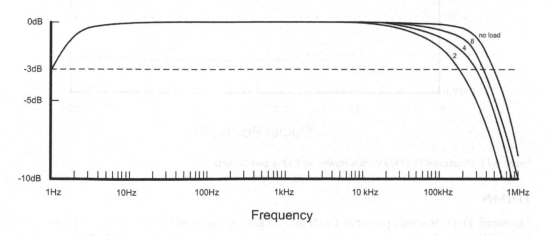

Figure 4.9 Frequency Response

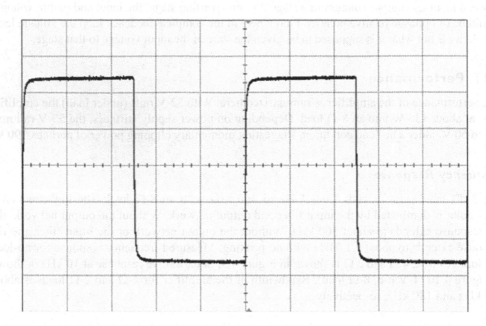

Figure 4.10 10-kHz Square Wave Response

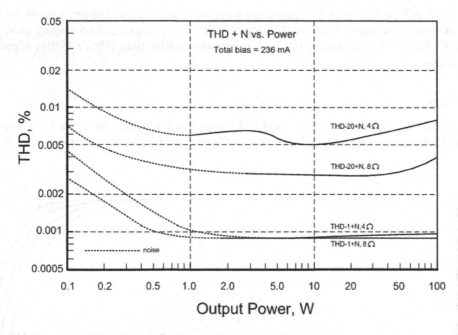

Figure 4.11 Measured THD+N Versus Power at 1 kHz and 20 kHz

THD+N

Measured THD+N versus power at 1 kHz and 20 kHz is shown in Figure 4.11 for 8-Ω and 4-Ω loads. The dotted curves show noise as opposed to crossover distortion. THD+N versus frequency at 10 W and 100 W is shown in Figure 4.12.

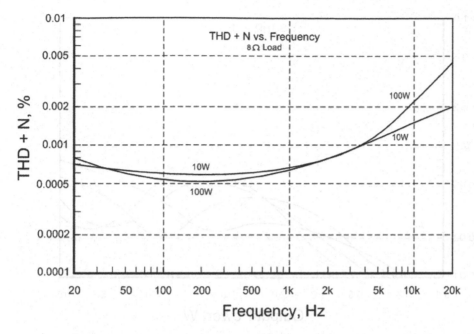

Figure 4.12 THD+N Versus Frequency at 10 W and 100 W with 8-Ω Load

Simulated and measured THD-1 at 100 W into 8 Ω are 0.00015% and 0.0005%, respectively. Simulated THD-20 and measured THD-20+N at 100 W into 8 W are 0.0034% and 0.004%, respectively. Measurement results are in reasonable agreement with simulation results, and this is encouraging. The simulated results do not include noise, so measured THD+N is higher than simulated THD at lower power levels. THD-1 without noise was also measured by connecting a spectrum analyzer to the distortion analyzer output. It measured less than 0.001% for all power levels up to 100 W into 8 Ω. The Distortion Magnifier discussed in Chapter 26 was used to make some of these measurements [5].

The THD versus frequency plot in Figure 4.12 shows the decreasing reduction of THD by negative feedback as frequency rises. The distortion at 100 W is rising at about 6 dB/octave between 10 kHz and 20 kHz, the same slope as the feedback loop gain is falling, indicating that dynamic HF distortion in the output stage is quite small.

Crossover Distortion

The output stage is usually the largest contributor to distortion in well-designed amplifiers. This is easily seen when one observes the increase in amplifier distortion with heavier loading. As discussed previously, the output stage quiescent bias current plays a significant role in the amount of crossover distortion created by an output stage. Biasing the output stage to the Oliver criteria of V_q = 26 mV across each emitter resistor is supposed to minimize crossover distortion in the ideal case [3]. Most amplifiers do not conform to the ideal case, however. Real-world factors like ohmic emitter and base resistance in the output transistors, and the resistance of base stopper resistors tend to reduce the value of V_q that minimizes crossover distortion. These resistances, as seen looking into the emitter, act as extensions of R_E, making the effective value of R_E larger and the optimum bias current smaller.

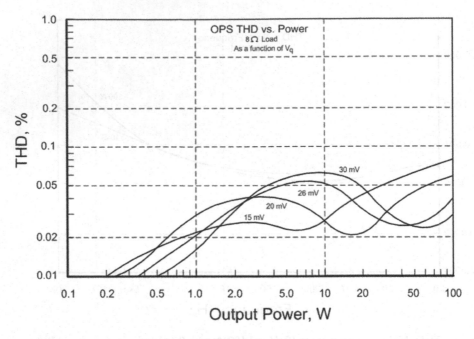

Figure 4.13 Output Stage THD-1 Versus Power for Different Bias Currents

Figure 4.13 shows measured output stage THD-1 as a function of power for four different values of V_q. Output stage distortion was measured by operating the amplifier with the feedback loop closed from the center-tap of the pre-driver emitter resistors (TP1). This technique exposes the output stage distortion, as long as the global negative feedback forces the distortion at TP1 to be much smaller. In order to avoid thermal effects, the amplifier was operated with no load at the chosen output voltage to allow the THD analyzer to stabilize. The chosen load was then applied briefly and the distortion measurement taken.

As expected, the optimum value of V_q is less than 26 mV. The distortion rises and then decreases as power is reduced from a high value. Here the optimum value for V_q is less than 20 mV. However, the amplifier was set to V_q = 20 mV to be on the safe side against under-bias under dynamic signal conditions. It is evident that crossover distortion is not overly sensitive to changes in bias current over a reasonable range, the maximum value changing by only a factor of 1.5 as V_q moves from 20 mV to 30 mV (corresponding to bias current per output pair moving from 91 mA to 136 mA). Maximum crossover distortion occurred at 3–7 W for V_q between 20 mV and 26 mV, and was no more than 1.6 times full-power distortion. With the chosen V_q setting of 20 mV, the output stage crossover distortion was a maximum of 0.04% at 3 W, while distortion at 100 W was 0.06%.

Noise

Input-referred noise of the amplifier is 9 nV/√Hz. This is respectable. A-weighted S/N is an excellent 100 dB with respect to 2.83 V output (1 W into 8 Ω). Un-weighted S/N with respect to 2.83 V is 97 dB in a 20 kHz measurement bandwidth. Input series resistor R4 and feedback shunt resistor R6, each 1 kΩ, are the largest noise contributors. Halving these resistors would reduce

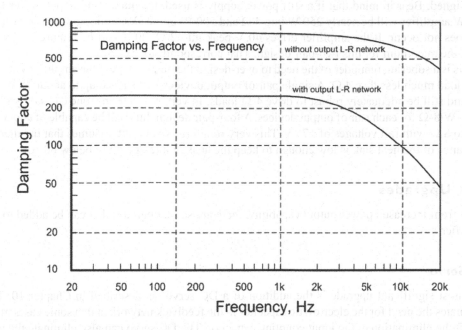

Figure 4.14 Damping Factor Versus Frequency

their contribution by 3 dB, but would reduce amplifier input impedance to about 13 kΩ unless some other circuit approaches were used.

Damping Factor

Damping factor (DF) as a function of frequency is shown in Figure 4.14 above. DF is over 200 from 20 Hz to 2 kHz and falls to 47 at 20 kHz. DF is limited primarily by the impedance of the output network. Absent the output network, DF is 800 at 1 kHz and 400 at 20 kHz. Output impedance at 1 kHz without the coil is 10 mΩ. The extremely low output impedance is due primarily to the use of an output Triple in combination with very high loop gain, especially at low frequencies. Loop gain is 92 dB at 20 Hz. More specifically, closed-loop output impedance is very low because open-loop output impedance is very low and is very much reduced by the high negative feedback loop gain. Open-loop output impedance is low because the shunt feedback of the Miller compensation keeps the output impedance of the VAS low, and that impedance is greatly reduced by the very high current gain of the triple-EF output stage.

4.12 Scaling

This amplifier is designed with transistors that can handle relatively high power supply rail voltages and the operation of the circuits is also not strongly dependent on rail voltages. As a result, the design can be scaled for different power output capabilities by selecting the appropriate power supply rail voltages and employing an adequate number of paralleled output transistor pairs. As shown, the two-pair design is adequate up to about 150 W with an 8-Ω load.

The maximum peak output current allowed by the current limiting circuit will be proportional to the number of output pairs, while the maximum V_{ce} that the output transistors will see is roughly proportional to the square root of the maximum power level for which the amplifier is designed. Bear in mind that if a stiff power supply is used, the maximum output power for a 125-W amplifier will be nearly 250 W into 4 Ω and 500 W into 2 Ω, assuming that current limiting does not occur. If the amplifier drives 40 V peak into 2 Ω, 20 A will be required, which is about the maximum for this two-pair design. That corresponds to 400 W.

This is a sobering reminder of the need to over-design the output stage if the amplifier will ever see a load much less than 4 Ω. A single pair of output devices will suffice up to about 75 W into 8 Ω and still be adequately robust to drive 4-Ω loads. In very rough terms, one can safely go up by 75-W/8-Ω for each pair of output devices. A four-pair design thus will be capable of about 300 W into 8 Ω with rail voltages of ±75 V. This very rough rule-of-thumb assumes that the thermal resistance of the heat sink is low enough to keep the heat sink temperature below 60 °C.

4.13 Upgrades

Apart from increased power output capability, there are some upgrades that can be added to this amplifier.

DC Servo

The most significant upgrade is the addition of a DC servo, as described in Chapter 10. This eliminates the need for the electrolytic capacitor in the feedback network and in some cases might permit the elimination of the input coupling capacitor. The DC servo can also eliminate the need for the feedback and input return resistances to be equal, since offset resulting from inequality will be compensated by the DC servo. This permits reduction of the feedback network impedance by at least a factor of 2 to about 500 Ω, reducing noise and reducing phase lag at the feedback input node.

Increased ULGF

This amplifier is capable of ULGF of greater than 1 MHz with adequate phase and gain margins. Compensation can be changed to allow a ULGF of at least 1.5 MHz, reducing distortion by perhaps 3 dB. ULGF can be increased by reducing the value of the Miller compensation capacitor (C9). However, it is unwise to go too small. The parasitic capacitance at the base of Q11 creates a capacitance voltage divider with C9 and increases the output impedance of the VAS at high frequencies. Instead, keep C9 the same and increase gm of the input stage. This can be done by increasing tail current and keeping the degeneration ratio the same by reducing the degeneration resistors in proportion. However, this increases input bias current.

Input LTP with Followers

One can decrease input bias current by adding emitter or source followers to the IPS without too much noise degradation. This allows increased LTP bias current and gm without an input bias current penalty. If JFETs are used, their drains can be connected to a −15-V supply to keep V_{ds} within their rated voltage. Adding either BJT or JFET followers allows the tail current of the LTP to be increased and its degeneration resistors decreased, so as to increase IPS gm without increasing input bias current. In fact, with decreased input bias current, one can reduce feedback network impedance and not reduce input impedance (and not suffer excessive input offset as a result). The ultimate implementation uses JFETs to completely eliminate input bias current. Of course, use of a DC servo also mitigates the effect of input bias current. These approaches are described in Chapters 9 and 10.

IPS *gm* can also be increased (and input bias current decreased) by the use of a BJT or JFET CFP LTP. In the case of a JFET CFP, there may be an issue, or opportunity for creativity, in keeping V_{ds} adequately low without resort to cascoding the IPS.

Advanced Feedback Compensation

More sophisticated feedback compensation schemes can be easily applied to further reduce distortion, as described in Chapter 11. Examples include 2-pole compensation (TPC) and transitional Miller compensation (TMC).

Inrush Control

If the power supply is equipped with an NTC inrush control device, some effective resistance will be added to the supply by that device, slightly reducing maximum power. The inrush device can be shorted by a relay once adequate rail voltages have been reached. In this design, a relay controlled by the same circuit as the speaker relay can be added to short the NTC resistor after turn-on. More on this can be found in Chapter 19.

Quasi-Boosted Supply Rails

A separate power supply rectifier and reservoir capacitors can be used for the front-end. This increases nominal available voltage under load, decreases rail drop under transient load, and can reduce ripple. It is a relatively inexpensive upgrade, as the added rectifier and reservoir capacitors can be relatively small.

References

1. Bart Locanthi, "Operational Amplifier Circuit for Hi-Fi," *Electronics World*, January 1967, pp. 39–41.
2. Bart Locanthi, "An Ultra-Low Distortion Direct-Current Amplifier," *Journal of the Audio Engineering Society*, vol. 15, no. 3, pp. 290–294, July 1967.
3. Barney M. Oliver, "Distortion in Complementary Pair Class B Amplifiers," *Hewlett Packard Journal*, February 1971, pp. 11–16.
4. Federal Trade Commission (FTC), "Power Output Claims for Amplifiers Utilized in Home Entertainment Products," CFR 16, Part 432, 1974.
5. Bob Cordell, "The Distortion Magnifier," *Linear Audio*, pp. 142–150, September 2010.

Chapter 5

Noise

5. INTRODUCTION

All passive and active circuits create noise, and noise is undesirable. Noise limits the dynamic range of a system. In order to design circuits for the lowest noise, the noise-generating mechanisms must be understood. There are numerous trade-offs, and the best low-noise design is usually very dependent on the particular application. More complete discussions on noise as it pertains to audio amplifiers and semiconductors can be found in these references [1–13].

Although the noise characteristics of a power amplifier are not as critical as those of a preamp, it is still important to achieve low noise because there is no volume control in the power amplifier to reduce noise from the input stage under normal listening conditions. This is particularly important when the power amplifier is used with high-efficiency loudspeakers. Here we will explore the ways in which noise is governed by the circuits and discuss ways to minimize it. There are three parts in this chapter. The first part covers noise in power amplifiers in general, without regard to the specific sources. The second part explains the device and circuit noise sources. The primary sources are thermal noise and shot noise. The last part of the chapter discusses how noise is created in amplifier circuits and how to evaluate it and minimize it.

5.1 Signal-to-Noise Ratio

Power amplifier noise is usually specified as being so many dB down from either the maximum output power or with respect to 1 W. The former number will be larger by 20 dB for a 100-W amplifier, so it is often the one that manufacturers like to cite. The noise referenced to 1 W into 8 Ω (or, equivalently 2.83 V RMS) is the one more often measured by reviewers.

The noise specification may be un-weighted or weighted. Un-weighted noise for an audio power amplifier will typically be specified over a full 20-kHz bandwidth (or sometimes up to 100 kHz or more). Weighted noise specifications take into account the ear's sensitivity to noise in different parts of the frequency spectrum. The most common one used is *A weighting*, whose frequency response weighting is illustrated in Figure 5.1. The A-weighted noise is measured by placing a filter with this frequency response in front of a true-RMS AC voltmeter. Notice that the weighting curve is up about +1.2 dB at 2 kHz, whereas it is down 3 dB at approximately 500 Hz and 10 kHz.

Consider a power amplifier with input-referred noise density of 15 nV/$\sqrt{\text{Hz}}$. Noise density will be discussed in more detail in Section 5.5. Think of a noiseless amplifier driven by a white noise source. That is what it means for the noise to be *input-referred*. All of the noise sources in the amplifier act together to produce the same noise result as the hypothetical noise generator

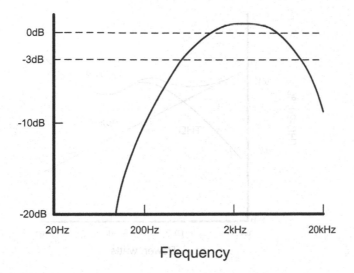

Figure 5.1 A-Weighting Frequency Response

mentioned above. Suffice it to say that 15 nV/√Hz is not difficult to achieve in a power amplifier design. Assume that the amplifier gain is 30. Output noise will be 450 nV/√Hz. The un-weighted noise in a 100 kHz bandwidth will be 316 times that, since there are 316 √Hz in a 100-kHz noise bandwidth. Output noise will be 142 µV. This is 86 dB below 2.83 V, so the un-weighted S/N is 86 dB. It is useful to note that this is the same amplitude as 0.005% THD at 1 W.

THD+N Measurements

Noise can introduce confusion into the traditional THD+N measurements made for power amplifiers [7]. Most traditional distortion analyzers report THD+N because they remove the fundamental and display what is left over, which is distortion harmonics and noise. This is also what most amplifier specification sheets and reviews report. Newer distortion analyzers can report THD by itself or THD+N.

Many THD+N plots for power amplifiers show a rise in THD+N at low power levels. This can be a result of crossover distortion, which is known to often rise at low power levels in power amplifiers. However, this can also be the result of the fixed output noise level rising in comparison to the falling output signal level. The plot is thus open to ambiguous interpretation.

Figure 5.2 is an illustrative plot that shows both THD and THD+N for a power amplifier as a function of power at a fundamental frequency of 20 kHz with an analyzer bandwidth of 100 kHz. The amplifier's un-weighted S/N is 86 dB in a 100 kHz bandwidth, as discussed above. The noise contribution is equivalent to 0.005% THD at a power level of 1 W. As expected, THD starts out very small at low power levels. At middle power levels it rises to a broad peak where crossover distortion has become prominent. THD then decreases and finally rises again as maximum power is approached. Notice, however, that the THD+N curve continues to rise as power is decreased, possibly giving the impression that significant crossover distortion is also present at very low power levels.

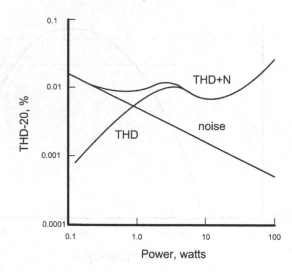

Figure 5.2 20 kHz THD and THD+N Versus Power

A key observation is that measured THD+N usually continues to rise as power level decreases, while THD by itself from crossover distortion often rises as power is decreased from maximum power and then largely disappears when output power falls to the point where the output stage enters its small class-A region (often less than 200 mW into 8 Ω).

5.2 A-Weighted Noise Specifications

The frequency response of the A-weighting curve is shown in Figure 5.1. It weights the noise in accordance with the human ear's perception of noise loudness. Note that it is up by 1.2 dB at 2 kHz. The A-weighting filter includes a second-order high-pass function, causing it to be down by 26 dB at 60 Hz. It also includes a third-order distributed pole low-pass function that causes it to be down by 10 dB at 20 kHz.

The A-weighted noise specification for an amplifier will usually be quite a bit better than the un-weighted noise because the A-weighted measurement tends to attenuate noise contributions at higher frequencies and hum contributions at lower frequencies. A very good amplifier might have an un-weighted signal-to-noise ratio of 90 dB with respect to 1 W into 8 Ω, while that same amplifier might have an A-weighted SNR of 105 dB with respect to 1 W. A fair amplifier might sport 65 dB and 80 dB SNR figures, respectively. The A-weighted number will sometimes be 10–20 dB better than the wideband un-weighted number.

Figure 5.3 shows the schematic of a simple A-weighting filter. The filter is entirely passive with the exception of input/output buffering and the necessary gain stage.

5.3 Noise Power and Noise Voltage

The noise arising from different sources is usually assumed to be uncorrelated. For this reason, it adds on a power basis. This means that noise voltage adds on an RMS basis as the square root of the sum of the squares of the various sources. Two noise sources each 10 μV RMS will add to

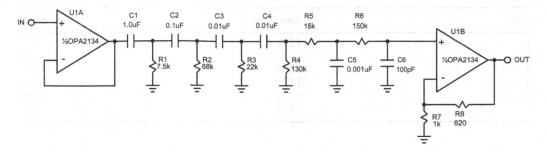

Figure 5.3 A-Weighting Noise Filter

14.1 µV RMS. Two noise sources, one 10 µV and the other 3 mV will sum to $\sqrt{(100 + 9)} = \sqrt{(109)} =$ 10.44 µV. This shows how a larger noise source will tend to dominate over a smaller noise source.

$$V_{noise\_total} = \sqrt{(V_1^2 + V_2^2)} \tag{5.1}$$

$$V_{noise\_total} = \sqrt{(10^2 + 3^2)} = \sqrt{109}$$

5.4 Noise Bandwidth

Most noise sources have a flat noise spectral density, meaning that there is the same amount of noise power in each Hz of frequency spectrum. This means that total noise power in a measurement is proportional to the bandwidth of the measurement being made. This gives rise to the concept of noise bandwidth.

A perfect brick-wall filter would have a noise bandwidth equal to its signal bandwidth. Because real filters roll off gradually, the noise bandwidth is slightly different than the 3-dB bandwidth of the filter (often slightly more). The term *equivalent noise bandwidth* (ENBW) is therefore often used to describe how much white noise will be passed by a real circuit [2]. For example, a 12.7-kHz single-pole low-pass filter has an *equivalent noise bandwidth (ENBW)* of 20 kHz. Conversely, the ENBW of a 20-kHz first-order LPF is 31.4 kHz. The ENBW of a single-pole roll-off is equal to 1.571 times the pole frequency.

Low-Pass Filter Noise Bandwidth

The noise bandwidth of several loss-pass filters of different order is shown in Table 5.1. As the order of the filter increases, the noise bandwidth tends to approach that of a brick-wall filter, as expected. The noise bandwidth of Chebyshev filters is not shown because it depends heavily on the amount of ripple for which the filter is designed. The numbers in the table are multipliers normalized to the 3-dB bandwidth of the filter.

A-Weighting Noise Bandwidth

The ENBW of the *A-weighting* function is 13.5 kHz. This means that a white-noise source passing through the A-weighting filter will result in about 4 dB less noise as compared to the same

Table 5.1 Low-Pass Filter Equivalent Noise Bandwidth

Order	Butterworth	Bessel
1	1.571	1.571
2	1.111	1.560
3	1.047	1.080
4	1.026	1.045
5	1.017	1.039
6	1.012	1.040

source passed through a first-order roll-off at 20 kHz. Compared to passing through a sharp 80-kHz roll-off, it will experience about 1/6 the noise bandwidth, so it will be smaller by a factor of $\sqrt{6} = 2.45$, or about 8 dB.

Consider the amplifier discussed earlier. It had 15 nV/$\sqrt{Hz}$ input noise and an un-weighted S/N of 86 dB in a 100-kHz bandwidth. Its output noise was 450 nV/$\sqrt{Hz}$. The A-weighted noise in the 13.5-kHz ENBW will only be 52 µV, which is 93 dB below 2.83 V.

All of these simplified noise discussions have assumed white noise sources and no contributions from hum, line harmonics and EMI. These latter noise sources do not necessarily add on an RMS basis. Only uncorrelated sources add on an RMS basis.

5.5 Noise Voltage Density and Spectrum

White noise has equal noise power in each Hz of bandwidth. If the number of Hz is doubled, the noise power will double, but the noise voltage will increase by only 3 dB or a factor of $\sqrt{2}$. Thus noise voltage increases as the square root of noise bandwidth, and noise voltage is expressed in nanovolts per root hertz (nV/$\sqrt{Hz}$). There are 141$\sqrt{Hz}$ in a 20-kHz bandwidth. A 100-nV/$\sqrt{Hz}$ noise source will produce 14.1 mV RMS in a 20-kHz measurement bandwidth.

The most common source of white noise is *Johnson noise*, sometimes also called *thermal noise*. It is typically created in a resistor. A 1-kΩ resistor creates about 4.16 nV/$\sqrt{Hz}$ of voltage noise at room temperature of 27 °C. It is valuable to keep this number in mind. This number goes up as the square root of resistance. It also goes up as the square root of temperature expressed in Kelvin (27 °C = 300 K). At 100 °C, this number becomes 4.64 nV/$\sqrt{Hz}$.

Pink Noise

So-called *pink noise* has the same noise power in each octave of bandwidth [9]. Pink noise is usually employed in certain acoustical measurements. Pink noise is created by passing white noise through a low-pass filter that has a roll-off slope of 3 dB per octave. Put another way, the noise power density of pink noise is inversely proportional to frequency. Indeed, so-called *1/f noise* is pink noise.

1/f Noise and Flicker Noise

Noise that increases at low frequencies, and whose power doubles for every decreased octave, is referred to as *1/f noise*. It is usually associated with circuit devices like transistors and diodes. Because it can extend to very low frequencies, it is sometimes called *flicker noise* [8, 9, 10]. It has the same spectral shape as pink noise. *1/f* noise does not have a predictable and mathematical basis to it, and is more often the result of imperfections in semiconductor devices. It is notable that many natural processes experience *1/f* noise.

The Color of Noise

As with white noise and pink noise, other noise having different spectral shapes are sometimes described by other colors [9].

Red noise has a power spectrum that falls off at 6 dB/octave. One can see that there is some connection between the nature of the noise and the colors in the light spectrum, the red end of the spectrum corresponding to lower frequencies.

Red noise is generated by white noise that is passed through a 6 dB/octave low-pass filter with a very low cutoff frequency. Put more accurately, it is generated by integrating white noise. It is heavier in the lower frequencies. Red noise is also called *Brown* noise, as it has statistics that are like those of Brownian motion.

The white noise present at the input to a phono preamp is converted approximately to red noise by the RIAA equalization that falls at a bit less than 6 dB per octave on average [11]. A condenser microphone capsule and its preamp also create red noise. A typical 30-pF condenser capsule is polarized to about 60 V through a 1-GΩ resistor, resulting in a low-frequency 3-dB point of about 5 Hz. The 4000-nV/$\sqrt{\text{Hz}}$ thermal noise of the 1-GΩ resistor is low-pass filtered at 5 Hz. The result is red noise whose amplitude is about 20 nV/$\sqrt{\text{Hz}}$ at 1 kHz.

Infrared noise falls off at 9 dB/octave. It can be created when $1/f$ noise is integrated. An example might be the open-loop noise of a DC servo integrator implemented with a JFET op amp that has significant $1/f$ noise. In practice, the infrared noise of a DC servo is quite small.

Grey noise is white noise that is weighted for constant audibility considering the human ear's varying sensitivity to different portions of the audible spectrum. Grey noise is white noise that has been passed through an inverse A-weighting filter. It has a higher spectral density at low frequencies and at high frequencies.

Blue noise is white noise that has been passed through a high-pass filter so that its power density increases at 3 dB/octave.

Violet noise is white noise whose power spectral density increases at 6 dB/octave. It is created by differentiation of white noise. A transmission system with a white noise source whose signal is subjected to 6 dB/octave pre-emphasis will create violet noise in the channel at frequencies above the pre-emphasis frequency breakpoint. The noise put into the channel will be converted back to white noise when the receive end subjects the signal to de-emphasis.

The Sound of Noise

All of the above noises can be generated electronically and then subjected to listening tests. Which noise sounds more pleasant for a given A-weighted S/N? Hiss at higher frequencies is generally less objectionable. Correlated noise, like that of mains frequency buzz, can be especially annoying. Keeping in mind the A-weighting curve and the ear's frequency-dependent sensitivity, two amplifiers with the same A-weighted S/N can have their respective noises sound different.

5.6 Relating Input Noise Density to Signal-to-Noise Ratio

Most of the noise in an amplifier is usually contributed by the input stage or other early stages. For this reason, the noise of an amplifier is often referred back to the input. Input-referred noise is equal to the measured output noise divided by the gain of the amplifier. A low-noise op amp might have input-referred noise of 2–3 nV/$\sqrt{\text{Hz}}$.

If a power amplifier has 10 nV/$\sqrt{\text{Hz}}$ of input-referred noise, what is its *signal-to-noise ratio (SNR)*? Assume that the SNR is in an un-weighted 20-kHz bandwidth and that it is referred to 2.83 V RMS out. Also assume that the amplifier has a voltage gain of 20. The output noise voltage will be 10 nV/$\sqrt{\text{Hz}}$ * 141$\sqrt{\text{Hz}}$ * 20 = 28.2 µV. This is 100,000 times smaller than 2.83 V, so the un-weighted SNR in a 20-kHz measurement noise bandwidth is 100 dB. If instead the un-weighted SNR is measured in an 80-kHz noise bandwidth, the SNR will be 6 dB worse, at 94 dB. In all of this, it is important to recognize that contributions of hum and its harmonics are not being considered here. Such low-frequency contributions can sometimes increase the un-weighted SNR by a significant amount. A-weighted SNR measurements largely remove the contributions of low-frequency noise.

Now consider a wideband un-weighted noise measurement of the same amplifier. Assume that the amplifier has a closed-loop bandwidth of 500 kHz with a single-pole roll-off. The ENBW will be 690 kHz ($831\sqrt{Hz}$), and the output noise will be 166 μV RMS. The SNR will be 17,000, corresponding to about 85 dB.

5.7 Amplifier Noise Sources

There are many sources of noise in a multi-stage amplifier. Every transistor and every resistor contributes noise, but some contributions are much larger than others [2]. The smaller contributors can often be ignored. Total noise is often referred to the input. The input-referred noise of a given noise source is divided by the gain ahead of it to determine its overall contribution to input-referred noise of the overall amplifier. If the second stage of an amplifier has input-referred noise of 10 nV/$\sqrt{Hz}$ and the first stage gain is 10, then the net noise contribution of the second stage will be 1 nV/$\sqrt{Hz}$ as referred to the input.

Input Stage

The input stage (IPS) is usually where most of the noise in an amplifier originates, and a designer first looks to the input stage for improvement of noise performance. A good power amplifier input stage may have input-referred noise less than 5–10 nV/$\sqrt{Hz}$. As we will see, however, there are numerous other noise sources that can make the amplifier's overall input-referred noise quite a bit larger.

VAS Noise

The input stage is not the only source of noise in an amplifier, even though it often dominates. The input stage load and later stages create noise, and their noise can be referred back to the input of the amplifier by the voltage gain that precedes them. For example, a VAS with input noise of 30 nV/$\sqrt{Hz}$ will contribute an amplifier input-referred noise component of 10 nV/$\sqrt{Hz}$ if the voltage gain of the input stage is only 3. That will increase the input-referred noise by 3 dB if the input stage noise is 10 nV/$\sqrt{Hz}$. The message here is that VAS noise cannot be ignored, and may even dominate the noise in some amplifier designs. This can happen because the VAS is not usually designed for low noise and input stage gain is sometimes quite small, sometimes as small as 1.5. This can happen in the case of a heavily degenerated differential input stage with a simple resistive load connected to a one-transistor VAS.

Power Supply Noise and PSRR

The power supply rails in any amplifier are often corrupted by numerous sources of noise. These may include random noise and other noises like power supply ripple, EMI and signal-dependent noise from the output stage. The power supply noise can get into the signal path as a result of the amplifier circuit's finite *power supply rejection ratio* (PSRR). If an input stage has 60 dB PSRR, then 1 mV of power supply noise in a 20-kHz ENBW will contribute 1 μV of input-referred noise. This alone corresponds to 7 nV/√Hz.

There are two important ways to control power supply noise. The first approach is to employ circuit topologies that have inherently high PSRR. One example is the use of a current mirror load on the input stage differential pair and a tail current source with high output impedance. Adding a cascode stage to the input pair can also increase PSRR by increasing the output imped-ance of the input pair. The ability of a circuit to reject power supply noise usually decreases as the frequency of the noise increases. In other words, PSRR degrades at high frequencies. Fortu-nately, it is often possible to do a more effective job of filtering the power supply rails at higher frequencies. Power supply noise and PSRR are discussed elsewhere in detail.

The second approach is to do a better job filtering the power supply rails. This is especially effective for power supply rails that provide power to the IPS and VAS circuits. Additional rail decoupling can be used, but it can eat into voltage headroom. A 120-Ω decoupling resistor for an IPS that draws 4 mA will drop 0.5 V. To get 20 dB of additional rejection at 60 Hz will require 100 μF. The same network providing 14 mA of filtered power to both the IPS and VAS will drop 1.7 V.

There is also power supply rectifier noise. This is largely a result of the fact that rectifiers do not turn off instantly due to stored charge that must be removed by reverse current flow. The turn-off process can be a local source of EMI. The time it takes to turn off the diode is usually referred to as the *recovery time*. Sharp current impulses and bursts of ringing with the power transformer inductance occur during turn-off. Fast-recovery and soft-recovery diodes can reduce this, along with rectifier snubbers that damp bursts of ringing during the switching intervals. Good discus-sions of this can be found in these references [12, 13].

Switchmode Power Supply Noise

Although power supply noise in conventional linear supplies has always been a concern (rip-ple, rectifier switching, transformer ringing, etc.), the noise from switch-mode power supplies (SMPS) has usually been thought to be worse. Nowadays, that need not be the case. Modern SMPS operate at much higher frequencies and create virtually no ripple. High frequencies are often easier to shield.

Having said that, SMPS are a potential local source of significant EMI. The good news is that they largely eliminate line-rate ripple and potentially large 60-Hz magnetic fields created by the power transformer. Noise that is magnetic in origin is insidious. Careful layout and shielding is imperative in amplifiers employing SMPS. The SMPS is often put in its own sub-enclosure; this is practical because of the small size of the SMPS.

Much has been made of EMI as a source of sound quality degradation in conventional ampli-fiers with linear power supplies, so one is inclined to suspect that EMI immunity is even more important when there is a local source of it right in the same enclosure. Amplifier circuits and layouts that are more robust to EMI will probably fare better in such designs. For example, amplifiers with JFET input stages may be preferred.

Capacitance Multiplier

A more efficient alternative to an R-C filter in the power supply rail is a capacitance multiplier. It is little more than an emitter follower that is fed a version of the power supply rail that has been filtered by a relatively high impedance R-C filter. It may drop as little as 0.6 V even for a moderate current load. A simple capacitance multiplier is shown in Figure 5.4. A small-signal

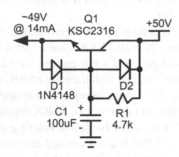

Figure 5.4 Capacitance Multiplier

transistor can be used because power dissipation is small. However, the transistor must be rated to withstand the full power supply rail voltage because the base will begin with 0 V when power is applied. The transistor operates at $V_{cb} = 0$ V. For this reason, input ripple should be less than 100 mV. With $I_{load} = 14$ mA and $\beta = 200$, base current will be 70 μA. A 4.7-kΩ decoupling resistor will drop an additional 350 mV for a total drop of almost 1 V. A 100-μF shunt capacitor will now provide 52 dB of attenuation at 120 Hz. The capacitance multiplier approach re-references the rail voltage to the local ground, and greatly decreases the amount of noise current dumped into the local ground by a conventional large bypass capacitor.

Diodes D1 and D2 protect the transistor in the event that the input rail voltage collapses quickly for some reason. If increased voltage drop is created by the addition of a shunt resistor across C1, a larger amount of power supply ripple can be tolerated on the input rail. Note that an amplifier having a 20,000-μF reservoir capacitor and drawing an average of 5 A at high power will have about 2 V peak-to-peak ripple on its incoming rails from the power supply. Care must be taken in the design of such capacitance multipliers that there is enough voltage margin to allow worst-case rail voltage dips that happen more quickly than C1 can discharge. In other words, in normal operation, D2 should never turn on.

Ground Loops and Noise Pickup

Avoidance of ground loops and other means of noise pickup is important in a power amplifier if best sound quality and S/N are to be achieved. The same best practices for grounding and interconnect often employed in low-level and line-level equipment should also be observed in the design of power amplifiers. Bill Whitlock of Jensen Transformers has published numerous excellent articles on this topic [14, 15, 16, 17].

Star Quad Cable

So-called *star quad* cable consists of four wires twisted together inside a shield with diagonally opposite wires connected together at both ends. Manufacturers of star quad cable include Belden, Mogami and Canare. This construction results in far superior immunity to externally induced noise from magnetic fields as compared to conventional shielded twisted pair cable [18, 19]. Magnetically induced noise is insidious and difficult to suppress. Cable shielding blocks noise from electric fields but does little for noise originating from magnetic fields. Twisted pairs,

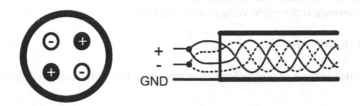

Figure 5.5 Star Quad Cable Construction

whether shielded or not, suppress noise induction from magnetic fields, but the geometry of star quad cable takes this to a much higher level, sometimes by as much as 20 dB [19].

Although most often associated with standard size microphone cable, it is available in miniature sizes as well, down to as small as 30-AWG conductors in a cable of 3/16 inch OD that is very flexible. These cables can be especially useful for short runs inside amplifiers for improved noise immunity from noisy magnetic fields (e.g., from transformers, mains wiring, power supply wiring, etc.). Figure 5.5 illustrates how star quad cable is constructed. In some instances, the star quad geometry has been emulated on printed wiring boards for improved noise immunity. This is accomplished with a particular arrangement of traces on multiple layers.

The concept of reciprocity means that star quad will also confine sources of noise passing through the cable. This means that it can be used to distribute power while suppressing radiation of magnetic noise. The rail voltages from the power supply in an audio amplifier can be routed to the amplifier board with 21-AWG star quad microphone cable. This results in 18-AWG equivalent wire size, since cross-sectional area doubles with each three increments in AWG. This is entirely adequate for the short runs found in such an application. The resistance of one foot of 18-AWG wire is only 0.05 Ω. The shield can be used for the ground.

Canare also makes star quad speaker cable. This cable is unshielded and is available in fairly heavy gauges. The Canare 4S8 consists of four 16-AWG conductors for an effective 13-AWG cable. Capacitance is only 145 pF/meter.

5.8 Thermal Noise

All resistors generate noise. This is referred to as *Johnson noise* or *thermal noise* [1, 2]. It is the most basic source of noise in electronic circuits. It is most often modeled as a noise voltage source in series with an ideal noiseless resistor. The noise power in a resistor is dependent on temperature.

$$P_n = 4kTB \text{ watts} = 1.66 \times 10^{-20} \text{ W/Hz} \tag{5.2}$$

where k = Boltzman's constant = 1.38×10^{-20} J/°K
T = temperature in Kelvin = 300 K @ 27 °C
B = equivalent noise bandwidth in Hertz

The open-circuit RMS noise voltage across a resistor of value R is simply

$$e_n = \sqrt{4kTRB} \tag{5.3}$$

$$en = 0.129 \text{ nV}/\sqrt{Hz} \text{ per } \sqrt{\Omega} \text{ (at 27°C)} \tag{5.4}$$

Noise voltage for a resistor thus increases as the square root of both bandwidth and resistance. A convenient reference is the noise voltage of a 1-kΩ resistor at 27 °C:

$$1\ k\Omega \Rightarrow 4.163\ nV/\sqrt{Hz}$$

From this the noise voltage of any resistance in any noise bandwidth can be estimated.

5.9 Shot Noise

Bipolar transistors generate a different kind of noise [1, 2]. This noise is related to the discreteness of current flow. This is called *shot noise* and is associated with the current flows in the collector and the base of the transistor. The collector shot noise current is usually referred back to the base as an equivalent input noise voltage in series with the base. It is referred back to the base as a voltage by dividing it by the transconductance of the transistor. Once again, the resulting input-referred noise is usually measured in nanovolts per root Hz.

The shot noise current is usually stated in picoamperes per root Hz (pA/$\sqrt{Hz}$) and has the RMS value of

$$I_{shot} = \sqrt{2qI_{dc}B} \tag{5.5}$$

$$I_{shot} = 0.57\ pA/\sqrt{Hz}/\sqrt{\mu A} \tag{5.6}$$

where $q = 1.6 \times 10^{-19}$ Coulombs per electron.
B = bandwidth in Hz.

It is easily seen that shot noise current increases as the square root of bandwidth and as the square root of current. A 1-mA collector current flow will have a shot noise component of 18 pA/$\sqrt{Hz}$.

$$1\ mA \Rightarrow 18\ pA/\sqrt{Hz}$$

5.10 Bipolar Transistor Noise

There are several sources of noise in the BJT. Collector shot noise current, when referred back to the input of the transistor by its transconductance represents an effective input noise voltage represented in nV/$\sqrt{Hz}$. Base shot noise current is simply an input noise current usually represented in pA/$\sqrt{Hz}$. The third significant noise source in a transistor is base resistance. This represents a thermal noise voltage source that adds to the effective input noise voltage created by collector shot noise. Base noise current flowing through the base resistance can also create a noise voltage component, but it is usually quite small.

BJT Input Noise Voltage

The transconductance of a BJT operating at 1 mA is 38.5 mS. Dividing the shot noise current by *gm* yields input-referred noise e_n = 0.47 nV/$\sqrt{Hz}$. From Equation 5.3 we can see that this is the voltage noise of a 13-Ω resistor.

At the same time, notice that *re'* ($1/gm$) for this transistor is 26 Ω. The noise voltage for a 26-Ω resistor is 0.66 nV/$\sqrt{Hz}$. The input-referred voltage noise (resulting from collector current shot noise) of a transistor is equal to the Johnson noise of a resistor whose value is half the value of *re'*. This is a very handy relationship.

Base resistance is the second major component of transistor voltage noise. A base resistance of 100 Ω will contribute 1.3 nV/√Hz. When added to the collector shot noise component of 0.47 nV/√Hz, we have 1.38 nV/√Hz input noise for a transistor operating at 1 mA. This demonstrates the importance of base resistance in BJT noise.

BJT Input Noise Current

The base current of a transistor also has a shot noise component measured in units of pA/√Hz. Recall that

$$I_{shot} = 0.57 \text{ pA}/\sqrt{Hz}/\sqrt{\mu A} \tag{5.7}$$

Consider a BJT biased at 1 mA and with β of 100. Base current will be 10 μA. This corresponds to 3.16 √μA. Input noise current will be 1.8 pA/√Hz. Put another way, base shot noise is collector shot noise divided by $\sqrt{\beta}$. It is always important to remember that shot noise goes down as the square root of the associated current.

If the base circuit includes source resistance, the base shot noise current will develop an equivalent noise voltage across that resistance. If the source impedance driving that transistor's base is 1 kΩ, then the input noise voltage due to input noise current will be 1.8 nV/√Hz.

Low-Noise Transistors

There are really two types of bipolar transistors that can be deemed *low-noise*. The difference relates to the type of application for which they are intended, depending on whether input current noise or input voltage noise is the greater concern. Transistors with very high current gain, like the 2N5089, have very low input-current noise due to their high current gain and the low collector current at which they can be operated. The 2N5089 has a typical current gain of over 400 for collector current as low as 10 μA. This makes it possible to achieve very low input noise current. When operated at 10 μA, its base current will be less than about 25 nA and its input current noise will be less than about 0.1 pA/√Hz. Of course, at this low operating current the 2N5089's voltage noise will be about 4.7 nV/√Hz.

For low impedance applications, transistor voltage noise is much more important. An example application here would be where the source impedance is 50 Ω. Here the source voltage noise is only 0.9 nV/√Hz. A transistor like the 2N4401 performs very well in this application because it has low base resistance on the order of r_{bb} = 15 Ω, whose thermal noise is only 0.5 nV/√Hz. If the 2N4401 is operated at 5 mA, its transconductance is about 200 mS (re' = 5 Ω) and its input noise voltage due to collector shot noise current will be about 0.2 nV/√Hz. The total will be about 0.54 nV/√Hz. Of course, with a minimum specified β of only 40, base current will be as high as 125 μA and base current noise will be as high as 6.4 pA/√Hz. In an actual circuit with 50-Ω source impedance, this input current noise will add 0.3 nV/√Hz as it flows through the source impedance. The NPN 2N4401 and the PNP 2N4403 have been popular choices for moving coil phono preamps.

For still lower voltage noise, one can consider the NPN ZTX851 and the PNP ZTX951, each with r_{bb} of only about 2 Ω. These are medium-power high current transistors in a TO-92 package. However, these transistors are characterized by rather high C_{cb} of 45 pF and 74 pF, respectively.

Noise Figure

Noise figure (NF) is the amount by which the inherent thermal noise of the source impedance for an amplifier is increased by the amplifier noise [1, 2]. NF is usually associated with the device used for the input stage of the amplifier, since that is where the major noise contribution of the amplifier often originates. As a simple example, consider a source impedance of 1 kΩ, which will have thermal noise of 4.2 nV/√Hz. If the associated amplifier has 4.2 nV/√Hz input voltage noise and no significant input current noise (like a JFET), then the noise figure for that amplifier will be 3 dB. If instead the voltage noise of the amplifier is 2 nV/√Hz, the total noise will be about 4.5 nV/√Hz and the noise figure will be about 1 dB. The 2N5089 datasheet shows a typical noise figure of 0.5 dB at 15 kΩ source impedance when operated at 10 μA.

Now consider the 2N4401 operating at 1 mA with a source impedance of 50 Ω. The source noise is 0.9 nV/√Hz. Voltage noise of the 2N4401 will be about 0.5 nV/√Hz. If the 2N4401 is operating at a typical β of 100, base current will be 10 μA and base current noise will be 1.8 pA/√Hz. Flowing through the 50-Ω source impedance, this will create 0.1 nV/√Hz of additional noise, which is negligible. Total noise with the transistor connected to the 50-Ω source impedance is 1.03 nV/√Hz. This corresponds to NF = 1.4 dB.

Optimum Source Impedance

If the source impedance for a transistor at a given operating point is high, input noise current will dominate the transistor's contribution to noise figure. Conversely, if the source impedance is low, input noise voltage will dominate obtainable NF. This suggests that there is an optimum value of source impedance for best NF, where input current and voltage noise contributions from the transistor are equal. The optimum source impedance for lowest NF for a given device is simply input voltage noise divided by input current noise.

$$Z_{opt} = e_{in}/i_{in} \tag{5.8}$$

For the 2N5089 operating at 10 μA, this works out to about 47 kΩ. For the 2N4401 operating at 5 mA, this works out to about 84 Ω.

5.11 JFET Noise

JFET noise results primarily from *thermal channel noise* [1, 20, 21, 22, 23]. That noise is modeled as an equivalent input resistor r_n whose resistance is equal to approximately $0.6/gm$ [24]. If we model the effect of gm as rs' (analogous to re' for a BJT), we have $r_n = 0.6 \, rs'$. This is remarkably similar to the equivalent voltage noise source for a BJT, which is the voltage noise of a resistor whose value is $0.5 \, re'$. The voltage noise of a BJT goes down as the square root of I_c because gm is proportional to I_c, and re' goes down linearly as well. However, the gm of a JFET increases as the square root of I_d. As a result, JFET input voltage noise goes down as the one-fourth power of I_d. The factor 0.6 is approximate, and SPICE modeling of some JFETs suggests that the number is closer to 0.67.

At $I_d = 0.5$ mA, gm for the LSK489 dual monolithic JFET is about 3 mS, corresponding to a resistance rs' of 333 Ω. Multiplying by the factor 0.67, we have an equivalent noise resistance r_n of 223 Ω, which has a noise voltage of 2.7 nV/√Hz [24].

At $I_d = 2.0$ mA, gm for the LSK389 dual monolithic JFET is about 20 mS, corresponding to a resistance rs' of 50 Ω. Multiplying by the factor 0.67, we have an equivalent noise resistance r_n of 33.5 Ω, which has a noise voltage of 0.73 nV/√Hz. JFET input voltage noise will also include a thermal noise contribution from ohmic gate and source resistance, but this is often negligible.

JFET Operation

The simplified equation below describes the DC operation of a JFET [1]. The term β is the transconductance coefficient of the JFET (not to be confused with BJT current gain).

$$I_d = \beta(V_{gs} - V_t)^2 \tag{5.9}$$

At $V_{gs} = 0$, we have I_{DSS}:

$$I_{DSS} = \beta(V_t)^2 \tag{5.10}$$

β can be seen from I_{DSS} and V_t to be:

$$\beta = I_{DSS}/(V_t)^2 \tag{5.11}$$

The operating transconductance gm is easily seen to be:

$$gm = 2 * \sqrt{\beta * I_d} \tag{5.12}$$

This last relationship is important, because transconductance of the JFET is what is most often of importance to circuit operation [24]. For a given transconductance parameter β, gm is largely independent of I_{DSS} and V_t, and goes up as the square root of drain current. This is in contrast to a bipolar transistor where gm is proportional to collector current. The value of β for JFETs ranges from about 1×10^{-3} to 100×10^{-3}. A typical low-noise JFET may have $\beta = 50 \times 10^{-3}$ and will have $gm = 20$ mS at an operating current of 2 mA. A still lower-noise JFET with $\beta = 81 \times 10^{-3}$ and operating at 4 mA will have $gm = 36$ mS.

JFET Noise Sources

In order to understand low-noise JFETs, it is helpful to briefly review the five major sources of noise in JFETs [1, 20, 21, 24].

* Thermal channel noise
* Gate shot noise current
* $1/f$ noise
* Generation-recombination (G-R) noise
* Impact-ionization noise

The first two sources of noise are largely fundamental to the device, while the remaining three sources are largely the result of device imperfections. Examples of such imperfections include lattice damage and charge traps. A major reduction in G-R noise is key to the LSK489's superior noise performance [24].

Thermal Channel Noise

Thermal channel noise, as discussed above, is akin to the Johnson noise of the resistance of the channel. However, it is important to recognize that the channel is not acting like a resistor in the saturation region where JFETs are usually operated. The channel is operating as a doped semiconductor whose conduction region is pinched off by surrounding depletion regions to the point where the current is self-limiting. Conduction is by majority carriers. The constant 0.67 in the equation where $r_n = 0.67/gm$ is largely empirical, and can vary with the individual device geometry. It is often a bit smaller than 0.67.

Gate Shot Noise Current

JFET input current noise results from the shot noise associated with the gate junction leakage current. Shot noise increases as the square root of DC current. A useful relationship is that $I_{shot} = 0.57$ pA/$\sqrt{Hz}$/$\sqrt{\mu A}$ [24]. Alternately, $I_{shot} = 0.57$ fA/$\sqrt{Hz}$/$\sqrt{pA}$. Gate shot noise is white, and has no $1/f$ component [3].

This noise is normally very small, on the order of fA/$\sqrt{Hz}$. It can usually be neglected. However, in extremely high-impedance circuits and/or at very high temperatures, this noise must be taken into account. Consider a circuit with 100-MΩ source impedance and a JFET with input noise current of 4 fA/$\sqrt{Hz}$ at 25 °C. The resulting voltage noise will be 400 nV/$\sqrt{Hz}$. Leakage current doubles every 10 °C, so at 65 °C the leakage current goes up by a factor of 16 and this noise contributor will go up by a factor of 4 to about 1600 nV/$\sqrt{Hz}$. For comparison, the Johnson noise of the resistive 100-MΩ source is about 1300 nV/$\sqrt{Hz}$.

1/f Noise

At very low frequencies the input noise power of a JFET amplifier rises as the inverse of frequency. That is why this noise is referred to as $1/f$ noise. When expressed as noise voltage, this means that the noise rises at a rate of 3 dB/octave as frequency decreases. In a good JFET, the $1/f$ spot-noise voltage at 10 Hz may be twice the spot noise at 1 kHz (up 6 dB) when expressed as nV/$\sqrt{Hz}$. The noise might typically be up by 3 dB at 40 Hz. $1/f$ noise is associated with imperfections in the fabrication process, such as imperfections in the crystal lattice [21]. Improved processing contributes to reduced $1/f$ noise. In fact, the amount of $1/f$ noise is sometimes considered as an indication of process quality.

By comparison, a good JFET IC op amp with input noise of 10 nV/$\sqrt{Hz}$ at 1 kHz may have its noise up by 3 dB at 100 Hz and the spot noise at 10 Hz might be up by 10 dB to 32 nV/$\sqrt{Hz}$. At 1 Hz that op amp may have spot noise on the order of 65 nV/$\sqrt{Hz}$.

Figure 5.6 illustrates a plot of voltage noise versus frequency. The $1/f$ corner frequency is defined as the frequency where the $1/f$ noise contribution equals the flat-band noise. Put another way, it is where the −3 dB/octave $1/f$ noise line intersects the flat-band noise line. At this frequency, the voltage noise will be up by 3 dB.

Generation-Recombination Noise

A less-known source of voltage noise results from carrier generation-recombination in the channel of the JFET. This is referred to as G-R noise [21]. This *excess noise* is governed by fluctuation in the number of carriers in the channel and the lifetime of the carriers. G-R noise manifests itself as drain current noise. When referred back to the input by the transconductance of the JFET, it is expressed as a voltage noise.

Like $1/f$ noise, G-R noise results from process imperfections that have created crystal lattice damage or charge-trap sites [21]. In contrast, however, G-R noise is not limited to low frequencies. In fact, it is flat up to fairly high frequencies, usually well above the audio band. The G-R noise power-spectral-density function is described in "Noise in Semiconductor Devices" [21] as:

$$S_{G\text{-}R(f)}/N^2 = [(\overline{\Delta N})^2/N^2] * [4\tau/(1 + (2\pi f\tau)^2]$$

(5.13)

where $(\overline{\Delta N})^2$ is the variance of the number of carriers N, and τ is the carrier lifetime.

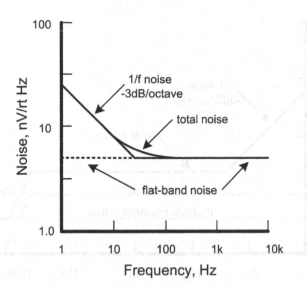

Figure 5.6 1/f Noise Voltage

Above a certain frequency, the *G-R* noise power decreases as the square of frequency. When expressed as noise voltage, this means that it decreases at 6 dB/octave. The point where the *G-R* noise is down 3 dB can be referred to as the *G-R* noise corner frequency. That frequency is governed by the carrier lifetime, and in fact is equal to the frequency corresponding to a time constant that is the same as the carrier lifetime [21]. We have,

$$f_{G-R} = 1/2\pi\tau \tag{5.14}$$

where τ is the carrier lifetime.

The 6 dB/octave high-frequency roll-off of *G-R* noise is only an approximation because there are normally numerous sites contributing to *G-R* noise and the associated carrier lifetimes may be different. As a result, the *G-R* noise corner frequency is poorly defined and the roll-off exhibits a slope that is usually less than 6 dB/octave over a wider range of frequencies. The inflection in the JFET's noise-versus-frequency curve may thus be somewhat indistinct. The important take-away here is that excess *G-R* noise can sometimes exceed the thermal channel noise contribution and thus dominate voltage noise performance of a JFET.

The idealized noise-spectral-density graph in Figure 5.7 illustrates how the three voltage noise contributors act to create the overall noise-versus-frequency curve for a JFET [24]. In the somewhat exaggerated case illustrated, *G-R* noise appears to dominate thermal channel noise, but it often just adds a bit to the existing thermal channel noise.

Noise improvements derive from process improvements that reduce device imperfections. Those imperfections create *G-R* noise and *1/f* noise. Such process imperfections include crystal lattice damage and charge trap sites. Put simply, most JFETs are not as quiet as they can be. Process improvements reduce both *1/f* noise and *G-R* noise.

Impact-Ionization Noise

An electron traveling in a strong electric field can be accelerated to the point where it has enough kinetic energy to knock another electron out of its valence band into the conduction band if it impacts an atom in the crystal lattice [21, 23, 25]. For convenience, the electron corresponding to

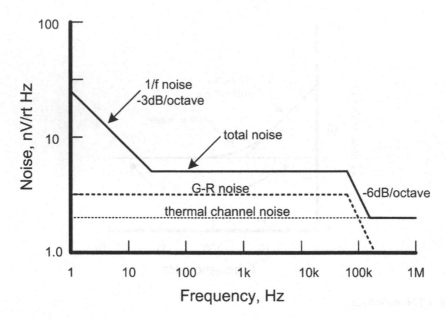

Figure 5.7 JFET Voltage Noise

normal current flow can be called a *seed* electron, since it starts the process. This collision creates a new hole-electron pair. This process is called impact ionization. The new hole and electron then act as additional charge carriers and add to the current flow. The new carriers may also be accelerated to the point where they create an impact ionization event, so the process may be multiplied. This is what is called an avalanche effect. Impact ionization often occurs in a *p-n* junction that is under a high reverse bias voltage that creates a large electric field.

In a JFET, the nominal gate junction saturation current corresponds to the flow of such seed electrons. Impact ionization in the gate-channel junction can create a current flow that results in excess gate leakage current. This leakage current includes a corresponding amount of shot noise current. The result is impact-ionization noise (IIN), sometimes also called avalanche noise.

A second source of IIN originates in the conducting channel of the JFET. Electrons passing along the channel encounter a high electric field due to the drain-source potential difference. The high electric field mainly exists at the drain end of the channel where the channel is pinched off when the JFET is operating in its saturation mode. These carriers are accelerated to a velocity where they can create impact ionization events in the channel and an avalanche multiplication effect.

The *seed* carriers for channel IIN are the carriers associated with the drain current flow, so channel IIN is proportional to drain current. The minority carriers created by impact ionization are swept across the *p-n* junction and create excess gate current. Thus, in an N-channel device, the channel impact-ionization current flows from the drain to the gate, just as it does for the junction impact-ionization current. The majority carriers created by impact ionization merely increase the drain current. The multiplicative effect of the avalanche action means that excess gate current grows exponentially with drain-gate voltage.

P-channel JFETs have much less impact ionization noise in the channel because the majority carriers are holes, which have lower mobility and less kinetic energy. However, gate junction IIN is still present in P-channel JFETs because electron majority carriers exist at the junction.

Impact-ionization noise can be minimized in JFET amplifiers by operating at lower drain-gate voltages and lower drain current. In fact, some circuit arrangements have been devised that keep the drain-gate voltage at the smallest value that will keep the JFET in its saturation region [26]. IIN from the channel usually is much larger than that from the drain-gate junction as long as the drain-gate voltage is not close to the breakdown voltage. Because impact ionization noise is primarily a gate noise current, circuits with low source impedance are much less susceptible to IIN. Conversely, circuits with very high source impedance, like condenser microphone preamplifiers, can be affected significantly by IIN. In most circuits, gate noise current due to impact ionization is insignificant, especially if the drain-gate voltage is less than about 10 V.

5.12 Op Amp Noise

Bipolar op amps have both input voltage noise and input current noise, since their input stage employs bipolar transistors [2, 6, 8, 9, 10, 22, 27, 28]. JFET op amps have input voltage noise but virtually no input noise current [22]. Table 5.2 shows some examples of op amp noise performance, beginning with the venerable μA741.

It is notable that several newer JFET op amps have achieved very low voltage noise levels that are competitive with BJT op amps. These include the LT1113 and the OPA827.

Noise Gain

Operational amplifiers are inevitably connected in negative feedback circuits that establish the gain of the circuit. These circuits often do not have the same gain for the signal and the op amp's noise. For this reason the term *noise gain* is used to describe the amount by which the op amp's input-referred noise is multiplied by the circuit to become output noise [8, 27]. For example, consider a unity gain inverter with 10-kΩ input and feedback resistors, as shown in Figure 5.8a. The gain for the signal is unity. However, with the input shorted, it is easy to see that the input and feedback resistors form a 2:1 attenuator from the output to the input. A noise input to the non-inverting terminal of the op amp will see a non-inverting gain of 2. If the input voltage noise of the op amp is 2.7 nV/√Hz, as with the LM4562, the output noise will be 5.4 nV/√Hz. The input noise of the overall amplifier will be 5.4 nV/√Hz because the output noise has been referred back to the input by dividing it by the gain of the stage, which is unity.

In Figure 5.8b an additional 10-kΩ resistor R3 is connected to a second input, as in a mixer application. This does not change the gain of the inverting path for the first input, but does increase the noise gain to 3 because of the increased attenuation in the feedback path. The input-referred noise now becomes 8.1 nV/√Hz.

Table 5.2 Examples of Op Amp Noise Performance

Op amp	e_n, nV/√Hz	i_n, pA/√Hz	Type
μA741	22	1.0	BJT
NE5534	3.5	0.4	BJT
LM4562	2.7	1.6	BJT
AD797	0.9	2.0	BJT
TL072	18	0.01	JFET
OPA134	8	0.003	JFET
ADA4627	6.1	0.003	JFET
LT1113	4.5	0.01	JFET
OPA827	4.0	0.002	JFET

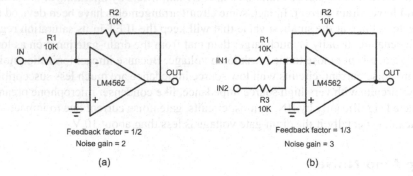

Feedback factor = 1/2
Noise gain = 2

(a)

Feedback factor = 1/3
Noise gain = 3

(b)

Figure 5.8 Noise Gain of an Inverter

By contrast, the noise gain of a unity-gain buffer is unity because there is no attenuation of the feedback signal. The main point here is that the noise gain of a circuit is usually not the same as the signal gain, and in fact is usually larger.

Input Noise Current

Operational amplifier input noise current will create noise voltage by flowing through the resistors in the input and feedback networks. Consider the inverter in Figure 5.8a with op amp input noise of 1.6 pA/√Hz, as with the LM4562. The noise current flows through the parallel combination of R1 and R2, creating a noise voltage of 8 nV/√Hz, which will be multiplied by the noise gain of 2 to 16 nV/√Hz. It will be referred back to the input by unity, resulting in input-referred noise of 16 nV/√Hz. Input noise current dominates input noise voltage in this example, and noise would be lower if R1 and R2 were made smaller. To put matters in perspective, however, bear in mind that R1 and R2 each contribute thermal noise of about 13 nV/√Hz.

If there is impedance in the non-inverting input circuit, noise will be contributed there as well. The noise currents of the inverting and non-inverting inputs of the op amp are considered as largely uncorrelated, so this additional noise source will add in an RMS fashion.

Input Noise Impedance

As with transistors, op amps have optimal input noise impedance values. This is simply the input voltage noise divided by the input current noise. For the LM4562, this is 2.3 nV/√Hz divided by 1.6 pA/√Hz, or 1.4 kΩ. This is merely the point where noise contributions from current and voltage are equal. If R1 and R2 were each made 2.8 kΩ, the circuit impedance would be 1.4 kΩ and would be at the optimum value. Because JFET op amps have virtually no input noise current, their input noise impedance is very high.

5.13 Noise Simulation

With an understanding of the basics of noise and the cause-effect relationships, noise analysis is best handled by SPICE simulations. As described in the LTspice® chapter, the noise contribution of any element can be evaluated by clicking on the circuit element. The base-current noise in a simulation will show up as a component of the transistor voltage noise contribution. This contribution is the sum of the transistor voltage noise and the transistor base current noise multiplied by the source impedance seen by the base. More information on using LTspice® for noise and other simulations can be found in Chapter 23 and in this reference [29].

Emitter Followers and Source Followers

The emitter follower and source follower circuits shown in Figure 5.9 are easy circuits to simulate for noise in order to obtain a better understanding of how sources of noise contribute to the total noise of a circuit.

The ideal current sources contribute no noise, so as shown the simulations focus only on the voltage noise sources in the transistors. Changing the current can reveal how the noise contributed by the transistor behaves as a function of current. If resistance is added to the input circuit, the effects of input current noise can be seen as well. This is especially useful in the case of the BJT.

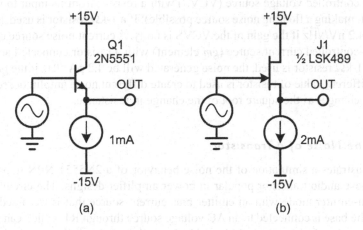

(a)   (b)

Figure 5.9 Emitter Follower and Source Follower

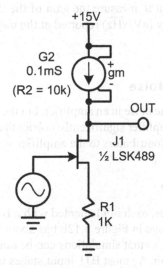

Figure 5.10 Simulation of a JFET with a Noiseless Load Resistor

Noiseless Resistors

Sometimes only a certain portion of a circuit needs to be analyzed for noise. Other elements, like load resistors, may get in the way. In these cases an ideal transconductance element can be used to create a *noiseless resistor*. The output of the *gm* element is connected back to its input, as shown in Figure 5.10. The arrangement acts as a two-terminal noiseless resistance whose value is 1/*gm*. Such a noiseless resistor can be used in the circuits of Figure 5.8 to isolate and evaluate input current noise.

The use of a noiseless load resistor in Figure 5.10 allows the simulation of the common source amplifier circuit noise apart from the effects of any noise contribution from the load.

Noise Sources

Sometimes it is useful to have a reference noise source in a simulation. This is easily achieved with a voltage-controlled voltage source (VCVS) with a resistor from its input to its ground (or reference point, making a floating noise source possible). If a 1-kΩ resistor is used, the noise generated will be 4.2 nV/$\sqrt{Hz}$ if the gain of the VCVS is unity. A current noise source can be created with a voltage-controlled current source (*gm* element) with a resistor connected across its input terminals. If a 1-kΩ resistor is used, the noise generated will be 4.2 pA/$\sqrt{Hz}$ if the *gm* value is set to 1 mS. If a different value of resistor is used to create different noise amplitude, remember that the noise level changes as the square root of the change in resistance.

Simulating the Noise of a Transistor

Figure 5.11 illustrates a simulation of the noise behavior of a 2N5551 NPN transistor. This is a general-purpose audio transistor popular in power amplifier designs. The circuit operates Q1 in the common-emitter mode with an emitter bias current source that is bypassed with a large capacitance. The base is connected to an AC voltage source through R1, which can be made zero or some chosen value. Load resistor R2 is an ideal noiseless resistor whose value is uncritical. It can be selected to provide a reasonable gain and a nominal voltage drop given the bias current chosen for the transistor.

An AC simulation is carried out to measure the gain of the stage. A noise simulation is then done and the noise voltage density (nV/$\sqrt{Hz}$) reported at the output is referred back to the input by dividing by the gain.

5.14 Amplifier Circuit Noise

The input stage often dominates the noise in an amplifier, but there are numerous other sources of noise that can make the overall amplifier significantly noisier than the input stage by itself would suggest. Here some of the major contributors to the amplifier system noise will be discussed.

Noise of a Degenerated LTP

Figure 5.12 shows two input stages, each implemented with a BJT differential pair. They are the same with the exception that the one in Figure 5.12b has about 10:1 emitter degeneration. Each has a 5-kΩ noiseless load resistor so that simulations can be done that focus solely on the noise performance of the differential pair. As most BJT input stages in power amplifiers are degenerated, the circuit in Figure 5.12b is more representative of expected noise performance. A question

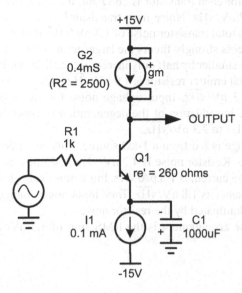

Figure 5.11 2N5551 Noise Simulation

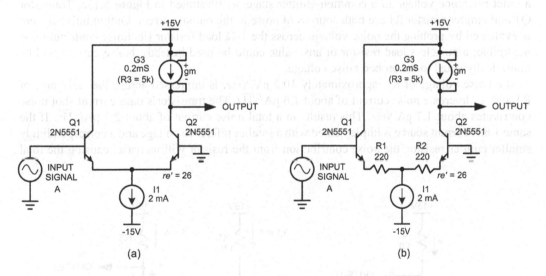

(a)   (b)

Figure 5.12 Differential LTP Input Stage

often asked is to what extent does emitter degeneration degrade the input noise performance of such an amplifier.

Each stage is biased with a tail current of 2 mA. The transconductance to the single-ended output has been reduced from about 20 mS in Figure 5.12a to about 2 mS in Figure 5.12b by the 10:1 emitter degeneration. Assume that the stage is fed from a voltage source. The voltage noise contributions of each 2N5551 transistor and its degeneration resistor will add to create the input noise because there are two of each effectively in series in forming the LTP. This is responsible for the usual 3-dB noise penalty when a differential pair is used instead of a single-ended stage.

Operating at 1 mA, *re'* for each transistor is 26 Ω and the transistor voltage noise is that of a 13-Ω resistor, so e_n = 0.47 nV/$\sqrt{Hz}$. Noise from transistor base resistance, about 90 Ω, is about 1.2 nV/$\sqrt{Hz}$. This results in total transistor noise of 1.3 nV/$\sqrt{Hz}$. It is always important to remember that the RMS sum process strongly favors the larger term at the expense of the smaller term. A second noise contributor smaller by half only increases the RMS-summed noise by about 12%.

The thermal noise of each emitter resistor is 1.9 nV/$\sqrt{Hz}$. Input-referred voltage noise of each half of the LTP is thus 2.3 nV/$\sqrt{Hz}$. Input voltage noise for the stage is 3 dB higher, at 3.3 nV/$\sqrt{Hz}$. At this point, the introduction of the degeneration resistors has increased input stage noise by about 5 dB from 1.3 to 2.3 nV/$\sqrt{Hz}$.

Now assume that the stage is fed from a 1-kΩ source. This could be from the amplifier feedback network, for example. Resistor noise is 4.2 nV/$\sqrt{Hz}$. Assume that transistor β is 100. Base current is 10 μA. Input noise current is 1.8 pA/$\sqrt{Hz}$. Input noise voltage due to input noise current flowing in the source resistance is 1.8 nV/$\sqrt{Hz}$. Total input noise voltage across the input impedance is thus 4.7 nV/$\sqrt{Hz}$, dominated by the resistor noise.

Total input noise for the arrangement is the RMS sum of 3.3 nV/$\sqrt{Hz}$ and 4.7 nV/$\sqrt{Hz}$, or 5.7 nV/$\sqrt{Hz}$.

Current Source Noise

BJT current sources can be made in numerous ways, but the most fundamental is to provide a quiet reference voltage to a common-emitter stage, as illustrated in Figure 5.13a. Transistor Q1 and emitter resistor R1 are both sources of noise in the output current. Output noise current is evaluated by probing the noise voltage across the 1-Ω load resistor (its noise contribution is negligible; a noiseless load resistor of any value could be used instead). Noise current will be numerically equal to the probed noise voltage.

The noise voltage of R1, approximately 10.2 nV/$\sqrt{Hz}$, is impressed across the resistance of R1 and *re'* to create noise current of about 1.6 pA/$\sqrt{Hz}$. The transistor's base current shot noise contributes about 1.7 pA/$\sqrt{Hz}$. This results in a total noise current of about 2.3 pA/$\sqrt{Hz}$. If the same 1 mA current source is implemented with a smaller reference voltage and a correspondingly smaller emitter resistor, the noise contribution from the resistor will increase, causing the total

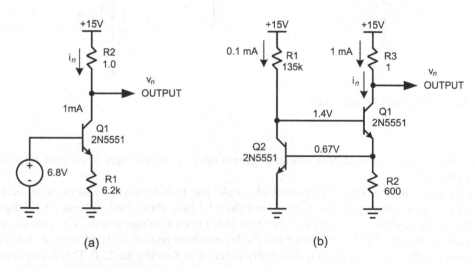

(a)   (b)

Figure 5.13 Noise in Current Sources

noise of the current source to increase. The contribution of the base shot noise will remain about the same, and in this case it will be dominated by the noise contributed by the emitter resistor.

The feedback current source shown in Figure 5.13b is a popular choice in power amplifiers. Its output current noise is influenced by the negative feedback provided by Q2. The feedback holds the base and emitter voltages of Q1 essentially constant, making them low-impedance points. The major noise contributor is the short-circuit current noise of R2, which is just its voltage noise divided by its resistance. This amounts to about 5.2 pA/√Hz. The voltage noise of Q2 operating at 100 µA is about 1.6 nV/√Hz. Impressed across R2, this contributes about 2.7 pA/√Hz. The third major contributor is the base shot noise current of Q1, which is about 2.4 pA/√Hz. These three contributors sum on an RMS basis to about 6.4 pA/√Hz.

The noise contribution of Q2 depends somewhat on the bias current supplied by R1, here 135 kΩ. If R1 is reduced, running Q2 at higher current, Q2's voltage noise will fall, but eventually Q2's base current noise will become a more significant contributor. A broad optimum for the bias current thus results. Here the optimum occurs at R1 = 30 kΩ, corresponding to bias current of 450 µA. This is rather high, and it should be kept in perspective that this noise contribution is not a major contributor to the total. Nevertheless, the bias current should probably never be set less than 1/10th of the current source value. The feedback current source is noisier than the conventional current source with a 6.8-V reference, but it requires much less voltage headroom.

Current Mirror Noise

A simple two-transistor current mirror is shown in Figure 5.14a. The sum of the noise voltages of Q1 and R1 is impressed across R1, creating a noise current in Q1. The same thing happens in Q2, adding another noise current. As the values of R1 and R2 go down, the noise current goes up. This means that current mirrors with smaller degeneration resistors will create more noise in their output current. The base current noises of Q1 and Q2 are also injected into the signal path.

A three-transistor current mirror is shown in Figure 5.14b. The base currents of Q1 and Q2 are supplied by emitter follower Q3. This reduces the contribution of base current noise from Q1 and Q2.

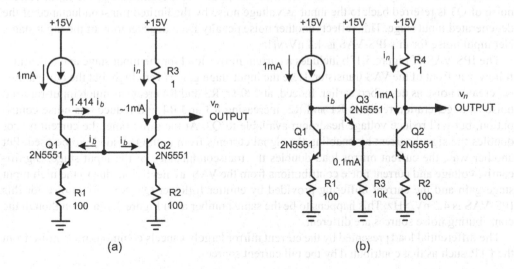

(a)   (b)

Figure 5.14 Noise in Current Mirrors

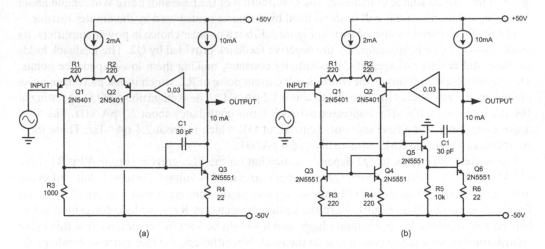

Figure 5.15 IPS-VAS Noise

IPS-VAS Noise

Two simple IPS-VAS circuits are shown in Figure 5.15. The negative feedback loop is closed with a buffer with gain of 0.03 so that noise can be evaluated in a circuit with gain of 33. The noise performance of an amplifier is often dominated by the input stage like the degenerated differential pair shown in Figure 5.12. This is because the gain of the input stage is often much greater than unity, meaning that noise created in later stages is divided by that gain when referred back to the input. If the input-referred noise of the VAS is 10 nV/√Hz and the gain of the input stage is 5, then the VAS noise contribution to the amplifier's input-referred noise is only 2 nV/√Hz. Unfortunately, things are usually not this simple.

In the crude IPS-VAS of Figure 5.15a, the gain of the input stage is only about 2. The 4-nV/√Hz of R3 is referred back to the input as 2 nV/√Hz. The input voltage noise of the VAS transistor is also divided only by 2 when it is referred to the input. More significantly, the input current noise of Q3 is referred back to the input as voltage noise by the limited transconductance of the degenerated input stage. This reflects another noise penalty from degeneration of the input stage. Net input noise for this IPS-VAS is 4.3 nV/√Hz.

The IPS-VAS of Figure 5.15b includes a current mirror load for the input stage and an emitter follower in front of the VAS transistor. Here the input stage gain is very high, but the current mirror creates noise as described earlier. Indeed, at 220-Ω, R3 and R4 create as much input-referred noise as degeneration resistors R1 and R2. Increasing R3 and R4 will reduce this noise contribution, but will reduce voltage headroom available to Q3. At the same time, the current mirror doubles the signal current by enabling the signal currents from both Q1 and Q2 to be used. Put another way, the current mirror load doubles the transconductance of the input stage. Significantly, voltage and current noise contributions from the VAS are negligible due to the high input stage gain and the current buffering provided by emitter follower Q5. Net input noise for this IPS-VAS is 4.3 nV/√Hz. This happens to be the same number as in Figure 5.15a, even though the contributing noise sources are different.

The differential load presented by the current mirror largely cancels common-mode noise from the LTP, such as that contributed by the tail current source.

Paralleled Devices

If two transistors or circuit blocks with current output are connected in parallel, the signal will sum on a voltage basis and the noise will sum on a power basis. This is because the signal contributions are correlated while the noise contributions are uncorrelated. A net improvement of 3 dB in the S/N ratio will thus result. If four such circuit blocks were connected in parallel a net 6 dB improvement would result. An additional 3 dB is gained each time the number of paralleled blocks is doubled.

Two low-noise dual monolithic LSK389 JFETs are connected in parallel to gain a 3-dB noise improvement in Figure 5.16 [11]. Matching is often a problem in connecting JFET transistors in parallel. Tight offset voltage matching is available in these dual monolithic devices, but this still leaves the issue of differences in I_{DSS} or threshold voltages between the two chips. Fortunately, we saw in Equation 5.11 that JFETs from the same process have about the same transconductance for a given operating current, independent of their individual I_{DSS} or threshold voltage. This means that both pairs will contribute about the same transconductance in the circuit as long as the tail current for each pair is the same. This is the reason why each pair has its own current source. The 100-Ω drain resistors isolate the JFETs to prevent HF instability. They add no noise. The input-referred noise of this circuit is about 1 nV/$\sqrt{\text{Hz}}$.

5.15 Excess Resistor Noise

Resistors can create more noise than just the theoretical thermal noise. This is called *excess resistor noise* [7]. It usually results from current flow in the resistor, such as when it has a DC voltage across it. It can result from the resistor material itself or from contacts between the resistance material and the leads. Carbon film resistors have the worst excess noise, while metal film

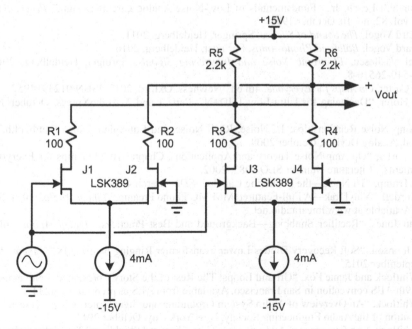

Figure 5.16 JFET Differential Pairs in Parallel

resistors have very little excess noise. Thick film resistors fall in between. All else remaining equal, a resistor of larger wattage will create less excess noise. Use metal film resistors throughout the amplifier and excess resistor noise will not be a problem.

5.16 Zener and LED Noise

Zener diodes are often used as voltage references, sometimes as part of a current source. Zener diodes can be noisy, and in fact they have sometimes been used to build noise sources. Zener diodes rated at less than about 6.2 V operate in the Zener breakdown mode, while higher-voltage Zener diodes (e.g., 12 V) actually work in the avalanche breakdown mode. The latter are quieter as a result of the different breakdown mechanism. Zener diodes should always be thoroughly bypassed if they are used in a noise-critical location. It is important to know that the effective source impedance of the Zener diode is its dynamic resistance, which can be quite low. The size of the bypass capacitor must be sufficiently large to take account of this reality. For a Zener diode having dynamic resistance of 100 Ω and bypassed with 100 μF, its noise is low-pass filtered beginning at 16 Hz. Alternatively, in low-noise applications, some resistance can be placed between the Zener diode and the filter capacitance, forming a low-pass filter.

Similarly, LEDs have often been used as a convenient low-voltage reference. LEDs also create voltage noise that originates from their shot noise mechanism. LEDs should be bypassed if they are used in a noise-critical location. Typical forward voltage for a Red LED is about 1.8 V and its voltage noise is on the order of 2.5 nV/$\sqrt{Hz}$. A Green LED is similar. Blue LEDs operated at 1 mA have a forward voltage of about 3.3 V and their voltage noise can be as high as 30 nV/$\sqrt{Hz}$.

References

1. Paul Horowitz and Winfield Hill, *The Art of Electronics*, 3rd ed., Cambridge University Press, Cambridge, 2015, ISBN 0521809266.
2. W. Marshall Leach, Jr., "Fundamentals of Low-Noise Analog Circuit Design," *Proceedings of the IEEE*, vol. 82, no. 10, October 1994.
3. Burkhard Vogel, *The Sound of Silence*, Springer, Heidelberg, 2011.
4. Burkhard Vogel, *Balanced Phono-Amps*, Springer, Heidelberg, 2016.
5. Gabriel Vasilescu, *Electronic Noise and Interfering Signals*, Springer, Heidelberg, 2010, ISBN 978-3-540-26510-8.
6. Bruce Carter, *Op Amps for Everyone*, 4th ed., Newness, Oxford, 2013, ISBN 0123914957.
7. Bruce Hofer, "Designing for Ultra-Low THD+N," Parts 1 and 2, AudioXpress, October–November 2013.
8. "Op Amp Noise Relationships: 1/f Noise, RMS Noise, and Equivalent Noise Bandwidth," MT-048 Tutorial, Analog Devices, October 2008.
9. Bruce Carter, "Op Amp Noise Theory and Applications, Chapter 10, Op Amps for Everyone, Texas Instruments," Literature Number SLOA082, 2002.
10. Bruce Trump, "1/f Noise—the Flickering Candle," *EDN*, March 4, 2013.
11. Bob Cordell, "VinylTrak—A Full-Featured MM/MC Phono Preamp," *Linear Audio*, vol. 4, September 2012. Available at www.linearaudio.net.
12. Morgan Jones, "Rectifier Snubbing—Background and Best Practices," *Linear Audio*, vol. 5, April 2013.
13. Mark Johnson, "Soft Recovery Diodes Lower Transformer Ringing by 10–20X," *Linear Audio*, vol. 10, September 2015.
14. Bill Whitlock and Jamie Fox, "Ground Loops: The Rest of the Story," presented 5 November 2010 at the 129th AES convention in San Francisco. Available from AES as preprint no. 8234.
15. Bill Whitlock, "An Overview of Audio System Grounding and Signal Interfacing," Tutorial T5, 135th Convention of the Audio Engineering Society, New York City, October 2013.
16. Bill Whitlock, "An Overview of Audio System Grounding and Shielding," Tutorial T2, Convention of the Audio Engineering Society, Star-Quad, p. 126.
17. Bill Whitlock, "Design of High Performance Audio Interfaces," Jensen Transformers, Inc.

18. Belden, How Starquad Works. Available at www.belden.com/blog/broadcastav/How-Starquad-Works.cfm.
19. Canare, "The Star Quad Story." Available at www.canare.com/UploadedDocuments/Cat11_p35.pdf and www.canare.com/UploadedDocuments/Star%20Quad%20Cable.pdf.
20. J. W. Haslett and F. N. Trofimenkoff, "Thermal Noise in Field-Effect Devices," *Proceedings of IEE*, vol. 116, no. 11, November 1969.
21. Alicja Konczakowska and Bogdan M. Wilamowski, "Noise in Semiconductor Devices," Chapter 11 in *The Industrial Electronics Handbook*, vol. 1, Fundamentals of Industrial Electronics, 2nd ed., CRC Press, Boca Raton, FL, 2011.
22. "Low-Noise JFETs—Superior Performance to Bipolars," AN106, Siliconix, March 1997.
23. Ken Yamagouchi and Shojiro Asai, "Excess Gate Current Analysis of Junction Gate FET's by Two Dimensional Computer Simulation," *IEEE Transactions on Electronic Devices*, vol. ED-25, no. 3, March 1978.
24. Bob Cordell, "Linear Systems LSK489 Application Note," February 2013. Available at www.linear-systems.com.
25. J. C. J. Paasschens and R. de Kort, "Modeling the Excess Noise Due to Avalanche Multiplication in (Hetero-Junction) Bipolar Transistors," 2003, Philips Research Laboratories.
26. James R. Butler, "JFET Differential Amplifier Stage with Method for Controlling Input Current," US Patent 4538115, August 27, 1985, assigned to Precision Monolithics Inc.
27. Walter G. Jung, *Op Amp Applications Handbook*, Elsevier, Newnes, Burlington, MA, 2005, ISBN 0-7506-7844-5, chapter 1.
28. Glen Brisebois, "Op Amp Selection Guide for Optimum Noise Performance," Design Note 355, Linear Technology Corporation.
29. Gilles Brocard, *The LTSPICE® IV Simulator*, 1st ed., Wurth Elektronik, Waldenburg, Germany, 2011, ISBN 978-3-89929-258-9.

Negative Feedback Compensation and Slew Rate

6. INTRODUCTION

In Chapters 1 and 3 the basic concepts of negative feedback were discussed briefly so that the basic amplifier design concepts could be understood. Negative feedback is fundamental to the vast majority of audio power amplifiers and its optimal use requires more understanding. Much of that will be provided in this chapter.

6.1 How Negative Feedback Works

Negative feedback was invented in 1927 by Harold Black to reduce distortions and better control gain and frequency response in telephone amplifiers [1]. Figure 6.1 shows a simplified block diagram of a negative feedback amplifier. The basic amplifier has a forward gain A_{ol}. This is called the *open-loop gain* because it is the gain that the overall amplifier would have from input to output if there were no negative feedback.

A portion of the output is fed back to the input with a negative polarity. The fraction governing how much of the output is fed back is referred to as *beta* (β). The negative feedback *loop gain* is the product of A_{ol} and β. The overall gain of the closed-loop amplifier is called the *closed-loop gain* and is designated as A_{cl}. The action of the negative feedback opposes the input signal and makes the closed-loop gain smaller than the open-loop gain, often by a large factor.

The output of the subtractor at the input of the amplifier is called the *error signal*. It is the difference between the input signal v_{in} and the divided-down replica of the output signal. The error signal, when multiplied by the open-loop gain of the amplifier, becomes the output signal. As the gain A_{ol} becomes large, it can be seen that the error signal will necessarily become small, meaning that the output signal will become close to the value v_{in}/β. If A_{ol} is very large and β is 0.05, it is easy to see that the closed-loop gain A_{cl} will approach 20. The important thing to notice here is that the closed-loop gain in this case has been determined by β and not by the open-loop gain A_{ol}. Since β is usually set by passive components like resistors, the closed-loop gain has been stabilized by the use of negative feedback. Because distortion can often be viewed as a signal-dependent variation in amplifier gain, it can be seen that the application of negative feedback also reduces distortion.

The closed-loop gain for finite values of open-loop gain is shown in Equation 6.1. As an example, if A_{ol} is 100 and β is 0.05, then the product $A_{ol}\beta = 5$ and the closed-loop gain (A_{cl}) will be equal to $100/(1 + 5) = 100/6 = 16.7$. This is slightly less than the closed-loop gain of 20 that would result for an infinite value of A_{ol}. If A_{ol} were 2000, A_{cl} would be $2000/(1 + 100) = 2000/101 = 19.8$. Here we see that if $A_{ol}\beta = 100$, the error in gain from the ideal value is about 1%. The product $A_{ol}\beta$ is called the *loop gain* because it is the gain around the feedback loop. The net gain around the

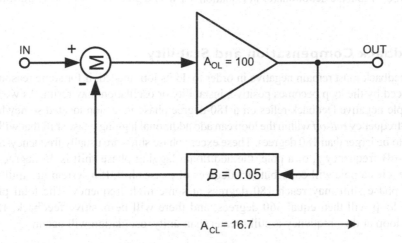

Figure 6.1 Block Diagram of a Negative Feedback Amplifier

feedback loop is a negative number when the negative sign at the input subtractor is taken into account.

$$A_{cl} = A_{ol}/(1 + A_{ol}\beta) \tag{6.1}$$

Notice that if the sign of $A_{ol}\beta$ is negative for some reason, the closed-loop gain will become greater than the open-loop gain. The closed-loop gain will ultimately go to infinity, and oscillation will result if the product $A_{ol}\beta$ reaches -1. The potential for positive feedback effects is of central importance to feedback compensation and stability.

Negative feedback is said to have a phase of 180 degrees, corresponding to a net inversion as the signal circles the loop. Positive feedback has a phase of zero degrees, which is the same as 360 degrees.

6.2 Input-Referred Feedback Analysis

A useful way to look at the action of negative feedback is to view the circuit from the input, essentially answering the question, "What input is required to produce a given output?" Viewing negative feedback this way essentially breaks the loop. We will find later that viewing distortion in an *input-referred* way can also be helpful and lend insight.

In the example above with $A_{ol} = 100$ and $\beta = 0.05$, assume that there is 1.0 V at the output of the amplifier. The amount of signal fed back will be 0.05 V. The error signal driving the forward gain path will need to be 0.01 V to drive the forward amplifier. An additional 0.05 V must be supplied by the input signal to overcome the voltage being fed back. The input must therefore be 0.06 V. This corresponds to a gain of 16.7 as calculated above in Equation 6.1.

Suppose that instead $\beta = -0.009$. This corresponds to positive feedback, which reinforces the input signal in driving the forward amplifier. With an output of 1.0 V, the feedback would be -0.009 V. The drive required for the forward amplifier to produce 1.0 V is only 0.01 V, of which 0.009 V will be supplied by the positive feedback. The required input is then only 0.001 V. The closed-loop gain has been enhanced to 1000 by the presence of the positive feedback. If β were -0.01, the product

$A_{ol}\beta$ would be -1 and the denominator in Equation 6.1 would go to zero, implying infinite gain and oscillation.

6.3 Feedback Compensation and Stability

Negative feedback must remain negative in order to do its job. Indeed, if for some reason the feedback produced by the loop becomes positive, instability or oscillation may result. As we have seen above, simple negative feedback relies on a 180-degree phase inversion located somewhere in the loop. High-frequency *roll-off* within the loop can add additional lagging phase shift that will cause the loop phase to be larger than 180 degrees. These excess phase shifts are usually frequency-dependent.

At the 3-dB frequency f_p of a pole, the additional lagging phase shift is 45 degrees. At very high frequencies a pole will contribute 90 degrees of phase shift. If a system has multiple poles this added phase shift may reach 180 degrees at some high frequency. The total phase shift around the loop will then equal 360 degrees, and there will be positive feedback. If the gain around the loop at this frequency is still greater than unity, oscillation will occur.

Poles and Zeros

The concepts of *poles* and *zeros* is fairly simple, and yet central to the understanding of feedback compensation. Ordinary circuits that create high-frequency and low-frequency roll-offs contain poles and zeros, respectively (or some combination of them). Poles introduce lagging phase shift (negative), and zeros introduce leading phase shift (positive).

Figure 6.2 illustrates three simple circuits. The first one implements a simple pole, with a 3-dB roll-off at the pole frequency $f_p = 1/2\pi R1C1$. This is just a simple first-order low-pass filter. The second circuit implements a zero at $f_z = 0$ Hz, causing the gain to rise as frequency increases, resulting in a high-pass filter. The gain can't get any larger than unity, so there is a pole to level off the frequency response at the frequency $f_p = 1/2\pi R1C1$.

The third circuit implements a *pole-zero pair*, with a resulting frequency response that begins at a lower value and increases to unity at higher frequencies. In this case the zero is not at 0 Hz, but rather at a finite frequency $f_z = 1/2\pi R1C1$. Once again, the gain cannot get any larger than unity, so there is a pole at a higher frequency $f_p = 1/2\pi R_P C1$, where R_P is the resistance of the parallel combination of R1 and R2.

Figure 6.2 also shows *Bode plots* of the three circuits. The first circuit forms a low-pass filter. Its frequency response falls at frequencies above the pole frequency f_p at a rate of 6 dB per octave

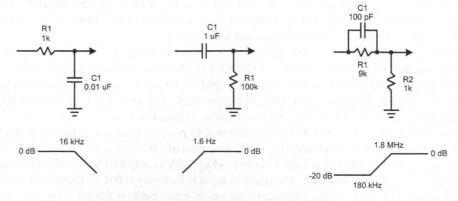

Figure 6.2 R-C Circuits Implementing Poles and Zeros

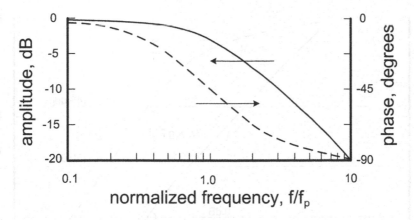

Figure 6.3 Gain and Phase for a Pole at Frequencies Far from the Pole Frequency

or 20 dB per decade. The phase lag (negative values of phase) increases to −45 degrees at the pole frequency and eventually increases to 90 degrees at high frequencies.

The second circuit forms a high-pass filter. Its frequency response is unity at high frequencies and falls to −3 dB at the frequency of the pole. At this frequency the phase shift is leading at +45 degrees. As frequency goes lower, its response falls off at 6 dB per octave and the phase lead eventually increases to +90 degrees.

The third circuit has a rising slope in frequency response beginning at f_z and continuing until f_p. Its gain is equal to $R2/(R1 + R2)$ at low frequencies and rises to unity at high frequencies. This circuit creates a leading phase shift that is at its maximum at the geometric mean of the pole and zero frequencies. At very high and very low frequencies its phase shift approaches 0 degrees.

Figure 6.3 shows the actual gain and phase for a pole at normalized frequencies f/f_p extending far away from the pole. This can be very useful in estimating the impact of multiple additional poles. It illustrates the way that the pole's phase contribution asymptotes to −90 degrees at frequencies far above the pole, while the attenuation from the pole continues to increase. The phase lag for a pole is about 26.6 degrees from its asymptotic value one octave on either side of the pole frequency.

Phase and Gain Margin

If the accumulation of lagging phase shift causes negative feedback to become positive feedback at some frequencies, instability can result. The nominal phase shift around a simple negative feedback loop is 180 degrees as a result of the inversion. Additional lagging phase shift will be introduced by high-frequency roll-offs created by feedback compensation networks and unwanted poles in the circuit. If the total phase shift in the loop reaches 360 degrees at the point where the magnitude of the loop gain is unity, oscillation will result. Central to stability analysis in the frequency domain are the concepts of *phase margin* and *gain margin*. They describe how much margin a circuit has against instability.

The concept of phase margin describes how close we are to 360 degrees of phase shift at the point in frequency where the loop gain has fallen to unity. This frequency is often referred to as the gain crossover frequency f_c. It is also referred to as the Unity-Loop-Gain-Frequency (ULGF). If the phase shift accumulates to a value greater than 360 degrees at a higher frequency than this where there is less than unity gain, oscillation will not result.

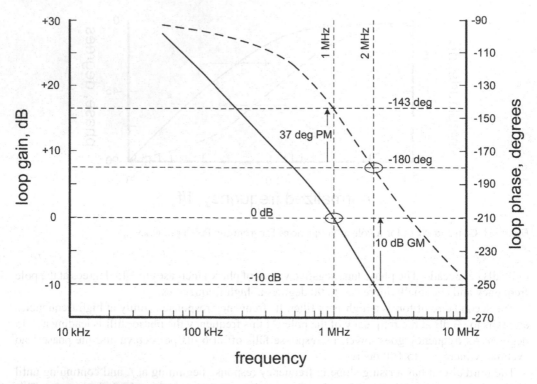

Figure 6.4 Plot Showing Phase Margin and Gain Margin

Gain margin refers to how much margin we have against the gain becoming unity at a frequency where the loop phase shift has accumulated to a total of 360 degrees. This is of particular concern because component tolerances can cause a nominal design to have higher gain under some conditions.

Figure 6.4 illustrates the concepts of phase margin and gain margin. The Y axis on the left is loop gain in dB. The Y axis on the right is lagging phase shift in the product $A_{ol}\beta$. When this lagging phase reaches 180 degrees, oscillation may result. The X axis is frequency.

The hypothetical amplifier for the plot has a gain crossover frequency of 1 MHz. This is where the loop gain has fallen to 0 dB. Loop gain rises 6 dB per octave as frequency is reduced, reaching about 40 dB at 10 kHz. Lagging phase shift at low frequencies is 90 degrees. Two poles in the open-loop response are located at 2 MHz. Each contributes 26.6 degrees at 1 MHz, so total lagging phase shift at 1 MHz is 143 degrees. This corresponds to a phase margin of 37 degrees. Loop gain is down to −10 dB at 2 MHz where loop phase shift is 180 degrees. Gain margin is thus 10 dB.

Stability against oscillation is not the only reason why adequate phase and gain margin is needed. Amplifiers with inadequate phase or gain margin usually have poor transient response and undesirable peaking in their frequency response.

Gain and Phase Variation

Component tolerances are not the only contributors to gain and phase variation. Changes in the operating points of active elements like transistors can cause gain and phase changes. One example is the change in collector-base capacitance of a transistor with signal voltage. Another example is change in transistor speed with operating current. Changes in the impedance of the load connected to the amplifier can also affect loop gain and phase.

6.4 Feedback Compensation Principles

The simplest way to compensate a negative feedback loop is to roll off the loop gain at a frequency low enough that the lagging phase shift accumulated in the loop is well less than 180 degrees at the frequency where loop gain has fallen below unity. This approach recognizes that as frequency increases in a real amplifier, more extraneous poles come into play to create extra lagging phase shift.

Dominant Pole Compensation

Dominant pole compensation is the basis for most compensation approaches. The strategy is to reduce loop gain with frequency while accumulating as little lagging phase beyond 90 degrees as possible. The key to achieving this is that a single pole continues to attenuate with frequency while its contribution to lagging phase shift is limited to 90 degrees. For this reason a feedback circuit with a single pole in its loop can never become unstable.

In real circuits there will be many poles, but if the roll-off behavior is strongly dominated by a single pole, the stability criteria will be more straightforward to meet. The effects of all other poles can then be lumped together as so-called *excess phase*. In this case, at the frequency where the gain around the loop has fallen to 0 dB, the phase lag of the dominant pole will be 90 degrees. This means that there is an additional 90 degrees to play with before hitting instability. You might allocate 40 degrees for excess phase from all other sources and keep the remaining 50 degrees in your pocket as phase margin. You then choose the maximum gain crossover frequency f_c as the frequency where the total amount of excess phase does not exceed 40 degrees.

Excess Phase

Excess phase is usually thought of as additional lagging phase shift that would not have been expected based on the amount of high-frequency amplitude roll-off. A pure delay is a very good example of excess phase. If a signal runs through 10 ft. of coaxial cable, it will encounter very little loss, but it will experience a time delay due to the speed of light in the cable. This will be on the order of 1.5 ns per foot, or 15 ns at 67 MHz, corresponding to about 360 degrees of phase lag.

More often, excess phase is not truly time delay, but rather the accumulation of phase shift from many high-frequency poles, sometimes called *parasitic poles*. Even together, these poles may not create much attenuation, but the amount of phase shift they create can accumulate to an amount sufficient to reduce phase margin in a negative feedback circuit. Excess phase in the amplifier output stage is an important contributor. It results from multiple far-out poles due in part to base resistance and AC beta roll-off of the driver and output transistors. Output stage excess phase can vary significantly with load and current/voltage conditions.

Excess phase can also originate at the feedback input to the LTP input stage. The resistance of the feedback network can form a pole against the input capacitance of the LTP. This input capacitance can be a combination of base-emitter capacitance due to finite f_T and the base-collector capacitance of the transistor. This latter capacitance can be multiplied by the Miller effect if the input stage has voltage gain to its collector at high frequencies (often not the case). This phase lag can be reduced by employing a smaller value of impedance for the feedback network. Alternatively, a very small capacitor placed across the feedback resistor can also reduce this effect by forming a capacitance voltage divider with the input capacitance of the stage. If the input capacitance were 5 pF and the gain of the amplifier were 20, this capacitor would have to be very small, on the order of 0.25 pF. A slightly larger value of capacitor will provide some phase lead, which may actually help phase margin a bit. Great caution is needed here to avoid using too much capacitance; it can reduce gain margin and provide a sneak path for EMI from the amplifier output back to the input stage.

Lag Compensation

Lag compensation is perhaps the simplest form of frequency compensation, but is usually not the best. It simply involves the introduction of a low-frequency pole by placing a shunt capacitor from the output of the VAS to ground. Unfortunately, there is also a pole in the base circuit of the VAS at a low enough frequency to add a large amount of excess phase. This will often force you to set the gain crossover frequency fairly low. You may have two poles *running together* over a significant frequency range before the loop gain falls to 0 dB.

Consider the amplifier of Figure 6.5 and ignore compensating components C1, C2 and R6 for the moment. The uncompensated amplifier has over 60 dB of loop gain at DC, and it does not fall to 0 dB until 6.2 MHz, at which point the accumulated phase lag is about 260 degrees, 80 degrees more than is needed for oscillation. Choose a conservative goal of 1 MHz for f_c. Gain of the uncompensated amplifier is about 40 dB at 1 MHz, so the compensation must get rid of 40 dB of gain at that frequency.

A large 0.02-μF shunt capacitor C2 loading the VAS collector node will provide lag compensation that brings the gain down by the required amount. C2 creates the dominant pole. However,

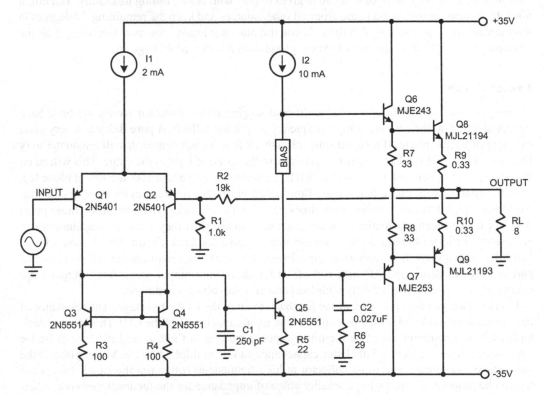

Figure 6.5 An Amplifier Where Lag Compensation Is Employed

the pole at the base of Q3 is also contributing phase lag, causing the total phase lag at 1 MHz to be 142 degrees, leaving only 38 degrees of margin against oscillation.

The pole at the base of Q3, at about 1 MHz, is heavily dependent on the f_T of Q3. A 250-pF shunt capacitor C1 can be added to the base circuit to bring that pole down to 200 kHz and stabilize it against f_T variations in Q3. That adds attenuation at high frequencies, bringing f_c down to about 513 kHz. C2 can now be reduced to 5200 pF to bring f_c back up to about 1 MHz. However, phase margin has actually decreased to only 8 degrees because both poles are contributing more phase lag—mainly as a result of moving the pole at the base of Q3 to a lower frequency.

Resistor R6 can now be inserted in series with C2 to create a zero at 200 kHz to cancel the pole at the base of Q3. The result is an approximate single-pole roll-off up to beyond 2 MHz. However, the increased gain resulting from canceling the pole at the base of Q3 has now pushed f_c up to 3.6 MHz. If the R-C network impedance is now reduced by changing C2 to 0.027 μF and R6 to 29 Ω, while keeping the frequency of the zero at 200 kHz, f_c (ULGF) will be returned to about 1 MHz. Total phase lag at 1 MHz is now only 112 degrees and phase margin is 68 degrees. Phase lag does not reach 180 degrees until 4.5 MHz, where loop gain has fallen to −16 dB.

As described here, this may seem like a tedious and iterative process, but the steps describe the tactics and component interactions for implementing one form of lag compensation. Lag compensation of this type is somewhat of a brute-force technique. With 10 mA available from the VAS to charge C1, the slew rate of this circuit is limited to only 0.36 V/us.

Miller Compensation

Miller compensation takes advantage of the *Miller effect*, wherein the effective capacitance seen at the input of an amplifier stage, due to a feedback capacitance from output to input, is approximately equal to that feedback capacitance multiplied by the voltage gain of the stage. The resulting large effective input capacitance pushes the pole frequency at the input node of Q5 down to a much lower frequency. At the same time, the shunt feedback resulting from the feedback capacitance lowers the impedance at the output node of the stage, pushing the pole at the collector to a higher frequency. The poles at the input and output nodes of the stage are thus pushed apart in frequency. This phenomenon is referred to as *pole splitting*. The first pole is thus in control of the phase response of the stage over a very wide frequency range. Over this frequency range, the phase lag will be a nearly constant 90 degrees.

Miller compensation thus uses local feedback to roll off the high-frequency response of the amplifier. The gain that is "thrown away" acts to linearize the VAS with shunt feedback. The amplifier in Figure 6.6 employs Miller compensation. Capacitor C1 is the so-called *Miller compensation capacitor* C_M. It stabilizes the global negative feedback loop by rolling off the high-frequency gain of the amplifier so that the gain around the feedback loop falls below unity before enough phase lag builds up to cause instability. At high frequencies the combined gain of the input stage and the VAS is equal to the product of the transconductance of the IPS and the impedance of C1. At high frequencies the gain is thus dominated by C1. Since the impedance of C1 is inversely related to frequency, the gain set by it will decrease at 6 dB per octave as frequency increases. The transconductance gm is just the inverse of the total LTP emitter resistance R_{LTP}.

The capacitor controls the high-frequency AC gain of the VAS by forming a shunt feedback loop around the VAS transistor. At higher frequencies, virtually all the signal current from the LTP input stage flows through C1. This creates a voltage drop across C1 that equates to the output voltage of the VAS. At this point the VAS is acting like a so-called *Miller integrator*, where the output voltage is the integral of the input current.

While at low frequencies the gain of the combined input stage and VAS is set by the product of the individual voltage gains of those two stages, at higher frequencies that combined gain is set

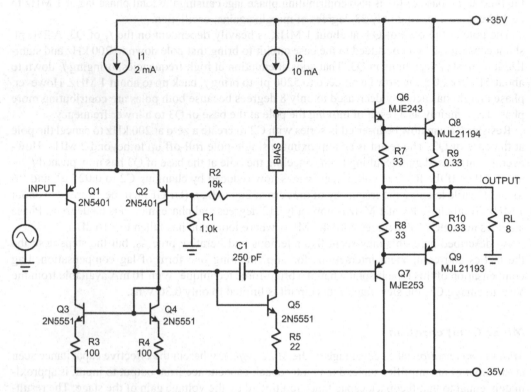

Figure 6.6 A Miller-Compensated Amplifier

by the ratio of the impedance of C1 to R_{LTP}. The frequency at which the gain set by R_{LTP} and C1 becomes smaller than the DC gain is where the roll-off of the amplifier's open-loop frequency response begins. R_{LTP} for this design is about 52 Ω. C1 has been chosen to be 250 pF in order to establish f_c at 500 kHz.

Assume for the moment an operating frequency of 20 kHz. At this frequency the reactance of C1 is $1/(2\pi * 20 \text{ kHz} * 250 \text{ pF}) = 3180$ Ω. If all signal current provided by the LTP passes through the capacitor, then the gain of the combined input and VAS stage at this frequency is $3180/52 = 61$. This is considerably less than the low-frequency forward gain of the amplifier. This means that the capacitor is dominating the gain at this frequency. Falling at 6 dB per octave, A_{ol} will become 26 dB at 500 kHz. Since A_{cl} is also 26 dB, the gain crossover frequency occurs at 500 kHz.

The peak signal current that the LTP can deliver to C1 is about 2 mA in either direction. This puts the amplifier slew rate at a symmetrical 8 V/μs, superior to the amplifier with lag compensation. Degenerating the input stage and reducing its transconductance allows the choice of a smaller value for C1 to achieve the same gain crossover frequency, improving slew rate.

An optional resistor can be placed in series with C1 to create a zero that can be used to cancel a pole elsewhere in the circuit or cancel some excess phase. A 160-Ω resistor placed in series with C1 will create a zero at 8 MHz, providing about 14 degrees of phase lead at 1 MHz.

6.5 Evaluating Loop Gain

In order to meet the stability target, the gain around the feedback loop must be estimated. There are a number of ways to do this, both in simulation and with laboratory measurements.

Breaking the Loop

The most obvious way to estimate or evaluate the loop gain is to break the loop. A stimulus signal is applied at the input side of the loop. The frequency and phase response is then measured at the output side of the loop. Means must be used to maintain proper biasing and DC levels in the amplifier when this is done. This is not always practical in the real world where very large amounts of gain may be involved.

There is also a caveat: the loading of the output side of the break will often not be identical after the loop is broken, and this will cause some error. If the source on the output side of the break is of very low impedance compared to the load seen looking into the input side of the break, the error will be quite small. This will often be the case in a power amplifier where the loop is broken at the input to the feedback network.

In SPICE simulations the loop can be kept closed at DC by connecting an extremely large inductor across the break. It is possible to employ a 1-GH inductor. The source signal is then applied to the feedback network through an AC coupling capacitor. The low-frequency corner of this coupling capacitor against the input resistance of the feedback network will determine how low in frequency the results of this method will be accurate. The loop gain is then inferred by viewing the signal at the output of the amplifier. This is illustrated in Figure 6.7.

In the laboratory, the loop can be kept closed at DC by connecting a non-inverting auxiliary DC servo circuit from the output of the amplifier to the feedback input of the IPS (see Chapter 10 on DC servos). The low-level test signal from the signal generator can then be applied to the input of the feedback network. This technique is illustrated in Figure 6.8. Caution should be observed here in light of the very high gain that may be encountered from the signal generator to the output of the amplifier. The accuracy of measurement is limited at low frequencies by the feedback through the DC servo. For example, in this arrangement, it will limit open-loop gain readings to about 76 dB at 10 Hz. R5 places a zero in the integrator at 1.6 Hz to prevent low frequency instability in the event that the open-loop bandwidth of the amplifier is less than about 10 Hz.

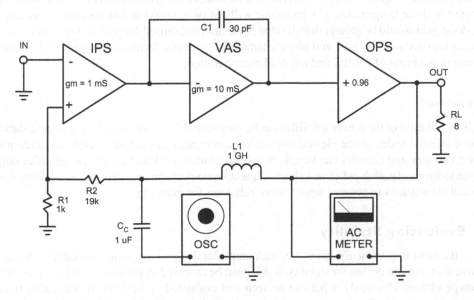

Figure 6.7 Breaking the Loop with a Large Inductor

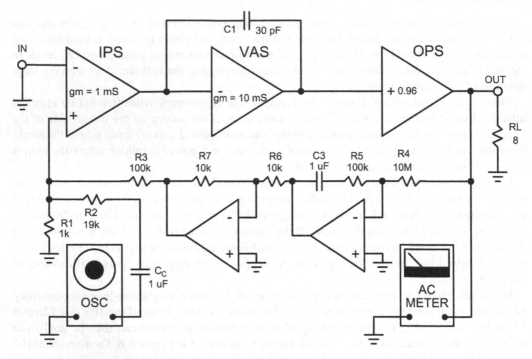

Figure 6.8 Breaking the Loop Using a Servo for DC Loop Closure

Exposing Open-Loop Gain

Exposing the forward gain A_{ol} by setting closed-loop gain very high is an accurate way to esti-mate the high-frequency open-loop gain roll-off and phase shift. The forward gain is *exposed* by increasing the closed-loop gain by a factor of 100. This decreases the loop gain by a sufficient amount over the frequency range of interest that feedback effects do not affect the gain from input to output at these frequencies. The technique will not be accurate at low frequencies where the open-loop gain would be greater than 100 times the nominal closed-loop gain. This approach will not take into account the gain and phase characteristics of the feedback network. This includes the pole at the input of the IPS and any lead compensation.

Simulation

SPICE simulation of the power amplifier can be very valuable in assessing loop gain and stability because internal nodes can be viewed, impractical component values can be used and functions of probed voltages and currents can be calculated and plotted, such as the ratio of amplifier output voltage to forward path input error voltage. Time domain performance can also be evaluated with transient simulations to observe square wave behavior, for example.

6.6 Evaluating Stability

One of the most important aspects of feedback amplifier design is assessing its stability. Although it is true that one can design for stability, it still must be assessed in simulation and/or in the actual prototype circuit. Obviously, what can be seen and evaluated is different in simulation than in the real circuit. Each has advantages where the other may be a bit blind. A proper assessment of stability in the real world using the real circuit is a must, but for those who can simulate, it is also desirable and beneficial to assess stability in simulation.

Feedback stability can often be inferred from viewing the closed-loop frequency response and looking for peaking. It is especially important in these tests to bypass any input low-pass filters in the amplifier. Peaking of the closed-loop response by more than about 1 dB just prior to final roll-off is a danger sign. However, apparent flatness of the closed-loop frequency response is not always sufficient evidence of adequate stability. Transient response must always be checked as well; this is best done with a transient simulation using a square wave source.

Instability can be either local or in the global feedback loop. Local instability can originate from a local feedback loop or from local circuit instability such as an emitter follower driving a capacitive load. Such local instabilities can often occur in output stages.

Probing Internal Nodes in Simulation

In assessing stability with AC simulations, it is important to look for evidence of peaking (or sometimes sharp dips) at nodes internal to the circuit. Figure 6.9 shows a block diagram of a simplified power amplifier with some suggested probing points. The feedback input $P1$ of the IPS should represent a unity-gain follower amplifier stage with respect to the amplifier input signal. Probe this point and look for overshoot, ringing and peaking. This may show behavior that is masked at the output of the amplifier by high-frequency roll-offs. If this is done in a real amplifier, the IPS feedback input should be probed with a high-impedance low-capacitance probe. This test should be done with no amplifier input filters in place.

Stability of local loops can be assessed with an AC simulation in a relatively non-invasive way by injecting an AC current at a chosen node and observing the resulting signal voltage. This procedure reveals the impedance of the node as a function of frequency. At frequencies where there is instability, the impedance will rise markedly. In simulation, the current injection is often conveniently carried out with a voltage source in series with a 10-MΩ resistor. The probing locations shown in Figure 6.9 are examples of where this technique can be applied in simulation. If the stability probe is placed at the input node $P2$ of a Miller-compensated VAS, the results may uncover a local instability. The same can be said for probing at $P3$. Such local oscillations can be in the 20- to 200-MHz range, and can easily be overlooked.

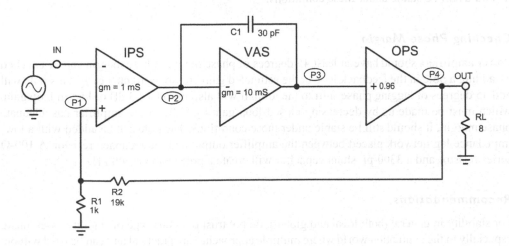

Figure 6.9 Probing Points for Stability Evaluation

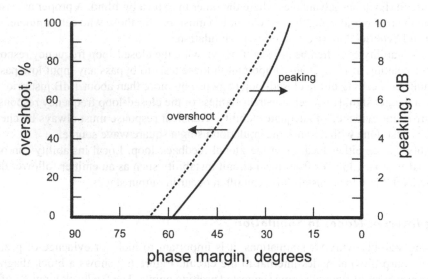

Figure 6.10 Square Wave Overshoot and Frequency Response Peaking as a Function of Phase Margin

Figure 6.10 is a plot of square-wave overshoot (left) and frequency response peaking (right) as a function of phase margin on the *X* axis from 90 degrees down to zero degrees. This data was obtained by simulating an amplifier with a dominant pole and four parasitic poles at a high frequency above the gain crossover frequency. As the frequency of the parasitic poles was reduced, the phase margin, square-wave overshoot and frequency response peaking were recorded and plotted. These numbers will not be accurate for all amplifier designs and roll-off profiles, but are helpfully representative.

Checking Gain Margin

Power amplifiers should have at least 6 dB of gain margin. Gain margin can be checked by reducing the closed-loop gain by 6 dB. In a power amplifier with a gain of 20, this would mean reducing the feedback resistor to change the gain to 10. If the amplifier has adequate gain margin, it should still be stable under these conditions.

Checking Phase Margin

Power amplifiers should have at least 45 degrees of phase margin. Phase margin can be checked by adding a pole in the feedback loop at the estimated gain crossover frequency f_c. This pole will add 45 degrees of lagging phase shift to the loop. It will also introduce 3 dB of loss in loop gain, which must be made up by decreasing closed-loop gain by 3 dB. If the amplifier has adequate phase margin, it should still be stable under these conditions. The pole can be added with a low-impedance lag network placed between the amplifier output and the feedback resistor. A 100-Ω series resistor and a 3300-pF shunt capacitor will create a pole at about 500 kHz.

Recommendations

For stability in general (both local and global), do not trust only one type of stability assessment, especially in the simulation world where multiple approaches are practical and can be tried without great effort. At the breadboard level, don't assume that there is adequate stability just because the circuit does not oscillate. Carefully evaluate frequency response and square-wave response.

Operate the circuit with lighter compensation to assess margin against instability. This can be accomplished by reducing the value of the Miller compensation capacitor by perhaps a factor of 2. Operate the circuit under large-signal conditions, recognizing that transistor parameters change with operating point and that instability at operating points other than quiescent can appear. Operate power amplifiers into difficult and diverse loads to see if instability can be provoked.

6.7 Compensation Loop Stability

The loop formed by the Miller compensation capacitor is itself a feedback loop and must also obey the rules for stability. This is often a very tight, wideband loop and will not need compensation for stability. Connecting the Miller capacitor from the collector to the base of a simple one-transistor VAS is a good example. If more complexity is added to the compensation loop, stability of this loop may have to be evaluated carefully. For example, the use of a cascoded Darlington VAS places three transistors in the compensation loop, all capable of contributing to excess phase (see the amplifier in Figure 3.12). If the Miller capacitor is connected to the output of a pre-driver emitter follower instead of to the VAS collector node, a further opportunity for instability is introduced. This will be touched on again in Chapter 9 where more complex compensation approaches are discussed.

Figure 6.11 illustrates an amplifier segment where all of these additions to a feedback compensation loop have been made. These latter arrangements make the feedback compensation

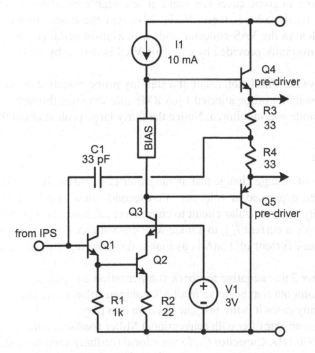

Figure 6.11 Example of a "Longer" Miller Compensation Loop

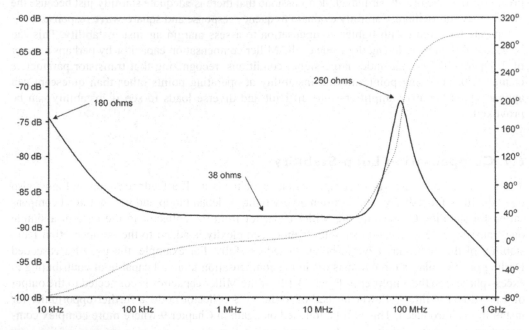

Figure 6.12 Stability Probe Result at the Collector of the VAS

loop less "local" and in some cases can make it less stable by allowing the introduction of excess phase shift by the added stages. Notice also that the connection of C_M to the output of the pre-drivers leaves the VAS collector node as a high-impedance point, without even the benefit of loading normally provided by C_M. This VAS is driven by an IPS loaded with a current mirror.

Figure 6.12 shows the simulation result of a stability probe placed at the collector of the VAS in this circuit. The stability probe injected 1 μA RMS into the node through a 1-MΩ resistor and the voltage on the node was monitored. Notice the fairly large peak at about 90 MHz.

6.8 Slew Rate

The maximum rate of voltage change that an amplifier can achieve is usually referred to as the *slew rate*. It is often expressed in volts per microsecond. Slew rate in an amplifier is usually limited by the ability of a particular circuit to charge a capacitance at a given rate. Indeed, for a circuit that can supply a current I_{max} to a node with capacitance C on it, the slew rate is simply I_{max}/C. If a maximum current of 1 mA is available for charging 100 pF, the slew rate will be limited to 10 V/μs.

We saw in Chapter 3 that negative feedback compensation can place a limitation on slew rate. It is important to point out that negative feedback compensation is not the only thing that limits slew rate, but in many cases it is the first one to come into play.

Figure 6.13 shows an amplifier with conventional Miller feedback compensation. Closed-loop gain is 20 and $f_c = 500$ kHz. Capacitor C_M forms a local feedback loop around the VAS that turns the VAS into an integrator. This means that the VAS will have a straight 6 dB/octave roll-off over a very large frequency range, from very low frequencies to frequencies above the gain crossover

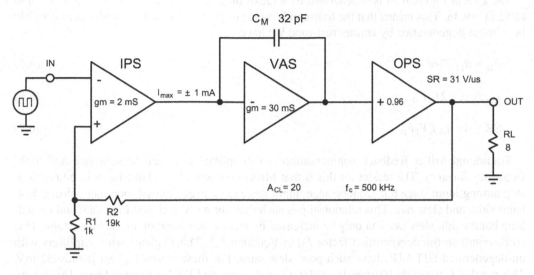

Figure 6.13 Traditional Miller Feedback Compensation

frequency. Once C_M is determined and we know the maximum peak signal current output of the input stage I_{max}, the slew rate is SR $= I_{max}/C_M$.

Calculating the Required Miller Capacitance

The required Miller compensation capacitance is calculated to yield the desired gain crossover frequency f_c. This calculation depends on the transconductance of the input stage and the chosen closed-loop gain for the amplifier. C_M is chosen so that $A_{ol} = A_{cl}$ at f_c. If transconductance of the input stage is gm, then

$$A_{ol} = gm/2\pi C_M f_c \tag{6.2}$$

Setting $A_{ol} = A_{cl}$, we have

$$C_M = gm/(2\pi A_{cl} f_c) \tag{6.3}$$

For $gm = 2$ mS, $A_{cl} = 20$, and $f_c = 500$ kHz, we have $C_M = 32$ pF.

Slew Rate

The slew rate of the amplifier is simply SR $= I_{max}/C_M$. Substituting Equation 6.3 for C_M and doing some rearrangement, we have

$$SR = 2\pi A_{cl} f_c(I_{max}/gm) \tag{6.4}$$

Notice that I_{max}/gm is a key parameter of the input stage that determines achievable slew rate. It has the units of volts. For Equation 6.4, when $I_{max} = 1$ mA and $gm = 2$ mS, $I_{max}/gm = 0.52$ V. Interestingly, $gm = I_c/V_T$ for a BJT and $gm = I_{tail}/2V_T$ for an un-degenerated differential pair. We thus have $I_{max}/gm = V_T = 52$ mV for an un-degenerated differential pair.

The LTP of Figure 6.13 is degenerated by a factor of $R_{LTP}/2re'$, or $F_D = 2(220 + 26)/52 = 490$ $\Omega/52 \ \Omega = 9.46$. This means that the transconductance or gain is 9.46 times smaller than it would be without degeneration by emitter resistors. We have

$$F_D = R_{LTP}/2re' \tag{6.5}$$

$$I_{max}/gm = 2V_T F_D \tag{6.6}$$

$$SR = 4\pi A_{cl} f_c V_T F_D \tag{6.7}$$

Traditional Miller feedback compensation is sub-optimal in regard to slew rate and high-frequency linearity. The reason for this is that Miller compensation establishes a fixed relationship among input stage transconductance, input stage tail current, closed-loop gain, closed-loop bandwidth and slew rate. This relationship is such that, for a given closed-loop gain and closed-loop bandwidth, slew rate can only be increased by adding degeneration to the input stage. This corresponds to the degeneration factor F_D in Equation 6.7. This explains why amplifiers with un-degenerated BJT LTPs have such poor slew rates. For these stages I_{max}/gm is only 52 mV. This number is typically 10 times larger for an un-degenerated JFET differential pair. This is why many designers do not degenerate JFET input stages. In Chapter 11 we'll discuss a compensation technique that breaks the relationships of Equations 6.4 and 6.7.

Reference

1. Harold Black, U.S. patent 2,102,671, issued 1937.

Chapter 7

Amplifier Classes, Output Stages and Efficiency

7. INTRODUCTION

The output stage of a power amplifier has perhaps the greatest influence on performance and cost. It is also challenged by its interface to the real world, where difficult loads, high voltages and high currents may exist. The output stage must also operate at high power levels, often at elevated temperatures. Indeed, there is often a trade-off between heat generation and sound quality. The class A output stage is perhaps the best example of this.

This chapter serves as an overview and introduction to output stages. Detailed design and nuances will be covered in Chapter 12. BJT output stages will be discussed here, while MOSFET output stages will be discussed in Chapter 14. Many of the principles and technical challenges are the same for MOSFET power amplifiers.

Several important issues involving output stages and different topologies and classes will be discussed. This chapter will begin with a review of the popular class AB emitter follower (EF) output stage and the basics of crossover distortion. The latter will be discussed in much greater detail in Chapter 12. The *complementary feedback pair* (CFP) output stage is an alternative to the EF output stage, and its merits and shortcomings will be considered. Output stage efficiency and power dissipation will be covered, as well as output stages that operate more efficiently.

7.1 Class A, AB and B Operation

The output transistors in a push-pull class A power amplifier remain in conduction throughout the entire cycle of the audio signal, always contributing transconductance to the output stage signal path. In contrast, the output transistors in a class B design remain on for only one-half of the signal cycle. When the output stage is sourcing current to the load, the top transistor is on. When the output stage is sinking current from the load, the bottom transistor is on. There is thus an abrupt transition from the top transistor to the bottom transistor as the output current goes through zero.

The formal definition of classes A and B is in terms of the so-called *conduction angle*. The conduction angle for class A is 360 degrees (meaning all of the cycle), while that for class B is 180 degrees. More accurately, the definition should really be the angle over which the transistor contributes transconductance to the output stage and signal current to the output. This precludes many so-called *non-switching* amplifiers from being called class A. Such amplifiers include bias arrangements that prevent the power transistor from completely turning off when it otherwise would.

Most power amplifiers are designed to have some overlap of conduction between the top and bottom output transistors. This smooths out the crossover region as the output current goes through zero. For small output signal currents, the output transistors are in the overlap zone and the output stage effectively operates in class A. These amplifiers are called *class AB amplifiers* because they possess some of the characteristics and advantages of both class A amplifiers and class B amplifiers. Most push-pull vacuum tube amplifiers operate in class AB mode. Class AB output stages have a conduction angle that is greater than 180 degrees, although sometimes only slightly so.

There is some semantic controversy in the definition of class B and class AB output stages. This arises partly because transistors do not turn on and off abruptly, so the definition of 180-degree conduction is fuzzy. There is a gray region between class B and class AB. I view class B as an amplifier that is under-biased. I also say that a class AB output stage has transitioned into its class B region when it exits its class A region.

I will use the term *class AB* to describe the optimally biased output stage, as it has important historical origins in push-pull vacuum tube amplifiers. Optimally biased class AB output stages can have a very substantial quiescent current when multiple output pairs and low-value emitter resistors are used. They have a class A region that extends to double the value of the output stage quiescent bias current.

7.2 The Complementary Emitter Follower Output Stage

The complementary emitter follower is the workhorse for the majority of power amplifiers that employ BJT output stages. Figure 7.1 shows simplified views of the Darlington and *Triple* emitter follower (EF) output stages. In each case the bias spreader is split to show how it would be driven from a voltage source. In a real amplifier the bias spreader (usually a V_{be} multiplier) is driven from a pair of VAS transistor collectors.

The output stage in Figure 7.1a is a class AB complementary Darlington arrangement comprising emitter follower drivers Q1 and Q2 followed by output devices Q3 and Q4. Emitter resistors R1 and R2 set the idle current of the drivers at about 50 mA. The output stage emitter resistors R3 and R4 provide thermal bias stability and also play a role in controlling crossover

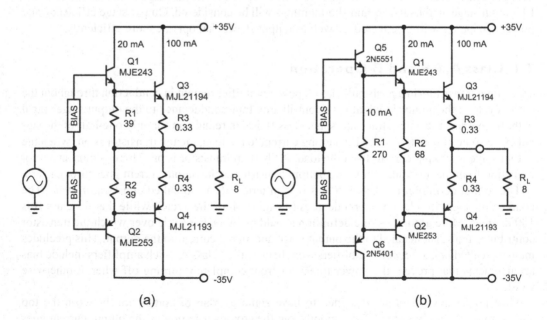

Figure 7.1 (a) Darlington Stage (b) Triple Class AB Output Stage

distortion. The output stage provides a voltage gain of slightly less than unity. Its main role is to buffer the output of the VAS with a large current gain. If driver transistor β is assumed to be 100 and output transistor β is assumed to be 50, the combined current gain of the output stage is 5000. When driving an 8-Ω load, the impedance seen by the VAS looking into the output stage will be about 40 kΩ.

The triple emitter follower (*Triple*) shown in Figure 7.1b provides much higher current gain and better buffering of the VAS. This output stage was popularized by Bart Locanthi, and is also known as the *Locanthi T circuit* [1, 2]. If the added pre-driver transistors Q5 and Q6 have betas of 100, the current gain of this output stage will be approximately 500,000, and when driving an 8-Ω load, the VAS will see a very light load of about 4 MΩ. The pre-driver and driver stages operate in class A in the Locanthi T circuit.

Returning to the Darlington output stage of Figure 7.1a, Q1 and Q3 conduct on positive half-cycles and transport the signal to the output node by sourcing current into the load. On negative half-cycles, Q2 and Q4 conduct current and transport the signal to the output node by sinking current from the load. When there is no signal, a small quiescent bias current of approximately 100 mA flows from the top NPN output transistor through the bottom PNP output transistor. A key observation is that the signal takes a different path through the output stage on positive and negative half-cycles. If the voltage or current gain of the top and bottom parts of the output stage is different, distortion will result. Moreover, the *splice point* where the signal current passes through zero and crosses from one path to the other is usually nonlinear, and this leads to so-called *crossover distortion*.

Output Stage Voltage Gain

The voltage gain of the output stage is determined by the voltage divider formed by the output stage emitter follower output impedance and the loudspeaker load impedance. The output impedance of each half of the output stage is the sum of the output transistor's *re'* and R_E. This is illustrated in Figure 7.2.

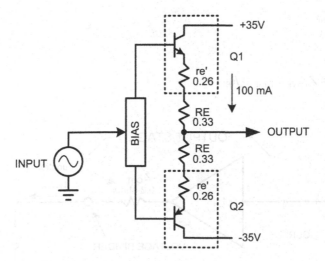

Figure 7.2 Output Stage Impedance

Since the two halves of the output stage act in parallel when they are both active in the quiescent state and under small-signal conditions, the net output impedance will be about half that of each side.

$$Z_{\text{out(small signal)}} \approx (re'_{\text{quiescent}} + R_E)/2 \tag{7.1}$$

If the output stage is biased at 100 mA, then re' of each output transistor will be about 0.26 Ω. The summed resistance for each side will then be $0.26 + 0.33 = 0.59$ Ω. Both output halves being in parallel will then result in a net output impedance of about 0.3 W. Because voltage gain is being calculated, these figures assume that the output stage is being driven by a voltage source. If the load impedance is 8 Ω, the voltage gain of the output stage will be $8/(8 + 0.3) = 0.96$. If instead the load impedance is 4 Ω, the gain of the output stage will fall to 0.93. The voltage divider action governing the output stage gain is illustrated in Figure 7.3.

Bear in mind that the small-signal gain of the output stage has been calculated at its quiescent bias current. The value of re' for each of the output transistors will change as transistor currents increase or decrease, giving rise to complex changes in the output stage gain. Moreover, for larger signal current swings, only half of the output stage is active. The output impedance under those conditions will be approximately $re' + R_E$ rather than half that amount. These changes in incremental output stage gain as a function of output signal current cause what is called *static crossover distortion*.

$$Z_{\text{out(large signal)}} \approx re'_{\text{high current}} + R_E \approx R_E \tag{7.2}$$

At high current, re' becomes very small. At $I_c = 1$ A, re' is just 0.026 Ω, much smaller than a typical value of R_E. At 10 A, re' is theoretically just 0.0026 Ω. That is why for large signals $Z_{\text{out}} \approx R_E$.

The Optimal Class AB Bias Condition

If R_E is chosen so that

$$R_E = re'_{\text{quiescent}} \tag{7.3}$$

then

$$Z_{\text{out(large signal)}} \approx Z_{\text{out(small signal)}} \approx R_E \tag{7.4}$$

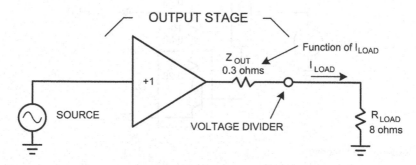

Figure 7.3 Amplifier Output Stage Gain

and crossover distortion is minimized. Oliver showed this mathematically [3].

We have

$$re'_{\text{quiescent}} = V_T/I_q \tag{7.5}$$

$$R_E = V_T/I_q \tag{7.6}$$

$$V_q = V_{R_{E\_quiescent}} = R_E I_q = V_T \tag{7.7}$$

Note that $V_T = kT/q$. The voltage V_q is the voltage that appears across each emitter resistor when the condition in Equation 7.3 is met and the class AB output stage is optimally biased. The optimal quiescent bias current $I_q = V_T/R_E$. Here that number is 79 mA.

When Oliver's condition is met, under quiescent conditions the output impedance of each half of the output stage is $re' + R_E = 2R_E$, so the parallel combination is once again equal to R_E. Thus, the nominal output impedance of the output stage when voltage driven is R_E for an optimally biased class AB stage. Satisfying the Oliver condition is about the best one can do to minimize crossover variation in net output impedance of the stage [3]. All of this is valid for ideal BJTs. Real-world BJTs have some effective intrinsic ohmic emitter resistance that must be counted as part of R_E. This ohmic resistance is approximately the sum of the ohmic base resistance R_b divided by β of the transistor, plus the ohmic emitter resistance R_e. If $R_b = 3\ \Omega$ and $\beta = 100$, and R_e is 0.05 Ω, this added resistance will be 0.08 Ω. This will reduce slightly V_q when the stage is optimally biased.

Equalizing the large-signal and small-signal output stage gains is only a compromise solution and does not eliminate static crossover distortion because the equality does not hold at intermediate values of output current as the signal passes through the crossover region. This variation in output stage gain as a function of output current is illustrated in Figure 7.4. This data corresponds

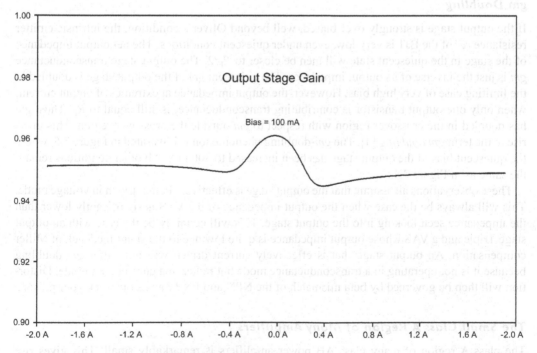

Figure 7.4 Output Stage Gain Versus Output Current

to the simple amplifier output stage of Figure 7.2 with a quiescent bias current of 100 mA and employing 0.33-Ω emitter resistors. The MJL21193/21194 output pair was used for the simulation. The theoretical optimum bias current for this arrangement is 79 mA (placing 26 mV across each R_E), so this represents a very slightly over-biased condition. As a result the output stage gain is slightly higher in the crossover region, evidencing slight *gm* doubling [4]. The slight asymmetry on the left and right sides of the plot results from differences in the NPN and PNP output transistors.

Output Stage Bias Current

The quiescent bias current I_q of the output stage plays a critical role in controlling crossover distortion. It is important that the right amount of bias current flows through the output stage, from top to bottom, when the output stage is not delivering any current to the load. Notice that together the two driver and two output transistors require at least four V_{be} voltage drops from the base of Q1 to the base of Q2 to begin to turn on. Any additional drop across the output emitter resistors will increase the required bias-spreading voltage.

The required bias voltage for the output stage is developed across the bias spreader, which is usually a V_{be} multiplier. The objective of the bias spreader design is temperature stability of the output stage quiescent current. The temperature coefficient of the voltage produced by the V_{be} multiplier should match that of the base-emitter junction voltages of the driver and output transistors. Since the V_{be} of a transistor decreases 2.2 mV/°C, it is important for thermal bias stability that these junction drops track one another. The output transistors will usually heat up the most. Because they are mounted on a heat sink, the V_{be} multiplier transistor is often mounted on the heat sink so that it is exposed to the same approximate temperature. This approach is only an approximation, because the drivers are often not mounted on the heat sink and the temperature of the heat sink changes more slowly than that of the power transistor junctions.

gm Doubling

If the output stage is strongly over-biased, well beyond Oliver's condition, the intrinsic emitter resistance *re'* of the BJT is very low, even under quiescent conditions. The net output impedance of the stage in the quiescent state will then be closer to $R_E/2$. The output stage transconductance *gm* is just the inverse of its output impedance. The quiescent *gm* of the output stage is doubled in the limiting case of very high bias. However, the output impedance at extremes of output current, when only one output transistor is contributing transconductance, is still equal to R_E. Thus, *gm* has doubled in the crossover region with respect to *gm* outside the crossover region. This gives rise to the term *gm doubling* [4]. The *gm*-doubling phenomenon is illustrated in Figure 7.5, where the quiescent bias of the output stage has been increased to 300 mA. All other conditions remain the same as in Figure 7.4.

These observations all assume that the output stage is effectively being driven in voltage mode. This will always be the case when the output impedance of the VAS is significantly lower than the impedance seen looking into the output stage. This will certainly be the case with an output stage Triple and a VAS whose output impedance is quite low due to the shunt feedback of Miller compensation. An output stage that is effectively current driven will not suffer *gm* doubling because it is not operating in a transconductance mode but rather in a current gain mode. Distortion will then be governed by beta mismatch of the NPN and PNP transistors. Pick your poison.

The Small Class A Region of Many Amplifiers

The class A region of many class AB power amplifiers is remarkably small. This gives rise to crossover distortion at very low signal levels. Consider an amplifier with 0.33-Ω emitter

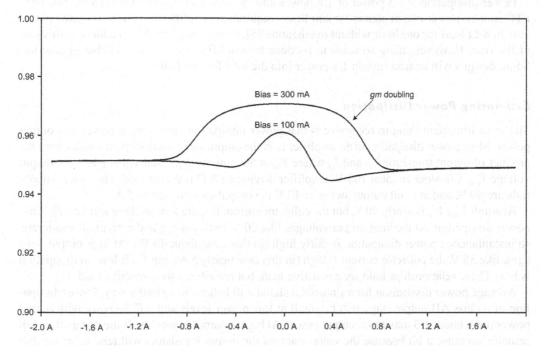

Figure 7.5 The *gm*-Doubling Effect in an Over-Biased Class AB Output Stage

resistors. With 26 mV across each R_E, the bias current will be 79 mA. The peak current output at the edge of the class A region will be twice this amount, or 158 mA. With a 4-Ω load, this will correspond to 0.63 V peak, for a power level of 50 mW.

If the quiescent bias is set well above the optimum described by Oliver, there will be a large conduction overlap region and a correspondingly large class A region. This will lead to lower distortion at small power levels, but result in *gm*-doubling distortion when signal amplitudes are sufficient to cause the output stage to exit the class A region.

7.3 Output Stage Efficiency

Even apart from the electric bill, output stage efficiency is very important because any input power that is not converted to output power is converted to heat. This directly affects the cost of the amplifier because increased heat production requires more expensive heat sinks.

Heat Versus Sound Quality

Minimizing heat dissipation is where any designer will have some difficult decisions to make. There are some unavoidable trade-offs to be made between heat and sound quality. This does not mean that one must choose class A or even heavily over-biased class AB, but such trade-offs with sound quality will exist.

The heat trade-off is especially so with respect to quiescent power dissipation in each output stage. For a 100-W amplifier, quiescent dissipation can range from about 10 W for a mediocre BJT output stage to over 30 W for an excellent BJT output stage to perhaps 45 W for a high performance MOSFET output stage.

Power dissipation at 1/3 power or 1/8 power into the load must also be considered. The 1974 FTC rule on power output claims for amplifiers required an amplifier to sustain 1/3 rated power into an 8-Ω load for one hour without overheating [5]. Some designers do not adhere to this part of the rule. However, doing so tends to produce an amplifier that is more reliable in practice. Some designs will instead sustain 1/8 power into the load for one hour.

Estimating Power Dissipation

The most important thing to recognize is that power dissipation equals input power less output power. Most power dissipation in an amplifier is in the output stage transistors. It results from the product of output transistor V_{ce} and I_c, where V_{ce} is the power supply rail voltage less the output voltage V_{out}. Consider an ideal 100-W amplifier driving an 8-Ω resistive load. The power supply rails are 40 V, and at a full output swing of 40 V the output current will be 5 A.

At small V_{out}, V_{ce} is nearly 40 V, but the collector current is quite low, so there is relatively little power dissipation. At medium output voltages like 20 V, both voltage and current are moderate, so instantaneous power dissipation is fairly high (in this case about 40 W). At high output voltages, like 38 V, the collector current is high (in this case nearly 5 A), but V_{ce} is low, so dissipation is low. These relationships hold for a resistive load, but are altered for a reactive load.

Average power dissipation for a sinusoidal signal will behave in a similar way. Power dissipation of a class AB output stage will be small at low power levels and will increase with output power up to about 1/3 maximum output power. At higher output power levels the dissipation will actually decrease a bit because the voltage across the output transistors will tend to be smaller when they are conducting the greatest current (at the peaks of the sine wave). When operated at 1/3 power, dissipation for an ideal amplifier is equal to about 40% of its clipping power. This number is closer to 46% for a real-world amplifier. When operated at full power, a real-world amplifier will dissipate power equal to about 37% of its clipping power; the ideal amplifier will dissipate about 26%. These figures do not include power dissipation in the drivers or any other part of the power amplifier preceding the output stage.

Interestingly, the power dissipation for a class A amplifier tends to decrease with increases in output power. This is so because the input power is constant while the output power is increasing; this leaves less power to dissipate.

Estimating the Input Power

Consider a 100-W, 8-Ω BJT design that uses a single pair of idealized output transistors with 0.5-Ω emitter resistors. The output of this idealized design can swing all the way to the 40-V rails. This is often the model used to estimate output stage power dissipation as a function of output power, ignoring many realities. This example also illustrates how low the quiescent bias current can be in such a class AB output stage. There were in fact many BJT output stages designed with 0.5-Ω emitter resistors. The optimum idle bias current in this design is only 52 mA, resulting in a quiescent power dissipation of 4 W.

At full power into an 8-Ω load, the output voltage is 40 V peak and the output current is 5 A peak. The current drawn from each supply rail is a half-wave-rectified waveform with a peak current of 5 A. The average of a half-sine is about 63% and the duty cycle is 50%, so the average rail current is 0.5 * 0.63 * 5 A = 1.58 A. Average input power for two 40-V rails is then 126 W. With 100-W output, the power dissipation is 26 W. The same calculation at 1/3 power of 33 W results in an average rail current of 0.91 A. This corresponds to an input power of 73 W, leaving a power dissipation of 40 W.

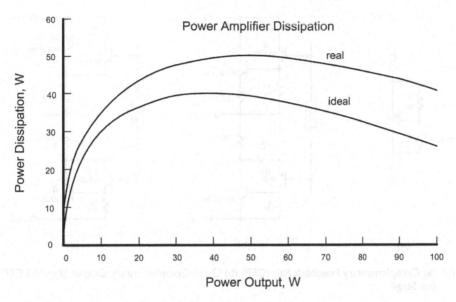

Figure 7.6 Power Dissipation as a Function of Output Power

An Example

A 100-W/8-Ω power amplifier was simulated to illustrate the power dissipation of a real amplifier design as compared with one whose dissipation is estimated with ideal output transistors and output voltage swings that extend to the rails. The amplifier employed an output Triple with ±48 V rails and two output pairs with 0.33-Ω emitter resistors.

Figure 7.6 is a plot of amplifier power dissipation as a function of output power for the 100-W amplifier when it is driving an 8-Ω resistive load. Only the power dissipation for the output stage is shown. The second curve represents an idealized 100-W amplifier with 40-V rails.

7.4 Complementary Feedback Pair Output Stages

The emitter follower output stage is simple and straightforward, but there also exist output stages based on common-emitter operation of the output transistors. In some situations, for example, they may be able to swing closer to the power supply rail. A *complementary feedback pair* is illustrated in Figure 7.7a. This circuit, called a CFP, can be used in small signal and power stages as well. It consists of an NPN transistor feeding a PNP transistor in a tight feedback loop. In many ways it acts like an NPN transistor with some improved properties. The CFP in Figure 7.7a is configured as an emitter follower. It has a current gain that is the product of the β of the two transistors, so it shares the buffering capability of the Darlington connection. However, because of the local feedback, it has very low output impedance. Put another way, each CFP in a push-pull CFP output stage has very high transconductance. The local feedback also makes each CFP by itself very linear, but this does not necessarily lead to a more linear combination of halves in a class B or class AB push-pull arrangement due to the behavior in the crossover region.

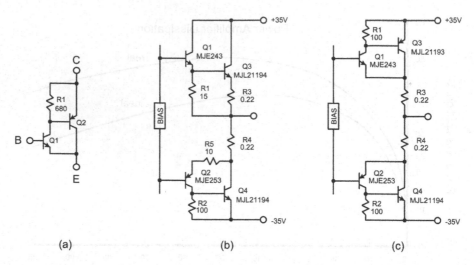

Figure 7.7 (a) Complementary Feedback Pair (CFP) (b) Quasi-Complementary Output Stage (c) CFP Output Stage

The Quasi-Complementary Output Stage

During the 1960s and early 1970s there were few if any good silicon PNP power transistors, so the complementary output stages that we take for granted nowadays were not common. In order to make an output stage with the properties of a complementary push-pull output stage, an NPN power transistor was configured with a PNP driver in a CFP arrangement that emulated a PNP power transistor. This was the basis of the quasi-complementary output stage illustrated in Figure 7.7b. This output stage suffered from its fundamental asymmetry. Different mechanisms govern the dynamic output impedance of the top and bottom parts of the stage, and the Oliver condition for minimizing crossover distortion is virtually impossible to meet. The quasi-complementary output stage is only mentioned here for historical purposes, and by way of introduction of the use of the CFP in modern true complementary output stages.

The CFP Output Stage

A *complementary feedback pair* (CFP) output stage is shown in Figure 7.7c. Here both output devices are operated in the common-emitter (CE) mode rather than as emitter followers. Some designers advocate the CFP output stage because it has a high degree of local feedback that linearizes each half of the output stage [4]. This can result in very low output impedance for each half of the output stage and thus, presumably, reduced crossover distortion. Each of the upper and lower CFP stages acts like a super emitter follower and may exhibit very high transconductance. This depends on the design and biasing details of the CFP. Assuming that emitter resistors (R_E) are still used between each stage and the output node, the output impedance may be very low and largely dominated by the R_E value. This may not be desirable.

Biasing and Thermal Stability

Bias stability in the CFP output stage will tend to be better than that in an emitter follower output stage. This is because the bias in the CFP is mainly dependent on the driver transistor, which is subject to less heating. The bias current is less influenced by the changing power dissipation of the output transistor, allowing bias to be set with greater precision and stability. For this reason it is possible to obtain adequate bias stability with smaller values of R_E in the CFP arrangement than would be permitted in the emitter follower arrangement. The use of smaller R_E can in principle lead to a reduced amount of crossover distortion because the net output impedance of the output stage is smaller, so dynamic variations in it will have less effect on incremental gain. However, the tendency to gm doubling in the crossover region can make this benefit elusive.

Optimum Class AB Bias Point and gm Doubling

The output impedance of the CFP is much lower than that of the emitter follower as a result of the local feedback. This means that the output transistor's re' is no longer a major determinant of the output impedance of each half. The output impedance is much more fully determined by the emitter resistor R_E. This makes it more difficult to avoid gm doubling and the crossover distortion that it brings.

The optimum quiescent voltage V_q across each emitter resistor in an EF output stage is ideally 26 mV. This makes the product of gm and R_E equal to unity, in accordance with Oliver [3]. The optimum V_q across the analogous R_E output resistor in the CFP output stage is on the order of 3.1 to 7.2 mV, depending on where R_E falls in the range of 0.47 Ω to 0.1 Ω [4]. This corresponds to output transistor bias currents on the order of 15 mA to 31 mA. This is a direct result of the higher gm of the CFP stages for a given operating current. The output transistors are thus operating in a starved mode in order to get the net stage transconductance down to a point where gm doubling will not be too bad. Indeed, when there are multiple devices in parallel, each has only a fraction of this current. The f_T of the output transistors is quite low under these conditions.

The greater bias precision and stability of the CFP provides little net value because it is really needed to keep crossover distortion down because crossover distortion in the CFP output stage is more sensitive to the bias setting. The high transconductance of each half of the output stage causes the crossover region for the CFP to be much narrower and more abrupt. This implies higher-order crossover distortion products.

Although the claim may be true that as a single stage the CFP is more linear than a double EF, this does not hold for a class AB stage consisting of a complementary CFP pair due to the gm doubling that will occur under most realistic conditions.

As an aside, the complementary CFP output stage can provide very low distortion in a class A output stage where both halves are on all of the time.

High-Frequency Stability

The second major concern is one of high-frequency stability. The CFP can be notoriously difficult to stabilize under all conditions. This is because of the larger amount of local feedback resulting from the feedback loop in the CFP. Its loop gain is a function of the load impedance and of the β, f_T, and C_{cb} of the output transistors. These stages may also be more vulnerable to destabilizing influences like capacitive loads. Some sort of local feedback compensation network is often needed with the CFP.

CFP stages tend to be unstable with capacitive loading. This includes the case where a CFP might be used as a driver. If a CFP is used as a driver for an emitter follower output pair, and a speed-up capacitor is connected between the bases of the output devices to help turn them off, this arrangement will tend to be unstable.

Turn-Off Issues in CFP Output Stages

The conventional CFP output stage has no problem turning on, as the driver transistor can pull quite a bit of current from the base in the turn-on direction. The CFP does, however, suffer from limited ability to turn off quickly. Turn-off current is supplied by R1 and R2 in Figure 7.7c, where only 7 mA of turn-off current is available. The voltage across these resistors can never be more than the V_{be} of the output transistor. This is in contrast to the emitter follower output stage where current flowing in the opposite half of the output stage provides an increased voltage drop across the driver emitter resistor in the Locanthi T circuit [1]. There are few options for improvement of turn-off in the CFP output stage. For example, the *speed-up* capacitor sometimes used in emitter follower output stages cannot be used. As a result, the CFP is often slower to switch off and may be more prone to high-frequency switching distortion and common-mode conduction (*shoot-through*). The driver transistors in the CFP output stage do not operate in class A. This is a further disadvantage for this arrangement.

Miller Effect in the CFP Output Stage

The collector-base capacitance in the CE-operated output devices creates a Miller effect in the CFP. The small-signal effect is to reduce CFP bandwidth and to partially compensate its feedback loop. The large-signal effect is to cause high-frequency distortion due to the nonlinearity of C_{cb} of the output transistor. Moreover, there is also the large-signal action of the Miller effect that opposes turn-off. Consider the case where total C_{cb} is 500 pF and the output voltage slew rate is 50 V/μs. This current will be 25 mA, more than what is often run through the base-emitter resistor.

CFP Triples

As with simple Darlington output stages, the conventional two-transistor CFP does not have adequate current gain to enable the really high performance achievable by lightening the load on the VAS. As with emitter follower Triples, there are CFP *Triples* [6, 7]. Although there are many ways to incorporate three transistors into a CFP, the safest and most straightforward is to simply precede the CFP with an emitter follower pre-driver. It adds 2 V_{be} to the required amount of bias spread and does not change any of the stability or transconductance characteristics of the CFP.

CFP Degeneration

The CFP output stages described thus far do not include emitter degeneration in the driver or output transistors. This is partially responsible for the high transconductance of the CFP. Modest emitter degeneration in each transistor can reduce CFP loop gain and provide smaller transconductance for a given amount of bias current. A typical design might employ a resistor of 10 Ω to 100 Ω in the emitter of the driver transistor. The output transistor might also be degenerated, often with a resistor of the same value as R_E. The use of paralleled output transistors requires emitter degeneration of the CFP output transistors anyway.

7.5 Stacked Output Stages

Some output stages are designed with two transistors in series to share the voltage swing required to produce the output signal [8–12]. This reduces the maximum rated voltage requirement for the

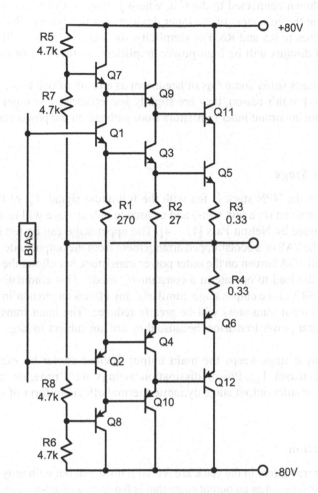

Figure 7.8 Stacked Output Stage Design

power transistors, but more importantly it increases the available *safe operating area* (SOA). This is especially significant because safe area in bipolar transistors often decreases rapidly at higher voltage due to secondary-breakdown effects. However, modern transistors have higher voltage ratings and are less prone to secondary breakdown. As a result, the use of stacked output stages is much less common than it used to be.

Figure 7.8 illustrates a simple stacked output stage design where two *Locanthi Triples* are stacked atop one another. A popular example of the stacked output stage is the *Double Barreled Amplifier* by Marshall Leach (also referred to as the *Leach Superamp*) [8]. The inner transistors drive the load in a conventional fashion, while the outer transistors provide a signal voltage to the collectors of the inner transistors. The outer stage of the stacked arrangement is driven in a bootstrapped arrangement with a signal derived from the output. The bootstrap signal is divided in half by R5 and R7 on the top half so that inner and outer parts of the output stage divide the voltage swing equally.

Resistor R5 is shown connected to the rail, where garbage will unnecessarily get into the circuit and appear at the collectors of the inner devices. For this reason, R-C filters should be placed in the rail lines to R5 and R6. For simplicity, the stage is shown with only one output pair. Most stacked designs will be high-power amplifiers, so that two or more pairs will be used.

Stacked output stages suffer some loss of headroom as a result of two power transistors being connected in series. For this reason, they are slightly less efficient. The outer transistors in the stack shield the more important inner transistors from garbage on the power rail and also reduce the Early effect.

Cascode Output Stage

If the upper stage in the NPN stack is fed with the full audio signal, V_{ce} of the bottom output transistors will be constant (at a few volts) and a cascode output stage will result. This architecture has been discussed by Nelson Pass [13, 14]. The upper stage can be fed in a feed-forward arrangement from the VAS or a bootstrapped arrangement from the output node. The arrangement places nearly the full SOA burden on the outer power transistors, but allows the power transistors that actually drive the load to operate in a constant V_{ce} mode. This eliminates the Early effect and improves the PSRR of the output stage. Similarly, any effects due to nonlinear base-collector capacitance of the output transistors will be greatly reduced. The inner transistors driving the load can also be faster power transistors because they are not subject to large voltage and SOA requirements.

The cascode output stage keeps the main output devices cooler by exposing them to a relatively small, constant V_{ce}. Power dissipation swings with program material are thus smaller, leading to smaller output stage dynamic thermal effects (a form of so-called *memory distortion*).

Soft Rail Regulation

If instead the upper transistors in the stack are fed with no signal but with only a filtered version of the rail voltage, this becomes an output stage that is fed from a quasi-regulated supply voltage. The filtered rail tracks the available rail voltage. This is what I refer to as *soft rail regulation*. It is little more than feeding the output stage through a pass transistor that is configured as a capacitance multiplier.

The advantage of this scheme is that the output transistor is shielded from the noise and ripple on the high-current main rail power supply. This imparts a very high effective PSRR to the output stage. It also prevents the nasty pass-though of power supply ripple to the amplifier output when the amplifier clips. As with other stacked output stages, some operating headroom and operating efficiency are lost. The pass transistor can also function as a fast electronic circuit breaker if appropriate control circuitry is included.

7.6 Classes G and H

The quest for higher efficiency has led to the development of other types of output stages, the better-known variants being *class G* and *class H*. The class G and class H output stage configurations are designed to reduce output stage power dissipation by effectively changing the rail voltage applied to a conventional class AB output stage as a function of the instantaneous signal amplitude. These amplifiers are popular in professional audio applications where the sonic penalties sometimes associated with these designs are not such a big problem.

Conflicting Terminology

In class G, the rail voltage is elevated in a linear way to a higher voltage once the peak signal output level requires it. Class G was popularized in 1977 by Hitachi with the introduction of its Dynaharmony line of amplifiers, an example being the HCA-8300 [15]. Class H is very much like class G, but the power supply for the output stage is switched abruptly to higher levels when the signal waveform exceeds a certain threshold.

The U.S. and Japanese naming conventions for these classes are sometimes reversed, as explained in Reference 15. The remainder of this section will focus on class G, in accordance with the Japanese nomenclature, where the supply voltages to the output stage are increased in a linear fashion when required.

Class G Operation

Figure 7.9 illustrates how the rail voltages change as a function of the signal for a class G amplifier. At low power levels the output stage operates from a fixed intermediate rail voltage supplied through *commutating diodes*. As the signal increases and the headroom for the output stage becomes small, the rail is lifted by a set of power transistors above in a fashion that is linear with the signal so as to provide a constant amount of headroom as the signal increases further. The upper power transistor is connected to the high-voltage rail. In a sense, the bottom output stage transistor is transformed into a cascoded output stage for the higher signal levels, since under these high-level conditions its collector is moving with the signal.

Figure 7.10 shows a simplified schematic of a class G output stage. This design is very similar to the stacked output stage of Figure 7.8, except that R7 and R8 have been replaced by bypassed Zener diodes D5 and D6 to provide a full bootstrapped signal swing to the outer Triples. Diodes D1 and D2 are the commutating diodes. They supply the rail current from the low-voltage power supply when the signal swing is small. When the signal swing becomes large enough to reduce the V_{ce} of Q5 or Q6 to a small value, outer transistors Q11 and Q12 lift the rail, maintaining a fixed minimum V_{ce} across Q5 and Q6.

The breakdown voltage of D5 and D6 establishes the minimum V_{ce} that Q5 and Q6 will see. The relevant voltage is between the collector of Q5 and the output node and must be set to take into account the voltage drop across the emitter resistors under high-current conditions. At the same time, the Zener breakdown voltage must not be so large that the base of Q7 is driven beyond the high-voltage rail. A typical value for the Zener voltage might be 6.8 V. The amplifier output

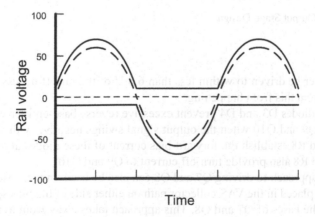

Figure 7.9 Rail Voltage as a Function of the Signal in a Class G Output Stage

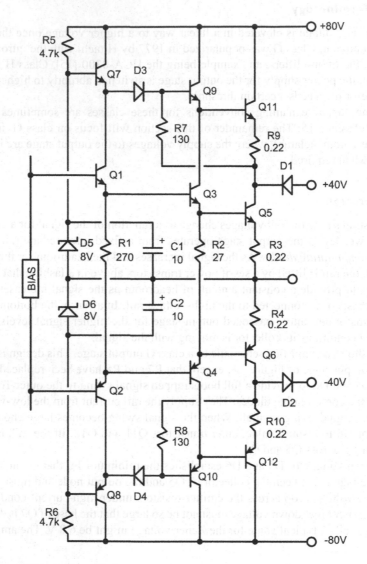

Figure 7.10 Class G Output Stage Design

therefore must never be driven to within less than 6.8 V of the rail. If necessary, Baker clamps can be used to prevent this from happening.

Driver isolation diodes D3 and D4 prevent excessive reverse base-emitter voltage from being applied to drivers Q9 and Q10 when the output signal swings negative with respect to the low-voltage rail. R7 and R8 establish the forward-bias current of these diodes at about 10 mA when they are on. R7 and R8 also provide turn-off current to Q9 and Q10.

A feed-forward approach to driving Q7 and Q8 can also be used. In that case Zener diodes D5 and D6 are instead placed in the VAS collector path on either side of the bias spreader to provide offset voltages to the bases of Q7 and Q8. This approach takes away from available VAS head-room, so boosted VAS rails are recommended.

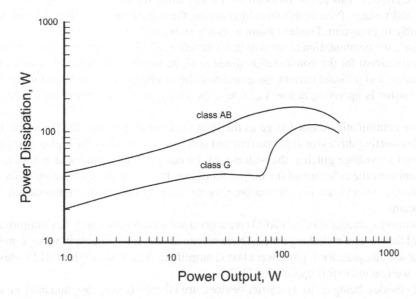

Figure 7.11 Class G Power Dissipation as a Function of Output Power

Class G Efficiency

Figure 7.11 is a plot of power dissipation as a function of output power for a 250-W class G output stage. The power dissipation for a comparable class AB design is also plotted.

Choice of Intermediate Rail Voltage

The choice of the intermediate rail voltage determines the signal amplitude at which the amplifier transitions its current draw from the low-voltage supply to the high-voltage supply. As such, it affects the power dissipation profile of the amplifier as a function of the signal level. If sine wave signals were the governing criteria, one might adjust the low-voltage rail so that the two peaks in power dissipation had the same maximum power dissipation. However, real program material has a higher crest factor than sine waves; this means that it spends relatively less time at high amplitudes. This argues for choosing a lower intermediate rail voltage to reduce power dissipation for conditions of small signal amplitude.

Headroom Considerations

Class G requires additional rail voltage headroom because of the presence of the outer stage and the need for headroom for the inner stage. This decreases the value-added contributed by class G for lower-power amplifiers. Class G may also benefit from boosted driver rails that allow the outer transistors to swing closer to the rails while maintaining adequate headroom for the VAS.

Rail Commutation Diode Speed

Turn-off time of the rail commutation diodes is a problem that creates a performance challenge for class G amplifiers. The commutation diode provides the intermediate rail voltage to the output transistor under low-swing conditions. When a signal peak occurs, the high-rail output transistor pulls the collector node of the main output transistor to a higher voltage, following the signal as its amplitude increases. This action turns off the commutating diode at a fairly high rate-of-change of current and voltage. Prior to this switching action, the commutation diode will have been conducting fairly high current. Diodes cannot instantly stop conducting current.

As a result, the commutation diode will resist the change in its voltage to the reverse direction. The turn-off current for the commutation diode must be supplied from the high-side transistor. This can cause a significant current spike and possibly a voltage spike at a time when the main output transistor is operating at low V_{ce}, where its ability to provide power supply rejection is minimal.

When the commutation diode lets go as the signal goes more positive, the high-side transistor has been conducting this extra diode current and still wants to conduct the higher current. It will create a positive voltage glitch at the collector of the main output transistor as it tries to put this extra current into the collector of the output transistor. It may in theory have no other place to go than into the output load. For this reason, reverse recovery time of the commutation diodes is very important.

Fast-recovery epitaxial diodes (FRED) are a good solution because they have minimal charge storage and fast reverse switching times. These silicon diodes are available in voltage and current ratings that are adequate for high-power class G amplifiers. The Vishay HEXFRED® devices are especially well suited to this application [16].

Schottky diodes, being majority-carrier devices, are inherently very fast, and have no minority carrier charge storage. Note that the voltage rating of the commutation diodes needs only to be greater than the maximum difference of the low-voltage and high-voltage rails.

The Transition to Cascode Operation

By now it should be apparent that the class G output stage is just another type of stacked output stage that operates in two modes. For smaller signals, it operates as a conventional class AB output stage powered with the intermediate rail voltage. At higher signal levels, when the commutation diode is turned off, the output stage acts like a cascode as described in Section 7.6. As observed by Self, [4] Early effect is present in the former case but not in the latter case. This means that transistor current gain will vary with the signal in the former case but not in the latter case. If changes in current gain alter the effective load impedance seen by the VAS, distortion will result. Similarly, if current gain variations alter the voltage gain of the output stage, distortion will also result. For this reason it is important to design the class G output stage so that beta variations have a minimal effect on its circuit behavior. Use of the Triple-based class G output stage discussed above is very effective to this end.

Safe Operating Area

The stacked nature of the class G output stage reduces the maximum required SOA on the output transistors as compared to a straight class AB design. This happens because the inner transistors are never subjected to a V_{ce} greater than the intermediate rail voltage plus the main rail voltage. The outer transistors are never subjected to V_{ce} greater than the difference between the main rail voltage and the intermediate rail voltage.

7.7 Class D

No chapter on output stages would be complete without discussion of class D amplifiers, whose popularity and performance increase daily. An entire book could be written on class D, not to mention a chapter. Here we touch on it very briefly for completeness, but defer any meaningful discussion to Chapters 33–36.

Class D amplifiers are popular as a result of the never-ending quest for efficiency. They rely on the fact that an on-off switch (a MOSFET transistor) cannot dissipate any power if it has no voltage across it or if it has no current passing through it. The simple class D output stage uses high-speed solid-state switches to connect the positive and negative supply rails to the output node alternately, with different duty cycles, such that the average value applied to the output is the desired value corresponding to the signal. This is referred to as *pulse width modulation* (PWM). The alternation between positive and negative connection of the supply rails to the output node is at a high frequency well above the audio band.

When the output signal is to have large positive amplitude, the positive supply rail is connected to the output most of the time (i.e., with a high duty cycle). When the average output signal should be zero, the pulse widths of the positive and negative connections of the rails to the output node will be equal, corresponding to a 50% duty cycle.

The way in which the input signal is converted into a PWM representation and the way in which the output switches are driven is where most of the complexity and innovation lie. Approaches range from simple analog techniques to sophisticated high-speed DSP-based techniques, some of which rely on *Sigma-Delta* techniques borrowed from the land of A/D and D/A converters.

Achieving low distortion and high sound quality is a special challenge for class D amplifiers, but great progress has been made. One example of such a challenge is the poor power supply rejection inherent in the simple class D approach. If the power supply rail is connected to the output node through a switch, it is easy to see that there will be very little power supply rejection in the circuit.

References

1. Bart N. Locanthi, "Operational Amplifier Circuit for Hi-Fi," *Electronics World*, pp. 39–41, January 1967.
2. Bart N. Locanthi, "An Ultra-Low Distortion Direct-Current Amplifier," *Journal of the Audio Engineering Society*, vol. 15, no. 3, pp. 290–294, July 1967.
3. Barney M. Oliver, "Distortion in Complementary Pair Class B Amplifiers," *Hewlett Packard Journal*, pp. 11–16, February 1971.
4. Douglas Self, *Audio Power Amplifier Design Handbook*, 5th ed., Focal Press, Burlington, MA, 2009.
5. Federal Trade Commission (FTC), "Power Output Claims for Amplifiers Utilized in Home Entertainment Products," CFR 16, part 432, 1974.
6. Dan Meyer, "Build a Four-Channel Power Amplifier," *Radio Electronics*, pp. 39–42, March 1973 and pp. 62–68, April 1973.
7. Peter Williams, "Voltage Following," *Wireless World*, pp. 295–298, September 1968.
8. W. Marshall Leach, Jr., "Double Barreled Amplifier," *Audio*, vol. 64, no. 4, pp. 36–51, April 1980.
9. Dan Meyer, "Tigersuarus: Build This 250-Watt Hi-Fi Amplifier," *Radio Electronics*, pp. 43–47, December 1973.
10. James Bongiorno, "High-Voltage Amp Design," *Audio*, pp. 30–36, February 1974.
11. "SAE Mk III CM Basic Amplifier" (review), *Audio*, pp. 61–65, January 1975 and p. 69, August 1975.
12. "G.A.S. Ampzilla Basic Power Amplifier" (review), *Audio*, pp. 57–60, September 1975.
13. Nelson Pass, "Cascode Amplifier Design," Pass Labs. Available at www.passdiy.com.
14. Nelson Pass, "Cascode Amp Design," *Audio*, pp. 52–59, March 1978.
15. Ben Duncan, *High Performance Audio Power Amplifiers*, Newnes, Oxford, 1996.
16. HFA25PB60 HEXFRED® Ultrafast Soft Recovery 25A Diode, Vishay. Available at www.vishay.com.

Chapter 8

Summary of Amplifier Design Considerations

8. INTRODUCTION

This chapter is intended to summarize the numerous considerations that should be addressed in designing a power amplifier. Now that basic power amplifier design principles have been covered, it is a good time to go over this before getting into deeper design considerations. This will also help put into perspective those later discussions. All of these topics will be covered in much greater detail in later chapters.

8.1 Power and Loads

This may seem obvious, but one first needs to decide what the rated power of the amplifier should be and into how low an impedance load it will be operated, taking into account the fact that rated power will usually be a function of the load impedance. Nominal power is usually stated into 8 Ω, but what power will be needed into 4 Ω or even 2 Ω? An ideal amplifier doubles its power as the load is halved. How close to this ideal should this amplifier come? Will this amplifier need to be able to operate continuously into a load impedance of 4 Ω or even 2 Ω? What are the expectations for testing? Will the amplifier be expected to operate at 1/3 power with both channels driven into a 2-Ω load for an extended period?

Worst-Case Loads

Loudspeakers are anything but a resistive load. What will the criteria be for the worst-case load? What minimum impedance will be designed for? What phase angle of the load impedance will be tolerated? How much direct load capacitance will be allowed without stability problems?

Peak Output Current

High-output-current capability is often associated with good-sounding amplifiers. At minimum, it is good insurance against the unpredictability of difficult speaker loads that are driven by unusual program material. It is not difficult to concoct scenarios where the peak load current into a reactive load can be twice that into a resistive load of the same impedance. An amplifier rated at 100 W into an 8-Ω load will attempt to produce peak currents of 40-V/2-Ω = 20 A into a resistive 2-Ω load. If you double this for a difficult reactive load, you get 40 A for a nominal 100-W amplifier. The numbers get ugly fast. This is, of course, an extreme example, and a well-designed amplifier will go into well-behaved current limiting before such extremely high currents are reached. Moreover, power supply sag will prevent such extreme output current values to be reached in practice.

Slew Rate

The required slew rate for an amplifier can sometimes be a point of controversy, but some minimum guidelines can put things into perspective and prove useful. Much of it amounts to how much margin is desired against the maximum slew rate likely to occur in program material. To be generous, assume it is that of a full-power sinusoid at 20 kHz. A decent margin to add on top of that is to require the amplifier to have sufficient slew rate to support a full-power sinusoid at 50 kHz.

The required slew rate for a sinusoid at 20 kHz is 0.125 V/μs per peak volt of output. This means that a 100-W amplifier producing 40 V peak requires a minimum slew rate capability of 5 V/μs. Adding the margin for full-power capability at 50 kHz raises this number to 12.5 V/μs. Required slew rate tends to increase as the square root of power, so a corresponding 400-W amplifier would require a minimum slew rate capability of 25 V/μs. In reality, for very high quality reproduction, substantially higher slew rate capabilities are desirable and not difficult to achieve. Unfortunately, high slew rate alone does not guarantee low values of high-frequency distortion.

8.2 Sizing the Power Supply

The power supply design will have the greatest influence on achievable output power, on both a short-term basis and a long-term basis. All unregulated power supplies sag under a heavy load. A stiffer power supply sags by a smaller amount. The amount by which a power supply sags under a given load is referred to as its regulation.

When the amplifier is operating under a no-load condition, the power supply must only deliver the bias currents, perhaps on the order of 100 to 200 mA per channel for a modest-sized power amplifier. When the amplifier is delivering full power into an 8-Ω load, the power supply will have to deliver considerably more current, on the order of amperes. Still more current will have to be delivered when a 4-Ω or 2-Ω load is being driven to rated power.

Average Power Supply Current

A 100-W amplifier driving a resistive 8-Ω load will put out 40 V peak at a current of 5 A peak. The average current drawn from each rail will be about 63% (2/π) of the peak for half of the cycle for a class AB output stage. This comes to about 1.6 A average per rail (ignoring bias current and IPS-VAS operating current). Note that with 45-V rails this corresponds to input power of 144 W and thus an output stage power dissipation of 44 W.

Average power supply current will increase as the square root of rated power and the inverse of the load resistance. The 100-W amplifier (with a perfect power supply that does not sag) will produce 400 W into a 2-Ω load, and its average power supply rail current will be 6.4 A. Input power will be 576 W, and power dissipation will be 176 W.

Effect of Power Supply Sag

Every power supply has an effective source resistance, although that resistance may be somewhat nonlinear, changing with the load current (it is often higher for light loads). This has two consequences. First, under no-signal conditions the supply voltage will rise, sometimes considerably. This means that component voltage ratings must exceed this higher voltage, including under conditions of maximum line voltage of perhaps 130 V. This also applies to no-signal power dissipation values. Second, the power supply will sag further when maximum power is delivered into loads less than 8 Ω. Consider a supply with effective resistance of 2 Ω for each rail that is sized to deliver 44 V at a load current of 1.8 A when the amplifier is delivering 100 W into an 8 Ω load, 1.6 A of which is for the power output and 0.2 A is for bias currents. Assume also that the amplifier needs 4 V of headroom from the peak output voltage to the rails. Assume this number is independent of load current for simplicity (it is not). Into a 4-Ω load, the amplifier will be able to deliver roughly 36 V peak for a power level on the order of 162 W. The power supply will have sagged to about 40 V. Into a 2-Ω load, the amplifier will be able to deliver roughly 33.5 V peak for about 280 W, with rails having sagged to less than 37 V.

Sizing the Power Transformer

Power transformers are usually rated in *volt-amperes* (VA) delivered as AC into a resistive load with a given degree of sag (regulation) from no-load to full-load. It is not always easy to correlate this to the amount of current that can be delivered as DC by a capacitor-input rectifier into a load with a given amount of regulation.

One of the biggest issues in sizing the power transformer is the desired degree of stiffness of the resulting power supply. Some designers will prefer a power supply with higher compliance (less regulation), resulting in greater dynamic headroom. The legendary *Phase Linear 700* is a good example of this philosophy. In more recent times, it appears that designers of very high-quality amplifiers prefer a stiff power supply, eschewing dynamic headroom for greater output current capability and greater ability to handle smaller load impedances. Stiffer power supplies may produce better sound quality as a result of less program-dependent rail voltage variations.

As a matter of reference, the 250-VA toroidal transformer used in the amplifier in Chapter 4 has a primary resistance of 1.3 Ω and an end-to-end secondary resistance of 0.9 Ω. It can provide approximately ±50-V rails at a load current of 2.5 A.

Sizing the Reservoir Capacitors

Large reservoir capacitors are always a good thing, but they are expensive. There is no objective way to state the required size, but a look at power supply ripple as a function of reservoir capacitance and output current can provide useful perspective.

The 100-W/8-Ω amplifier driving 280 W into a 2-Ω load will consume an average rail current of about 5.5 A. A gross approximation of the peak-to-peak ripple voltage is arrived at by assuming that the rail voltage decays over one half-cycle (8.3 ms) in the amount of $V = I * T/C$. If the 100-W amplifier has reservoir capacitors of only 10,000 µF, then the ripple under these conditions (5.5 A average load current) becomes 4.5 V peak-to-peak. It is easy to see why some better power amplifiers have 100,000 µF of reservoir capacitance on each rail. Apart from ripple reduction, large reservoir capacitance also reduces short-term power supply sag for large signal transients. In very rough terms, 100,000 µF against an effective power supply resistance of 2 W with yield a time constant of 200 ms. Put differently, a 20 Hz, 25-ms half-cycle burst of 40 V peak into 2 Ω, corresponding average current of 12.6 A over the half-cycle, will discharge a 100,000 µF capacitor by only 3 V.

The use of large reservoir capacitors, however, increases inrush current at turn-on, and may require the use of inrush control circuits in the power supply. These issues are discussed in Chapter 19.

8.3 Sizing the Output Stage

The output stage must be able to handle the maximum anticipated power supply rail voltages that will occur under no-load conditions with high mains voltage conditions. It is easy for a designer to reason that a 100-W/8-Ω amplifier requires 45-V rails and thus the output stage will see no more than 90 V. This is far, far from safe. In addition to the usual safety margins, one must take into account power supply regulation (the rails will rise under no-load conditions) and high mains voltages. With 20% regulation and 130-V mains, the rail voltages may rise to over 60 V if the nominal mains voltage is 120 V. This is a disadvantage of using power supplies that sag a lot.

Even more importantly, the amount of available safe operating area in the output stage must be adequate to handle the most difficult load conditions anticipated, taking into account the kind of output stage protection to be employed. This will influence the number of output pairs needed. More aggressive protection circuits will allow the use of a smaller output stage, but will likely interfere with audio performance under some conditions. How rarely do we want the protection circuit to have to engage? Will we employ a *V-I* protection circuit at all?

Number of Output Pairs

The required number of output pairs depends on power dissipation, needed safe operating area, and desired maximum output current. Even if the latter two considerations are ignored for now, a conservative look at power dissipation can lend some insight.

The 100-W/8-Ω amplifier driving 400 W into a 2-Ω load will dissipate 176 W in its output stage. Let's arbitrarily argue that the output stage should be sized to withstand this full-power dissipation indefinitely for purposes of full-power bench testing into a 2-Ω load. Let's further assume that the heat sink temperature is allowed to reach 70 °C under these conditions and that junction temperature must not exceed 150 °C. There is thus an 80 °C rise allowed to the junctions for 176 W. This amounts to a net thermal resistance from heat sink to junction of 0.45 °C/W spread out over the total number of output transistors including their heat sink insulators.

Consider two output pairs comprising four transistors. This means that each transistor is allowed 1.8 °C/W from junction to heat sink. Assume the insulators have a thermal resistance of 0.5 °C/W. This means that the junction-to-case thermal resistance of each transistor must be no greater than 1.3 °C/W. This corresponds to a transistor with a power dissipation rating of about 96 W. Thus two output pairs are adequate for this output stage power dissipation criteria. For simplicity, assume that the required number of output pairs increases in proportion to the rated amplifier power.

For 150-W output transistors, the amplifier could possibly go to 150-W/8-Ω under these dissipation criteria. A rule-of-thumb thus emerges: divide the rated power by 75 and round up to the next integer. This is the minimum number of output pairs recommended to satisfy this thermal criterion. A 400-W amplifier would thus require six pairs of output devices.

It is important to remember that this rule-of-thumb does not take into account SOA requirements and possible output current requirements. It also does not take into account operating at 1/3 power into a 2-Ω load, where output stage dissipation will be higher. Power transistors are not as expensive as they used to be, so it is easier to be generous in this regard. A greater number of output transistors relaxes thermal concerns, improves thermal stability and makes the amplifier performance less vulnerable to beta-droop effects.

8.4 Sizing the Heat Sink

Getting rid of the heat in a power amplifier is one of the most important design considerations and it can be a large factor in the cost of the amplifier. The size of the heat sink will largely be determined by the nominal power rating of the amplifier in combination with the lowest anticipated load impedance. How long the amplifier will have to be able to be run safely into that load will also influence the required size of the heat sink. A key point here is that the maximum power dissipation for a typical class AB output stage occurs at an output power that is approximately 1/3 rated power. It is important to note that other types of output stages, such as class A, class G and class H, may have very different heat sinking requirements.

A Simple Guideline

A fairly conservative approach to estimating the required heat sink thermal resistance can be arrived at by assuming that the heat sink temperature shall not exceed 60 °C when the amplifier is operated at 1/3 power into an 8-Ω load. The highest temperature of an object that you can touch without excessive pain is about 60 °C. The FTC metric of 1/3 power into an 8-Ω load for one hour may seem conservative, but remember that things get worse rapidly when the amplifier is called on to deliver its power into the many loudspeakers that exhibit much lower impedances than 8 Ω [1].

A 100-W/8-Ω amplifier will dissipate about 50 W in its output stage at 1/3 rated power when driving an 8-Ω load. If the ambient temperature is 25 °C, then a rise of 35 °C will be allowed. This translates to a required heat sink thermal resistance of 0.7 °C/W. Amplifiers with higher power ratings will require a correspondingly smaller thermal resistance to ambient.

Amplifiers designed to handle today's often highly compressed program material at high levels while driving low-impedance speaker loads should be designed with significantly lower heat sink thermal resistance, especially if they are expected to operate in hot environments.

8.5 Protecting the Amplifier and Loudspeaker

One of the tougher practical design considerations for a power amplifier is protection. There are two major aspects here. The first, and most important, is that the amplifier should not fail in such a way that it will destroy the expensive loudspeakers to which it is connected. The second is that the amplifier should not self-destruct when driving a difficult load at high power levels or when it is subjected to a short circuit at the output. Finally, it is desirable that when the protection circuits engage, they do so in a way that is benign and that causes as little damage to the audio signal as possible. Some protection circuits can create high-amplitude spikes that are very audible and can even damage tweeters. Some jokingly call these *tweeter eaters*.

Loudspeaker Protection

It is very important to protect loudspeakers from high DC voltages at the output of the amplifier. An output stage that fails will often do so by shorting the output of the amplifier to one of the rails. Speaker fuses are often used, but they can be unreliable and introduce low-frequency distortion. Active circuits that sense a DC level at the output of the amplifier can be used to open a loudspeaker relay or disable the power supply to the output stage. An alternative is to *crowbar* the output of the amplifier to ground with a TRIAC that is fired when DC is sensed.

Short-Circuit Protection

Short-circuit protection is the most fundamental form of protection for the amplifier itself. With a rugged output stage, a loudspeaker fuse or rail fuses or circuit breakers may be sufficient. How long it takes them to act is the key here. Current-limiting and active amplifier shut down circuits can also play an important role in protecting the amplifier against short circuits.

Safe Area Protection

SOA protection is much more sophisticated than short-circuit protection or current-limiting. Its purpose is specifically to protect the output transistors from unsafe combinations of voltage and current that may cause the output transistors to fail. Safe area protection is often associated with *V-I* combinations that occur when complex load impedances are being driven. Safe area protection, often synonymous with *V-I limiting*, has a dubious reputation for interfering with sound quality, and some designers dispense with it. If the output stage is made big enough, *V-I* limiting can sometimes be avoided without incurring undue risk to the output transistors.

8.6 Power and Ground Distribution

The schematic may not tell all when it comes to power and ground distribution. Even well-designed amplifiers can sound bad if they are implemented with poor power and distribution. The key here is to understand the nature of all the current flows in the power and ground circuits of the amplifier, and to recognize that no wires have zero impedance. Moreover, current flowing in any wire will create a magnetic field that may induce noise or distortion into neighboring circuits.

When Ground Is Not Ground

Ground is whatever single reference point you pick it to be. Ground is not ground when current flows from the point being considered to the reference point. This reality is simply due to the finite impedance present in any wire or piece of interconnect. The best recommendation is to follow the currents and to understand the nature of those currents.

The well-known star grounding approach seeks to keep the currents of one circuit from flowing through the ground line of another circuit. In such a case, a small-signal reference ground will have very little current flowing in it and will remain at pretty much the same potential as the reference ground. Life is not that simple, and most power amplifier ground topologies are only an approximation to star grounding at best. A key point to understand is that most of the currents that are important to avoid are AC currents. Bypass capacitors can often destroy the integrity of a star ground arrangement by providing another (unintended) path through which these AC currents can flow.

Ground Loops

Unintended ground loops can make great antennas. They don't just pick up and introduce hum. They can serve as an entryway for EMI. They can also serve to pick up and distribute voltages resulting from nonlinear magnetic fields in the amplifier. These include magnetic fields associated with power supply rectification and signal currents.

Nonlinear Power Supply Currents

The class AB output stage creates nasty half-wave-rectified nonlinear currents of large amplitude in the power rails. These currents can cause unwanted nonlinear voltage drops in power lines and grounds through which they are allowed to flow. They can also create AC magnetic fields that induce nonlinear voltages into nearby circuits [2].

Current Flows Through the Shortest Path

Make this reality work for you rather than against you. Manage impedances and wiring topology so as to force currents to circulate locally when possible.

8.7 Other Considerations

There is a myriad of additional design choices and implementation considerations that come into play in the design of an amplifier. Some could be classified as feature choices, whereas others could be classified as performance targets.

Output Stage Bias and Thermal Stability

Special attention needs to be paid to how the output stage quiescent bias will be set and how its variation over environmental and program conditions will be minimized. This will help minimize crossover distortion. At the same time, it is very important that the output stage never be allowed to enter a condition of thermal runaway. There can be conditions where the output stage can run away before the heat sink even knows what happened. Much more on this important topic will be discussed in Chapter 17.

Output Node Catch Diodes

The output node of the amplifier must never be allowed to go beyond the power supply rails by more than a diode drop. Such high-voltage excursions can occur as the result of an inductive speaker load whose current has suddenly been interrupted or limited. Such inductive kicks can damage output transistors, speaker relay contacts and tweeters. For this reason, silicon diodes are wired in a reverse-biased manner from the output node to each of the power supply rails. They will conduct if the output voltage attempts to go beyond the rail voltage. For amplifiers that incorporate speaker relays, it is best to include such catch diodes on both sides of the relay.

Protection of Speaker Relay Contacts

Speaker relay contacts can be more vulnerable to damage than one might think, often as a result of arcing when they open. Such damage can result in pitted contacts that will impair the sound quality. It is good practice to connect the speaker terminals to the *swinger* of the relay and the *NC* contacts to ground, so that the speaker terminals are shorted when the relay opens. The output catch diodes mentioned above can help by avoiding voltages greater than the rail voltage from appearing across the contacts during the opening transition interval from the *NO* contacts to the *NC* contacts. This interval will be on the order of 2 ms for the speaker relay specified for the amplifier in Chapter 4 (this is not the same as the release time of the relay).

Physical Design and Layout

One should always be aware of the surroundings of the path of a sensitive signal or of an inductive component. For example, input and feedback lines should be kept away from the power supply and the main rail lines to the output stage. If the amplifier uses an output coil, that coil should not be near ferrous material or low-level circuits.

The Feedback Path

The negative feedback should be tapped from the output after the high-side and low-side currents of the class AB output stage have been merged where the emitter resistors are joined. The signal should remain a high-level, low-impedance signal until it reaches the physical location of the input stage. In other words, most of the feedback network should reside in close proximity to the input stage. This reduces the effect of any corrupting influences on its way from the output to the input stage.

References

1. Federal Trade Commission (FTC), "Power Output Claims for Amplifiers Utilized in Home Entertainment Products," CFR 16, Part 432, 1974.
2. Edward M. Cherry, "A New Distortion Mechanism in Class B Amplifiers," *Journal of the Audio Engineering Society*, vol. 29, no. 5, May 1981.

Part 2

Advanced Power Amplifier Design Techniques

Part 2 delves deeply into the design of advanced power amplifiers with state-of-the art performance. Advanced input, VAS and output stages are discussed in depth, as are DC servos for DC offset control. A complete chapter is devoted to advanced forms of negative feedback compensation that can provide high slew rate and low distortion without compromising stability. Crossover distortion, one of the most problematic distortions in power amplifiers, is studied in depth in Chapter 12. Both static and dynamic crossover distortions are covered. Additional aspects of output stage design are covered in Chapter 13. Part 2 also includes a detailed treatment of MOSFET output stages and error correction techniques. The Part closes with a discussion of other sources of distortion that are less well known.

Chapter 9

Input and VAS Circuits

9. INTRODUCTION

The amplifier that was evolved in Chapter 3 served as a good platform for amplifier design understanding, but it did not include significant sophistication of the input stage (IPS) and voltage amplifier stage (VAS) circuits. Rather, it started with the most basic IPS-VAS and evolved it in a linear way to achieve much-improved performance. Although the end result was quite good, there are many ways to skin a cat and achieve further improved performance. Moreover, the analysis of the IPS-VAS was fairly superficial. For example, there was little discussion of noise.

9.1 Single-Ended IPS-VAS

The single-ended IPS-VAS was discussed at length in Chapter 3 where a basic amplifier was evolved to a high-performance amplifier. Most of the evolution in the design took place in the IPS-VAS. It is referred to as single-ended because the VAS is single-ended with a current source load. Later in this chapter we will focus on designs that include a push-pull VAS for improved performance.

The IPS-VAS shown in Figure 9.1 is similar to the simple IPS-VAS that was used as a starting point in Chapter 3. It is provided with ±45 V rails that correspond to an amplifier capable of delivering about 100 W into an 8-Ω load. This IPS-VAS includes emitter degeneration and is arranged with output stage pre-drivers and drivers as if a Triple EF was being used for the output stage. The output stage is not present, and the feedback is taken from a center tap on the driver emitter bias resistor. This allows distortion of the IPS-VAS to be evaluated absent the distortion of an output stage.

The pair of 234-Ω emitter degeneration resistors implements 10:1 degeneration of the input differential pair by increasing the total emitter-to-emitter resistance R_{LTP} from 52 Ω to 520 Ω. This reduces its transconductance by a factor of 10.

Recall the relationship described in Chapter 2 for Miller compensation:

$$C_{Miller} = 1/(2\pi f_c * R_{LTP} * A_{cl})$$

where A_{cl} is the closed-loop gain, R_{LTP} is the total emitter-to-emitter LTP resistance including re', and f_c is the desired gain crossover frequency for the negative feedback loop. Setting f_c to 500 kHz and closed-loop gain to 20, we have

$$C1 = C_{Miller} = 0.159 / (500kHz * 520\Omega * 20) = 30.6 \text{ pF}$$

By this calculation C1 must be about 30 pF.

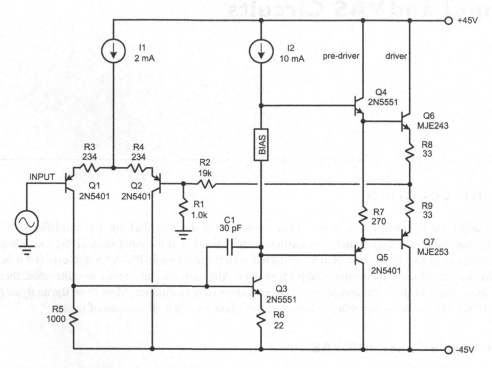

Figure 9.1 A Simple Single-Ended IPS-VAS

Notice that ±1 mA is available from the input stage to charge and discharge C1. This results in an achievable slew rate of 1 mA/30 pF = 33 V/μs. This is a respectable value of slew rate for an audio power amplifier of modest power level.

Improved Single-Ended IPS-VAS

The IPS-VAS shown in Figure 9.2 is very much like the one shown in Chapter 3 (Figure 3.12) that was evolved to a high-performance level. The major improvements made to that design included a current mirror load on the IPS and a Darlington-cascode VAS with current limiting. This combination of improvements greatly increased the open-loop gain while virtually eliminating the Early effect in the VAS.

Figure 9.3 plots simulated 20-kHz THD as a function of output voltage level corresponding to power into an 8-Ω load for the IPS-VAS circuits of Figures 9.1 and 9.2. One can see the great improvement in performance achieved by merely adding a few small-signal transistors to the circuit. Bear in mind that this distortion does not include output stage distortion. Isolating the IPS-VAS distortion is the best way to compare different designs of this portion of an amplifier.

Shortcomings of the Single-Ended IPS-VAS

The single-ended IPS-VAS is asymmetrical. With a 10-mA quiescent current, the single-ended VAS can never source more than 10 mA to the output stage. However, it can sink an amount of current that is limited only by whatever current limiting is built into the VAS. The transconductance of the single VAS transistor varies in accordance with the amplitude and polarity of the

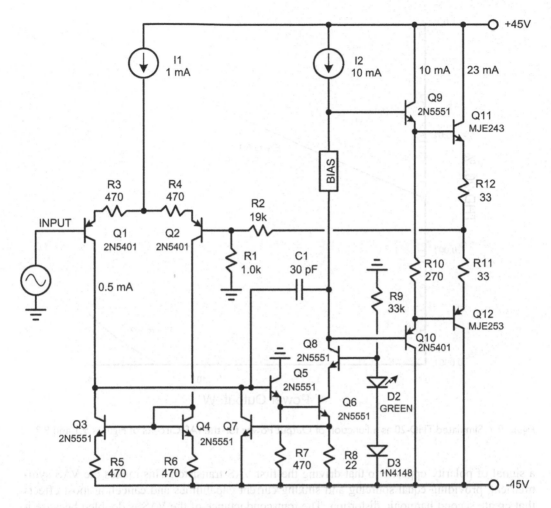

Figure 9.2 An Improved Single-Ended IPS-VAS

output current. When the VAS is sourcing high current to the output stage, the VAS transistor is operating at a low collector current, and its transconductance is correspondingly low. When the VAS is sinking high current from the output stage, the VAS transistor is operating at high current and has high transconductance. Such signal-dependent changes in transconductance lead to open-loop gain variations that are dependent on the signal in such a way as to cause second harmonic distortion. The degeneration of the VAS mitigates this problem but does not eliminate it. Output stages with high current gain, such as triples, minimize this source of distortion in the single-ended VAS.

A VAS design in which the Early effect can play a significant role will suffer second harmonic distortion from the Early effect as well, since the current gain and output impedance of the VAS will depend on the output signal voltage. Emitter degeneration reduces Early effect distortion, as does the use of VAS transistors with high Early voltage. Cascoding the VAS is another approach to minimizing this distortion.

Perhaps the single biggest improvement that can be made to the VAS is to make it push-pull, replacing the current source load with a second common emitter VAS transistor that is driven with

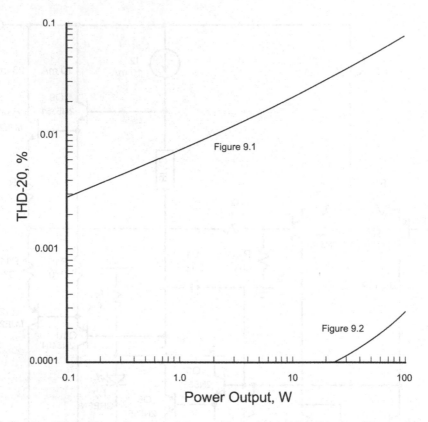

Figure 9.3 Simulated THD-20 as a Function of Output Power for the VAS Circuits of Figures 9.1 and 9.2

a signal of polarity opposite to that driving the first VAS transistor. This makes the VAS symmetrical, providing equal sourcing and sinking current capabilities and canceling most effects that create second harmonic distortion. The transconductance of the VAS is doubled because it becomes the sum of the transconductances of the positive and negative VAS transistors. When one transistor's *gm* is high, the *gm* of the other one is low. For a given quiescent current, the maximum available VAS output drive current is doubled.

Many of the IPS-VAS variations that will be seen in this chapter simply reflect different approaches to delivering the necessary drive signal to the added complementary VAS transistor.

Opportunities for Further Improvement

Many variations on the IPS and VAS are possible. Some provide improved symmetry and performance, while others bring functional features like Baker clamps to control clipping behavior. Others simply represent alternative approaches to the IPS-VAS that some believe sound better or are more immune to things like EMI ingress.

Input Stage Stress

Input stage distortion cannot be ignored. Some IPS-VAS architectures will increase or decrease the susceptibility of the amplifier to distortion that originates in the IPS. Perhaps the best-known effect is high-frequency distortion caused by increased error signal input amplitude at high frequencies. This distortion is associated with *transient intermodulation distortion* (TIM) and *slewing-induced distortion* (SID) [1, 2, 3, 4]. Input stage distortion is reduced, but not eliminated, by input stage degeneration. Degeneration reduces input stage transconductance, and this in turn reduces the size of the Miller compensating capacitor for a given amplifier unity-loop-gain-frequency (ULGF). The smaller compensating capacitor then requires less signal current from the input stage for a given output voltage slew rate. Less input stage distortion is then the result.

The size of the error signal presented to the input stage is a measure of input stage stress that results in input stage distortion. For a given output signal amplitude, the input stage stress is inversely proportional to the open-loop gain for a sinusoidal signal. In a typical amplifier design with dominant pole compensation, the open-loop gain is smaller at high frequencies (like 20 kHz), leading to greater input stage stress. At low frequencies, the open-loop gain is substantially higher, leading to substantially reduced input stage stress. Amplifiers with wide open-loop band-width have the same open-loop gain at low and high audio frequencies and thus place just as much stress on the input stage at low frequencies as they do at high frequencies.

Amplifiers with low amounts of negative feedback (and thus low open-loop gain) across the audio band subject the input stage to correspondingly greater input stage stress. Amplifiers with no global negative feedback have open-loop gain simply equal to the amplifier gain; this means that the full amplitude of the line level input signal is applied to the input stage, creating the greatest stress and demanding the design of an input stage that is very linear up to high input signal levels. This requires an input stage with very high dynamic range.

There is one caveat to the above observations. The error signal applied to the input stage in amplifiers with high open-loop gain at low frequencies is essentially a differentiated version of the input signal. If the amplifier is driven with a square wave, the peak error signal under some conditions can approach the peak-to-peak value of the input signal, implying an error signal swing that could in theory be twice as large as the case where open-loop gain was uniform across the bandwidth to which the square wave is limited.

9.2 JFET Input Stages

A popular alternative to the BJT input differential pair is the JFET input pair. This choice has advantages and disadvantages. Many believe that the sound is better while others believe that its superior resistance to input EMI is important. JFETs usually have increased input referred voltage noise, but in power amplifier applications this is not a serious issue due to the line-level signal voltages involved. Moreover, JFETs have virtually no input current noise. When a BJT LTP is degenerated to the same transconductance as a JFET (to help slew rate, for example), the noise contributed by the emitter degeneration resistors will often increase the input voltage noise of the BJT stage to be similar to that of the JFET stage.

JFET Noise

As mentioned above, input-referred noise for a JFET input stage can be a bit higher than that for a BJT stage that has emitter degeneration to make the transconductances of the stages the same. Consider an input stage using a Linear Integrated Systems LSK389 monolithic dual matched N-channel JFET with tail current of 2 mA and degenerated with 160-Ω source resistors to total *gm* of 4 mS [5]. Compare it to a BJT input stage biased at the same current and degenerated to the same *gm* with 220-Ω emitter resistors. Simulations show that input-referred noise for the JFET and BJT stages is 2.7 nV/$\sqrt{\text{Hz}}$ and 2.9 nV/$\sqrt{\text{Hz}}$ for the JFET and BJT stages, respectively. In both cases, the degeneration resistors are larger noise contributors than the active devices.

An alternative to consider is the LSK489, which has lower transconductance and capacitance than the LSK389 [5]. If an un-degenerated LSK489 is operated at a tail current of 4 mA, input-referred noise will be 2.8 nV/√Hz, comparable to the BJT stage. However, the stage trans-conductance will be only 3.1 mS, down 2.2 dB from that of the BJT stage.

In a complete power amplifier, the noise of the input stage active devices may not dominate, anyway. If the BJT 2N5551 and JFET LSK489 LTPs described above see 500-Ω source imped-ances on both negative and positive inputs, and they are loaded with BJT current mirrors with 220-Ω degeneration resistors, input-referred noise for the BJT and JFET stages is 6.2 nV/√Hz and 6.5 nV/√Hz, respectively. Authors who complain about the larger input-referred noise volt-ages of JFETs as compared to BJTs appear to be missing the bigger picture.

JFET Distortion

In the strictest sense, the *gm* of a JFET LTP is not as linear as that of a BJT LTP degenerated to the same value of transconductance. However, the cutoff characteristic of a BJT pair is much sharper. Figure 9.4 shows a comparison of the differential pair transfer characteristic for BJT and JFET input pairs that have the same small-signal transconductance. For the JFET differential pair, the device is the LSK489 operating with a tail current of 2 mA and with *gm* of about 2.0 mS for each transistor [5]. The BJT LTP operates at a tail current of 1 mA and is degenerated by a factor of 10 with 470-Ω emitter resistors to bring its transconductance down to the same 2 mS. Both LTPs are loaded with a current mirror.

Figure 9.5 shows simulated THD-1 as a function of signal level for the same two input stages. Bear in mind that both stages have the same small-signal transconductance and that the JFET stage is biased at twice the current as the BJT stage. The bottom two traces show the sum of fifth- and seventh-order harmonics for these stages, giving an idea of the relative levels of the less benign higher-order harmonics.

In an actual amplifier application, both stages will be required to deliver the same amount of signal current to the same value of compensation capacitor for a given output signal slew rate. The JFET will ultimately be able to deliver twice the peak current, however. This may seem unfair to some, but operating a JFET input stage at twice the tail current of a degenerated BJT stage is quite reasonable, given that the JFET stage has no input bias current and virtually no input current noise. Under these conditions, the JFET has only slightly more distortion than the

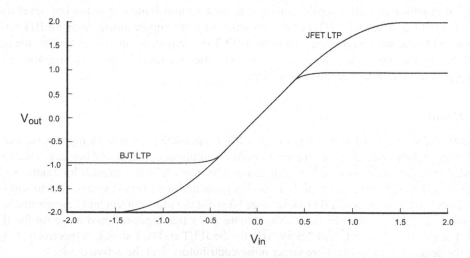

Figure 9.4 Transfer Characteristics for BJT and JFET Differential Pairs with the Same Transconductance

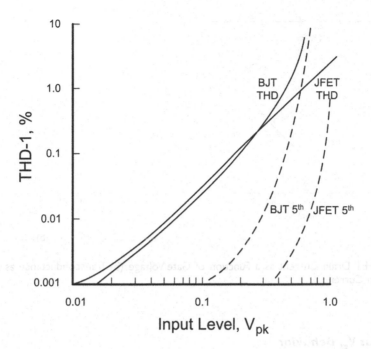

Figure 9.5 THD-1 for the Input Stages of Figure 9.4

BJT for low and medium signal levels, and lower THD at higher signal levels. Moreover, the JFET stage has drastically lower fifth-order harmonic distortion for all signal levels. With twice the tail current and using the same value of compensation capacitor, the JFET input stage will be able to deliver about twice the slew rate as the BJT input stage, all else remaining equal.

Very interestingly, if we compare a BJT LTP with a higher-gm LSK389 JFET pair, and both are run at the same 2-mA tail current, with the BJT degenerated to the same gm as the LSK389, we find that THD-1 is very similar (within ±16%) up to 150 mV peak input. At this amplitude, THD-1 is about 2.5%. Once again, however, the higher-order fifth harmonic is substantially lower (less than half) for the JFET stage at all levels. Some authors' concerns about lower trans-conductance of JFETs and smaller opportunity for degeneration of JFETs leading to higher distortion as compared to BJTs appear to be misplaced.

JFET Input Current and Voltage Offset

The JFET has the advantage of extremely high input impedance and essentially no input bias current, but sometimes its higher input offset voltage detracts from the advantage of no input bias current. The dual monolithic LSK489 has input offset voltage of less than ±20 mV [5]. In a power amplifier with gain of unity at DC, this means only ±20 mV of offset at the speaker terminals, quite acceptable. Of course, less expensive discrete JFET pairs can be matched by selection as well. Because JFETs do not need a relatively low-value return resistor for bias, they simplify the DC offset design of the input stage, even when DC servos are employed. The ability to employ large-value return resistors results in an amplifier that can have inherently higher input imped-ance and employ a much smaller input coupling capacitor.

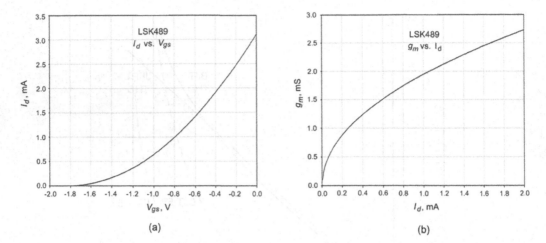

Figure 9.6 (a) JFET Drain Current as a Function of Gate Voltage (b) Transconductance as a Function of Drain Current

JFET I_d Versus V_{gs} Behavior

Figure 9.6a shows how drain current changes as a function of gate voltage in the saturation region; Figure 9.6b illustrates how transconductance changes as a function of drain current in the same region. The device is one-half of an LSK489 dual JFET [5]. Threshold voltage for this device is nominally about −1.8 V.

The JFET I-V characteristic (I_d versus V_{gs}) obeys a square law, rather than the exponential law applicable to BJTs.

JFET RFI Immunity

When operating in the quiescent state, the gate-source junction of the JFET is typically reverse-biased by a volt or two. In contrast, the base-emitter junction of the BJT is forward-biased by about 600 mV. This means that the latter is more prone to rectification effects at its base-emitter junction, making the un-degenerated BJT far more prone to RFI pickup, demodulation and inter-modulation. A degenerated BJT is less prone to this effect because it has a higher signal overload voltage as a result of the emitter degeneration.

Voltage Ratings

JFETs often do not have as high a voltage rating as BJTs, and this sometimes requires that JFET input stages be cascoded. As examples, the LSK389 has maximum V_{ds} of 40 V, while the LSK489 has maximum V_{ds} of 60 V [5]. The latter can be used directly in smaller amplifiers whose rails will never exceed 60 V. Otherwise, most of the voltage drop required in an IPS can be accommodated by the addition of BJT cascode transistors. Figure 9.7 illustrates a JFET IPS that can handle the rail voltages encountered in a typical power amplifier. Each LSK489 JFET is operated at a drain current of 1 mA, and its drain is loaded by a BJT cascode whose bases are biased at +15 V. Cascoding the JFET input devices also minimizes their power dissipation.

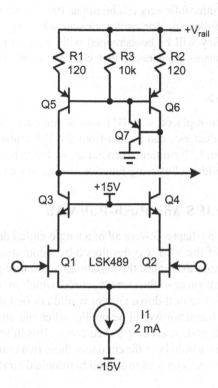

Figure 9.7 A Cascoded JFET Input Stage

JFET Input Pairs and Matching

The threshold voltages among discrete JFETs can be quite different, so it is very desirable to employ monolithic dual-matched pair devices. While discrete P-channel JFETs are readily available, dual-matched pairs like the Toshiba 2SK389/2SJ09 are scarce and largely no longer in production. That is why the IPS-VAS of Figure 9.7 is upside down from the BJT version of Figure 9.2. Dual monolithic N-channel JFET pairs are better performing and much more readily available. However, the release of the P-channel dual monolithic LSJ689 in 2016 has greatly improved matters [5]. The LSJ689 is the P-channel complement of the LSK489.

Some IPS-VAS circuits use a full-complementary IPS-VAS with two pairs of JFET LTPs, one N-channel and one P-channel. Until the LSJ689 became available, the scarcity of P-channel matched pairs made this kind of IPS difficult to build in practice. In some cases, P-channel-matched pairs may not be well matched to the N-channel complements for transconductance at a given operating current. Selective degeneration of one pair can mitigate this issue.

9.3 Buffered Input Stages

Sometimes it is desirable to operate the IPS at higher current to improve slew rate and achieve greater transconductance with a given percentage of degeneration. However, this can magnify problems associated with input bias current of a BJT IPS, especially in amplifiers that do not employ a DC servo. Buffering both sides of the IPS with emitter followers or source followers can mitigate this problem and improve performance. IPS bandwidth may be increased and input current noise will be decreased. Bipolar emitter followers can be run at 100 µA, while 500 µA might be a good operating current for source followers. The emitter or source bias resistors can be returned to the tail of the IPS LTP, where they will be bootstrapped with the common-mode signal. If voltage breakdown of the buffer transistors is a concern, they can be powered from a lower supply voltage.

9.4 CFP Input Stages

If the BJT IPS transistors are replaced with BJT complementary feedback pairs or JFET-BJT CFPs, much higher transconductance can be had from the IPS transistors for a given operating current. The linearization from LTP emitter degeneration will then be much greater, reducing IPS distortion. The CFP also provides a buffering function similar to that described above.

9.5 Complementary IPS and Push-Pull VAS

The VAS designs illustrated in Chapter 3 were all of a single-ended drive nature, where the large voltage swing at the output of the VAS was developed by a common-emitter stage loaded by a constant current source. In those examples, the maximum amount of pull-up current on the VAS output node was limited to the value of the current source, which in turn was equal to the quiescent bias current of the VAS. The pull-down current would not be limited in the same way. The collector current of the VAS transistor would be smaller when the output node was being pulled up and larger when the output node was being pulled down. This in turn meant that the transconductance of the VAS transistor would be different under these two conditions. This asymmetrical structure and behavior of the VAS can lead to second harmonic distortion and other performance impairments.

The push-pull VAS addresses these limitations by employing active common-emitter VAS transistors of each polarity, one pulling up and one pulling down. Such a symmetrical architecture has many advantages. Twice the peak drive current is available for a given VAS quiescent current and the symmetry suppresses the creation of second harmonic distortion.

The key challenge with the push-pull VAS is how to drive the two common-emitter VAS transistors, one referenced to the positive rail and the other referenced to the negative rail. In the approaches described in this section, two input pairs are employed, one a PNP LTP and one an NPN LTP. P-channel and N-channel JFET LTPs can be used also. The PNP LTP provides drive to the NPN VAS transistor on the bottom while the NPN LTP drives the PNP VAS transistor on the top. A simplified version of an IPS-VAS using this approach is illustrated in Figure 9.8 [6, 7].

Each LTP is powered by its own tail current source. Twin Miller compensation capacitors are employed; one feeds back to the input of each polarity of VAS transistor. Many amplifiers have been designed with arrangements like this one. It is an elegant way to drive a push-pull VAS and, if nothing else, has appealing symmetry on the schematic. Many of the usual IPS-VAS improvements can be applied to the arrangement of Figure 9.8, such as Darlington-connected VAS transistors.

Complementary IPS with Current Mirrors

We saw in Chapter 3 the great improvement that resulted from employing current mirror loads on the IPS LTP. Such an approach is illustrated in Figure 9.9. Here there is a current mirror on top driving the PNP VAS transistor and another current mirror on the bottom driving the NPN VAS transistor. Both current mirrors enable the input stage to provide high gain, especially given that both of the VAS transistors employ a Darlington connection. Look at the circuit and see if you can calculate the VAS quiescent current from the components and currents shown. You can't. That is the problem with this circuit. The VAS quiescent current is indeterminate. This is not a practical and reliable circuit. Some means must be introduced to establish with some reliability the VAS

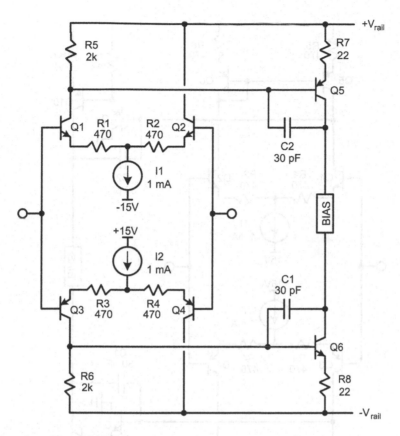

Figure 9.8 Complementary IPS-VAS Circuit

bias current. Such a means will cause the output voltage of the current mirror to be at a defined level when the LTP is in balance.

One approach to solving this problem is illustrated in Figure 9.10. First notice that each current mirror has had a *helper transistor* added to it (Q13, Q14). This is little more than an emitter follower that supplies the base current for the current mirror transistors, rather than having it drained from the incoming current. This improves DC balance of the current mirror and greatly decreases the influence of transistor beta on its operation.

Notice also that the helper transistor separates the voltage at the current mirror input (collector of Q6) by one additional V_{be} from the rail. This puts that voltage 2 V_{be} plus the drop across R6 away from the rail. This happens to be the same voltage drop from the rail as exists at the input to the Darlington VAS transistor if V_{be} drops are the same and amount of degeneration voltage drop is the same.

Resistors R14 and R16, connected across the collectors of Q1, Q2 and Q3, Q4 accomplish the remainder of solving the problem. When the LTP is in balance, no current flows through R14 or R16; the voltage on both sides is the same. The voltage at the emitter of the VAS transistor then becomes approximately the same as the voltages at the emitters of the current mirror transistors, which are set by the LTP tail current. It is easily seen that the quiescent current of the VAS is now established so that it depends directly on the tail current in the LTPs.

The price being paid here is a slight reduction in the gain of the input stage. Q1 and Q2 are differentially loaded by R14 rather than by high indeterminate impedance set by transistor betas. The effective load resistance is half the value of R14, since the current flowing through R14 is

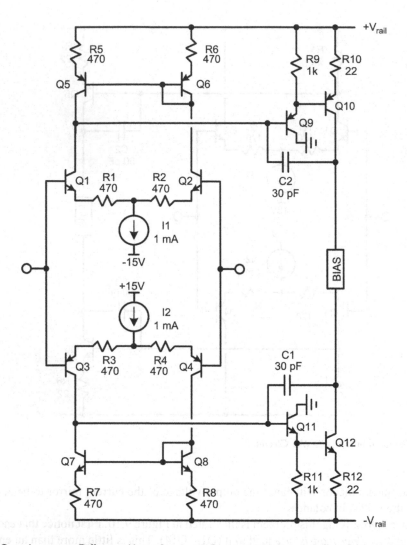

Figure 9.9 Complementary Differential Input Stage with Current Mirror Loads Suffers Quiescent Bias Instability

re-circulated through the current mirror to further oppose voltage change at the output of the current mirror. In the same way, R16 provides a well-determined load for Q3 and Q4. The value of the differential shunt resistors can be made quite high if well-balanced circuitry is used with precision resistors.

This circuit is sensitive to differences in the tail currents for the NPN and PNP LTPs. It can be seen that any such difference will attempt to set a different quiescent current for the top and bottom halves of the VAS. This cannot be, so an input offset voltage will be created for the overall stage that forces the top and bottom VAS transistors to operate at the same current.

Complementary IPS with JFETs

Many fine amplifiers have been made with the complementary IPS using JFETs, as illustrated in Figure 9.11. All of the same approaches and caveats apply. The JFET LTPs are cascoded to keep the drain voltages low enough to meet their V_{ds} ratings.

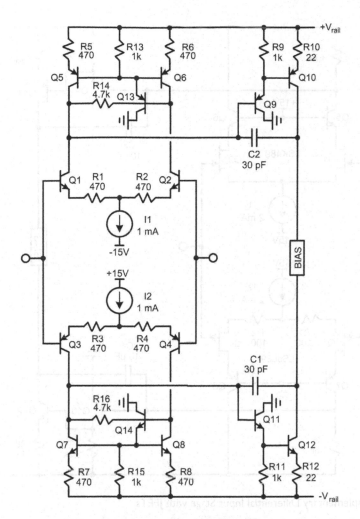

Figure 9.10 Complementary Differential Input Stage with Stabilized Current Mirror Loads

There are two concerns that apply with complementary JFET input stages. First and foremost, P-channel dual-matched JFETs are difficult to find. Most went out of production many years ago. The availability of the Linear Systems LSJ689 as of 2016 has once again made such fully complementary JFET input stages practical [5]. Second, the transconductance of the P-channel pair does not match that of the N-channel pair at the same operating current.

Complementary IPS circuits do not perform optimally if the transconductances of the top and bottom halves are not well matched; each half will have different gain, and the top and bottom parts of the VAS will tend to fight each other. This can result in second harmonic distortion. This is especially so if twin Miller compensation capacitors are used as shown. Each compensation capacitor creates a shunt feedback loop that makes the output impedance of its half of the VAS low. Two low-impedance sources connected in parallel (the top and bottom halves of the VAS) will fight each other unless each one is trying to put out the same voltage as the other one. If the transconductances of the top and bottom LTPs are not the same, this condition will not be satisfied. Source degeneration resistors R1 and R2 on the P-channel devices in Figure 9.11 help to equalize the LTP transconductances.

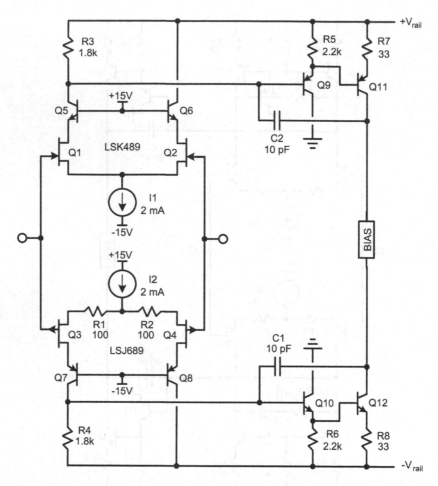

Figure 9.11 Complementary Differential Input Stage with JFETs

The LTP tail currents are set by current sources I1 and I2, here 2 mA each. The tail currents also determine the VAS bias current. For a given design, the values of R3 and R4 help establish the desired VAS bias current, here 10 mA. Component values are also chosen so that the JFETs in each LTP are operating at the same current under normal conditions. Simulation of this circuit is strongly recommended for choosing component values. Other versions of this circuit can be implemented with current mirrors in place of R3 and R4, as was shown in Figure 9.10.

The compensation capacitors C1 and C2 are set to provide a 900-kHz ULGF for a closed-loop gain of 20. The input-referred voltage noise for this IPS is less than 4 nV/$\sqrt{\text{Hz}}$.

Floating Complementary JFET IPS

An elegant complementary JFET IPS popularized by John Curl floats the complementary differential pairs. An arrangement like this is shown in Figure 9.12. This design takes advantage of the depletion mode biasing of the JFETs to create what is effectively a floating common current source for the N-channel and P-channel LTPs. The sum of the V_{gs} voltages of the top and bottom JFET pairs at the chosen operating current is forced to appear across bias resistor R9 connecting

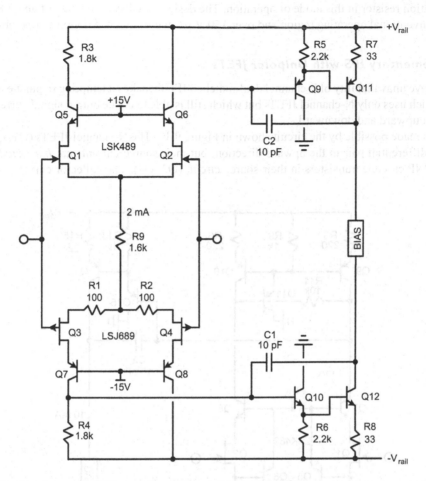

Figure 9.12 Complementary Differential Input Stage with Floating JFETs

the two pairs, thus establishing the tail currents, which will be equal. Here the tail current is set to 2 mA, as it was in Figure 9.11.

The tail current in this design will depend directly on the threshold voltages of the JFET pairs, so some selection or adjustment of R9 may be necessary to arrive at the desired tail current in order to accommodate JFET V_{gs} spreads. In this arrangement, the tail currents also determine the VAS bias current. Once the circuit is built, R9 is adjusted to establish the desired VAS bias current, here 10 mA. VAS bias current can be measured by probing the voltage across VAS emitter resistor R7. In a given design, the values of R3 and R4, here 1.8 kΩ, are chosen so that the JFETs in each LTP are operating at the same current. Simulation of this circuit is strongly recommended for choosing component values.

This complementary IPS can provide unusually high slew rate. If the LTPs go into clipping due to a large error voltage, the diagonally opposite input transistors will continue to conduct at a higher current by passing increased current through tail resistor R9. Under these quasi-class B IPS operating conditions, the net transconductance of the IPS smoothly falls to about 50% of its small-signal value. This also tends to reduce the closed-loop bandwidth by about 50% under these high-slew-rate conditions. The degree of transconductance reduction in this operating regime may be somewhat different for other designs, but 50% will be typical. R9 acts as a degeneration resistor in this mode of operation. The design here has a slew rate of about 80 V/μs in the conventional operating region and over 150 V/μs in the extended region of operation.

Complementary IPS with Unipolar JFETs

The relative unavailability of P-channel dual-matched JFETs makes it tempting to pursue an input stage which uses only N-channel JFETs but which still is able to create output signal currents that face both upward and downward.

This is made possible by the circuit shown in Figure 9.13. The N-channel JFETs (Q1, Q2) still act as a differential pair in the upward direction, but their source currents are *harvested* with a pair of PNP cascode transistors in their source circuit (Q3, Q4). The collector currents of these

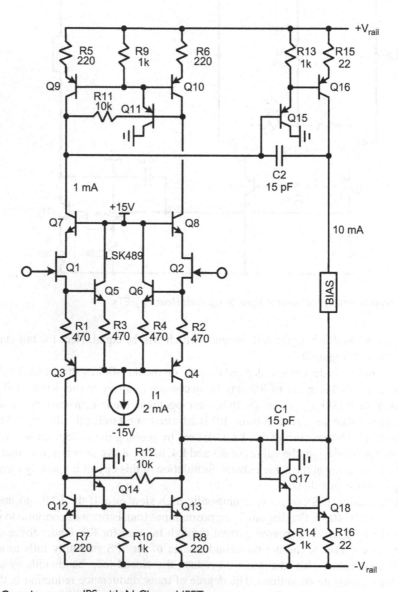

Figure 9.13 Complementary IPS with N-Channel JFETs

cascode transistors flow downward to a current mirror to drive the NPN VAS transistor. Resistors R1 and R2 provide the degeneration. If one assumes that the base line connecting Q3 and Q4 floats, it is easy to see that the differential input voltage creates a current that flows through the loop created by Q1, R1, Q3, Q4, R2 and Q2.

The key to making this arrangement work is to get the right bias currents to flow without interfering with common-mode rejection. This involves properly generating the voltage at the interconnected bases of Q3 and Q4. This voltage must float with the common-mode input voltage and be offset downward by the proper amount to establish the desired bias current. Transistors Q5 and Q6 perform this function.

Q5 and Q6 are emitter followers whose output is summed by R3 and R4 to form a replica of the common-mode voltage. This replica is applied to the bases of Q3 and Q4. Current source I1 provides the necessary pull-down bias current for the emitter followers. Notice that the V_{be} drops of Q3 and Q5 cancel each other, as do those of Q4 and Q6. As a result, the voltage drops across R1 and R3 are equal if all V_{be} drops are equal. Likewise, the voltage drop across R2 is the same as that across R4. The DC voltage drops across R1 and R2 set the bias current, and the drop across R3 and R4 is controlled by current source I1. If R3 and R4 equal R1 and R2, then the bias current flowing in Q1 will be equal to one-half of I1. Current I1 is the "tail current" of the arrangement.

The arrangement is completed with upper cascode transistors Q7 and Q8. Current mirrors at the top and bottom (Q9, Q10, Q11 and Q12, Q13, Q14) provide the loading. R11 and R12 provide differential loading on the current mirror outputs to provide VAS quiescent current stability.

9.6 Unipolar Input Stage and Push-Pull VAS

The relative scarcity of complementary JFET pairs also makes it desirable to have designs that require only a conventional LTP of one polarity (usually N-channel JFET) while still being able to drive a push-pull VAS. The key to achieving this is how to *turn around* the signal drive for the opposite side of the VAS. Tom Holman's *APT-1* power amplifier is a good example of this design approach (wherein NPN BJT devices were used) [8, 9].

Figure 9.14 shows such an approach where a current mirror (Q5, Q6) is used to generate the drive signal for the NPN VAS transistor. The signal current from one side of the LTP output is reflected down to the negative rail where it is used to drive the NPN VAS transistor. A cascode transistor is placed in the path of the reflected current so as to reduce quiescent voltage on Q6. This architecture does not appear to lend itself well to the use of LTP current mirror loads in the way that they were employed in some earlier arrangements.

Differential Pair VAS with Current Mirror

A slightly different approach is illustrated in Figure 9.15. Here the PNP VAS is actually a differential pair. It is fed from an IPS with a current mirror load, so high gain and good DC balance is achieved in the IPS. The IPS current mirror has a "helper" transistor (Q7), and sometimes stability is improved by the addition of about 470 pF from base to emitter of Q7.

The VAS LTP is fed from both outputs of the IPS, but because of the current mirror arrangement only the IPS output from Q3 has significant voltage movement. The IPS output from Q4 acts as a voltage reference for the other input of the VAS LTP. Diodes D1 and D2 prevent the current mirror transistors from saturating when the amplifier clips. The PNP VAS LTP (Q10, Q11) is preceded by NPN emitter followers Q8 and Q9. Their V_{be} drops cancel those of the LTP transistors. The LTP is powered by twin current sources and its gain is set by R8 connected between the emitters of Q10 and Q11. The twin current source arrangement helps conserve voltage headroom because the VAS bias current does not flow through the emitter degeneration resistance.

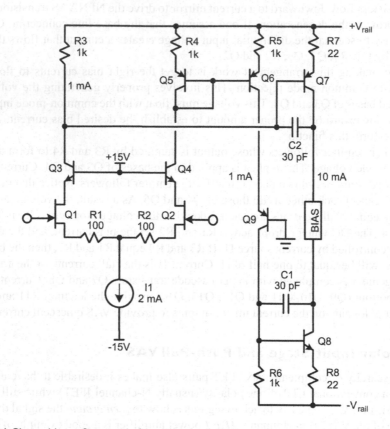

Figure 9.14 N-Channel Input Stage with Complementary VAS

The bottom half of the VAS is formed by a current mirror (Q13, Q14, Q15). A cascode (Q12) is situated in the path of the current from Q10 to the current mirror to share the voltage drop that spans both supply rails. Conventional Miller compensation is provided by C1 connected from the collector of Q11 to the base of Q9. A further advantage of this design is that the VAS is naturally current limited; it will never sink or source more than the total amount of its LTP tail current.

IPS with Differential Current Mirror Load

A more advanced form of unipolar IPS-VAS front-end is shown in Figure 9.16 [10]. The key advantage of this circuit is that it allows the use of a balanced current mirror structure to load the input stage. The differential current mirror exhibits very high impedance in the differential mode, but rather low impedance in the common mode. It thus provides some additional common-mode rejection. The primary elements of the differential current mirror are current sources Q5 and Q6. Emitter followers Q7 and Q8 jointly act as the current mirror helper transistors, creating and feeding back a common-mode voltage to control current sources Q5 and Q6. The emitter followers also buffer the differential signal before application to the VAS LTP.

The differential current mirror establishes a well-defined common-mode voltage level to be applied to the VAS LTP. This, combined with the differential drive of the LTP, allows the use of a simple resistor tail for Q9 and Q10. The VAS in Figure 9.16 also employs cascodes Q11 and Q12, suppressing the Early effect and allowing the use of fast, low-voltage transistors for Q9

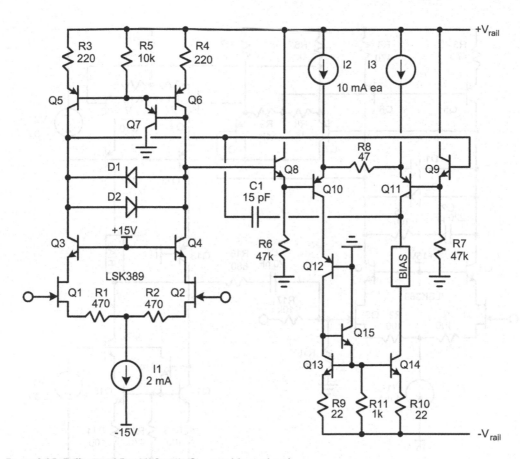

Figure 9.15 Differential Pair VAS with Current Mirror Load

and Q10. The bias voltage provided for the cascodes is also used to power emitter followers Q7 and Q8, again allowing the use of fast, low-voltage transistors. A similar approach is used for the cascoded current mirror Q14–Q17.

Diodes D1 and D2 clamp the IPS differential output voltage to prevent saturation of Q5 and Q6 when the amplifier clips. This IPS-VAS is best used with *Miller input compensation* (MIC) as implemented by C1 and R18. MIC will be described in Chapter 11. The series R-C network across the IPS collectors (C2 and R5) provides some lag-lead frequency compensation for the high-impedance intermediate nodes between the IPS and VAS. This compensates the local MIC feedback loop. In some MIC implementations an additional series R-C network is required from the VAS output to ground.

9.7 Input Common-Mode Distortion

In the customary non-inverting amplifier configuration the full line-level input signal is applied to the input stage as a common-mode signal. This can create distortion in a number of ways. Since this distortion makes itself apparent as an effective input-referred differential signal at the input stage, negative feedback cannot reduce it.

One source of common-mode distortion can originate in the tail current source. A cascoded current source can reduce this contributor. Another source can arise from the Early effect in the

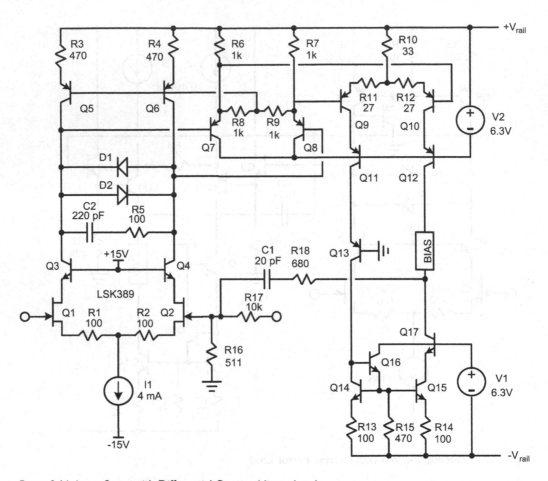

Figure 9.16 Input Stage with Differential Current Mirror Load

LTP transistors. At high frequencies the nonlinear collector-base capacitance of the LTP transistors can also introduce common-mode distortion, especially if the impedances of the networks driving the LTP are not very low. An IPS with a single-ended output and not loaded by a current mirror is more vulnerable to common-mode distortion. In some cases JFETs may be more susceptible to common-mode distortion than BJT devices.

Input common-mode distortion can be reduced with a *driven cascode*. Figure 9.17 shows how an input stage cascode can be driven with a replica of the common-mode signal, making the collector-base voltage of the LTP transistors constant, independent of signal. The replica signal is created by passing the amplifier output signal through a second feedback network comprising R3 and R4. Other approaches to generating the common-mode signal for driving the cascode are also possible. For example, the tail signal of the LTP can be buffered and level shifted as needed.

9.8 Early Effect

As discussed in Chapters 2 and 3, the current gain of a transistor is mildly dependent on the collector voltage. This leads to a finite output resistance in a common-emitter stage, and unfortunately this resistance is nonlinear. It also leads to base current being a function of collector voltage. This can impair power supply rejection in an input stage by introducing common-mode

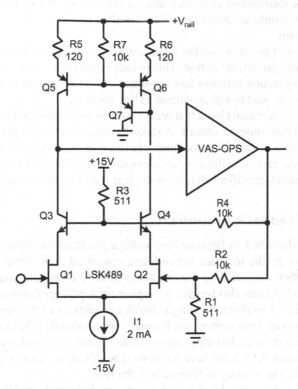

Figure 9.17 N-Channel Input Stage with Driven Cascode

input base current that will flow through the source impedances on either side of the input pair. Having equal source impedances on both sides can reduce this effect.

As discussed in Chapter 3, the use of a cascode VAS greatly reduces Early effect, but does not eliminate it completely. The usual cascode creates a fixed potential at the collector of the common emitter stage, so that signal voltage at the base still modulates V_{bc} (but to a much smaller extent). Some circuit techniques bootstrap the base voltage of the cascode with a replica of the degenerated CE transistor's emitter voltage to reduce this remaining effect. However, these arrangements are sometimes unstable. Since Early effect is essentially a modulation of transistor current gain, preceding a CE stage with an emitter follower also substantially reduces the consequences of Early effect in the CE transistor. This is a significant benefit that results from the use of a 2T VAS. Using such a VAS with 10:1 emitter degeneration and a transistor with high Early voltage can often obviate the need for adding a cascode to the VAS.

There is also the Early effect at the input to the driver or pre-driver transistors of the output stage. This is because the base-collector voltage of the driver transistor is modulated by the signal, and thus its beta is influenced via its Early effect. The bias current flowing through the driver collector, even if constant, will result in a changing driver base current being pulled from the VAS output node. The shunt feedback provided by the usual Miller compensation will reduce the output impedance of the VAS, and this will also reduce the influence of the Early effect.

9.9 Baker Clamps

When amplifiers are overdriven, they will clip. How cleanly they clip can have an effect on the sonic performance of the amplifier. How often amplifiers clip depends on many factors, not the least of which include loudspeaker efficiency and the crest factor of the program material being played. In some cases, amplifiers may clip more often than we think. When they do, it is important that they clip cleanly.

In the VAS circuits we have seen thus far, if the amplifier is driven to clipping, the VAS transistor will almost certainly go into saturation. This occurs when the collector voltage goes so low that the base-collector junction becomes forward-biased. Transistors tend to be slow to come out of saturation, and this can lead to a phenomenon called *sticking*.

A Baker clamp is a diode-based circuit that prevents the signal excursion from going far enough to allow the protected transistors to saturate. A Baker clamp can be as simple as two diodes connected from the top and bottom VAS output nodes to fixed voltage references. The APT-1 power amplifier was one of the earliest audio designs to employ Baker clamps [8, 9]. Baker clamps and related circuitry to control amplifier behavior in the real world are discussed in Chapter 21.

9.10 Current Feedback Amplifiers

All of the amplifiers described so far have been *voltage feedback amplifiers* (VFA); the output voltage is scaled down by the feedback network and compared to the input voltage to form an error voltage that is then amplified by the open-loop gain to create the output signal. *Current feedback amplifiers* (CFA) can also be used to implement a push-pull power amplifier [11, 12, 13, 14, 15]. In a CFA, the feedback dividing network connects to a low-impedance node in the input stage, and delivers an *error current* that is usually proportional to the difference of the input voltage and the divided value of the output signal (if the feedback network were not connected to the low-impedance node). A CFA can have the same type of output stage as a VFA and often the VAS is similar as well. The primary difference is in the IPS.

Figure 9.18 shows a simple model of a CFA power amplifier that employs a complementary VAS. It consists of a unity-gain input buffer, a pair of current mirrors that mirror the positive and negative supply currents of the buffer and a unity-gain output buffer that can be a conventional audio power amplifier output stage. The feedback network divides the output signal in the usual way to establish the closed-loop gain, but is typically of low-impedance design. Here the gain is

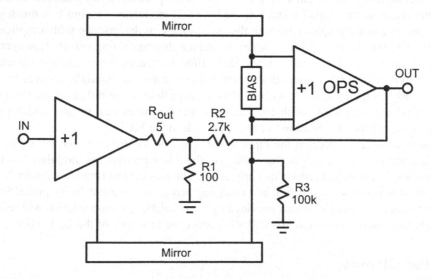

Figure 9.18 A Simple Current Feedback Amplifier

set to be 28. As in a VFA, the feedback network produces a voltage that is the same as what the input voltage to the amplifier should be. However, in the CFA the feedback network's signal is connected to the low-impedance output node of the input buffer.

The input buffer forces the input signal to appear at this node. The difference between the input voltage forced on this node and the Thévenin voltage of the feedback network causes a current to flow in accordance with the Thévenin impedance of the feedback network. This is thus an *error current*. If the input buffer has high input impedance and low output impedance, this error current must flow equally in what are normally the power supply lines for the input buffer. It has nowhere else to go. The input buffer could be something as simple as a push-pull emitter follower.

The error currents are reflected by the current mirrors and summed at a high-impedance node; together, the current mirrors have implemented the VAS. For convenience of analysis, the VAS output node is loaded with resistance R3, but in most designs the resistor is not present and the DC gain of the VAS is very high and poorly defined. A signal current I_{fb} delivered into the input buffer's output node will appear as a voltage equal to $2I_{fb} * R3$ at the VAS output node. The forward path gain of the amplifier is thus described as a *transimpedance*. From a feedback loop perspective, all of the current in feedback resistor R2 flows as a current I_{fb} into the input buffer. The loop gain is thus the ratio of the open-loop transimpedance of the amplifier to the feedback resistance (or impedance). If R3 is 100 kΩ and R2 is 2.7 kΩ, then loop gain is ideally 74 (37 dB). Notice that the value of R1 plays no role in affecting the loop gain if the output impedance of the input buffer is ideally zero. Here it is illustrated as 5 Ω, so 95% of the current through R2 flows through the buffer.

Figure 9.19 shows a simplified circuit implementation of a CFA power amplifier. The input stage is a unity-gain diamond buffer. This buffer has high input impedance, low output impedance and ideally zero input-output DC voltage offset. 3T current mirrors with 20:1 degeneration are used to provide high output impedance in implementing the VAS. These current mirrors have a current gain of 2. With Q3 and Q4 operating at 5 mA, Q6 and Q9 operate at 10 mA. Feedback current sources with high output impedance are used to bias the diamond buffer input stage. A stiff CFP V_{be} multiplier is used to bias a Locanthi T output Triple. Output offset can be controlled by a DC servo that injects a DC correction current into the output of the input buffer.

Miller Input Compensation (MIC) is provided by a single capacitor from the VAS output node to the feedback node. A Zobel network shunts the VAS output node to improve stability. Closed-loop bandwidth is about 2 MHz. The main compensation capacitor is the primary limitation on slew rate. The current mirrors can deliver 20 mA peak to the VAS output node while the input stage remains in class A, so if C1 = C2 = 20 pF, slew rate will be limited to 500 V/μs, assuming that all IPS-VAS stages remain in class A. Traversing the feedback loop, notice that the input stage looks like a cascode to the feedback signal. The feedback path thus looks largely like a cascode, a common emitter stage and an emitter follower, making for a fast loop.

If the Triple has current gain of 125,000 (β = 50³), its input impedance will be 500,000 with a 4-Ω load. With R6 = 2.7 kΩ and current gain of 2 in the mirrors, loop gain would be about 740 (57 dB), were it not for Early effect in the VAS and pre-driver transistors. CFAs do tend to have lower loop gain than VFAs because they have only one voltage amplifying stage. Other approaches, such as using a 2T common emitter VAS in place of the current mirrors, can provide higher loop gain at some expense in speed.

Performance and Ease of Design Comparison with VFAs

In reality, CFAs are more difficult to design than VFAs for a given amplifier performance level. Managing DC offset is more difficult because the different V_{be} of NPN and PNP transistors causes substantial input-output offset in the diamond buffer. Simulation of the simple CFA in

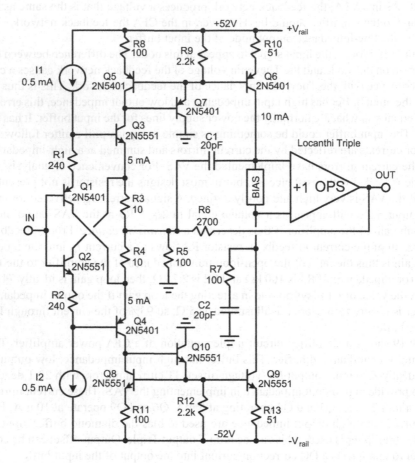

Figure 9.19 A Simplified CFA Power Amplifier

Figure 9.19 shows a DC offset of −0.9 V. PSRR of many CFA designs is inferior to that of a VFA. The promised higher maximum ULGF for CFAs compared to a well-designed VFA can in reality be illusory. The CFA can usually offer very high slew rate in comparison to a VFA, but 500 V/μs is simply not necessary for an audio amplifier to achieve very low amounts of THD-20 and CCIF intermodulation distortion even at full power. In any case, 300 V/μs was achieved with MIC in the VFA of Reference 10. The design of Figure 9.19 achieved 0.012% THD-1 and THD-20 and a ULGF of 2 MHz using the same Triple output stage as the BC-1 of Chapter 4.

9.11 Example IPS-VAS

Figure 9.20 shows an example of an IPS-VAS that was employed in a lateral MOSFET power amplifier described in Chapter 14 [16]. The IPS-VAS is a full complementary JFET design using a floating tail, similar to the one described in Figure 9.12. Although designed for use in a Lateral MOSFET amplifier, this IPS-VAS can be used with any type of output stage with the appropriate bias spreader. All bipolar transistors in the design are NPN 2N5551 or PNP 2N5401 unless noted.

Dual monolithic N-channel LSK489 and P-channel LSJ689 JFETs are used for the IPS. Q1–Q4 form the floating full-complementary IPS, which operates at a tail current of 4 mA as determined by R3 and R4. R4 is required to trim the tail current to its design value because of the variability of JFET threshold voltages. Source degeneration resistors R5 and R6 decrease

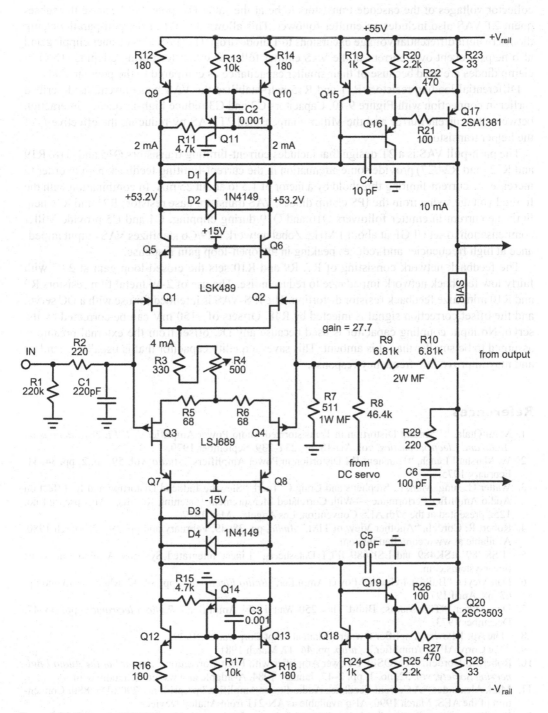

Figure 9.20 Example IPS-VAS

the transconductance of the P-channel LTP to equal that of the N-channel LTP. The JFET LTPs are cascoded to allow for higher rail voltages than the JFETs would otherwise support.

Each LTP is loaded by a current mirror, which includes an EF "helper" transistor to supply the base current for the mirror transistors. The V_{be} voltage drop of the helper causes the differential collector voltages of the cascode transistors to be at the same DC potential because the subsequent 2T VAS also includes an emitter follower. This allows the use of the anti-parallel clamp diodes to limit differential voltage excursions to 1 diode drop. This provides cleaner clipping and also helps prevent over-current in the VAS emitter follower transistor during clipping. 1N4149 clamp diodes are used because of their smaller capacitance as compared to the popular 1N4148.

Differential loading resistors R11 and R15 help stabilize the VAS bias current, as described earlier in connection with Figure 9.10. Capacitors C2 and C3 reduce high-frequency interaction between the current mirror and the Miller-compensated 2T VAS by reducing the effective f_T of the helper transistors.

The push-pull VAS is a 2T design that includes current-limiting transistors Q15 and Q18. R19 and R22 (and R24/27) provide some attenuation in the current-limiting feedback path in order to increase the current-limiting threshold by a factor of 1.5 to about 25 mA. In combination with the limited voltage swing from the IPS clamp diodes, VAS transistor base resistors R21 and R26 help limit the current in emitter followers Q16 and Q19 during clipping. C4 and C5 provide Miller compensation to set ULGF at about 1 MHz. Zobel network R29/C6 stabilizes VAS output impedance at high frequencies and reduces peaking in the open-loop gain response.

The feedback network consisting of R7, R9 and R10 sets the closed-loop gain at 27.7 with fairly low feedback network impedance to reduce noise. The use of 2-W metal film resistors R9 and R10 minimize feedback resistor distortion. The IPS-VAS is intended for use with a DC servo, and the offset correction signal is injected by R48. Offsets of ±150 mV can be corrected by the servo. No input coupling capacitor is used because any DC offset from the external preamp is assumed to be smaller than this amount. This saves a quality capacitor that is usually redundant and may improve low-frequency response.

References

1. Matti Otala, "Transient Distortion in Transistorized Audio Power Amplifiers," *IEEE Transactions on Audio and Electroacoustics*, vol. AU-18, pp. 234–239, September 1970.
2. W. Marshall Leach, "Transient IM Distortion in Power Amplifiers," *Audio*, vol. 59, no. 2, pp. 34–41, February 1975.
3. Walter G. Jung, Mark L. Stephens and Craig C. Todd, "Slewing Induced Distortion and Its Effect on Audio Amplifier Performance—With Correlated Measurement Listening Results," AES preprint no. 1252 presented at the 57th AES Convention, Los Angeles, May 1977.
4. Robert R. Cordell, "Another View of TIM," *Audio*, pp. 38–49, February and pp. 39–42, March 1980. Available at www.cordellaudio.com.
5. "LSK389, LSK489 and LSK689 JFET Datasheets," Linear Integrated Systems. Available at www.linearsystems.com.
6. Dan Meyer, "Build a 4-channel Power Amplifier," *Radio Electronics*, pp. 39–42, March 1973 and pp. 62–68, April 1973.
7. Dan Meyer, "Tigersaurus: Build This 250-Watt Hi-Fi Amplifier," *Radio Electronics*, pp. 43–47, December 1973.
8. "The Apt 1 Power Amplifier Owner's Manual," Apt Corporation, 1979.
9. "Apt Corp. APT 1 Amplifier," *Audio*, pp. 44–47, March 1981.
10. Robert R. Cordell, "A MOSFET Power Amplifier with Error Correction," *Journal of the Audio Engineering Society*, vol. 32, no. 1, pp. 2–17, January 1984. Available at www.cordellaudio.com.
11. Mark Alexander, "A Current Feedback Audio Power Amplifier," preprint no. 2902-D5, 88th Convention of the AES, March 1990. Also available as AN-211 from Analog Devices.
12. Erik Barnes, "Current Feedback Amplifiers-1," *Analog Dialog*, vol. 30, July 1996. Available at www.analog.com/analog-dialogue/articles/current-feedback-amplifiers-1.
13. Erik Barnes, "Current Feedback Amplifiers-II," *Analog Dialog*, vol. 30, October 1996. Available at www.analog.com/analog-dialogue/articles/current-feedback-amplifiers-II.
14. Andrew C. Russel, "CFA vs. VFA: A Short Primer for the Uninitiated," January 2014, V2.00. Available at www.hifisonix.com.
15. Hans Palouda, "Current Feedback Amplifiers," TI/National Semiconductor, AN-597, June 1989.
16. Presented at Burning Amp 2016. Available at www.youtube.com/watch?v=V7-27fDgqco.

Chapter 10

DC Servos

10. INTRODUCTION

Virtually all of the discussion on power amplifiers thus far has ignored the reality of AC coupling and DC offset. The primary focus of this chapter is *DC servos*. They make up a separate global feedback loop that acts at DC to control amplifier output offset. However, it is also important to understand the origins and potential magnitudes of DC offsets in amplifiers of conventional design. Some approaches to reducing offsets in conventional designs are also discussed, and some of them are relevant to designs incorporating DC servos.

Figure 10.1 shows a simple amplifier with the usual DC blocking arrangements. The VAS and output stage are shown as an amplifier symbol. Coupling capacitor C1 blocks any DC from the source. Input return resistor R1 biases the LTP input node at about 0 V. With a 20-kΩ input return resistor, a 5-µF coupling capacitor is required to push the input cutoff frequency below 2 Hz. This capacitor should be of very high quality [1].

The negative feedback network includes a 20-kΩ feedback resistor (R3) and a 1.05-kΩ feedback shunt resistor (R2). These resistors set the closed-loop gain at 20. The network also includes an electrolytic capacitor (C2) in the shunt leg that allows the DC gain of the amplifier to fall to unity at DC. This prevents LTP input voltage offsets from being multiplied by the closed-loop gain of the amplifier.

The electrolytic capacitor introduces a second low-frequency roll-off with a time constant approximately equal to $R2 * C2$. The value of $C2$ must be 100 µF to push this second roll-off frequency to below 2 Hz. This is why the capacitor usually must be an electrolytic type, preferably non-polarized. The back-to-back diodes across the electrolytic prevent it from being subjected to excessive voltages in the event that the output of the amplifier becomes stuck to one rail or is otherwise driven to a large voltage for a significant period of time. These diodes also prevent the input stage from being subjected to damaging high voltage in the event that the output is stuck at one rail. A 100-W sinusoid at 2 Hz would attempt to place about 1.4 V peak across C2. Under these conditions the diodes would become forward-biased and cause distortion. At 20 Hz, about 140 mV peak will appear across the diodes, and less than 0.001% distortion will result. However, distortion due to any conduction in these diodes rises steeply as frequency decreases. Sometimes the diodes are left out for this reason. In other cases two diodes are put in series or two Zener diodes are used in anti-series. It is also true that the nominal voltage at the base of Q2 resulting from base current flow in R3 can place a 200 mV forward bias across D2 even under quiescent conditions.

The electrolytic capacitor is problematic because it is in the signal path. Electrolytic capacitors can be notoriously nonlinear and will cause distortion. Any signal voltage appearing across the electrolytic will be distorted. At minimum, capacitance is nonlinear with voltage across the capacitor and ESR is nonlinear with current through the capacitor. Distortion from the capacitor

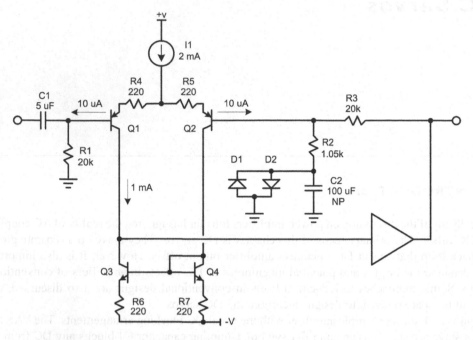

Figure 10.1 Conventional Amplifier DC Block Arrangement

rises at lower frequencies as a larger signal voltage appears across the capacitor. A very high-quality non-polarized electrolytic should be used for C2. NP electrolytic capacitors with higher voltage ratings tend to introduce less distortion. Capacitor distortion is covered more deeply in Chapter 16. Some designers bypass C2 with a quality film capacitor. This can improve the sound, but the higher impedance of the film capacitor over much of the audio band limits its effectiveness in reducing distortion from C2.

An alternative to the electrolytic capacitor is a large film capacitor in combination with higher-circuit impedances in the feedback network. If R2 is set to 5 kΩ and R3 is set to 95 kΩ, then a 22-μF film capacitor will yield a low-frequency corner below 2 Hz. This is a costly and bulky solution. The high impedance in the feedback network invites noise and DC offset impairments.

We will see that the single biggest reason for employing a DC servo is the elimination of the electrolytic capacitor. The DC servo is a smaller and less expensive solution that provides much higher sound quality and performance.

10.1 Origins and Consequences of DC Offset

Input offset voltage of the LTP can be less than 1 mV for dual monolithic BJTs or can be more than 20 mV for unmatched discrete BJTs. A JFET LTP implemented with dual monolithic JFETs can be had with input offset voltages less than 10 mV without great difficulty. As long as these input voltage offsets are not multiplied by a DC closed-loop gain greater than unity, satisfactory amplifier output offsets are achievable. Amplifier DC offsets greater than about 50 mV should be avoided.

Input Bias Current

Input offset voltage of the differential pair is not the only source of DC offset in the amplifier. A more serious problem with DC offset occurs in amplifiers with BJT input stages as a result of base current. Importantly, JFET input stages do not suffer this problem.

The PNP LTP transistors in Figure 10.1 are each biased at 1 mA. If transistor beta is 100, base current will be 10 μA. The base current on the input side of the LTP flows through input return resistor R1, resulting in a small positive voltage at the base of Q1. If the base current is 10 μA, the base of Q1 will be at +0.2 V.

Assume that the betas of the LTP transistors are the same, so that 10 μA also flows from Q2 through feedback resistor R3. If the feedback resistor is the same value as the input return resistor, 0.2 V will be dropped across it as well. In this case the output voltage of the amplifier will be zero, as desired (assuming no input offset voltage in the LTP). DC offset due to base current will largely be canceled if the betas of the transistors in the LTP are reasonably matched. If the betas of the LTP transistors are mismatched by 10%, then a net offset of 20 mV will occur. Once again, this is acceptable if it is only multiplied by unity at the amplifier output. Many amplifiers are designed with this approach.

If the amplifier design has R1 and R3 at substantially different values, offset from input bias current can become serious. This could happen if the designer wished to have amplifier input impedance larger than 20 kΩ (which would also allow C1 to be smaller) without compromising noise by increasing the impedance of the feedback network. This is illustrated in simplified form in Figure 10.2. Here the entire forward path of the amplifier is abstracted by an amplifier symbol with the understanding that the input stage is a PNP LTP that sources about 10 μA from each of its input terminals. The diodes are also removed for clarity. A 50 kΩ input return resistor combined with a 20 kΩ feedback resistor will result in a net offset of about 300 mV. This would be completely unacceptable.

A similar problem arises if a DC coupled low impedance feedback network is employed, as in the case when a DC servo is being used. In this case, if R1 is set to 50 kΩ and the electrolytic

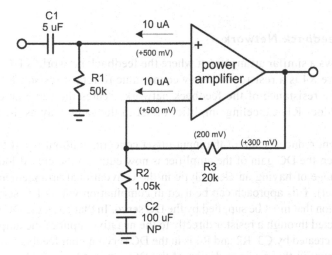

Figure 10.2 Conventional Amplifier DC Bias Arrangement with Higher Input Impedance and Worse DC Offset Performance

capacitor is removed, an input offset of +500 mV will result at the base of Q1. This large offset will have to be opposed and canceled by the DC servo.

Conflicting Impedance Requirements

In amplifiers with AC coupling at the input, the input impedance to ground is simply that of the return resistor R1. This resistor sets the input impedance of the amplifier and also sets the size of the coupling capacitor required to achieve a given low-frequency cutoff. For these reasons, it is desirable for this resistor to be a large value, perhaps on the order of 50 kΩ.

Unfortunately, for reasons of noise and distortion, the feedback resistor R3 should be of a lower resistance than a value like 50 kΩ. Bear in mind that the AC impedance feeding the LTP on the input side is quite low because the input coupling capacitor acts like a short circuit at audio frequencies. Ideally, the AC impedance on the feedback network side should also be low in order to minimize noise. There is thus a conflict between the DC and AC requirements at the input stage. It is also notable that when the feedback network impedance is reduced, the value of the electrolytic capacitor in the feedback shunt path must be increased accordingly.

Bypassed Equalizing Resistor

Some designers have sought to mitigate the impedance conflict by using a slightly different arrangement on the feedback side, as shown in Figure 10.3a. R4 is added in series with the input of the LTP and bypassed. The sum of R3 and R4 is made equal to that of R1, returning DC balance to the arrangement. Here the feedback network can be of arbitrarily low impedance, subject to power dissipation in the network and to the needed size of C2. Bypass capacitor C3 has very little signal current flowing in it because of the high impedance seen looking into the LTP, so this capacitor need not be as large a value as the coupling capacitor used on the input side. It is not expensive to use a high-quality polypropylene capacitor here. Over the years, numerous other techniques have been used to deal with the input offset issues and the size of the capacitor that serves the role of C2 [2–6].

DC-Coupled Feedback Network

Figure 10.3b shows a similar arrangement where the feedback network is DC coupled and C2 is eliminated. Here R4 is a resistor of nearly equal value to return resistor R1. The combination of R4 and the resistance of the feedback network creates the same amount of positive voltage offset as does R1, canceling amplifier offset in the same way as the arrangement of Figure 10.3a.

This arrangement reduces some of the input offset concerns without resort to an electrolytic capacitor. However, the DC gain of the amplifier is now equal to the closed-loop gain, so a big part of the advantage of having an electrolytic in the conventional arrangement is lost (due to input voltage offset). This approach can be used in combination with a DC servo to reduce the amount of correction that must be supplied by the DC servo. In that case, the DC servo correction signal can be injected through a resistor directly to the negative input of the amplifier. However, the pole-zero pair created by C3, R2 and R4 is in the DC servo global feedback path, so this must be taken into account in the dynamic design of the DC servo. Alternatively, the DC servo correction can be applied to the junction of R2 and R3 in the conventional fashion without concern about the pole-zero pair introduced in the arrangement above.

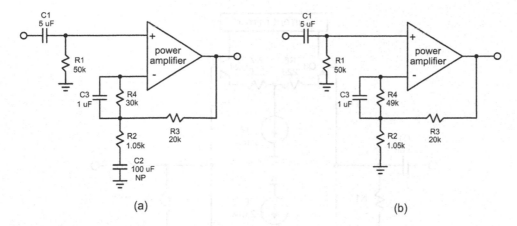

Figure 10.3 (a) Amplifier with Bypassed Equalizing Resistor (b) Amplifier with DC-Coupled Feedback Network

Complementary Input Stages

Many amplifiers with BJT input stages employ the popular complementary differential input stage architecture shown in Figure 10.4. This approach reduces the input bias current problem because the base current flowing in the PNP transistors is of opposite sign to that of the NPN pair, resulting in some cancellation. This reduces the offset problem to the extent that the betas of the PNP and NPN input pairs are well matched. However, if betas are not matched between the NPN and PNP pairs, there may be as much as a 2:1 difference in beta, reducing the advantage gained by this arrangement to a factor of only two compared to a unipolar input stage.

If the NPN and PNP betas are matched to within 10%, this architecture will reduce the offset problem by a factor of 10 as compared to the situation where only a single NPN differential pair is used. Matching incurs extra cost, however. If the net input bias current is 1 μA in Figure 10.4, then output offset will be a tolerable 30 mV.

DC Trim Pots

Regardless of which of the approaches described above is used, some amplifiers end up using a DC trimmer potentiometer to achieve low output offset. This adds cost, and its effectiveness is sometimes temperature-dependent as a result of transistor beta temperature dependence. The correction current can be injected on either side of the input through a large-value resistor. It has been pointed out that R4 in Figures 10.3a and b can be implemented as a trimmer to control DC offset in a relatively temperature-independent way as long as the betas of the input pair of transistors change in the same way. It is a clever use of the BJT input bias current.

JFET Input Stages

Although JFET input stages typically start off with greater input voltage offset than BJT stages (assuming both are dual matched pairs), it should be very apparent by now that their absence of input bias current makes them superior in terms of overall amplifier DC offset. They free

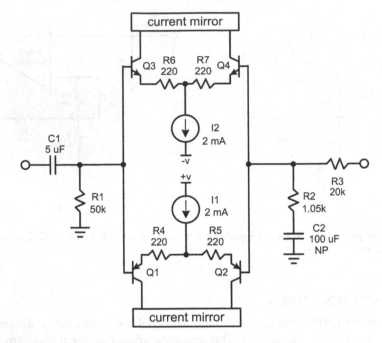

Figure 10.4 The Complementary Input Differential Pair

the designer from trying to balance the DC resistances on the input and feedback sides of the input stage.

10.2 DC Servo Basics

The concept of a DC servo is quite simple. The average DC level at the output is extracted by a low-pass filter, amplified and fed back to the feedback side of the input stage. This drives the output DC value to zero or a very small value. This permits the use of a low-impedance DC-coupled feedback network while retaining high amplifier input impedance. In practice, an integrator is almost always used to provide both the low-pass filtering function and the gain. This is illustrated in Figure 10.5 where the input pair of the amplifier is implemented with JFETs. Without the DC servo, output offset would be 200 mV with a JFET offset of 10 mV.

The amplifier output is applied to a conventional inverting integrator followed by a unity-gain inverter to provide the proper feedback polarity. The integrator input resistor R4 is chosen to be 1 MΩ, while the integrator capacitor C3 is set to 1 µF. The integrator is usually implemented with a JFET op amp to avoid integrator offsets created by input bias current. The integrator inputs are protected from excessive input voltages by diodes D1 and D2. The servo output from the inverter is applied to the feedback input of the amplifier input stage. This effectively creates an auxiliary feedback loop that is active at DC and very low frequencies.

If there exists a small positive average DC value at the amplifier output, integrator capacitor C2 will charge by the current sourced to it through R5, driving the output of the integrator negative. The output of the inverter will go positive and source current to the feedback input to drive the feedback input in a positive direction. This in turn will drive the output of the amplifier

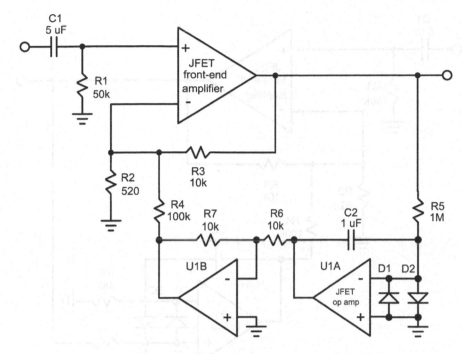

Figure 10.5 A Power Amplifier with a Simple DC Servo Using an Inverting Integrator

negative. The very high DC gain of the integrator forces the output of the amplifier to essentially zero. In practice, it forces the output voltage to equal the input offset voltage of the integrator op amp in the absence of input bias current. This will typically be less than ±10 mV for a JFET op amp.

On paper the complexity of the amplifier is higher, but cost and space for the same quality is lower. The only major components are a dual op amp and a film integrating capacitor. The output of the servo drives the amplifier's feedback input node through a fairly high-value resistor (R4) because it needs only to inject enough correction current to overcome the maximum anticipated input-referred offset error. The large resistor tends to reduce the ability of the servo and its op amp to adversely impact sound quality via noise or distortion in the servo. As discussed below, caution is required to avoid setting R4 too high.

The servo provides increased negative feedback as frequency goes lower. As such it does indeed introduce a high-pass filter function into the audio path, but so did the simple electrolytic in the feedback return leg. With the DC servo, however, you have now removed an evil 100-μF electrolytic that would have been bad for the sound even if bypassed by a smaller film capacitor.

DC Servo Architectures

Figure 10.6 shows a non-inverting integrator that requires only one op amp. It is like a single op-amp differential amplifier but with the feedback and shunt resistors replaced with capacitors to make it into an integrator. It requires two capacitors, and this is a disadvantage.

I prefer DC servos that employ a dual op amp and only require a single integrating capacitor, perhaps on the order of 1 μF. It is quite economical to employ a high-quality 1-μF film capacitor. Dual op amps that are of high quality are also relatively inexpensive.

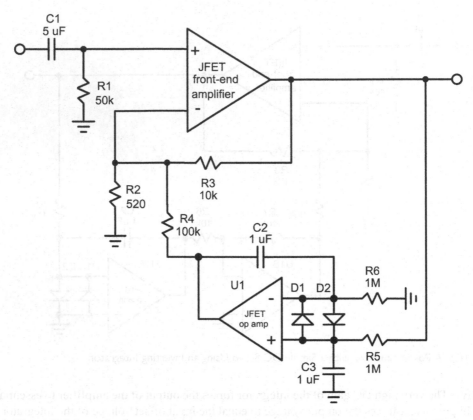

Figure 10.6 A DC Servo Implemented with a Non-Inverting Integrator Requiring Only One Op Amp but Two Integrator Capacitors

Setting the Low-Frequency Corner

Referring to Figure 10.5, consider the way in which the DC servo establishes the LF 3-dB point for the overall amplifier, excluding the high-pass filter at 0.6 Hz formed by the input coupling capacitor. The LF corner will occur at the frequency where the amount of feedback provided through the servo circuit equals that provided through the feedback network. Because R4 is 10 times the value of feedback resistor R3, and the gain of inverter U1B is unity, this will occur at the frequency where the integrator gain is ten. This in turn will be at the frequency where the impedance of the capacitor is 10 times that of R5, which in this case is 0.015 Hz. The gain of the servo to the output of the inverter is only 0.1 at 1.5 Hz. This means that a 1.5-Hz, 40-V peak test signal at the output of the amplifier (100-W/8-Ω) will produce a signal of 4 V peak at the output of the DC servo op amps. They will have more than adequate margin against clipping, even at this very low frequency, if they are powered from ±15-V supplies.

Amount of Offset to Be Corrected

If a DC servo is used with a low-impedance DC-coupled feedback network, then the DC servo must correct almost the full amount of offset created by input bias current flowing in the input return resistor. Consider the case above where a single NPN differential pair is used with transistors biased at 1 mA and having beta equal to 100. Input bias current will be 10 μA, dropping 500 mV across input return resistor R1 and 100 mV across feedback resistor R2, for a difference of 400 mV. The DC servo must be able to create a full −0.4 V of compensating offset on the feedback side of the LTP. This large amount of required correction is due to the choice of having a high amplifier input impedance of 50 kΩ while having a low feedback resistor value of only 10 kΩ, all in combination with input bias current of 10 μA. The high input impedance enabled a nice low 0.6-Hz input corner frequency using only 5 μF for the coupling capacitor (small enough to use a quality film capacitor instead of resorting to an electrolytic). This also lightens the load on the source preamplifier, which probably has its own output coupling capacitor. The low impedance of the feedback network helped keep noise down. Here, there is an advantage in using JFETs for the LTP because the input offset due to base current is eliminated.

This amount of compensating offset is substantial. This offset needs to be taken into account in determining how much correction range the servo will need to be able to apply. In the example of Figure 10.5, if the op amp can provide ±14 V, the servo will be able to provide only ±70 mV of compensating offset. This is inadequate. This would require that R4 be reduced to less than 20 kΩ in order to provide adequate control range. Any noise and distortion at the output of U1B would then be attenuated by only 6 dB to the output of the amplifier. This underscores the fact that the DC servo *is* in the signal path and that its audio quality cannot be ignored.

The trick shown in Figure 10.3b can help. It adds a DC balancing resistor between the feedback shunt resistor and the input of the BJT input stage. The value of this resistor is chosen to be that of the return resistor less the DC resistance of the feedback network. The added resistance will equalize the offset due to input bias current if betas for the input pair are the same. This resistor can be bypassed with a 1-μF film capacitor, since signal current flow in it is very small. This will reduce the required DC servo control range when BJT input stages are used. If beta for the input pair is matched to ±10%, then an increased current of 1 μA flowing in one of the 50-kΩ arms results in a 50 mV offset that can be accommodated by the circuit of Figure 10.5.

Servo Control Range

My philosophy in applying a DC servo is that it should do as little as possible to the amplifier circuit, and its effect should be as subtle as possible. The DC servo is there to eliminate the *natural* offset of the amplifier that would exist if a capacitor in the feedback return leg were not used. The servo should not be more powerful than is necessary to do this job with some margin. This philosophy argues for the use of a servo injection resistor with the highest value that is consistent with meeting all servo performance requirements. The 100-kΩ injection resistor in Figure 10.5 meets this objective, but limits servo control range to only ±70 mV.

Servo Clipping

The signal voltage at the output of the DC servo will increase as frequency decreases. If the amplifier is subjected to full-power subsonic signals during testing, it is possible that the output of the DC servo will clip. For this reason, the gain of the DC servo at low frequencies must be kept in mind. If that gain is set too high so as to allow a larger value of injection resistor, then the clipping point might be reached at frequencies that are not sufficiently below the audio band. The tendency to clip is also governed by the chosen low-frequency cutoff for which the servo is designed. Servos designed to yield a higher cutoff frequency will tend to clip at a higher frequency. A good rule of thumb is that the servo should never clip at a frequency higher than 10 Hz when the amplifier is operating at full power.

The DC servo described above having a 0.015-Hz bandwidth will produce a peak signal swing of 4 V when the amplifier is producing a 40-V peak at 1.6 Hz (corresponding to 100-W/8-Ω). A 400-W/8-Ω amplifier will only produce an 8-V peak output from the servo at 1.6 Hz, indicating that this servo design is adequate for such an amplifier, but there is a caveat.

Servo Headroom

If the servo op amp can produce ±14 V, it will be able to counteract ±70 mV of offset at the input. If a BJT input stage is being used with 10 µA of input bias current flowing through a 50-kΩ return resistor, and even using the offset balancing circuit of Figure 10.3b with ±10% beta matching, ±50 mV of correction is needed. The servo can handle this, but it will exhibit a constant output voltage of up to ±10 V under quiescent conditions. This asymmetry will eat into the clipping headroom of the servo on low-frequency signals, leaving only 4 V on the negative side, just enough to handle 40 V peak at the amplifier output at 1.6 Hz. This is an important consideration in the design of the servo.

The JFET Advantage

While some prefer JFET input pairs for reasons of sound quality and EMI resistance, there is another reason that makes them attractive. The JFET input stage does not suffer from DC offset caused by input bias current, and that makes the job of the DC servo much easier.

If an amplifier is to have reasonably high input impedance, its input return resistor must be large (at least 20 kΩ). BJT input bias current flowing through this resistor will cause a far larger offset than the input voltage offset of the input pair or of input-referred offset from the VAS stage. If a JFET input stage is used instead, there is no DC offset from input bias current, and the servo need only compensate for the 5–15 mV of offset of a dual-matched monolithic JFET pair. Correcting ±10 mV of JFET input offset in the circuit of Figure 10.5 requires the servo op amp to produce only ±2 V. The bottom line here is that the servo must typically work much harder when used in a typical DC servo arrangement with BJT input pairs as opposed to JFET input pairs. Authors who complain about the larger input offset voltages of JFETs as compared to BJTs appear to be missing the bigger picture. In fairness to BJTs, a monolithic JFET pair may cost a few dollars, while a pair of discrete BJTs costs a few pennies. Alternatively, matching a pair of discrete JFETs adds some labor cost.

10.3 The Servo Is in the Signal Path

The DC servo in Figure 10.5 is in the feedback path with modest gain at low frequencies. As such, the DC servo is in the signal path of the amplifier, and its performance can affect sound quality. The fact that it is injecting a signal at the input stage gives it opportunity to inject noise and distortion into the signal path. This can influence the quality of the audio signal, but is still better than having an electrolytic in the signal path.

For this reason, audio-grade op amps should be used for the DC servo's integrator and inverter. Pay attention to the noise and class B crossover distortion created by the op amp and consider pulling its output down to the −15-V rail with a resistor. This will force its output stage to operate in class A [7]. Design the servo as if it were part of a quality IC-based preamp. Provide a good clean power supply to the op amp and use a quality integrating capacitor (e.g., polypropylene film) [1]. The capacitor can be of slightly lower quality than those in the main signal path of the amplifier only because its output is attenuated before being applied to the input circuit.

Servo Op Amp Distortion and Noise

If the output of the servo is attenuated by 100:1 before application to the amplifier input and the amplifier has a closed-loop gain of 20, then the attenuation of the servo output to the output of the amplifier is only 5:1, or 14 dB. Put differently, if the servo injection resistor is the same value as the feedback resistor, then there is unity gain from the output of the DC servo to the output of the amplifier.

There is thus flat gain (albeit usually less than unity) from the output of the servo to the output of the amplifier. This will depend on the ratio of the amplifier feedback resistor to the servo injection resistor. In some designs, this gain can be near unity. Given the fact that the output of the servo will be reproducing a strongly low-pass-filtered version of the output signal at some amplitude, it is possible for the servo op amp to create distortion that will make its way to the output of the amplifier. Indeed, if the servo op amp's output is hovering around zero, it could be experiencing some crossover distortion from the class B output stage of the op amp. As mentioned above, this can be minimized by using a pull-down resistor on the output of the op amp to force its output stage into class A operation [7].

Noise created by the integrator and inverter op amps will be transported to the amplifier output with the gain mentioned above. The noise will also be influenced by the size of the integrator input resistor and the size of the integrator capacitor. Using a large input resistor, like 1 MΩ, allows the use of a much smaller integrator capacitor, sometimes as small as 0.1 µF. This takes up less space and is less expensive to obtain in a high-quality capacitor. The price paid is increased servo noise. This noise increases at low frequencies, however, so its sonic effect is limited. It can be thought of as being akin to *1/f* noise. If high impedances are used in the integrator, it is especially important to employ a JFET op amp for the integrator for two reasons. First, input base current of a bipolar op amp will create an undesirable DC offset voltage across the integrator input resistor. Second, the input noise current of a BJT op amp will cause significant noise with such a high-impedance source. The JFET op amp should be a low-noise design.

Beware of well-intentioned efforts to reduce servo noise and distortion. Some can result in instability, some can result in frequency response anomalies and some can result in servo clipping. The very best way to reduce injected servo noise and distortion is to use high-quality parts and audio design practices in the servo.

Adding a Second Pole

The integrator in the DC servo may not be a perfect integrator at all frequencies. This can lead to some high-frequency program material or interference sneaking through the integrator. This leakage can be reduced by adding a passive low-pass filter at the output of the servo, as shown in Figure 10.7. This can be done by splitting the servo injection resistor and taking a capacitor from the junction to ground. This is implemented by R8 and C3, placing a pole at about 16 Hz. The idea is to further keep noise and distortion from the servo op amp out of the signal path.

However, this technique can have subtle effects on the low-frequency response if it is not implemented with care. This is because the frequency-dependent impedance seen looking back into the servo injection network acts as though it is in parallel with the main negative feedback network shunt resistor to ground (R2). The low-frequency response step created in the design of Figure 10.7 is only 0.005 dB. The added capacitor is also somewhat in the effective signal path, so its quality matters.

A different approach is shown in Figure 10.8. It can be implemented in the case of a servo with an inverting integrator followed by an inverter, as in Figure 10.5. Some capacitance can be put across the inverter's feedback resistor, giving the inverter a low-pass response. Here the second pole is implemented by C3 working against R7, producing a pole at 1.6 Hz. This approach

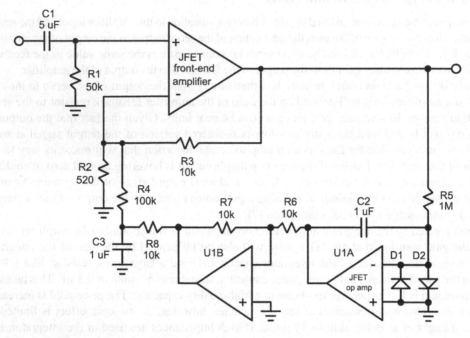

Figure 10.7 DC Servo with Additional Low-Pass Filter

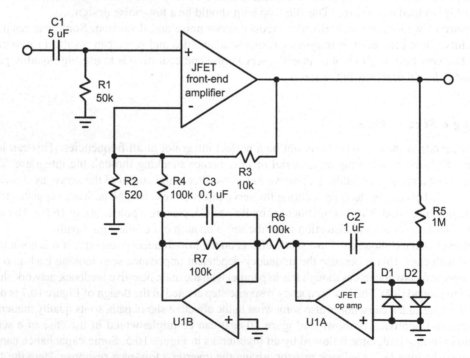

Figure 10.8 DC Servo with LPF Implemented with a Capacitor Around the Inverter

still leaves the door open a crack for noise and high-frequency sneak-through from the inverter, however.

Adding a second pole to a feedback loop usually invites instability or frequency response peaking if it is not done carefully. Figure 10.9 shows the frequency response of a poorly designed servo. In order to avoid this, the added pole should be well above the servo bandwidth frequency, perhaps by a decade. This constraint is easily satisfied in both of the designs above where the servo bandwidth is only 0.016 Hz.

10.4 DC Offset Detection and Protection

The servo's integrator output provides a convenient monitoring point for the DC health of the amplifier. If DC persists at a certain voltage level over a certain period of time, the servo integrator will build up a large voltage to try to counteract the DC offset. This voltage can be fed to a window detector to open the speaker relay or otherwise engage protection circuits. The window detector can be constructed from two comparators, each fed an appropriate threshold voltage.

As discussed earlier, there may often be a considerable servo correction voltage being delivered to the amplifier, especially in the case of amplifiers with BJT input stages. This generally means that the window detector should just be set to a range that will detect when the servo is near clipping. Alternatively, D1 or D2 will conduct when there is excessive DC offset that cannot be corrected by the DC servo. In principle, D1 and D2 can be replaced with transistors. Their collector current can be sensed as an indicator of excessive offset. The potentially significant time delay incurred when using the DC servo for offset detection must be considered, however. If the amplifier output goes to 40 V DC, 40 μA will be sourced to the integrator of Figure 10.5, and the servo will clip in about 350 ms.

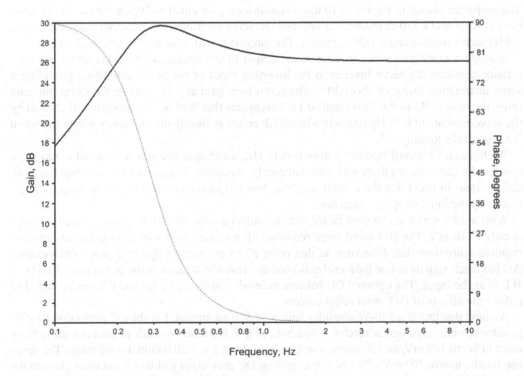

Figure 10.9 Frequency Response with a Poorly Designed DC Servo

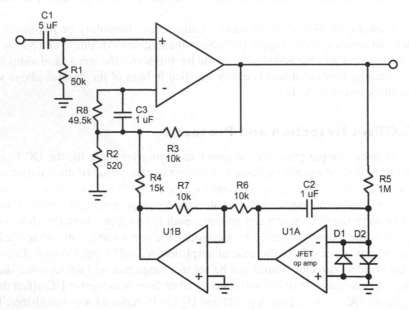

Figure 10.10 Example Power Amplifier with a DC Servo

10.5 DC Servo Example

The amplifier shown in Figure 10.10 has a closed-loop gain equal to 20, set with a 10-kΩ feedback resistor and a 520-Ω feedback shunt. The DC servo uses an inverting servo integrator with a 1-MΩ series resistor and a 1-μF capacitor. The integrator will have a gain of 0.16 at 1 Hz.

Assume a unity gain servo inverter at the output of the integrator. A 15-kΩ servo injection resistor connects the servo inverter to the inverting input of the power amplifier, providing a servo attenuation factor of about 30:1. The servo loop gain at 1 Hz will be the integrator gain times the ratio of R3 to R4. This equals 0.11. This means that the low-frequency cutoff created by the servo is at about 0.11 Hz (the servo-based LF cutoff is simply the frequency where the servo loop gain falls to unity).

We have an LF cutoff frequency of only 0.11 Hz, not simply because we wanted to really go low, but because the numbers will work out nicely, especially in terms of servo signal-handling ability. Bear in mind that the overall amplifier low-frequency cutoff will still be largely determined by the input coupling capacitor.

With a 30:1 servo attenuation factor and op amps capable of ±14 V, servo correction range is only ±470 mV. The BJT input stage requiring 10 μA combined with a 50-kΩ return resistor requires more than that. Moreover, at that point all of the servo's dynamic range for handling AC feedback signals at low frequencies is used up. This servo requires the use of high-β BJTs or JFETs at the input. The optional DC balance network consisting of R8 and C3 can be added to reduce the effects of BJT input offset current.

Assume that this is a 100-W amplifier with a 40-V peak output. Let the DC protection be triggered when the servo output reaches a threshold of 6 V. If the servo needs to correct a fairly large input offset of 100 mV, the servo output only goes to about 3 V, well within its safe range. The servo can handle nearly 200 mV offset before triggering DC protection and is still not even close to the output limits of the integrator op amp. However, LF signal swings have yet to be taken into account.

With a full-power (40-V peak) 10-Hz sinusoid produced by the amplifier, the sinusoidal output of the servo integrator will only be about a 0.6-V peak. It will be 10 times this if the input signal is at 1 Hz and will just trigger the DC protection (in the absence of offset correction). If the amplifier fails shorted to a rail and produces 40 V of DC, the servo integrator output will rise to the 6-V protection trigger point within 0.25 second.

If we want to reduce servo influence at the expense of offset that can be handled, we can simply increase the servo injection resistance from 15 kΩ to 50 kΩ. This will reduce by 10 dB the opportunity for servo noise and distortion to get into the signal path. This will also reduce the LF cutoff to 0.03 Hz. As long as the amplifier design and architecture does not cause a turn-on settling issue that must be corrected, there is likely no problem with this lower LF cutoff frequency. The peak voltage at the output of the integrator will increase by nearly a factor of 3 for a given amount of offset correction. This may increase the chances of triggering DC protection circuits. This approach is best suited to a JFET input LTP so that base current offsets are absent.

If for some reason we want to go to a higher LF cutoff frequency, we must increase servo loop gain. We can do this by reducing the 15-kΩ servo injection resistance at the expense of greater opportunity for servo garbage to get into the signal path. Alternatively, we can achieve the same result by increasing the integrator gain. Changing to a 0.1-μF integrator capacitor will increase the integrator gain by a factor of 10 and bring the LF cutoff up to 1 Hz. Unfortunately, the servo is now more easily overloaded by large low-frequency signals at the amplifier output. A full-amplitude 10-Hz signal at the output of the amplifier will now cause triggering of the DC protection. This is just a design decision and it may be OK. Many audiophiles would not want a high-level signal at 10 Hz to drive their loudspeakers anyway. There are many possible variations to this kind of servo, but this is a good illustration of typical operation and trade-offs.

10.6 Eliminating the Input Coupling Capacitor

A system signal path often contains more coupling capacitors than necessary. There will usually be one at the output of the preamplifier, and yet another one at the input of the power amplifier. If the power amplifier has a DC servo, the input coupling capacitor of the amplifier can be bypassed, knowing that the output of the preamp is probably at or very close to zero. This is no more risky than using a DC-coupled power amplifier. Any small DC offset present at the input of the amplifier will be handled by the DC servo. In this case, an overall improvement in low-frequency transient response will have been had by the use of a servo. Removal of the input coupling capacitor is strictly a choice in regard to managing risk. Many DC-coupled power amplifiers have the option of switching in a blocking capacitor at the input. This is a good idea that can be applied in the general case if one uses a servo in the power amplifier.

10.7 DC Servo Design Issues and Nuances

It is easy for even experienced designers to fall into a trap when it comes to servo design. Just because the servo may not need to do much correction in some designs, do not make the mistake of making the servo strength too small. If you do, the integrator gain will be necessarily higher, all else remaining equal, and the servo will be vulnerable to clipping on large subsonic signals.

Servo Start-Up Transients

The servo will often be designed to have a very low cutoff frequency. This reflects the fact that it should act largely as a long-term automatic screwdriver adjustment. The integrator capacitor will start out at zero on power-up, so initially there will be no servo correction. This means that the power-on transient will include the uncorrected offset of the amplifier, which will often be dominated by the offset induced by base current in the case of a BJT input stage. A power-on delay implemented with a speaker relay will eliminate this.

Low-Frequency Testing of Amplifiers Employing Servos

Full-power testing of power amplifiers at frequencies well below 20 Hz may cause problems with servo clipping or with servo-based DC protection in some designs. This is because the servo is merely an integrator fed with the output signal of the amplifier. The integrator gain increases as frequency decreases. Consider an amplifier producing a peak output of 45 V. At a frequency where the integrator gain has risen to 1/3, the signal at the output of the integrator will be 15 V peak, enough to clip the integrator op amp.

In any amplifier, but especially one using a DC servo, do not overlook THD testing at low frequencies, like 20 Hz or 50 Hz. It is sometimes too easy to become complacent about amplifier distortion at low frequencies because of the high amount of negative feedback normally available to reduce distortion at low frequencies.

Simulation

It is wise to conduct a SPICE simulation of all DC servo designs to verify behavior. This is especially the case when verifying the LF amplifier corner and any frequency response peaking that might occur. This can be done with an AC simulation.

The issue of unbalanced LTP return resistances and the resulting need for correcting servo action can be evaluated with a DC simulation. Static offset at the output of the servo should always be evaluated. Bear in mind that the current gain of all devices of the same type will be the same by default in a SPICE simulation.

References

1. Walter G. Jung and Richard Marsh, "Picking Capacitors," *Audio*, pp. 50–63, March 1980.
2. James Bongiorno, "High Voltage Amp Design," *Audio*, pp. 30–36, February 1974.
3. "SAE MKIIICM Basic Amplifier," *Audio*, pp. 61–65, January 1975 and p. 69, August 1975.
4. "G.A.S. Ampzilla Basic Power Amplifier," *Audio*, pp. 57–60, September 1975.
5. "DB Systems Model DB-6 Stereo Power Amplifier," *Audio*, pp. 102–108, January 1980.
6. W. Marshall Leach, "Build a Double Barreled Amplifier," *Audio*, pp. 36–51, April 1980 and pp. 58–60, May.
7. Gionvanni Stocchino, "Reducing Op-amp Crossover Distortion," *Wireless World*, p. 35, April 1984.

Chapter 11

Advanced Forms of Feedback Compensation

11. INTRODUCTION

The conventional Miller compensation described in Chapter 6 is by far the most widely used form of frequency compensation for amplifiers employing negative feedback. It is simple and reliable. However, as we will see here, there are alternatives to, and variants of, Miller compensation that can provide substantial performance improvements. This chapter takes up where Chapter 6 left off.

The following are some of the reasons for choosing advanced forms of compensation:

- Improved slew rate
- Greater amounts of in-band negative feedback
- Increased negative feedback at high frequencies
- Addition of local feedback around the output stage
- Inclusion of the input stage in local feedback loops

Miller compensation sets a strict relationship that limits slew rate in accordance with closed-loop gain, input stage transconductance and gain crossover frequency. Some forms of advanced compensation allow this limitation to be avoided. A greater amount of in-band negative feedback reduces in-band IM products, even if it does not reduce further the amount of higher-harmonic high-frequency out-of-band distortion. Some advanced compensation techniques provide greater in-band feedback without a corresponding increase in high-frequency feedback that might compromise stability.

Increased feedback at high audio frequencies like 20 kHz is very desirable because some circuits create more distortion at high frequencies, like crossover distortion. Compensation techniques that permit greater negative feedback at 20 kHz without requiring a higher gain crossover frequency are valuable. The output stage is a major contributor to distortion, and some advanced compensation techniques permit the introduction of additional local negative feedback around the output stage to reduce its distortion. The input stage is a source of distortion in many amplifiers, yet global negative feedback is not fully effective in reducing some forms of input stage distortion, such as common-mode distortion. Some advanced compensation techniques enclose the input stage in a local feedback loop to increase its dynamic range and reduce such distortion.

11.1 Understanding Stability Issues

The general concepts of feedback stability were addressed in Chapter 6, and we will review them only briefly here. The main point to take away is that the gain and phase response of the feedback loop must be tailored in such a way that adequate gain margin and phase margin are achieved under all conditions.

The most significant influence on stability is the choice of the gain crossover frequency, f_c. This is often referred to as the *unity-loop-gain-frequency* (ULGF). If f_c is chosen too high, the accumulation of *excess phase* with increasing frequency will reduce phase margin to the point where instability is likely. If it is chosen too low, slew rate will be reduced and high-frequency distortion will be increased as a result of there being less feedback at frequencies like 20 kHz.

The open-loop gain and phase will often be a function of signal voltage and current. For example, f_T droop in the output transistors can reduce phase margin under conditions of high current. For this reason it is important that the feedback compensation be sufficiently conservative that adequate phase margin is preserved over all signal swing conditions.

The load placed on an amplifier can have a profound influence on stability because it affects the open-loop gain and phase of the amplifier. Even though the open-loop output impedance is generally thought to be small at high frequencies, declining VAS transistor AC beta at high frequencies can cause it to increase. Heavier loads in general tend to exacerbate the creation of excess phase in the output stage. This is especially so for capacitive loads whose impedance becomes small at high frequencies.

Some feedback compensation schemes are stable only for a range of open-loop gain and phase characteristics, and even become unstable if open-loop gain decreases. These arrangements are referred to as having conditional stability. If an amplifier breaks into oscillation when it is turned off, this is an example of conditional stability.

Finally, some advanced feedback compensation schemes involve feedback loops themselves that may extend over more than one stage in the open-loop amplifier. The stability of these local compensation loops must also be assured. These loops often involve much higher gain crossover frequencies. The conventional Miller compensation scheme involves a shunt feedback loop of very wide bandwidth, but it is usually very stable because it only encircles one or two transistors.

Dominant Pole Compensation

Most conventional feedback amplifiers depend on frequency compensation in which a single-pole roll-off dominates the gain and phase characteristics of the loop gain. If there really is only one pole in the loop, the phase shift around the loop at high frequencies will be 90 degrees up to extremely high frequencies. This means that the phase margin will be 90 degrees and that the gain margin will be an unlimited number of decibels. As explained in Chapter 6, the non-dominant poles at higher frequencies ultimately contribute excess phase that increases with frequency, detracting from phase margin and leading to instability if the gain crossover frequency is set too high. The capacitor that establishes the dominant pole is sometimes referred to as C_{dom}.

11.2 Miller Compensation

Miller compensation is the most common form of dominant pole compensation employed in audio amplifiers. It is named in connection with the Miller effect and Miller integrators. As explained in Chapter 6, Miller compensation is implemented by a negative feedback process and, as a result, enjoys inherent advantages. In particular, it makes use of all of the gain for distortion reduction. The shunt feedback process in Miller compensation increases local feedback around the VAS as it decreases global feedback with increasing frequency. This acts to further linearize the VAS at high frequencies. Shunt feedback by its nature decreases the output impedance of an amplifier stage. This means that the output impedance of the Miller-compensated VAS decreases with frequency, making the VAS gain and phase more immune to loading effects from the output stage.

Figure 11.1 shows an amplifier with Miller compensation. Throughout this chapter the IPS will be assumed to be an LTP that is simply characterized by a transconductance of 4 mS. In the

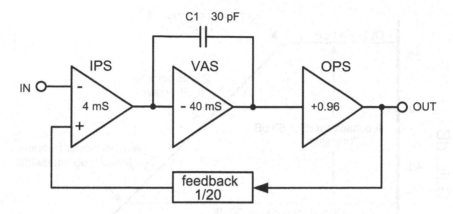

Figure 11.1 A Miller-Compensated Amplifier

arrangement of Figure 11.1, all of the signal current produced by the LTP will be assumed to flow into C1. Capacitor C1 provides the dominant pole, so here it serves the function of C_{dom}.

Since C1 provides compensation by acting as a Miller integrator capacitor, we refer to it as C_M. The VAS will be assumed to be a common-emitter stage characterized by a transconductance of 40 mS. The OPS will be assumed to be an emitter follower with nearly unity voltage gain and high current gain. The VAS in such a design is essentially a Miller integrator that is characterized by a 6 dB per octave roll-off over a very wide range of frequencies. If the loop gain is 0 dB at the 1-MHz gain crossover frequency, it will be 40 dB at 10 kHz and 60 dB at 1 kHz, assuming that there is enough DC gain in the real circuit to support this amount of loop gain.

As explained in Chapter 2, the value of C1 in combination with the IPS transconductance determines the open-loop gain as a function of frequency. The open-loop gain is simply $gm * X_{C1}$, where X_{C1} is the impedance of C1 at a given frequency. The gain crossover frequency f_c is determined by the open-loop gain A_{ol} and the closed-loop gain A_{cl}. The frequency where A_{ol} falls to equal A_{cl} is the gain crossover frequency f_c, (ULGF). These relationships are illustrated by the Bode plot in Figure 11.2. It follows that the value of C1 to set f_c is

$$C1 = gm/\left(2\pi * A_{cl} * f_c\right) \qquad (11.1)$$

The circuit of Figure 11.1 is designed to have $f_c = 1$ MHz, making the required value for C1 equal to 30 pF.

Pole Splitting

The output of the IPS is a high-impedance point. The output of the VAS is also a high-impedance point. In the absence of Miller compensation shunt feedback, both of these nodes can form poles. Both of those poles can lie at a similar frequency. Such a two-pole situation in the open-loop amplifier will greatly jeopardize stability because both poles will contribute nearly 90 degrees of lagging phase shift at frequencies well above the pole frequencies but probably below the gain crossover frequency.

The shunt feedback provided by the Miller compensation causes what is called *pole splitting* to occur. Of the two poles that would have been present without Miller feedback, one pole is moved far down in frequency and the other pole is moved far up in frequency. This is why the process is called *pole splitting*; the frequencies of the poles are split to become far away from each other.

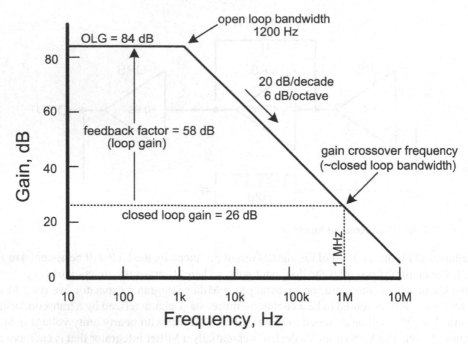

Figure 11.2 Bode Diagram for the Amplifier of Figure 11.1

The pole that is pushed to a high frequency is usually pushed to a frequency that lies well above the gain crossover frequency. The result is an amplifier stage characterized by a 6-dB per octave roll-off that begins at a very low frequency and continues over many decades to a very high frequency, without the lagging phase contribution exceeding 90 degrees.

Limitation on Slew Rate

As explained in Chapter 6, conventional Miller compensation places a limit on slew rate (SR) that is closely bound to the gain crossover frequency and the closed-loop gain of the amplifier. The equation for slew rate is

$$SR = 2\pi f_c A_{cl} I_{max} /gm \tag{11.2}$$

The term I_{max}/gm is the ratio of the maximum output current of the IPS to the transconductance of the IPS. Once closed-loop gain and gain crossover frequency are chosen, slew rate is determined by the ratio I_{max}/gm. This number is 52 mV for an un-degenerated BJT differential pair. The number is about 10 times larger for an LTP that has 10:1 emitter degeneration. This is why it is so important to incorporate emitter degeneration into BJT input stages. The number is also about 10 times larger for an un-degenerated JFET differential pair.

An amplifier with $f_c = 1$ MHz and $A_{cl} = 20$ will have a slew rate of only 6.6 V/μs with an un-degenerated BJT LTP. If the LTP is degenerated by a factor of 10, slew rate will rise to a respectable 66 V/μs. Some of the advanced compensation schemes to be discussed in this chapter break that relationship, allowing higher slew rate to be achieved without requiring higher gain crossover frequencies or excessive amounts of IPS degeneration.

Distortion Reduction as a Free Side Benefit

Feedback compensation involves throwing open-loop gain away at high frequencies. It would be nice if this *thrown-away* gain could be put to good use. This is exactly what Miller compensation does. It exchanges high-frequency global feedback for high-frequency local feedback. Miller compensation reduces high-frequency open-loop gain through the use of local feedback that increases with frequency. This local shunt feedback serves to reduce distortion in the VAS. Virtually all of the advanced forms of compensation to be discussed here operate in the same way, linearizing the VAS while reducing gain as frequency increases.

VAS Output Impedance

The output impedance of the VAS is reduced by the effect of the shunt feedback, making the VAS less affected by nonlinearities of the output stage load. The VAS output impedance is estimated by injecting a small voltage at the output node of the VAS and calculating how much change in VAS current will result (with the amplifier operating open-loop). This current will be related to the voltage attenuation ratio of the Miller feedback and the transconductance of the VAS stage.

Consider an amplifier with a Darlington VAS operated at 10 mA and having 10:1 emitter degeneration. Its effective emitter resistance will be approximately 26 Ω and its gm will be about 38 mS. Assume that C_M = 15 pF and that there is a 3.4-kΩ IPS load resistor shunting the input to the VAS. At 1 kHz the impedance of C_M is about 10.6 MΩ. The Miller feedback attenuation ratio will be approximately 3.4 kΩ/10.6 MΩ = 0.00032.

A 1-V change forced on the VAS collector will thus result in a 0.32-mV change at the input to the VAS, creating a 12-mA change as a result of the 38-mS transconductance. The VAS output impedance is thus 1 V/12 μA = 83 kΩ at 1 kHz. This value assumes that there are no other effects causing the output impedance to be lower, such as the Early effect or output stage loading. The estimated output impedance will decrease with frequency as the impedance of C_M decreases with frequency. At 20 kHz the VAS output impedance is about 4 kΩ.

Now assume that the Darlington VAS is driven from an LTP with a current mirror load, as shown in Figure 11.3. The impedance at the input of the VAS will be much higher, on the order of 40 kΩ. At 1 kHz and with a C_M = 30 pF and X_{CM} = 5.3 MΩ, the Miller feedback attenuation factor will be 40 kΩ/5.3 MΩ = 0.0075 and the VAS output impedance will be Z_{out} = 1/(0.0075 * 38 mS) = 3.5 kΩ. This is a remarkable and important reduction from the example above. At 20 kHz the impedance of C_M is about 265 kΩ and the impedance ratio of the divider elements is about 0.15. VAS output impedance will be on the order of 175 Ω.

At very high frequencies, where C_M begins to look like a short, the VAS output impedance will begin to level off at close to 26 Ω, which is 1/gm of the VAS. In these estimates the capacitance at the input node of the VAS is assumed to be small compared to C_M. Capacitance to ground at that node will tend to increase VAS output impedance because it will form a capacitance voltage divider with C_M for the shunt feedback.

The Feed-Forward Zero

At very high frequencies where C_M is essentially a short circuit, it can be seen that the VAS is no longer inverting and that the voltage gain of the VAS is gm_1/gm_2 (non-inverting), where gm_1 is the transconductance of the IPS and gm_2 is the transconductance of the VAS. This means that there is a *right-half-plane* (RHP) zero in the transfer function of the VAS created by feed-forward of current from the IPS through C_M. This zero can reduce phase margin of the global feedback loop. It can be eliminated by placing a resistance in series with C_M whose value is equal to or greater

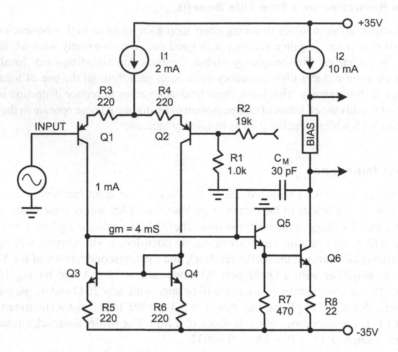

Figure 11.3 Simple IPS-VAS with a Current Mirror LTP Load

than $1/gm_2$. The zero is located at the frequency where the reactance of C_M equals $1/gm_2$. In the case of $C_M = 30$ pF and $gm_2 = 38$ mS, the zero is at about 200 MHz and is of little concern.

Inserting a Zero to Cancel or Mitigate a Pole

Increasing the resistance in series with the Miller capacitance to a larger value than $1/gm_2$ creates an ordinary zero that can cancel other parasitic poles in the circuit. If this zero is placed well above f_c at a frequency where the first parasitic pole is thought to be located, phase margin can be improved. If the first parasitic pole is thought to lie at 5 MHz and $C_M = 30$ pF, a value for R2 of 1 kΩ might be appropriate. Caution must be exercised with this technique to avoid compromising gain margin. Bear in mind that parasitic poles can move around with operating point, especially with voltage and current operating points in the output stage, and with capacitive loading.

Power Supply Rejection

The Miller compensation capacitor straddles nodes that are referenced to the power supply rail and ground, respectively. Power supply noise thus creates noise current through C_M, resulting in injection of noise into the signal path. This is a potential disadvantage of conventional Miller compensation and calls for extra care in keeping the power supply to the IPS-VAS quiet. In some alternative compensation approaches both ends of the compensation capacitor are referenced to signal ground, removing this potential source of power supply noise ingress.

Buffered Miller Feedback Pick-Off Point

In some cases it is advantageous to tap the Miller feedback from the pre-driver or driver instead of from the collector of the VAS. This reduces the load of the compensating capacitor on the VAS and (for what it is worth) includes the pre-driver in the local feedback loop formed by the Miller compensation. Although it may seem of limited benefit for conventional Miller compensation, it can be more valuable for other compensation schemes that will be discussed.

A key issue with buffered compensation loop pick-off is its effect on pole-splitting and compensation loop stability. The compensation loop now encloses more circuitry and the VAS high-impedance collector circuit is not directly part of that loop and is allowed to have very high impedance. It is tempting to argue that the buffered and un-buffered schemes must behave identically because the buffer has unity gain, so that the signal picked off is essentially identical in both cases. However, the loading of the VAS collector node is not the same, allowing the gain of the loop so formed to be much higher, especially at very high frequencies. As discussed in Chapter 6, the stability of the compensation loop itself must always be considered.

If the VAS high-impedance node is loaded with a replica of what it would have seen, it can be argued that the behaviors may be the same. What is the net benefit, then? One can argue that the replica load need not be as heavy as the load that the actual compensation network would have placed on it; it only needs to be heavy enough to adequately limit the high-frequency loop gain of the compensation loop.

Indeed, one could even argue that the light replica load could be brought back to the input of the VAS, so that at very high frequencies the architecture would devolve to that of a conventional Miller compensation approach.

The Summing Node Pole

The resistance of the conventional feedback network in combination with the capacitance at the input of the LTP forms a pole that should not be neglected. If the resistance of the feedback network at the summing node is 1 kΩ and the capacitance seen looking into the LTP is 5 pF, a pole will be formed at about 32 MHz. This is not too bad, but the effective input capacitance of the LTP may be more than 5 pF in some cases. For example, if a JFET complementary differential pair input stage employs the 2SK389/2SK170 pair, each with typical gate-drain capacitance of 6 pF, the total capacitance seen will be 12 pF (even ignoring impedance effects of gate-source capacitance). Moreover, in the open loop, the IPS has significant gain, so the Miller effect created by C_{bc} or C_{gd} of the IPS transistors may create a more significant pole. This will not be an issue if the input stage is cascoded. In any case these considerations argue for the use of a low-impedance feedback network.

Another pole can sneak into the summing operation if the source impedance to the amplifier input side of the LTP is significant at high frequencies. For this reason it is usually a good idea to position the usual input LPF shunt capacitor electrically close to the input base of the LTP (I always separate it by at least 100 Ω for the sake of HF stability, however).

Sometimes a small capacitor is placed across the feedback resistor as part of the compensation scheme. This capacitor creates a zero that contributes leading phase shift that will oppose some of the lagging phase shift in the forward path. That is why this is often called a "lead" capacitor. The leading phase shift that it introduces can improve phase margin. If this capacitor in combination with the input capacitance of the IPS is thought of as implementing a capacitance voltage divider, the summing node pole can be largely eliminated and the response of the feedback network can be made flat to many tens of MHz. This usually requires a capacitor of quite small value, sometimes less than 0.6 pF. SPICE simulations should be used to evaluate the response of the feedback network *in situ* in a loop gain simulation. Such simulations show that the situation is not as simple as a capacitance voltage divider because the effective input impedance is not usually well approximated by a simple capacitance, especially when emitter degeneration is used.

In the amplifier of Chapter 4, use of a 0.6 pF capacitor across the feedback network flattened the feedback network response to 20 MHz, but allowed the response to increase by 10 dB at 100 MHz, inviting EMI ingress to the input stage from the external speaker cable. Placing a 15-kΩ resistor in series with the capacitor made the network response flat within ±1 dB to 100 MHz. This improved phase margin by 5 degrees at 1 MHz, and would have permitted ULGF to be increased to 2 MHz with 67 degrees of phase margin and 8 dB of gain margin. Increasing the capacitor to 2 pF buys 15 degrees of leading phase shift at 2 MHz, but increases loop gain by 5 dB at 6 MHz, eating into gain margin.

11.3 Miller Input Compensation

Some amplifiers incorporate what is called *input compensation*, as illustrated in Figure 11.4. This is usually in the form of dominant-pole lag compensation placed across the input of the IPS. This form of compensation has the advantage that it does not suffer the bond between slew rate and gain crossover frequency that conventional Miller compensation introduces. Unfortunately, it can have detrimental effects on input-referred noise and input impedances. It also does not provide the benefit of pole-splitting.

Notice that this approach allows there to be poles at fairly low frequencies at both the input and output of the VAS. R3 can be used to insert a zero that will cancel one of those poles. Other measures may have to be taken to push the remaining pole to a high enough frequency. This kind of input compensation is not recommended for audio amplifiers.

Combining the Best of Input and Miller Compensation

Miller input compensation (MIC) implements input compensation by means of negative feedback to the input stage [1]. It provides many advantages analogous to those that Miller compensation provides over simple shunt lag compensation. Figure 11.5 illustrates an amplifier with Miller input compensation. Instead of routing the compensation capacitor back to the input of the VAS, it is routed all the way back to the input of the IPS. This encloses the input stage in the wideband compensation loop, reducing its distortion and increasing its dynamic range. For this reason, it breaks the relationship between gain crossover frequency and slew rate.

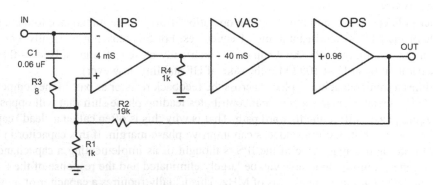

Figure 11.4 Amplifier with Input Compensation

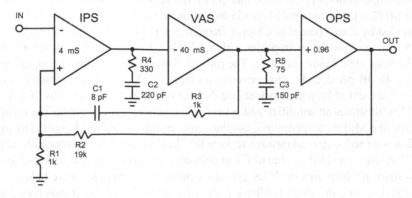

Figure 11.5 Amplifier with Miller Input Compensation

The 50-W amplifier in Reference 1 achieved a slew rate of 300 V/μs using this compensation technique. The gain crossover frequency f_c in this scheme is the frequency where $X_{C1} = R2$. This frequency is set to 1 MHz in Figure 11.5.

Compensating the Compensation Loop

The Miller compensation loop in Figure 11.5 spans several stages. These stages are implemented with fairly fast small-signal transistors, so the gain crossover frequency of the compensation loop can be made fairly high, perhaps on the order of 10–20 MHz. Nevertheless, that loop itself must be compensated. There still exists the high-impedance intermediate node at the input of the VAS where a pole can contribute instability to the local loop (which is not so local anymore). For this reason the series R-C network consisting of C2 and R4 is added at this intermediate node.

In some designs a second series R-C network (C3, R5) is added shunting the output node of the VAS. This reduces the proportion of current at high frequencies that can pass through C1 to the IPS input node, reducing loop gain of the compensation loop. Compensating the compensation loop can be difficult and requires a good deal of experimentation and simulation.

Notice that R3 places a zero in the open-loop roll-off at a frequency above f_c that can cancel some excess phase in the output stage. However, it also transforms the compensation loop to a flat-gain amplifier at higher frequencies. That gain will be R3/(R1 + R3).

11.4 Two-Pole Compensation

Bode showed that optimal compensation might be had with a 9-dB per octave roll-off (30 dB per decade) instead of a 6-dB per octave roll-off (20 dB per decade). In this case, the phase margin would still be a respectable 45° and yet the steeper roll-off would permit much higher gain in the audio band for a given gain crossover frequency. Consider an amplifier with a 2-MHz gain crossover frequency. There are two decades between 20 kHz and 2 MHz. With a conventional 20 dB per decade loop gain roll-off, the amplifier will have 40 dB of negative feedback at 20 kHz. With a 30 dB per decade roll-off, the amplifier enjoys fully 60 dB of negative feedback at 20 kHz. This implies potentially 10 times less distortion at 20 kHz. In practice, such a loop gain roll-off can be approximated with numerous pole-zero networks for compensation. However, compensation with a constant 9 dB per octave roll-off results in bad peaking and poor transient response.

Two-pole compensation (TPC) can be thought of as a very simple and crude approximation to a steeper roll-off [2, 3]. A second pole is put in the dominant pole roll-off characteristic at 0 Hz. The pole is canceled by a zero placed at a higher frequency f_z (usually well below the gain crossover frequency). For example, in an amplifier with a 1-MHz ULGF, one might insert this 12 dB per octave roll-off segment below 200 kHz. The roll-off characteristic in this region will be steeper, approaching 40 dB per decade. This provides an extra 20 dB of roll-off between 20 kHz and 200 kHz. The amount of loop gain available at 20 kHz will thus be increased by 20 dB.

Figure 11.6 illustrates an amplifier with a two-pole compensation network. The compensation is essentially like Miller compensation, but the compensation capacitor is split. The junction of C1 and C2 is returned to ground through resistor R1. To first order, the series combination of C1 and C2 will be the same as the value of C1 in conventional Miller compensation to arrive at the same f_c. Resistor R1 increases the VAS gain at frequencies at least an octave below f_c by introducing attenuation into the shunt feedback path. This network can sometimes create additional loading on the VAS that might not be desired, and this is an example of where the compensation feedback might instead be tapped off from the pre-driver emitter follower [2]. Bear in mind that C1 and C2 need not have the same value, and in fact in some designs C1 might be significantly larger than C2.

Figure 11.7 shows a Bode plot of loop gain for Miller and TPC compensated amplifiers. Both amplifiers have f_c of 1 MHz. Frequency f_z is the location of the TPC zero where the roll-off slope returns to a 6 dB/octave slope. The amplifier in this example has a maximum loop gain of 80 dB (open-loop gain at DC is 106 dB). Loop gain is enhanced by TPC at frequencies below f_z. Here loop gain enhancement (LGE) is about 17 dB at 20 kHz (LGE_{20}).

As the TPC zero is placed closer to the gain crossover frequency f_c, LGE increases and the phase margin decreases. Dymond [3] has shown that the zero in the TPC loop gain is located at:

$$f_z = 1/2\pi R1(C1+C2) \tag{11.3}$$

For the special case where C1 = C2, $f_z = 1/8\pi R1C_M$, where C_M is what would have been the Miller compensating capacitance with conventional compensation. This corresponds to 1/4 the corner frequency of R1 and C_M. In the example circuit where C1 = C2 = 60 pF, the zero is located at f_z = 133 kHz. The range from 20 kHz to 133 kHz is a factor of 6.65:1 in frequency, and over

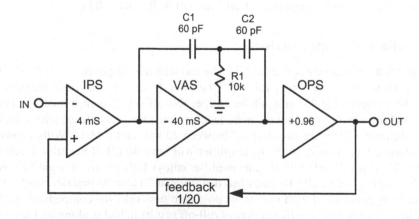

Figure 11.6 An Amplifier with Two-Pole Compensation

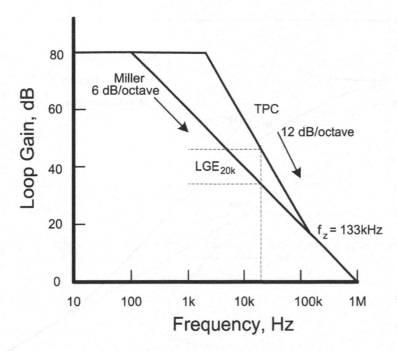

Figure 11.7 Bode Plot for the Amplifier with Two-Pole Compensation

this range LGE increases at 6 dB/octave as frequency falls. Therefore, LGE at 20 kHz is about 16.5 dB. The gain crossover frequency will be at:

$$f_c = gm/2\pi\, C_{series} A_{cl} \tag{11.4}$$

where C_{series} = C1 * C2/(C1 + C2), as shown by Dymond [3]. If C1 >> C2, the value of C2 will approach that of C_M, and the VAS loading will be about the same as for Miller compensation.

There are many approaches to picking C1, C2 and R1. I recommend doing it using SPICE simulation of the amplifier, starting with C1 = C2 = 2 C_M of a conventional Miller compensation scheme. This keeps the gain crossover frequency f_c about the same. Then add R1, starting with a high value that has little effect. Decrease R1 until closed-loop gain peaking of about 1 dB is evident. This gets you in the ballpark. Check the loop gain characteristic, phase margin and square-wave overshoot. Experiment with component values from there. This is definitely a process of iteration.

Figure 11.8 shows the loop gain for the amplifier of Figure 11.6 with the curve labeled *TPC*. Loop gain for conventional Miller compensation is also shown for comparison. Almost 17 dB of additional loop gain is available at 20 kHz. A further increase in loop gain is achieved at frequencies below 20 kHz. However, notice the large peak in loop gain just above 1 kHz. This happens with conventional TPC as a result of the steep slope in the local compensation feedback loop caused by the series combination of C1 and C2. Such an anomaly in the open-loop gain of the amplifier in the middle of the audio band can be undesirable.

Taming the In-Band Loop Gain Peak and Phase Transition

The peak in loop gain is due to the sharp intersection of the 12 dB/octave loop gain curve in TPC with the flat portion of the curve at low frequencies. The peak is accompanied by a rapid change

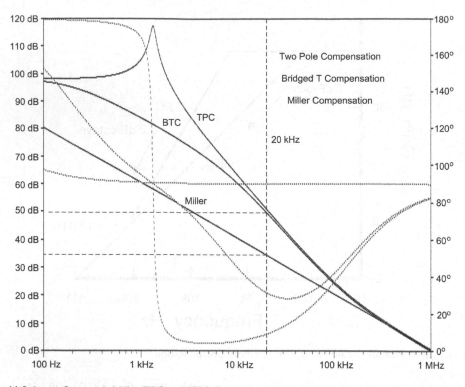

Figure 11.8 Loop Gain with Miller, TPC and BTC Compensation

in loop phase from 90° toward 180°. The peak can be removed by bridging the compensation network with a small capacitor C3, as shown in Figure 11.9a. This reduces LGE at 20 kHz by only a very small amount [2]. The addition of C3 to the circuit eliminates the anomaly by limiting the increase in compensated VAS gain to about 20 dB. It does so by bridging the combination of C1 and C2. Notice that C3, with a value of 2 pF, is smaller than the series combination of C1 and C2 by a factor of 10; this is why the gain step is limited to about 20 dB. I call this *bridged T compensation* (BTC). The loop gain curve labeled BTC in Figure 11.8 shows none of the gain peaking anomaly associated with conventional two-pole compensation. The phase transition is much smaller and smoother.

Capacitor C3 may be inconveniently small (~2 pF). Dymond has shown that the same peak-mitigating effect can be obtained by putting a larger capacitor C3 in series with R1, as shown in Figure 11.9b [3]. Placing C3 in series with R1 reduces loop gain peaking by bringing the complex poles together and to ultimately become real. In the examples here, this capacitor will be on the order of 2200 pF to achieve the same response as in Figure 11.9a. Both of these measures effectively limit the maximum amount of low-frequency LGE to a reasonable value that just eliminates the peak. These techniques also reduce the amount by which the phase shift falls toward 180° just above the audio band.

Conditional Stability

The open-loop phase of conventional Miller-compensated amplifiers is a constant 90 degrees (absent excess phase). The open-loop phase of a TPC design dips more negative at frequencies

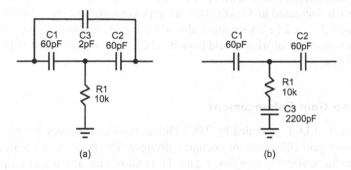

Figure 11.9 TPC Networks to Tame the In-Band Peak

between the added pole and zero. This implies a very small phase margin in this frequency region as the total phase lag approaches 180°. However, as long as the loop gain is well above unity where this phase dip occurs, there is little degradation of stability. If for some reason the forward gain is substantially reduced so that the gain crossover frequency falls to the frequency of maximum dip, then the phase margin at the reduced gain crossover frequency could be much smaller. One example of such an open-loop gain reduction occurs when power is turned off and the circuits become starved as rail voltages fall. Such a gain reduction can also occur as clipping or slew rate limiting is approached. It is therefore best if the phase at the dip does not go closer than about 20° to the 180° point.

Frequency Response Peaking and Overshoot

It is fundamental to TPC that there will be some closed-loop frequency response peaking at high frequencies above the audio band. There will also be some corresponding square-wave overshoot. In general, more aggressive TPC will cause greater frequency response peaking. TPC based on the suggested values above and with LGE_{20} equal to about 17 dB will yield a frequency response peak of about 0.8 dB at 231 kHz and square-wave overshoot of about 15% in an amplifier with no excess phase. The usual input low-pass filtering in the amplifier will reduce these effects.

Designing the Compensation Network

The basic design of the TPC compensation network can be very simple. Design the amplifier with conventional Miller compensation with compensation capacitor C_M to achieve the desired f_c (e.g., 30 pF for 1 MHz). Replace C_M with C1 and C2, each of which is double the value of C_M (e.g., 60 pF). So far you have the same conventional Miller compensation. Add shunt resistor R1 from the junction of the capacitors to ground. Choose the value of R1 to set how much in-band loop gain enhancement will be delivered. A smaller value of R1 will result in a greater increase in negative feedback loop gain at 20 kHz, corresponding to higher LGE_{20}. An example value for R1 might be 10 kΩ, corresponding to a corner frequency with the parallel combination of C1 and C2 at 133 kHz, or about $f_c/7.5$. The TPC zero will be at this frequency [3]. This is quite far below f_c, so the reduction in phase margin will be minimal. This network is illustrated in Figure 11.10a.

This is a very simple baseline design approach. Many different ratios of C2 to C1 can be used to arrive at a series combination that still equals the value of C_M. Larger ratios will require a smaller value of R1 if f_z is to be kept the same. For example, if C1 = 5 C2, where C1 = 180 pF and C2 = 36 pF, and C_{series} remains at 30 pF, C1 + C2 will equal 216 pF and f_z will move down to 74 kHz unless R1 is decreased to 5.6 kΩ. This network is shown in Figure 11.10b.

Some advocate C1 >> C2 [3]. A smaller value of C2 then becomes necessary because C2 will represent the dominant part of what would have been C_M. This should result in less VAS loading by the TPC network.

Degree of Loop Gain Enhancement

R1 sets the amount of LGE provided by TPC. This defines how aggressive the TPC is. As R1 decreases, the loop gain enhancement becomes stronger. The price for increased in-band loop gain is usually reduced stability margins. Figure 11.11 shows the amount of loop gain enhancement at 20 kHz as a function of R1 when C1 = C2 = 60 pF for a TPC design with a nominal f_c of 1 MHz. Also shown is the phase margin for this idealized design with no extra poles in the circuit. The further below f_c that f_z is, the less reduction in phase margin occurs, but then there is less LGE at 20 kHz.

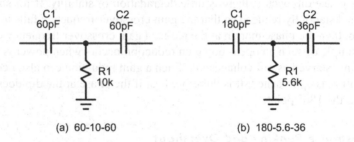

(a) 60-10-60   (b) 180-5.6-36

Figure 11.10 Alternative TPC Networks

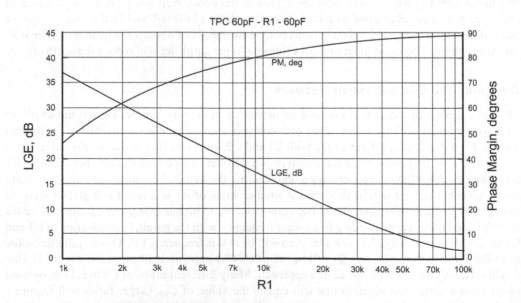

Figure 11.11 LGE at 20kHz and Phase Margin as a Function of R1

Effect on Distortion Spectrum

Distortion is reduced in accordance with the amount of loop gain provided by negative feedback. With Miller compensation, loop gain falls off at 6 dB/octave, so the amount of distortion reduction decreases as frequency increases. If there is 34 dB of feedback at 20 kHz, there will only be about 28 dB of feedback at the second harmonic (40 kHz) and 24 dB at the third harmonic (60 kHz). The higher harmonics are reduced less than the lower harmonics, so the shape of the harmonic distortion spectrum is changed. The loop gain for TPC will drop off nearly twice as fast (12 dB/octave) as frequency increases in the frequency range below f_z. This means that if LGE is 16 dB at 20 kHz (50-dB NFB at 20 kHz), it may be as little as 10 dB at the second harmonic, resulting in only 38 dB of NFB at 40 kHz. There may be only 30 dB of NFB at the third harmonic.

It is especially important to recognize that the reduction in THD-20 will not be as much as LGE at 20 kHz, since LGE is smaller at the higher harmonic frequencies. For a 60-10-60 TPC design, LGE at the various harmonics is shown in Table 11.1 for 20-kHz and 1-kHz fundamentals.

Consider an amplifier wherein THD is dominated by H3. LGE is only 7.5 dB at 60 kHz for 60-10-60 compensation, so THD-20 might be reduced by only 7.5 dB by using TPC.

Fortunately, the real culprit in sound quality impairment from high-frequency distortion is generally understood to be intermodulation distortion products that fall in the audio band. Such distortion is best evaluated with the 19 + 20 kHz CCIF IM test. The IM products in the audio band will be reduced by the loop gain at those frequencies. TPC provides at least as much distortion reduction at those frequencies as at 20 kHz. This means that the reduction in THD-20 when using TPC may underestimate the improvement in sound quality.

Input Stage Effort and VAS Loading

The more signal current the IPS must deliver for given amplitude at the output of the VAS, the greater distortion will be created in the IPS. This input stage effort as a function of frequency is well defined for Miller compensation, but is different for TPC. Indeed, it is smaller by the amount of LGE at a given frequency. For a given output amplitude, less current makes its way back through the TPC network to the input of the VAS, since a portion of that current flows through R1 to ground.

IPS effort is simply the current through C1, which for audio frequencies Diamond [3] has shown to be:

$$i_{C1} \approx V_{out}\, \omega^2\, R1 * C1 * C2 \tag{11.5}$$

where radian frequency $\omega = 2\pi f$.

The IPS will not be required to work as hard at audio frequencies. This is consistent with the fact that open-loop gain is higher, and increasing as the square of frequency reduction. Input stage effort has nothing to do with the C1/C2 ratio and everything to do with LGE. For the

Table 11.1 Loop Gain Enhancement

Frequency	HD-20	HD-1
Fundamental	16.6 dB	37.0 dB
Second	10.7 dB	36.5 dB
Third	7.5 dB	33.1 dB
Fourth	5.4 dB	30.6 dB
Fifth	3.9 dB	28.6 dB

60-10-60 case, it is less by 17 dB at 20 kHz. Amplifiers that suffer significant IPS distortion may benefit more from TPC.

VAS loading by the compensation network increases required VAS signal current swings at high frequencies and can limit the slew rate achievable by the VAS. A 100-pF load on a single-ended VAS with a bias current of 10 mA will limit slew rate to 100 V/μs. This is where a push-pull VAS has a strong advantage, being able to provide twice the amount of slew rate for a given amount of bias current.

While the loading on the VAS output node by Miller compensation is just equal to C_M, for TPC it is different. It is the current flowing through C2. The signal current drawn from the VAS by the TPC compensation network at audio frequencies is approximately:

$$i_{C2} \approx V_{out} \, \omega C2 \qquad\qquad (11.6)$$

when C1 >> C2 [3]. This current may be smaller when C1 is not significantly larger than C1, as in the case where C1 = C2. Compared to Miller compensation, VAS effort is a bit higher for TPC because some current flows to ground through R1. It is higher by about 1.7 dB at 20 kHz for a 60-10-60 arrangement.

11.5 Transitional Miller Compensation

Cherry proposed a form of Miller compensation that included the amplifier output stage in the local feedback loop formed by Miller compensation [4]. This was done by connecting one end of the Miller capacitor to the output of the output stage instead of to the output of the VAS, as shown in Figure 11.12. This can be called *inclusive Miller compensation* (IMC). The idea was to extend the distortion-reducing benefit of the Miller compensation loop to the output stage. Unfortunately, the output stage is often the largest source of excess phase shift in an amplifier, and this connection often leads to unacceptable risk of instability.

Baxandall proposed a compromise approach that was later dubbed *transitional Miller compensation* (TMC) by Stuart [2, 5]. Figure 11.13 shows an amplifier that employs TMC. Its topology resembles that of TPC, with a pair of series-connected compensation capacitors and a resistor at their junction. The resistor is connected to the output node instead of to ground. At very high frequencies, the two capacitors dominate the resistor and create an ordinary Miller compensation loop that has an effective C_M that is approximately the value of the two capacitors in series.

At low frequencies, the resistor is more conductive than C2, and the output stage is effectively enclosed within the local feedback compensation loop. TMC is well behaved and does not create the frequency response peaking and overshoot that necessarily accompany TPC. Moreover, TMC

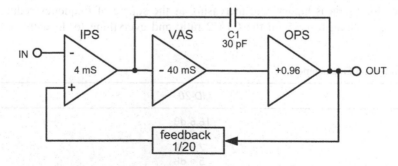

Figure 11.12 Amplifier with Inclusive (Cherry) Miller Compensation

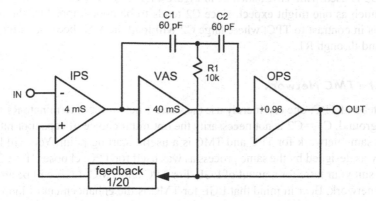

Figure 11.13 Amplifier with Transitional Miller Compensation

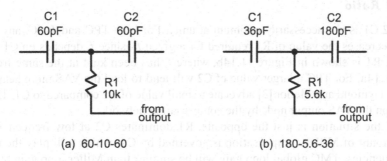

(a) 60-10-60   (b) 180-5.6-36

Figure 11.14 Alternative TMC Networks

reduces distortion by putting more local feedback around the output stage instead of by merely increasing global feedback in a certain frequency band. The global negative feedback rolls off at 6 dB per octave in the TMC scheme. This is the same as with conventional Miller compensation. TMC is very effective in reducing amplifier distortion because it introduces some local feedback around the output stage, where feedback is often needed most. Because the Miller compensation tap-off *transitions* from the output stage to the VAS at high frequencies, it does not suffer the instabilities introduced by Cherry compensation.

The total loop gain enclosing the output stage in TMC is important. It includes both the global loop gain and the local loop gain introduced by the TMC arrangement. The amount of the total loop gain is similar to the global loop gain of TPC if the same network is used. For this reason, we define the LGE for TMC as the enhancement of the *total* loop gain enclosing the output stage. For convenience, we will use the term *OPSLG* to denote total loop gain enclosing the output stage, as the output stage loop gain includes contributions from both the global loop and the compensation loop around it.

It is notable that global loop gain for TMC is smaller at audio frequencies than with conventional Miller compensation or TPC with the 60-10-60 component values shown in Figure 11.14a. This is because R1 dominates C2 at low frequencies and makes the effective Miller compensation capacitance equal to C1 instead of the series combination of C1 and C2 [2]. The loop gain reduction will thus be about 6 dB for C1 = C2. This effect is greatly reduced if C2 is chosen significantly larger than C1.

If a large C2 is used with TMC, such as in Figure 11.14b, where C2 = 5 C1, it does not load the VAS as much as one might expect, since C2 tends to be bootstrapped by the output signal via R1. This is in contrast to TPC, where large C2 will load the VAS because of the current that flows to ground through R1.

Designing the TMC Network

For TMC instead of TPC, one can merely use the same T network but terminate R1 to the output instead of to ground. C1 = C2 is not necessarily the optimum choice of capacitor ratio for TMC, but using the same network for TPC and TMC is a useful starting point. You will have a TMC network that was designed by the same process as was used for TPC: choose C1 = C2 = 2 C_M and choose R1 to suit your taste for amount of LGE. For TPC and TMC, LGE will be about the same for a given T network. Bear in mind that LGE for TMC is the enhancement of loop gain around the output stage, and this is what determines the reduction in output stage distortion.

The C2/C1 Ratio

The ratio C2/C1 is not necessarily optimum at unity. For both TPC and TMC, any ratio other than unity decreases the value of R1 required for a given f_z, since f_z depends on C1 + C2. This reduction of R1 is shown in Figure 11.14b, where f_z has been kept at the same frequency as in Figure 11.14a. For TPC a larger value of C2 will tend to load the VAS more heavily at high frequencies. Dymond and Mellor [3] advocate a small value of C2 compared to C1. This reduces the loading on the VAS output node by the compensation network.

For TMC the situation is just the opposite. R1 dominates C2 at low frequencies and the effective amount of Miller compensation is governed by C1, which will play the role of C_M. At low frequencies, TMC global loop gain will be smaller than Miller loop gain by the factor C2/(C1 + C2). For this reason we want to keep C1 small. For a given f_c, this means that C2 should be large. We do not want or need to give up 6 dB of global loop gain using TMC with C1 = C2.

The benefit of having C2 >> C1 for TMC seems very clear. At minimum, it removes the closed-loop frequency response droop that is seen with C1 = C2, and ultimately moves the closed-loop frequency response toward that of a Miller-compensated amplifier, as expected in the first place. It also greatly reduces the reduction in global loop gain in the audio band seen with TPC when C2 = C1.

Degree of Loop Gain Enhancement

R1 sets the amount of LGE provided by TMC, just as it did for TPC. Its value defines how aggressive the TMC is. As R1 decreases, the aggressiveness increases. The price for increased in-band loop gain is usually a reduced stability margin. TMC does not provide any loop gain enhancement for the IPS.

As discussed earlier for TPC, the application of TMC affects the shape of the distortion spectrum, since LGE decreases with frequency. Output stage distortion dominates in most well-designed amplifiers. In this case, the LGE of either TMC or TPC will largely govern the distortion reduction for the amplifier, since the LGE is applied to the output stage in both arrangements. If the TMC and TPC are designed to provide the same LGE magnitude with frequency, then both approaches will improve the overall amplifier in about the same way under these conditions. The difference is that TMC does not change the global feedback loop from a first-order loop to a second-order loop. As a result, gain margin and phase margin for the global loop are not significantly reduced from the case of straight Miller compensation. Moreover, the closed-loop peaking and overshoot introduced by TPC are not present with TMC.

TMC Loop Stability

Most power amplifiers have one or more local feedback loops in addition to the global feedback loop. Indeed, the conventional Miller-compensated VAS is an example of such a local loop. The stability of all the local loops in an amplifier must be considered. The local loop formed around the output stage by TMC is one such loop. The output stage is enclosed by this loop and the global loop. Both loops contribute to the total loop gain around the output stage (OPSLG). For Miller and TPC, OPSLG is the same as the global loop gain. For TMC, this loop gain includes the contributions from the local loop formed by the TMC network plus the global loop gain.

Although not intuitive, TPC and TMC amplifiers implemented with the same compensation network will exhibit very similar output stage loop gain characteristics. TPC and TMC are different forms of compensation, but beyond the *T* shape of the compensation network, and despite the different place where R1 is connected, they exhibit similar OPSLG gain and phase, and *apparent* degradation of stability margins with increased compensation aggressiveness. Aggressive TPC and TMC both exhibit small phase margins when OPSLG is evaluated.

However, the TMC loop around the output stage encloses less of the amplifier than for TPC. Therefore, TMC may present a better PM in a real amplifier due to fewer sources of excess phase in the loop, such as that from the input stage. It is notable that the peak in loop gain seen with TPC is also present in the total loop gain that encloses the output stage with TMC, and that the same remedies apply.

As with conventional compensation, loop gain can be evaluated by breaking the loop in the right place. For TMC OPSLG, the loop is broken right at the output of the OPS. This captures the contributions of the global and local loops. The loop can be broken in simulation by placing a very large inductor in series with the signal path so as to provide feedback only at DC. A test input signal is then injected through a capacitor on the far side of the inductor. This is shown in Figure 11.15.

If the OPSLG for TMC is found to have very small phase margin, how will it affect the closed-loop behavior of the amplifier and will it pose a real stability risk? Or does the benign-looking bootstrapping nature of R1 in TMC cause a false suggestion of instability when there is none?

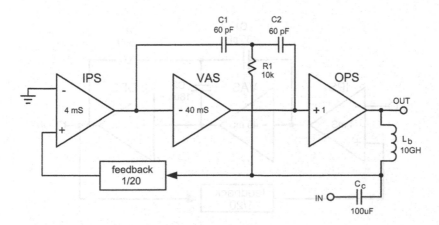

Figure 11.15 Breaking the TMC Loop to Evaluate Output Stage Loop Gain

An OPSLG gain and phase plot that indicates small phase margin does not appear to correspond to small stability margin for the amplifier. A TMC network was designed that was very aggressive, resulting in an output stage loop gain phase margin of about 2 degrees. When the loop was closed and sine wave and square wave signals were applied to the amplifier, there were no signs of frequency response peaking or of square wave overshoot, much less ringing. This was done in an idealized amplifier built with VCVS elements. The same arrangement, when connected for TPC, showed gross evidence of instability. Similar results were seen when the idealized amplifier included excess phase that reduced Miller-compensated phase margin to 65° at f_c equal to 1 MHz.

Closed-loop stability was also evaluated by plotting output stage sensitivity versus frequency. Any instability will be evidenced by peaking in the sensitivity function. The TMC arrangement showed little or no peaking while the TPC arrangement showed gross peaking. The tentative conclusion is that TMC is more stable than TPC for a given amount of LGE, in spite of the fact that apparent OPSLG phase margin is the same in both cases of TPC and TMC. Small OPSLG phase margin in the TMC arrangement appears to give a false or over-pessimistic indication of instability. Nevertheless, it is wise to not depend too heavily on this observation when selecting the degree of aggressiveness for a TMC design.

TMC as a Form of Cherry Compensation

Recall that Cherry compensation encloses the output stage within the Miller compensation loop, but this minor loop is generally unstable because of the excess phase introduced by the output stage and the failure to split the pole at the output of the VAS. TMC can be viewed as a form of Cherry compensation wherein the minor loop has itself been compensated. Indeed, in this case the compensation is provided by C2. This is seen if the TMC arrangement is redrawn showing R1 as the feedback resistor and C2 as a shunt capacitance to the input node of the output stage, where C2 is large compared to C1. This is illustrated in Figure 11.16. Here C1 acts largely as C_M while R1 and C2 compensate the Cherry loop with leading feed-forward compensation around the output stage. R1 closes the Cherry compensation loop. C2 can start out small and be increased in size until sufficient stability is attained. This suggests an entirely different way of choosing the values for the TMC network. With C2 >> C1, the combination of C2 and R1 acts like a crossover network whose output impedance is small compared to the impedance of C1.

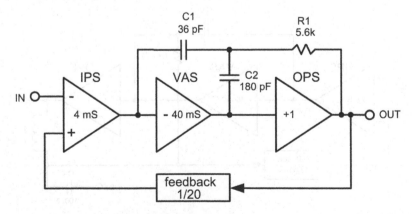

Figure 11.16 TMC Re-Drawn as Compensated Cherry Compensation

Input Stage Effort and VAS Loading

The in-band signal current required of the input stage depends on the amount of gain in the amplifier after the input stage. In contrast to TPC, TMC does not decrease IPS effort, since it does not increase VAS transimpedance or open-loop gain. Indeed, in situations where TMC reduces open-loop gain, as where C1 = C2, IPS effort will be increased somewhat. This is not desirable and is one of the reasons why TMC is best implemented with C2 >> C1, such as with a 5:1 ratio, as shown in Figure 11.14b. In this case, IPS effort with TMC is only about 1.5 dB greater than that with Miller compensation.

VAS loading by the TMC network is a bit smaller than that for Miller compensation because the output signal bootstraps R1. At 20 kHz VAS effort is 0.7 dB less than that for Miller compensation in the 60-10-60 arrangement.

Feeding the T Network from the Driver

The *T* compensation network can be fed from the driver or pre-driver instead of from the high-impedance output of the VAS. The buffering provided by the driver can circumvent the VAS loading issue and permit larger values of C2 and smaller values of R1. This can be advantageous for both TPC and TMC. For TMC, the use of lower impedances for C2 and R1 can make the "crossover" action of C2 and R1 more ideal and less interactive with C1. Such an arrangement is shown in Figure 11.17, where C2 is 25 times the value of C1. The series R-C network R2-C3 loading the VAS improves stability in some amplifiers.

Transitional Miller Input Compensation (TMIC)

Miller Input Compensation (MIC) has been discussed earlier in this chapter and in this reference [1]. In it, C_M is fed back to the input stage feedback node instead of to the input of the VAS. This encloses the IPS in the compensation loop and can greatly increase slew rate. MIC can be combined with TMC to reap the advantages of both techniques. The TMC network is merely returned to the feedback node of the IPS instead of the VAS input. This is illustrated in Figure 11.18. R4-C4 and R5-C5 compensate the compensation loop [1].

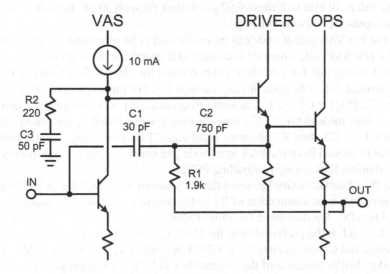

Figure 11.17 Connecting the TMC Compensation Network to the Driver

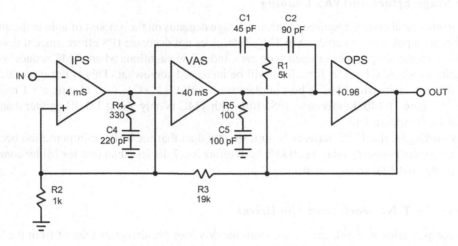

Figure 11.18 Transitional Miller Input Compensation (TMIC)

Power Supply Rejection

Conventional Miller compensation degrades PSRR to the VAS rail because the input side of C_M is referred to the rail, while the output side is referred to signal (ground). Assume that no signal is present and that the VAS output node, with no signal on it, is at ground. Recognize that global feedback keeps that node at ground. The VAS input node must move with the rail, else large signal current would flow. Rail noise thus causes current flow in C_M. In order to keep the VAS output node at zero, the IPS must supply error current to the VAS input node to keep this node from moving as a result of the current in C_M. Thus, the current flowing in C_M due to rail noise must be supplied by the IPS, and this means that there is input-referred (rail) noise at the input to the IPS.

Put simply, any current noise flowing from ground reference through the compensation network to the rail-referenced input node of the VAS will create input-referred noise. For example, 10 mV on the rail at 10 kHz will cause 0.02 μA to flow through 30 pF. With IPS $gm = 4$ mS, this will result in input-referred noise of 50 μV.

The fact that the VAS output node can be considered to be at ground simplifies the analysis, especially for TPC and TMC where R1 is connected to ground or to the ground-referenced output node. It is easy to see that for TPC there are two paths for current flow to ground, one through C2 and one through R1, with both flowing through C1. For purposes here, R1 is effectively in parallel with C2. If C1 = C2 = 2 C_M, at audio frequencies R1 will dominate C2 and the current flow will be largely through C1 = 2 C_M, thus showing a worse PSRR as compared to Miller compensation. If C1 >> C2, then R1 becomes small and C1 becomes large, significantly reducing the impedance to ground from the VAS input node and causing more current to flow for a given amount of rail noise, thus strongly degrading PSRR.

For TMC, the situation is similar, since the VAS output node and the amplifier output node are both at ground, so the connection of R1 to the output node for TMC makes no difference as compared to TPC. We thus have the same PSRR behavior for C1 = C2 = 2 C_M. However, for C2 >> C1, which is the preferred ratio for TMC, C1 now dominates the control of the noise current flowing, and C1 approaches C_M as C2/C1 becomes larger. Thus, for TMC, PSRR is not quite as good as Miller (because of the current flow in R1), but it approaches that of Miller for C2 >> C1.

11.6 A Vertical MOSFET TMC Amplifier Example

Figure 11.19 shows a simplified schematic of the 50-W amplifier that was built and measured to evaluate TMC in a real amplifier [2]. It is a vertical MOSFET power amplifier with a cascoded JFET input stage and a single-ended VAS. The single pair of vertical MOSFETs is biased at 200 mA. THD+N at 20 kHz is 0.016% at 50 watts into 8 Ω when conventional Miller compensation is used to set f_c to 1 MHz (here with 100 pF in accordance with IPS gm of about 12 mS).

IPS, VAS and TMC

The input stage employs a dual monolithic LSK389 that is cascoded to a current mirror load. The LSK389 is degenerated with 100 Ω on each source. The cascode is a *driven cascode*, with a replica of the feedback signal feeding its bases. This arrangement bootstraps the drains of the JFETs, largely eliminating the effect of C_{gd} and minimizing common-mode distortion. The feedback network employs 10-kΩ and 511-Ω resistors to establish a closed-loop gain of 20.6. The 10-kΩ feedback resistor is a 2-W metal film type to minimize signal-dependent heating and other sources of resistor distortion. The VAS is a conventional Darlington single-ended arrangement with a 10-mA current-source load. The amplifier is compensated with a 120-1-620 TMC network, providing LGE_{20} of 17 dB.

Output Stage and Temperature Compensation

Two V_{be} multipliers are connected in series to form the bias spreader, with one of the transistors mounted on the heat sink. Vertical MOSFET amplifiers are usually temperature stabilized properly when about half of the bias spread is from a V_{be} multiplier with its transistor on the heat sink.

The output stage uses a single pair of vertical MOSFETs, each with 0.33-Ω source resistors. Source resistors R44 and R45 are not for temperature stability but rather for distortion reduction in the output stage (this will be discussed in Chapter 14 on MOSFET output stages). 100-Ω gate stopper resistors are used in combination with gate Zobel networks to stabilize the MOSFETs without the use of higher-resistance gate stopper resistors [1]. The amplifier is rated at 50 W with 50-V rails. This lower-than-normal power for 50-V rails is due to the required turn-on voltage of the vertical MOSFETs (about 4 V) in combination with the use of ordinary emitter follower drivers. Higher power can be had if boosted rails are used for the IPS and VAS. For power levels over about 75 W, a second output pair with its own gate stoppers, gate Zobels and source resistors can be added.

DC Servo and Power Supply

The amplifier incorporates a DC servo that employs a dual JFET op amp to implement an inverting integrator and an inverter. This is much like the arrangement shown in Figure 10.5 in the DC servo chapter. The ±15-V power supply is implemented with Zener diodes and dropping resistors fed from the main rails.

The main power supply rails are immediately decoupled with 1000-μF capacitors on the amplifier board. Each rail is then decoupled on the way back to the IPS-VAS with a series resistor and another filter capacitor.

Performance

Figure 11.20 shows THD+N versus frequency for the amplifier when it is delivering 50 watts into 8 Ω. With 120-1-620 TMC (C2 = 5.2 C1), THD+N at 20 kHz drops from 0.016% to 0.0055% for an improvement of about 10 dB. THD at 10 kHz drops from 0.009% to 0.0015% with TMC, for

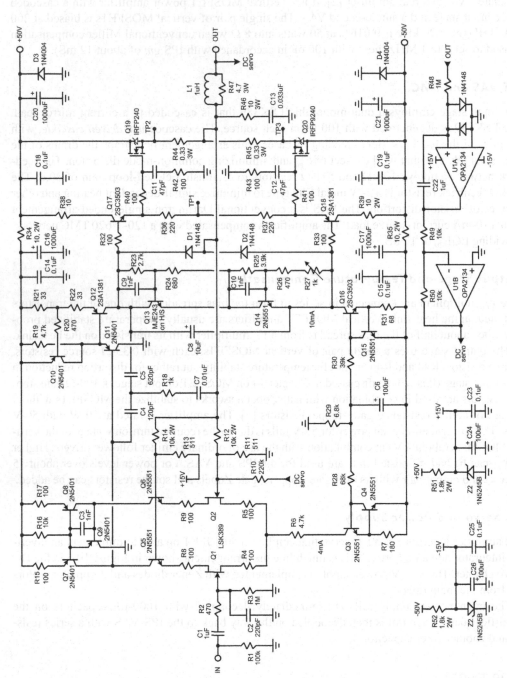

Figure 11.19 A Vertical MOSFET TMC Amplifier

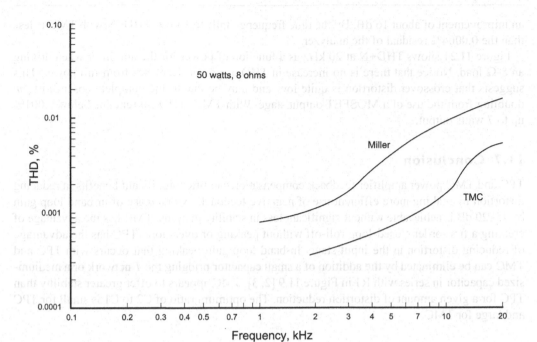

Figure 11.20 THD+N Versus Frequency at 50 W into 8 Ω

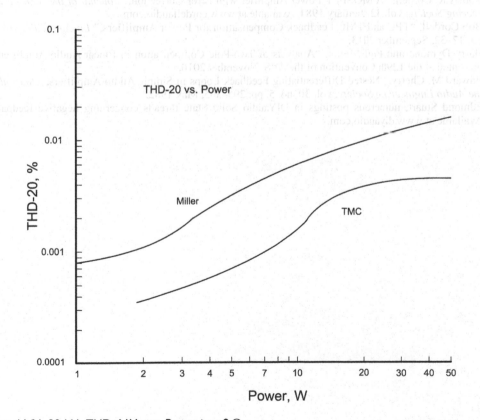

Figure 11.21 20-kHz THD+N Versus Power into 8 Ω

an improvement of about 16 dB. By the time frequency falls to 2 kHz, THD+N with TMC is less than the 0.0004% residual of the analyzer.

Figure 11.21 shows THD+N at 20 kHz as a function of power for the amplifier when driving an 8-Ω load. Notice that there is no increase in THD as power decreases from full power. This suggests that crossover distortion is quite low, and may be due to the complete absence of *gm* doubling from the use of a MOSFET output stage. With TMC, THD-20 remains below 0.001% up to 7 watts output.

11.7 Conclusion

TPC and TMC power amplifier feedback compensation can offer significant benefits in reducing distortion by making more efficient use of negative feedback. An increase of in-band loop gain by 15–20 dB is achievable without significant loss in stability margins. TMC has the advantage of creating a first-order closed-loop roll-off without peaking or overshoot. TPC has the advantage of reducing distortion in the input stage. In-band loop gain peaking that occurs with TPC and TMC can be eliminated by the addition of a small capacitor bridging the *T* network or a medium-sized capacitor in series with R1 in Figure 11.9 [2, 3]. TMC appears to offer greater stability than TPC for a given amount of distortion reduction. The optimum ratio of C2 to C1 is small for TPC and large for TMC.

References

1. Robert R. Cordell, "A MOSFET Power Amplifier with Error Correction," *Journal of the Audio Engineering Society*, vol. 32, January 1984. Available at www.cordellaudio.com.
2. Bob Cordell, "TPC and TMC Feedback Compensation for Power Amplifiers," *Linear Audio*, vol. 6, pp. 17–32, September 2013.
3. Harry Dymond and Phil Mellor, "Analysis of Two-Pole Compensation in Linear Audio Amplifiers," presented at the 129th Convention of the AES, November 2010.
4. Edward M. Cherry, "Nested Differentiating Feedback Loops in Simple Audio Amplifiers," *Journal of the Audio Engineering Society*, vol. 30, no. 5, pp. 295–305, May 1982.
5. Edmond Stuart, numerous postings in DIYaudio Solid State threads concerning negative feedback. Available at www.diyaudio.com.

Chapter 12

Output Stage Design and Crossover Distortion

12. INTRODUCTION

The output stage is in many ways the most important part of a power amplifier. It is surely the most difficult section in which to reduce distortion when all reasonable measures have been taken to reduce distortion in the input and VAS stages. There is also a lot more money tied up in the output stage.

The output stage discussion was begun in Chapter 7. Concepts of crossover distortion were introduced and other classes and topologies were covered. Here we focus on the emitter follower (EF) output stage. The CFP output stage was discussed and was found to have some serious challenges and shortcomings. However, many concepts discussed here also apply to CFP output stages.

12.1 The Class AB Output Stage

The class AB output stage is the workhorse of most audio power amplifiers. This stage, shown in simple form in Figure 12.1, was examined in some detail in Chapters 3 and 7, where a simple power amplifier was evolved to a fairly high-performing design of conventional topology.

The output stage of Figure 12.1 is the classic Locanthi T circuit, [1, 2] which will also be referred to here as a Triple EF or simply a Triple. This is my preferred BJT output stage, as it provides far higher performance than a simple Darlington output stage (a double EF).

It will also be assumed throughout this chapter that the output stage is being driven in voltage mode. This loosely means that the output impedance of the VAS is much lower than the input impedance of the output stage. Bear in mind that the output impedance of the VAS is often fairly low, especially at high frequencies, due to the shunt feedback created by Miller compensation. Assuming pre-driver and driver transistor beta of 100 and output transistor beta of 50, the current gain of the Triple is 500,000, meaning that its input impedance is about 1 MΩ even when driving a 2-Ω load.

The voltage gain of the output stage is determined by the voltage divider formed by the output stage emitter follower output impedance and the loudspeaker load impedance. The output impedance of each half of the output stage is approximately the sum of the dynamic emitter resistance re' and the external emitter resistance R_E. The output impedance of the stage is thus approximately equal to R_E when $re' = R_E$. For simplicity, we are here ignoring the ohmic component of the output transistor emitter resistance, often on the order of 0.03–0.1 Ω.

Class B or Class AB?

In a class B amplifier, the top and bottom halves each conduct for one half-cycle and the hand-off from one side to the other is abrupt. In a class AB amplifier, the stage behaves largely like a

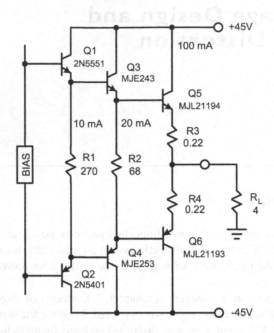

Figure 12.1 A Simple Class AB Output Stage Using a Triple

class B amplifier, but some overlap of conduction is permitted, so that the transition of contribution to the output signal is a bit more smooth and broad. In the most widely used output stage, a small *optimum* bias is applied, called the *quiescent bias* I_q. In the region of $\pm 2\ I_q$ the stage is essentially in class A, with both transistors contributing transconductance. This does not mean it is linear, as Figure 12.2 demonstrates. This chapter is focused exclusively on class AB output stages. Other authors have described this mode of operation as class B, and this is really just a matter of semantics [3].

12.2 Static Crossover Distortion

When the contribution to the output signal is transitioned from one half to the other half of the output stage, the effective output impedance of the output stage may change. This changes its gain into the load and causes static crossover distortion. This phenomenon was discussed at length in Chapter 7.

If the gain of the output stage is plotted as a function of the current being driven into the load, a so-called wingspread plot results [3]. Such a plot is shown in Figure 12.2 for the Triple output stage of Figure 12.1 biased at optimum quiescent current and at values of half and twice optimum current. Barney Oliver of HP showed in 1971 that the optimum class AB bias point was that where *re'* of the output devices equals the value of the emitter resistors. This amounts to biasing the stage so that 26 mV (kT/q) appears across the emitter resistors under the quiescent bias condition [4]. In reality, the optimum value is a bit less than 26 mV because a few mV must be accounted for in the effective ohmic emitter resistance of the power transistor, which is being ignored here.

The output stage in Figure 12.1 employs 0.22-Ω emitter resistors and its theoretical optimum bias is 118 mA. Ideally the gain of the output stage is unity, while a typical gain is on the order of 0.95 (95%) when driving a 4-Ω load. The peak-to-peak gain variation for the optimum-bias case in Figure 12.2 is about 1.1%.

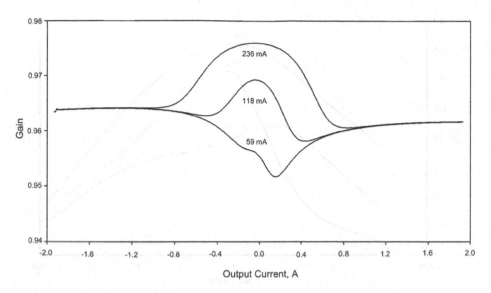

Figure 12.2 Wingspread Plot Illustrating the Crossover Distortion Mechanism

If a class AB output stage is over-biased, *gm* doubling can occur. Suppose that the quiescent bias current of the output stage is sufficiently large that the intrinsic emitter resistance (*re'* = 1/*gm*) of the top and bottom transistors is much smaller than the resistance of the emitter resistors R_E. Under these conditions, the output impedance of the output stage will approach $R_E/2$ when no current is flowing into the load. The effective transconductance of the output stage will be $2/R_E$. If the output voltage swings positive to produce a current into the load greater than twice its quiescent current, then only the top transistor will be conducting and the output stage impedance will be equal to R_E. Under these conditions the output stage transconductance is simply $1/R_E$. Thus we find that the effective transconductance of the output stage has nearly doubled in the *crossover* region where both top and bottom halves of the output stage conduct.

Evidence of *gm* doubling can be seen in Figure 12.2 for the over-biased case. If the nominal gain at high current was 0.94 and the gain rose to 0.97 in the crossover region (half the loss from unity), this would represent full *gm* doubling, which can only be approached in practice.

So-called *gm* doubling can happen when a conventional class AB design is unintentionally over-biased or it can happen in class AB designs that are deliberately over-biased to achieve a larger crossover region. The *gm*-doubling phenomenon also occurs in class A amplifiers when they are called on to deliver more than twice their quiescent current into the load. This will often happen in class A designs when they drive low-impedance loads to substantial power levels, where they effectively transition into class AB operation.

Because crossover distortion is largely a function of output current, it will always be larger for loads that have lower impedance. Virtually no crossover distortion will be produced when an amplifier is under a no-load condition.

Crossover Distortion as a Function of Signal Level

Crossover distortion is small at very small signal levels where the class AB output stage is operating within its class A region, extending to peak output current equal to about twice I_q. Crossover

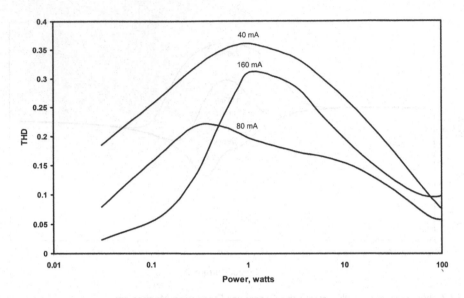

Figure 12.3 Crossover Distortion Versus Power Level with 4-Ω Load

distortion also tends to be low at high output levels where the crossover region of nonlinearity is only a small fraction of the signal swing. Crossover distortion is usually highest at low power levels. Figure 12.3 shows crossover distortion of the output stage of Figure 12.1 as a function of power level when driving 4 Ω. It can be seen to peak at about 0.5 W when bias is optimal at 80 mA. The other two curves show crossover distortion when I_q is one-half and twice the optimum value. It is notable here that the optimum bias is 80 mA, somewhat less than the theoretical value of 118 mA. This corresponds to $V_q = 18$ mV across each emitter resistor. The smaller V_q than 26 mV reflects the fact that there is some ohmic resistance contributions to effective R_E from the power transistors. The point of maximum crossover distortion occurs at a higher power level when the stage is over-biased, as would be expected. In any case, crossover distortion is usually worst at typical listening levels. Bias adjustment to minimize crossover distortion should be done at low power levels with a 4-Ω load, where crossover distortion is worst. Doing the adjustment at a higher frequency, like 20 kHz, helps because the crossover distortion at high frequencies is not reduced as much by negative feedback.

12.3　Optimum Bias and Bias Stability

Static crossover distortion is minimized when the variation in output impedance through the crossover region is minimized. This variation occurs as a result of the changes in transconductance with current of the upper and lower output transistors. The variation is thus a function of net output current, not net output voltage.

Assume ideal output transistors, each with an emitter resistor R_E. The emitter resistor serves to stabilize the bias current by providing a resistance across which a voltage drop will establish the bias current. However, it also plays a critical role in determining crossover distortion.

The bias spreader is usually implemented with a V_{be} multiplier. Part or all of the V_{be} multiplier circuit is usually mounted on the output transistor heat sink so that long-term output stage temperature can be tracked, providing a measure of temperature compensation. In Chapter 17 the use of more sophisticated output transistors that include built-in temperature-sensing diodes will be discussed. These permit the design of bias spreaders that track internal output transistor temperature changes much more quickly and accurately.

Setting the Bias

The output stage bias is inevitably a function of heat sink temperature and junction temperature of the output transistors. It is also a function of the temperatures of the pre-driver and driver transistors. Under what conditions should the quiescent bias be set? Should it be set while the amplifier is at idle after it has stabilized? Should it be set by a distortion measurement? If so, at what power level into what load impedance? In what way is the degree of thermal compensation established in the amplifier design? This will affect the biasing process as well. These issues will be addressed in Chapter 17.

Bias Stability

Bias stability is important to class AB output stages because crossover distortion is sensitive to the quiescent bias current I_q. Figure 12.4 shows a plot of simulated crossover distortion as a function of quiescent bias current for the simple class AB output stage of Figure 12.1 driven from a voltage source. The stage is driving a 4-Ω load. Shown on the plot are THD-1 and the amplitudes of the individual odd harmonics up to the seventh harmonic. The even harmonics are of less interest. The power level is set to be 1 W. Notice that the different harmonics do not share the same minimum. These curves show that it is much safer to err on the high side when setting the bias current.

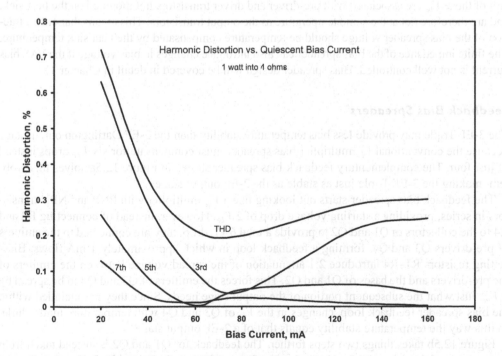

Figure 12.4 Crossover Distortion Versus Quiescent Bias Current

Stability of the quiescent bias current can be a challenge because of the temperature variations in the output transistors and because it only takes a few millivolts to produce a 10% change in quiescent bias current.

Dynamic Bias Stability

Dynamic bias stability refers to the behavior of the output stage quiescent bias as a function of output transistor junction temperature changes caused by real program material. This does not show up under ordinary continuous sine wave testing on the bench. Dynamic bias tracking error can lead to conditions of over-bias or under-bias after the average power of the program material changes. This problem will be discussed in detail in Chapter 17.

The Bias Spreader

The V_{be} multiplier and minor variations on it are almost universally employed for the biasing of output stages. However, there are some challenges in designing it to provide good thermal bias stability and temperature tracking of the output transistors. Tracking is important because just 10 mV of error represents a large fraction of the ideal 26 mV that should appear across each emitter resistor. Tracking is made difficult because of the numerous time constants involved, particularly the very slow time constant of the heat sink on which the temperature sensing transistor or diode is usually mounted. The temperature of the power transistor junction can change much more quickly than that of the temperature-tracking device.

The ideal 52-mV that appears emitter-to-emitter at the output devices is the result of a subtraction of multiple junction drops from the bias spreader voltage. In the case of an output Triple, a spreader voltage on the order of nearly 4 V is required, from which 6 V_{be} will be subtracted to arrive at the 52-mV number. This process introduces increased sensitivity and opportunity for error. Often, several of those V_{be} are associated with pre-driver and driver transistors not mounted on the heat sink, and are therefore not at the same temperature as the output transistors. This means that only a fraction of the bias spreader voltage should be temperature compensated by the heat sink temperature. The finite impedance of the bias spreader can also introduce changes in bias voltage if the VAS bias current is not well controlled. Bias spreader design will be covered in detail in Chapter 17.

Feedback Bias Spreaders

The 3-EF Triple may provide less bias temperature stability than the 2-EF Darlington output stage because the conventional V_{be} multiplier bias spreader must compensate for six V_{be} drops instead of just four. The complementary feedback bias spreader shown in Figure 12.5a solves this problem, making the 3-EF Triple just as stable as the 2-EF output stage.

The feedback bias spreader starts out looking like a V_{be} multiplier with PNP and NPN transistors in series, providing a starting voltage drop of 2 V_{be}. However, instead of connecting R3 and R4 to the collectors of Q1 and Q2 to provide a total 6 V_{be} drop, they are connected to the emitters of pre-drivers Q3 and Q4, forming a feedback loop in which approximately 1 mA flows. Bias-setting resistors R1–R4 introduce 2:1 attenuation of the spread voltage between the emitters of the pre-drivers and the bases of Q1 and Q2. This forces the emitters of Q3 and Q4 to be spread by 4 V_{be}, just what the subsequent portion of the output stage needs. Since they are included within the bias spreader feedback loop, changes in the V_{be} of Q3 and Q4 with temperature matter little. In this way the temperature stability equals that of a 2-EF output stage.

Figure 12.5b takes things two steps further. The feedback for Q1 and Q2 is instead taken from the emitters of the drivers, Q5 and Q6, forcing these emitters to be spread by 2 V_{be}, just what the

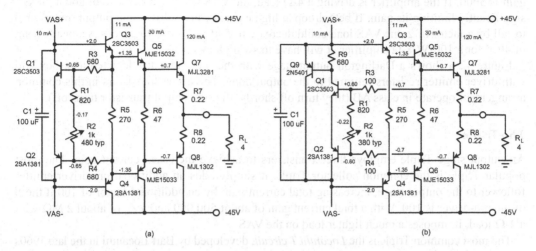

(a)

Figure 12.5 Feedback Bias Spreaders

pair of output transistors needs. This makes the bias stability of the Triple potentially *superior* to that of the Double using a conventional V_{be} multiplier. Approximately 1 mA flows in the feedback loop, creating a drop of 100 mV across each of R3 and R4. Because the base voltages of Q1 and Q2 differ by 2 V_{be}, very little attenuation of the DC bias feedback voltage is needed between the emitters of the drivers and the bases of Q1 and Q2. This limits the amount of downward bias control range allowed by adjustment of R2. Smaller V_{be} drops in Q1 and Q2 as compared with those in the output transistors (plus 2 R_E) permit more adjustment range, and are therefore desirable. This is accomplished by operating Q1 and Q2 at 1 mA instead of 10 mA, a decrease by a factor of 10, corresponding to a 60-mV reduction in V_{be} for each transistor. The addition of Q9 and R9 transforms the feedback bias spreader into a CFP-based design, enabling the reduction in current flowing in Q1 and Q2 and also reducing the impedance of the bias spreader. If for some reason adjustment range is still insufficient, a small current can be fed into the bases of Q1 and Q2 by resistors connected from these bases to the emitters of Q3 and Q4. 1.5-kΩ resistors will flow about 400 µA into the base nodes, decreasing spread at the power transistors further by 40 mV on each side.

One of the bias spreader transistors will typically be mounted on the heat sink or one of the output transistors for temperature compensation.

12.4 Output Stage Driver Circuits

So far we have discussed output stages with simple or ideal drivers. In reality, the driver must be considered an integral part of the output stage. The two most common driver arrangements for emitter follower (common-collector) output stages were discussed briefly in Chapters 3 and 7. These are the Darlington and the Triple, with one and two EF stages preceding the output transistors, respectively.

Darlington Output Stage

The Darlington output stage (2 EF) consists of two emitter followers in tandem. It produces approximately unity gain and a current gain that is the product of the betas of the driver and output transistors. If the beta of the driver is 100 and that of the output transistor is 50, then the current gain is 5000. If the amplifier is driving a 4-Ω load, the VAS will see a net load of about 20 kΩ, significantly limiting its gain. If beta droop at high current causes driver and output transistor β to fall by a factor of 2, the VAS load could decrease to 5 kΩ. If the VAS produces a peak swing of 40 V for a 100-W/8-Ω amplifier, it will have to swing 8 mA peak.

Figure 12.6 shows a Darlington output stage with the bias spreader depicted as a box and with driver emitter resistors returned to the output node. The driver transistors in this common arrangement operate in class AB (they turn off shortly after the output transistor turns off).

The Triple

An output stage Triple employs three transistors to achieve increased current gain. The most popular Triple is the common-collector Triple. It simply adds an additional pre-driver emitter follower to the output stage, increasing total current gain by an additional factor of 100 if the β of the pre-driver is 100. With a total current gain of about 500,000 and a Z_{in} of about 2 MΩ with a 4-Ω load, it imposes a much lighter load on the VAS.

The most common Triple is the *Locanthi T circuit*, developed by Bart Locanthi in the late 1960s [1, 2]. This is the output stage shown in Figure 12.1. The pre-driver and driver emitter resistors are each returned to the emitter of the complementary device. This allows for improved turn-off of the subsequent device by reverse bias. It also increases the input impedance of the output stage because the emitter resistors have essentially no signal across them. The pre-driver and driver stages in this arrangement operate in class A. This stage requires a bias spread of about six V_{be} (about 4 V).

The Diamond Driver

The *diamond driver* [5, 6] is a four-transistor arrangement in which the first emitter followers are folded, or upside-down, as shown in Figure 12.7a. It is a more complex circuit because it

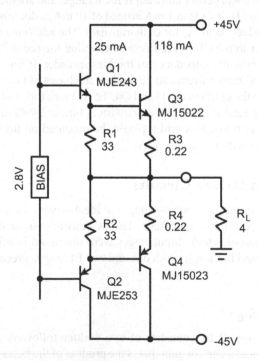

Figure 12.6 A Simple Darlington Output Stage

requires a pair of current sources to bias the folded emitter follower. An output stage driven by a diamond driver can be considered as another form of Triple, which we will call a *diamond buffer Triple* (DBT).

The pre-driver in the DBT adds one V_{be} of VAS headroom, while the pre-driver EF in the conventional Triple subtracts one V_{be} of headroom. Thus, the DBT provides 2 V_{be} more headroom to the VAS. The current sources that bias the pre-drivers can have their collectors swing very close to the rail, and their output impedance can drop somewhat without introducing signal distortion because their load is the low output impedance of the folded EF pre-driver.

The V_{be} drops of the pre-driver and driver cancel, leaving the bias spreader to provide only the 2 V_{be} spread required by the output transistors. If the pre-driver and driver transistors are bolted together, the temperature dependence of their V_{be} drops will largely cancel. A TO-126 pre-driver can be bolted to a TO-220 driver, separated by an insulator if the TO-126 device is not self-insulated. The bias spreader then need only correct for the temperature dependence of the output transistor base-emitter junctions. Less opportunity for error is introduced into the driver chain than in the conventional triple EF arrangement. The drivers and pre-drivers all can be mounted together on an aluminum bar that spans across the amplifier PWB so they are all at the same temperature and there is some modest common heat sinking.

The current source feeding the folded EF pre-driver limits the amount of current that can be driven into the base of the driver transistor (unlike the situation with the conventional Triple). This can reduce the chances of catastrophic failure under some conditions like output short circuits.

A bootstrapped version of the DBT is shown in Figure 12.7b. Here the collector voltages of the folded emitter followers move with the signal because they are bootstrapped by the signals at the emitters of the opposite driver transistors. This mitigates the nonlinear effects of their collector-base capacitances on the VAS output node. The Early effect in the pre-drivers is also greatly reduced. This arrangement permits the use of fast, low-voltage transistors for the pre-driver.

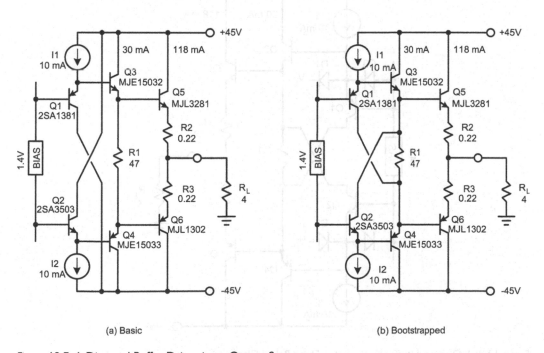

(a) Basic   (b) Bootstrapped

Figure 12.7 A Diamond Buffer Driver in an Output Stage

Power dissipation in the pre-driver is also much smaller. Most of the dissipation that was once in the pre-driver is now in the current source, whose signal characteristics have little influence on the signal path.

If a diode is connected from the emitter of the folded EF to a fixed voltage, a Baker clamp is conveniently formed whose capacitance effects are buffered from the VAS by the pre-driver. Such a connection is shown in Figure 12.8. D1 and D2 are connected to positive and negative clipping voltages. Additional diodes (D3 and D4) are needed from base to emitter of the pre-driver to prevent excessive reverse biasing of the base-emitter junction when the clamping action takes effect. These diodes also prevent the VAS transistors from saturating. D3 and D4 float with the signal, so their capacitance effects are essentially eliminated. This diode can be a small, fast diode. The arrangement is here referred to as a *flying Baker clamp*.

Diamond Buffer Quad

In some cases additional buffering in the output stage might be desirable, as when a large multi-pair output stage employs a large driver transistor with substantial capacitance like an MJL3281/MJL1302 pair. Here the *diamond buffer quad* (DBQ), shown in Figure 12.9a, may be a good choice. It consists of a diamond buffer followed by a 2-EF Darlington output stage. The input transistor of the diamond buffer is thus a pre-pre-driver and the output transistor of the diamond buffer is thus the pre-driver. Since there is nominally no net DC shift passing through the diamond buffer, the bias spreader must supply a total of 4 V_{be}, just as with a simple 2-EF output stage. The two transistors in the diamond buffer can be bolted together to facilitate their temperature tracking and cancel their V_{be} temperature dependencies. Simulations show

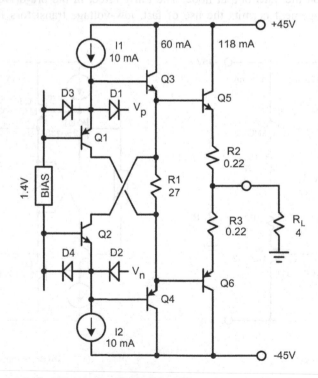

Figure 12.8 A Diamond Buffer Triple with a Flying Baker Clamp

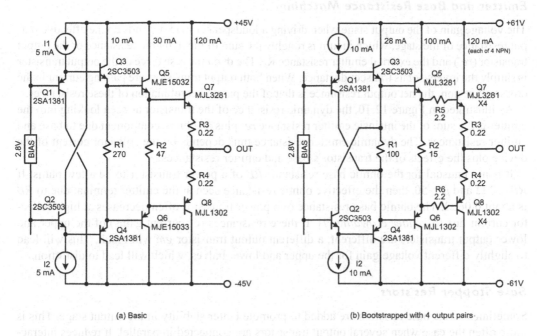

Figure 12.9 A Diamond Buffer Quad

that the DBQ has good HF stability, similar to that of a 4-EF stage but without the bias stability shortcomings. Figure 12.9b shows a bootstrapped DBQ designed for use with a four-pair output stage.

An alternative DBQ places the diamond buffer directly ahead of the output transistor, with a simple EF ahead of the diamond buffer. This is essentially a DBT preceded by an emitter follower. It allows the two higher-power drivers in the diamond buffer to be bolted together and have their temperature coefficients cancel, but eliminates the ability to bootstrap the first EF.

12.5 Output Transistor Matching Considerations

The audio signal passes through the top output transistors on the positive half-cycle and through the bottom transistors on the negative half-cycle. The signal takes two different paths through the output stage depending on the direction of net current flow to the load. If the behavior of the top and bottom halves is not the same, even-order distortion will result.

Beta Matching

If the current gain of the upper half of the output stage is different than that of the lower half, the load on the VAS will be different on the positive and negative half-cycles of signal current swing, creating second harmonic distortion. This effect is greatly reduced by the use of a Triple. In such a case, if the VAS output impedance is low at higher frequencies (where output stage distortion counts most), the output transistor will behave as if it is being driven by a voltage source and β mismatch will matter much less.

Emitter and Base Resistance Matching

The voltage gain of the output stage when driving a loudspeaker load depends on the effective output resistance of the stage, which in turn is roughly the sum of the dynamic resistance of the output transistor (re') and the external emitter resistance R_E. The dynamic resistance of the output transistor is simply the inverse of its transconductance. When both output transistors are passing current in the crossover region, the net output resistance is that of the parallel combination of these resistances.

As illustrated in Figure 12.10, the dynamic resistance of the transistor as seen looking into the emitter is the sum of the intrinsic emitter resistance re' plus an ohmic component due to base and emitter resistances. The dynamic emitter resistance thus depends on the collector current of the device plus the effects of the transistor's base and emitter resistances.

It is not unusual for the ohmic base resistance RB of a power transistor to be a few ohms. If $RB = 3\ \Omega$ and $\beta = 50$, then the effective ohmic resistance seen at the emitter terminal due to RB is about $0.06\ \Omega$. The ohmic base resistance of a power transistor often decreases at high collector current due to *emitter crowding* [7]. If these resistances (or current gains) of the upper and lower output transistors are different, a different output transistor gm will result. This will lead to slightly different voltage gain for the upper and lower halves, which will lead to distortion.

Base Stopper Resistors

Sometimes series base resistors are added to promote better stability in the output stage. This is more often the case when several output transistors are connected in parallel. It reduces interactions among the paralleled output transistors that can lead to instability or oscillation.

The base stopper resistors *stop* the fall of impedance seen looking into the base of the output transistor as frequency increases. They place a minimum value on this impedance and prevent it from going negative under conditions of capacitive loading at the emitter. They *stop* oscillation. Small-signal simulations may not show the instability that is of concern because the ohmic base resistance of the transistor may not have been reduced by emitter crowding under the quiescent conditions of the small-signal simulation. The base stopper resistors also tame parasitic oscillations by reducing the Q of resonant circuits involving the base terminal of the transistor and associated inductances and capacitances.

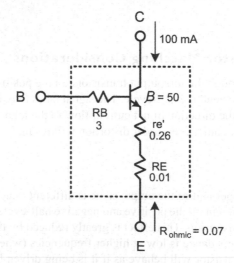

Figure 12.10 Dynamic Emitter Resistance Components

If base stopper resistors are employed, the voltage drop across them can be a significant function of base current, and this will lead to dependence of output stage gain (and bias) on the power transistor β and matching of β among the NPN and PNP output transistors with their individual base stopper resistors. Base stopper resistors should not be made larger than necessary. In many cases as little as 2 W is sufficient.

The base stopper resistor is just a further extension of RB. It will thus contribute to effective emitter resistance and rob transconductance in the same way as RB. By itself, a 5-Ω base stopper resistor with an output transistor whose $\beta = 50$ will add a full 0.1 Ω to the transistor's ohmic component of emitter resistance. This ohmic addition can influence the effective value of emitter resistance and alter the optimum quiescent bias current required for minimum crossover distortion. Conversely, if an output stage was designed for optimum bias with $R_E = 0.22$ Ω for an ideal transistor, the actual value of R_E with the real transistor might have to be smaller. These effects can be seen in Figures 12.3 and 12.4 where the optimum bias current for the output stage is less than the theoretical optimum value.

In some cases a base stopper resistor at the driver base is also helpful to stability. This is because the driver is an emitter follower and it sees a capacitive load, namely, the sum of all of the base-collector capacitances of the paralleled output transistors. Each output transistor can contribute on the order of 600 pF of C_{cb} when V_{cb} is small, as when the output is near the rails. This capacitance is a significant function of V_{cb}, and may only be 200 pF or less under quiescent conditions. Small-signal AC simulations for stability carried out under quiescent conditions may thus be optimistic.

12.6 Dynamic Crossover Distortion

Transistors do not turn off instantaneously. This is especially true of power transistors because they are slower than small-signal transistors. As the signal current goes through crossover in a class AB output stage at high frequencies, the rate of change of current can be substantial, and it can be difficult for the transistor turning off to do so fast enough.

The failure to turn off quickly and cleanly can result in a brief period when both transistors are conducting together (not just the quiescent bias current). This is variously referred to as *common-mode conduction, totem-pole conduction, conduction overlap, cross-conduction* or *shoot-through conduction*. The resulting impairment to output stage behavior is called *dynamic crossover distortion*. Figure 12.11 shows the simulated collector current waveforms of the power transistors in a class AB output stage being driven hard at 20 kHz. Notice the separation of the two waveforms at the crossover point. It is well in excess of the quiescent bias current. This is evidence of cross-conduction. Notice also the sharper edge of the current waveform of the transistor turning off.

Cross-conduction results when the transistor turning on must conduct additional current to overcome that of the transistor that is turning off too slowly. Obviously, any hope of an optimal-biased class AB crossover transition goes out the window under these conditions. In many amplifiers the power supply current will rise at high frequencies when driving a load to high levels. This is a symptom of cross-conduction. In some cases cross-conduction can lead to destruction of the amplifier.

Transistor Turn-Off Current Requirements

Consider a 100-W/8-Ω amplifier delivering 200 W into a 4-Ω load at 20 kHz. Voltage slew rate will be 5 V/μs and the corresponding current slew rate will be 1.25 A/μs. If the f_T of the transistor is known and the rate at which it needs to turn off (in amperes per microsecond) is known, then

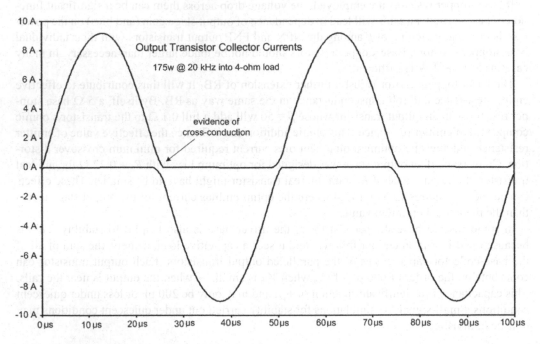

Figure 12.11 Output Transistor Current Waveform

the base current required to be pulled out of the base can be calculated. For a BJT, the hybrid-pi input capacitance is approximately

$$C_{\pi} = gm/\left(2\pi * f_T\right) \qquad (12.1)$$

For a respectable 4-MHz power BJT (MJL21193/4) biased at 150 mA, $C_{\pi} = 0.24$ µF. At higher collector current C_{π} will be much larger. At only 1 A, C_{π} will be a whopping 1.6 µF! The rate of voltage change at the base is

$$dV_{be}/dt = I_b/C_{\pi} \qquad (12.2)$$

where I_b is the base discharge current. In a conventional output stage with an EF driver, that current can be no more than the driver bias current. Substituting for C_{π}, we have

$$dV_{be}/dt = I_b * 2\pi * f_T/gm \qquad (12.3)$$

The collector current slew rate dI_c/dt is merely $gm * dV_{be}/dt$, so we have, re-arranging,

$$dI_c/dt = 2\pi * f_T * I_b \qquad (12.4)$$

Required base turn-off current is then

$$I_b = \text{ISR}/\left(2\pi * f_T\right) \qquad (12.5)$$

where ISR is the output current slew rate.

Notice that I_c and gm are not in this simple equation. Notice also that the number of paralleled output devices does not appear in this equation. If the number of devices is doubled, current per device gets halved and C_π per device gets halved, so total charge that must be removed from all of the bases for a given ISR remains the same. However, effects of collector-base capacitance have been ignored in this simple estimate. Current flowing through the collector-base capacitance when the signal voltage is changing rapidly will subtract from the available base turn-off current.

An Example BJT Power Transistor

For a conventional power transistor with $f_T = 4$ MHz and a healthy $I_b = 50$ mA, the achievable current slew rate is only 1.2 A/μs. This is just equal to the current slew rate for the amplifier example above, where 200 W was being driven into a 4-Ω load at 20 kHz. To put this in perspective, bear in mind that the peak output voltage is 40 V and the peak output current is 10 A.

As a sanity check, consider the case where the transistor is turning off and its operating current is 1 A. This will be at a point in the waveform where the current rate of change is near its maximum value of 1.2 A/μs. As mentioned above, C_π will be 1.6 μF at this point. The voltage rate of change across the base-emitter junction will be $I_b/C_\pi = 0.031$ V/μs. Intrinsic transconductance of the transistor will be about 40 S. The rate of change in the transistor collector current will be about 40 * 0.031 V/μs = 1.24 A/μs.

Figure 12.12 shows the collector current of the driver transistor under these conditions with the exception that driver bias current has been reduced to 20 mA. When the collector current of the emitter follower driver transistor goes to zero, it has lost control of the output transistor and is unable to make it turn off faster.

Because a push-pull class AB output stage is assumed, the opposite power transistor may be turning on and contributing to the total demanded output current slew rate. It is therefore arguable that the transistor turning off only needs to be turning off at a rate of about 0.6 A/μs in order to achieve the total current rate of change of 1.2 A/μs demanded by 40 V peak at 20 kHz into a 4-Ω load. If the PNP power transistor must turn on in order to decrease the current being sourced into a resistive load while the output voltage is still positive, distortion will surely be created, however.

The situation is better for BJT *ring emitter transistors* (RETs) or devices with similar advanced structures that provide higher f_T. The ON Semiconductor MJL3281 device is a good example. Even at $I_q = 150$ mA, these devices still manage a respectable f_T on the order of 20 MHz. This means that with a base discharge current of 50 mA, each device is capable of turning off at a rate of about 6 A/μs.

Figure 12.13 shows power transistor f_T as a function of collector current for a nominal 4-MHz transistor and a RET. The Y-axis values for the MJL3281 are 10 times those for the slower MJL21194. Notice the decline of f_T for both devices at low and high collector current. The decline of f_T at high current is due to the *Kirk effect* [8].

Turning Off the Transistor Under Conditions of Beta Droop and f_T Droop

An additional concern is turning off the output power transistor in the vicinity of its maximum current at the peak of a sinusoid (when driving a resistive load). This is different than the other challenge of turning off the transistor during high di/dt in the crossover region.

Here the problem occurs when the transistor β and f_T become very low at high power under conditions of low V_{ce} and high I_c. A lot of current flows into the base as the peak amplitude of the signal is approached. This high base current is necessary to turn on the transistor to the high demanded current level in light of the decreased β at low V_{ce}.

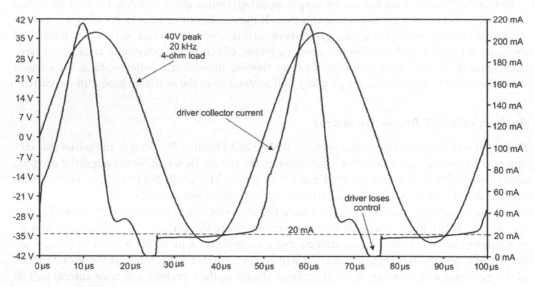

Figure 12.12 Driver Collector Current Losing Control of the Output Transistor

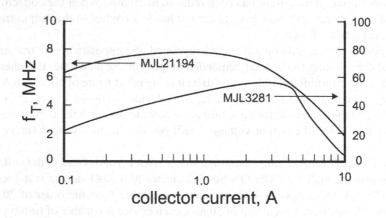

Figure 12.13 Power Transistor f_T Versus Collector Current

Just after the peak, the rate of change of collector current becomes reversed, and turn-off of the transistor must begin. All of that charge in the base must be pulled back out of the base. This tends to be a bigger problem at high frequencies.

The Role of Collector-Base Capacitance

Consider once again a 100-W/8-Ω amplifier delivering full power of 200 W into a 4-Ω load at 20 kHz. Voltage slew rate (VSR) will be 5 V/μs and the corresponding current slew rate will be 1.25 A/μs. The slew rate of 5 V/μs will cause a current to flow in the output transistor collector-base capacitance that will tend to oppose transistor turn-off.

For C_{cb} of 500 pF, this current will be VSR * C_{cb} = 5 V/μs * 500 pF = 2.5 mA. This is small compared to the amount of reverse base current needed for turn-off (here 50 mA for f_T = 4 MHz), but bear in mind that an output stage employing n paralleled output pairs will suffer a loss in turn-off current of n times 2.5 mA in this example. This current's proportion of the total required turn-off current will increase with the number of paralleled pairs because the total amount of minority carrier charge in the output transistor(s) will remain about the same as more devices are added.

Higher slew rates delivered into lighter loads will increase the relative proportion of this component of turn-off current even further. Consider a 200-W/8-Ω amplifier with four output pairs delivering 20 V/μs to an 8-Ω load. I_{Ccb} will be 40 mA. Under these conditions ISR = 2.5 A/μs. If transistor f_T is 20 MHz, I_b required for sweep out of stored charge carriers is only I_b = ISR/$2\pi f_T$ = 20 mA, which is less by a factor of 2 than the current through C_{cb}. Finally, it is important to bear in mind that C_{cb} becomes much larger at signal amplitudes near clipping, where V_{cb} is small.

The Speedup Capacitor

We have seen that the emitter follower driver is much more effective in turning on the output transistor than in turning it off quickly. Of course, the driver transistor on the opposite side would be in a good position to turn off the output transistor. Figure 12.14 shows an output Triple that employs a *speedup capacitor*. The capacitor works by resisting voltage change from base to base of the output transistors. This allows the complementary driver to contribute in pulling current out of the base of the transistor being turned off. Under AC conditions, it effectively provides push-pull drive for the bases of the output transistors.

A number of amplifiers use the speedup capacitor and it does reduce cross-conduction current and high-frequency distortion on sinusoidal tests. The amount of the capacitance must be

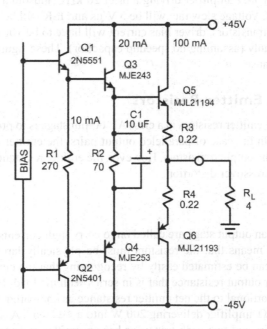

Figure 12.14 A Triple with a Speedup Capacitor

considered. If it is to provide turn-off current to the base of one of the power transistors, it will accumulate an increased voltage and it will begin to allow a larger spread voltage between the bases of the output devices, corresponding to the increased common-mode current. If the speedup capacitor is small, it may be ineffective in drawing charge out of the base of the power transistor because it will charge up quickly. If it is large, and does its job, it may leave the output stage in an over-biased state just after the cross-conduction event. This may not be a bad thing, given the alternative.

Suppose that a 10-μF speedup capacitor is asked to deliver 60 mA of turn-off current over an interval of 10 μs (corresponding to 1/5 cycle at 20 kHz). The capacitor will acquire a charge of 60 mV. If each R_E is 0.22 Ω, this added V_{spread} will correspond to increased bias current of 136 mA. Note that this is on the order of typical I_q.

Suppose a 10-μF speedup capacitor is being used with a driver biased at only 20 mA, as shown in Figure 12.14. The driver emitter resistor is 70 Ω. The time constant of 70 Ω and 10 μF is 700 μs. More importantly, the 20-mA of available driver discharge current will require 30 μs to discharge the speedup capacitor back to the nominal spread voltage, corresponding to 0.3 cycles at 10 kHz. During this time one or both driver transistors will be conducting less than its normal class A bias current. Because the speedup capacitor stores charge, it can have a lasting effect on the behavior of the output stage after the transistor base discharge event. The use of a larger speedup capacitor will do a better job of sweeping out stored minority carriers, but it will lengthen the hangover period. Pick your poison. The speedup capacitor may not be the panacea that some think it to be. Caution is always required when a capacitor is used in a nonlinear circuit. I do not use a speedup capacitor in my designs, and instead use at least 30 mA of driver bias current in combination with high-f_T output transistors like the MJL3281/1302 (whose f_T is typically 20–50 MHz).

Current Slew Rate Requirements

Consider a beefy 100-W/8-Ω amplifier driving a brief 20 kHz sinusoidal burst into a 2-Ω load, corresponding to 400 W. Voltage slew rate will be 5 V/μs and ISR will be 2.5 A/μs. If the amplifier uses 4-MHz power transistors, driver bias current will have to be 100 mA in order to handle this transient signal cleanly (assuming no speedup capacitor). These figures may seem extreme, but they are cause for pause.

12.7 The Output Emitter Resistors

The main purpose of the emitter resistors in a class AB output stage is to properly help set a stable quiescent bias current. In the case of paralleled output pairs, the emitter resistors also promote current sharing among the output transistors. However, the emitter resistance also plays an important role in controlling crossover distortion.

Power Dissipation

The emitter resistors in an output stage are called on to carry high currents, and so must be sized accordingly. This often means that the resistors must be physically large and well ventilated. The power dissipation can be estimated easily by recognizing that the emitter resistors together account for an amplifier output resistance that is in series with the load. Power dissipated in the emitter resistors is proportional to the net emitter resistance as compared to the load resistance.

Consider a 100-W/8-Ω amplifier delivering 200 W into a 4-Ω load. Assume that the amplifier employs two output pairs and that each transistor has an emitter resistor of 0.2 Ω. The output resistance to be used in the estimate is then the output resistance of the output stage, which is 0.1 Ω. If 200 W is being dissipated in a 4-Ω load, then 5 W will be dissipated in the emitter resistors. Because there are four of them, each one will be called on to dissipate 1.25 W. It is important to stress that this is average power dissipation over a cycle.

The peak current delivered to the load is 10 A. This means that each emitter resistor must be capable of handling peak current of 5 A. In some cases, this will govern the required size of the emitter resistors. The peak voltage drop across the emitter resistors will be 1 V, and the peak power dissipation for each resistor will be 5 W.

Inductance

It is usually recommended that the output stage emitter resistors be non-inductive. However, there is little evidence that anyone has taken the trouble to measure the inductance of ordinary wire-wound emitter resistors and to quantify their effect. I am also unaware of any good technical analysis of emitter resistor inductance and its actual effect on output stage behavior. However, given the switching nature of a class AB output stage, one would instinctively assume that very low inductance in the emitter resistors is important.

The inductance of several typical inductive power resistors in the usual range of values used for emitter resistors was measured. The results indicate that conventional (inductive) wire-wound resistors in the 0.1 Ω to 0.5 Ω range commonly used as emitter resistors in power output stages can be expected to have between 16 nH and 70 nH of parasitic inductance. Inductance appears to be smaller with smaller resistances, as expected. A safe worst-case estimate for typical modern output stages would be 120 nH for a 0.33-Ω wire-wound resistor. The axial and radial versions of the 0.33-Ω resistor had nearly identical values of inductance. At the other extreme, a pair of 0.68-Ω metal oxide 2-W resistors in parallel had a very low inductance of 25 nH. This is little more than that of a 1-inch trace.

One thing that must be considered in estimating the impact of small amounts of inductance in the emitter resistors is the expected current rate of change (ISR) in the resistor and the resulting inductive component of the voltage drop. The rate of change will be greatest near the crossover region, where the power transistor is in the process of turning on or turning off. Using ISR = 2.5 A/μs from the example above, each resistor will see 1.25 A/μs. If resistor inductance is 100 nH, we have an inductive voltage drop of 125 mV. This is not insignificant.

Paralleled Emitter Resistors

One approach to reducing inductance and increasing power dissipation without resort to large wire-wound resistors is to parallel several smaller power resistors to achieve the necessary resistance and power dissipation. Two 3-W metal oxide film (MOF) resistors can be connected in parallel to obtain a suitable non-inductive emitter resistor. The penalty here is an increase in occupied board space unless the resistors are stacked.

12.8 Output Networks

Emitter followers are the basis of most output stages. They can be picky about their load when it comes to local high-frequency stability. Emitter followers can become unstable if they are lightly loaded at high frequencies or if they are called on to drive a capacitive load [9].

A capacitive load can also introduce another pole into the output stage frequency response, possibly destabilizing the global negative feedback loop. For these reasons, most amplifiers incorporate an output network that controls the impedance seen by the output stage and isolates the load from the output stage at high frequencies. This helps to make the amplifier stable with the great variety of unknown loads it may be presented with by speaker cables and loudspeakers.

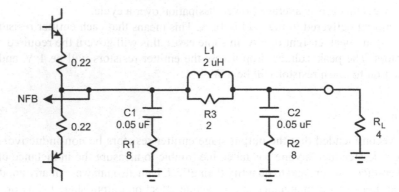

Figure 12.15 An Output Network Arrangement

Figure 12.15 illustrates a simplified output stage with an output network and a load. The network includes a series R-C Zobel network to ground on the input side and a parallel R-L network in series with the signal path to the load. For generality, a second Zobel network is shown on the output side of the network for variants that will be discussed later.

The Zobel Network

The series combination of R1 and C1 shunting the output node is called a Zobel network. The purpose of the Zobel network is to ensure that the emitter follower output stage sees at least some resistive loading out to very high frequencies. This is important if the amplifier has no load or if the loudspeaker load becomes inductive at high frequencies.

Power dissipation in the Zobel network's resistor must be considered. At minimum, the resistor in the Zobel network should have a sufficient power dissipation rating to withstand continuous operation of the amplifier at full power at 20 kHz. Consider a 100-W/8-Ω amplifier with a typical Zobel network consisting of 0.05 µF and 8 Ω. This combination will be resistive at frequencies above about 400 kHz. The impedance of the capacitor will be about 159 Ω at 20 kHz. If the output voltage is 28 V RMS, the current in the network will be 176 mA. The power dissipation in the resistor will be 0.25 W. However, in the unfortunate event that the amplifier breaks into a high-power oscillation at ultrasonic frequencies, a small Zobel network resistor will likely get fried. For this reason, Zobel network resistors are often very oversized.

Because the function of the Zobel network is to maintain a resistive load, it is important that the resistor be non-inductive. The load presented by the Zobel network should remain resistive up to at least the f_T of the power transistors. An 8-Ω wire-wound resistor with 500 nH inductance will become inductive at frequencies above 2.5 MHz. This argues against the use of wire-wound resistors in the Zobel network unless they are non-inductive. Metal oxide resistors with adequate dissipation are a better choice.

Distributed Zobel Networks

The Zobel network is often built with fairly large components in order to dissipate the power that will be present under high-frequency conditions, especially sine wave testing. Such big components are not usually very good out to very high frequencies.

An attractive alternative is to use multiple smaller Zobel networks in parallel, where each Zobel network is placed very close to one output transistor pair. Smaller components can be used;

they can be less inductive, and they can more effectively damp out high-frequency resonances. Such an approach provides a better high-frequency shunt path for output stages employing multiple output pairs in parallel. This can be more important when fast output transistors are being used. An output stage employing four pairs might include four Zobel networks, each consisting of a 33-Ω, 2-W metal oxide resistor and a 0.01-μF polypropylene film or COG ceramic capacitor.

The Series L-R Network

Most amplifiers include a small inductor in series with the output, usually with inductance between 0.5 μH and 5 μH. At high frequencies the impedance of the inductor increases and isolates the output stage and the global feedback take-off point from capacitive loads. The inductor is shunted by a small resistor (1–10 Ω) that helps damp out resonances that the inductor might have in combination with load capacitance.

A combination of 1 μH and 2 Ω might be employed in a high-performance amplifier. At very high frequencies the output stage will never see load impedance less than 2 Ω, even if the output is shunted by a large capacitance with low ESR. The impedance of the 1-μH inductor is 2 Ω at about 3 MHz. At frequencies above 3 MHz the load seen by the output stage will be substantially resistive regardless of what kind of load is connected to the amplifier output. The coil will reduce the amplifier's damping factor at high frequencies. The impedance of a 1-μH inductor at 20 kHz is about 0.13 Ω, implying a damping factor of 62 at 20 kHz.

Details of the amplifier output stage and global feedback compensation will govern how small an inductance can be used without risk of instability when driving difficult high-frequency loads. Amplifiers with low open-loop output impedance at high frequencies will usually permit the use of a smaller inductor.

The Effect of the Coil on Sound Quality

Some high-end amplifier designers claim that the presence of the output coil degrades the sound. It is hard to justify this solely on the basis of its impact on frequency and phase response when small inductances are used. A 1-μH output coil feeding a 4-Ω speaker load will cause a frequency response droop of less than 0.01 dB at 20 kHz. However, some amplifiers employ a 5-μH inductor in parallel with a 5-Ω resistor. Such a combination reduces the damping factor to only 13 at 20 kHz.

The coil must always be implemented as an air-core coil for best sound quality [10, 11]. Coils implemented with steel or ferrites will suffer nonlinearity, in some cases due to the approach of core saturation. Fortunately, air-core coils with values on the order of only 1 μH are quite small. The coil should be kept away from magnetic materials like steel for the same reason. The coil should also be kept away from devices or circuitry sensitive to radiation from it or from which it can pick up radiation, such as power wires carrying half-wave-rectified signal currents. The coil should be solid and not springy; no opportunity for vibration should exist. The high current flowing through the coil and the resultant magnetic fields could cause electromechanical movement that could create nonlinearity in the inductance. The coil should be held together by coil dope. Silicone or heat shrink tubing will also suffice.

Variations on the Networks

In some cases the shunt Zobel network will be placed on the output (downstream) side of the output L-R network, as shown with R2 and C2 present in Figure 12.15 and with R1 and C1 absent. This will normally work adequately, since at high frequencies the resistor across the coil will act as part of the series resistance of the Zobel network, ensuring loading of the output stage that extends to high frequencies.

Sometimes, the series resistance of the Zobel network in this position (R2) will be eliminated, allowing the shunt resistor across the output coil to effectively take on this duty as well. Placing the Zobel network on the downstream side of the coil can help with physical design, getting it off of the amplifier printed wiring board and perhaps out at the speaker terminals. However, this can also compromise the integrity of the load it provides for the output stage at very high frequencies due to the increased wiring inductance in the path to the Zobel network.

There may also be concerns about where the Zobel is returned to ground (local to the output stage or local to the speaker return). One advantage of the downstream Zobel network is that it provides some degree of low-impedance high-frequency termination directly at the speaker terminals, possibly reducing the opportunity of EMI ingress from the external world via the speaker cables.

The Pi Output Network

A further variation on output networks combines the advantages of the two approaches above. In this case a Zobel network is placed on both sides of the output coil as shown in Figure 12.15 when all of the components are present. The upstream Zobel network provides a low-inductance load for the output stage to very high frequencies and allows high-frequency currents to circulate local to the output stage. The downstream Zobel network provides a good resistive termination right at the speaker terminals at high frequencies, helping to reduce RFI ingress and damp resonances with, or reflections from, the speaker cables. Once again, in some cases the downstream Zobel is implemented without a series resistor, reducing it to merely a shunt capacitor.

Eliminating the Output Coil

Some designers do not employ an output coil in an attempt to eliminate its perceived influence on sound quality. This incurs some risk of instability when driving unusually capacitive loads with little effective series resistance.

In seeking to do without the output coil, three things must be considered.

- Local output stage stability
- Global negative feedback loop stability
- How bad a capacitive load one is willing to tolerate

The likelihood of trouble operating without an output coil can be reduced by employing a large number of output devices operating in parallel so as to greatly reduce effective output impedance. In some cases, the judicious use of base stopper resistors can also help. Finally, amplifiers that have a stiff, high-speed output stage and do not have a high gain crossover frequency for their negative feedback may better tolerate the absence of a coil. I believe that the risk of eliminating the output coil does not justify the perceived gain as long as the value of the output coil is not greater than 1 μH.

12.9 Output Stage Frequency Response and Stability

As mentioned in Chapter 6, the excess phase shift introduced into the negative feedback loop by poles in the output stage is one of the main limitations on how high a gain crossover frequency can be chosen. The frequency locations of the numerous poles in an output stage are governed by several factors, including f_T of the driver and output devices, base resistances and collector-base

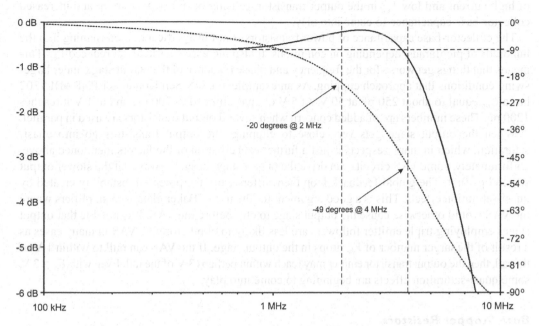

Figure 12.16 Gain and Phase of the Output Triple Driving a 4-Ω Load and 0.01 µF

capacitances. Device bond wire inductances and external stray inductances can also play a role in influencing the frequency and phase response of the output stage. Finally, the load as seen by the output stage through the output network can have a large influence.

Figure 12.16 illustrates the gain and phase of the Triple EF output stage when it is driving a 4-Ω load with 0.01 µF in parallel. The output stage employs a single output pair (MJL21193/4) with $R_E = 0.22$ Ω and includes the output network of Figure 12.15. Gain and phase are simulated at the feedback take-off point upstream of the output network. The Triple is driven from a source impedance of 100 Ω.

Excess phase at 2 MHz is 20 degrees. Notice that there is no amplitude loss at 2 MHz; response is actually up by 0.4 dB at this frequency. The addition of 2-Ω base stopper resistors eliminates the peak and increases phase lag to 26 degrees. The use of 5-Ω base stopper resistors increases phase lag to 35 degrees.

Variation with Operating Point

The f_T of output transistors varies with current. It gets smaller at low currents. This can happen at the quiescent bias current value. The f_T will often peak in the neighborhood of 1 A to 3 A. At high currents the f_T also droops, as does β. This effect was evident in Figure 12.13, where the f_T of the MJL21194 was shown as a function of collector current for V_{ce} of 10 V.

The f_T of output transistors also droops at lower collector-base voltages. This can be especially troublesome under conditions of output swings near clipping, where current is high and collector-base voltage is low, creating a double whammy working against f_T. The f_T of the MJL21194 drops by about 10% when V_{ce} falls from 10 V to 5 V. Parasitic oscillations can result if too much excess phase shift results from the reduced f_T.

It is wise to simulate the amplifier with output transistors whose models are operating at the minimum expected f_T and maximum C_{cb}. This can be accomplished by doing an AC simulation with a large DC offset while driving a load resistance. This will simultaneously create conditions of high current and low V_{ce} in the output transistor, causing beta droop, f_T droop and increased collector-base capacitance to come into play.

The collector-base capacitance of power transistors is not insignificant, often running into the hundreds of picofarads, depending on conditions. It also increases markedly at reduced V_{ce}. This means that things get worse for the frequency and phase response of the output stage under large-swing conditions that approach clipping. As an example, the ON Semiconductor PNP MJL1302 has a C_{ob} equal to about 250 pF at 50 V. At 5 V C_{ob} has climbed to 700 pF, and at 1 V it reaches 1200 pF. These numbers are of added concern when several output transistors are used in parallel.

When the output stage gets very close to clipping, the output transistors go into quasi-saturation, which in many respects is just a further exacerbation of the factors mentioned above. Unfortunately, some VAS circuits can drive the output stage in such a way that the slower output stage clips first. The global feedback loop then suffers from the potential instability created by quasi-saturation effects. This is a good argument for the use of Baker clamps in amplifiers where the VAS would otherwise cause the output stage to clip before the VAS. It is notable that output stages employing triple emitter followers are less likely to clip before the VAS in many cases as a result of the larger number of V_{be} drops in the output stage. If the VAS can pull to within 1 V of the rail, then the output transistor emitter may reach within perhaps 3 V of the rail. Even with $V_{ce} = 3$ V, some quasi-saturation effects are beginning to come into play.

Base Stopper Resistors

Base stopper resistors are sometimes added in series with the base of the output transistors to improve local output stage stability. This is more often seen in output stages where several devices are connected in parallel. Recognizing that an output stage emitter follower can have negative input impedance under some conditions (such as capacitive loading), the addition of a base series resistance can bring that negative impedance back to positive impedance. The base stopper resistor also maintains a lower limit on the total value of base resistance in light of reduced intrinsic base resistance of the output transistor under conditions of emitter crowding at high currents. Base stopper resistors are more important when fast output transistors are employed.

The use of base stopper resistors is not without a price. They add to the total base resistance of the output transistor and ultimately contribute to a reduction in the transconductance of the output transistor at high frequencies as the β_{ac} of the device decreases with increases in frequency. As explained earlier, the base stopper resistors can also influence the optimum class AB bias point by altering the ohmic resistance seen looking into the emitter of the output transistor.

Load Capacitance

Because the output stage has non-zero output impedance, placing any capacitive load on it will introduce a high-frequency pole and consequent frequency response roll-off at high frequencies. It is important to recognize that the open-loop output impedance increases at high frequencies. The bottom line here is that feedback loop stability must be assessed with worst-case load impedances in place.

Excess Phase

The concept of *excess phase* recognizes that a circuit like an output stage may have many poles at higher frequencies, each adding some small amount of phase shift at frequencies in the range of the gain crossover frequency. These phase contributions add up, and sometimes are not accompanied by very much amplitude roll-off. For this reason, it is convenient to lump these parasitic high-frequency pole effects into what is called *excess phase*. A pole at 10 MHz will create a loss of only 0.04 dB at 1 MHz, but will introduce nearly 6 degrees of phase shift. Seven poles at 10 MHz will introduce almost 45 degrees of phase shift (the same as a single pole at the target frequency), but will introduce only 0.3 dB of loss, 1/10 that of a single pole at 1 MHz. In the vicinity of 1 MHz, the effect of these far-out poles is much like that of a constant time delay on the order of 125 ns.

Gain and Phase Margin

We have seen that the output stage plays a major role in establishing how high the gain crossover frequency can be made while retaining adequate loop stability. The gain and phase response of the output stage can vary considerably over different operating conditions. For this reason, it is important to design the power amplifier with adequately conservative feedback compensation so as to achieve generous gain and phase margins. It is crucial to avoid amplifier parasitic oscillations under all conditions of load and signal excursion, including clipping.

Stabilizing the Triple

The Triple has been strongly advocated throughout this book for BJT output stages. However, there is a dark side to the Triple. It tends to be more prone to local parasitic oscillations than the Darlington.

Emitter followers are prone to oscillate when they drive a capacitive load because their falling AC beta with frequency transforms the capacitive load impedance at the emitter into a negative resistance as seen at the base. Frequency-dependent negative resistance (FDNR) is always an invitation to oscillation. Bear in mind that the base-collector capacitance of a subsequent emitter follower can look like a capacitive load to an emitter follower. This is partly why the inclusion of a collector resistor or base resistor sometimes improves stability. The numerous terminal inductances and inter-element capacitances like C_{bc} and C_{be} further contribute to instability by creating resonances and forming Colpitts and Hartley oscillator topologies. The formation of these oscillator topologies is illustrated in Figure 14.6 in the discussion of MOSFET power transistor stability. While a BJT does not have significant collector-emitter capacitance, analogous to C_{ds} in Figure 14.6a, any capacitance from the output node of the output stage to ground will have the same effect. The Triple is more difficult to stabilize because it involves more EF stages that interact with each other.

Figure 12.17 shows an output Triple with the numerous parasitic inductances that may be found in transistors and their wiring. Bond wire inductance is usually on the order of 10–15 nH, while a 1-inch copper trace contributes about 20 nH. For purposes of simulation and illustration, every base terminal and every emitter terminal has a 35-nH inductor in series with it. Every collector terminal has 10 nH in series with it. Notice also the inclusion of some common rail inductance. This can allow the formation of a feedback path from the collector of the output transistor back to the driver or pre-driver. Assume that there is no such thing as a solid AC ground at high frequencies. Transistor base-collector capacitances are also shown, since they often play a key role in forming oscillator topologies.

Figure 12.17 also shows some circuit precautions that suppress instability. Both the output transistors and drivers incorporate base stopper resistors. Placing small base stopper resistors in the bases of the driver transistors in a Triple may provide a stability advantage often overlooked by designers. Emitter followers that are isolated from each other by a resistance equal to at least $4/gm$ of the driving stage are sometimes more stable (my rule of thumb). The use of a base stopper for the driver should always be simulated to see whether it improves or degrades stability in

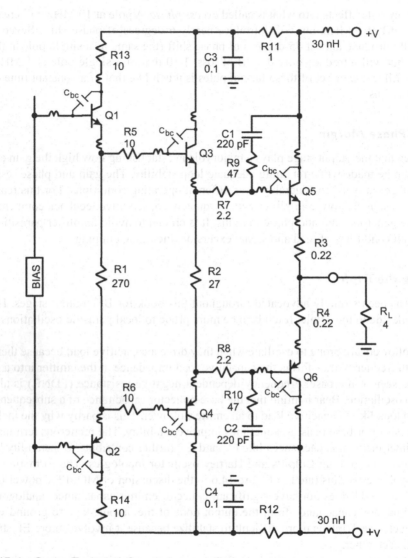

Figure 12.17 An Output Triple with Stray Inductances Shown

a particular design. Small resistances in series with EF collectors can also help stability. This is shown for the pre-driver transistors.

Base stopper resistors prevent the formation of negative impedances and damp resonances. Other circuit approaches that damp resonances and kill oscillator Q are also helpful. Base-collector Zobel networks on the output transistors help damp resonances formed by C_{bc} and terminal inductances. Shown also is a high-frequency R-C filter in the rail between the output transistors and the preceding driver stages. This acts in two ways. First, it breaks the feedback path via the common rail at high frequencies. Second, it acts as a Zobel network on the supply rail to dampen resonances there. This R-C filter is an opportunity for stabilization that many designers overlook.

The inclusion of ferrite beads in collector or base leads is a popular and effective practice for stabilizing circuits. Ferrite beads provide impedance that rises from zero at DC to some resistive value (often 10–100 Ω) at frequencies in the MHz range. The resistive component of their imped-ance serves to damp resonances. However, their use is generally frowned on in audiophile circles because of their potential for introducing nonlinearity.

Good layout is important, but it alone is often not enough. Designers should strive for a layout-robust circuit design. Even the best, tightest layout cannot completely eliminate inductances and the formation of oscillator topologies. Many of the precautions shown in Figure 12.17 are very inexpensive to implement. However, it is also true that not all of them together are necessary in a given design.

12.10 Sizing the Output Stage

The output stage is where the money is in an amplifier (with the possible exception of the power supply). It includes expensive power transistors and expensive heat sinks.

Power Dissipation

Power amplifier output stages are not 100% efficient. This means that power is wasted in the form of heat. The power dissipation in an output stage is simply the power delivered to the output stage from the supply rails less the amount of power being delivered to the load. It was shown in Chapter 7 that the maximum power dissipation in an output stage does not occur when the amplifier is producing maximum output, but rather when it is producing a lesser amount, sometimes on the order of 1/3 rated power. Choosing the number of output pairs to provide adequate static power dissipation is only the first step, and is usually not sufficient to guarantee reliable operation. Sizing the output stage for power dissipation will be discussed in detail in Chapter 17.

Safe Operating Area

Power transistors cannot dissipate rated power under all conditions, even when that power is properly de-rated for worst-case temperature conditions. This reduced capability is usually evidenced at higher operating voltages and is attributed to an effect called *secondary breakdown*. One cause of secondary breakdown is thinning of the base at high voltage. At higher voltages the collector-base junction depletion region widens and causes the base region to become thinner. This can make the device more prone to uneven heat distribution across the die and lead to hot spots. Because the temperature coefficient of current as a function of voltage is positive for BJTs, the hot spots will conduct more current, making them still hotter and eventually leading to localized thermal runaway. Power transistors must always be operated within their *safe operating area* (SOA).

The key thing to bear in mind with SOA is that resistive loads do not tax the output stage nearly as much as reactive loads. Because load currents and output voltages are out of phase with a reactive load, it is possible for one half of the output stage to be sourcing significant current to the load when it has a high voltage across it. Safe operating area is both an amplifier protection issue and an output stage sizing issue. SOA and protection circuits are discussed in detail in Chapter 18.

12.11 Delivering High Current

While the focus on amplifier clipping is most often on voltage clipping, it is equally important that amplifiers do not clip by failing to deliver the amount of output current demanded by the load. Such a failure can happen as a result of the output stage not having enough current gain (especially when beta droop occurs at high current) or it can happen as a result of protection circuits limiting the amount of output current to protect the output transistors from a safe area violation or a short circuit.

Driving Low-Impedance Loads

When you design an amplifier, you must choose the lowest load impedance into which the amplifier will be able to operate satisfactorily. This is the simplest aspect of determining how much output current must be delivered. Given a value of resistive load impedance and maximum output voltage swing, the peak output current can be calculated. We saw examples of such calculations in Chapter 1. A 100-W/8-Ω amplifier driving a 4-Ω resistive load to 40 V peak can be called on to deliver 10 A. Ignoring power supply sag, the amplifier could be confronted with the task of delivering 20 A into a 2-Ω resistive load. Things get worse, however.

Loudspeaker Peak Current Requirements

The model of a simple loudspeaker is largely that of a parallel resonance. The resonance is electromechanical. It is created by the compliance of the suspension and the mass of the cone. As such, this resonance stores energy, often in the form of cone velocity. The cone velocity creates counter-emf; the loudspeaker acts like an electric generator. The counter-emf usually opposes current flow in the voice coil. The impedance of the loudspeaker rises at the resonant frequency. This would cause most to think that the resonance effect would result in a lighter load on the amplifier and less peak load current. However, if the phase of the driving signal becomes opposite that of the emf created by the velocity of the cone, much larger currents can result [12, 13]. Under some conditions the current required to drive a loudspeaker load can be twice that required to drive an equivalent resistive load. This phenomenon is discussed in detail in Chapter 22.

Beta Droop

The current gain of power transistors tends to fall off rapidly above a certain collector current. This is called *beta droop*. While a power transistor might have a beta of 80 at 1 A, the beta might fall to 20 or less at a collector current of 10 A. This makes it much harder for the amplifier to deliver high current to the load. This can also endanger the driver transistor because it will be asked to supply very high base current to the output transistor. Figure 12.18 illustrates the β versus

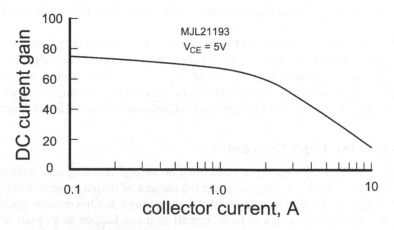

Figure 12.18 Beta Versus I_c for the MJL21193

I_c characteristic for the PNP MJL21193 power transistor at a collector-emitter voltage of 5 V. Beta drops to about 17 at 10 A.

Consider the 100-W/8-Ω amplifier above when it must deliver 20 A on rare transient occasions into a 2-Ω resistive load. A conservatively designed 100-W/8-Ω amplifier would include two pairs of MJL21193/21194 output devices. Each pair would be called on to produce 10 A on a transient basis. Current gain for the PNP MJL21193 can be as low as 20 at 10 A. The driver transistor will thus be called on to deliver 1 A. Some driver transistors may already be experiencing some beta droop themselves at 1 A. The beta of the PNP MJE253 falls to about 35 at 1 A. In a simple Darlington output stage, the VAS would then be called on to deliver 29 mA, more than many VAS designs are able to produce (surely without distortion). This illustration underlines the importance of employing an output Triple if high current is to be delivered into difficult loads by the amplifier. It also argues for larger driver transistors like the MJE15032/33 pair.

Newer transistors often do not suffer as much beta droop at 10 A. The current gain of the MJL3281/1302 is 60 at 10 A and falls to about 35 at 15 A. Delivery of high current into difficult loads is one area in which vertical power MOSFET output stages excel, since they do not suffer from beta droop.

f_T Droop

The speed of a power transistor also degrades at high current. This is referred to as f_T droop. It is caused by the *Kirk effect* [8]. The MJE21193 has an f_T of about 7 MHz at a collector current of 1 A, but this falls to less than 2 MHz at 10 A with V_{ce} less than 10 V. This means that the amplifier is operating with slower output transistors when it is delivering high current into the load. This can compromise feedback loop stability. It is another good reason to parallel output transistors.

Safe Operating Area of the Driver

The SOA of the driver transistor can also be taxed when the amplifier is called on to deliver high current, especially in light of the fact that the β of the power transistors falls at high currents. Under these conditions, the driver will still try to supply the increased base current required by the output transistor in order to drive the load. In some cases the failure of the output stage begins with the driver transistor entering second breakdown. The shorted driver transistor may then take out the power transistor. It is not always possible to know which came first in such a failure situation when all you have left is a pile of melted silicon.

Bear in mind that the driver in a Locanthi Triple operates in class A, and conducts over the full cycle. This can tax SOA in an amplifier with a generous driver bias current of 50 mA and 70 V rails. The driver must be able to handle 50 mA at nearly 140 V. The MJE15032/33 drivers can handle about 150 mA at 140 V.

The message here is that the drivers should be generously sized and that output protection circuits should be designed with protection of the drivers in mind as well. If an amplifier is built with numerous output pairs to increase output SOA, the SOA of the drivers must also be increased. In some cases, a single driver transistor of the same type as the output transistor is a good choice. Indeed, the f_T of RET output transistors makes them plenty fast enough to serve as drivers in an output Triple arrangement. The caveat here is that the pre-driver bias current must be sufficient to drive the larger collector-base capacitance of the driver transistor.

12.12 Driving Paralleled Output Stages

Most power amplifiers rated at 100 W and above into 8 Ω employ paralleled output transistors to achieve adequate power dissipation, current handling, safe area and current gain when driving heavy loads. This can introduce some technical challenges.

Output Transistor Current Sharing

A hotter transistor wants to conduct more current. This makes it still hotter, causing it to draw even more current. Even if this process does not lead to thermal runaway, it can easily cause very uneven sharing of load current among the multiple transistors in a paralleled output stage.

Beta droop may mitigate this to some extent, but this will apply less where the high transistor power dissipation is a result of high voltage rather than high current. For example, a power transistor conducting 3 A with 30 V across it will be dissipating 90 W, but it will be operating near its peak beta. Output transistors with matched betas will tend to share current more effectively. Increased emitter resistance will also help significantly.

Output Transistor Capacitances

Some larger amplifiers connect as many as four to ten pairs of output transistors in parallel. It is especially important to bear in mind that the collector-base capacitance of all of these transistors adds up and can become a significant sum. While C_{cb} under normal collector-base voltage conditions is often on the order of only 200 pF, this value can increase to nearly 1000 pF per device under large-swing conditions where the collector-base voltage becomes small. Such a large capacitive load can have small-signal and large-signal effects on the performance of the output stage. In a small-signal sense, the bandwidth of the output stage can decrease and global feedback stability may be reduced. Under large-signal high-frequency conditions, the driver may not be able to discharge this capacitance quickly enough.

The base-emitter capacitances of paralleled output devices will also add up, but the result of this is reduced by the bootstrapping effect of the output emitter followers. Output stages with more paralleled devices will often tend to have gain that is closer to unity, so the increased bootstrapping effect tends to offset the effect of the larger number of base-emitter junctions connected in parallel.

12.13 Advanced Output Transistors

The ON Semiconductor MJL3281/1302 complementary pair of output transistors is a good example of modern output transistors that combine high speed, good beta linearity and respectable SOA, and their electrical characteristics are worth mentioning here [14]. They are made with an advanced high-speed perforated-emitter process that puts them in the general class of ring emitter transistors (RETs). This means that they have high f_T and good safe operating area.

The ON Semiconductor datasheet for the MJL3281/1302 complementary pair of output transistors shows typical β to be quite flat at about 100 for most operating currents, falling to about 70 at 9 A [10]. Peak f_T is about 50 MHz at about 3 A, but is still about 20 MHz at 100 mA. At about 6 A, f_T has fallen to about 20 MHz. These transistors suffer much less beta droop and f_T droop than most other BJT power transistors. Collector-base capacitance is about 600 pF at $V_{cb} = 10$V. Safe operating area for the device is almost 100 W at 100 V.

The NJL3281/1302 ThermalTrak™ versions of these output transistors include a temperature-tracking diode in the package for much-improved thermal bias tracking [15]. These will be discussed at length in Chapter 17.

References

1. Bart N. Locanthi, "Operational Amplifier Circuit for Hi-Fi," *Electronics World*, pp. 39–41, January 1967.
2. Bart N. Locanthi, "An Ultra-low Distortion Direct-current Amplifier," *Journal of the Audio Engineering Society*, vol. 15, no. 3, pp. 290–294, July 1967.

3. Douglas Self, *Audio Power Amplifier Design Handbook*, 5th ed., Focal Press, Burlington, MA, 2009.
4. Barney M. Oliver, "Distortion in Complementary Pair Class B Amplifiers," *Hewlett Packard Journal*, pp. 11–16, February 1971.
5. Walt Jung, "Op-Amp Audio—Realizing High Performance Buffers, Part 2," *Electronic Design*, October 1, 1998.
6. Jim Williams, "High Speed Amplifier Techniques," Linear Technology Corporation, vol. AN-47, August 1991.
7. Giuseppe Massobrio and Antognetti Paolo, *Semiconductor Device Modeling with SPICE*, 2nd ed., McGraw-Hill, New York, 1993.
8. C. T. Kirk, "A Theory of Transistor Cut-off Frequency, f_T, Falloff at High Current Density," *IEEE Transactions on Electron Devices*, vol. ED-9, March 1964.
9. A. Neville Thiele, "Load Circuit Stabilizing Networks for Audio Amplifiers," *Journal of the Audio Engineering Society*, vol. 24, no. 1, pp. 20–23, January and February 1976.
10. A. Neville Thiele, "Air-Cored Inductors for Audio," *Journal of the Audio Engineering Society*, vol. 24, no. 5, pp. 374–378, June 1976.
11. A. Neville Thiele, "Air-Cored Inductors for Audio—A Postscript," *Journal of the Audio Engineering Society*, vol. 24, no. 10, pp. 830–832, December 1976.
12. Matti Otala, and J. Lammasniemi, "Intermodulation Distortion in the Amplifier Loudspeaker Interface," 59th Convention of the Audio Engineering Society, preprint no. 1336, February 1978.
13. Robert R. Cordell, "Open-Loop Output Impedance and Interface Intermodulation Distortion in Audio Power Amplifiers," 64th Convention of the Audio Engineering Society, preprint no. 1537, 1982. Available at www.cordellaudio.com.
14. ON Semiconductor datasheets for MJL3281D and MJL1302D, June 2006. Available at www.onsemi.com.
15. Mark Busier, "ThermalTrak™ Audio Output Transistors," ON Semiconductor, AND8196/D, February 2005. Available at www.onsemi.com.

Chapter 13

Output Stages II

13. INTRODUCTION

Chapter 13 is a second chapter dedicated to the topic of output stages. Output stages are usually the single biggest source of distortion in audio power amplifiers, and so warrant additional treatment.

13.1 VAS Output Impedance and Stability

VAS output impedance is important to output stage stability. The Miller-compensated VAS is a feedback circuit with complex output impedance, and it feeds an output stage that has complex input impedance. This can cause local instability and unforeseen interactions between the stages. Evaluating output stage stability in isolation, driven from a voltage source or from simple resistive source impedance, will not always tell the whole story. Taking steps to control the VAS output impedance can improve output stage stability and make it more valid to evaluate output stage stability in isolation with a resistive source. A good objective is to keep the VAS output impedance below a few hundred ohms from 10 kHz up to several tens of MHz, with only small dips and peaks due to reactive elements. Most output stages perform well when driven by a low-resistance source.

The shunt feedback provided by the Miller compensation capacitor will accomplish most of this goal up to several MHz if that local loop itself is very stable. However, being a shunt feedback circuit, its output impedance may begin to rise at higher frequencies as the amount of local feedback decreases, indicating an inductive component in its impedance. The addition of a shunt R-C Zobel network to ground at the VAS output node will take over the responsibility of keeping the impedance low and resistive to higher frequencies. Typical values for such a Zobel network are 100 pF and 100 Ω. If clipping occurs, the output impedance of the VAS may increase because the Miller feedback is no longer working. This might allow the output stage to become unstable under clipping conditions. This is a further reason to add a Zobel network at the output of the VAS.

VAS output impedance can be evaluated by simulation in the same manner as open loop gain is measured. One opens the feedback loop and closes it at DC with a very large inductor. An AC current is then injected at the VAS output node and the resulting voltage is measured in order to infer impedance. This measurement is best done with the output stage replaced by an ideal unity gain buffer to eliminate the influence of output stage loading. This approach properly takes into account any high-frequency interactions among the input stage, current mirror load and Miller-compensated VAS. Figure 13.1 shows the VAS output impedance for the BC-1 amplifier of Chapter 4 without and with the Zobel network. Notice how the Zobel network flattens the impedance curve out to beyond 50 MHz, with Z_{out} staying between 58 and 63 Ω from 40 kHz to 50 MHz.

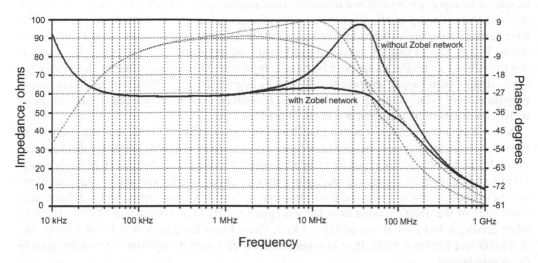

Figure 13.1 VAS Output Impedance Without and with Zobel Network

The VAS mid-band output impedance can depend on the ratio of C_M to the amount of capacitance C_{in} at the VAS input. C_M and C_{in} form a voltage divider that affects the shunt feedback loop gain. Without any such attenuation, VAS output impedance would be on the order of $1/gm$ of the VAS. C_{in} includes IPS and current mirror collector capacitances and VAS input capacitance. The larger the C_{in} capacitance, the higher the mid-band VAS output impedance.

VAS Design

The desirability of driving the output stage with low impedance underscores the importance of the VAS design. In a Miller-compensated VAS it can helpful to operate it with a larger Miller capacitor and perhaps at a higher current to keep its output impedance down up to very high frequencies. It is especially important that a more complex VAS design with more transistors in its local loop not have peaks in its output impedance at high frequencies. Running the VAS at higher current also allows the use of more stabilizing shunt capacitance in the VAS Zobel network without seriously compromising slew rate.

Double EF Stability

Three output stages were simulated to evaluate stability issues: a double EF, a tripe EF and a quad EF. Each output stage was evaluated for frequency and phase response, and for input impedance. The output stages employed MJE15032/33 drivers biased at 30 mA and MJL3281/MJL1302 output transistors biased at 100 mA with $R_E = 0.22\ \Omega$. Where applicable, pre-drivers and pre-pre-drivers used 2SC3503/2SA1381 transistors biased at 10 mA. The output stages were equipped with 5-Ω base stopper resistors and Zobel networks (10 Ω and 0.02 µF). They were evaluated with 2-, 4- and 8-Ω loads. They were driven through a 100-Ω source impedance, consistent with the VAS design recommendations above.

The 2-EF output stage shows a +3 dB peak around 8 MHz with an 8-Ω load. The peak was smaller for smaller load impedances. The output transistor can transform a resistive load into a capacitance at its base at frequencies above f_β. The preceding EF driver thus sees a capacitive load. When an emitter follower sees a capacitive load, it can transform it into a negative resistance as seen at its input, giving rise to a frequency response bump. If the driver and output transistors have 50 MHz f_T and β of 50, f_β is at 1MHz. This means that at 10 MHz the phase angles of the base currents have been shifted by nearly the full 90 degrees. Moreover, at 10 MHz, β of each device is only 5. β^2 is thus 25, causing the reflected load impedance of 8 Ω to possibly appear as negative 200 Ω. In reality, things are a bit more complex, but this simple example illustrates what kinds of things can happen and how stability can be degraded at higher frequencies. Many output transistors have f_T only in the 5–15 MHz range. Moreover, f_T of the output transistors can droop under conditions of high current as a result of the Kirk effect.

A worse situation can result with a capacitive load, even when an output coil shunted by 2 Ω is used. At high frequencies, the load impedance can drop to as low as 2 Ω resistive. In the above example, the reflected load impedance could become something like −50 Ω at 10 MHz. It is useful to simulate/measure the input impedance of the output stage under various loading conditions when evaluating stability issues and possible interactions with VAS output impedance. Simulated magnitude of the double's input impedance dropped to 700 Ω at 1 MHz and 18 Ω at 10 MHz when driving a 4-Ω load. It was 60 kΩ at 1 kHz. Phase lag of the double was 17° at 1 MHz, 58° at 2 MHz and 125° at 4 MHz. Bear in mind that fairly fast output stage transistors were used in these simulations.

Triple EF Stability

Even though the Triple has one more stage, it is far superior to the double in terms of bandwidth, phase lag and input impedance. The Triple had a +1.4 dB peak at 8 MHz when driving an 8-Ω load, and less when driving a smaller load impedance. Phase lag was 8° at 1 MHz, 16° at 2 MHz and 36° at 4 MHz when driving a 2-Ω load. The input impedance when driving a 4-Ω load was 32 kΩ at 1 MHz and 252 Ω at 10 MHz. Input impedance was 2.6 MΩ at 1 kHz, and partially influenced by Early effect in the pre-driver. With three stages, there is more opportunity for the impedance seen at the input of the Triple to pass through a large phase angle range, since the ultimate phase angle of the current at the input can theoretically reach 270 degrees. This gives rise to the possibility of negative input impedance, but that should not be a problem if the source impedance is less than 100 Ω resistive to tens of MHz. For this reason, stray inductances in the signal path should be kept small.

Quad EF Stability

Even though the quad has yet another added stage, it appears to be relatively stable. This is partly because it provides the subsequent Triple with low source impedance and its first transistor can have fairly high f_T. The quad had a +0.02-dB peak at 5 MHz when driving an 8-Ω load, and none when driving a smaller load impedance. Phase lag was 8° at 1 MHz, 17° at 2 MHz and 37° at 4 MHz when driving a 2-Ω load. The input impedance when driving a 4-Ω load was 36 kΩ at 1 MHz and 4 kΩ at 10 MHz. Input impedance was 2.6 MΩ at 1 kHz, and mainly influenced by Early effect in the pre-driver. With four stages, there is more opportunity for the impedance seen at the input of the quad to pass through a large phase angle range, since the ultimate phase angle of the current at the input can theoretically approach 360 degrees. This gives rise to the possibility of negative input impedance, but that should not be a problem because the input impedance is high. Nevertheless, stray inductances in the signal path should be kept small.

Good Practices for Output Stage Stability

Parasitic interactions are usually the biggest threat to output stage stability. Progressive power supply decoupling should be used to virtually eliminate feedback at high frequencies to the earlier stages in the output stage. This also enhances power supply rejection in earlier amplifier stages. The output Zobel network should be placed very close to the output transistors and should employ a non-inductive resistor. I prefer metal oxide film (MOF) resistors to non-inductive wire-wound resistors, even if it requires more than one MOF resistor to achieve adequate dissipation capability. Multiple distributed Zobel networks, each with higher impedance, are also an attractive option.

13.2 Complementary Feedback Pair (CFP)

The CFP output stage was discussed briefly in Chapter 7. Here we have a few more details to discuss. As mentioned previously, its major shortcomings include limited turn-off current for the power transistors, increased tendency to oscillation and inability to meet the Oliver condition for biasing to achieve low crossover distortion [1].

The first use of the CFP output stage appears to have been by Dan Meyer, beginning with his Universal Tiger amplifier in 1970 [2]. A typical CFP output stage is shown in Figure 13.2a [3]. It has a 100-Ω driver load resistor, setting the driver bias current at about 6 mA. There are no degeneration resistors in the driver or output transistor circuits. The output stage employs 0.1-Ω R_E resistors. These values are unusually low, and cause output transistor bias current to be quite sensitive to bias spreader voltage, somewhat offsetting the bias stability advantage of the CFP. Under operating current conditions normal for an EF stage, the CFP output stage suffers strong gm doubling because of its very low Z_{out}. The effect of this is mitigated a bit by the fact that the gm is very high in the first place due to the local feedback in the CFP stage and use of the 0.1-Ω R_E resistors.

Because of its local feedback, the CFP output stage can have a very low output impedance before its R_E resistor (analogous to R_E in an EF output stage). This can lead to gm doubling. In order to reduce the gm doubling and achieve an acceptable wingspread characteristic, the output stage is starved of quiescent bias current. The output transistor bias current is made quite small so as to reduce the local feedback loop gain and increase output impedance of the CFP. This lifts the intrinsic output impedance of the stage to nearly 0.1 Ω so that it approximates R_E, making

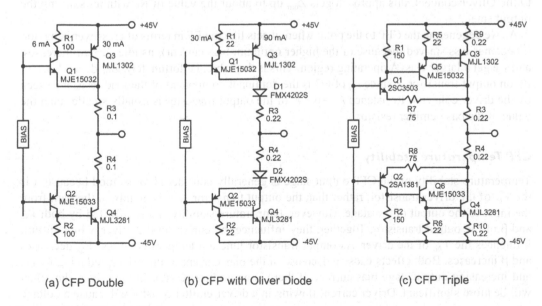

(a) CFP Double   (b) CFP with Oliver Diode   (c) CFP Triple

Figure 13.2 CFP Output Stages

quiescent transconductance the same as that at high current swing. This is the same functional criteria as for EF outputs, but it is a poor approximation to meeting the Oliver condition, as is done in the EF output stage by setting of the voltage across R_E to be about 26 mV [1]. In such a CFP, the proper bias may only correspond to a few mV across R_E. It appears that there is no simple procedure for setting the bias of the CFP for minimal crossover distortion like the Oliver condition. It is unclear that one can just adjust for some known voltage drop across R_E.

The output transistor is starved to as little as 30 mA in the quiescent state [3]. This means that the class A region is very small and that the handoff from one side of the output stage to the other is very abrupt. This can lead to higher order distortions. The loop gain of the CFP is often a significant function of output current and transistor β, so as signal currents change the bandwidth and frequency response peaking of the output stage change. The current available to turn off the output transistor in Figure 13.2a is only 6 mA. This is simply not enough to turn off the output transistor with enough speed when output current rate of change is large.

CFP Output Impedance and Crossover Distortion

The output impedance of the CFP can be estimated by applying a small voltage to its output. The current that flows can be used to estimate Z_{out}. Transconductance of the driver transistor will determine how much current flows into the parallel combination of its load resistor and r_{in} of the output transistor. At 6 mA, gm of the driver is about 0.25 S. At 30 mA and $\beta = 50$, re' of the output transistor is 0.83 Ω and r_{in} is 42 Ω. The effective driver load resistance is thus about 30 Ω, so voltage gain of the input stage is 30 Ω * 0.25 S = 7.5. If 1 mV is forced at the output, 7.5 mV will be delivered to the base of the output transistor. With output transistor gm of 1.2 S, the change in output current will be 9 mA. Output impedance of the CFP is thus 1/9 = 0.11 Ω, on the same order as the 0.1-Ω R_E.

If instead driver current was 30 mA and output current was 90 mA (typical for a good EF output stage), the same estimates would yield CFP output impedance on the order of 0.029 Ω, and severe gm doubling would occur in a narrow notch at crossover. If 2.2 Ω is put in series with the driver emitter and 0.1 Ω is put in series with the output emitter, estimated output impedance becomes about 0.12 Ω, close to the 0.1 Ω R_E. Although still a poor approximation to the Oliver concept, this approach gets Z_{out} up to about the value of R_E without starving the output stage.

As we degenerate the CFP to the point where it acts like the EF in terms of crossover behavior, it becomes less starved (because of the higher optimum bias current), its idle current increases, and we get a larger class A operating region. This is desirable. Unfortunately, Z_{out} depends heavily on output transistor β, since r_{in} of Q3 is the dominant component of the load resistance seen by the driver collector. Resistance $r_\pi = \beta * r_e$ of the output transistor is usually smaller than the value of the base-emitter resistor.

CFP Temperature Stability

Temperature stability of the CFP output stage is generally considered to be good because it is the V_{be} of the driver transistor, rather than the output transistor, that is mainly in the path from the input to the output of the stage. However, temperature stability is also affected by both V_{be} and β of the output transistor. Together, they influence the current in the driver, which in turn influences the V_{be} of the driver. As output transistor junction temperature rises, V_{be} decreases and β increases. Both effects cause a decrease in the bias current of the driver, reducing its V_{be} and increasing output stage bias current. If there is an emitter resistor in the driver, the effect will be more significant. Driver current flowing in a driver emitter resistor will cause a voltage drop that must be made up by the bias spreader. This further sensitizes the CFP bias current to driver current.

The conventional starved CFP will run quite cool. It will run much cooler than an optimally biased EF. This may seem like a good thing, but it is not. The cooler quiescent operation means that there will be greater program-dependent dissipation and temperature variation, which are never good. Amplifiers usually work best in thermal equilibrium. Any bias temperature stability shortcomings will only be made worse.

CFP with Oliver Diode

The CFP in Figure 13.2b has a fast silicon diode in series with it and $R_E = 0.22$ Ω. If the CFP is run with 30 mA driver current and 90 mA output transistor current for a total of 120 mA, it has high *gm* and output impedance below 0.05 Ω (if driven by a voltage source). The Oliver diode then takes the role of the exponential nonlinearity that enables establishing compliance with the Oliver condition, where the output stage is biased to put 26 mV across R_E [1]. This is especially advantageous, since this diode operates at low dissipation and temperature, and can be a fast-recovery rectifier diode. Temperature stability should be good, but there will be some power dissipation in the diode, so this will influence temperature stability. The diode can be mounted on the board or the heat sink, but that decision will affect how temperature compensation should be implemented. The diode should have low ohmic resistance, preferably less than 0.1 Ω. The widespread use of such diodes in switching power supplies makes such diodes readily available. This technique should be used with a CFP Triple, as described below.

CFP Triple

A common version of a CFP Triple is shown in Figure 13.2c [4]. Here the output transistor is made into a Darlington. This greatly reduces the required bias current for the first transistor in the CFP, which now becomes a pre-driver. This adds the driver's V_{be} to the voltage impressed across the pre-driver's load resistor, allowing it to be larger to enable more gain. This Triple greatly reduces the influence of output transistor β on CFP operation. The circuit shown operates the pre-driver at 10 mA and includes emitter degeneration in the pre-driver. This limits and stabilizes the gain of the pre-driver to be a bit less than R1/R7 = 2.0. Emitter degeneration has also been added to the output transistor, helping to stabilize transconductance. Z_{out} is now on the order of $(R9 + r_e'_{Q3})$ * R7/R1. With Q3 biased at $I_q = 120$ mA, $r_e'_{Q3}$ is about 0.22 Ω, and Z_{out} is about 0.22 Ω in the quiescent state if β is high. Z_{out} will decrease with increased current in Q3, as desired, as $r_e'_{Q3}$ decreases with current.

HF Stability

The CFP involves a local feedback loop whose stability must be considered. This feedback loop is also characterized by its closed-loop frequency response, whose phase lag can add to the phase lag of the output stage.

The CFP can have a peak in its response if no resistor is used in the emitter of the first transistor. However, adding even 10 Ω helps to flatten the response. Notice that the bias current of the CFP input transistor (pre-driver) flows through the 10-Ω degeneration resistor, creating a voltage drop of 100 mV if the pre-driver is biased at 10 mA. This means that the bias spreader must provide an additional 100 mV on each side of the output stage. Use of a 75-Ω resistor, as in Figure 13.2c, will result in a 750-mV drop, increasing required bias spread. The CFP Triple above provides two opportunities for local feedback compensation. A lead capacitor can be placed across R7 or a lag-lead R-C network can be placed across R1.

13.3 CFP Output Stages with Gain

While the great majority of output stages have unity gain, some output stage arrangements can be configured to have some gain, often in the range of 1.1 to 3 [2, 5]. This gain can provide a headroom advantage to the VAS, since its output will not have to swing the full amplitude of the output signal. In some cases, it can obviate the need for, or advantage of, boosted rails for the VAS and IPS. The gain of 3 is quite large, and permits the use of lower rail voltages that are well regulated to power the IPS and VAS. This can be useful in high-power designs. It also permits the use of low-voltage transistors in the IPS-VAS.

Virtually all output stages with gain use a variant of the CFP output stage [2]. It is straight-forward to obtain gain from a single-ended CFP stage by attenuating the feedback voltage from the output to the emitter of the CFP input transistor, as illustrated in Figure 13.3a. It is a little less straightforward to create a push-pull output stage with gain, since there are two CFP stages involved, and they must smoothly transfer the responsibility of delivering output current from one to the other at the crossover point, and both halves should accurately have the same gain. The figure shows a two-transistor CFP with gain equal to 1.5. Each half has a 20-Ω feedback resistor and a 40-Ω shunt resistor to ground.

The driver transistor operates at 6 mA and the output transistor operates at 30 mA, so they are somewhat starved as described in the previous section. The 6-mA driver current flows through R5 and R7 in the top half and creates a DC offset that modifies the amount of bias spread voltage needed for the output stage. Because the 6-mA driver bias current depends on V_{be} of the power transistor, some bias dependence on output transistor junction temperature is introduced that was rather insignificant in the simple unity gain CFP stage. Smaller resistor values for R5 and R7 reduce this effect, but they will dissipate a significant amount of power as they become smaller. With the values shown, the driver bias current will introduce about 80 mV of additional required bias spreading voltage on each side. This voltage will have a negative temperature coefficient similar to that of a V_{be}. Together these four resistors will add a 30-Ω load to the output of the

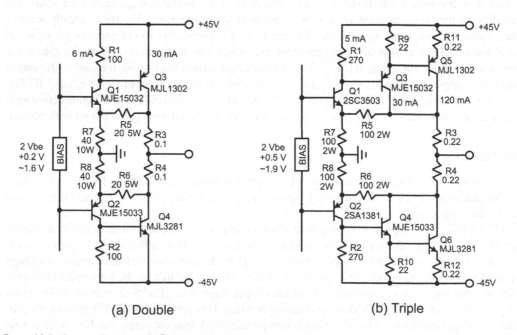

(a) Double (b) Triple

Figure 13.3 Output Stages with Gain

amplifier and will dissipate 27 W when the amplifier is delivering 100 W to an 8-Ω load. That is a high price to pay for that gain. Even with the relatively low resistor values shown, loop gain in the CFP is significantly reduced and Z_{out} of the stage is higher than desired.

Clipping Behavior

Under negative clipping conditions, the VAS output and the amplifier output will swing close to the negative supply rail, which here is at −45 V. If the output stage is designed for a gain of 1.5, the CFP feedback network, with 1.5:1 attenuation, will take the emitter of the top CFP input transistor to about −30 V, putting about 15-V of reverse V_{be} on the input transistor, which may be rated for as little as 5 V and rarely greater than 10 V. Clipping can thus cause the allowed reverse breakdown voltage to be exceeded, causing reverse base-emitter conduction, leading to permanent β degradation or even device failure. If sufficiently lower voltage rails are used for the VAS this problem can be avoided. If the gain of the output stage is held to only 1.1, then output voltage of −45 V will swing the CFP input transistor emitter to about −41 V. If the VAS output goes to −45 V, then reverse base-emitter voltage will be limited to about 4 V.

CFP Triple with Gain

A CFP Triple with a Darlington-connected output transistor is shown in Figure 13.3b [4, 5]. It is configured for a gain of 2. The pre-driver current can be small, here 5 mA, because of the Darlington current gain. At the same time, 30 mA of turn-off current is available. Pre-driver load resistor R1 can be larger because it has 2 V_{be} across it and because the input impedance of the Darlington is over 500 Ω for $\beta = 50$. The feedback resistors can be larger, here 100 Ω, for the same reasons, while maintaining enough CFP loop gain to achieve Z_{out} of about 0.22 Ω. As a result, dissipation in the feedback resistors is much lower. The pre-driver current of 5 mA flowing through the net feedback network resistance of 50 Ω will add about 250 mV to the required bias spread for each half of the output stage. For smaller gains on the order of 1.1, the numbers get even better, where a 22-Ω feedback resistor might be used with a 200-Ω shunt resistor. Such a small gain is enough to give the VAS some useful headroom relief, in this case about 4 V.

A CFP output stage Triple with gain can also be arranged as a two-transistor CFP followed by an output emitter follower. This arrangement has also been used, but is not a true CFP output stage since the actual output comes from an emitter follower.

Two CFP Stages in Parallel

If push and pull CFP output stages are implemented, each individually with a gain of 2, there may be problems with them fighting each other. This may happen because two circuits, each with low output impedance, are being connected in parallel and may have slightly different gains. If there is a 1% mismatch in CFP stage gain with a nominal gain of 2, and the output amplitude is 40 V peak, one will want to produce 40.1 V and the other will want to produce 39.9 V, for a difference of 0.2 V. If $R_E = 0.22$ Ω, then as much as 455 mA peak AC current might flow between the top and bottom portions of the output stage. This can't be good.

The Bottom Line

Unity-gain CFP output stages bring with them enough problems. Configuring them for gain makes matters worse. Output stages with gain seem not to be worth the trouble.

13.4 Bryston Output Stage

Bryston, Ltd. developed a novel output stage that employs both PNP and NPN output transistors in each half of the output stage [6, 7]. The simplified output stage, shown in Figure 13.4, is a variant of a CFP output stage with gain. Each half of the stage is like a combination of a CFP and an emitter follower stage. It can be viewed as a CFP Triple output stage that has an emitter follower added to it to contribute output current. The PNP and NPN power transistors in each half contribute equally to output current and transconductance. A key feature of this design is that the driver transistor contributes driving base current to the CFP transistor from its emitter and to the EF transistor from its collector.

Each power transistor has the same-value R_E in its emitter. As a result, ideally, both output transistors conduct the same current and deliver it to the load. This assumes that the NPN and PNP output transistor V_{be} and β are the same. We thus get the benefit of a two-pair output stage wherein the driver does double-duty, with each of its collector and emitter driving only one output transistor. This means that the driver needs to contribute only half the signal current compared to a conventional 2-EF stage for a given amount of output power.

Output Transistor Matching

Base current of the power transistors is a significant portion of the current flowing in the 22-Ω driver emitter and collector transistors. If the power transistors have the same β, no imbalance is created and the power transistor contributions to output current are equal. For this reason, the Bryston output stage requires β matching between the NPN and PNP power transistors. This can be a challenge, and may be difficult to maintain under high current conditions where β droop may come into play. Even if the PNP and NPN power transistors are β-matched at normal currents, they might not have matched β droop characteristics. If there is β mismatch between the PNP and NPN output transistors, they will not share the load current equally.

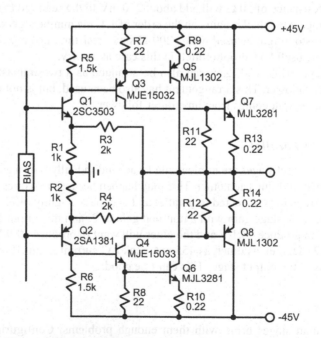

Figure 13.4 Simplified Bryston Output Stage

Gain and Biasing

The Bryston CFP feedback network comprises a 2-kΩ feedback resistor and a 1-kΩ shunt resistor, resulting in an output stage gain of 3. The collector load resistor of the first CFP transistor is 1.5 kΩ. The load resistor has about 2 V_{be} across it, implying that this transistor is operating at about 1.5 mA when driver base current is included. This current must pass through the 667-Ω feedback network resistance, causing a drop of about 1 V. This voltage drop is dependent on V_{be} and β of the driver and output transistors. It increases the required bias spreader voltage by about 1 V for each side of the output stage. The low operating current of the CFP input transistor permits the use of higher-value CFP feedback resistors that will not dissipate a lot of power.

13.5 ThermalTrak™ Output Stage

Thermal stability is an important issue for emitter-follower output stages because the V_{be} of the output transistor changes with the substantial junction temperature changes that the power transistor experiences. Without some kind of temperature compensation based on output transistor temperature, thermal runaway can occur, since as the transistor gets hotter, V_{be} decreases and current will increase. Good tracking of the power transistor temperature is also important in having the quiescent bias current meet the Oliver condition for minimizing static crossover distortion [1].

Temperature compensation is usually achieved with a diode or transistor mounted on the heat sink or on one of the output transistors, but really good tracking is made difficult by long thermal time constants and thermal attenuation between the actual transistor junction and the case or heat sink. ON Semiconductor makes a line of output transistors called ThermalTrak™ to address this issue [8, 9]. These five-leaded power transistors incorporate a silicon diode mounted inside the transistor package on the same header as the power transistor chip. The two leads of that diode are brought out and can be connected as part of the bias spreader, providing superior thermal tracking. These transistors and the related circuits for using them are described in Chapter 17. A ThermalTrak™ power amplifier with measured performance is described in Chapter 17.

13.6 Class A Output Stage

Class A amplifiers are associated with very high performance because the output transistors both stay on throughout the entire cycle and both contribute nearly equally to output stage transconductance, eliminating crossover distortion and minimizing many other sources of distortion. The price paid is very high power dissipation, since the quiescent current must usually be one-half the peak output current that can be delivered while operating in the class A mode. A 100-W push-pull class A amplifier must deliver peak current of 5 A into an 8-Ω load, meaning that its quiescent current I_q must be at least 2.5 A. If the amplifier employs ±50-V rails, quiescent power dissipation is a whopping 250 W. This same amplifier can only deliver 50 W into 4 Ω and 25 W into 2 Ω while remaining in class A. In order to keep the heat sink temperature below 50 °C (uncomfortable to the touch), a temperature rise of only 25 °C above ambient is allowed. For dissipation of 250 W, this requires a large heat sink with thermal resistance of less than 0.1 °C/W.

Output transistor junction temperatures will also be constantly high, leading to the need for more paralleled output pairs than one might think for a 100-W amplifier. If there are three pairs in the example above, each transistor will have to dissipate 42 W. The TO-264 200 W MJL3281 has 0.625 °C/W thermal resistance junction to case. The insulator will add another 0.5 °C/W at minimum, for a total of 1.125 °C/W. The junction temperature will be elevated by 47 °C above the 50 °C heat sink temperature, for a nominal junction temperature of 97 °C. Maximum operating junction temperature for this device is 150 °C. As explained in Chapter 7, power dissipation in class A amplifiers goes down as the output power goes up. In practice, this is of little help because the power level on well-recorded music is very low most of the time.

Transition to Class AB

At higher power levels and into lower-impedance loads, where peak load current exceeds $2\,I_q$, the push-pull class A amplifier will naturally transition to class AB—a very richly biased class AB, with gm doubling. This is not the end of the world because it will occur at high power levels where any gm doubling distortion will be less audible.

Biasing Circuits

All biasing circuits for class A amplifiers must be designed to work properly when the amplifier transitions into class AB. This eliminates seemingly clever bias schemes that take advantage of the fact that the sum of the top and bottom currents in class A operation is constant.

Bias stability is important in class A to avoid thermal runaway. At the same time, a Triple is desirable for best class A linearity, but the bias stability of a 3-EF stage is usually not as good as that of a 2-EF stage. If a 3-EF is used, a feedback bias spreader should be used, taking feedback from the emitters of the drivers, so only the V_{be} of the output transistors can be a source of bias drift. If a diamond buffer Triple is employed, a feedback bias spreader can also be used, but it is less necessary because the pre-driver and driver V_{be} drops cancel each other if the pre-driver and driver transistors are bolted together.

13.7 Crossover Displacement (Class XD™) Output Stage

Douglas Self patented an output stage that moves the point of crossover away from the center of the output swing, where low-power signals are most affected. The idea is that the output stage operates effectively in class A for small signals up to a certain power level. He called this a Crossover Displacement (Class XD™) output stage [3].

The technique borrows from an old idea used with operational amplifiers that had fairly poor class B output stages. Op amps of the time had much better top-half NPN output stages than bottom-half output stages. Bias was also usually kept low to keep power dissipation down. An external resistor was used as a pull-down to the negative supply, forcing a quiescent current to flow in the top half of the output stage, keeping it in single-ended class A operation for signals that did not require output current in excess of the pull-down current. In some cases, an external current source was used instead of a resistor to keep the pull-down current constant and independent of output voltage. This corresponds to connecting a single-ended class A output stage in parallel with a class B stage.

Power Dissipation

The Class XD™ output stage simply employs a larger pull-down current source whose amplitude is controlled by the voltage of the output signal, as shown in Figure 13.5. This adds significant power dissipation to the output stage, depending on how large the displacement of the crossover (and hence size of the class A region) is made. With a typical pull-down current of 1 A, the crossover point is offset from zero by 1 A [3]. The wingspread is shifted to the left by 1 A. This corresponds to a single-ended class A region of 4 W when driving an 8-Ω load. In a 100-W power amplifier with 50-V rails under no-signal conditions, this adds 100 W to the quiescent dissipation. This can greatly increase heat sink cost. For comparison, a conventional 100-W class AB output stage with one pair of output transistors dissipates only 13 W under quiescent conditions and 58 W at 1/3 rated power.

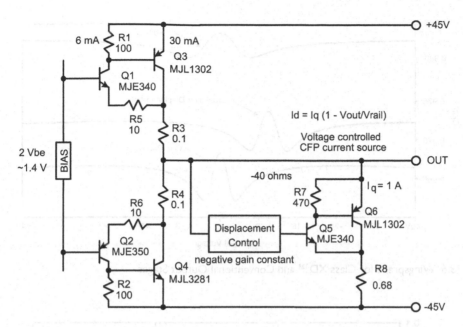

Figure 13.5 Crossover Displacement Output Stage

The pull-down current is modulated as: $I_d = I_q (1 - V_{out}/V_{rail})$. Modulating I_d reduces the burden of I_d for large positive output voltage swings. At 100 W, it reduces peak current in the top transistor from 6 A to 5 A. Modulation increases the size of the class A region, but by an insignificant amount. With ±50 V rails and $I_q = 1$ A, the I_d modulation causes the displacement circuit to act as a negative resistance load of −50 Ω. It is unclear if this negative resistance characteristic presents a problem under any operating conditions. Unfortunately, the effectiveness of this approach depends heavily on the impedance of the load being driven. With 1-A pull-down, the class A region extends to only 2 W for a 4-Ω load.

Wingspread Comparisons

Figure 13.6 compares the wingspread for the Class XD™ stage to a conventional one-pair output stage. The output stage in this case is an emitter follower type with $R_E = 0.22$ Ω. The crossover wiggle of the XD stage is displaced to the left by about 8 V (1 A), as expected. The gain is slightly higher due to the modulation of the pull-down current. This also makes the wiggle slightly smaller. The wingspread for the conventional two-pair output stage is also shown for perspective. This costs an extra output transistor, but does not have high power dissipation. It does not shift the crossover point away from zero. It has higher gain because there are two pairs contributing transconductance. For the same reason, the wiggle is of smaller amplitude, is less sharp and is spread over a larger span of signal swing.

Biasing and Temperature Asymmetry

The Class XD™ arrangement makes for an asymmetrical output stage, with the bottom output transistor operating cold (largely off) and the upper and pull-down transistors operating quite hot. Temperature changes with signal are different for the two output transistors. This can interfere with achieving and maintaining the optimum class AB bias at the displaced crossover point

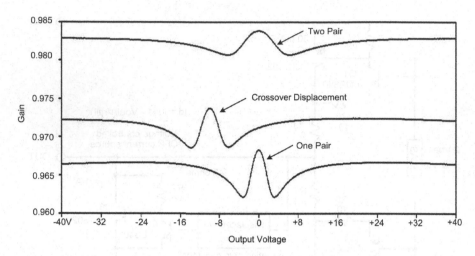

Figure 13.6 Wingspread for Class XD™ and Conventional Output Stages

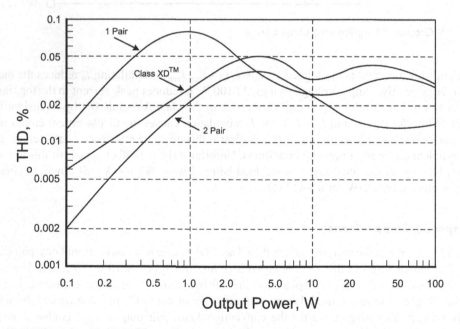

Figure 13.7 THD-1 for Class XD™ and Conventional Output Stages

(Oliver condition) [1, 10]. It is unclear how the bias is set for the optimum in the Class XD™ amplifier. It is likely that it is set to satisfy the Oliver condition when the displacer is disabled. However, junction temperatures are different when the displacer is enabled. The top output transistor is dissipating 50 W while the bottom one is dissipating 0 W. With 1 °C/W thermal resistance from junction to heat sink, the junction temperatures under quiescent conditions will differ by 50 °C.

Performance and Cost Comparisons

Figure 13.7 shows simulated output stage THD-1 for conventional one-pair, conventional two-pair and Class XD™ output stages. All are EF output stages with 0.22-Ω R_E and driven by a voltage source. The Class XD™ stage is significantly better than the conventional one-pair stage up to 6 W, and is actually a bit worse at higher power levels. The conventional two-pair output stage is actually superior to the Class XD™ stage at all power levels. The extra cost is one output transistor, but that is more than offset by the smaller cost of the heat sink.

13.8 DoubleCross™ Output Stage

The DoubleCross™ (Class DX™) output stage implements a different kind of crossover displacement that is symmetrical, eliminating the shortcomings of the Class XD™ output stage. It features two crossover points displaced equally from zero. A fully push-pull class A region extends from one crossover point to the other. The approach usually requires the use of an even number of output transistor pairs.

The simplest way to understand operation of the DoubleCross™ is to visualize two separate output stages whose quiescent output voltages are offset from one another by a spreading voltage V_{DX}. Their outputs are then connected together, as illustrated in Figure 13.8a. A quiescent *tensioning* current I_q then flows that is approximately equal to V_{DX} divided by the sum of open-loop output impedances of the two output stages. Each output stage is biased to the Oliver condition when operated individually [1]. A wingspread with two sets of wiggles spaced away from zero by $\pm 2 I_q$ results. Figure 13.8b is a rearrangement showing how a main bias spreader can feed Q1 and Q4, while individual bias offset voltages can be used to drive Q2 and Q3. These offset voltages will be referred to as *retarding voltages* (V_r) in Figure 13.8c. Notice that R1 and R4 are reduced to 0.18 Ω in Figure 13.8b and c. This will be explained momentarily.

A second way of looking at the DoubleCross™ stage is to visualize a conventional output pair that is over-biased so that it has a large class A region. This is illustrated in Figure 13.8c, where Q1 and Q4 implement this first pair. This portion of the circuit will suffer from *gm* doubling. The *gm* doubling occurs in the region of the crossover point because both halves of the output stage are contributing a substantial fraction of their full transconductance (e.g., $1/R_E$) under these conditions. Elimination of the *gm* doubling requires the replacement of the *gm* lost when Q1 or Q4 turns off as the output current exceeds twice I_q and the class A region is exited. *Transconductance replacement* is accomplished by a second pair of transconductance replacement transistors, Q2 and Q3. Their turn-on is retarded so that one of them turns on as one of the first pair of transistors is turning off. The amount by which their turn-on is retarded is carefully set so that the transistor turning off and the transistor turning on meet the Oliver condition. Emitter resistors R1 and R4 have been reduced to 0.18 Ω in Figure 13.8b and c to compensate for incomplete *gm* doubling in the first pair of output transistors. In Figure 13.8c, V_r is achieved by creating a voltage drop across the base stopper resistors by flowing about 43 mA through them via R9. This also provides the driver bias current.

Wingspread Comparisons

Figure 13.9 shows the wingspread for the DoubleCross™ output stage of Figure 13.8a driving an 8-Ω load. Also shown for comparison are the wingspreads for a conventionally biased two-pair output stage and for a Class XD™ circuit. All of the stages use $R_E = 0.22$ Ω. I_q for the DoubleCross™ stage is set to 0.5 A, while that for the Class XD™ stage is set to 1 A. I_q for each pair in the

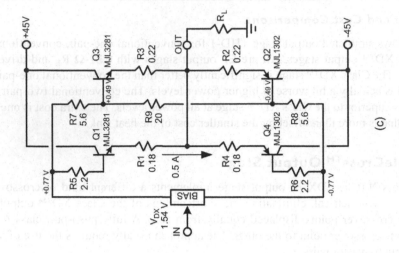

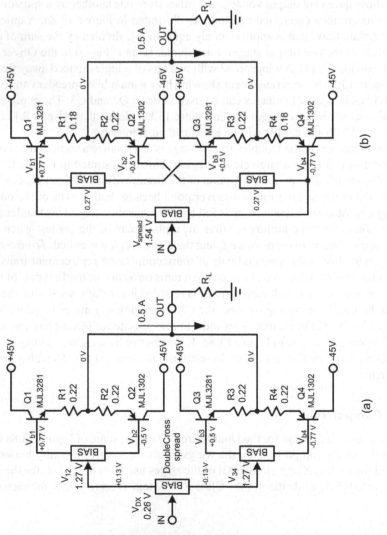

Figure 13.8 The DoubleCross™ Output Stage

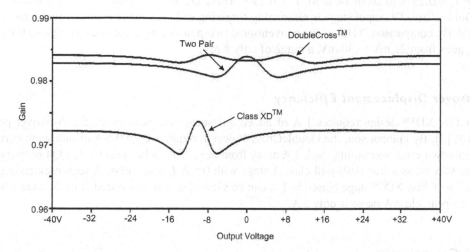

Figure 13.9 DoubleCross™ Wingspread

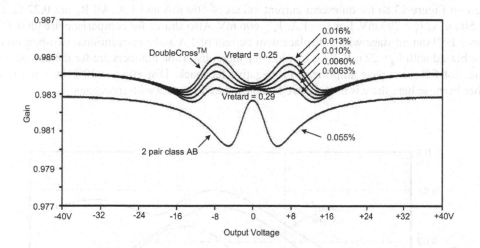

Figure 13.10 Wingspread as a Function of Retarding Voltage

conventional stage is 120 mA. The nominal gain of the former two stages is closer to unity than that for the Class XD™ stage because there are two output pairs contributing transconductance instead of one. More important is the greatly reduced variation of gain for the DoubleCross™ output stage, with the displaced crossover wiggles visible in the vicinity of ±2 I_q. Peak-to-peak gain wiggle for the DoubleCross™, conventional and Class XD™ circuits is 0.0015, 0.0031 and 0.0052, respectively.

Wingspread as a function of retarding voltage V_r is shown in Figure 13.10 for the circuit of Figure 13.8b. Also shown for reference is the wingspread for a conventional two-pair output stage. Average gain for the DoubleCross™ stage is slightly higher due to $R_E = 0.18\ \Omega$ for Q1 and Q4. The appearance of the five wingspreads changes as V_r goes from 0.25 to 0.29 V, and maximum THD of the output stage changes from 0.016% to 0.006% for any power level up to 25 W. This is the power range in which crossover distortion is most likely to dominate. The power level at which crossover distortion was greatest was different for each value of V_r, moving from about 6

W for $V_r = 0.25$ V to about 14 W for $V_r = 0.29$ V. These are all very good distortion numbers and the DoubleCross™ output stage is reasonably forgiving of bias voltage changes over a range of 40 mV. By comparison, THD of the conventional two-pair output stage goes as high as 0.055% as V_q goes from 22 mV to 30 mV, a range of only 8 mV.

Crossover Displacement Efficiency

The Class XD™ design requires 1 A of added quiescent current to move the crossover point to −1 A [3]. By comparison, the DoubleCross™ design requires only 0.5 A of quiescent current to create two crossover points, each 1 A away from zero. This is because the Q1/Q4 output pair can be viewed as a true push-pull class A stage with 0.5 A I_q and a class A region extending to ±1 A. The Class XD™ stage biased at 1 A can be viewed as a single-ended class A stage whose peak-to-peak class A range is only 2 A.

THD Comparisons

Figure 13.11 shows simulated THD-1 as a function of power for DoubleCross™ output stages based on Figure 13.8a for quiescent current values of 500 mA and 1 A. All R_E are 0.22 Ω. For $I_q = 500$ mA, $V_r = 293$ mV. For $I_q = 1$ A, $V_r = 606$ mV. Also shown for comparison are plots for a Class XD™ output stage with a displacement current of 1 A and a conventional two-pair output stage biased with $V_q = 22$ mV. Bear in mind that these distortion numbers are for the output stage alone, not for a complete amplifier with negative feedback. Distortion at 20 kHz is not much higher because here there is no negative feedback that decreases with frequency.

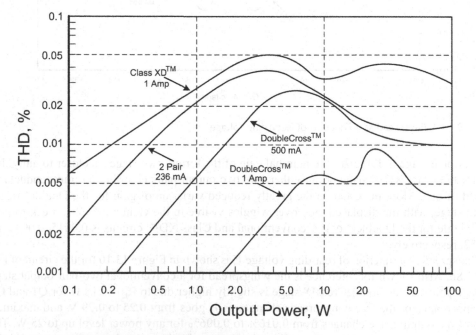

Figure 13.11 Simulated DoubleCross™ Output Stage THD-1 Versus Power

A DoubleCross™ Output Stage with Dual Drivers

Figure 13.12 shows a DoubleCross™ output stage built with a Triple arrangement. The 2.2-Ω base stopper resistors are not shown. Each output pair has its own set of drivers. This allows the retarding voltage V_r for the second pair to be established in the emitter circuit of the shared pre-drivers. The retarding voltage is dropped across R6 and R7, and the magnitude of V_r is set by R5. The conventional bias spreader is responsible for establishing a temperature-stable quiescent bias current I_q, here about 500 mA. For a given I_q, the magnitude of V_r is responsible for establishing the best crossover transition through the displaced crossover points that minimizes *gm* variation. The value of I_q is not critical to distortion performance as long as the value of V_r is appropriate. Adjustment of the bias spreader voltage does, however, influence the value of V_r. Although in some designs it may not be necessary, adjustment of R5 to achieve a specified amount of current in the second pair of transistors under quiescent conditions will lead to the lowest distortion. Such an adjustment will take account of V_{be} differences between the transistors of the first pair and the second pair. Fortunately, the DoubleCross™ output stage has a broad distortion minimum as a function of V_r.

Isothermal Output Stage

The DoubleCross™ output stage will have significant power dissipation in the quiescent state. This means that as average power is increased, average dissipation will not increase too much, so the operating temperature of the heat sink will remain relatively constant and quite warm. Biasing the output stage at a similar dissipation as 1/8 of full power is probably a good compromise for achieving constant dissipation, recognizing that excursions to 1/3 power and above will be rare. Doing so will result in a toasty output stage with minimal temperature variation and less

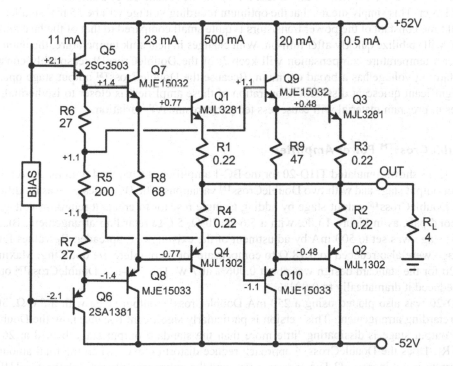

Figure 13.12 A 3-EF DoubleCross™ Output Stage

thermally induced distortion [10]. A DoubleCross™ output stage with ±50 V rails and I_q = 500 mA dissipates about 50 W in the output transistors with no signal. Such an amplifier might be conservatively rated at 100 W.

As power output rises to 12.5 W (1/8 power) dissipation will increase from 50 W to about 55 W. As output power increases to 54 W, dissipation will reach a maximum of 66 W. Dissipation will then decrease to 59 W as rated power of 100 W is reached. Power dissipation is relatively constant, especially in the range below 1/8 power, where the amplifier will spend most of its time. For comparison, the same amplifier with a conventional two-pair output stage will reach maximum power dissipation of 63 W at output power of 55 W. This is greater than the usually quoted value of 1/3 rated power due to the conservative value for rated power, taking into account power supply sag and line voltage variations. From quiescent to 1/8 power, dissipation of the conventional output stage rises from 20 W to 45 W, an increase of 125%. Over that same range, dissipation of the DoubleCross™ output stage increases by only 10%.

Retarding Voltage Temperature Compensation

All of the simulations presented have assumed that the first and second stage output transistors are at the same junction temperature. This is obviously not the case in the real world, since the second stage transistors operate with very little quiescent dissipation. While they will see a similar heat sink temperature, the junctions of the first-stage transistors will be at a higher temperature due to the thermal resistance from junction to case and from case to heat sink (θ_{js}). Fortunately, this ΔT_j is reasonably predictable. In a 500-mA DoubleCross™ design with ±50-V rails, each first-stage power transistor will dissipate 25 W, and dissipation in the second stage transistors will be negligible.

With θ_{js} = 1 °C/W, ΔT_j will be 25 °C and the V_{be} of the second stage transistors will be about 55 mV larger. This simply means that the optimum retarding voltage will be 55 mV smaller. The thermal time constant of the power transistors is quite small compared to that of the heat sink, so this ΔT_j will stabilize quickly after turn-on. With changes in heat sink temperature, the main bias spreader's temperature compensation will keep I_q of the DoubleCross™ reasonably constant. The retarding voltage has a broad optimum. Because the DoubleCross™ output stage operates with significant quiescent dissipation, operation of these amplifiers is closer to isothermal, and changes in program material will cause less temperature and ΔT_j variation.

A DoubleCross™ Power Amplifier

Figure 13.13 shows simulated THD-20 for the BC-1 amplifier of Chapter 4 with its conventional two-pair output stage and with two DoubleCross™ variations. The BC-1 design was modified to have a DoubleCross™ output stage by adding a single resistor to create a retarding voltage for the second pair, as in Figure 13.8c, with a 5.6-Ω, 20-Ω, 5.6-Ω retarding arrangement. Bias for the first stage was set to 500 mA by adjustment of the existing bias spreader. R_E values for the first stage were also reduced to 0.18 Ω to compensate for incomplete gm doubling. Maximum THD-20 for the standard design was about 0.006% at 7 W. The 500-mA DoubleCross™ output stage reduced it dramatically to 0.001%.

THD-20 was also plotted using a 250 mA DoubleCross™ output stage with a 5.6-Ω, 30-Ω, 5.6-Ω retarding arrangement. This version is particularly significant because now the DoubleCross™ output stage is dissipating little more than the standard output stage biased at 26 mV across R_E. Does the DoubleCross™ approach reduce distortion even when the total amount of bias current is not increased? The answer is yes, and the improvement is still dramatic. THD-20 at 7 W is reduced from 0.006% to 0.0015%. Lest the added advantage of biasing at 500 mA be

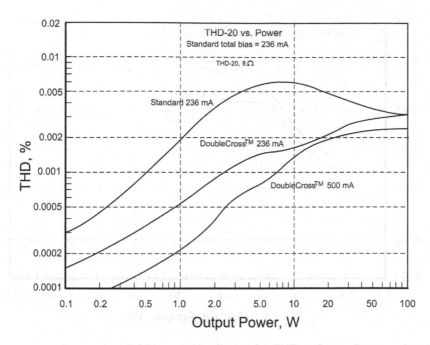

Figure 13.13 Simulated THD-20 for a DoubleCross™ Power Amplifier

under-stated by this 3-dB difference in THD-20 reductions, note that the factors of improvement compared to the conventional output stage at 2 watts are 9.2:1 and 4.1:1, respectively.

DoubleCross™ Power Amplifier Measurements

The BC-1 amplifier of Chapter 4 was modified to have a DoubleCross™ output stage with separate drivers for each output pair as shown in Figure 13.12. In this alternate arrangement, the secondary drivers are driven from taps on the pre-driver emitter resistors to establish the retarding voltage V_r. Resistor R5 of Figure 3.12 was made adjustable to enable optimization of the retarding voltage. All of the emitter resistors remained at 0.22 Ω and bias for the main pair was set to 236 mA, double the current for the Oliver condition. This setting allows comparison of DX effectiveness at identical power dissipation to that of the conventional design.

Figure 13.14 shows measured output stage THD-1 for the BC-1 amplifier of Chapter 4 with its conventional two-pair output stage and with the DoubleCross™ output stage. The base-to-base voltage spread for the main output pair at the driver bases was 2.28 V, while the spread at the secondary driver bases was 1.90 V. The difference in these voltages indicates that the DX retarding voltage V_r was 0.19 V. The value of V_r was adjusted empirically, based on measurements of output stage distortion. Under this condition, the quiescent current in the secondary output pair was only 3 mA.

Performance of the DoubleCross™ output stage is significantly improved even when total power dissipation has not been increased. Notably, distortion was not a strong function of V_r. Notice the reduced amount of distortion at power levels below 10 W, indicating a relative absence of crossover distortion. At 5 W output into an 8-Ω load, THD-1 decreased from 0.05% to 0.023% with the DX output stage.

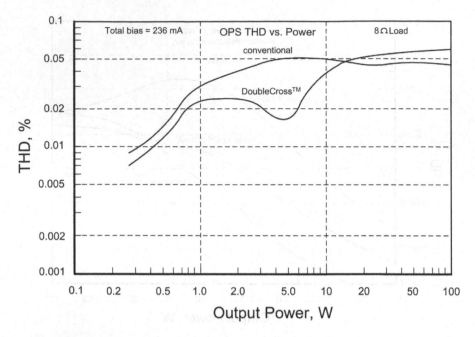

Figure 13.14 Measured Output Stage THD-1 for the DoubleCross™ BC-1

A second measurement was made with increased total power dissipation. Quiescent current of the main output pair was increased to 354 mA, 1.5 times the conventional bias current. A further reduction in distortion resulted. With the increased DX quiescent current, THD-1 decreased further to 0.014% at 8 W. These reductions in output stage distortion translate to reductions in overall closed-loop distortion at both 1 kHz and 20 kHz because output stage distortion is the main contributor.

There is a caveat to operation of the DX output stage at higher total bias currents than in a conventional output stage. The main DX output pair bears virtually all of the quiescent power dissipation, at least twice that of each pair in a conventional two-pair output stage. While this is not generally a problem under quiescent thermal conditions, it can be of concern when the amplifier is first turned on. This is because the power transistor junctions heat up much more quickly than can be tracked by the V_{be} multiplier, resulting in much higher bias current for the first 30 seconds or so. This can also happen with conventional output stages, but is a stronger phenomenon with the more heavily biased DX main pair. In the design measured here, initial bias peaked at about twice the nominal value just after turn-on. While not a problem here, this is a cautionary note for more aggressive designs. This issue can be greatly mitigated by the use of ThermalTrak™ output transistors, as described in Chapter 17. The addition of a second paralleled main pair to share the quiescent power dissipation can also reduce the amount of bias current peaking following turn-on.

13.9 Sliding Bias and Non-Switching Output Stages

A common concern with class B and class AB output stages is that the transistors in one half of the output stage switch off, usually at a small value of output current. There are two concerns. First, static crossover distortion results because the transconductance of the output stage changes as the signal goes through the crossover region. Second, switching spikes may occur as the slow power transistors turn off. Many approaches to non-switching output stages have been implemented in the past, mostly during the late 1970s and early 1980s [11–19].

There are two types of non-switching output stages. In both cases, neither transistor turns off; both transistors are always conducting some amount of current. In the first type, circuitry is merely applied to force the transistors to always conduct at least a small somewhat fixed amount of current. Transconductance of one transistor will still go to zero, but its current will not. That current will usually not be a function of the signal and the transistor will not be contributing to the transconductance of the output stage. These are often called some form of a class A output stage because both transistors are on for the full cycle. However this is not really class A, because each transistor is only really contributing to the output signal for a bit more than half the cycle. Using the term *class A* in labeling these stages is misleading. This approach will do little or nothing to avoid static crossover distortion, but it may smooth slightly the switch-off event. This will be referred to as *Type 1*.

The second type of non-switching output stage, sometimes called *sliding bias*, not only keeps both transistors on at all times, but enables them both to contribute at least some transconductance to the output signal at all times. Although the contribution of one will fall to a smaller value as the output current swings in one direction, the hand-off of transconductance contribution from one transistor to the other will be smooth and take place over a large part of the cycle. This will be referred to as *Type 2*.

Type 1 Non-Switching Output Stage

If a non-switching arrangement prevents turn-off so that one side becomes a constant current source of some value, the *gm* will still abruptly go to zero, but both output transistors will always stand ready to begin their increased current as the signal demands. In such a case, the base-emitter voltage will never go to zero, much less negative; it will never fall to less than one V_{be}, where that V_{be} corresponds to the defined off-state idle current. Does the prevention of a substantial negative swing in V_{be} help matters, either in a slow-speed sense or a high-speed sense? SPICE simulations suggest that it does not change matters very much. This approach, by itself, does nothing to reduce static or dynamic crossover distortion.

Type 2 Non-Switching Output Stage

Turn-off can be prevented in a fairly conventional output stage by dynamically increasing the bias spread when one side of the output would otherwise turn off. The total voltage across the emitter resistors increases when the amount of current delivered to the load increases beyond twice I_q. The bias spread must be increased if both transistors are to remain on. If a dynamic increase in the required bias spread is implemented properly, the transition region can be made smoother and the decrease in transconductance of the device turning off can be made slower. For example, if a circuit monitored the voltage from emitter to emitter of the top and bottom output transistors, and somehow increased the voltage of the bias spreader when that voltage began to increase, and by about the same amount, the transistor turning off would not completely turn off. It would still have signal driving it, and would still contribute some transconductance. Such an arrangement constitutes feedback to the bias spreader.

Tanaka Bias Circuit

This circuit works on the principle that the increased voltage from base to output of the heavily conducting transistor is caused to increase the bias spread voltage by the same amount.

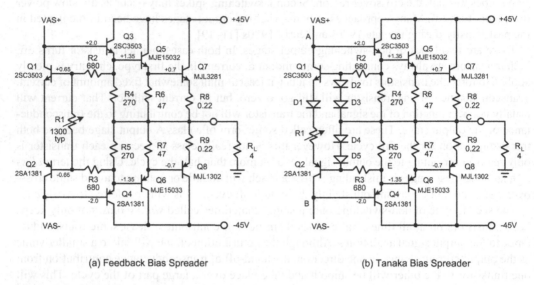

(a) Feedback Bias Spreader   (b) Tanaka Bias Spreader

Figure 13.15 Tanaka Biasing Circuit

Consider the complementary feedback bias spreader shown in Figure 13.15a. Its spreading transistors get feedback from the emitters of the pre-driver transistors in the triple-EF output stage. That feedback is transmitted via a 2:1 resistor attenuator so that the spread voltage between the emitters of the pre-drivers is 4 V_{be}, as needed for the driver and output transistors. The spread will be 2 V_{be} plus the sum of the drops across R2 and R3. Because R1 has 2 V_{be} across it, R2 and R3 will each have 1 V_{be} across them. The key is that increased voltage drop across R2 and R3 causes increased spreading voltage.

In the Tanaka bias circuit of Figure 13.15b, two pairs of diodes are placed between the bases of the spreader transistors and the output node [16]. Normally the diodes are conducting only a small amount of current, as the base nodes across which they are connected are set up to span only 4 V_{be}, and resistors R2 and R3 are in series with them (assume the diode drops are equal to a V_{be} and that R1 has about one V_{be} across it). However, as current increases in the top output transistor Q7, the voltage from the pre-driver emitter to the output node will necessarily increase. This will increase the current in D2 and D3, causing more current to be conducted through R2. The greater voltage drop across R2 will cause increased bias spreader voltage. That increase in spreader voltage will keep the bottom output transistor Q8 turned on. Careful choice of diodes and transistors can enable this scheme to work.

Geometric Mean Biasing

In geometric mean biasing, the voltages across the output emitter resistors R_E are monitored and controlled in such a way that neither can ever go to zero. A feedback control circuit adjusts the bias spread so that the product of the currents in the two R_E resistors is made to be a constant. Under quiescent conditions the currents in the top and bottom R_E might each be 100 mA. The product of the currents is 10,000 mA^2. If a positive output signal causes the top transistor to conduct 200 mA, the bottom transistor will conduct 50 mA. If the top transistor conducts 1 A, the bottom transistor will conduct 10 mA. The currents in both transistors are always non-zero, and the current in the transistor turning off decreases very gradually as the current in the conducting transistor increases. Both transistors always contribute some transconductance to the output stage.

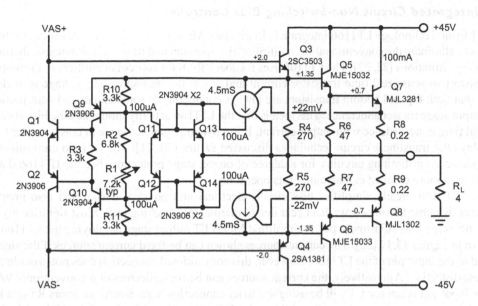

Figure 13.16 Geometric Mean Biasing

A so-called *translinear* circuit can be arranged to do this [20, 21]. Translinear circuits take advantage of the fact that V_{be} of a transistor changes as the log of the collector current and the collector current changes as the anti-log (exponential) of base-emitter voltage. Functions like squaring and multiplication can be done in the log domain by addition. A geometric mean bias circuit based on a translinear cell (Q1–Q4) is depicted in simplified form in Figure 13.16.

A current of 4.5 μA per millivolt across R_E is forced through diode-connected transistors Q13 and Q14, creating a base-to-base voltage of two V_{be}, each of which is proportional to the log of the currents in the R_E resistors. This same voltage is impressed across the bases of Q11 and Q12, causing them to conduct in accordance with the anti-log of the sum of the log V_{be} voltages of Q13 and Q14. If the output transistors are conducting 100 mA in the quiescent state, Q11–Q14 will be conducting 100 μA each.

The current through Q11 and Q12, in combination with the additional 100 μA flowing through R1 and R2, creates 1 V_{be} to turn on Q9 and Q10 so as to conduct 400 μA through R3 to turn on Q1 and Q2 so as to conduct the nominal 10 mA flowing through the VAS. The resulting feedback loop through the output stage forces the product of the currents in the R_E resistors to be a constant value set by the bias currents in the translinear cell. A BJT output stage controlled by a geometric mean bias spreader can theoretically have constant transconductance, independent of output current, and thus have a flat wingspread and no static crossover distortion.

13.10 LT1166 Output Stage

The LT1166 bias control IC implements a non-switching output stage wherein both transistors remain on throughout the whole cycle and always contribute some transconductance to the output stage. Unlike most bias spreader circuits, the LT1166 requires no bias adjustment. The LT1166 is designed for MOSFET output stages, but can be used with BJT output stages if enough bias spread voltage is available for proper operation (4 V).

An Integrated Circuit Non-Switching Bias Controller

The Linear Technology LT1166 integrated circuit class AB bias controller is a clever device that seeks to eliminate the conventional V_{be} multiplier bias spreader and its attendant dynamic thermal tracking limitations [22, 23]. The LT1166 uses feedback from the sources or emitters of the output transistors to sense real-time transistor current and adjust the bias-spreading voltage in such a way that both top and bottom transistors are always conducting to some extent and contributing to output stage transconductance. This means that the LT1166 actually modulates the bias spread in real time in accordance with output current. It is a dynamic geometric mean bias spreader that employs the translinear circuit techniques discussed earlier [20, 21]. The LT1166 conveniently includes current-limiting circuitry for a degree of output stage protection. See the LT1166 datasheet for a more detailed explanation of its operation.

The LT1166 greatly simplifies bias circuit electrical and physical design and also greatly reduces dynamic biasing errors that occur due to thermal mis-tracking [10]. Most significantly, it eliminates the bias adjustment pot. A simplified MOSFET output stage employing the LT1166 is shown in Figure 13.17 [22]. The current sources shown can be fixed current sources if the signal is fed to the input pin of the LT1166. However, this conventional connection does not provide the lowest distortion. Alternatively, the current sources can be the collectors of a conventional VAS if the input pin is not used. I will be using the latter connection here. Source resistors R1 and R2 sense the currents in output transistors Q1 and Q2. The key functionality of the LT1166 is that it servos the bias spread so that the product of V1 and V2 remains constant at 0.0004 V². This means that neither output transistor ever fully turns off. The product of $I1$ and $I2$ is held constant.

When no output current is flowing, $V1$ and $V2$ are each 20 mV. The product of these two 0.02 V numbers is 0.0004 V². If R1 and R2 are each set to 0.1 Ω, then the quiescent bias current I_q of the MOSFETs will be 200 mA. If Q1 is conducting 5 A during a large positive output voltage swing V_{R1} will be 0.5 V and V_{R2} will be 0.0008 V and I_{Q1} will be 0.008 A, or 8 mA. Q2 will still be contributing about 31 mS to output stage transconductance. A very smooth transfer of transconductance from one transistor to the other thus results.

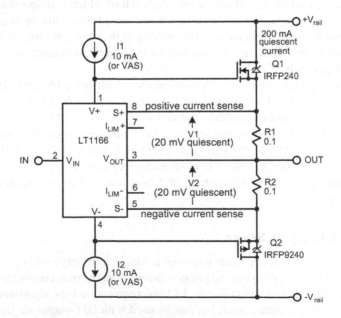

Figure 13.17 A Simplified MOSFET Output Stage Using the LT1166

It is easy to see how the servo action of the LT1166 sets the output stage bias current to a known stable value by this feedback process. It is important to understand that this process is achieved by dynamically modulating the bias spread voltage as a function of signal current. The bias spread voltage will increase as output current increases from zero in either direction. This in turn means that the signal across the bias spreader is quite distorted with third-order distortion products. This also means that the bias spreader formed by the LT1166 should not be bypassed (V_{be} multiplier bias spreaders usually are bypassed).

A simplified amplifier employing the LT1166 is shown in Figure 13.18. There are several ways in which the LT1166 can be used, but here we focus on its use as a shunt-regulating bias spreader, analogous to a V_{be} multiplier. In this arrangement the signal does not actually pass through the LT1166 on its way to the output stage. This arrangement provides a much lower distortion solution than a circuit that applies the input signal to the V_{in} pin. The V_{in} pin of the LT1166 is left floating.

The LT1166 floats with the signal at the VAS output and is powered by the VAS bias current. It requires a minimum bias-spreading voltage of 4 V and minimum bias current of 4 mA for proper operation. It is especially well suited for use with vertical power MOSFETs because of their higher gate-to-gate voltage spread, from which the LT1166 is powered. However, the device also works well with bipolar output stages if they are designed to require a sufficiently high bias spreader voltage. A Triple that is designed to require sufficient bias spread at its pre-driver inputs under worst-case conditions (like high temperature) will work if a bit of extra DC drop is introduced into its forward path.

The IRFP240/9240 MOSFET output pair requires a total bias spread of about 7–8 V under quiescent conditions, so that voltage is what appears across the LT1166 between V_{top} and V_{bot}. Source resistors R14 and R15 are set to 0.15 Ω, resulting in I_q = 133 mA. R9, R10, C4 and C5 control the current-limiting process of the LT1166. Current limiting begins when the $I+$ or $I-$ pin

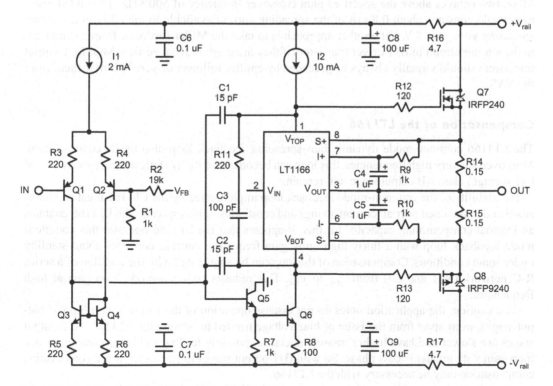

Figure 13.18 A Simplified Amplifier Using the LT1166

voltage differs from the output pin voltage by more than 1.3 V. With the 0.15-Ω source resistors shown, the current limit will be about 8.7 A. V_{top} and V_{bot} are limited to no more than a 12-V difference with respect to the output node by internal clamps in the LT1166 to protect the device. This also limits gate drive to the MOSFETs in the event of a fault.

Compensation of the Amplifier

The voltage across the bias spreader varies with signal current in a nonlinear way; it increases when the signal goes positive or negative. It is important that this nonlinear *common-mode* spreading voltage does not make its way into the signal path and introduce distortion. This is particularly important at high frequencies. To first order, changes in the bias spread will not cause distortion. For this reason, we want the center point of the bias spread to remain a linear representation of the signal. This impacts the way in which Miller feedback compensation must be applied to the VAS. If the Miller capacitor is just connected to one end of the LT1166 dynamic bias spreader, distortion will result because this node contains the signal plus half the nonlinear spreading voltage. Instead, Miller capacitors of equal value (C1 and C2) should be connected from each end of the dynamic bias spreader back to the input of the VAS. This will force the virtual center tap of the bias spreader to be the desired linear representation of the signal.

Alternatively, a center tap can be formed with two precision resistors of equal value connected in series across the dynamic bias spreader. A single Miller capacitor can then be connected from this point back to the input of the VAS. This arrangement will introduce a zero into the Miller compensation roll-off, which may actually be helpful in some designs. In the design of Figure 13.18, a pair of 5-kΩ resistors would result in an effective source resistance of 2.5 kΩ, which in combination with the single 30-pF Miller capacitor would introduce a zero at about 2 MHz, two octaves above the specified gain crossover frequency of 500 kHz. The added resistors would consume about 0.8 mA of the spreading current available to the LT1166 if the bias spreading voltage is 8 V. Yet another approach is to take the Miller feedback from a center tap in the emitter circuit of BJT driver transistors if they are used. Indeed, vertical MOSFET output transistors should virtually always be preceded by emitter follower drivers to buffer them from the VAS.

Compensation of the LT1166

The LT1166 common-mode dynamic bias-spreading feedback loop also needs compensation. Moreover, for very high frequencies, this loop can become essentially static and a properly biased (on average) class AB output stage will remain.

The stability of the common-mode feedback loop implemented by the LT1166 must be understood and maintained over all current swings and conditions of bias spread. The LT1166 contains an internal compensating capacitor for this. It appears that the LT1166 operates this common-mode feedback loop with a fairly high unity gain frequency, causing concerns about stability under some conditions. Compensation of the loop can be augmented with the addition of a series R-C network (R11 and C3) from V_{top} to V_{bot}. This reduces common-mode loop gain at high frequencies.

As a caution, the application notes do not discuss operation of the LT1166 with bipolar output stages, even apart from the issue of bias voltage needed to operate the LT1166. BJT output stages are slower and have higher transconductance, possibly leading to higher common-mode loop gain with greater excess phase. So with BJT output stages in particular, more conservative compensation may be necessary with the LT1166.

A Non-Switching Amplifier

By its very way of biasing the output stage, the LT1166 provides a non-switching amplifier, since neither output transistor ever turns off. There will be no abrupt hand-off of signal handling from one transistor to the other. The transistor that would normally be off and reverse-biased will always be conducting at least slightly, always at the ready to contribute transconductance to the output stage. The beauty of the LT1166 is that it provides correct, accurate automatic bias without trimming and eliminates the effects of thermal stability variation due to dynamic junction temperature changes. Finally, there is reason to believe that the dynamic bias-spreading action tends to reduce MOSFET crossover distortion due to better control of the class AB transition and appropriate modulation of the bias. Notice that the bias current in the output stage is effectively increased as the signal output current departs from zero.

13.11 Measuring Output Stage Distortion

In simulation and in the lab it is useful to measure output stage distortion *in situ*. This is because the output stage is usually the dominant source of distortion in well-designed power amplifiers. It is easier to achieve extremely low distortion in designing an IPS-VAS than it is for an output stage. It is important to measure output stage distortion in isolation because global negative feedback will obscure information.

There are two ways to measure output stage distortion in isolation within a working power amplifier. Both involve being able to tap a version of the signal prior to the output stage. This can be accomplished by creating a center-tap on a pre-driver or driver emitter resistor that connects the positive and negative halves of the output stage.

In the first approach, the global feedback loop is closed from this center tap instead of from the output of the amplifier. We now have an open-loop output stage being driven by a very low distortion version of the signal. Distortion at the output of the amplifier is then measured. This is a very good representation of the output stage distortion as long as the distortion of the signal feeding the output stage, as measured at the driver center tap, is at least 10 times lower than that of the output stage.

In the second approach, the amplifier is operated with the global feedback loop in its normal connection, and distortion is measured at the aforementioned driver center tap. Here we will see the actual inverse of the output stage distortion, since the distortion in this drive signal is the compensating distortion that must be fed to the output stage to obtain very low distortion at the output of the amplifier. This is a valid representation of the open-loop output stage distortion as long as the distortion of the amplifier output is at least 10 times lower than the measured distortion at the driver center tap. The waveform of the distortion at the center tap will be inverted from that which would be measured in the actual output stage or in the first method.

13.12 Setting the Bias

It has been mentioned earlier that the optimum bias to satisfy the Oliver condition for minimum static crossover distortion is not necessarily the point at which the voltage across R_E (V_q) is set to 26 mV. This is because there is ohmic resistance seen looking into the emitter of the power transistor. This includes internal emitter resistance, internal base resistance divided by β and external base stopper resistance divided by β. In theory, some of that 26 mV should essentially appear across that effective ohmic resistance seen looking into the emitter. As a result, the optimum value of V_q seen across R_E is usually less than 26 mV, often more like 20 mV.

Bias can be adjusted by finding the point of minimal distortion of the output stage by itself at 1 kHz using a distortion analyzer and using the technique described above. Since output stage distortion measured by that technique has not been reduced by negative feedback, it is a larger value that can be seen by a less sensitive distortion analyzer. In fact, in the absence of a distortion analyzer, one can use a passive or active twin-T notch filter to make the measurement at 1 kHz. Since crossover distortion is greatest at lower power levels, this adjustment should not be done at full power, but rather at something like 1/10 rated power. In fact, one can make a first adjustment at 1/10 power and then vary the power to find the point of greatest distortion in that region to carry out subsequent optimization. The amplifier should first be powered for at least 30 minutes to let its temperatures stabilize. Then the bias adjustment should be made quickly with the amplifier briefly excited to the test power level for each change of the adjustment. In practice, this can be accomplished by operating the amplifier initially with no load. The load is then briefly connected and the THD measurement is made. This approach has the advantage that the THD analyzer has already auto-tuned to the signal and has settled.

References

1. Barney M. Oliver, "Distortion in Complementary Pair Class B Amplifiers," *Hewlett Packard Journal*, pp. 11–16, February 1971.
2. Dan Meyer, "Assembling a Universal Tiger," *Popular Electronics*, October 1970.
3. Douglas Self, *Audio Power Amplifier Design*, Focal Press, Burlington, MA, 2013.
4. Dan Meyer, "Build a Four-Channel Power Amplifier," *Radio Electronics*, March 1973, pp. 39–42 and April 1973, pp. 62–68.
5. Dan Meyer, "Tigersuarus: Build This 250-Watt Hi-Fi Amplifier," *Radio Electronics*, December 1973, pp. 43–47.
6. Christopher W. Russell, *A Novel Output Stage with Superior Performance*, Bryston Ltd., Peterborough, Ontario, Canada.
7. Bryston 4B-SST power amplifier.
8. ON Semiconductor datasheets for NJL3281D and NJL1302D, June 2006. Available at www.onsemi.com.
9. Mark Busier, "ThermalTrak™ Audio Output Transistors," ON Semiconductor, AND8196/D, February 2005. Available at www.onsemi.com.
10. T. Sato et al., "Amplifier Transient Crossover Distortion Resulting from Temperature Change of Output Power Transistors," preprint no. 1896, 72nd Convention of the AES, October 1982.
11. Ben Duncan, *High Performance Audio Power Amplifiers*, Newnes, Oxford, 1996.
12. Peter Blomley, "New Approach to Class B Amplifier Design," *Wireless World*, February and March 1971.
13. Nelson Pass, "Active Bias Circuit for Operating Push-Pull Amplifiers in Class A Mode," U.S. Patent 3,995,228, November 1976.
14. N. Sano, T. Hayashi and H. Ogawa, "A High Efficiency Class A Audio Amplifier," preprint no. 1382, 61st AES Convention, November 1978 (the Technics SE-A1).
15. Y. Kawanabe, "Non-Switching Amplifier," preprint no. 1421, 61st AES Convention, November 1978.
16. Nico M. Visch, "A Novel Class B Output," *Wireless World*, April 1975.
17. Susumo Tanaka, "New Biasing Circuit for Class B Operation," *Journal of the Audio Engineering Society*, March 1981.
18. Erik Margan, "Crossover Distortion in Class B Amplifiers," *Electronics World & Wireless World*, July 1987.
19. Marcel van de Gevel, "Audio Power with a New Loop," *Electronics World & Wireless World*, February 1996.
20. Barrie Gilbert, "Translinear Circuits: A Proposed Classification," *Electronics Letters*, January 9, 1975.
21. E. Seevinck and R. J. Wiegerink, "Generalized Translinear Principle," *IEEE Journal Solid-Sate Circuits*, vol. 26, no. 8, August 1991.
22. Linear Technology Corporation Design Note 126, "The LT1166: Power Output Stage Automatic Bias System Control IC." Available at www.linear.com.
23. Linear Technology Corporation/Analog Devices, Inc. datasheet "LT1166 Power Output Stage Automatic Bias System." Available at www.analog.com.

Chapter 14

MOSFET Power Amplifiers

14. INTRODUCTION

While the majority of audio power amplifiers employ *bipolar junction transistors* (BJT) for their output stage, the power MOSFET presents an alternative with some significant advantages. These include high speed, freedom from secondary breakdown and ease of driving them to very high currents.

Figure 14.1 shows a simple lateral MOSFET amplifier using the same kind of IPS-VAS that was discussed earlier in connection with BJT amplifiers. This design is much like that of Figure 3.8, the main difference being that Q10 and Q11 have been replaced by lateral power MOSFETs (2SK1056 and 2SJ160) [1]. The V_{be} multiplier is set to a slightly higher bias spreading voltage to accommodate the larger V_{gs} of the MOSFETs as compared to V_{be} of BJT output transistors. The MOSFETs are typically biased at a quiescent current of about 150 mA, where their V_{gs} is about 0.7 V. The V_{be} multiplier is usually not mounted on the heat sink with lateral MOSFET power amplifiers because the temperature coefficient of drain current for a given gate voltage for a lateral MOSFET at the typical bias current is nearly zero. At higher currents, the temperature coefficient becomes negative, promoting good temperature stability.

Notice also the absence of source resistors in this design. Many MOSFET power amplifiers do not employ source resistors because they are not very effective in promoting bias stability or current sharing among paralleled devices (unless they are of an unreasonably large value). Source resistors are also not required as part of the overall crossover distortion management scheme. In designs with parallel-connected MOSFETs, some matching of devices is required.

MOSFETs do not require DC current drive into the gate, so some designers do not employ drivers (here Q8 and Q9) and instead drive the MOSFETs directly from the high-impedance VAS output. This is not usually a good practice in my opinion because the MOSFETs do need drive current to charge and discharge their input capacitances at high frequencies.

The MOSFET power amplifier also employs gate stopper resistors (R17 and R18) in series with the MOSFET gates. These resistors are often required with power MOSFETs to prevent parasitic oscillations. Values between 100 Ω and 500 Ω are often used.

The design of Figure 14.1 would be changed very little if vertical power MOSFETs were employed instead of the lateral MOSFETs. Vertical MOSFETs typically require a higher V_{gs} turn-on voltage, so the bias spreading voltage will be larger. Some vertical MOSFETs, like the 2SJ201/2SK1530, have a V_{gs} of about 1.7 V at the typical quiescent bias current of 150 mA, whereas others, like the IRFP240/9240, have a V_{gs} on the order of 4 V at 150 mA [2, 3].

The V_{be} multiplier is still used as the bias spreader, but with vertical MOSFETs some or all of the V_{be} multiplier arrangement will be mounted on the heat sink for bias temperature compensation, as is done with BJT designs. This is so because vertical MOSFETs have a negative temperature coefficient of gate voltage at typical values of quiescent current (on the order of 150 mA).

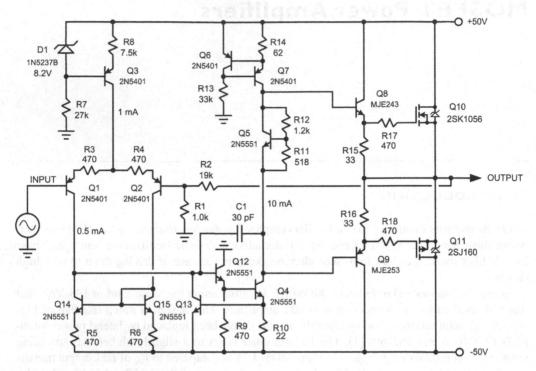

Figure 14.1 A Simple Lateral MOSFET Power Amplifier

14.1 MOSFET Types and Characteristics

Lateral MOSFETs were the first to be applied to power amplifiers in widespread production, perhaps the best known being the Hafler MOSFET amplifiers, [4] which employed Hitachi lateral MOSFETs like the Hitachi 2SJ56/2SK176. Other popular lateral MOSFETs included the Hitachi 2SJ49/2SK134 complementary pair [5]. Unfortunately, these devices are no longer available.

Vertical MOSFETs matured at a later point in time, their development spurred by their use in switching power supplies. They are characterized by higher transconductance and higher peak current capability.

Some have argued that vertical power MOSFETs were designed for switching applications and are not suitable for use in analog audio circuits. This is simply not true. What matters are the actual device characteristics, not the primary application for which the parts were made. It is true that the driving application often influences the optimization of certain characteristics over others, but the effect of this on other applications (good or bad) is usually just coincidental.

Lateral MOSFET Structure

The modern power MOSFET is made possible by many of the same advanced techniques that are employed in MOS large-scale integrated circuits, including fine-line photolithography, self-aligned polysilicon gates and ion implantation. Two planar structures, one lateral MOSFET and one vertical DMOS, are currently the most suitable devices for audio applications. Both are available in complementary pairs, offer suitable current and voltage ratings and are realized with a cellular structure that provides the equivalent of thousands of small-geometry MOSFETs connected in parallel.

The structure of the lateral power MOSFET is illustrated in Figure 14.2a. The N-channel device shown is similar to small-signal MOSFETs found in integrated circuits, except that a lightly doped n-type drift region is placed between the gate and the $n+$ drain contact to increase the drain-to-source breakdown voltage by decreasing the gradient of the electric field. Current flows laterally from drain to source when a positive bias on the silicon gate inverts the p-type body region to form a conducting n-type channel. Note that the arrow in Figure 14.2a illustrates the direction of carrier flow rather than conventional current flow. The device is fabricated by a self-aligned process where the source and drain diffusions are made using the previously formed gate as part of the mask. Alignment of the gate with the source and drain diffusions thus occurs naturally, and the channel length is equal to the gate length less the sum of the out-diffusion distances of the source and drain regions under the gate. Small gate structures are required to realize the short channels needed to achieve high transconductance and low *on* resistance.

While providing high breakdown voltage, the lightly doped drift region tends to increase *on* resistance. This partly explains why higher voltage power MOSFETs tend to have higher *on* resistance. A further disadvantage of this structure is that all of the source, gate and drain interconnect lies on the surface, resulting in fairly large chip area for a given amount of active channel area, which in turn limits transconductance per unit area. Series gate resistance also tends to be fairly high (about 40 Ω) as a result, limiting maximum device speed. Lateral power MOSFETs have been widely used in audio amplifiers. Examples of this structure are the Hitachi 2SK-134 (N-channel) and 2SJ-49 (P-channel). Desirable features of these devices include a threshold voltage of only a few tenths of a volt and a zero temperature coefficient of drain current versus gate voltage at a drain current of about 100 mA, providing good bias stability.

Vertical MOSFET Structure

A more advanced power MOSFET design is the vertical DMOS structure illustrated in Figure 14.2b. When a positive gate bias inverts the p-type body region into a conducting N-channel, current initially flows vertically from the drain contact on the back of the chip through the lightly doped n-type drift region to the channel, where it then flows laterally through the channel to the source contact. The double-diffused structure begins with an n-type wafer that includes a lightly doped epitaxial layer. The p-type body region and the $n+$ source contact are then diffused into the wafer in that order. Because both diffusions use the same mask edge on either side of the gate, channel length is the difference of the out-diffusion distances of the body and source regions.

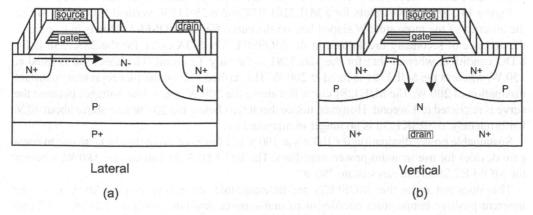

Lateral

(a)

Vertical

(b)

Figure 14.2 Structures of N-Channel Lateral and Vertical Power MOSFETs

As a result, short channels are easily realized without heavy dependence on photolithographic resolution. Short channels permit high transconductance and low *on* resistance. The geometry and dimensions of the *n*-type drift region are such that its effective resistance can be much smaller than that of the drift region for the lateral devices. This also aids in achieving low *on* resistance while retaining high voltage capability.

The vertical DMOS structure is much more compact and area efficient than the lateral structure because the source metallization covers the entire surface; the polysilicon gate interconnect is buried under the source metallization. Also, each gate provides two channels, one on each side. The amount of active channel area for a given chip area is thus higher than for the lateral geometry. The fact that source metallization areas can occupy virtually an entire side of the chip leads to high current capability. Finally, the length of the gate can be greater in this structure because it does not directly control channel length. This feature, combined with the compact structure, results in lower series gate resistance (about 6 Ω) and higher speed. Because of its many advantages, the planar vertical DMOS structure is the main-line power MOSFET technology. Examples of this cellular structure are the International Rectifier IRFP240 (N-channel) and IRFP9240 (P-channel).

14.2 MOSFET Advantages and Disadvantages

Power MOSFETs enjoy several advantages over BJTs when used in power amplifier output stages. The best known is the ease with which they can be driven. While BJTs require input current to drive the base, MOSFETs have an almost infinite input resistance at DC. They only require input current to charge and discharge their internal capacitances. In a sense, the DC current gain of a MOSFET is virtually infinite.

This also means that the MOSFET is free from the problem of beta droop at high current that BJT power transistors suffer. As seen in earlier chapters, the beta of a BJT can droop as low as 20 in some cases at current approaching 10 A. This makes them much more difficult to drive at high current and creates distortion. The f_T of BJTs also droops at high current, whereas the equivalent f_T of power MOSFETs actually tends to increase at high current.

Freedom from Second Breakdown and Device Protection

MOSFETs are generally free from the second breakdown phenomenon that haunts BJT power transistors at high voltage. The safe operating area for a MOSFET is usually bounded by constant power dissipation. The point of device destruction is primarily a function of peak die temperature.

Figure 14.3 shows SOA plots for a MJL3281 BJT and a 2SK1530 vertical MOSFET. Notice the absence of the more steeply sloped line on the curve for the MOSFET. This is indicative of the absence of secondary breakdown in the MOSFET. The SOA curve for the 2SK1530 is for a DC condition, whereas that for the MJL3281 is for only 1 second. The 2SK1530 is rated at 150 W, whereas the MJL3281 is rated at 200 W. The dashed line on the plot represents constant dissipation of 200 W. The MJL3281 curve lies above the 200 W line at low voltages because the curve is restricted to 1 second. However, notice that it falls below the 200 W line above about 40 V. Unfortunately, the 2SK1530 is no longer in manufacture.

Sustainable power dissipation for 100 ms at 100 V is a figure of merit that I like to use to compare devices for use in audio power amplifiers. The BJT MJL3281 can sustain 180 W, whereas the MOSFET 2SK1530 can sustain 290 W.

This does not mean that MOSFETs are indestructible, especially vertical MOSFETs. The inherent positive temperature coefficient of drain-source resistance over a wide range of current helps protect laterals and makes them quite resistant to burnout. The region of positive

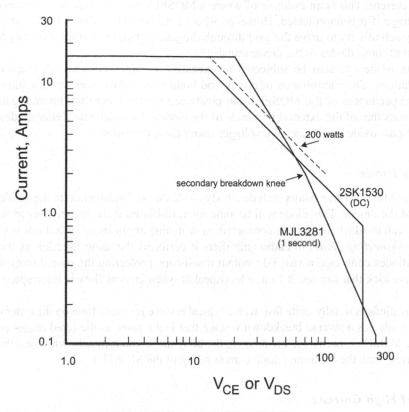

Figure 14.3 SOA Comparison for BJT MJL3281 and MOSFET 2SK1530

temperature coefficient of gate voltage and drain resistance for vertical MOSFETs begins at a much higher current level and does less to protect them from thermal runaway and localized hot spots on the die.

Because vertical power MOSFETs are inherently capable of very high current, they usually require protection from short circuits. The smaller maximum current capability of lateral MOS-FETs, combined with their negative temperature coefficient of drain current, means that lateral MOSFETs often do not require active short-circuit protection and that they will survive a short circuit until a fuse has time to blow.

Fragile Gate Oxide

The thin gate oxide used to form the gate in MOSFETs often has a breakdown voltage of only about 20 V. Without protection, this is a serious source of vulnerability to destruction for MOSFETs. If this voltage is exceeded, a pinhole may develop in the gate oxide causing instant destruction of the device. Most lateral MOSFETs actually incorporate internal gate-source Zener protection diodes. Many vertical MOSFET power amplifier designs employ external Zener diodes to protect the gate from excessive voltages of either polarity. In other cases, it is possible to design driver circuits that are inherently unable to create excessive gate-source voltages.

In some amplifier designs, if the output is shorted or if the output stage is otherwise unable to supply the current demanded by the load, the driver circuitry will attempt to drive the output MOSFETs very hard, to excessive gate voltages, in an attempt to get the output stage to produce very high current. This is an example of where a MOSFET may be exposed to excessive gate-source voltage if it is unprotected. However, where gate protection Zener diodes are used, the driver may actually try to drive the load through the gate protection diodes, possibly leading to destruction of those diodes or the driver transistor.

The gate oxide can also be subjected to excessive voltages during high-frequency parasitic oscillations. The combination of circuit and bond wire inductances and the internal inter-electrode capacitances of the MOSFET can create resonances where the internal gate voltage swing exceeds that of the external terminals of the device. External gate Zener diodes may not protect the gate oxide from excessive voltages under these conditions.

The Body Diode

Most power MOSFET transistors include a body diode that is fundamental to the semiconductor structure of the device. This diode will become forward-biased if the source ever tries to swing beyond the rail to which the drain is connected. In switching applications, this diode is sometimes called a *freewheeling diode*. In audio amplifiers it performs the same function as the external rail catch diodes connected across BJT output transistors, protecting the output transistors from the inductive kick that can result from a loudspeaker when current flow is interrupted for some reason.

The body diode is usually quite fast, with a typical reverse recovery time on the order of 100 ns. The body diode has a reverse breakdown voltage that is the same as the rated drain-source voltage for the MOSFET. In reverse breakdown, the body diode can typically withstand the same or greater current than the maximum drain current rating of the MOSFET.

Supply of High Current

Lateral power MOSFETs are limited in their ability to deliver high current by their comparatively high drain-source resistance, and yet compete reasonably with BJTs overall. Vertical MOSFETs, on the other hand, are able to conduct very high current on a transient basis, largely because of their much lower $R_{ds(on)}$. A single vertical power MOSFET can conduct over 30 A on a transient basis and needs almost no drive current to do it. Figure 14.4 shows drain current as a function of gate voltage for typical lateral and vertical power MOSFETs.

Notice that the forward bias required for 7 A is 8.3 V for the 2SK1056 lateral device, whereas it is 2.8 V for the 2SK1530 vertical device and 5.6 V for the IRFP240. This is in spite of the fact that the vertical devices start out requiring a higher V_{gs} to turn on. The difference in V_{gs} from $I_d =$ 150 mA to $I_d = 7$ A is 7.6 V for the lateral and about 1.2 V for the two vertical devices. The lateral device is incapable of delivering 10 A, whereas the 2SK1530 and IRFP240 vertical devices easily deliver 12 A and 50 A, respectively.

Table 14.1 shows the current that typically can be delivered from each device with strong forward gate drive and a V_{ds} of only 5 V. This is relevant because one may wish to evaluate maximum amplifier current delivery with a voltage drop from rail to output of no more than 5 V. If two output pairs are used to build a 100-W/8-Ω amplifier that is expected to be able to deliver 10 A into a 4-Ω load, the lateral devices are marginal by this criterion.

It is also notable in Figure 14.4 that the slope of I_d versus V_{gs} for the two vertical devices is very nearly 600 mV per decade over the range of interest from 300 mA to 3A. That is the slope of the dashed line drawn for reference between the curves of the two devices. Recall that the analogous slope for BJTs is 60 mV per decade. This implies that the *gm* of the vertical MOSFETs is about 1/10 that of a BJT at a given current. It also suggests that their behavior in this region is more exponential than square law.

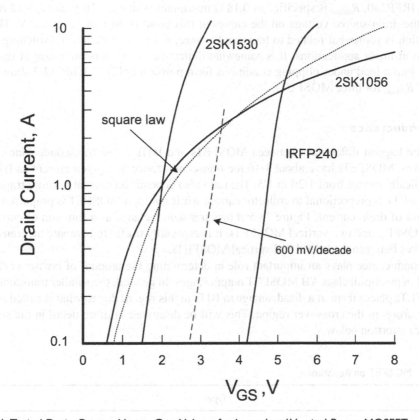

Figure 14.4 Typical Drain Current Versus Gate Voltage for Lateral and Vertical Power MOSFETs

Table 14.1 Maximum Drain Current at V_{ds} = 5 V

Device	Type	I_{max}, A
2SK1056	Lateral	6
2SK1530	Vertical	12
IRFP240	Vertical	35

In contrast, look at the I_d versus V_{gs} curve for the lateral 2SK1056. It is similar in shape to the square-law curve plotted for reference with dotted lines. One is tempted to argue that the lateral device is more of a square-law device in the region of interest than the vertical devices. Those who argue that the sound of lateral MOSFETs is superior to that of vertical MOSFETs might find comfort in this observation.

The Role of $R_{ds(on)}$

Power MOSFETs are often characterized by the parameter $R_{ds(on)}$. This is the effective value of resistance from drain to source when the MOSFET gate is strongly forward-biased. It is essentially the slope of the I_d versus V_{ds} curve of the output characteristic curve in the linear region for high gate drive.

For an IRFP240, $R_{ds(on)}$ is specified as 0.18 Ω maximum with V_{gs} = 10 V and I_d = 12 A, implying that the drain-source voltage on the curves at this point is no more than 2.2 V. The $R_{ds(on)}$ specification is somewhat related to transconductance, but is of more use in switching applications than in linear applications. It is somewhat indicative of the maximum output that can be provided into a load under clipping conditions for a power amplifier. Table 14.2 shows typical values of $R_{ds(on)}$ for three MOSFETs.

Transconductance

One of the biggest differences between MOSFETs and BJTs is the transconductance. In very rough terms, MOSFETs have about 1/10 the transconductance at a given current as BJTs. This factor typically ranges from 1/20 to 1/5. The ratio also depends on current because transconductance for a BJT is proportional to collector current while *gm* for a MOSFET is proportional to the square root of drain current. Figure 14.5 shows transconductance as a function of current for a lateral MOSFET and two vertical MOSFETs. It is especially useful to compare the transconductance curves between the lateral and vertical MOSFETs.

Transconductance plays an important role in determining the amount of crossover distortion produced in push-pull class AB MOSFET output stages. In general, the smaller transconductance of MOSFETs places them at a disadvantage to BJTs in this regard due to what is called *transconductance droop* in the crossover region. This will be discussed in more detail in the section on crossover distortion below.

Table 14.2 MOSFET *on* Resistance

Device	Type	$R_{ds(on)}$, Ω
2SK1056	Lateral	0.55
2SK1530	Vertical	0.40
IRFP240	Vertical	0.18

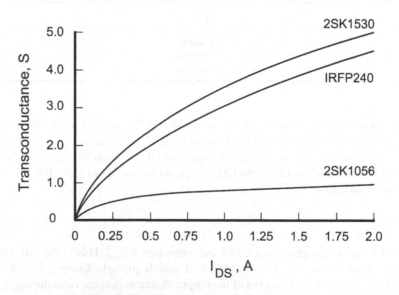

Figure 14.5 Transconductance Versus Drain Current for 2SK1056, 2SK1530 and IRFP240

Table 14.3 MOSFET f_T

Device	Type	gm, S	C_{in} pF	f_T, MHz
2SK1056	Lateral	1.0	620	250
2SK1530	Vertical	2.5	1100	360
IRFP240	Vertical	3.6	1300	440

High Speed

The speed of a MOSFET can be characterized by equivalent f_T in the same way as a BJT. In the case of the MOSFET, the f_T is

$$f_T = gm/2\pi C_{in}$$

where C_{in} is the sum of the gate-source and gate-drain capacitances. The capacitance C_{in} does not generally increase significantly as current is increased (as is the case for the diffusion capacitance of a BJT). Therefore, the equivalent f_T of a MOSFET continues to rise with increased current as *gm* increases. As is often the case for BJTs, this estimate is made at 1 MHz (the estimate of f_T for a BJT is often taken as the value of AC beta at 1 MHz). The f_T is estimated at a current of 1 A for three MOSFETs in Table 14.3.

The transconductance of the IRFP240 is about 1 S at the typical class AB bias current of 150 mA. Its f_T under these conditions is about 120 MHz.

Transconductance Frequency Response

Transconductance can also be a function of frequency for MOSFETs as a result of gate resistance working against the gate-source capacitance. This is a special problem for lateral MOSFETs because they have higher gate resistances. Gate resistance for a lateral MOSFET is often on the order of 40 Ω, while for vertical MOSFETs it is on the order of 6 Ω or less. The transconductance for a typical lateral power MOSFET is down 3 dB at 2.5 MHz. This means that there will be excess phase created by this device that is not explained fully by its f_T.

The different device structure of the vertical MOSFETs results in much lower gate series resistance. For this reason the f_T estimates are more "real" for the vertical devices. The influence of gate series resistance on effective f_T of the MOSFET underlines the potential deleterious effect of large gate stopper resistors used with MOSFETs.

MOSFET Disadvantages and Caveats

MOSFETs have their disadvantages in comparison to BJTs as well. They are summarized here and will be discussed in later sections.

- Lower transconductance
- Higher required drive voltage
- Greater tendency to high-frequency oscillations
- Higher price in some cases, especially with laterals
- Fragile gate structure

MOSFETs Versus Bipolar Transistors

BJTs have been the workhorse for audio power amplifiers since their inception as solid-state designs, and it will likely stay that way. BJTs have improved greatly over the years and are quite rugged and fast. Power MOSFETs didn't arrive until the late 1970s. At the time they were quite a bit more expensive than BJTs, but were arguably faster and more rugged.

MOSFET class AB biasing tends to be simpler and less critical, and the bias point is more stable with temperature. MOSFETs do not have an optimum class AB bias current. Instead, they operate better with greater bias as long as thermal objectives are met. For this reason, typical MOSFET power amplifiers operate at higher bias current per output pair and have a larger class A region of operation for small signals. However, their lower transconductance tends to result in higher measured values of static crossover distortion. Because of their high speed, MOSFET amplifiers are less prone to dynamic crossover distortion as a result of switch-off characteristics.

MOSFETs do not suffer from beta droop and f_T droop at high currents and are generally able to handle high peak currents better than BJTs. This is less so for lateral MOSFETs. The ease with which MOSFETs are driven also means that there is less stress on driver transistors when the amplifier is delivering high current to the load.

MOSFETs are a bit more prone to high-frequency parasitic oscillations as a result of their inherently higher speed. For this reason, circuit design and layout can require more care than for BJT designs.

14.3 Lateral Versus Vertical Power MOSFETs

As a result of the structural differences, lateral devices have much higher $R_{ds(on)}$ than vertical devices and are not as adept at conducting high currents. Their gate structure also includes higher resistance, and this tends to limit their speed and performance at high frequencies. Although lateral MOSFETs are slower than vertical MOSFETs, they do tend to have reduced gate-source and gate-drain capacitances. Moreover, the gate-drain capacitance does not tend to rise substantially at low reverse gate-drain voltages, as it does in vertical devices.

The reduced turn-on voltage for lateral MOSFETs is initially appealing, but when you look at the forward bias required for high currents you quickly discover that any perceived advantage here disappears. This is a result of their smaller transconductance.

While technically not as high performing as vertical MOSFETs, the sound of lateral MOSFETs is preferred by some listeners as being more tube-like. Prices for vertical MOSFETs have fallen over the last 35 years, but the prices for lateral MOSFETs have risen and availability has decreased sharply.

14.4 Parasitic Oscillations

Vertical MOSFETs are about 10 times as fast as RET BJTs, with equivalent f_T in the range of 100–500 MHz. Lateral MOSFETs are quite fast as well. For this reason, MOSFETs are more prone to parasitic oscillations due to stray inductances and capacitances in the surrounding circuitry. They are also sufficiently fast that their own internal bond wire inductance can come into play in parasitic oscillations.

Gate Stopper Resistors

The standard approach to controlling parasitic oscillations in MOSFETs is to employ gate stopper resistors in series with the gate of the device. These resistors are placed very close to the gate terminal of the MOSFET. The resistors tend to kill the speed of the MOSFETs and to damp out resonant circuits that may be formed by inductance operating in conjunction with gate capacitances.

Unfortunately, the practice of using gate stopper resistors discards one of the big advantages of the power MOSFETs, namely their speed. These resistors are often on the order of 100–500 Ω. It is easy to see that such a resistance up against a capacitance on the order of 1000 pF will result in a frequency roll-off in the vicinity of the low megahertz region. This will create excess phase in the output stage and limit the gain crossover frequency that can be employed for global negative feedback.

Parasitic oscillations become even more likely when power MOSFETs are connected in parallel, since more complex high-frequency interactions among the devices are then made possible [7]. Inevitably, individual gate stopper resistors, placed very close to the gate terminals, are necessary.

Origin of Parasitic Oscillations

A simplified analysis of the origin of such parasitic oscillations can provide insight into how to eliminate them with less brute force than simply the use of large-value gate stopper resistors. Some of the high-speed benefits of the MOSFETs can then be retained. The high-frequency parasitic oscillations are often caused by the inadvertent formation of oscillator topologies by the various inductances and capacitances present in the circuit. The Hartley and Colpitts oscillator topologies shown in Figure 14.6 are often the source for the parasitic oscillations [6]. The R-C Zobel network that is usually placed across the load helps damp and kill the effect of C_{ds} in Figure 14.6a, but this presumes that the Zobel is placed very close to the source and has very little inductance. To this end, the Zobel can actually be connected from source to drain of the MOSFET.

The parasitic oscillations can often occur at quite high frequencies, from 25 MHz to 250 MHz. In some cases these oscillator topologies can be formed even in the absence of external inductances in the printed wiring board layout.

MOSFET Internal Inductances

The MOSFET internal bond wire and packaging inductances can play an important role in the formation of parasitic oscillators. The IRFP240 datasheet provides typical values of drain and gate inductances, as measured at the device leads 0.25 inch from the package [3]. The drain inductance is given as 5 nH, whereas the source inductance is given as 13 nH. The former is smaller because

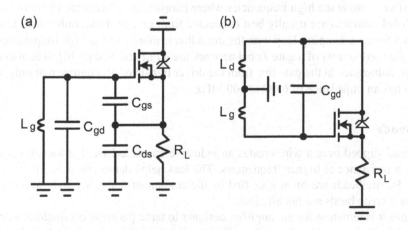

Figure 14.6 (a) Colpitts and (b) Hartley Oscillator Topologies

it does not involve a bond wire; the drain is in direct electrical contact with the package heat spreader. The gate inductance is not given, but it can be assumed to be approximately the same as the source inductance, since it includes a bond wire as well. For comparison, the inductance of a 1-inch trace on a printed wiring board is on the order of 20 nH.

It is instructive to calculate the resonant frequency of 13 nH with 1000 pF. It is 44 MHz. This frequency range is where such parasitic oscillations often occur. The reactance of each of these elements is 3.6 Ω at the resonant frequency of 44 MHz.

MOSFET Output Capacitance

The device structure of the vertical MOSFET creates a junction capacitance from drain to source called C_{ds}. This capacitance is a function of drain-source voltage. For an IRFP240, it is on the order of 1000 pF at 1 V, falling to 500 pF at 10 V and falling still further to 200 pF at 50 V.

This capacitance represents a direct capacitive load on the source-follower output stage and can play a role in creating parasitic oscillations. This can be especially true when the output swings close to the rail, where C_{ds} becomes large. The total amount of this capacitance will also accumulate in designs where multiple output pairs are connected in parallel.

Fortunately, the amplifier's output Zobel network will place a fairly low resistance in parallel with this capacitance at high frequencies, helping to kill the Q of resonant circuits formed by this capacitance. This underlines the importance of having the drain power supply rail well bypassed locally to the circuit ground at high frequencies through very little stray inductance. It also suggests that the Zobel network should be located close to the output transistors. Indeed, a high-frequency Zobel network with smaller resistance and capacitance values can be connected directly from source to drain at each MOSFET. Such a network might comprise a 1000 pF capacitor and a non-inductive 3.3-Ω resistor. This approach reduces the influence of stray inductance in the path from the supply rail to ground.

Gate Zobel Networks

Rather than just killing the high-speed performance of the MOSFET, it is better to reduce the Q of the circuits involved in the oscillator topologies and damp out the oscillations. This can be done with series R-C Zobel networks like those used to shunt the output of power amplifiers [6]. These circuits look largely like open circuits at audio frequencies, but transition to a fairly low-value resistance at high frequencies. When placed in the gate circuit of a MOSFET, these networks provide a resistive shunt at the high frequencies where parasitic oscillators are likely to be formed.

These Zobel networks are usually best connected from gate to drain, rather than from gate to ground. This forms a damping loop with the drain that is more local at high frequencies. Typical values for the components of a gate Zobel network are 47 Ω and 100 pF [6]. It is also important to minimize inductance in the gate line from the driver transistor. Recognize that only 100 nH of inductance has an impedance of 63 Ω at 100 MHz.

Ferrite Beads

A ferrite bead slipped over a wire creates an inductive characteristic that transitions to what is effectively a resistance at higher frequencies. The loss helps damp resonance that could cause instability. Ferrite beads are often specified by the amount of resistance they introduce at high frequencies. Ferrite beads are not all alike!

Sometimes it is tempting for an amplifier designer to tame parasitic oscillations with strategically placed ferrite beads. These are often effective, but are frowned on in high-end audio circles as affecting the sound quality in a negative way. Ferrite beads are fundamentally nonlinear. In fact, significant current should never be passed through a wire on which a ferrite bead is placed because the bead will become magnetically saturated and its inductive effect will be defeated. Ferrite bead datasheets usually specify how much current can be passed through the wire without causing the impedance of the bead to be decreased by more than some percentage.

Paralleled MOSFETs

Many power amplifiers of medium to high power require the use of multiple output pairs in parallel. The connection of MOSFETs in parallel increases the opportunity for parasitic oscillations, especially if the gates of the paralleled devices are not isolated at very high frequencies [7]. This means that individual gate stopper resistors are required for each device. It is also best that each gate have its own gate Zobel network if they are used.

Spotting Parasitic Oscillations

It is important to recognize that such parasitic oscillations may be difficult to spot with ordinary lab test equipment. It is necessary to have an oscilloscope with a bandwidth of at least 100 MHz in order to see some of these parasitic oscillations. Moreover, these oscillations will often show up only as brief bursts. This happens because the capacitances and transconductance of the MOS-FET are a function of its operating point. It may be that the conditions for a parasitic oscillation are right only at certain places on a sine wave. Because gate-drain capacitance of MOSFETs increases dramatically under conditions of low V_{dg}, it is especially important to be watchful for parasitic oscillations at high output amplitudes at and near clipping.

The presence of parasitic oscillations will sometimes show up as increased harmonic distortion, but I have often seen this to be a subtle effect that can be dismissed as being merely the natural harmonic distortion created by the amplifier. Subtle increases in THD caused by parasitic oscillations are more likely to be noticed in an amplifier design that normally produces very low amounts of THD.

14.5 Biasing Power MOSFETs

Biasing power MOSFETs for class AB operation is different from that for BJTs in two ways. First, MOSFETs require greater forward bias at the gate than BJTs do at the base. Lateral MOS-FETs typically require on the order of 0.7 V, but vertical MOSFETs can require up to 4 V. Some vertical MOSFETs like the 2SK1530 require only about 1.7 V, however. If MOSFETs requiring 4 V forward bias are combined with emitter follower drivers, the total bias spreading voltage will be on the order of 9.2 V. This compares to a typical bias spreading voltage of about 4.0 V for a BJT output Triple. This difference in required bias spreading voltage is not a problem for the traditional V_{be} multiplier or minor variants of it, but it does imply that the driver circuitry will require more voltage headroom in a MOSFET design. This is sometimes dealt with through the use of boosted power supplies for the circuits preceding the output stage [6].

TC$_{Vgs}$ Crossover Current

The temperature coefficient of V_{gs} for MOSFETs is different from that for BJTs as well. Lateral power MOSFETs are characterized by a V_{gs} temperature coefficient that is nearly zero at typical quiescent bias currents on the order of 150 mA. This is illustrated in Figure 14.7a, where the I_d versus V_{gs} transfer characteristic for a 2SK1056 lateral MOSFET is illustrated for junction temperatures of 25 °C and 75 °C. It is easy to see that the temperature coefficient of V_{gs} (TC_{Vgs}) is zero at the point where the curves cross. At lower currents, the drain current is higher at high

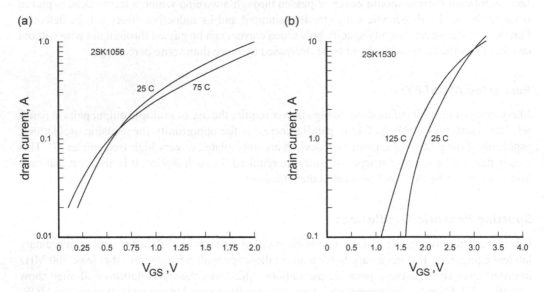

Figure 14.7 I_d Versus V_{gs} at Different Temperatures for (a) Lateral and (b) Vertical MOSFETs

temperatures, indicating a negative TC of V_{gs}. At currents above the crossover point, drain current is smaller at high temperatures, indicating a positive TC of V_{gs}. Such a positive TC of V_{gs} makes for greater temperature stability, since higher junction temperature will result in smaller drain current and reduced power dissipation.

Because TC_{Vgs} for the lateral MOSFET is approximately zero at the quiescent bias level, the traditional V_{be} multiplier bias spreader need not be mounted on the heat sink. Notice also that the typical value of V_{gs} at the quiescent current for a lateral MOSFET is not very different than that for a BJT, so the V_{be} multiplier design will be very similar.

The situation is different for vertical MOSFETs. While the temperature coefficient of V_{be} for a BJT is about -2.2 mV/°C, the temperature coefficient for a 2SK1530 vertical power MOSFET is on the order of -4 mV/°C at a typical quiescent bias current of 150 mA. The TC_{Vgs} behavior of a 2SK1530 vertical MOSFET is illustrated in Figure 14.7b. Notice that there is still a current at which TC_{Vgs} is zero, but that it occurs at a very high current of about 8 A. The TC_{Vgs} crossover point for the IRFP240 (not shown) is at about 15 A and its TC_{Vgs} at 150 mA is about -6 mV/°C. This means that bias spreaders for vertical MOSFET output stages must incorporate some temperature compensation based on heat sink temperature.

Bias Spreaders for MOSFET Output Stages

Consider the IRFP240 vertical MOSFET. The -6 mV/°C TC_{Vgs} at the typical bias point of 150 mA may sound worse than for a BJT, but it is actually a much smaller percentage of the typical forward bias number (4 V for the vertical MOSFET compared to 0.7 V for the BJT). This means that a traditional one-transistor V_{be} multiplier with its transistor mounted on the heat sink will provide too much temperature compensation for a MOSFET design. In rough terms, it will provide -2.2 mV/°C multiplied by the ratio of V_{be} to V_{gs}, or about -14 mV/°C for the output device. This is about twice as much as needed. For this reason, a modified version of the V_{be} multiplier, with only a portion of its temperature sensing circuitry on the heat sink, should be used for vertical MOSFET output stages.

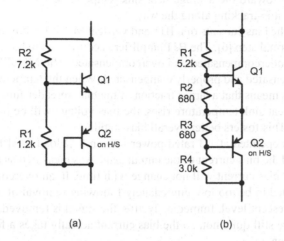

Figure 14.8 Bias Spreader V_{be} Multipliers for Vertical MOSFETs

The V_{be} multipliers illustrated in Figure 14.8 can be used. The first bias spreader just adds a diode to the emitter circuit of the V_{be} multiplier. Only the diode is then mounted on the heat sink. The V_{be} multiplier is set to provide a spread of about 14 V_{be} (9.2 V), of which about half is compensated. The diode should be implemented as a diode-connected transistor. The second bias spreader consists of two V_{be} multipliers in series. The transistor for one of them is mounted on the heat sink. It also creates a spread of 14 V_{be}, of which 40% is compensated with the heat sink temperature. This permits significant freedom in picking the proportionality of the temperature compensation. The TC of the board-mounted V_{be} multiplier transistor Q1 responds to ambient temperature, which may or may not be desirable in a given design. It accounts for about 5.7 V of the spread in Figure 14.8b.

If desired, the Q1 V_{be} multiplier could be replaced with a Zener diode or an LED-based voltage drop. Some ambient compensation should probably be retained for BJT driver transistors, however. As such, one could put a green LED in series with Q1's emitter in (b) and adjust the values of R1 and R2 accordingly.

Dynamic Thermal Bias Stability

Mounting the V_{be} multiplier bias spreader (or a portion of it) on the heat sink provides for long-term bias stability against temperature variations by providing a form of biasing feedback. This feedback will be very slow acting, since its action is governed by the long thermal time constant of the heat sink. This is true for both BJT and vertical MOSFET designs. As the temperature of the heat sink changes with program material over time, its temperature can lag what the temperature should really be for proper bias. As a result, bias may be too high or too low after the average power of the program material changes.

Imagine running the amplifier at 1/3 power for an extended period of time, with the average heat sink temperature reaching 60 °C and the portion of the heat sink under the power transistor reaching 65 °C (heat sinks are not isothermal). The transistor package may reach 75 °C and the transistor junction may reach 90 °C. If the temperature compensation was keeping the quiescent bias at its nominal value, it was compensating for quite a bit more than just the temperature rise of the heat sink. When the drive signal is removed, the junction, case and local heat sink temperatures all move toward the average heat sink temperature with different time constants. This results in serious mis-tracking along the way.

Figure 14.9 shows the bias currents of a BJT and vertical MOSFET power amplifier as a function of time after a thermal step [6]. The BJT amplifier is configured to have the bias temperature compensation set to undercompensated and overcompensated. The MOSFET amplifier is set to have the bias uncompensated and properly compensated. When the temperature compensation is overcompensated, this means that a larger fraction of the bias spreader function is located on the heat sink. When the heat sink temperature rises, the bias voltage will be made to decrease a bit more than necessary. This fosters better overall bias stability.

Each amplifier was operated at 1/3 rated power into an 8-Ω load for 10 minutes. The signal was then removed and the bias current of the output stage was measured as a function of time. In an ideal amplifier, the bias current will not change with time. In an overcompensated amplifier, the bias can be expected to be too low immediately following removal of the signal and slowly rise to its nominal quiescent level. Immediately after the signal is removed, however, the output devices themselves are still quite hot, so the bias current actually takes a few seconds to fall to the under-bias condition.

Figure 14.9 shows dramatically how much more thermally stable the MOSFET design is. The serious under-bias that is evident in the BJT amplifier following removal of the signal can lead to crossover distortion that might not be revealed in static bench tests.

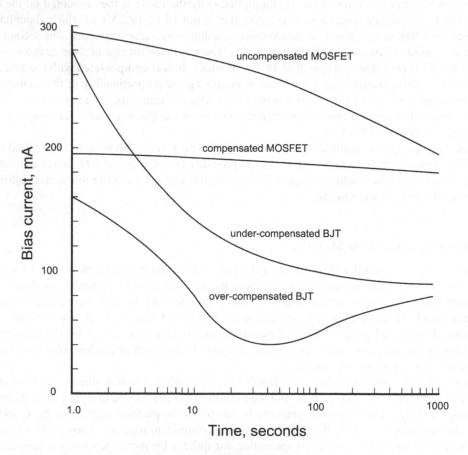

Figure 14.9 Bias Current Versus Time for BJT and MOSFET Amplifiers

14.6 Crossover Distortion

Crossover distortion is one of the most insidious distortions in class AB power amplifiers. It occurs at fairly low signal levels and often contains a high-order distortion spectrum that is more dissonant and difficult to remove with negative feedback. It is a result of the changing gain of the output stage as the signal current delivered to the load goes through zero (the crossover). We have seen how it originates in BJT output stages and the measures that must be taken to minimize it. MOSFET output stages are also subject to crossover distortion.

Static crossover distortion in BJT output stages is a result of the output impedance changing as the output current goes through zero. The output impedance forms a voltage divider with the load impedance; as a result the gain of the output stage changes. The lower the value of the output impedance, the smaller the crossover distortion will be for a given percentage change in the output impedance.

The same is true for power MOSFET output stages, but the output impedance is generally quite a bit higher for a given amount of bias current. This is because the transconductance of a MOSFET is much smaller than that of a BJT. The transconductance of a BJT at $I_c = 100$ mA is about 4 S. The transconductance of an IRFP240 biased at $I_d = 150$ mA is about 1 S. As a result, the sum of the transconductances of the upper and lower MOSFETs dips in the crossover region. This is referred to as *transconductance droop* [6].

Transconductance Droop

Typical transconductance characteristics as a function of drain current were shown in Figure 14.5. A class AB output stage comprises both an N-channel and a P-channel MOSFET. The transconductance of the source follower push-pull stage is thus the sum of the transconductances of the N- and P-channel devices. The output impedance of the push-pull output stage is simply the inverse of the sum of the transconductances.

The typical transconductance of different MOSFETs is shown in Table 14.4 for drain current values of 150 mA, 1 A and 5 A.

Since both transistors are *on* at crossover, we should compare twice the *gm* of each transistor at the quiescent current with the *gm* of one transistor at 1 A. This helps to get an idea of the degree of transconductance droop at modest power levels. For the lateral device the transconductances are 0.8 S and 1.0 S, respectively, for a 20% transconductance droop at crossover. This corresponds to the output impedance changing from 1.25 Ω to 1.0 Ω. With an 8-Ω load this corresponds to output stage gains of 0.86 and 0.89, respectively.

The vertical devices exhibit increased transconductance, but also a greater amount of increase in transconductance from quiescent current to 1 A. The comparison for the 2SK1530 is 1.4 S and 2.5 S, corresponding to output impedances of 0.7 Ω and 0.4 Ω, respectively. With an 8-Ω load this corresponds to output stage gains of 0.92 and 0.95, respectively.

The comparison for the IRFP9240 is 2.0 S and 3.6 S, corresponding to output impedances of 0.5 Ω and 0.28 Ω, respectively. With an 8-Ω load this corresponds to output stage gains of 0.94 and 0.97, respectively.

Table 14.4 Transconductances of Several MOSFETs

Device	Type	$I_d = 150$ mA	$I_d = 1$ A	$I_d = 5$ A
2SK1056	Lateral	0.4 S	1.0 S	1.2 S
2SK1530	Vertical	0.7 S	2.5 S	4.5 S
IRFP240	Vertical	1.0 S	3.6 S	7.6 S

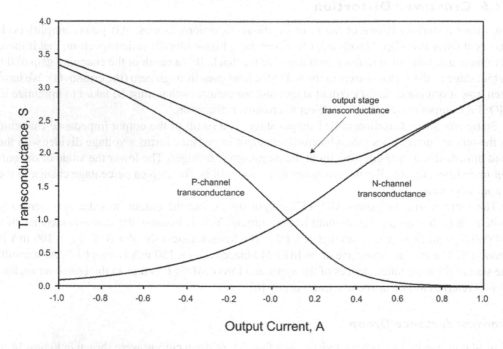

Figure 14.10 Illustration of Transconductance Droop

Figure 14.10 shows the transconductances of a complementary pair of vertical MOSFETs, the IRFP240 and the IRFP9240, as a function of output current from a class AB output stage. The curve in the center is the sum of the transconductances. It is easily seen that the total transconductance droops in the crossover region where the magnitude of the output current is small [6]. This corresponds to smaller gain in the output stage and thus static crossover distortion.

Absence of gm Doubling

When a BJT output stage is over-biased, its transconductance can be substantially larger in the crossover region than elsewhere because both transistors are on and contributing transconductance. The different transfer characteristics of MOSFETs, particularly vertical MOSFETs, make it almost impossible for *gm* doubling to occur. The MOSFETs would have to be operating at very high quiescent bias current for the *gm* at crossover to be substantially larger than that at output currents outside the crossover region. In general, more quiescent bias current is better for MOS-FET output stages as long as no thermal problem results. If source resistors of significant value are used with MOSFET output stages, the possibility of *gm* doubling may arise, however. In some ways *gm* doubling is the opposite of transconductance droop.

Use of Source Resistors

Source resistors do not help very much with setting and equalizing bias current among paralleled MOSFETs, so they are often not used. However, sometimes they can be used to reduce crossover distortion in two ways. First, they can be used to reduce transconductance somewhat at high currents. This may reduce crossover distortion by reducing the relative amount of transconductance droop. Source resistors will have a larger effect on reducing transconductance at high current than

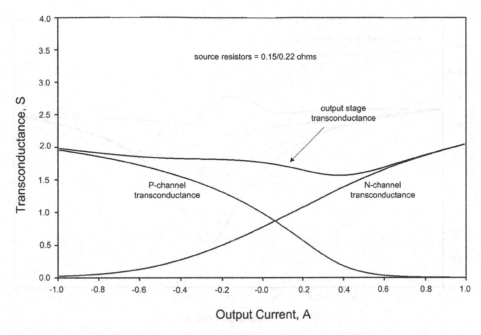

Figure 14.11 Transconductance Versus Current When Using Source Resistors

at the low current in the crossover region. Figure 14.11 is similar to Figure 14.10, but includes source resistors to reduce transconductance variations.

Second, the use of different source resistances for the N- and P-channel MOSFETs may reduce crossover distortion slightly by improving the transconductance match of the devices. This has also been employed in the circuit used for Figure 14.11. The source resistor for the N-channel device is 0.15 Ω while the source resistor for the P-channel device is 0.22 Ω.

Wingspread Simulations

A *wingspread* simulation for a MOSFET output stage is shown below in Figure 14.12. It is a plot of output stage gain as a function of output current. It was carried out using EKV models created by the author for the IRFP240 and IRFP9240. These models are thought to be reasonable approximations to the behavior of the devices. The simulation is shown for three bias levels: 100 mA, 150 mA and 250 mA. It is readily apparent that the higher bias yields reduced crossover distortion and that there is no evidence of *gm* doubling. For comparison, the wingspread for a BJT output stage is also shown on the plot. The BJT output stage employs 0.22-Ω emitter resistors and is optimally biased at 120 mA. The MOSFET output stages include 0.06-Ω source resistors to make the BJT and MOSFET output stage gains equal at the output current extremes.

Memory Distortion

Memory distortion in output stages results when the history of output stage thermal variations with program material causes bias errors that can lead to increased crossover distortion [8]. This can be a significant concern with BJT output stages because of their sensitivity of bias point to temperature. As mentioned above, MOSFET output stages are inherently more temperature stable than BJT output stages, so they are less prone to memory distortion.

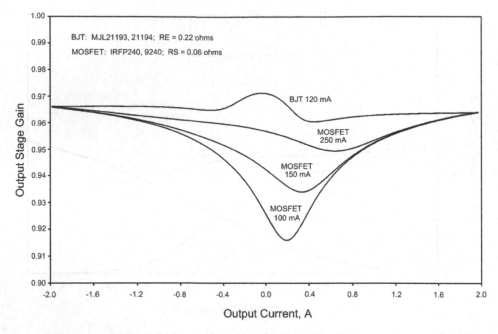

Figure 14.12 Wingspread Simulations for MOSFET and BJT Output Stages

14.7 Driving Power MOSFETs

The high input impedance of power MOSFETs makes them much easier to drive than BJT output stages, but with a few caveats. In essence, they have almost infinite DC beta. This means that they are not plagued by beta droop when driving high currents.

The high DC input impedance of MOSFETs can make it tempting to drive them directly from the VAS, and indeed some amplifier designers do just that. This is generally not a good idea because the input capacitance of the power MOSFETs is not insignificant, so the AC input impedance at higher frequencies definitely presents a loading effect on the VAS—and a nonlinear one at that.

Driving the Gate Capacitance

The gate capacitance of the MOSFET has two components: gate-source capacitance C_{gs} and gate-drain capacitance C_{gd}. As with BJTs in an emitter follower configuration, MOSFETs in a source follower configuration have their gate-source capacitance bootstrapped by the output signal. In contrast, the gate-drain capacitance is not bootstrapped. C_{gd} will sometimes dominate the effective input capacitance of the MOSFET output stage, especially when the output is near the rail and C_{gd} is large.

Gate-Source Capacitance

Since the gate-source voltage must change in order for the current of the MOSFET to change, the gate-source capacitance must be charged and discharged. C_{gs} is usually between 500 pF and 1500 pF and is relatively constant with changes in V_{gs} for vertical MOSFETs. While this capacitance can seem significant, its effective value is much smaller because it is bootstrapped by the output signal. The current through the capacitance is only in proportion to the signal voltage across C_{gs}. If the gain of the output stage when driving a load is 0.9 and C_{gs} is 1200 pF, the apparent load presented to the driver by the gate-source capacitance is only 120 pF.

Gate-Drain Capacitance

The drain voltage for a source follower output stage is at a fixed potential, while the gate voltage is changing at approximately the voltage rate of change of the amplifier output. The current required to charge and discharge the gate-drain capacitance C_{gd} is thus proportional to the rate of change of the output voltage.

The gate-drain capacitance for MOSFETs is usually small, but there is a major exception that is cause for concern. As shown by the plot of C_{gd} versus V_{dg} for the IRFP240 in Figure 14.13, the gate-drain capacitance becomes much larger for vertical MOSFETs when the drain-to-gate voltage becomes small. At $V_{dg} = 50$ V, C_{gd} is only 50 pF. This climbs to 300 pF at 10 V and over 1200 pF at 1 V. C_{gd} can thus be quite large when the amplifier output is driven close to clipping. Fortunately, the voltage rate of change of program material is usually not at its maximum near clipping; it is more often greatest at smaller output voltage magnitudes. However, a square wave test signal will often exhibit its maximum rate of change just as its voltage begins to move away from the rail.

Required Drive Current Versus Slew Rate

The drivers for a MOSFET output stage must be capable of providing adequate source and sink current to support the rated slew rate of the amplifier. The power MOSFET will often be driven with a BJT emitter follower. In such a case, the maximum available turn-off current is limited to the bias current of the driver stage. As with bipolar designs, if the turn-off drive current is not sufficient, there may be significant totem pole conduction through the output stage.

Consider an IRFP240/9240 output stage driving a 4-Ω load to 40 V peak with a slew rate of 100 V/µs. Output stage transconductance will be assumed to be at least 2 S and output stage

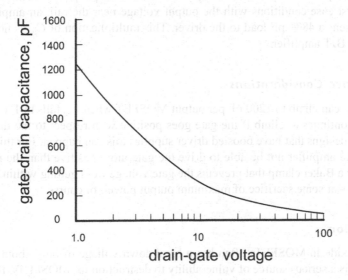

Figure 14.13 C_{gd} Versus V_{ds} for IRFP240

gain will be at least 0.89. C_{gs} will be bootstrapped by a factor of about 10. If C_{gs} = 1200 pF, its apparent value will be about 120 pF. Even if C_{gd} is only 200 pF (the value at V_{dg} = 20 V), total capacitance that must be driven will be on the order of 320 pF. Required driver current will be 32 mA. Other demanding situations can be imagined where a high rate of change is required when the signal amplitude is near the rail, where C_{gd} is much higher. The message here is not to be complacent about required drive current for both turn-on and turn-off of MOSFETs under demanding dynamic signal conditions.

Excess Phase at Signal Peaks

The significant increase in C_{gd} when the signal is close to the rail can lead to degraded feedback loop stability in the vicinity of large signal peaks. The C_{gd} capacitance working against the source impedance feeding the MOSFET forms a pole whose frequency will decrease under these conditions. This is of special concern because C_{gd} will be usually working against a gate stopper resistance. Even if a relatively small gate stopper resistance of 47 Ω is used with an IRFP240 and C_{gd} increases to 1200 pF, that pole will be at about 3 MHz.

Driving Multiple Output Pairs

Many amplifiers employ multiple output pairs. This clearly can lead to increased capacitive loading on the driver. The paralleling action affects the C_{gs} and C_{gd} loading differently, however.

The apparent gate-source capacitance seen by the driver does not necessarily increase as expected. Recall that the gate-source capacitance is bootstrapped by the output signal. The closer the gain of the output stage is to unity, the more effective is the bootstrapping. If the added output pairs are each biased at the same quiescent current as the first, their presence will increase the total transconductance of the output stage and drive the output stage gain closer to unity, increasing the bootstrapping effect. In principle, the increased bootstrapping effect will offset the effect of the increase in total gate-source capacitance. Apparent input capacitance of the output stage due to C_{gs} will remain about the same.

The situation is not so rosy for C_{gd}. The gate-drain capacitance adds up with the introduction of multiple output pairs as expected. An amplifier with two output pairs will have twice as much C_{gd}. Under worst-case conditions with the output voltage near the rail, an amplifier with four pairs could present a 4800-pF load to the driver. This multiplication of C_{gd} is no different than that for C_{cb} in a BJT amplifier.

Maximum Drive Considerations

The fact that C_{gd} can climb to 1200 pF per output MOSFET when V_{dg} falls to 1 V is bad enough. However, C_{gd} continues to climb if the gate goes positive with respect to the drain voltage. In some amplifier designs that have boosted driver supplies this can happen. For this reason, I recommend that the amplifier not be able to drive the gate more positive than the rail. Better yet, one can employ a Baker clamp that prevents the gate voltage from getting within less than a few volts of the rail—at some sacrifice of maximum output power, of course.

Gate Protection

The thin gate oxide in MOSFETs often has a breakdown voltage of only about 20 V. Without protection, this is a serious source of vulnerability to destruction for MOSFETs. If this voltage is exceeded, the gate oxide may break down, causing instant destruction of the device. Many lateral

MOSFETs actually incorporate internal gate Zener protection diodes, often with a breakdown of 15 V. Many vertical MOSFET power amplifier designs employ external Zener diodes to protect the gate from excessive voltages of either polarity. In other cases, it is possible to design driver circuits that inherently are unable to create excessive gate-source voltages. These will be discussed later in this chapter.

If the amplifier output is shorted or if the output stage is otherwise unable to supply the current demanded by the load, the driver may attempt to drive the MOSFETs to excessive gate voltages, in an attempt to get the output stage to produce very high current. This is an example of where a MOSFET may be exposed to excessive gate-source voltage if it is unprotected. However, even in the case where gate protection Zener diodes are used, the driver may actually try to drive the load through the gate stopper resistors and protection diodes, possibly leading to destruction of those diodes or the driver transistor (depending on the value of the gate stopper resistors).

The gate oxide can also be subjected to excessive voltages during high-frequency parasitic oscillations. The combination of bond wire inductances and the internal capacitances of the MOSFET can create resonances by which the internal gate voltage swing exceeds that at the external terminals of the device. External gate Zener diodes may not protect the gate oxide from excessive voltages under these conditions.

Flying Catch Diodes

Figure 14.14 shows a driver circuit that incorporates so-called flying catch diodes. These diodes connect from each end of the bias spreader to the output node of the amplifier. They are normally reverse-biased by the voltages at the ends of the bias spreader. These diodes will become forward-biased when the gate drive voltage exceeds the magnitude of the bias spreader voltage. For example, if a vertical MOSFET is being used that requires $V_{gs} = 4$ V for a bias current of 150 mA, the maximum gate drive to that MOSFET will be limited to about 8 V by the catch diodes. This protects the gate from dangerous overdrive voltages without resort to gate Zener diodes. The diodes are referred to as flying catch diodes because both ends of the diode move with the signal.

Natural Current Limiting

There is a second benefit from the use of the flying catch diodes. If the gate drive voltage is prevented from exceeding approximately twice the bias value, the maximum current that the MOSFET can deliver will be naturally limited. In the case of the IRFP240 the maximum current will be limited to approximately 30 A. The corresponding number for a 2SK1530 is about 12 A. We refer to this as *natural current limiting*.

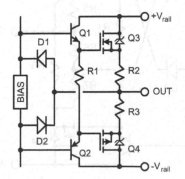

Figure 14.14 A MOSFET Driver Circuit Employing Flying Catch Diodes

If a smaller current limit is desired, a small source resistor can be added as shown in Figure 14.14. The voltage drop across the source resistor will subtract from the available gate drive, reducing the current limit. If a 0.1-Ω source resistor is used and 15 A flows, the available gate drive will be reduced by 1.5 V to 6.5 V. This is approximately the amount of gate drive required for the IRFP240 to conduct 15 A.

If for some unrelated reason a larger value of source resistor is used, the flying catch diodes can be connected to a voltage divider across the larger source resistor to retain the higher value of current limit. This is illustrated in Figure 14.15.

Short-Circuit Protection

The flying catch diodes provide yet another opportunity in the form of short-circuit protection. As mentioned above, the gate drive is limited to twice the bias spreading voltage. If the bias spreader is collapsed (shorted out), the output stage will be turned off and disabled because no gate drive will be available to either MOSFET. The VAS signal current will be dumped harmlessly into the output node. One approach to collapsing the bias spreader is to simply place a TRIAC across it. The TRIAC can be triggered by an optocoupler from circuitry that detects a short-circuit condition. The amplifier then will be latched in a disabled state until power is cycled. Such behavior might not be desirable for a pro audio environment, but it is perfectly acceptable for home audio use.

Folded Drivers

Figure 14.16 illustrates an alternative way of driving a MOSFET output stage, referred to here as a *folded driver*. Because the MOSFETs require essentially no DC current to drive them, it is feasible to use an inverted driver EF that acts to turn off the MOSFET rather than turn it on. Turn-on current is then provided by a current source. The current must be able to provide adequate

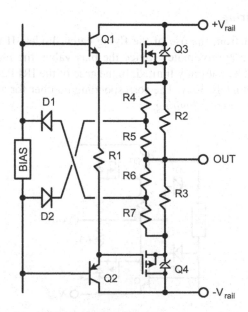

Figure 14.15 Adjustable Natural Current Limiting with Flying Catch Diodes

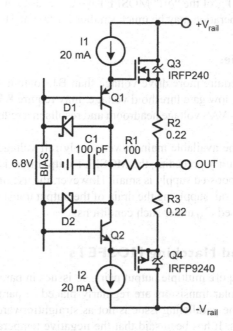

Figure 14.16 A MOSFET Output Stage with Folded Drivers

turn-on slew rate for the device (as opposed to adequate turn-off slew rate as in the conventional approach).

Notice that the collectors of the driver transistors are connected to the output node of the power amplifier. The unused drive current is thus dumped harmlessly into the output node. This means that there is low voltage across these transistors, allowing fast devices to be used. The signal from the output node is filtered somewhat at very high frequencies to avoid the possibility of feedback effects that might come into play at very high frequencies.

This approach has some advantages. First, the amount of gate drive is current limited. Second, the driver transistor can be a small, fast device if its collector is connected to the output node. The driver transistor controls the MOSFET by turning it off, shunting current from the current source to the output node. The driver transistor also draws charge out of the gate-source capacitance by shunting gate current right back to the source. Third, the current source, which bears virtually all of the driver power dissipation, need not have exceptional signal-handling characteristics, since the signal does not pass through it. Fourth, the bootstrapping of the driver collectors greatly reduces the effects of their nonlinear collector-base capacitance and Early effect on loading of the VAS. The folded driver transistors also dissipate less power and experience less temperature rise. This helps overall output stage thermal bias stability.

The base-collector junction of the driver transistor also serves as the flying catch diode. In many cases the driver transistors can be allowed to saturate under natural current-limiting conditions without causing sticking. If the driver transistors must be kept out of saturation under these conditions, Shottky flying catch diodes can be incorporated as shown.

Notice that this circuit will not allow V_{gs} to go below zero for either output device. This is not a problem with these vertical MOSFETs because they have a fairly large turn-on voltage typically on the order of 4 V, and they quickly go to several amperes of conduction with a couple of additional volts. With the circuit shown, by the time V_{gs} for Q3 would need to go to zero or below, Q4 will be turned on with about 8 V of forward V_{gs}, corresponding to something on the order of 20 A.

Under normal conditions, V_{gs} of the "off" MOSFET never needs to go below zero. However, the margins for this kind of operation may be much smaller for lateral MOSFET implementations.

Boosted Driver Supplies

MOSFET output stages require more drive voltage than BJT output stages. Even lateral MOS-FETs, with their relatively low gate threshold voltage, may require 8 V or more to drive them to high currents. This reduces VAS voltage headroom and results in a reduced power output capability for given rail voltages.

Better use is made of the available main power supply rail voltage if a boosted supply voltage is provided to the circuitry preceding the MOSFET output transistors [6, 9]. The amount of current required from the boosted supply is small. However, the use of boosted supplies to allow drive of the gate above the rail supplying the drain of the output transistor may be unwise in light of the dramatically increased C_{gd} under such conditions.

14.8 Paralleling and Matching MOSFETs

Most power amplifiers require multiple output pairs, so issues in paralleling of MOSFETs must be considered. While bipolar transistors are regularly placed in parallel with small individual emitter ballast resistors, the paralleling issue is not as straightforward for power MOSFETs, at least in linear applications. It has been said that the negative temperature coefficients of trans-conductance and *on* resistance of MOSFETs act to suppress current hogging by one transistor, thus permitting easy paralleling of MOSFETs without ballast resistors. This is somewhat true for lateral MOSFETs where some degree of simple matching has been employed [4]. This also appears to be true for hard-switching applications using vertical MOSFETs where the paralleled devices are all fully turned on together (i.e., channels fully enhanced by forward gate voltage) so that current and dissipation imbalances are only a result of mismatched *on* resistance.

However, the issue is more complex for linear, and especially low-distortion, applications because the operating region of interest is not the fully turned-on region, but rather the linear region wherein drain current at a specified gate voltage is important. Specifically, recognizing that the gate threshold voltage specification for vertical power MOSFETs is 2–4 V for the IRFP240, for example, you'll find that an examination of the gate transfer characteristics quickly shows that a very serious current imbalance can exist unless gate threshold voltages among paralleled devices are reasonably matched. It is also apparent that reasonable temperature differentials will not adequately reduce the imbalance. This is especially true if a common heat sink is employed. Because of the size of the worst-case threshold voltage differentials possible, the use of source ballast resistors is not a reasonable approach to achieving balance.

It thus appears that, for high-quality audio applications where paralleled devices are necessary, both threshold voltage and transconductance of paralleled devices should be matched. Matching that guarantees that all devices are carrying ±50% of their nominal current share in the quiescent bias state and ±25% of their share at high currents is probably adequate. For the IRFP240, V_{gs} matching of ±0.1 V at 150 mA and ±0.25 V at 4 A will typically satisfy this matching criterion [6].

Fortunately, this is not as difficult as it sounds. With the manufacturing consistency in place for modern MOSFETs, most of the devices in a tube will be like peas in a pod. In many cases they will all have come from the same wafer. Moreover, if their threshold voltages are well matched, their transconductances will likely be well matched. For this reason, one can often get by with just matching V_{gs} at 150 mA.

Matching requirements can be relaxed a bit by the use of source resistors. However, the amount of resistance needed to do much good in this regard is so high as to waste voltage headroom under high-current conditions. Source resistances on the order of less than 0.5 Ω can be used to reduce distortion by adjusting wingspread symmetry at high current, as in Figure 14.11 (0.15 Ω and 0.22 Ω), but will do little for matching paralleled devices under quiescent bias conditions. Their voltage drop, at a typical bias current of 150 mA, is less than 75 mV. They will tend to reduce current hogging among paralleled devices under high-current conditions, however. It is useful to note that the 4 S conductance of a 0.25-Ω source resistor is about the same as the transconductance of an IRFP240 operating at 2 A, as indicated in Figure 14.5. This is the point at which the help with matching and thermal stability of the source resistor becomes significant. The price paid is an additional 2.5 V of required gate drive when the MOSFET is delivering 10 A.

14.9 Simulating MOSFET Power Amplifiers

Simulation of power amplifiers with SPICE can be very valuable, and this includes simulation of crossover distortion. However, extra caution is required in simulating MOSFET output stages, especially in regard to crossover distortion. The reason for this is that most SPICE models for power MOSFETs do not properly model the low-current region, sometimes called the weak inversion region. The MOSFET is generally thought of as a square-law device.

$$I_d = K(V_{gs} - V_t)^2 \qquad (14.1)$$

The simple square-law equation for drain current goes to zero at the threshold voltage, causing a discontinuity in transconductance. This is simply not accurate for MOSFETs. In fact, at low currents, the MOSFET characteristic transitions to an exponential law that is much like that followed by BJTs, but with far different coefficients. This is referred to as *subthreshold* conduction.

Because the simulation of crossover distortion involves behavior of the devices at low currents (150 mA is considered to be in the transition region of the models), the normal SPICE models will give misleading results. There are better models for MOSFETs, one of which is called the EKV model [10]. However, EKV model parameters for power MOSFETs are extremely rare. The EKV model is discussed in depth in Chapter 24. With datasheet information and perhaps some low-current measurements, you can create an EKV model for a MOSFET and install it into LTspice®.

The LTspice® VDMOS model now incorporates a parameter called KSUBTHRES that allows the modeling of subthreshold conduction. This usually makes it unnecessary to resort to using the EKV model. Using the VDMOS model also allows one to take advantage of its excellent modeling of MOSFET gate-drain capacitance. The KSUBTHRES VDMOS model parameter is covered in Chapter 24.

High-Frequency Simulations

The gate-drain capacitance is highly nonlinear in MOSFETs and can become quite large when V_{dg} is small. This effect is not modeled well in most models available from manufacturers. In fact, many of their models are optimized for switching applications where different approaches to modeling C_{gs} may be satisfactory. Fortunately, the VDMOS model in LTspice® does a good job of modeling C_{gd} that is equally well suited to linear applications. The VDMOS model should be used when the frequency response and gate drive current requirements of the MOSFET output stage matter the most.

14.10 A Lateral MOSFET Power Amplifier Design

Figure 14.17 shows the bias spreader and output stage of a lateral MOSFET power amplifier that is a redesign of the Hafler DH-220. The full complementary JFET input stage and push-pull VAS were discussed in Chapter 9, where they were shown in Figure 9.20. The amplifier, here dubbed the DH-220C, uses the 2SK134/2SJ49 lateral devices, heat sinks, chassis and power supply from the original amplifier [11]. The amplifier employs two output pairs and is capable of delivering 110-W/8-Ω using the original Hafler power supplies. Each output pair is biased at 200 mA for low crossover distortion.

Input Stage, VAS and DC Servo

The schematic of the IPS and VAS is shown in Figure 9.20. It includes a full-complementary JFET input stage with a 4 mA floating tail, and employs the LSK489 and LSJ689 dual monolithic JFETs. The input stage is cascoded and is loaded with current mirrors. The push-pull VAS is a 2T design that incorporates current limiting. It is biased at 10 mA. The amplifier uses a DC servo as shown in Figure 10.5. The servo can correct up to 150 mV of offset, and this obviates the need for an input coupling capacitor. Closed loop gain is set to 27.7 and conventional Miller compensation sets the ULGF to about 1 MHz.

Bias Spreader and Output Stage

The bias spreader is a CFP design that typically supplies a spread of about 2.7 V to bias the output lateral MOSFETs and their BJT emitter follower drivers. The CFP bias spreader provides a stiff, low-impedance voltage drop that is advantageous for full complementary designs like this,

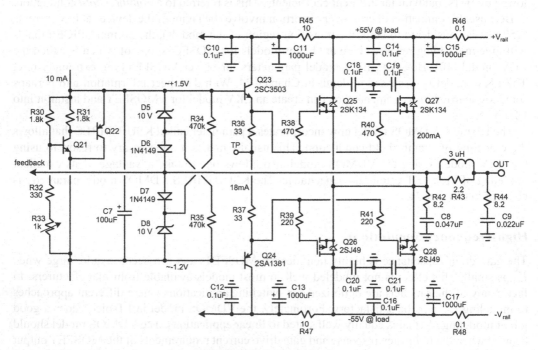

Figure 14.17 A High-Performance Lateral MOSFET Power Amplifier

wherein the VAS bias current can be somewhat more variable than in single-ended designs where the VAS bias current is set by a fixed current source. 10-V Zener diodes D5 and D8 provide gate protection and current limiting for the output MOSFETs by limiting their gate voltage with respect to the output rail to about ±10 V. R34 and R35 reverse-bias the Zener diodes by about 2.7 V under normal signal conditions to reduce their capacitance.

The output transistors are driven by emitter followers that are biased at a healthy 18 mA so as to be able to drive the capacitances of the output MOSFETs with adequate turn-off current even under conditions of fast turn-off. The driver bias resistors provide a center-tap test point that allows measurement of output stage distortion alone or closure of the feedback loop without the output stage for testing purposes. Gate stopper resistors are included with the same values as used in the original Hafler design.

Performance

Figure 14.18 shows measured THD+N versus power output into 8 W. Using a test power supply that delivered 52 V under load, the amplifier clipped at over 100 W and delivered 90 W into an 8-Ω load with THD+N of 0.007% at 1 kHz and THD+N of 0.02% at 20 kHz. There was no rise of distortion at low power that would be indicative of crossover distortion. THD-1 was below 0.0011% in the critical power range between 1 W and 10 W. In that same range, THD-20 was below 0.007%. Use of the DH-220's power supply increases maximum output power to over 120 W.

Figure 14.19 shows measured THD+N versus frequency at 1 W, 10 W and 75 W into 8 Ω. THD+N remains below 0.007% up to 20 kHz at all of the power levels, and rises gently at less than 6 dB/octave above 1 kHz, where it hovers around 0.001%, as feedback loop gain decreases, as expected.

Figure 14.20 shows measured THD+N versus power of the output stage by itself at 1 kHz for no-load, 8-Ω and 4-Ω loads. This measurement is of interest because the output stage is almost

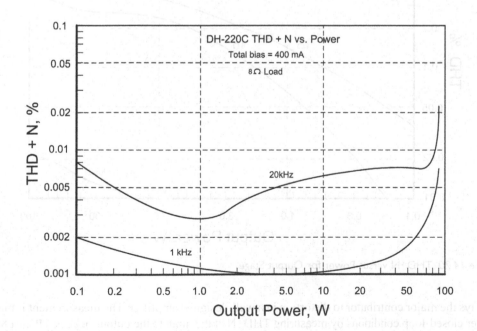

Figure 14.18 THD+N Versus Power

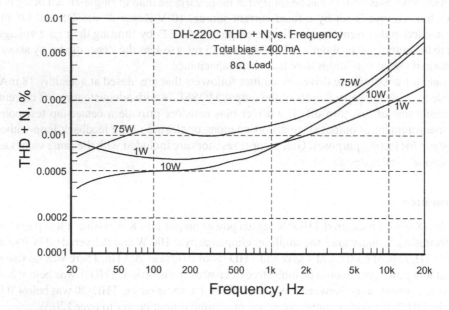

Figure 14.19 THD+N Versus Frequency

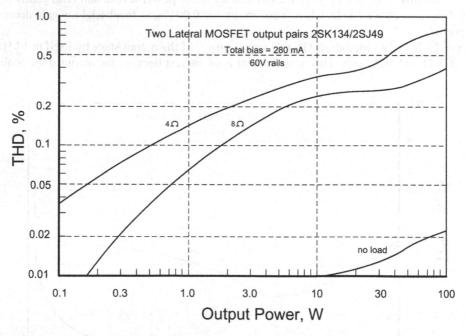

Figure 14.20 THD+N versus Power for Output Stage

always the major contributor to distortion in a well-designed amplifier. The measurement is made under closed-loop conditions by measuring THD+N at the input to the output stage at TP1. At 8 W into 8 Ω, the spectrum had second harmonic at −54 dB, third at −58 dB, fifth at −74 dB and all other harmonics lying below −80 dB. The relative absence of high-order harmonics testifies to the smooth transfer characteristic of the lateral MOSFET output stage.

14.11 A Vertical MOSFET Power Amplifier Design

A good example of a vertical MOSFET power amplifier was presented in Chapter 11, Figure 11.19, where TMC feedback compensation was discussed [12]. It employed a single pair of IRFP240/9240 MOSFET output devices biased at 200 mA and had a fairly straightforward JFET input stage with a single-ended VAS. With ordinary Miller compensation and ULGF of 1 MHz, the amplifier achieved THD+N of only 0.015% at 20 kHz at its rated full power of 50 W into 8 Ω ($\pm$50-V rails). With TMC, THD-20 at 50 W fell to 0.0045%.

References

1. Renesas datasheet for 2SK1056/2SJ160.
2. Toshiba datasheet for 2SK1530/2SJ201.
3. Vishay datasheet for IRFP240/IRFP9240.
4. David Hafler Company, DH220 Power Amplifier.
5. "Hitachi Power MOSFET Data Book HLN600," Hitachi America, Ltd., 1983.
6. Robert R. Cordell, "A MOSFET Power Amplifier with Error Correction," *Journal of the Audio Engineering Society*, vol. 32, no. 1, pp. 2–17, January 1984. Available at www.cordellaudio.com.
7. Edward S. Oxner, "Analyzing and Controlling the Tendency for Oscillation of Parallel Power MOSFETs," *Proceedings Powercon*, vol. 10, 1983.
8. T. Sato, K. Higashiyama and H. Jiko, "Amplifier Transient Crossover Distortion Resulting from Temperature Changes in the Output Power Transistors," presented at the 72nd Convention of the Audio Engineering Society, *Journal of the Audio Engineering Society* (Abstracts), vol. 30, pp. 949–950, December 1982, preprint no. 1896.
9. Bart Locanthi, "Operational Amplifier Circuit for Hi-Fi," *Electronics World*, January 1967, pp. 39–41.
10. Christian C. Enz and Eric A. Vittoz, *Charge-based MOS Transistor Modeling*, Wiley, New York, 2006.
11. Presented at Burning Amp 2016. Available at www.youtube.com/watch?v=V7-27fDgqco.
12. Bob Cordell, "TPC and TMC Feedback Compensation for Power Amplifiers," *Linear Audio*, vol. 6, pp. 17–32, September 2013.

Chapter 15

Error Correction

15. INTRODUCTION

Conventional negative feedback is not the only way to reduce distortion. Various error-correction techniques can be used in place of, or in connection with, negative feedback [1, 2, 3]. In this chapter we'll examine some of these techniques and study one of them in detail.

In virtually any well-designed power amplifier the output stage ultimately limits performance. It is here where both high voltages and large current swings are present, necessitating larger, more rugged devices that tend to be slower and less linear over their required operating range. The performance-limiting nature of the output stage is especially evident in class AB designs where the signals being handled by each half of the output stage have highly nonlinear half-wave-rectified waveforms and where crossover distortion is easily generated. In contrast, it is not difficult or prohibitively expensive to design front-end circuitry of exceptional linearity.

Overall negative feedback greatly improves amplifier performance, but it becomes progressively less effective as the frequency or speed of the errors being corrected increases. High-frequency crossover distortion is a good example. The philosophy here is based on the observation that only the output stage needs extra error correction and that such local error correction can be less complex and more effective.

While the power MOSFET has many advantages, it was pointed out in Chapter 14 that the lower transconductance of the MOSFET will result in moderate crossover distortion unless rather high bias currents are chosen. This effect was shown in Figure 14.10 where the individual and summed transconductances of both halves of a class AB MOSFET output stage were plotted as a function of net output current. The output stage transconductance is smaller in the crossover region, a phenomenon dubbed *transconductance droop* [2]. At a bias current of 150 mA and a load of 8 Ω, transconductance droop can result in open-loop output stage harmonic distortion on the order of 1% [2]. Mismatch in the transconductance characteristics of the top and bottom output devices also contributes to crossover distortion.

The primary focus of this chapter will be the application of error correction to MOSFET power amplifiers. However, output stage error correction can also be applied with advantage to BJT output stages, and that will be covered in this chapter as well.

15.1 Feed-Forward Error Correction

Figure 15.1 depicts a simple *feed-forward error correction* circuit. The idea is to calculate the error and then feed it forward and subtract it from the output. No loop is formed, so there are no stability concerns. The output stage is assumed to be a unity-gain stage with error injected. All forms of output stage error, including gain less than unity, are represented by the source $\varepsilon(x)$. The error is subtracted from the unity-gain signal path to signify nominal gain less than unity. Summer

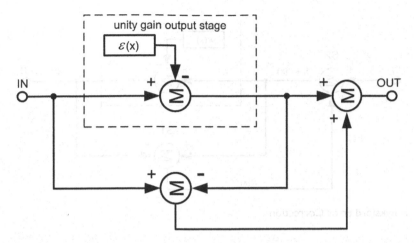

Figure 15.1 Feed-Forward Error Correction

S1 compares the input and output of the output stage and recovers $\varepsilon(x)$. Summer S2 adds $\varepsilon(x)$ back into the signal path to restore the signal to its original input value.

The problem with this kind of error correction is the difficulty with which an additional error signal can be passively added in a precise way to the output signal after the output stage. Making such a passive adding circuit at high power levels is difficult. It is even more difficult to do it in such a way as to not seriously compromise the output impedance. Any departures from the ideal signal will not be corrected after the summing network.

Reduced Effectiveness at High Frequencies

The cancellation of distortion with feed-forward requires that the error signal have just the right amplitude and phase. In principle, these can be adjusted with potentiometers, but this adds to cost and difficulty of setup. It is also generally the case that it becomes progressively more difficult to maintain the proper phase relationship as the frequency increases. Ultimately, this means that the amount of achievable distortion reduction through error correction decreases with increasing frequency, just as does the distortion reduction afforded by negative feedback.

15.2 Hawksford Error Correction

In 1980 Hawksford introduced a form of error correction that did not depend on a passive addition of the error signal to the output signal after the output stage [1]. Instead, as illustrated below, it fed back the error signal in a particular way. This will be referred to as *Hawksford error correction* (HEC). This technique is illustrated in Figure 15.2. As in Figure 15.1, the output stage is modeled as having exactly unity gain with an error voltage $\varepsilon(x)$ added. This error represents any departure from unity gain, whether it is a linear departure due to less than unity gain, a distortion due to transconductance nonlinearity or injected errors like power supply ripple. A differential amplifier, represented by summer S1, subtracts the output from the input of the power stage to arrive at $\varepsilon(x)$. This error signal is then added to the input of the power stage by summer S2 to provide that distorted input which is required for an undistorted output. Note that this is an error-cancellation technique like feed-forward as opposed to an error-reduction technique like negative feedback.

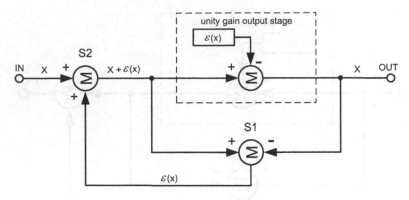

Figure 15.2 Hawksford Error Correction

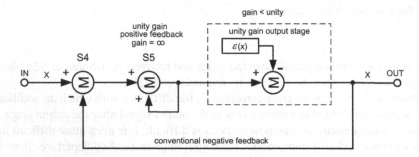

Figure 15.3 HEC Viewed as Negative Feedback Whose Open-Loop Gain Is Enhanced by an Embedded Positive Feedback Loop

This technique is in a sense like the dual of feed-forward. It is less expensive because the point of summation is a low-power internal amplifier node. Like feed-forward, HEC tends to become less effective at very high frequencies because the required phase and amplitude matching for error cancellation becomes progressively more difficult to maintain. It can be shown that the required amplitude and phase matching is indeed identical to those for feed-forward. HEC also tends to become less effective at very high frequencies because, being a feedback loop (albeit not a traditional negative feedback loop), it requires some amount of frequency compensation for stability, which detracts from the phase and amplitude matching.

A Specialized Form of Negative Feedback

If the HEC circuit is redrawn as in Figure 15.3, it is apparent that there is a summer S5 with unity-gain positive feedback, around which there is a negative feedback loop (S4). In essence, the unity-gain positive feedback stage provides infinite forward gain and thus infinite negative feedback loop gain.

If HEC is just another form of negative feedback, then what are its advantages? Those include simplicity of implementation and a tight local feedback loop. The negative feedback view of HEC also underlines the need for some form of frequency compensation of the loop. In the implementation to be discussed below, the forward path is compensated by feed-forward shunt capacitance that does not introduce a pole into the forward path of the output stage [2].

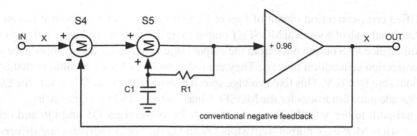

Figure 15.4 Compensating the Error-Correction Loop

Frequency Compensation

Because HEC can be seen as a negative feedback loop with a high-gain stage formed by embedded positive feedback, it needs compensation just like any other negative feedback loop. A crude approach to compensating the *error correction* (EC) loop is illustrated in Figure 15.4, where the positive feedback view of the EC circuit is repeated, but with a compensating R-C circuit in the positive feedback path.

It is easy to see that this network kills the loop gain of the EC circuit at high frequencies without placing a pole in the forward signal path. This helps avoid introducing additional phase lag into the global feedback loop of the main amplifier. This simplified example only reduces the EC loop gain to unity, so the practical implementation to be described is a bit different.

Effect on Output Impedance

The error-correction circuit seeks to force the gain of the output stage to unity under all conditions. Its output will then become load-independent. Increased loading will not cause the output to fall. This is equivalent to the output impedance being zero. As a result, the open-loop output impedance of an amplifier with EC is extremely low, even in the absence of global negative feedback. When global negative feedback is applied, the output impedance is driven even lower; this results in an extremely high damping factor out to quite high frequencies. The high-frequency damping factor for such an amplifier is largely dependent on the impedance of the output coil.

15.3 Error Correction for MOSFET Output Stages

The MOSFET output stage is an ideal candidate for the use of error correction. First, it profits greatly from error correction because the EC greatly mitigates transconductance droop. Second, the power MOSFET output stage is very fast and can provide the large bandwidth that allows HEC to work well. There are two caveats with respect to output stage speed. First, lateral MOSFETs are a bit slower than verticals because they have inherently higher distributed internal gate resistance [2]. Second, the necessary use of gate stopper resistors, as described in Chapter 14, tends to slow down the MOSFET output stage, so their value should be made as small as possible while providing adequate stability. In "A MOSFET Power Amplifier with Error Correction," [2] the gate stopper resistors were only 47 Ω. This low value of resistance was made possible by the use of gate Zobel networks that killed the Q of any resonance local to the MOSFETs.

Simplified Error-Correction Circuit

The simplified error-correction circuit of Figure 15.5 shows how an error amplifier can be interposed in the signal path of a vertical MOSFET output stage [2]. Emitter followers Q1 and Q2 isolate the high-impedance VAS output node from the output stage and provide a low-impedance signal for the error-correction summation process. They are fed from the VAS with voltages that are spread from nominal zero by ±11 V. This fixed voltage spread provided by Zener D1 powers the EC circuits and provides adequate headroom for the MOSFET bias voltages and V_{gs} signal swing.

The signal path to the MOSFETs passes through EF pre-drivers Q5 and Q6 and drivers Q7 and Q8 to vertical MOSFET output transistors Q9 and Q10. The pre-drivers and drivers provide a high-current drive capability for the MOSFET gates and isolate the error-correction summing nodes from the MOSFET gate loads. Note that Q5 and Q6 can be fast, inexpensive small-signal transistors because their collectors are driven with the signals from Q1 and Q2.

Complementary transistors Q3 and Q4 make up the error amplifier, where the error-correction action takes place. The base and emitter circuits of Q3 and Q4 implement the function of summer S1 in Figure 15.2. The signal from the error amplifier is injected into the forward path by error current flowing through series resistors R1 and R2. The collector nodes of Q3 and Q4 play the role of summer S2 in Figure 15.2.

The error amplifier made up of Q3 and Q4 is a differential amplifier formed by complementary transistors. The inverting input is applied to the bases while the non-inverting input is applied to the emitters. Emitter resistors R3 and R4 control the loop gain of the bias loops, provide emitter degeneration for the error amplifier and improve stability.

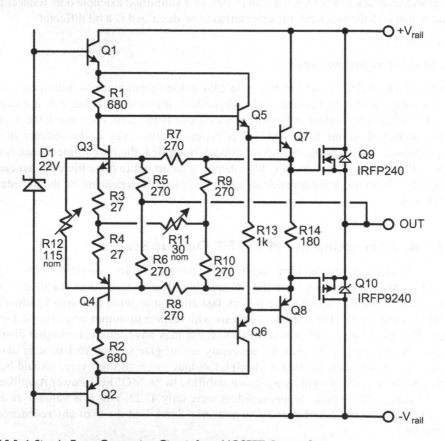

Figure 15.5 A Simple Error-Correction Circuit for a MOSFET Output Stage

In the context of Figure 15.2, MOSFETs Q9 and Q10 make up the output stage. The signal at their gates corresponds to the input signal supplied to the output stage by summer S2 in Figure 15.2. This signal makes its way back to the non-inverting input of the error amplifier through R9, R10 and R11. A simple circuit analysis shows that the feedback loop formed from the emitters of Q7 and Q8 through the error amplifier and the summing node load resistors R1 and R2 has non-inverting gain of unity. This corresponds to the behavior illustrated in Figures 15.2 and 15.3. The error correction condition is optimized when this gain is exactly unity, as optionally trimmed by R11. Adjustment of R11 nulls the distortion.

Transistors Q3 and Q4 also act as a V_{be} multiplier to implement the bias spreader for the MOS-FET output stage. Q3 and Q4, in conjunction with bias adjust resistor R12, control the DC voltage drop across R1 and R2. They thus set the bias for the MOSFETs by means of a V_{be}-referenced feedback loop that also includes Q5, Q6, Q7 and Q8. The V_{be} drops of Q3 and Q4 together are multiplied by the combination of resistors R5–R8. Transistor Q3 is mounted on the heat sink to provide thermal feedback. Thus, about half of the total bias spread is compensated with the temperature of the heat sink. Q3 is preferably a fast TO-92 transistor embedded in a hole in the heat sink. The wiring to this transistor should be fairly short so as to avoid parasitic oscillations.

Error-Correction Circuit Operating Voltage

The error amplifier circuit floats with the signal and receives its operating voltage from the fixed bias spread provided to the EC circuit. The bias spread applied to the output transistors also plays a role in establishing the available operating voltage for the EC circuit. In the circuit of Figure 15.5, a fixed bias spreader provides the initial bias spread with which the error amplifier works. It is easy to see that the error-correction circuit requires additional headroom from the VAS. Because the error-correction circuit demands that the output stage have unity gain, it must provide all of the gate drive required by the MOSFETs to drive the load, even to very high currents.

Error-Correction Circuit Clipping

Under very high current demands by the load, the error-correction circuit may not be able to provide all of the gate drive demanded by the output transistors. Under these conditions, one of the error amplifier transistors will go into cutoff. When this happens, any additional signal swing and gate drive must be provided by the VAS, meaning that the output stage gain for such large signal swings will fall somewhat below unity. Under these conditions, however, the MOSFETs are in a high-current state and their transconductance is quite high, so output stage gain will be fairly close to unity even without the action of the error-correction circuit. It is also possible that the error amplifier can clip by going into saturation. This should be avoided.

15.4 Stability and Compensation

Because error correction is a form of negative feedback, it must be stabilized in some way at high frequencies. As with conventional negative feedback, this is done by controlling the gain and phase in the local feedback loop so that stability is preserved.

Stability Considerations

It is important that the compensation of the local feedback loop formed by the error-correction scheme not interfere with the open-loop gain and phase of the amplifier. Otherwise, the gain crossover frequency of the global feedback loop will have to be reduced and the distortion reduction will be reduced. It is not desirable to achieve distortion reduction by error correction at the expense of distortion reduction by global negative feedback. The goal is to have an output stage with error correction that has little or no additional excess phase as compared to the same output stage without error correction.

The EC loop is active to fairly high frequencies, so its stability may be significantly affected by loading at the output of the amplifier. For this reason, it is important that the usual L-R network be placed in series with the amplifier output to isolate the output node from the load at high frequencies. It is also important to employ very good high-frequency layout and design techniques in the output stage when EC is being used.

Frequency Compensation Approach

Compensation of the error correction loop requires that the gain around that loop be reduced with frequency to achieve a gain crossover frequency, just as with an ordinary feedback loop. Because the EC loop is local, this gain crossover frequency can typically be at a higher frequency than that used for the global feedback loop.

Figure 15.6 shows the addition of frequency compensation components to the basic EC circuit of Figure 15.5. The primary reduction of EC loop gain with frequency is achieved with the series R-C networks that shunt the error amplifier collector load resistors R1 and R2. Clearly, if this impedance goes to zero, there will be no error-correction action and no error-correction loop gain. The important feature of this arrangement is that when it reduces gain of the EC loop, it does so by shunting the input signal around the error amplifier load resistor in a feed-forward fashion to the next emitter follower in the signal path. As such, this frequency compensation does not add phase lag to the overall open-loop amplifier.

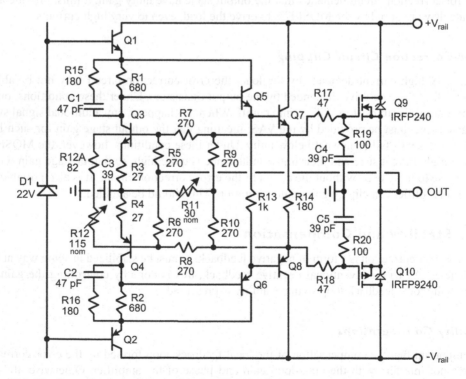

Figure 15.6 The Error-Correction Circuit Used in "A MOSFET Power Amplifier with Error Correction"[2]

C1 and C2 provide the primary means of frequency compensation by shunting R1 and R2. Resistors R15 and R16 in series with the compensation capacitors effectively place a zero in the loop of the EC circuit to improve phase margin. It is notable that at very high frequencies the collector-base capacitances of Q5 and Q6 create a further feed-forward bypass of the signal. As expected, the action of C1 and C2 at higher frequencies reduces the effectiveness of the error correction. However, in Reference 2 the error correction still reduced THD by almost 30 dB even at 20 kHz.

The remaining element of the EC compensation scheme is the single capacitor C3 connected from the error amplifier emitter circuit to ground. This capacitor acts as a positive feedback loop spoiler. It upsets the balance of the EC circuit at high frequencies in such a way that the feedback loop gain of the inner positive feedback loop is decreased from unity, so that the forward gain contributed by the positive feedback loop falls with increasing frequency.

The gate stopper resistors (R17 and R18) and gate Zobel networks (R19, R20, C4 and C5) used in Reference 2 for output stage stability are shown for completeness.

Simulation of Effective Gain Crossover Frequency

Estimating the effective gain crossover frequency of the error-correction circuit is not an easy matter. However, SPICE simulation can be used to evaluate it. SPICE simulation can also be used to evaluate the stability of the local error-correction feedback loop by viewing the frequency response and square-wave response of the error-corrected output stage by itself.

Figure 15.7 illustrates one way in which the EC loop can be broken in order to evaluate its loop gain. Recall that the EC circuit can be viewed as a negative feedback loop with an embedded high-gain stage formed by unity positive feedback as illustrated in Figure 15.2. The loop is broken in Figure 15.7 between the amplifier output (point A) and the input to the error amplifier

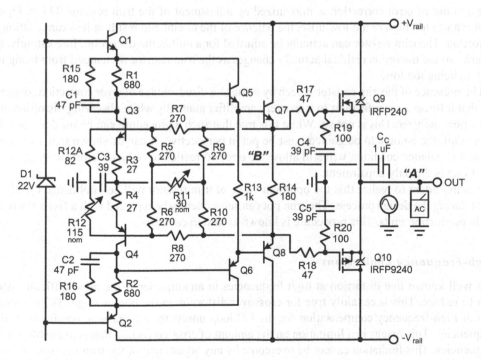

Figure 15.7 Breaking the Loop for Stability Analysis

(point *B*). The impedance at the output of the amplifier is much lower than the impedance at the input of the error amplifier. This means that the loop can be broken with little loss of accuracy by inserting an AC voltage source at the error amplifier side of the break and viewing the signal at the output of the amplifier. DC stability of the loop is not affected by breaking the loop in this way, so a large inductor is not needed to close this loop at DC. An AC test signal is merely injected into point *B* through a coupling capacitor.

Effect on the Global Feedback Loop

It is important that the error correction circuit does not detract significantly from the phase margin and gain margin of the global negative feedback loop. Fortunately, when the EC loop gain diminishes and EC action is no longer in play, the output stage defaults to a conventional output stage. In principle, this output stage then exhibits no more excess phase than a conventional output stage of the same design.

To this end, the gain and phase of the EC output stage by itself should be evaluated by simulation with and without the error correction enabled. Error correction is easily disabled by removal of R11.

15.5 Performance and Design Issues

There are some performance-limiting issues and design nuances in the EC circuit shown in Figure 15.6 that will be discussed here. In some cases understanding these can point the way to improvements that can be made to the error-correction circuit.

Trimming

The amount of error correction is maximized by adjustment of the trim resistor R11 in Figure 15.6. Values too high or too low upset the balance of the circuit and result in less cancellation of distortion. The trim resistor can actually be adjusted for a null in the distortion. Interestingly, the polarity of the distortion residual actually changes as the trim resistor is changed from being too high to being too low.

The presence of the trim resistor is seen by some as a disadvantage of error correction, suggesting that it forces a manufacturer to trim every amplifier manually while observing distortion on a distortion analyzer. This is not so. While it is true that such a procedure can be used to optimize the circuit, the amount to be gained must be put in perspective. It can be shown that a 1% error in the EC balance condition will still allow the error correction to reduce distortion by about 40 dB if that is the only impairment.

It is important to realize that the optimum value of trim resistor will be established by adjustment during the design process. The trim pot can subsequently be replaced with a fixed 1% resistor in production units. This procedure is known as *design centering*.

High-Frequency Limitations

It is well known that distortion at high frequencies in an amplifier is the most difficult distortion to reduce. This is certainly true for crossover distortion in the output stage. The necessary evil of high-frequency compensation for the EC loop upsets the EC balance condition at high frequencies. This results in a limitation on the amount of error correction that is available at high frequencies. This limitation cannot be overcome by any adjustment of the trim resistor, and usually dominates over trim resistor inaccuracies at high frequencies.

Given that high frequencies are where error correction is most needed, it is of relatively little value for the trim resistor to be substantially more optimal than that which achieves a similar degree of error correction at low frequencies. This is a further reason for arguing that a trim pot is superfluous for an EC circuit built with 1% resistors.

Nonlinearity in the Error Amplifier

The error-correction circuit sees any departure of the output stage from unity gain as an error, whether its origin is from a linear or nonlinear process. The EC will attempt to make the net gain of the output stage unity, regardless of load and (ideally) regardless of frequency.

This means that all of the gate voltage increase required by the MOSFETs for production of signal current must be supplied by the EC circuit. Where high signal currents are involved, the difference in gate voltage drive from the quiescent state to the high-current state can be several volts. The production of a signal swing of several volts by an amplifier as simple as the error amplifier in Figure 15.6 can incur some distortion. The main cause of distortion is the dynamic emitter resistance of Q3 and Q4. The value of re' changes with error current and creates distortion.

Table 15.1 shows typical V_{gs} at 150 mA and 5A for three different power MOSFETs. Also shown is the difference in V_{gs} for these two operating currents. This shows how much gate voltage swing is required when a single output pair is delivering 100 W into an 8-Ω load. The lateral MOSFET begins to turn on with a low value of V_{gs} but has a high value at 5 A due to its relatively low transconductance. This means that it must actually be driven harder than the vertical devices, and this suggests higher nonlinearity in the error amplifier.

Headroom and Clipping

The simple error amplifier will produce useful output only as long as the collector current of Q3 and Q4 is non-zero, so there must always be a sufficient voltage drop across the error amplifier load resistors R1 and R2. Increased gate drive for the output transistor is produced by reduction of the error amplifier transistor collector current. If the output MOSFET requires more gate drive than is available, that transistor will go into cutoff. Similarly, the error amplifier transistor on the opposite side should never saturate. This means that under maximum conditions of error correction, the emitter current of the error amplifier in the path of the output transistor that is off must not exceed the maximum collector current available through the load resistor. The available headroom is largely determined by the amount of fixed spreader voltage supplied by Q1 and Q2.

Boosted Rails

The added operating headroom required by the error-correction circuitry takes the form of increased fixed spreading voltage at the VAS. The fixed spreading voltage in the design of Reference 2 was 22 V. This means that the available VAS voltage swing will be seriously limited if the VAS is powered from the same rail voltage as the output transistors. For this reason, it is

Table 15.1 V_{gs} for Three MOSFETs

Device	Type	V_{gs} @ 0.15A	V_{gs} @ 5A	ΔV_{gs}
IRFP240	Vertical	4.2	5.5	1.3
2SK1530	Vertical	1.6	2.8	1.2
2SK1056	Lateral	0.7	5.7	5.0

advantageous to employ boosted supply rails for all circuits preceding the output transistors [2]. This was also recommended in the previous chapter for conventional amplifiers employing MOSFETs because of their significant gate voltage drive requirements. It is doubly important here. It is still desirable to employ Baker clamps to limit VAS swing in light of the higher VAS supply voltage. Be mindful that with EC, the output stage has virtually unity gain.

Use with Low-V_{gs} MOSFETs

The EC circuit is partly powered from the bias spread provided to the output transistors. For output stages employing MOSFETs like the IRFP240/9240, this voltage spread is adequate because these devices have threshold voltages on the order of 4 V. This will not be the case for lateral MOSFETs with threshold voltages less than 1 V and may not be the case for other types of vertical MOSFETs with threshold voltages as low as 1.5 V (e.g., Toshiba 2SK1530/2SJ201). V_{gs} for the lateral 2SK1056/2SJ160 at a quiescent current of 100 mA can range from 0.2 V to 1.5 V. V_{gs} for a vertical 2SK1530 at 100 mA can range from 0.8 V to 2.8 V. The high percentage range of V_{gs} variation for such low-threshold devices can present other challenges for driver design, such as variation in the driver quiescent current when that current is set by a resistor connected between emitter follower driver transistors.

The EC circuit can be used with such low-threshold voltage MOSFETs, but some modification of the circuit is required to present the EC circuit with sufficient operating voltage. A simple approach is to introduce an additional voltage drop ahead of each of the output MOSFET gates. This might be as simple as introducing a diode or green LED in series with the emitters of the driver transistors, keeping the feedback to the EC error amplifier connected to the driver emitters while feeding the output transistors from the inner ends of the diodes or LEDs. However, if the drivers are biased at 50 mA, this will be too much current for ordinary LEDs.

An alternative approach is shown in Figure 15.8. Additional pre-driver transistors Q11 and Q12 are introduced into the signal path, and their outputs are used for the feedback to the error

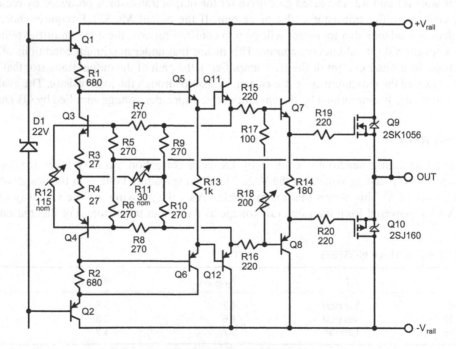

Figure 15.8 An EC Circuit Suitable for Use with Low-V_{gs} MOSFETs

amplifier. This introduces an additional V_{be} of drop in the path to the drivers. Further reduced voltage spread for the driver and output transistors is obtained from an attenuator formed by R15, R16, R17 and R18. Shunt resistor R18 can be trimmed in order to accommodate variation in the threshold voltages of the MOSFETs so that proper operating voltages for the EC circuit can be obtained. This approach also encloses the driver transistor in the error-correction loop, reducing distortion that it introduces. Notice, however, that the quiescent bias current of the driver transistors is still subject to considerable variation depending on the actual MOSFET output transistor threshold voltages. This is a potential issue for any amplifier that uses EFs to drive MOSFETs with a wide range of threshold voltages.

Error Correction for BJT Output Stages

Although Hawksford first described HEC in the context of BJT output stages, it has seen less application in BJT output stages. For one thing, the biasing arrangement is not necessarily advantageous. Second, BJT output stages do not need error correction as much, due to their higher transconductance. Finally, error correction does best when employed in connection with a very fast output stage, and even RET BJTs are not as fast as MOSFETs.

The circuit of Figure 15.8 can be used with BJT output transistors with little or no modification, where the situation and solution are really no different from those for lateral MOSFETs, whose threshold voltages are not too dissimilar to the V_{be} of a BJT. Note, however, that the BJT output stage does not have such a wide range in turn-on voltages as the MOSFETs, so some of those problems of variability are smaller. The turn-on voltage will always be nearly 0.6 V. For this reason, trimmer R18 can be eliminated.

One way in which EC can be used to advantage in a BJT design is to allow the BJT output stage to be somewhat over-biased. It will then incur some crossover distortion due to *gm* doubling. That distortion can be greatly reduced by error correction. The net advantage here is that the BJT amplifier can be over-biased and enjoy a larger class A operating region, while being more immune to the effects of dynamic bias errors caused by temperature swings with program material.

15.6 Circuit Refinements and Nuances

The simple error-correction circuit described thus far and used in Reference 2 is very effective in reducing output stage distortion. The circuit adds only four transistors to what would otherwise be an output Triple (BJT or MOSFET). Given its simplicity, the circuit works amazingly well, in some cases providing 30 dB of distortion reduction at 20 kHz [2]. The MOSFET amplifier in Reference 2 achieved THD below 0.001% at 20 kHz. A Distortion Magnifier circuit had to be used to measure distortion this low [2, 4]. Nevertheless, these simple circuits have some limitations, and improvements can be made at little relative increase in parts cost.

Complementary Error Amplifier

Because the error amplifier consists of an NPN-PNP pair, it operates in a complementary fashion. When the signal current increases in one device, it decreases in the other device. This provides a desirable degree of second harmonic distortion reduction of the same type that accompanies the use of a conventional differential pair.

The assumption behind this distortion mitigation is that the output signals of the two error amplifier transistors both have the same effect on the amplifier output signal. Unfortunately, this is not always the case with the simple EC circuit described earlier. When one of the output transistors is in the off state during class AB operation, the output of one of the error amplifier

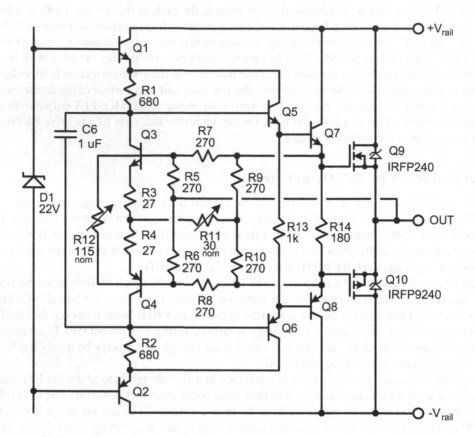

Figure 15.9 C6 Forces Complementary Behavior of the Error Amplifier

transistors is largely ignored. This is because the drive signal paths for the N-channel and P-channel MOSFETs are independent.

Ideally, the signal voltages at the collectors of the two error amplifier transistors will be identical. For this reason, it is allowable to couple these two collectors together with a capacitor. When this is done, both collector currents will be forced to exert the same influence on the output signal, regardless of the *on* or *off* state of the associated output transistors. In effect, the signal contributions of the two transistors are merged immediately rather than at the output node of the amplifier. This simple arrangement is illustrated in Figure 15.9, where C6 provides an AC short circuit between the collectors of Q3 and Q4, forcing them to act as a differential pair even though they are of opposite polarity. C6 need not be large, since its most important contribution is at high frequencies where the need for error correction is greater. A value of 1 μF is adequate.

CFP Error Amplifier

The nonlinearity of the error amplifier is caused by signal-dependent changes of *re'*. If each transistor in the error amplifier is replaced with a *complementary feedback pair* (CFP), this effect will be greatly reduced. The CFP is implemented by the addition of Q11 and Q12 to the error amplifier as shown in Figure 15.10. The bias current of the input transistor of the CFP can be set at about 1/10

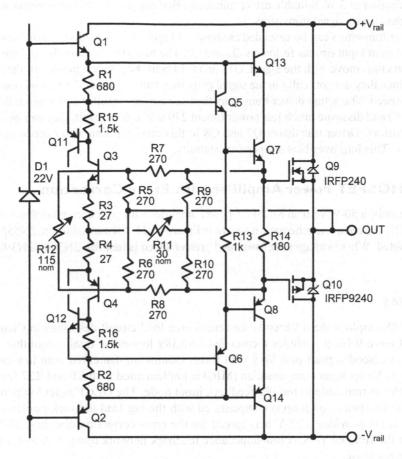

Figure 15.10 Use of a CFP-Based Error Amplifier and Cascoded Drivers

that of the CFP overall. The effective transconductance of the error amplifier transistors is greatly increased by the use of the CFP. This virtually eliminates the nonlinearity caused by *re'* variation.

A further advantage of the CFP error amplifier is that the effective beta of the CFP is much larger than that of a single transistor. This reduces the load placed on the resistive summing network and further improves linearity. A cost associated with the CFP is voltage headroom. The CFP will add at least 1 V_{be} of required supply voltage to each driver rail.

Cascoded Drivers

All of the transistors in the error correction circuitry should be as fast as possible. This allows for a higher local loop gain crossover frequency and lighter EC compensation, which will lead to increased EC effectiveness at high frequencies.

The driver transistors (Q7, Q8) in the arrangements described so far must dissipate considerable power if their bias current is sufficient to quickly turn off the output MOSFETs. Bias current of 50 mA and quiescent V_{ce} of 60 V are not uncommon for these drivers. This corresponds to power dissipation of 3 W. Suitable driver transistors that can handle this dissipation will usually be slower than small-signal transistors.

The driver transistors can be cascoded as shown in Figure 15.10. Cascode transistors Q13 and Q14 are fed from input emitter followers Q1 and Q2. The bases of the cascodes and the collectors of the drivers thus move with the signal. Q13 and Q14 bear the greatest portion of the power dissipation. Since they are not really in the signal path, they can be slower devices without affecting EC circuit speed. The actual driver transistors then see a much smaller collector-emitter voltage (about 5.4 V) and dissipate much less power (about 270 mW). As a result, they can be fast, small-signal transistors. Driver transistors Q7 and Q8 in this configuration also operate at fairly low temperature. This improves bias temperature stability.

15.7 A MOSFET Power Amplifier with Error Correction

Here we describe a 50-W vertical MOSFET power amplifier with error correction that was built and measured [2]. The complete schematic is shown in Figure 15.11. All transistors are 2N5551/2N5401 except as noted. Where voltages are lower and greater speed is helpful, 2N3904/2N3906 devices are used.

IPS and VAS

The IPS-VAS employs the differential current source load circuit described in Chapter 9 and shown in Figure 9.16. It includes a cascoded LSK389 low-noise dual monolithic JFET IPS coupled to a cascoded push-pull VAS with Baker clamps implemented with low-capacitance diodes D5–8. Miller input compensation (MIC) is implemented with C6 and R27 feeding back from the VAS output node to the IPS feedback input node. The ULGF is set to approximately 2 MHz. The compensation loop is compensated with the lag-lead network implemented with C5 and R11. D4 provides a 22-V bias spread for the error correction circuitry. 50-V boosted rails power the IPS and VAS. A low-impedance feedback network using 2-W metal film resistors minimizes noise.

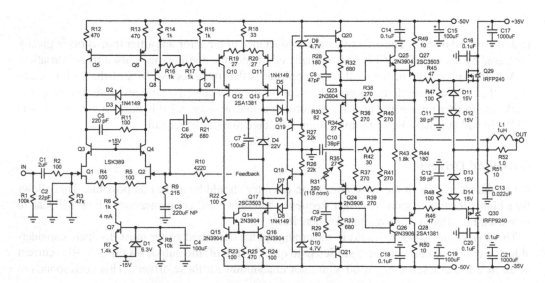

Figure 15.11 The 50-W MOSFET Power Amplifier with Error Correction

Error Correction and Output Stage

The error correction circuit and output stage is a straightforward practical implementation of the circuit described earlier in Figure 15.6 with the exception that here the error correction circuits are powered from the 50-V boosted rails. The output stage is powered from 35-V rails, which provide the amplifier with a power output capability in excess of 50 W.

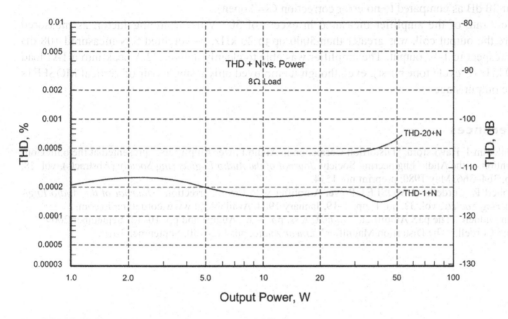

Figure 15.12 THD+N Versus Power at 1 kHz and 20 kHz

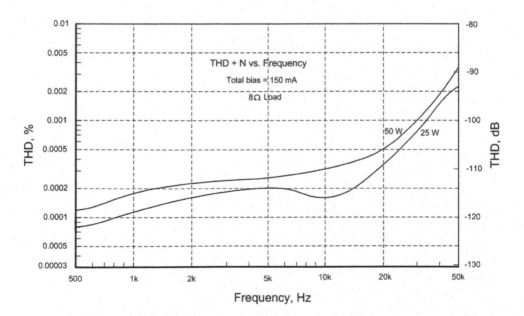

Figure 15.13 THD+N Versus Frequency at 25 W and 50 W

Performance

Figure 15.12 shows measured THD+N versus power at 1 KHz and 20 kHz when driving an 8-Ω load. THD-20 remains below 0.0007% at all power levels up to 50 W.

Figure 15.13 shows THD+N versus frequency for 25 W and 50 W driven into an 8-Ω load. A distortion magnification instrument in combination with a sensitive THD analyzer was required to make these low-distortion measurements [4]. The error correction circuit reduced THD-20 by about 30 dB as compared to no error correction (R42 open).

Slew rate of the amplifier measured in excess of 300 V/us. Damping factor, as measured before the output coil, was greater than 5000 up to 20 kHz. A-weighted S/N measured 108 dB with respect to 1-W output. The amplifier was able to cleanly deliver 22 A peak into a 1-Ω load at 20 kHz (2-cycle tone burst), even though it employed only a single pair of vertical MOSFETs in the output stage.

References

1. Malcolm J. Hawksford, "Distortion Correction in Audio Power Amplifiers," presented at the 65th Convention of the Audio Engineering Society, *Journal of the Audio Engineering Society* (Abstracts), vol. 28, pp. 364–366, May 1980; preprint no. 1574.
2. Robert R. Cordell, "A MOSFET Power Amplifier with Error Correction," *Journal of the Audio Engineering Society*, vol. 32, no. 1, pp. 2–19, January 1984. Available at www.cordellaudio.com.
3. Jan Didden, "The paX Amplifier," *AudioXpress*, pp. 6–15, August and pp. 10–15, September 2009.
4. Bob Cordell, "The Distortion Magnifier," *Linear Audio*, pp. 142–150, September 2010.

Chapter 16

Other Sources of Distortion

16. INTRODUCTION

The discussions on distortion have so far focused on well-understood distortions that are caused by nonlinearities largely in semiconductor devices. Examples are the nonlinear relationships of current to base or gate voltage, Early effect wherein transistor current gain is a function of collector voltage or junction capacitance that is a function of reverse bias.

These kinds of nonlinearity produce many types of distortions. Harmonic and CCIF IM distortion are just different ways of measuring the symptoms of such nonlinearities. This is also the case with many other distortions, such as TIM [1]. Here we focus on other sources of distortion that must be considered as well. In many cases, these distortions result from the way in which the circuits are implemented, either as a result of other component imperfections or as layout-related matters.

In many cases, minimizing these distortions requires a great amount of attention to detail, such as layout, ground topology and power supply bypassing. It is a shame that in some cases these sources of distortion can ruin an otherwise finely designed amplifier. Although some of these sources of distortion have been touched on in other chapters, they can often easily be overlooked. For that reason, each will be discussed here.

16.1 Distortion Mechanisms

Nonlinear distortion occurs as a result of a variety of sources, but these sources often share many of the same distortion-creating mechanisms. In the most general terms, distortion occurs when one or more parameters of an active or passive device change as a function of the signal. Examples include changing transistor current gain as a function of signal voltage and changing junction capacitance as a function of signal voltage. If a resistance changes as a result of signal-dependent temperature, this will also create nonlinear distortion.

Another trait of nonlinear distortion is that a small-signal characteristic of the amplifier changes as a function of signal excursion. The most common example is when the incremental gain of the amplifier changes over the signal swing. If the gain is higher on positive swings and lower on negative swings, this corresponds to second-order distortion. If the gain is smaller for large signal deviations from zero, this corresponds to third-order distortion. If instead the frequency response or phase response of the amplifier changes with signal, this is also a form of nonlinear distortion. Anything that changes as a function of signal is a likely source of distortion.

16.2 Early Effect Distortion

The current gain of a transistor is mildly dependent on the collector-base voltage. This is referred to as the Early effect. To the extent that the gain of an amplifier stage is dependent on the current gain of the transistor, the Early effect can cause distortion. The Early effect was discussed in Chapter 2. Early effect distortion is minimized by employing cascodes (which reduce signal-dependent collector-base voltage changes) or by designing stages whose gain is less dependent on transistor current gain. A common-emitter stage preceded by an emitter follower suffers less Early effect distortion because beta changes in the common emitter (CE) transistor have less influence.

16.3 Junction Capacitance Distortion

The capacitance of p-n junctions in transistors and diodes is a function of voltage. This means that the frequency response of a gain stage may change with signal voltage. This in turn means that the gain at high frequencies can be modulated. Whenever gain changes as a function of signal voltage, this is a nonlinearity that can cause distortion.

A good example of this distortion is the Miller effect in a gain stage. A larger value of collector-base capacitance will reduce high-frequency gain of the stage. In some amplifiers with simple Miller feedback compensation, the collector-base capacitance of the VAS transistor forms a part of the total compensation capacitance. The gain crossover frequency then becomes a function of signal, resulting in small in-band gain and phase changes that are signal-dependent [2]. Consider the collector-base capacitance of the 2N4401 NPN small-signal transistor. It is 9 pF at 0 V, 4.5 pF at 5 V and 1.5 pF at 50 V.

It is also important to recognize that the gate of a JFET forms a reverse-biased p-n junction with the source-drain channel of the JFET. This junction will also be subject to modulation of its capacitance by the signal voltage.

MOSFET Gate Capacitance Nonlinearity

Just as with p-n junctions, the gate capacitance in a MOSFET can also change with voltage, so the same sort of nonlinearity and distortion can occur. Although the capacitor formed by a MOSFET gate has an insulator as its dielectric, there still exists a depletion region in the doped silicon underneath the gate. Increased amounts of reverse bias on the gate increase the width of this depletion region and effectively move the plate further away, decreasing capacitance.

The most serious example of this is the increased gate-drain capacitance C_{gd} of a vertical power MOSFET when the reverse bias of the gate with respect to the drain becomes small. Most of the time, the reverse bias from gate to drain is large. When the output signal is at zero and the gate is at a nominal forward bias voltage of a couple of volts, the reverse bias will be nearly equal to the rail voltage. When the audio signal swings to a large value (approaching clipping), the amount of this reverse bias will decrease to a low value, sometimes close to zero. When this happens, C_{gd} may increase very substantially. It can literally increase from 50 pF to over 500 pF as the drain-source voltage approaches zero.

16.4 Grounding Distortion

There exist very distorted half-wave-rectified signal currents in class AB output stages. Current flows from the positive rail on positive half-cycles and from the negative rail on negative half-cycles. Sometimes these nonlinear currents can make their way into the grounding circuit, causing nonlinear voltage drops across the finite resistance of ground traces.

A good example is the bypassing of the main rails of the output stage. Those bypass capacitors will be carrying substantial amounts of the nonlinear currents. The better those capacitors are, the greater percentage of that current they will carry. That current has to be returned to ground somewhere. If that current is dumped into a ground that signal circuits depend on, distortion will be coupled into the signal path. This is one reason why star grounds and other features of grounding architecture are so important. It is very important that bypass capacitors not form AC ground loops that can destroy the star grounding architecture at high frequencies. Always remember that signal currents (including nonlinear signal currents) will tend to take the path of least impedance. Sometimes the strategic placement of a small resistance will effectively break a ground loop and make a great improvement.

16.5 Power Rail Distortion

All of the stages in a power amplifier have some amount of *power supply rejection ratio* (PSRR). This describes the degree to which the stage can suppress the influence of noise and distortion on the power supply lines on the signal output of the stage. The power supply rail will often contain a mixture of rectifier ripple, noise, hum and distortion components. This is especially true of the high-current main rails that supply the output stage. If any of this hash gets into a signal stage, it will make its way to the amplifier output. Nonlinear signals on power supply rails will thus result in amplifier distortion.

There are two basic ways to reduce distortion from the power rails. The first is to minimize the garbage appearing on the power rails. This can be done by employing generous R-C filtering in the power supply lines or by regulating the power supply lines. Sometimes a *soft* form of regulation is implemented with a pass transistor acting as an emitter follower for the power supply voltage. This will often take the form of a *capacitance multiplier*. Such an arrangement is shown in Figure 16.1. A nice feature of this arrangement is that the filtering capacitors C4 and C5 can have their ground referenced to the ground of the circuit that will be employing the resulting power rail. D1 and D2 protect the transistor in the event that the main rail voltage falls below the regulated rail voltage. R4 isolates emitter follower Q1 from the capacitive load.

The second approach is to design the individual amplifier stages to have higher PSRR out to higher frequencies. In many cases the PSRR of an amplifier stage will fall with increasing frequency. Differential, push-pull and cascode circuits often exhibit better power supply rejection up to higher frequencies.

Output Stage Power Supply Rejection

Although the output stage is usually just an emitter follower or source follower with some drivers in front of it, it is not immune to power supply garbage. This is largely because of the Early effect in the output transistors (or its analogous effect in MOSFETs). This is a weak effect in an emitter follower or source follower arrangement, but the amplitude of the garbage on the rails at this point is often quite large. This is of special concern when the output swings close to the rails. This effect is reduced in designs employing cascoded output stages. Splitting the power supply

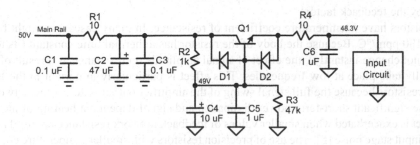

Figure 16.1 Filtering the Power Supply with a Capacitance Multiplier

reservoir capacitors and separating them by as little as 0.1 Ω can make a big improvement here. In addition to providing a second filtering pole at high frequencies, the series resistor reduces the effect of finite ESR/ESL in the second capacitor.

16.6 Input Common-Mode Distortion

An often-neglected source of distortion results from the common-mode signal swing on the input differential pair of a non-inverting power amplifier. The common-mode voltage is simply equal to the input signal swing, which may be on the order of a couple of volts.

The distortion can result from the collector-base voltages of the LTP changing with signal, causing beta to change via the Early effect. It can also result from changes in the collector-base junction capacitances as a result of their voltage dependence on V_{cb}. Nonlinear output impedance of the tail current source may also create common-mode distortion. Techniques for minimizing input common-mode distortion were discussed in Chapter 9.

An input stage that has better common-mode rejection will tend to have smaller common-mode distortion. Conversely, steps taken to reduce common-mode distortion will often improve common-mode rejection. Noise on power supply rails is often seen as common-mode noise by the input stage. For this reason, input stages with good common-mode rejection will tend to have better PSRR.

Testing for Common-Mode Distortion

One way to test for the presence of common-mode distortion is to drive the amplifier through a resistor connected to the inverting input, with no signal applied to the conventional non-inverting input. This forces the amplifier to operate in the inverting mode, where there is no common-mode signal swing on the input LTP. If distortion is significantly reduced under these conditions, it is likely that there is common-mode distortion.

16.7 Resistor Distortion

Some resistors can change their value slightly as a function of the voltage across them or the current through them [3, 4]. These effects are quite small and very difficult to measure in most resistors of reasonable quality. In some cases carbon composition resistors will show this effect. Sometimes the interface between the material of the resistor element and the resistor leads can develop some nonlinear voltage drop.

The feedback network resistors are especially important because they play a direct role in setting the gain of the amplifier. There is nearly a one-to-one dependence of gain on resistor value; a 1% change in resistance will cause a 1% change in gain. The sensitivity of gain to change in the value of other resistors in the open-loop portions of the amplifier is usually much less (it is reduced by the feedback factor).

All resistors have a temperature coefficient of resistance. In many cases this might be on the order of 100 ppm/°C. Because the body of the resistor has a thermal time constant (its temperature will not change instantly), the effect of signal amplitude on resistance as a result of heating will usually take place at low frequencies. This effect is particularly important in the feedback network resistors because the full signal swing of the amplifier output is across the network. This may cause significant short-term power dissipation and signal-dependent heating of the resistor. This effect is exacerbated when smaller values of feedback network resistance are used in pursuit of lower input stage noise [5]. The use of precision resistors with smaller temperature coefficients and higher power dissipation is very helpful.

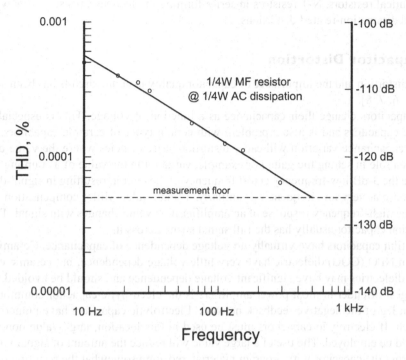

Figure 16.2 THD of 1/4 W Resistor Versus Frequency at 1/4 W Dissipation

Suppose a resistor's temperature swings by 10 °C peak to peak. If the resistor has a TC of 100 ppm/°C, then the resistance will swing by 0.1% peak-to-peak, and distortion on the order of 0.05% may result. Suppose we have a feedback resistance of 4 kΩ and the amplifier output swings by 40 V peak. This will result in a peak current of 10 mA and peak power dissipation of 400 mW. Almost all of this power dissipation is in the series feedback resistor. It is easy to see how this could swing the temperature of the resistor by several degrees Centigrade at sufficiently low frequencies. A 6.81-kΩ 1/4 watt metal film resistor subjected to 40 volts DC dissipated 230 mW and its resistance changed by 0.06%.

Figure 16.2 shows the measured distortion of a 1/4 W metal film resistor as a function of frequency. The 1-kΩ resistor was subjected to an average power dissipation of 1/4 W. Distortion was measured by forming two 20-dB attenuators. One employed the resistor under test in the series arm and the other employed a 2-W version of the same resistor in its series arm. The difference of the outputs of the two attenuators was then amplified and analyzed. This technique virtually eliminates the influence of distortion from the driving amplifier and enhances the sensitivity of the spectrum analyzer used for measurement. Resistor distortion was primarily of third order and decreased with frequency, as expected. The key thing to keep in mind is that the resistor distortion in Figure 16.2 is mainly as a result of the temperature coefficient of resistance in combination with signal-dependent swings in power dissipation.

The lesson here is to employ resistors of larger power dissipation in the feedback network, especially in the series feedback resistor. In some cases series-parallel combinations can be employed to achieve higher total power dissipation. Series combinations are superior because they also reduce the voltage swing on each resistor, mitigating voltage-dependent resistor distortions. Although often not practical, a nearly distortionless N:1 feedback divider can be made with N identical resistors, N-1 resistors in series forming the feedback resistor. This will cancel voltage and dissipation-related distortions.

16.8 Capacitor Distortion

Capacitor distortion and the importance of capacitor quality in audio circuits has been studied by many [3, 4, 6, 7, 8].

Some capacitors change their capacitance as a function of voltage. This is especially so for electrolytic capacitors and is also a problem with certain types of ceramic capacitors. Voltage-dependent capacitance variation will cause distortion at frequencies where the value of capacitance plays a role in setting the gain. For example, variation in the value of a coupling capacitor will change the 3-dB low-frequency cutoff frequency of the circuit, resulting in signal-dependent gain and phase at very low frequencies. Similarly, a small-value Miller compensation capacitor will alter the high-frequency response of an amplifier if its value changes with signal. The Miller compensation capacitor usually has the full signal swing across it.

Quality film capacitors have virtually no voltage dependence of capacitance. Ceramic capacitors with an NPO (COG) dielectric have very little voltage dependence, but ceramic capacitors with other dielectrics may have significant voltage dependence and should be avoided.

The biggest offender in most power amplifiers is the electrolytic capacitor commonly found in the shunt leg of the negative feedback network. Electrolytic capacitors have numerous forms of distortion. If electrolytic capacitors must be used in this location, large-value non-polarized types should be employed. The use of a large value will reduce the amount of signal voltage that appears across the capacitor with program material, reducing somewhat the resulting distortion.

If conventional polarized electrolytic capacitors must be used, placing them in series instead of in parallel can be better for reducing distortions. This is not quite equivalent to using an NP electrolytic, but polarized capacitors may be more readily available [7]. Of course, one must bear in mind that placing capacitors in series reduces the total value of capacitance.

These capacitors should be bypassed with a film capacitor of at least 1 μF. The low ESR of the film capacitor will effectively short out the electrolytic capacitor at very high frequencies, limiting the damage done by the electrolytic to lower frequencies. This will also limit EMI developing across the electrolytic due to its ESR/ESL. Expectations for improvements in the audio band must be tempered by the realization that the impedance of a 1-μF capacitor at 20 kHz is still 8 Ω.

Back-to-back silicon diodes will often be placed in parallel across the electrolytic capacitor to protect it and the input stage from a large DC voltage if the amplifier output goes to the rail. At very low frequencies some signal voltage will appear across the capacitor and these diodes depending on the cutoff frequency established by the capacitor and feedback network. Some distortion at these frequencies may result from small amounts of conduction by the diodes if peak signal voltages of more than 200 mV or so appear across the diodes. Use of a larger value of capacitance obviously reduces this effect. If thought to be necessary, each diode can be replaced by two diodes in series. Alternatively, the diode combination can be replaced by two Zener diodes in anti-series. Silicon transistors whose collector and base are connected together can also be used. This takes advantage of the reverse base-emitter breakdown of the transistors (typically less than 6 V).

Curve (a) in Figure 16.3 shows the THD of the voltage across a 100-μF, 16-V polarized electrolytic capacitor as a function of frequency. The capacitor is driven by a 5-kΩ resistor with a 20-V RMS input, corresponding to a 50-W amplifier with a low-value feedback resistor. Signal current in the capacitor is 4 mA RMS. THD reaches almost 0.01% at 20 Hz. Bear in mind that this is the distortion of the voltage across the capacitor; the resulting amplifier distortion will be smaller. The signal voltage across the capacitor at 20 Hz is about 300 mV. The frequency response of an amplifier built with this arrangement would be down by 3 dB at about 6 Hz.

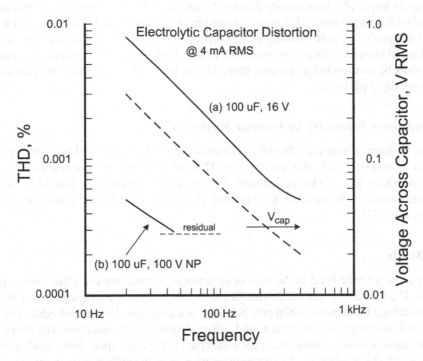

Figure 16.3 THD for a 100-µF Non-Polarized Electrolytic Capacitor

Curve (b) in Figure 16.3 shows the result when a 100-µF, 100-V audio-grade NP electrolytic capacitor is substituted. The capacitor is designed for use in loudspeaker crossovers and is not expensive. Its distortion is about 25 dB less than that of the low-voltage polarized capacitor. If you must use an electrolytic capacitor in the feedback network, use an NP capacitor with a substantial voltage rating.

16.9 Inductor and Magnetic Distortions

The use of inductors is frowned upon in audio power amplifiers, so few, if any, are found in the signal path. The usual series output coil is a widespread exception. The absence of inductors does not mean that there are no nonlinear magnetic effects, so other magnetic effects will be examined here as well.

Magnetic Core Distortion

Magnetic materials have a magnetization curve that is nonlinear. The degree to which a magnetic material can be magnetized is ultimately limited. In an inductor or a transformer, this is referred to as *core saturation*. If the magnetization of the core approaches the saturation point, the inductance decreases. The simplest example of this is when an inductor is carrying DC current. As the current increases, the inductance will decrease. In general, the inductance of coils with ferrous cores is nonlinear. Great caution should be used if SMD inductors or ferrite beads are used for filtering or EMI suppression in power supply circuits, as their effectiveness decreases greatly when passing modest amounts of current. Discrete ferrite beads or cores should be made to have zero net DC current flowing through them by passing all of the supply leads through the same bead or core.

A serious example of a nonlinearity that can be introduced into an amplifier is the use of an output coil with a ferrite core. This might be done to obtain adequate inductance with a smaller coil. The loudspeaker currents flowing in the output coil may cause saturation effects in the coil that will lead to increased distortion, especially at high frequencies. For this reason, output coils should always be constructed as air-core coils. This is not a serious problem because the value needed is usually 5 µH or less.

Distortion from Proximity to Ferrous Materials

A more subtle form of magnetic distortion can occur when the output coil is in close proximity to a magnetic material, such as a steel chassis. This can influence the inductance and can also introduce nonlinear losses into the inductor. For this reason output coils should be kept away from ferrous materials. Interposing an aluminum plate between coils and ferrous materials can reduce this effect [3].

Ferrite Beads

The insertion of a ferrite bead in the base or collector of a transistor can often tame a parasitic oscillation. The ferrite bead acts somewhat like an inductor at middle frequencies and then experiences loss at high frequencies, ultimately behaving like a resistor. The loss introduced by the ferrite bead tends to damp resonant circuits and reduce the tendency to oscillate. The ferrite bead is a nonlinear passive device whose use in the signal path is frowned upon. How much measurable distortion it introduces when used in a transistor base or in the collector of an emitter follower is debatable. The safest approach is to avoid the use of ferrite beads. As mentioned above, the effectiveness of ferrite beads, especially those in SMD packages, can be greatly reduced if DC current is passed through them. Always consult the datasheet when using these devices.

16.10 Magnetic Induction Distortion

There are very distorted signals running around in class AB amplifiers. The biggest source of these is the nonlinear half-wave-rectified current in the output stage. Current flows from the positive rail on positive half-cycles and from the negative rail on negative half-cycles. The magnetic fields created by these currents can induce nonlinear voltages in nearby signal lines, coupling distortion into the signal path [9].

Minimizing Magnetic Induction Distortion

The most important step in reducing magnetic induction distortion is to reduce the emission of nonlinear magnetic fields and then to keep as much distance from them to signal-carrying wires as possible. Twisting the positive and negative rail supply wires together can be quite helpful in reducing the emissions from these lines. This is effective because the sum of the currents (and thus the magnetic fields) of the positive and negative rail wires is a linear representation of the amplifier's output current signal. Bypassing the positive rail directly to the negative rail with low-ESR capacitors right at the output stage can improve matters by forcing fast-changing currents to circulate locally rather than through the wires back to the power supply.

The physical arrangement of the PNP and NPN output transistors on the heat sink can also have a significant effect on nonlinear magnetic fields emitted by output stage wiring. Interleaving the PNP and NPN power transistors on the heat sink is superior to grouping all of the PNP transistors on one end and all of the NPN transistors on the other end.

Arranging the emitting and pickup wires to be at right angles is also helpful. Reducing the susceptibility of signal lines to induction by magnetic fields can be accomplished by twisting the leads together with ground lines. The input and feedback signals are the most vulnerable to magnetic induction distortion.

16.11 Fuse, Relay and Connector Distortion

Many amplifiers employ fuses or relays in the output signal path for protection or muting. These devices can cause distortion because of the high currents flowing through them.

Fuse Distortion

When high current passes through a fuse, it heats up the fuse element by causing a small voltage drop that leads to power dissipation and temperature rise. The fuse element must therefore be resistive for this process to take place. The fuse blows when the temperature rises to the melting point of the fuse element. The fuse element has a positive temperature coefficient of resistance, so the process is accelerated as more current flows through the fuse.

At low frequencies the audio signal can heat up and cool down the fuse element within a single cycle, causing the resistance of the fuse to vary as a function of the signal amplitude. This leads to distortion because the attenuation of the fuse resistance against the load impedance changes as a function of signal swing [10]. Fuses are often undersized with respect to the peak audio current they may be called on to pass, recognizing that a smaller fuse will provide relatively more protection; with normal audio signals, such high currents are brief events much shorter than the time constant of the fuse element. The cold resistance of a 2-A 3AG fuse was measured to be 78 mΩ, while its resistance when passing 2 A DC was 113 mΩ. This represents a 45% increase in fuse resistance. Note that even a constant 78 mΩ will limit damping factor to 100 right off the bat.

The distortion of a fuse can be measured by looking at the voltage across the fuse with a sinusoidal signal current passing through it. The fuse under test is put in the ground leg of an 8-Ω load resistor so that the signal voltage across the fuse can be easily analyzed. This technique largely takes the distortion of the driving source out of the picture. Figure 16.4a is a plot of fuse distortion versus frequency when a 2-A fast-blow 3AG fuse is passing a 2-A RMS sine wave signal. As expected, fuse distortion increases dramatically at low frequencies. Signal voltage across the fuse was 250 mV. Amplifier THD (due to the fuse) is calculated by normalizing the fuse distortion voltage to the amplifier output voltage. The resulting amplifier distortion is shown in Figure 16.4b. Amplifier distortion is lower than fuse distortion by a factor of 64 because of the small voltage across the fuse compared to the total signal voltage. At 20 Hz, amplifier distortion due to the fuse is calculated to be 0.0033%.

Relay Distortion

All contacts have some resistance. Sometimes that resistance can change as a function of the current flowing through them. This gives rise to distortion. When a relay is placed in series with an amplifier output line, the contact resistance will form a voltage divider with the load impedance, causing a tiny amount of attenuation. If the contact resistance varies with the signal, this attenuation will vary with the signal and distortion will result.

Not all relay contacts are alike, and some are much less prone to distortion than others. The tendency to distortion will depend on the contact metallurgy, the condition of the contacts, the contact area and the contact pressure. High-current relays will often produce less distortion because the starting contact resistance is smaller and the current density in the contact is smaller.

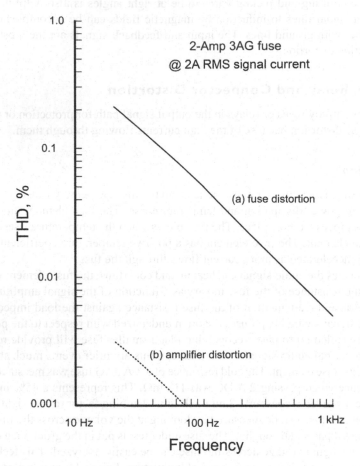

Figure 16.4 THD Versus Frequency for a 2-A Fuse Passing 2 A RMS

However, it is unwise to generalize that high-current relays (like automotive relays) will always yield lower distortion.

The distortion produced by a relay can be measured in the same way as that of a fuse. The output of a power amplifier is passed through an 8-Ω resistor that is connected to ground through the closed contacts of the relay under test. The voltage drop across the relay can be fed to a distortion analyzer or a spectrum analyzer. In some cases that voltage may have to be amplified before being applied to the input of the analyzer.

The test setup is shown in Figure 16.5. A 16-V, 1-kHz test signal is applied to an 8-Ω resistor, producing a current of 2 A RMS. This corresponds to a power level of 32 W. The relay under test is placed in the ground leg of the resistor. A spectrum analyzer is connected across the closed relay contacts. With this setup, the analyzer's response at 1 kHz is reflective of the resistance of the relay contacts, while harmonics in the spectrum represent distortion. Six relays were tested. The good news is that one of the relays was exceptionally good; the bad news is that some were surprisingly bad. This means that relays for this kind of application need to be selected carefully.

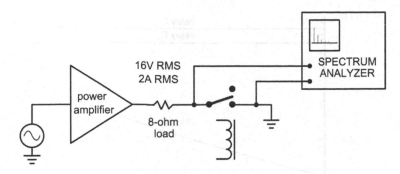

Figure 16.5 Test Setup for Measuring Relay Distortion

The magnitude of the fundamental displayed on the spectrum analyzer was used to infer the contact resistance by dividing it by 2 A RMS. The in-service amplifier distortion attributable to the relay was calculated by normalizing the magnitude of each distortion component of the relay voltage to the 16-V amplifier output operating level. This test procedure does a good job of isolating relay distortion from amplifier distortion.

Relays 1, 3 and 5 were of medium performance, with resistance ranging from 25 mΩ down to 1.3 mΩ, and third-order distortion on the analyzer ranging from −56 dB through −64 dB, corresponding to 0.001% (−100 dB) to 0.0004% (−108 dB) of added amplifier distortion. Third-order distortion predominated on most relays.

Relays 2 and 6 were notably bad, each with 25-mΩ contact resistance and third-order distortion as bad as −50 dB on the analyzer, corresponding to in-service third-order distortion of 0.002% (−94 dB). Relay 6 had a spectrum characterized by quite a bit of grunge above the analyzer noise floor.

Relay 4 was the clear winner. This was an 80 A automotive relay. It had a third-order distortion reading of −91 dB on the analyzer, corresponding to 0.00002% (−135 dB).

Although Relay 4 was the only "automotive" relay in the test, it would be wrong to conclude that an automotive relay will always be better than others for an audio application. It would also be premature to conclude that the higher current rating is responsible for the enhanced performance, on the basis of the variability among the other relays that were rated at 30 A.

There is no evidence that contact resistance is a good predictor of relay distortion performance. It is tempting to speculate that the contact surface chemistry plays a large role in the distortion performance; two different sets of contacts might exhibit the same resistance, but with different amounts of nonlinearity in that resistance.

Self has reported some frequency-dependence of relay distortion, speculating that soft iron in the swinger circuit path that is part of the magnetic circuit may be responsible for increased distortion at high frequencies [11]. A double-pole relay is less likely to be constructed this way, since there are two swingers. The plot in Figure 16.6 shows distortion of five relays as a function of frequency from 1 kHz to 16 kHz. This was a different group of relays than those discussed above. These five relays are labeled with the letters A—E to distinguish those from the group above.

The distortion was measured with the arrangement of Figure 16.5 with a 2-A signal flowing through the contacts. The THD shown is that of the small fundamental signal that appears directly across the contacts. Contact resistance ranged from 1.3–11 mΩ, corresponding to signal levels of 2.6–22 mV. In each case, the fundamental signal appearing across the contacts was amplified as needed to satisfy the input level requirements of the distortion analyzer.

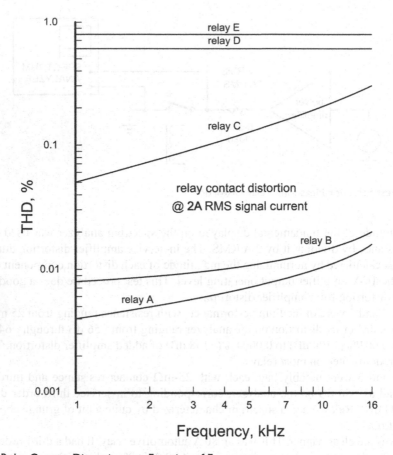

Figure 16.6 Relay Contact Distortion as a Function of Frequency

These are believed to be representative relays, but it is not known what material is used in these relays. Three relays exhibit frequency-dependence and three do not. With one exception, third-harmonic distortion dominated the measurements. Second-harmonic distortion dominated for Relay A. Distortion voltage across the contacts at 16 kHz for the three best relays was below 16 μV RMS, corresponding to 0.0001% distortion at the 16-V RMS operating level of the amplifier. Distortion does increase with frequency in some relays, but is not a problem if it remains very low at high frequencies.

The spread in net distortion among these relays was a remarkable 33 dB at 16 kHz, but the worst relays had distortion that was flat with frequency. Pick your speaker relays wisely.

Because distortion can be a function of the condition of the contacts, the performance of a new relay may not be representative of that relay after it is worn or abused. For example, if the relay is frequently opened or closed when there is a large signal present, this can cause some pitting of the contacts that may lead to increased distortion.

Fuse or relay distortion can be reduced if the device is enclosed within the global negative feedback loop. Approaches to doing this are discussed in Chapter 18. The key consideration is what happens to the feedback loop when the device opens.

Connector Distortion

As with relays, any connector consists of a pair of metal-metal contacts. As such, the interface will exhibit contact resistance which may be nonlinear and which can be a source of distortion. This will be especially the case with loudspeaker connections (banana plugs in particular). Some combinations of alloys and/or plating can act as semiconductors, and should be avoided.

Distortion in banana plugs was measured using the same kind of testing that was applied to relay contacts. For this simple spot test, an ordinary banana plug and jack combination was compared with an expensive one. For the ordinary pair, the jack was gold-plated in its interior, but the plug was merely nickel-plated. In the more expensive pair, both were of very substantial construction and fully gold-plated.

Interestingly, the banana plugs showed more second harmonic than third, and the ordinary plug combination was slightly better than the expensive one in this regard. Both plug combinations show a connector resistance of about 13 mΩ. All normalized distortion spectra were between −142 dB (0.000004%) and −135 dB (0.000016%). These are very encouraging numbers, but they are based on a very small sample. Both sets of connectors were brand new. I think it is very important that we err on the conservative side with any kind of connectors, especially ones that may become loose with time.

16.12 Load-Induced Distortion

Don't overlook the possibility of distortion artifacts induced by the load resistor when testing an amplifier. Large wire-wound resistors can sometimes be quite nonlinear as a result of the way in which the resistive element is attached to the terminals. The nonlinear load of such a resistor can cause falsely high distortion readings for amplifiers with very low distortion levels.

It is also possible to be fooled by distortion in a simulated speaker load. If a passive speaker simulator network uses an inductor with a ferrous core, the distorted current that flows in the load can induce distortion in the amplifier, especially if the amplifier does not have a very low damping factor.

16.13 EMI-Induced Distortion

Amplifiers are subject to *radio frequency interference* (RFI), more commonly referred to as *electromagnetic interference* (EMI). Electric drills, light dimmers, radio stations, wi-fi and cell phones are common sources of EMI. Amplifiers are susceptible to EMI through three conductive ports: the input port, the output port and the mains port. If EMI gets into an amplifier, it can show up as noise or buzzing which alone is objectionable. This is not distortion as such. It is additive noise. However, if the EMI disturbs amplifier stage operating points or intermodulates with the audio signal, then nonlinear distortion can result. Any EMI effect that is correlated with the audio signal will be perceived as nonlinear distortion. Distortion created by EMI is discussed in greater detail in Chapter 22.

16.14 Thermally Induced Distortion (Memory Distortion)

Whenever a temperature change can affect the gain or operating point of an amplifier, distortion may result when the program material contains low frequencies or when the short-term program power changes with time. This is sometimes called *memory distortion.*

This type of distortion was discussed in the cases of resistor and fuse distortion. If the gain of an amplifier stage changes with temperature, then it can also cause distortion at low frequencies or intermodulation with higher frequencies. Thermal distortion is often much lower at high frequencies because the thermal inertia of the affected elements prevents the temperature from changing too much with faster signal swings. However, it should be kept in mind that the local thermal time constant of a transistor junction can be on the order of milliseconds.

Memory distortion will not show up with conventional steady-state tests like THD-1 and THD-20. Memory distortions can sometimes be unmasked with an intermodulation distortion test where a large, very low frequency signal is added to a higher frequency signal and the resulting amplitude modulation of the higher frequency is measured. This is how the SMPTE IM test works, where frequencies of 60 and 7000 Hz are mixed in a 4:1 ratio. For thermal distortion, more sensitive results will be obtained if the low frequency is 20 Hz or even much lower. For such studies, I designed and built an IM analyzer that used coherent AM detection and whose low-frequency component could be chosen to be additional frequencies of 1 Hz, 5 Hz and 20 Hz [2]. Thermal distortion may also show up as increased THD at low frequencies like 20 Hz. These symptoms can be confused with distortion from electrolytic capacitors in feedback networks, however.

A special form of thermal distortion occurs in the output stage of a class AB power amplifier [12]. Here the critical quiescent bias current can be disturbed by junction temperature changes in the output transistors that are too fast to be tracked out by the bias temperature compensation. This will often result in quiescent bias errors that are dependent on the recent history of the signal. In some cases following a loud passage, the output stage may be temporarily under-biased after the program material returns to a low level; this will result in increased crossover distortion.

References

1. Robert R. Cordell, "A Fully In-band Multitone Test for Transient Intermodulation Distortion," *Journal of the Audio Engineering Society*, vol. 29, no. 9, pp. 578–586, September 1981. Available at www.cordellaudio.com.
2. Robert R. Cordell, "Phase Intermodulation Distortion—Instrumentation and Measurements," *Journal of the Audio Engineering Society*, vol. 31, no. 3, pp. 114–124, March 1983. Available at www.cordell audio.com.
3. Bruce Hofer, "Designing for Ultra-Low THD+N," *AudioXpress*, Parts 1 and 2, pp. 20–23, November 2013 and pp. 18–23, December 2013. Available at audioxpress.com.
4. Bruce Hofer, "The Ins and Outs of Audio." Available at www.aes-media.org/sections/uk/Conf2011/ Presentation_PDFs.
5. Robert R. Cordell, "A MOSFET Power Amplifier with Error Correction," *Journal of the Audio Engineering Society*, vol. 32, no. 1–2, pp. 2–17, January 1984. Available at www.cordellaudio.com.
6. Walt Jung and Richard Marsh, "Picking Capacitors," *Audio*, pp. 52–62, February 1980 and pp. 50–63, March.
7. Cyril Bateman, "Understanding Capacitors," *Electronics World*, pp. 998–1003, December 1997.
8. Cyril Bateman, "Capacitor Sound?" Electronics World, Parts 1–6, July 2002–March 2003.
9. Edward M. Cherry, "A New Distortion Mechanism in Class B Amplifiers," *Journal of the Audio Engineering Society*, vol. 29, no. 5, pp. 327–328, May 1981.
10. G. Randy Slone, *High-Power Audio Amplifier Construction Manual*, McGraw-Hill, New York, 1999.
11. Douglas Self, *Audio Power Amplifier Design Handbook*, 5th ed., Focal Press, Burlington, MA, 2009.
12. T. Sato, K. Higashiyama and H. Jiko, "Amplifier Transient Crossover Distortion Resulting From Temperature Changes in the Output Power Transistors," presented at the 72nd Convention of the Audio Engineering Society, *Journal of the Audio Engineering Society* (Abstracts), vol. 30, pp. 949–950, December 1982, preprint no. 1896.

Part 3

Real-World Design Considerations

Part 3 addresses those real-world design considerations that come into play when designing a high-performance amplifier that must operate in the real world. These include power supply design and grounding architectures, short-circuit and safe area protection and control of amplifier behavior when driving difficult loads. Switching power supplies are of growing importance in power amplifier design, and a full chapter is devoted to their design. Thermal design and the oft-overlooked matter of thermal stability are given special attention. The use of advanced power transistors that include internal temperature monitoring means for improved thermal stability is covered in depth.

The ways in which amplifiers misbehave often account for sonic differences. How amplifiers clip is an example. These issues are discussed in Chapter 21. The world is full of electromagnetic interference, and it is only getting worse. The challenges presented by EMI ingress and egress via the input, output and mains ports of the amplifier is one of the topics covered thoroughly in Chapter 22.

Chapter 17

Output Stage Thermal Design and Stability

17. INTRODUCTION

Audio power amplifiers are not all that efficient, so they tend to produce a lot of heat. This has several implications. First, the amplifier must provide adequate means to get rid of the heat without excessive temperature buildup and consequent loss of reliability or risk of destruction. This requires heat sinks, which add to cost, physical size and weight.

Another issue in thermal design is that of thermal stability. It can often be the case that when power transistors get hot, they tend to increase their current flow. This can lead to thermal runaway and destruction.

At higher temperatures the safe operating area of power transistors decreases. This means that protection circuitry must be more conservatively designed if the amplifier is to safely operate at higher temperatures. This in turn means that, for a given size or cost of output stage, the amplifier will be less able to supply high currents to difficult loads.

The effect of temperature variations on output stage bias levels is important to sound quality. The crossover distortion produced by a class AB output stage depends strongly on the output stage quiescent bias being at the right value. If poor temperature control and stability characterize the design, substantial increases in crossover distortion may occur either on a continuous or a transient basis. In some cases an amplifier that is run hot with high levels of program material will become under-biased after the high-amplitude program material is removed. This will leave the amplifier vulnerable to crossover distortion on subsequent low-level passages.

The thermal design of class AB output stages will be the primary focus of this chapter. Of course, many of the concepts presented here will be equally applicable to other types of output stages.

17.1 Power Dissipation Versus Power and Load

Although it might seem intuitive that a power amplifier produces the most heat when it is putting out the most power, this is not generally so. The top curve in Figure 17.1 shows that the power dissipation (wasted heat) for an idealized 100-W class A amplifier actually decreases as the output power increases. This is so because the current drawn from the output stage power rails by a class A output stage is constant, regardless of output power. The idealized class A amplifier must have ±40-V rails and an idle current of 2.5 A in order to remain in class A when driving an 8-Ω load to 100 W (40 V peak, 5 A peak). When the amplifier is at idle, all of this power is being dissipated as heat. When the amplifier is putting out its maximum power into the load, that maximum power is being dissipated in the load instead of in the amplifier. The laws of power conservation dictate that the amount of power being dissipated as heat in the amplifier will be less by the amount of output power. The important principle here is that power dissipation is simply average DC input power less the output power.

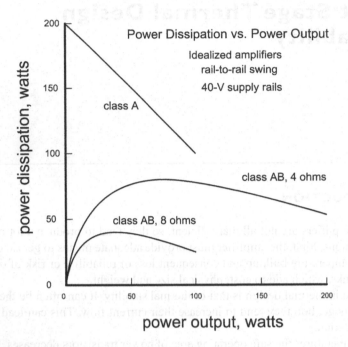

Figure 17.1 Power Dissipation of a 100-W/8-Ω Amplifier Versus Power Out (a) Class A (b) Class AB (c) Class AB Driving 4-Ω Load

The more interesting curves in Figure 17.1 are the curves for an idealized 100-W/8-Ω class B power amplifier delivering its power into 8-Ω and 4-Ω loads. Notice that the output stage power dissipation when driving an 8-Ω load is at its maximum value when the output power is at approximately 40% of its maximum value (actually $4/\pi^2 = 40.5\%$). The peak in power dissipation as a function of signal voltage occurs because power input to the amplifier increases linearly with signal voltage, whereas power being delivered by the amplifier increases as the square of the signal voltage. The 40% number is for the ideal case where there is no voltage drop from the rail to the peak output voltage into the load.

An optimally biased class AB amplifier will dissipate on the order of 10–11 W at idle due to bias current, adding somewhat to total power dissipation as a function of output power. This causes the point of maximum power dissipation for a class AB amplifier to occur at a slightly lower percentage of maximum power output, closer to 1/3 maximum power. The peak in real-world power dissipation occurring close to 1/3 maximum power dissipation explains why the FTC amplifier specification rule required that amplifiers be *preconditioned* at 1/3 their rated power into an 8-Ω load for 1 hour [1]. This seems a fairly brutal test for an amplifier until one realizes that often the same amplifier will be called on to deliver nearly its same output voltage into a 4-Ω load.

17.2 Thermal Design Concepts and Thermal Models

Thermal analysis is key to thermal design. Fortunately, there are electrical analogs to thermal "circuits" that allow many electrical engineering concepts (particularly Ohm's law) to be applied to thermal analysis.

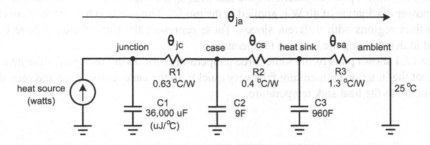

Figure 17.2 Thermal Model for a Power Transistor Mounted to a Heat Sink

Figure 17.2 illustrates an electrical model for thermal analysis. It is a very simple model of a power transistor mounted on a heat sink. The model is an electrical analog for the thermal behavior of the system and can actually be simulated with SPICE. Resistors represent thermal resistance and current sources represent heat sources. A 1-Ω resistor corresponds to a thermal resistance θ of 1 °C/W, while a 1-A current source corresponds to a power dissipation (or heat source) of 1 W. Node voltages correspond to temperatures in degrees Celsius relative to the ambient. Ground corresponds to the ambient temperature (usually 25 °C). Capacitances are the analog of thermal mass. A 1-W heat source flowing into a 1-Farad thermal capacitance will raise the temperature 1 °C in 1 second. One Farad of capacitance corresponds to 1 J/°C of thermal mass.

R1 in Figure 17.2 represents the thermal resistance from junction to case of the transistor, called θ_{jc}. Here it is 0.63 °C/W, representing a typical 200-W power transistor in a TO-264 package and having a maximum junction operating temperature of 150 °C. Shunt capacitor C1 represents the thermal mass of the die itself. The die has little physical size and mass, and its thermal mass is correspondingly very small. This means that the die temperature can change quickly, with a short-time constant. The value of C1 is 36,000 µF (36,000 µJ/°C). The thermal time constant τ_1 of R1 and C1 is about 23 ms. Voltage across R1 represents junction temperature rise above the case temperature. Voltage across C1 represents the junction temperature rise over ambient in degrees Celsius.

R2 models the thermal resistance of the insulator that separates the case of the transistor (usually the collector terminal) from the heat sink. This thermal resistance is designated θ_{cs}, meaning resistance from case to sink. This resistance depends heavily on the material and thickness of the insulator, and the area of contact. The most common material employed for insulators is mica. The value shown is 0.4 °C/W, representative of an insulator for the larger TO-264 transistor package. The insulator's thermal resistance will be higher for the smaller TO-3P or TO-247 packages. The thermal mass of the transistor package is represented by C2, here 9 Farads (9 J/°C). The thermal time constant τ_2 of R2 and C2 is about 3.6 seconds. The voltage across C2 represents the case temperature rise over ambient in degrees Celsius.

The heat sink is modeled by R3 and C3, here 1.3 °C/W and 960 F, respectively. The corresponding time constant τ_3 is about 1250 seconds (about 21 minutes). The voltage across C3 represents the heat sink temperature rise over ambient in degrees Celsius. The three thermal time constants are separated in time by more than an order of magnitude.

Temperature Versus Log Time Plots

Temperature changes can take place over very short time intervals or very long time intervals. For this reason it is best to plot temperature changes as a function of log time. The slope of the resulting curve is in degrees Celsius per decade.

Figure 17.3 shows plots of junction, case and heat sink temperature as a function of log time when a power dissipation of 40 W is applied to the model. The junction temperature curve shows three distinct regions with different slopes. These represent the three different time constants involved in determining the junction temperature.

Figure 17.4 shows plots of the same three temperatures following removal of the heat source. Notice that the junction temperature falls very quickly to the case temperature and then that falls fairly quickly to the heat sink temperature.

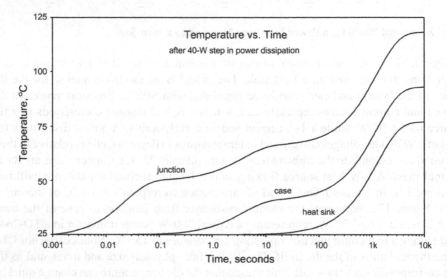

Figure 17.3 Junction, Case and Heat Sink Temperature Versus Log Time

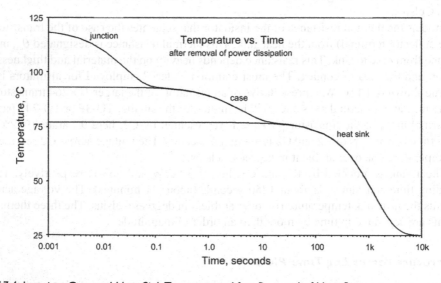

Figure 17.4 Junction, Case and Heat Sink Temperature After Removal of Heat Source

Thermal Attenuation

When the transistor is dissipating power it is easy to see that the series string of thermal resistances causes the junction temperature to be significantly higher than the heat sink temperature. There is so-called thermal attenuation from the junction to the heat sink. It means that during prolonged dissipation, the junction temperature remains significantly elevated over the heat sink temperature. In Figure 17.2, heat sink temperature rise over ambient will only be 56% of the junction temperature rise over ambient. This can be seen in Figure 17.4. However, when power dissipated by the transistor stops, the junction temperature falls to that of the heat sink within seconds. Heat sink temperature is not a good representation of junction temperature, even absent the time-dependent effects introduced by thermal mass and thermal time constants.

Lumped and Distributed Models

The models discussed above are highly simplified lumped models that do not accurately take into account the fact that heat travels as a diffusion process. For this reason, distributed models that include multiple R-C elements to model each part of the process are often used to obtain more accurate results.

Transient Thermal Impedance

Some power transistors are characterized by *transient thermal impedance* (TTI). This is often the case with MOSFETs, but is rarely the case with BJTs. The TTI curve is a valuable way of estimating the peak junction temperature when a power pulse is applied to the device. It recognizes the thermal mass of the parts of the transistor and the fact that heat diffuses through the device in a three-dimensional way. Because it takes distributed thermal mass into account, it is referred to as thermal impedance rather than a thermal resistance.

Figure 17.5 illustrates a TTI curve for an IRFP240 power MOSFET [2]. At long time intervals the value is simply the *DC* value of thermal resistance from junction to case. In this case that is

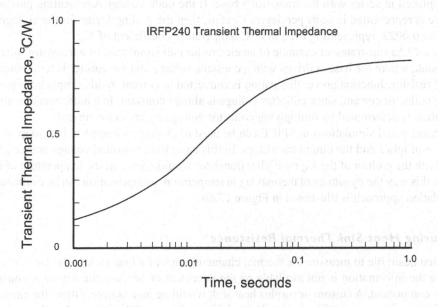

Figure 17.5 Transient Thermal Impedance Curve for an IRFP240 Power MOSFET

0.83 °C/W. The maximum junction temperature for this 150-W device is 150 °C. For shorter time intervals the effective impedance from junction to case (TTI) is smaller. This impedance can be used to infer the peak junction temperature reached at the end of a pulse of power of the given time duration.

For example, at 10 ms, the TTI is 0.42 °C/W, about half the DC value. If the case temperature is 70 °C, a rise of 80 °C is permitted at the junction with a 0.42 °C/W rate for 10 ms. The allowable transient power dissipation under these conditions is then 80°C/0.42°C/W = 190 W. The real action in the TTI curve is around 10 ms, implying that the time constant is of this order of magnitude. The thermal time constant appears to actually be about 17 ms.

The published operating boundary criterion for this device is simple: never let the peak junction temperature exceed 150 °C. This is actually very conservative. The maximum junction temperature in plastic-packaged devices is dictated by packaging reliability considerations. It is not the die failure temperature.

Thermal Simulation

Having the electrical analog to the thermal circuit makes it possible to arrange for SPICE simulations to predict thermal behavior and gain insight. Using current sources as heat sources and the kinds of thermal models described above, simulations can be carried out where node voltages represent temperatures.

Simulations can also be created that combine the thermal and electrical aspects into a single "electrothermal" simulation. Such simulations can be used to evaluate thermal stability. Bear in mind that this kind of simulation does not necessarily use the temperature models in SPICE, but instead uses controlled voltage sources to alter effective transistor V_{be} drops as a function of the temperature (which is represented as a voltage elsewhere). For example, if the voltage of a node in a simulation signifies transistor junction temperature above the reference 25 °C, a voltage source driven by a function of the voltage representing the transistor's junction temperature can be placed in series with the transistor's base. If the node voltage representing junction temperature is represented in volts per degree Celsius, then the scaling factor for the voltage source would be 0.0022, representing a 2.2 mV/°C temperature coefficient of V_{be}.

Figure 17.6a illustrates an example of an electrothermal simulation of a transistor mounted on a heat sink, where the base is driven with a constant voltage and the emitter is terminated with a 0.22-Ω resistor. Junction power dissipation is converted to current in this simple arrangement by scaling collector current, since collector voltage is almost constant. In a more accurate approach, dissipation is determined by multiplying collector voltage by collector current.

Electrothermal simulations in SPICE can be used to evaluate a complete bias loop, including the V_{be} multiplier and the output transistors. In this case, the controlled voltage source placed in series with the emitter of the V_{be} multiplier transistor would represent the temperature of the heat sink. In this way the dynamics of thermal lag in temperature compensation can be evaluated. Such a simulation approach is illustrated in Figure 17.6b.

Measuring Heat Sink Thermal Resistance

It is often desirable to measure the thermal characteristics of a heat sink in the lab. This may be because the information is not available on the datasheet or because the device is unusual or a custom component. A custom or surplus heat sink would be an example. Often, the chassis of an amplifier provides some of the heat sink capability and its contribution needs to be evaluated.

One of the most common approaches to measuring thermal resistance is to mount a metal-cased 50-W power resistor on the heat sink or other material in question and apply a known number of watts to the resistor. At the same time, an IC temperature sensor can be mounted to

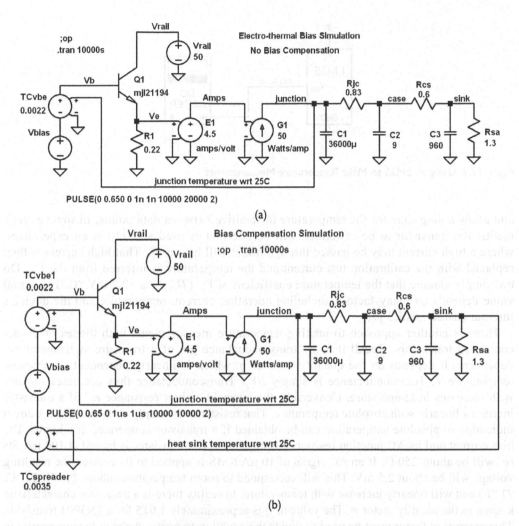

Figure 17.6 (a) Electrothermal Simulation of a Transistor Mounted on a Heat Sink (b) Transistor's Bias Is Controlled by a V_{be} Multiplier

the heat sink to monitor the temperature rise. The thermal resistance is then easily calculated. By observing the rate of rise of temperature, the thermal mass can also be inferred. Alternatively, you can raise the heat sink to a known high temperature, remove the heat source and record the time it takes for the temperature to fall by 63% of the difference between the ambient and the starting temperatures. This time value is then the time constant of the thermal capacitance and the known thermal resistance of the heat sink.

Figure 17.7 shows a simple schematic for using the three-terminal LM35 Centigrade tempera-ture sensor. The output reads directly in degrees Celsius with a linear 10 mV/°C [3] voltage. This handy device comes in a TO-92 package that can be placed in a 3/16-inch hole drilled in a heat sink and filled with some thermal grease. The LM35 is also available in a TO-220 package that can be mounted to the heat sink.

It is sometimes desirable to use a power transistor as a heat sensor in laboratory measure-ments. Here the V_{be} versus temperature characteristic of the transistor needs to be character-ized at a known current flow when the transistor is diode connected. This can be accomplished by putting the transistor and an accurate thermometer in an oven. Take several data points,

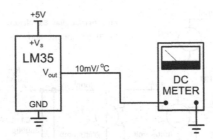

Figure 17.7 Using an LM35 to Make Temperature Measurements

and allow a long time for the temperature to stabilize between data points. In some experiments, the transistor to be calibrated will be one that is used actively in an experiment, where a high current may be passed through it that will heat it up. That high current is then replaced with the calibration test current and the temperature is inferred from the V_{be}. Do not simply assume that the temperature coefficient of V_{be} (TC_{Vbe}) is −2.2 mV/°C. The actual value depends on many factors, including operating current, temperature and the device's internal characteristics [4].

There is another approach to making temperature measurements with diodes or diode-connected transistors. Recall that the transconductance *gm* of a transistor is temperature-dependent. It depends on the quantity $V_T = kT/q$, which is linearly dependent on absolute temperature *T*. Transconductance is simply I/V_T. Transconductance thus decreases linearly with increases in temperature. Conversely, the intrinsic emitter resistance *re'* of a transistor increases linearly with absolute temperature. That resistance is simply equal to kT/qI. A direct indication of absolute temperature can be obtained if a transistor is operated at a known DC bias current and its AC junction resistance is measured. If a transistor is biased at 100 μA, its *re'* will be about 250 Ω. If an AC signal of 10 μA RMS is applied to its emitter, the resulting voltage will be about 2.5 mV. This will correspond to room temperature (about 295 K, 22 °C, 71 °F) and will linearly increase with temperature. In reality there is a transistor characteristic known as the ideality factor η. The value of η is approximately 1.015 for a 2N3904 transistor. This correction factor must be used to divide the equation to arrive at absolute temperature in degrees Kelvin. We then have

$$T = re'qI/k\eta \tag{17.1}$$

Another area sometimes in need of actual measurement is the effective resistance from case to heat sink for a power transistor, taking into account the insulator. In this case, the transistor can be mounted to a large heat sink using the insulator and thermal grease (if applicable). The transistor is then energized to heat up the arrangement with known power dissipation to a stable temperature. Temperature sensors mounted on the heat sink and on the case of the transistor can then be used to evaluate the temperature rise of the transistor case. Alternatively, the power transistor itself can be used as the case temperature sensor. The operating current is replaced with the transistor's temperature calibration test current and the case temperature is quickly inferred from V_{be}, before it has a chance to fall to the temperature of the heat sink.

A handheld infrared thermometer can also be a useful tool for thermal investigations. Most of them are equipped with laser targeting. Choose one with a good distance-to-spot ratio of at least 10:1. This means that a 1-inch area can be measured from a distance of 10 inches. This is important when trying to measure the case temperature of a power transistor. Also be aware that the color of the surface can influence temperature readings made by the infrared technique.

17.3 Transistor Power Ratings

A power transistor is rated at 150 W if its junction temperature remains below its maximum value when dissipating 150 W and its case is held at 25 °C. A transistor rated at 150 W with a typical maximum junction temperature of 150 °C will be allowed to have its junction temperature rise by 125 °C above the 25 °C ambient when it is dissipating 150 W. Its thermal resistance θ_{jc} from junction to case must then be 125/150 = 0.83 °C/W. This is a typical θ_{jc} for a TO-247 package, but θ_{jc} for the same package type can vary significantly depending on the die size of the actual transistor.

In most situations the transistor case must be electrically insulated from the heat sink with a mica insulator or an insulator made of some other material. This will add thermal resistance from case to heat sink (θ_{cs}) on the order of another 1 °C/W for a TO-247 package. This will raise the total thermal resistance from junction to heat sink in the example to about 1.8 °C/W. If this transistor is to dissipate 50 W, then the heat sink temperature must be held to no more than 150−1.8 * 50 = 60 °C. In practice we prefer that the heat sink temperature not exceed 60 °C, so the maximum power this transistor can dissipate on average will be about (150−60)/1.8 = 50 W.

Transistor Insulators

The thermal resistance of an insulator depends on its thickness, its area and its material. Thermal resistance will be proportional to thickness and inversely proportional to area. For a given thickness and area, a given insulator will be rated in °C/W. Insulators are available in many different materials, including Mica, Kapton, polyamide films (Sil-Pad®), silicone (Sil-Pad®) and ceramic [5, 6, 7].

The net thermal resistance of the insulator for a transistor depends on the contact area of the metal portion of the package of the transistor. For a TO-247 package, the area is about 183 mm², while for a TO-264 it is about 317 mm². As an interesting reference point, the thermal resistance of a TO-247 case to a heat sink without an insulator is about 0.25 °C/W when the interface is greased with thermal compound. Some numbers are shown in Table 17.1 for a typical *pad*-type insulating material.

There are numerous different insulator materials. Some require thermal grease to fill in the microscopic valleys (irregularities) in the mating metal surfaces while others perform this function themselves. We will assume typical thermal insulator θ_{cs} for TO-247 and TO-264 packages as 1.0 °C/W and 0.6 °C/W, respectively.

Table 17.1 Thermal Resistance of Transistor Insulators

Package	θ_{jc}	Mounting Area, mm²	Pad θ_{cs} °C/W	Total θ_{js} °C/W
TO-220	1.7	88	2.0	3.7
TO-247	0.83	183	0.96	1.79
TO-264	0.69	317	0.56	1.25
MT-200	0.63	620	0.29	0.92

17.4 Sizing the Heat Sink

One of the key design decisions is sizing the heat sink [8]. More is better from a performance and robustness point of view, but more is not the optimum solution in terms of cost, size and weight. Four key questions must be asked.

- What is the maximum power that will be dissipated?
- What is the maximum tolerable heat sink temperature?
- What is the maximum tolerable transistor junction temperature?
- What is the net thermal resistance from junction to the heat sink?

Depending on the numbers, the answers to either the second or third question will often dominate the decision about how big the heat sink must be. The thermal resistances from the junctions to the heat sink of all of the output transistors are effectively in parallel for purposes of calculating the net thermal resistance from the junctions to the heat sink.

Figure 17.2 illustrated the thermal resistance path from the transistor junction to the ambient. The first resistance encountered is from junction to case, designated θ_{jc}. The second is from the transistor case to the heat sink, θ_{cs}, which is dominated by the thermal resistance of the insulator. The third resistance is θ_{sa}, representing the thermal resistance of the heat sink to ambient air. The total thermal resistance from junction to ambient is θ_{ja}. The junction temperature rise above ambient is $P_j * \theta_{ja}$. Junction temperature T_j will then be $T_j = T_a + P_j\theta_{ja}$.

Output Stage Power Dissipation

Consider a 150-W amplifier design with two pairs of MJL21193/4 output transistors. These transistors have θ_{jc} of 0.63 °C/W and a 150 °C maximum junction temperature. The devices are in the larger TO-264 package, which has more surface area than the standard TO-3P package. An insulator thermal resistance of 0.6 °C/W will be assumed. The total thermal resistance to heat sink θ_{js} is then 1.23 °C/W for each device.

Output voltage at 150 W into 8 Ω will be 49 V peak, with a peak current of 6.1 A. Average current per rail will be 1.95 A. Assume that 5 V headroom is needed from the rail to the peak output voltage, making the required rail voltage 54 V at full load. For simplicity, we will assume a stiff power supply and nominal mains voltage. Power input to the amplifier at rated power is the product of 1.95 A and the rail-to-rail voltage of 108 V, which is 210 W. With 150 W being dissipated in the load, 60 W will be dissipated in the output stage.

The worst-case power dissipation will occur at an output power level near 1/3 rated power. Assume that the amplifier will support 1/3 rated power to the point where the thermal circuit breaker opens at 70 °C. At 50 W, output voltage will be 28 V peak and output current will be 3.5 A peak. Average power supply current will be 1.125 A per rail. Input power will be 121.5 W. With 50 W being dissipated in the load, about 72 W will be dissipated in the output stage. This comes to 18 W for each of the four output transistors, corresponding to a junction temperature rise of 22 °C with respect to the heat sink temperature. If the heat sink is at 70 °C, the junction temperature will be about 92 °C, well within the junction temperature rating (often 150 °C).

To keep things in perspective, bear in mind that if 0.22-Ω emitter resistors are used, optimal class AB bias current will total about 236 mA, resulting in quiescent power dissipation of about 25 W.

Required Heat Sink Thermal Resistance

If we assume a 25 °C ambient, the heat sink will be allowed to rise by 45 °C to its 70 °C thermal cutout value while dissipating 72 W. This requires a heat sink thermal resistance θ_{sa} of 0.625 °C/W.

Power Dissipation into Reactive Loads

Power dissipation for a given load impedance may be greater if that impedance is reactive. This is because the power factor is non-ideal for a reactive load; this means that the reactive portion of the load does not dissipate power. In the worst case of driving a purely reactive load, the load dissipates no power, but the power input to the output stage may be essentially the same. For example, the 150-W amplifier dissipates 72 W when driving an 8-Ω load at 1/3 power. However, when driving an 8-Ω purely reactive load, the power input is 210 W with no power being dissipated in the load, so the output stage dissipation is the full 210 W. This is far worse than the 1/3-power worst case for a resistive load. The example here is extreme and unlikely to ever occur in the real world. Fortunately, when the loudspeaker is most reactive, its impedance is much greater than 8 W. However, reviewers may test with a 2-μF load at 20 kHz, which is a 4-Ω reactive load, which is much worse. Testing with such a load should be avoided at substantial power levels for extended periods of time, and is best left to determining stability and transient response if it is to be done.

Transistor Junction Temperature

The transistor junction temperature should now be checked. Each transistor is dissipating 18 W with a thermal resistance from junction to heat sink θ_{js} of 1.23 °C/W. This results in a junction rise of 22 °C, for a junction temperature of 92 °C when the heat sink is at 70 °C. This leaves a comfortable margin of 58 °C to the rated maximum junction temperature of 150 °C. This allows for driving lower-impedance loads and for safe operating area allowance.

The junction temperature should also be checked when driving a 2-Ω load at 1/3 the maximum (2-Ω) power. This check is done when the heat sink temperature is 70 °C, the point at which the thermal breaker on the heat sink will open. This check is necessary because a reviewer may run the amplifier into a 2-Ω load until thermal shutdown occurs. It will be assumed that the power supply is very stiff and sags only 10% from its nominal 54 V under the 1/3-power condition at 2 Ω. The rail is thus at about 49 V. It is further assumed that increased headroom of 7 V is required from the rail. Under these conditions, the full-power output is 42 V peak, corresponding to 2-Ω *rated* power of 441 W, one-third of which is 147 W. Power input will be 374 W and dissipation will be 227 W, or 57 W per transistor. With θ_{js} = 1.23 °C/W, junction temperature rise will be 70 °C, resulting in T_j = 140 °C. This is a marginally safe junction temperature for this difficult testing condition.

The Thermal Breaker

A key element in protection of the amplifier is the thermal breaker. This sets an upper limit on operating heat sink temperature and actually makes maximum transistor junction temperature planning easier. A cutout temperature of 70 °C is a good choice. The device can be as simple as a mechanical thermal switch mounted on the heat sink that interrupts the mains supply. An alternative solution is to employ an IC temperature sensor like the National LM35 [3]. This can be arranged with an accurate threshold circuit so that it can open a mains relay or disable the amplifier in some other way until power is cycled. A second threshold can be incorporated to activate a warning LED when the heat sink temperature reaches 60 °C.

Sometimes a thermal breaker is embedded in the transformer windings. This is helpful for fire safety reasons, but does not necessarily guard against the heat sinks becoming dangerously hot to the touch or for the output transistors, since the power transformer may be relatively oversized and less prone to overheat.

The Finger Test

As a convenient rule of thumb, you would generally like the heat sink temperature to be less than 60 °C under strenuous operating conditions. It turns out that 60 °C is approximately the highest temperature at which you can keep your finger on a surface for at least 30 seconds. For example, you might wish to design the amplifier so that you can run it at 1/3 power into an 8-Ω load for 1 hour and still be able to put your finger on the heat sink for 30 seconds. For such an amplifier, you might also set the thermal breaker to operate at 70 °C. Any temperature over 70 °C is considered a burn hazard. At 70 °C you have time to remove your hand before a burn occurs.

The Heat Sink Is Not Isothermal

Aluminum has thermal resistance that cannot be ignored. This means that there will be thermal drops across the heat sink. In particular, the heat sink will be hotter in the vicinity of a power transistor. This can add to the effective thermal resistance of the heat sink as seen by the transistor, and so eat into thermal margins. For best thermal performance, space out the power transistors across the heat sink. The heat sink is hotter toward the top, so the thermal cutout sensor should be placed above the power transistors.

Sizing the Output Stage

Sizing the output stage is one of the most basic decisions in designing an amplifier if one is to achieve the desired performance and reliable operation. Much but not all of it involves thermal considerations. There are several constraints that must be obeyed. Often, one constraint will dominate and set the size of the output stage. These constraints include:

* Static power dissipation in the output stage
* Safe operating area
* Thermal bias stability
* Peak output current
* Beta droop (for BJTs)
* f_T droop (for BJTs)

This topic was discussed in Chapter 12, and will be touched on here in respect to the thermal considerations. Because things tend to scale, it is not unreasonable to make a recommendation of the number of output pairs needed for a given power output rating into 8 Ω. The amplifier in the above example was reasonable, from a junction temperature point of view, at one output pair per 75 W of rated power into 8 Ω. That amplifier benefited from the use of large TO-264 200 W power transistors. On the other hand, it was deliberately designed to withstand the punishment of operating at 1/3 power into a 2-Ω load until the thermal breaker opened.

A simple rule of thumb is to take the rated power into 8 Ω, divide by 75 and round up to the next integer number of output pairs. Recognize that this rule of thumb does not take SOA into account. That will be discussed in the next chapter. Depending on the type of protection employed, that consideration may trump worst-case average junction temperature considerations.

Table 17.2 provides a generalized rule of thumb for choosing the number of output pairs to use. This is based on a thermal breaker temperature of 70 °C and rated power into a resistive 8-Ω load. It assumes that a maximum junction temperature of 150 °C is reached by the time the thermal breaker opens, under conditions of driving 1/3 rated 2-Ω power into a 2-Ω load. The table assumes that this output stage dissipation is (227/150) * 75 ≈ 114 W for every 75 W of rated power into 8 Ω. The ratio 227/150 comes from the exercise above where the maximum power dissipation of a 150-W/8-Ω amplifier driving 2 Ω at 1/3 the 2-Ω maximum power was 227 W. This is thus used as a scaling factor.

Table 17.2 Minimum Required Number of Output Pairs

Package	P_{diss}, W	θ_{js}, °C/W	P_j, W	Pairs/75 W
TO-247	150	1.8	44	1.30
TO-264	200	1.2	67	0.85

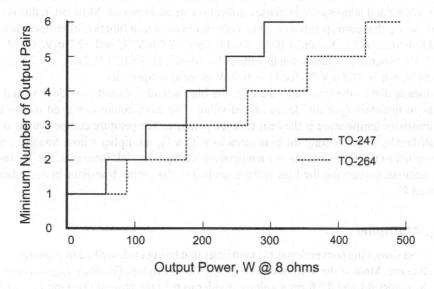

Figure 17.8 Recommended Number of Output Transistors Versus Rated Power into 8 Ω

The numbers are based on the generic estimate listed in the table for net thermal resistance from junction to heat sink θ_{js}, including the insulator. The column labeled P_j is the estimated allowable power dissipation per transistor with the heat sink at 70 °C and a rise of 80 °C to the rated 150 °C maximum junction temperature. The number of output pairs is simply the power dissipation of 114 W divided by twice the available power dissipation for each output device. Figure 17.8 illustrates how the recommended number of output pairs climbs with increased power rating into 8 Ω.

We can see that a 150-W amplifier misses the criteria if it is implemented with two pairs of TO-247 transistors (should have 2.6 pair). We can also see that a 200-W amplifier misses with two pairs of TO-264 devices (it should have 2.3 pair).

Modern output transistors are relatively inexpensive. For this reason, it is best to err on the side of conservatism when determining the number of output pairs to employ in an amplifier design. This approach also provides added SOA margin and may allow the use of less invasive protection circuits.

17.5 The Bias Spreader and Temperature Compensation

The quiescent bias current I_q in a class AB output stage plays an important role in controlling crossover distortion. This is especially true for BJT output stages where there exists an optimum value of bias. In theory, this optimum bias places 26 mV across each of the output transistor emitter resistors (the value is actually kT/q, which is 25.7 mV at 25 °C, and increases with temperature). As pointed out previously, the actual optimum value is a bit less due to ohmic resistance as seen looking into the emitter of the power transistor.

The bias current requirements are less precise for MOSFET output stages, where more bias current is usually better. Proper bias in the case of MOSFETs is still important if for no other reason than avoidance of overheating. In this section the focus will be on BJT output stages because their bias-setting requirements are more critical, but most of the concepts apply equally to MOSFET output stages.

As explained in earlier chapters, BJTs have a positive temperature coefficient of current. If V_{be} is held constant and temperature increases, collector current increases. More often this is stated in terms of the V_{be} that corresponds to a given collector current as a function of temperature. This is referred to here as TC_{Vbe}. For most BJTs, it is between -2.0 mV/°C and -2.2 mV/°C [4]. $TC_{Vbe} = (V_{bg} - V_{be})/T$, where V_{bg} is the bandgap voltage for silicon, about 1.24 V. TC_{Vbe} is -2.2 mV/°C for $V_{be} = 0.58$ V, and is -1.8 mV/°C for $V_{be} = 0.70$ V at room temperature.

This means that as the transistor heats up, the bias spreader should provide it with a smaller V_{be} so as to maintain I_q at the desired fixed value. The most commonly used measure of the output transistor temperature is the heat sink temperature. Temperature compensation is usually accomplished by implementing the bias spreader with a V_{be} multiplier whose transistor is physically attached to the heat sink. As the temperature of the heat sink increases, the V_{be} multiplier voltage decreases, reducing the bias voltage applied to the output transistor in accordance with its reduced V_{be}.

The V_{be} Multiplier

Figure 17.9a shows the conventional V_{be} multiplier that has been described in numerous places in earlier chapters. Most of the VAS bias current I_{VAS} flows through Q1. Here I_{VAS} is assumed to be 10 mA. Resistors R1 and R2 form a voltage divider to feed the base of Q1. One V_{be} must appear across R1 for Q1 to be turned on. The remainder of the bias-spreading voltage will appear across R2. If R1 is 700 Ω and R2 is 2100 Ω, then the total bias spread will be 4 V_{be} if base current in Q1 is negligible. The V_{be} multiplication factor is thus 4. About 1 mA will flow in the divider and about 9 mA will flow in Q1.

If Q1 has β equal to 100, then its base current will be 90 µA and an additional 189 mV will be added to the drop across R2 and to the total bias-spreading voltage. Because β is temperature-dependent, this constitutes a source of error in the V_{be} multiplier. However, β increases with temperature, decreasing this error current with temperature, so the net temperature effect of this error is mitigated a bit.

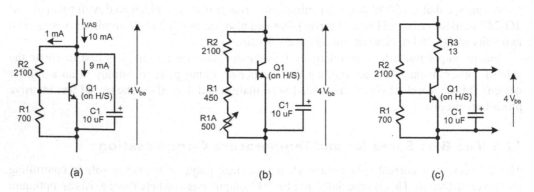

Figure 17.9 Conventional V_{be} Multiplier Bias Spreader

The usual bypass capacitor should be placed across the bias spreader to ensure stability and maintain low spreader impedance to high frequencies. Transistor Q1 will usually be mounted on the heat sink to temperature-compensate the junction drops of the output transistors. Sometimes Q1 will be implemented with a TO-126 device or something similar that is conveniently mounted to the heat sink. High-frequency stability precautions should be taken in light of the wiring inductance that may be encountered between the transistor on the heat sink and the remainder of the bias spreader on the circuit board. A 0.01-μF capacitor connected from collector to base of Q1 right at Q1 usually helps by converting Q1 into a diode-connected transistor at high frequencies.

Figure 17.9b shows a more practical version of the bias spreader where trimming resistor R1A has been added in series with R1 to provide for adjustment of the output stage quiescent bias I_q. If for some reason the trimming resistor fails open, the bias spreader fails safely by defaulting to a multiplication value of unity.

V_{be} Multiplier Impedance

The intrinsic emitter resistance re' of Q1 will be about 2.9 Ω and the dynamic impedance of the bias spreader will be greater than this by the V_{be} multiplication factor of 4, or about 12 Ω. This means that if I_{VAS} increases by 1 mA (10%), the bias spread will increase by 12 mV. This is a significant fraction of the 52 mV that should appear across the output stage emitter resistors for optimum class AB bias.

This error can be reduced by a simple compensation scheme that is sometimes used [9]. As shown in Figure 17.9c, a resistor is inserted in the collector circuit of the V_{be} multiplier. As I_{VAS} increases, total spreader voltage increases, but so does the drop across the collector resistor. The voltage drop across this resistor acts to reduce spreader voltage applied to the output stage. If the value of this resistor is the same as the DC impedance of the spreader, the effect will be nearly compensated. Because the impedance of the spreader is a function of bias current, this approach will be optimum at only one bias current level. Fortunately, the error curve is very broad and shallow.

If I_{VAS} is established by a feedback current source, I_{VAS} will depend on V_{be} of the control transistor, which will increase slightly with supply voltage as its bias current is increased, and will decrease a bit with ambient temperature. If supply voltage increases by 20%, I_{VAS} will increase by about 0.8%. If ambient temperature increases by 20 °C, I_{VAS} will decrease by about 0.6%. If the total I_{VAS} variation is 1.4% as a result, the bias spread will change by about 1.7 mV, which is not too bad. In a push-pull VAS, I_{VAS} is established by a different set of variables and may not be as stable.

Figure 17.10 shows a typical arrangement with a bias spreader like that of Figure 17.9 providing bias for an output stage Triple. Here the V_{be} multiplication factor is set to 6, with R2 increased to 3500 Ω. The spread is thus 6 V_{be}, with all V_{be} drops being controlled by the heat sink temperature if Q1 is mounted on the heat sink. This arrangement would work well in theory if the pre-driver and driver transistors were also mounted on the heat sink, placing them at the same temperature as the output transistors.

In some designs the driver transistors are mounted to the heat sink because they also dissipate moderate power, but the pre-drivers are often not mounted on the heat sink. In other designs neither the pre-drivers nor the drivers are mounted on the heat sink. Thus, not all of the 6 V_{be} supplied by the bias spreader should be compensated by heat sink temperature. If all of them are compensated, then the output stage may be overcompensated, meaning that when the heat sink is hot, the bias-spreading voltage will be too small and I_q will be too low. Depending on the details of the output stage and the transistor mountings, some fraction of the six V_{be} supplied by the bias spreader should be compensated by heat sink temperature.

If the pre-drivers and drivers are not mounted on the heat sink, then in principle only two of the V_{be} of spread should be sensitive to heat sink temperature. In practice, this will often result in

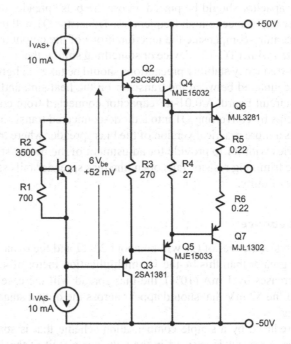

Figure 17.10 Conventional Bias Spreader Driving a Triple Output Stage

an undercompensated output stage because the actual junction temperature of the output transistor is higher than that of the heat sink, especially under conditions of significant power output. Recall that there is thermal attenuation between the junction and the heat sink. For this reason, somewhat more than the theoretical 2-V_{be} of compensated bias spread is appropriate. When bias stability and thermal bias modulation are discussed below, it will be seen that the situation is somewhat more complex.

V_{be} *Multiplier Variations*

The need to obtain different amounts of compensated and uncompensated bias spreading leads to the use of V_{be} multipliers where part of the multiplier is mounted on the heat sink and part is mounted on the circuit board and exposed to the board ambient. We will refer to these as split bias spreaders. Figure 17.11a shows a simple variant where diode-connected transistor Q2 is inserted in series with the emitter of Q1. Transistor Q2 is mounted remotely on the heat sink while Q1 is mounted on the board. To achieve the same 6-V_{be} drop, R1 is set to 1400 Ω and R2 is set to 2800 Ω. In this case half of the bias spread is compensated. This technically overcompensates the output stage, but that may be desirable in light of thermal attenuation from junction to heat sink. There will be situations where it is preferred to employ a real diode for temperature sensing instead of a diode-connected transistor, but the considerations are the same (see the discussion on ThermalTrak™ transistors in Section 17.8).

Figure 17.11b shows a different approach where Q2 is also configured as a V_{be} multiplier. This allows a completely flexible choice of the relative amount of compensation. The arrangement shown compensates 4 of the total spread of 6 V_{be}.

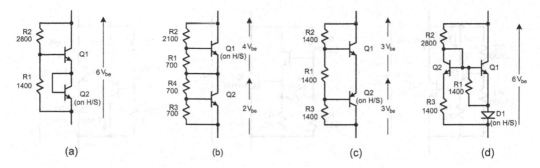

Figure 17.11 Four Different Split Bias Spreader Arrangements

Figure 17.11c is a complementary bias spreader employing an NPN and a PNP transistor with emitters connected. Here we assume that Q2 is mounted on the heat sink. This arrangement provides 50% compensation.

Figure 17.11d shows a bias spreader that employs a remote diode on the heat sink while also providing the flexibility of choosing the degree of compensation over a wide range. By means of diode-connected Q2, both R1 and R3 have 1 V_{be} across them (for convenience we will often refer to the voltage drop of a forward-biased diode like D1 as 1 V_{be}). With the values shown, the degree of compensation is about 50%. If R3 is made much larger than R1, compensation will be quite small because most of the influence will come from the uncompensated V_{be} of Q1. If R1 is made much larger than R3, compensation will be quite high because most of the influence comes from the compensated diode drop of D1. Notice in this case the V_{be} of Q2 largely cancels the influence of Q1 in the current path through R3.

Darlington Bias Spreaders

There is sometimes concern about the error introduced by the base current in a conventional V_{be} multiplier. Figure 17.12a shows a bias spreader that employs a Darlington arrangement. Base current drawn from the divider formed by R1 and R2 is reduced by a factor of about 9 with the values shown. This arrangement multiplies two V_{be} (those of Q1 and Q2) by a factor of 3 to arrive at the 6-V_{be} spreading voltage. If Q1 is placed on the heat sink, compensation will be about 50%.

Figure 17.12b shows how the addition of R4 enables different contributions of compensated and uncompensated bias spreading by the two transistors. Reducing R4 and increasing R1 decreases the proportion of spread contributed by Q1. The fact that either Q1 or Q2 can be the transistor placed on the heat sink makes a wide range of compensation available.

Figure 17.12c illustrates a folded Darlington bias spreader. As shown, the V_{be} drops of Q1 and Q2 cancel out, leaving the full spreading influence to D1. This is useful when it is preferred to remote a diode to the heat sink instead of a transistor.

CFP Bias Spreaders

The single transistor in the conventional V_{be} multiplier can be replaced with a *complementary feedback pair* (CFP). The local feedback in the CFP greatly increases the stiffness of the bias spreader (lower impedance across its terminals) and greatly reduces the influence of transistor beta on the action of the spreader.

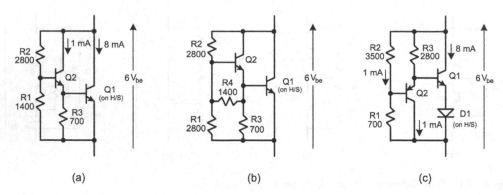

(a)   (b)   (c)

Figure 17.12 Darlington V_{be} Multipliers

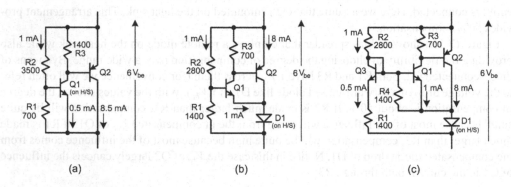

(a)   (b)   (c)

Figure 17.13 Three Variations of the CFP Bias Spreader

Figure 17.13a illustrates a straightforward CFP bias spreader that produces a 6-V_{be} spread. R1 and R2 perform the usual bias-setting function, with one V_{be} appearing across R1 and the remainder of the spread across R2. Q1 performs the primary function of the bias spreader, while R3 and Q2 complete the CFP. If the total spread is 6 V_{be}, 5 V_{be} will appear across R2. Although the VAS bias current is 10 mA, only 0.5 mA is flowing through Q1. This means that the base current of Q1 is much smaller than in the one-transistor designs. Errors caused by beta variations in Q1 will be much smaller.

Caution regarding high-frequency stability is always advised when using a CFP. In most cases the usual bypass capacitor across the spreader will be sufficient. A small resistor, here on the order of 50 Ω, can also be placed in series with the emitter of Q2 without seriously degrading performance. This reduces the loop gain of the CFP. Bias spreader impedance will be less than 2 Ω.

A remote temperature-sensing diode on the heat sink can be employed in the arrangement of Figure 17.13b. Here diode D1 is simply added in series with the emitter of the CFP. The voltage across R1 is now 2 V_{be}, only 1 V_{be} of which is sensitive to the heat sink temperature. This bias spreader is 50% compensated. The remote diode should be bypassed at the spreader to avoid potential instability caused by wiring inductance to the diode. Notice that placing D1 in series with the majority of the bias spreader current will increase the bias spreader impedance, taking away much of the impedance-reducing effect of the CFP arrangement. Other circuit arrangements may reduce or eliminate this disadvantage.

If an intermediate value of temperature compensation is desired, the arrangement of Figure 17.13c is useful. Here a diode is placed in series with R1 to cancel the voltage drop effect of Q1 so that only one V_{be} (that of D1) appears across R1. The added diode is conveniently implemented as transistor Q3 that can be identical to Q1. R4 is added to make the current flow in R2 dependent on a portion of the uncompensated V_{be} of Q1. R1 and R4 each see one junction drop. If the nominal values of R1 and R4 are set to be equal, the degree of compensation will be 50%. This circuit is analogous to that of Figure 17.11d.

Feedback Bias Spreaders

The feedback bias spreader incorporates feedback from the emitters of the pre-driver in a 3-EF output stage Triple. Consider a V_{be} multiplier that consists of an NPN and a PNP transistor in series, with a common base-to-base shunt resistor and collector-base transistors to the top and bottom of the bias spreader. If all three resistors are the same value (say 1500 Ω), we get a spread of 6 V_{be}, approximately what the Triple requires. Now connect the top and bottom resistors to the emitters of the pre-drivers instead of to the collectors of the bias spreader. The spread at the pre-driver emitters is down by 2 V_{be}, so the value of those two resistors should be cut in half. A spread of 4 V_{be} is now presented to the drivers, just what is needed. The pre-drivers have now been incorporated into the feedback loop of the bias spreader, and their V_{be} is far less influential. This makes the temperature stability of the 3-EF Triple just as good as that of a 2-EF double. The feedback bias spreader was first discussed in Chapter 12.3, Figure 12.5. It can also be used with other types of output stages, and many of the temperature compensation arrangements discussed above can be used with it.

Location of the Sensing Junction

When the V_{be} multiplier or a portion of it is mounted on the heat sink it should be as close as possible to the heat source. Self has advocated mounting the sensing junction on the case of one of the power transistors so that it can react more quickly and with less thermal attenuation. This is a useful approach, and a number of amplifier manufacturers have adopted it with various physical design approaches. Far superior results are obtained with the ThermalTrak™ line of output transistors by ON Semiconductor [10, 11, 12] to be discussed below in Section 17.8. These devices include a temperature-sensing diode inside the power transistor package right next to the transistor die.

Isothermal Bias Spreader and Driver Circuit

The Triple output stage generally provides superior performance in large part because of its much higher current gain and better isolation of the load from the VAS. However, it stacks to a voltage drop of 6 V_{be} that must be provided by the bias spreader. These 6 V_{be} all can come from different thermal environments and different conditions that affect the power dissipation and temperature of their corresponding transistors. As a result, thermal bias stability of the Triple can be more challenging than that of some other output stage designs. Here one approach to improving thermal stability of the Triple will be discussed.

Consider the case where the pre-drivers and drivers of a Triple output stage are mounted together on their own single heat sink on the circuit board. This heat sink could be as simple as an aluminum bar extending partly across the circuit board. It acts as a heat spreader. Assume also that Q1 of the V_{be} multiplier is mounted on this bar. These five transistors will then be essentially at the same temperature. The V_{be} of Q1 will track those of the pre-drivers and drivers very well.

Now assume that the bias spreader incorporates a remote sensing diode mounted on the main heat sink, using one of the bias spreader circuits like those discussed above. The arrangement can now accurately take into account the two groups of V_{be} drops in the desired proportion. This approach can significantly improve the bias stability of the Triple.

Thermal Attenuation Revisited

Consider again the 150-W power amplifier with 54-V rails discussed above. It uses two output pair wherein each transistor has thermal resistance from junction to heat sink of 1.1 °C/W (using the TO-264 package). The required heat sink was determined to have a thermal resistance to ambient of 0.65 °C/W based on an allowed rise to 70 °C when the output stage was dissipating 69 W (at 1/3 rated power into 8 Ω). This means that θ_{sa} for each output transistor is 2.6 °C/W. The thermal attenuation from junction to heat sink is then 2.6/(2.6 + 1.1) = 0.7. This means that if the transistor junction temperature rises by a sustained 20 °C, the heat sink will ultimately rise by 14 °C.

Suppose the designer decides to be conservative and sets the bias compensation so that the bias is still correct when the amplifier is dissipating 35 W with a continuous sine wave. Under these conditions the heat sink will have risen by 23 °C to 48 °C (assuming a 25 °C ambient). The junction temperature will have risen by 34 °C to 58 °C (remember, each of the four output transistors is dissipating only 8.75 W). The bias spreader will have to be designed to supply about 1.5 times the correction that would normally be created by the heat sink temperature rise. This is due to the thermal attenuation that is in effect when the transistor junction is dissipating continuous power. In rough terms, if the output stage needs two V_{be} drops of temperature-compensated bias in theory, it will in practice really need three V_{be} drops that are compensated. As a result of the 11 °C difference in rise between the junction and the heat sink, the bias spreader will have backed off the bias voltage by about 2.2 mV/°C times 11 °C or by 24 mV (as compared to the case where there was no thermal attenuation).

Now consider what happens when the signal is removed. Assuming that quiescent dissipation is small by comparison, the junction temperature "relaxes" back to the heat sink temperature with a time constant of less than 10 seconds. The heat sink temperature does not change significantly for quite some time. The bias spreader sees the same conditions and continues to supply a bias voltage that has been backed off by 24 mV to account for higher junction temperature rise than heat sink rise. However, there is no longer such an extra junction temperature rise. The output stage is now under-biased.

Setting the Bias and the Temperature Compensation

Under what conditions should you set the quiescent bias? In what way is the degree of thermal compensation established in the amplifier design? These are the questions to be addressed here. These two processes will be a bit interactive during the design process, but once the degree of thermal compensation is established and fixed by design, setting the bias will be fairly straightforward.

One procedure for designing the bias spreader for the proper amount of temperature compensation begins with a simple static bias adjustment. Assume for the moment that the bias spreader has been initially configured to provide a reasonable amount of temperature compensation. The bias is set to the low end of the range and the amplifier is turned on. While monitoring the emitter-to-emitter voltage, the bias is set for 2 V_q = 52 mV or whatever value has been determined to be optimal for the design. The amplifier is then allowed to reach thermal equilibrium over an extended period of time with no signal applied. The bias is retrimmed to the desired value during this time.

Bear in mind that if two output pair with 0.22-Ω emitter resistors are used, optimal class AB bias current will total about 236 mA, resulting in quiescent power dissipation of about 25 W.

This will raise the quiescent heat sink temperature by 16 °C to 41 °C if θ_{sa} of the heat sink is 0.65 °C/W.

A reasonable objective for temperature compensation is to have the quiescent bias current be the same when the heat sink is hot as when it is only warm (some may choose a slightly different criteria). Assuming that the quiescent bias has been set as above, the amplifier is then driven at 1/3 power into a load for 15 minutes. This heats up the heat sink. The input signal is then removed, and the bias voltage is checked about 20 seconds later. If the bias voltage is too high, the amplifier is undercompensated. If it is too low, the amplifier is overcompensated. The bias spreader is altered and the process is repeated.

A dynamic approach to bias setting using a THD measurement can be optionally used after the adjustments and design revisions above have been carried out. The bias setting can be trimmed based on actual harmonic distortion measurements taken at the amplifier power level where crossover distortion is estimated to be greatest. The peak in crossover distortion will occur at different percentages of full power for different designs, but will often be in the range of 3% of full power. The dynamic adjustment procedure is best applied while measuring THD-20 into a 4-Ω load. The bias adjustment is done after the amplifier has reached thermal equilibrium driving the load at the selected power level. In this case, the heat sink will be warmer than in the static approach. Suppose that the power amplifier described earlier is operated at 5 W into a 4-Ω load. Under these conditions it will dissipate about 49 W. This is a surprisingly high amount considering that the amplifier is delivering only 5 W. The heat sink temperature will rise to 57 °C if its θ_{sa} is 0.65 °C/W. The difference in quiescent heat sink temperature and the temperature at 5 W into 4 Ω will be about 57 °C −41 °C = 16 °C (recall that idle dissipation is assumed to be about 25 W).

There are many compromises in setting the temperature compensation and choosing the conditions for setting the bias. It is easy to say *Just set 2 V_q= 52 mV*, but the imperfect thermal dynamics of the output stage make that easier said than done. There is no right answer. Bias will be wrong under some conditions. The approach described above is just one of many possible choices.

Biasing Lateral Power MOSFETs

The biasing requirements for class AB output stages based on lateral power MOSFETs are much like those for BJT output stages, but in many ways are much more forgiving. The temperature coefficient of BJT V_{be} (TC_{Vbe}) for constant collector current is about −2.2 mV/°C. This is what gives rise to the temperature instability and potential for thermal runaway in BJT output transistors. In contrast, TC_{Vgs} for lateral power MOSFETs is negative at drain currents below about 100 mA and transitions from a negative value to a positive value as drain current increases. Conveniently, TC_{Vgs} passes through zero at a drain current typically lying somewhere between 100 mA and 300 mA. This is close to the typical bias current value used for a MOSFET output stage, so the result is a rather stable bias current over temperature. For this reason an ordinary V_{be} multiplier is used for the bias spreader with the bias spreader transistor mounted on the board instead of the heat sink. Figure 17.14 compares TC_{Vgs} for lateral and vertical MOSFETs.

Biasing Vertical Power MOSFETs

The biasing requirements for vertical power MOSFETs lie between those of BJTs and those of lateral MOSFETs. The TC_{Vgs} of gate voltage for vertical MOSFETs starts out negative at low drain currents and transitions to a positive value as drain current increases, as shown in Figure 17.14. This behavior is similar to that for lateral devices, but the transition through zero TC_{Vgs} occurs at a much higher drain current, usually on the order of several amperes. This level of drain current is

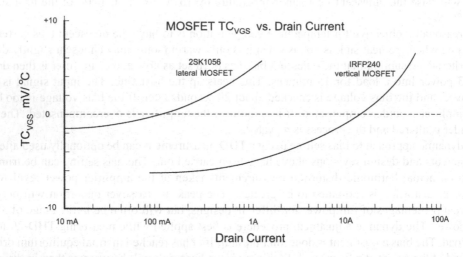

Figure 17.14 Temperature Coefficient of V_{gs} Versus Drain Current for Typical Lateral and Vertical MOSFETs

much higher than the typical 150 mA bias current. As a result, in the operating region for typical bias currents, TC_{Vgs} for vertical MOSFETs is negative (about −6 mV/°C).

Bias spreaders for vertical MOSFET output stages are much like those used for BJT output stages, with a portion of the spreading voltage compensated by heat sink temperature. The ratio of TC_{Vbe} to V_{be} is (−2.2 mV/°C)/0.7 V = 0.0031/°C, while the ratio of TC_{Vgs} to V_{gs} for an IRFP240 is about (−6 mV/°C)/4 V = 0.0015/°C. This means that the required percentage compensation for the bias spreader will typically be smaller for vertical MOSFET output stages. Vertical power MOSFETs tend to have significantly greater thermal stability than BJTs, and the penalty for being over-biased is smaller [13].

17.6 Thermal Bias Stability

The biasing schemes described above actually constitute a thermal feedback system. As the transistor junction heats up, it heats up the heat sink, which in turn heats up the bias-spreader-sensing element. This then reduces the V_{be} applied to the power transistor, reducing its current. The reduced current then results in reduced power dissipation that then results in reduced temperature. Clearly a negative feedback loop has been formed. As with any negative feedback system, stability must be considered. As can be seen from a thermal circuit equivalent like that in Figure 17.2, there are several points of thermal inertia separated by thermal resistances. This is equivalent to electrical poles in the feedback loop. However, the thermal inertia of the heat sink forms a very dominant pole.

In contrast, a conventional bias temperature compensation scheme is not predominantly a negative feedback system when output stage power dissipation is dominated by program material rather than quiescent bias current. There the change in bias point that the bias spreader is trying to manage is not the dominant influence on the temperature. The dominant influence on the temperature is the program material and the power dissipation it creates. In this situation, the bias temperature compensation scheme is really more of a sluggish feed-forward scheme. The loop gain of the negative feedback aspect of its operation is quite small.

These aspects of bias stability involving the bias spreader will be referred to as *global bias stability*.

Local Bias Stability

Many power amplifiers incorporate multiple pairs of output transistors, and current sharing among those transistors is a concern. The dynamic behavior of this current sharing is referred to as *local bias stability*. Larger values of emitter resistor R_E promote better current sharing. However, it is often possible to achieve lower crossover distortion with smaller emitter resistors. There is thus a trade-off between bias stability and crossover distortion.

The use of small-value emitter resistors can lead to current hogging by one of the output transistors (I consider 0.15 Ω or less to be small). This is a local positive feedback phenomenon. A transistor that is conducting more current (for whatever reason) will run hotter. The negative TC_{Vbe} will then cause the transistor's current to increase. The increased bias current, in combination with the rail voltage across the transistor, will cause an increase in the transistor's power dissipation. That transistor will then get hotter still. The key observation here is that an increase in bias current causes a further increase in bias current. In some cases this may result in a thermal runaway cycle. This behavior can be more pronounced in amplifiers using more output pairs.

Local thermal stability can be estimated by some simple mathematical considerations. For local thermal stability, the heat sink is assumed not to change in temperature. No matter how big the heat sink is, local thermal stability can be a concern.

Figure 17.15 illustrates the local positive thermal feedback loop formed by these effects. The open-loop gain A is set to unity; in the absence of these effects the change at the output would simply be the same as the change at the input. In Figure 17.15 the input is the bias current setting $I_{q,in}$. The output is the bias current result $I_{q,out}$. The analysis could also be carried out with other variables, such as temperature or power dissipation, as the input and output. The gain G of the circuit is $\Delta I_{q,out}/\Delta I_{q,in}$. You can think of $\Delta I_{q,in}$ as a disturbance. Remember, this analysis is for one output transistor with the temperature of the heat sink constant in the time frame of the analysis. The time constants in this thermal feedback circuit are much faster than that of the heat sink.

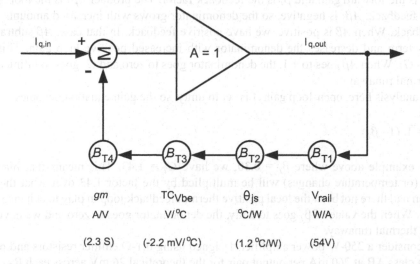

Figure 17.15 Local Positive Thermal Feedback Around the Output Transistor

The feedback path comprises four gain elements that together account for the thermal feedback factor β_T. The first element β_{T1} converts increased bias current to increased power dissipation. It is numerically equal to the rail voltage. The second element β_{T2} converts increased power dissipation to junction temperature rise. It corresponds to θ_{js}. The third element β_{T3} converts junction temperature rise to decreased V_{be}. It corresponds to TC_{Vbe} in millivolts per degree Celsius. The fourth element β_{T4} converts decreased V_{be} to increased bias current. It has units of amperes per volt and corresponds to transconductance gm. The feedback factor β_T is summarized by the equation below.

$$\beta_T = -TC_{Vbe} * gm * V_{rail} * \theta_{js} \tag{17.2}$$

In the equation above, β_T is considered to be a positive number for convenience. Assume that the negative sign is incorporated into the feedback summer. Remember that TC_{Vbe} is a negative number.

Consider a typical 150-W amplifier with 54-V rails and two pairs of TO-264 output transistors. Each transistor has $R_E = 0.22\ \Omega$ and is optimally biased at 118 mA, making its net transconductance, including R_E, equal to 2.3 S. Assume further that the thermal impedance from junction to heat sink for each transistor is $\theta_{js} = 1.1\ °C/W$.

$$\beta_T = 2.2\text{mV}/°C * 2.3\ S * 54\ V * 1.1°C/W = 0.30 \tag{17.3}$$

Recognizing that the unit of transconductance siemens has units of amperes per volt, we see that β_T is dimensionless, as expected.

Notice that the relevant thermal resistance is from junction to heat sink. In this analysis, it is assumed that the heat sink temperature remains constant over short periods of time and that there is no temperature-compensating effect occurring in the bias spreader. In the simple feedback system of Figure 17.15, closed-loop thermal gain is given by

$$G_T = A / (1 - A * \beta_T) \tag{17.4}$$

where A is the forward gain and β_T is the feedback factor. The product $A\beta_T$ is the loop gain. For negative feedback, $A\beta_T$ is negative, so the denominator grows with increased amounts of negative feedback. When $A\beta$ is positive, we have positive feedback. In that case, $A\beta$ subtracts from the unity term and decreases the denominator with increased positive loop gain. This in turn increases G_T. When $A\beta$ goes to $+1$, the denominator goes to zero and G_T goes to infinity, and we have thermal runaway.

In the analysis here, open-loop gain A is set to unity, so the gain equation becomes

$$G_T = 1 / (1 - \beta_T) \tag{17.5}$$

In the example above where $\beta_T = 0.30$, we have $G_T = 1.43$. This means that bias current changes (or temperature changes) will be multiplied by the factor 1.43 over what they would have been had there not been the local positive thermal feedback loop in play to enhance the thermal gain. When the value of β_T goes to unity, the denominator goes to zero and we have infinite gain and thermal runaway.

Now consider a 250-W power amplifier designed using 0.1-Ω emitter resistors and optimally biased in class AB at 260 mA per output pair for the theoretical 26 mV across each R_E. Ignoring the real-world effects of intrinsic base resistance and possible gate stopper resistance in calculating effective *gm* for each output transistor, we get *gm* = 5 S. Assume that the rail voltage is 70 V. Assume further that the output stage is built with 150-W TO-247 devices, with θ_{jc} = 0.83 °C/W and 0.8 °C/W insulator thermal resistance. Thermal resistance from junction to heat sink θ_{js} will thus be 1.6 °C/W.

We have

$$\beta_T = 2.2\text{mV/°C}*5\text{S}*70\text{V}*1.6\text{°C/W} = 1.23 \tag{17.6}$$

This amplifier is in deep trouble with respect to thermal bias stability, with a positive thermal feedback factor β_T greater than unity.

As a sanity check, we can "walk" around the loop starting with an assumed junction temperature increase of 10 °C. This will cause a 22-mV decrease in V_{be}, which will in turn, through *gm*, cause a 110 mA increase in bias current. With a rail voltage of 70 V, power dissipation in the transistor will increase by 7.7 W. With junction to heat sink resistance of 1.6 °C/W, this will result in a junction temperature rise of an additional 12.3 °C. Because this is greater than the starting assumption of 10 °C, there is greater than unity positive feedback and the transistor is off to the races.

This thermal stability analysis presents a strong case for employing more output devices in parallel with larger R_E. In the example above, if the number of output devices is doubled, each device can have $R_E = 0.22\ \Omega$ while keeping static crossover distortion about the same. This will reduce β_T by a factor of almost 2. Effective total θ_{js} will also be halved, further improving global bias stability. Additional advantages include increased SOA, reduced beta droop and reduced f_T droop. However, the driver circuits must be able to drive the increased collector base capacitance of the larger number of output transistors.

The evaluation of short-term local thermal stability by the calculation of β_T is of course a simplified approximation to the real word. There are other effects that are ignored, some of which are aggravating and some of which are mitigating. That *gm* increases with current is aggravating. That transistor current gain increases with temperature is aggravating. That base resistance decreases effective *gm* is mitigating.

The β_T analysis of thermal stability is very convenient and also applies to MOSFET output stages. Because the reaction time of the heat sink is so slow, local thermal runaway can happen in seconds.

Base Stopper Resistors and Thermal Bias Stability

Base stopper resistors are often placed in series with the bases of BJT output transistors to improve high-frequency stability. These resistors will create a DC voltage drop which is proportional to base current and which will factor into the bias current setting. This voltage drop will depend on transistor β and may be different among paralleled output transistors that are not matched for β. In some cases this can lead to current hogging. A transistor with higher β will conduct more current, get hotter and have its *β* increase further. This effect will be more pronounced with output stages where very low-value emitter resistors are used.

Consider an output transistor with β = 40, bias current of 170 mA and a 10-Ω base stopper resistor. Base current will be 4.3 mA and the DC drop across the base stopper resistor will be 43 mV. If another transistor in parallel with the first one has β = 60, the drop across its base stopper resistor will only be 28 mV. If the dynamic emitter resistance is 0.15 Ω and R_E = 0.15 Ω, then the difference of 15 mV will cause a difference in collector current of 50 mA. Beware of thermal instability that is exacerbated by the use of large base stopper resistors.

Measuring Thermal Bias Stability

Once an amplifier is built, it is desirable to evaluate its thermal bias stability. One certainly does not want it to show a tendency to destructive thermal runaway. However, even if that is not the case, one wants it to be thermally stable in regard to maintaining the optimum class AB bias current level under all static and dynamic temperature conditions.

Dynamic bias stability refers to the behavior of the output stage quiescent bias as a function of output transistor junction temperature changes caused by real program material. This does not show up under ordinary continuous sine wave testing on the bench. Dynamic bias mis-tracking can lead to conditions of over-bias or under-bias after the average power of the program material changes.

One way to evaluate dynamic bias stability is to place a meter across the emitter resistors of one of the output pairs, emitter to emitter. This is where one theoretically will see a 52-mV drop in an optimally biased class AB output stage (2 V_q). In practice, this voltage will often be somewhat less than 52 mV.

Measure 2 V_q when the amplifier has first been turned on. Measure it again after 1 hour (no signal). This will provide an indication of static bias temperature stability. Now run the amplifier at 1/3 rated power with an 8-Ω load until the heat sinks get too hot to touch (about 60 °C). Ignore the bias voltage reading while the amplifier is providing a signal to the load. Remove the signal and record the bias voltage as a function of time. Variations in 2 V_q are reflective of dynamic bias instability.

There is a caveat. In theory, the optimum bias places kT/q (~26 mV) across each of the output transistor emitter resistors. This means that when the transistors are hot, V_q should be larger, meaning that the bias current should indeed be higher. If junction temperature has increased by 20 °C under these conditions, V_q, and thus bias current, should be larger by about 7%.

Bias Stability of MOSFETs Versus BJTs

Figure 17.16 presents the results of such an exercise for four amplifier designs: an undercompensated bipolar, an overcompensated bipolar, an uncompensated MOSFET and a slightly overcompensated MOSFET design [13]. All of the amplifiers had identical rail voltages, power ratings (50 W) and heat sinks.

The first 10 seconds illustrate the effect of the faster power transistor thermal time constant, while the remaining time illustrates the heat sink time constant. Notice that both bipolar cases are actually very over-biased during and immediately following the high-dissipation "program" interval because the power transistor junctions run hotter than the heat sink due to thermal resistance from junction to heat sink. Overcompensation cannot reduce this effectively and will result in a seriously under-biased condition at other times.

In comparison, the compensated MOSFET design has much greater short-term and long-term thermal bias stability. Even the uncompensated MOSFET design has better thermal performance than the bipolar designs; this suggests that smaller vertical MOSFET amplifiers (say, below 50 W) with good heat sinking can probably be made without thermal feedback. TC_{VGS} for MOSFETs starts out negative, but becomes smaller as drain current is increased. It conveniently goes through zero for lateral MOSFETs at about 150 mA, giving these devices very good thermal stability, even without temperature compensation. TC_{VGS} does not go through zero until drain current reaches several A for vertical MOSFETs (those in Figure 17.16), so they do need temperature compensation.

17.7 Thermal Lag Distortion

Thermal lag distortion is probably the most insidious form of what some people call *memory distortion*. The pacing of the program material causes a modulation of the output stage power dissipation and temperature at subsonic to low audio frequencies, and this modulates the output

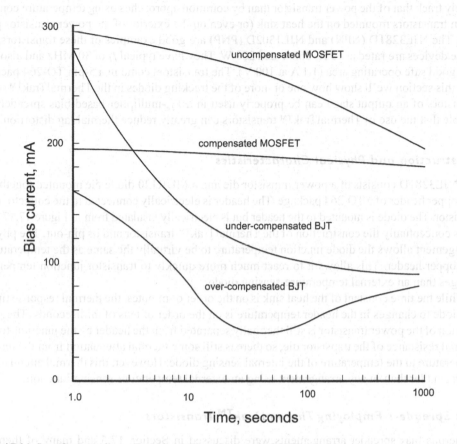

Figure 17.16 Bias Current Versus Time for BJT and Vertical MOSFET Amplifiers

stage bias [14]. As we saw above, the output stage bias control in a BJT amplifier cannot keep up with the changing power dissipation in the output stage, often leaving the output stage under-biased after cessation of a high-power program material interval.

Thermal lag distortion is most directly caused by the phenomenon of thermal attenuation and is exacerbated by the long time constant of the heat sink when the bias spreader temperature-sensing element is mounted on the heat sink. The use of ThermalTrak™ BJT output devices with their internal temperature-sensing diode greatly reduces thermal lag distortion. These transistors will be discussed in the next section.

MOSFET power amplifiers are far less susceptible to thermal lag distortion for two reasons. First, they are far more temperature stable than BJT-based amplifiers. Second, with most MOSFET power amplifiers, more bias is better, as opposed to the optimum value that must be observed with BJT class AB output stages. Thus, the performance of a MOSFET output stage is less affected by the minor bias current variations that occur with program material.

17.8 ThermalTrak™ Power Transistors

Many problems with output stage bias stability have been mitigated by the introduction of the ThermalTrak™ line of output transistors by ON Semiconductor® [10, 11, 12]. These transistors incorporate an electrically isolated tracking diode inside the transistor in close thermal contact with the output transistor die. This enables the junction temperature of the tracking diode to more closely track that of the power transistor than by common approaches using temperature compensation transistors mounted on the heat sink (or even on the exterior of the power transistor package). The NJL3281D (NPN) and NJL1302D (PNP) are good examples of these transistors [11]. These devices are rated at 15 A, 260 V and 200 W. They have typical f_T of 30 MHz and also have very good safe operating area (1.1 A at 100 V). The transistors come in a 5-pin TO-264 package.

In this section we'll show how one or more of the tracking diodes in the ThermalTrak™ output transistors of an output stage can be properly used in a V_{be}-multiplier-based bias spreader. It is notable that the use of ThermalTrak™ transistors can greatly reduce thermal lag distortion.

Construction and Physical Characteristics

The NJL3281D consists of a power transistor die and a MUR120 diode die mounted together on the copper header of a TO-264 package. The header is electrically connected to the collector of the transistor. The diode is mounted to the header but is electrically insulated from it. Figure 17.17 illustrates conceptually the construction of the ThermalTrak™ transistor and its pin-out. The physical arrangement allows the diode junction temperature to be virtually the same as the temperature of the copper header. This allows it to react much more quickly to transistor junction temperature changes than an external temperature-sensing diode.

While the time constant of the heat sink is on the order of minutes, the thermal response time of the diode to changes in the header temperature is on the order of tens of milliseconds. The actual junction of the power transistor is still thermally separated from the header by the junction-to-case thermal resistance of the transistor die, so there is still some thermal attenuation from the junction temperature to the temperature of the internal sensing diode. However, this thermal attenuation is much smaller than in any arrangement using an external temperature-sensing junction.

Bias Spreaders Employing ThermalTrak™ Transistors

Numerous bias spreader arrangements were discussed in Section 17.5 and many of them can be adapted for use with ThermalTrak™ transistors and their tracking diodes. The same principles apply, including those governing the choice of temperature compensation percentage. The bias spreaders employing remote sensing diodes, like those in Figures 17.11a and d, 17.12c and 17.13b and 17.11c are particularly good candidates.

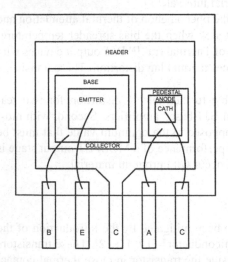 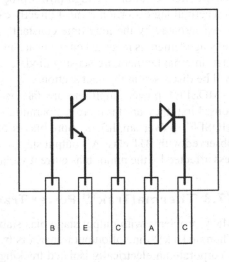

Figure 17.17 ThermalTrak™ Output Transistor Illustration

There is one important difference, however. It is important that at least one tracking diode from each of the NPN and PNP output transistors be included in the bias spreader. The reason for this is related to power transistor junction temperature changes at low frequencies. The reaction time of the tracking diodes is fairly fast, so it is desirable to take advantage of the complementary nature of the junction temperature changes of the NPN and PNP output transistors. For this reason the bias spreaders should be designed to employ two remote diodes.

The MUR120 is a 1 A SWITCHMODE™ power rectifier with a junction area that is large by comparison to those of the small-signal transistors usually employed in bias spreaders. Its forward junction drop is only about 600 mV at 25 °C at a junction current of 10 mA, which is a typical VAS bias current. This compares to about 750 mV for a small-signal transistor like the 2N5550 operating at 10 mA. This will affect somewhat the distribution of junction drops being multiplied by the V_{be} multiplier. Finally, the slope of the temperature coefficient of junction voltage for the MUR120 (TC_{TTD}) at 10 mA is about -1.7 mV/°C, while for the NJL3281 is about -2.1 mV/°C at its typical bias current of 118 mA ($R_E = 0.22\ \Omega$). This must also be taken into account in establishing the compensation percentage for the bias spreader. While the bias spreaders shown below have the ThermalTrak™ diodes in the emitter circuits, and consequently run the VAS current of about 10 mA through them, those sensing diodes can also be placed in series with the collector-base resistor R2. That operates them at lower current and increases their TC_{Vbe} [4, 15].

Figure 17.18a shows a simple bias spreader employing a conventional V_{be} multiplier that includes two of the internal diodes from a ThermalTrak™ output transistor, TTD1 and TTD2. The tracking diodes are simply placed in series with a conventional V_{be} multiplier. Q1 is not mounted on the heat sink and has no role in temperature compensation of the output transistors.

The bias spreader is designed for an output Triple and produces a nominal spread of about 4.2 V. The tracking diodes introduce the temperature effects of the output transistors, while the V_{be} multiplier transistor takes care of controlling that part of the bias spread necessary for the pre-driver and driver transistors. Sensitivity to the tracking diode $TC(S_{TTD})$ is unity. This bias spreader likely has too little R_{TC} to compensate the output transistors because TC_{TTD} is only about 80% of TC_{Vbe} for the output transistors. TC_{spread} for this bias spreader is only -3.4 mV/°C, while the output pair requires -4.2 mV/°C.

Figure 17.18b is a bias spreader that represents the opposite extreme. It encloses the two tracking diodes inside the V_{be} multiplier loop. As a result, TC_{TTD} is multiplied. The overall multiplier ratio in the spreader of Figure 17.18b is about 2.1; this is what is required to obtain the nominal spread of 4.2 V. As a result, TC_{spread} is about 7.1 mV/°C. S_{TTD} is about 2.1. This is more compensation than needed.

An intermediate solution is obviously needed. The spreader of Figure 17.18c is a tempting choice, but it still yields an estimated TC_{spread} of 6 mV/°C. The temperature coefficient is larger

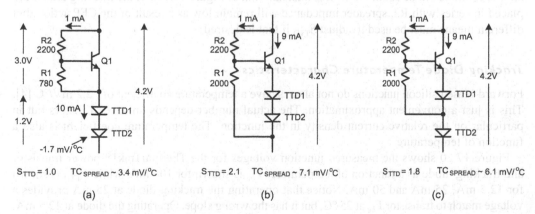

Figure 17.18 Three Bias Spreaders for ThermalTrak™ Output Transistors

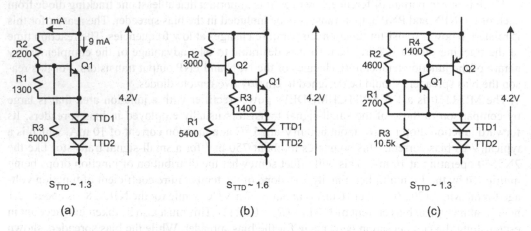

Figure 17.19 Bias Spreaders with Selectable Temperature Compensation

than one might expect because the multiplier ratio has increased, enhancing the influence of TTD1. A more significant concern is that the compensating influence of TTD1 and TTD2 is no longer equal; this degrades the balance brought by using one tracking diode from each of the top and bottom output transistors.

The bias spreader arrangement of Figure 17.19a is equally sensitive to TTD1 and TTD2 while providing a selectable value for TC_{spread}. R1 and R3 control the proportion of temperature coefficient introduced by the output transistor temperature. A larger ratio of R1/R3 provides increased S_{TTD}. With R1 = 1.3 kΩ and R3 = 5.0 kΩ, S_{TTD} is about 1.3, yielding TC_{spread} of about 4.4 mV/°C. This is just slightly more than the −4.2 mV/°C required for the output stage. However, slight overcompensation is often preferred.

Many other bias spreader arrangements can also be used with the ThermalTrak™ transistors. Figure 17.19b shows a Darlington bias spreader that yields a fixed TC_{spread} of about −5.3 mV/°C. It can be adjusted downward by adding a resistor (not shown) between the base of Q2 and the anode of TTD1 in similar fashion to what was done in Figure 17.19a.

Figure 17.19c shows a CFP bias spreader that performs analogously to the spreader of Figure 17.19a. Spreader impedance is reduced, but not by as much as with a conventional CFP bias spreader as described in Section 17.5. This is because the tracking diodes are outside the CFP loop in order to have most of the 10 mA VAS bias current flowing through them. Spreader impedance is estimated to be about 7.3 Ω, less than half that of Figure 17.19a. If the tracking diodes are placed in series with R2, spreader impedance will remain low as a result of the CFP action, but different means must be used to adjust S_{TTD} if that is desired.

Tracking Diode Temperature Characteristics

Forward-biased silicon junctions do not always have a temperature coefficient of −2.2 mV/°C [4]. This is just a convenient approximation. The actual number depends on several factors but in particular on the relative current density in the junction. The temperature coefficient is also a function of temperature.

Figure 17.20 shows the measured junction voltages for the ThermalTrak™ power transistor and tracking diode as a function of temperature. V_{be} is shown for 100 mA, while V_{TTD} is shown for 12.5 mA, 25 mA and 50 mA. Notice that operating the tracking diode at 25 mA provides a voltage match to transistor V_{be} at 25 °C, but it has the wrong slope. Operating the diode at 12.5 mA, near the typical VAS bias current level, results in reduced V_{TTD} and about the same slope as at

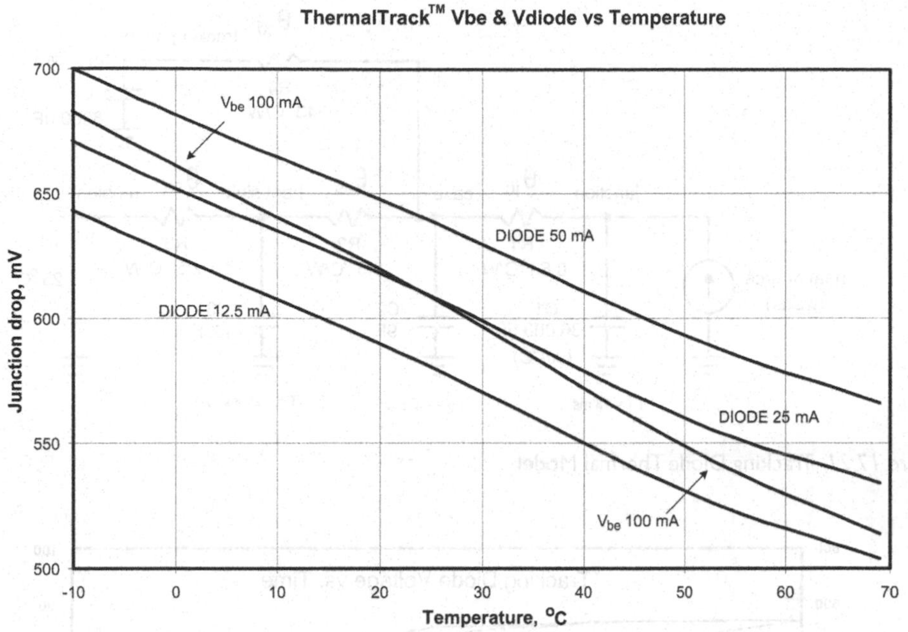

ThermalTrack™ Vbe & Vdiode vs Temperature

Figure 17.20 V_{be} and Tracking Diode Drop Versus Temperature for Different Tracking Diode Current

25 mA. There is really no advantage to operating the diode at higher current as long as the slight difference in bias voltage can be made up.

Thermal Model

Figure 17.21 shows a thermal model for the ThermalTrak™ transistor with emphasis on the action of the tracking diode. This model was arrived at by measurement of a ThermalTrak™ transistor under several different conditions, combined with SPICE simulation of the model. The current represents the heat source in watts while the R-C ladder represents the path of heat flow from the source to the heat sink, with voltage representing temperature above ambient temperature in degrees Celsius. R1 represents θ_{jc}. Here it is 0.63 °C/W, representing the 200-W power transistor in a TO-264 package and having a maximum junction operating temperature of 150 °C. Shunt capacitor C1 represents the thermal mass of the die itself. The capacitance of 36,000 μF has units of microjoules per degree Celsius. C2, at 9 F, represents the thermal mass of the copper header. R2 represents insulator thermal resistance $\theta_{cs} = 0.5$ °C/W. R3 corresponds to this transistor's share of a heat sink with $\theta_{sa} = 0.65$ °C/W. This transistor is assumed to be part of the 150-W amplifier described earlier in which there were two output pairs. Thus, θ_{sa} for this model is 2.6 °C/W. C3 = 480 F is this transistor's share of the heat sink thermal mass.

The tracking diode is affixed to the copper header with insulating epoxy that, combined with the smaller area of the tracking diode, results in thermal resistance $R4 = 43$ °C/W from header to diode. At the same time, the thermal mass of the diode, represented by C4, is only 8000 μF.

Tracking Diode Response Time

The tracking diode in a ThermalTrak™ transistor is mounted on the same heat spreader (header) as the die of the power transistor, putting it in intimate thermal contact with the power transistor. There is still thermal attenuation and thermal delay in this arrangement, but it is far less than

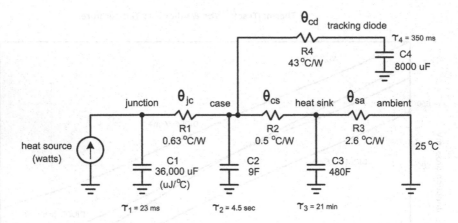

Figure 17.21 Tracking Diode Thermal Model

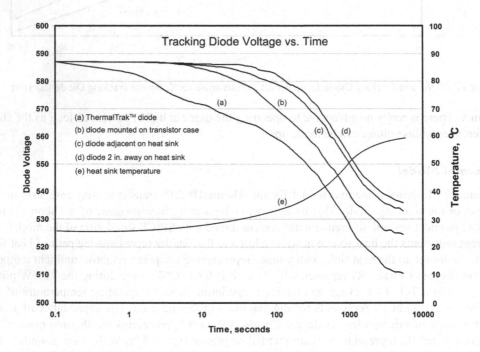

Figure 17.22 Measured Diode Thermal Response Time

that encountered by conventional temperature compensation arrangements employing tracking diodes external to the power transistor. The thermal time constant of the tracking diode is $\tau_4 = 350$ ms. It is useful to note that the thermal time constant for the transistor with respect to the header is $\tau_1 = 23$ ms. Both of these time constants are fairly small compared to the time constant of the copper header with respect to the heat sink, $\tau_2 = 4.5$ seconds. The time constant for the heat sink is $\tau_3 = 1250$ seconds, or about 21 minutes.

Figure 17.22 illustrates measured diode thermal response time for four diodes used with the same ThermalTrak™ power transistor. The first diode is the internal ThermalTrak™ diode. The second one is affixed to the transistor package surface with thermal grease and some mechanical pressure. The third diode is mounted on the heat sink right at the power transistor, and the fourth diode is attached to the heat sink 2 inches away from the power transistor. All four diodes were run at 10 mA. The three external diodes were ThermalTrak™ diodes in other ThermalTrak™ transistors whose BJT devices were left unconnected. The power transistor was then operated at 25 W.

The plot in Figure 17.22 shows the junction voltage of each diode as a function of logarithmic time in seconds. The ThermalTrak™ sense diode reacted much more quickly than the external diodes. The ThermalTrak™ diode voltage dropped 5 mV in 1.5 seconds. The package-mounted diode took 27 seconds to drop by 5 mV. The diode mounted on the heat sink directly adjacent to the power transistor required 50 seconds to drop by 5 mV. The diode mounted 2 inches from the power transistor required 105 seconds to drop by 5 mV. The curve in Figure 17.22e shows the heat sink temperature as measured close to the power transistor.

Thermal Attenuation

Thermal attenuation can also be seen in Figure 17.22 by looking at the final value of junction drop for the four diodes. The actual junction of the ThermalTrak™ transistor is still thermally separated from the header by the junction-to-case thermal resistance of the device, so there is still some thermal attenuation from the junction to the internal sensing diode. However, this attenuation is much smaller than that for the external sensing diodes. Notice that beyond about 200 seconds all four diodes generally track the changes in heat sink temperature. The difference in the near-final values at 90 minutes illustrates the effect of thermal attenuation. Relative to the ThermalTrak™ diode, the external diodes exhibit thermal attenuation factors of 0.82, 0.71 and 0.67.

Compensation of Pre-Driver and Driver

It is always important to bear in mind that not all V_{be} drops in an output stage are created equal. Those in the output transistors are subject to the greatest thermal variations, while those in the pre-driver and driver are subject to rather small temperature variations once they warm up and reach a stable temperature. This means that they should not be temperature compensated in the same way as the output transistors, if at all. A significant choice to be made is how to provide heat sinking for the drivers, which will each dissipate 2.5 W to 4 W when biased at 50 mA. The drivers can be mounted to the main heat sink or they can include their own heat sink. The pre-drivers usually do not require heat sinks, although in some situations it may be advantageous to mount them on the main heat sink as well.

Any driver or pre-driver transistors that are mounted on the heat sink should be temperature-compensated by a V_{be} multiplier (or portion thereof) that is also mounted on the heat sink. In principle, if both drivers and pre-drivers are mounted on the heat sink, Q1 of the bias spreaders of Figures 17.18 and 17.19 should also be mounted on the heat sink. Alternatively, the five devices can be mounted on a separate isothermal bar on the circuit board that serves as a heat sink.

Bias as a Function of Time

The effectiveness of temperature compensation using the ThermalTrak™ transistors was evaluated by measuring V_q as a function of time after an amplifier using the devices was allowed to cool down with no program material after having been run at high dissipation for 10 minutes.

The amplifier used for these tests employed an output Triple and two ThermalTrak™ output pair with nominal 55 V power rails. The amplifier included two bias spreaders, one of which was selected for each test. The conventional design used a split bias spreader like that in Figure 17.11b, with one transistor mounted to the heat sink and the other transistor mounted to a common heat spreader bar to which the pre-drivers and drivers were also mounted. The ThermalTrak™ bias spreader was implemented with the circuit of Figure 17.19a, with R1 = 820 Ω, R2 = 1 kΩ, R3 = 1.2 kΩ plus a 500 Ω trimmer potentiometer. The single transistor in this bias spreader was mounted on the common heat spreader bar used for the pre-drivers and drivers.

Component values for proper bias compensation were determined by measuring V_q as the temperature of the heat sink was raised and lowered. Compensation was adjusted so that the same value of V_q was obtained when the heat sink temperature was 60 degrees Celsius as when it was at room temperature. An LM35 Centigrade temperature sensor was mounted to the heat sink to measure heat sink temperature. The temperature of the heat sink was elevated by applying power to a pair of 50-W resistors mounted to the heat sink. A fan was used to cool down the heat sink in a timely fashion.

A dynamic bias stability test of the amplifier was then conducted. The output stage bias was adjusted for $V_q = 26$ mV in the quiescent state. The amplifier was then operated at 50 W (near 1/3 maximum power) for an extended period of time with the fan speed adjusted to obtain a final heat sink temperature of 60 °C. The input signal was then removed and V_q was measured as a function of time. Figure 17.23 shows the results of this test for both the conventional and ThermalTrak™ bias spreaders. It is quite apparent that during high-dissipation signal intervals the amplifier is seriously over-biased, with V_q in excess of 41 mV. This is due to transistor junction and package heating that is not taken into account by the bias spreader transistor that is mounted on the heat sink. This behavior is evidenced by the fairly rapid decay of the over-bias condition after the signal is removed. However, notice that it has a fairly long tail, crossing through 26 mV at about 9 seconds and then exhibiting some undershoot.

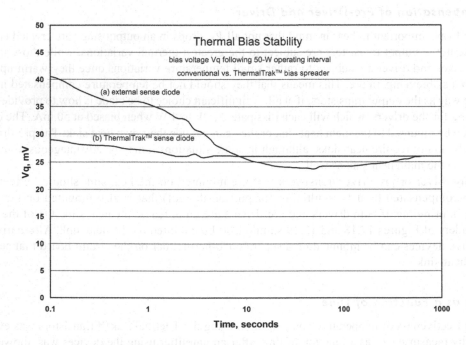

Figure 17.23 Bias (V_q) as a Function of Time After a Thermal Step

In stark contrast, when using the ThermalTrak™ bias spreader, V_q rises to only about 29 mV during the program interval and settles to 26 mV in little more than 1 second. This is dramatic evidence of the improvement gained by use of the ThermalTrak™ transistors.

THD as a Function of Bias Setting

Crossover distortion of the output stage was then measured as a function of the initial setting of V_q, which was set when the amplifier was in thermal equilibrium and passing no signal. Crossover distortion of the output stage was measured by closing the global feedback loop from a center-tap on the driver bias resistor that is connected between the emitters of the driver transistors in the Locanthi T circuit. This technique exposes the open-loop distortion of the output stage. Static crossover distortion was measured at 1 kHz at an operating level of 4.5 W while driving an 8-Ω load. As a sanity check, the distortion was measured under no-load conditions and was less than 0.001%. When the 8-Ω load was connected, the crossover distortion appeared as expected.

The output stage bias voltage V_q was adjusted to various values from 5 mV to 30 mV under stabilized quiescent conditions. The amplifier was then operated at 4.5-W and THD-1 was measured for each bias setting. The 4.5 W operating level is in an operating power range that is sensitive to crossover distortion. These measurements were conducted using both the conventional bias spreader and the one based on the ThermalTrak™ transistors. The results when using the conventional bias spreader are shown in Figure 17.24a.

It is especially notable that the minimum amount of crossover distortion when using the conventional bias spreader occurred at a bias setting of $V_q = 9$ mV, when in theory the number should be 26 mV. This suggests that the junction and case of the power transistor heat up (even at only 4.5 W output) and effectively increase the bias current while the signal is present. This tends to agree with the observation above that amplifiers are actually operating at increased bias current under signal conditions. In this case, the distortion residual could actually be seen to decrease with time after the signal was applied.

Equally notable is the depth of the crossover distortion minimum and the fact that the distortion percentage changes over a range of 5:1 as the bias setting is changed from the optimum value to an over-bias value. Bear in mind that this is the open-loop distortion of the output stage.

The results for the same amplifier using a bias spreader based on the ThermalTrak™ technology are shown in Figure 17.24b. Here we see the same crossover distortion behavior, but the bias voltage V_q that yields the minimum value of crossover distortion is at a higher value of 14 mV. Changes in the level of the distortion residual with time after the signal was applied were still observable, but to a much lesser degree. This suggests that the temperature compensation that I used was good but not optimal.

The reduced value of V_q for minimum crossover distortion may be partly attributable to output transistor base resistance effects that cause the optimum bias to occur at a lower bias current than the expected 26 mV. This amplifier employed 2.2-Ω base stopper resistors that also contribute to this phenomenon.

ThermalTrak™ Transistors as Part of a Monitoring and Protection Scheme

If some of the ThermalTrak™ diodes are not used as part of the bias spreader, they can be used as output transistor temperature monitors. In this way, they can function as over-temperature devices that might be more accurate and responsive than thermal cutouts mounted on the heat sink. There is, of course, a higher level of circuit complexity implied. Of course, in amplifiers with multiple output pairs, all but one pair can employ the non-ThermalTrak™ devices, the MJL3281 and MJL1302.

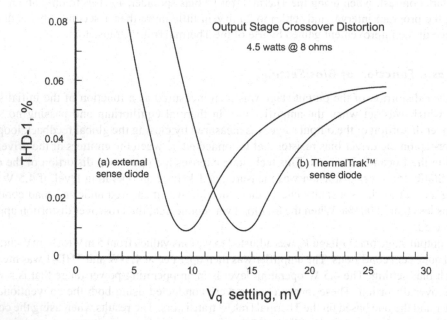

Figure 17.24 Output Stage Crossover Distortion Versus Initial Bias Setting V_q

17.9 A ThermalTrak™ Power Amplifier

The schematic of the ThermalTrak™ power amplifier built and measured for the data discussed above is shown in Figure 17.25. It is a modest no-frills design that provides good performance. It has an NPN BJT input stage loaded by a 3T current mirror. The 2T single-ended VAS operates at 10 mA and is loaded with a feedback current source. Closed loop gain is set to 19.6 by the feedback network, whose impedance is on the order of 530 Ω so as to keep noise low. Conventional Miller compensation sets ULGF to about 1 MHz via C3.

The bias spreader comprises a V_{be} multiplier with two ThermalTrak™ diodes in series with its emitter (DQ18 and DQ19 from Q18 and Q19). Q13 is mounted on a 3/4 × 3-1/2 × 1/8 inch aluminum heat spreader where the pre-driver and driver transistors are also mounted. Q15 thus provides temperature compensation for the pre-drivers and drivers. R23 helps control the proportion of temperature compensation between that for the devices on the heat spreader and that for the ThermalTrak™ output transistors.

The output stage employs two pair in a *Locanti Triple*, with 2.2-Ω base stoppers and 0.22-Ω emitter resistors. The second pair of devices, Q19 and Q20, are the non-ThermalTrak™ MJL3281 and MJL1302 devices. The design does not include the usual input LPF, output coil and short-circuit protection.

Performance

Figure 17.26 shows THD+N as a function of output power into 8 Ω and 4 Ω at 1 kHz and 20 kHz for the amplifier.

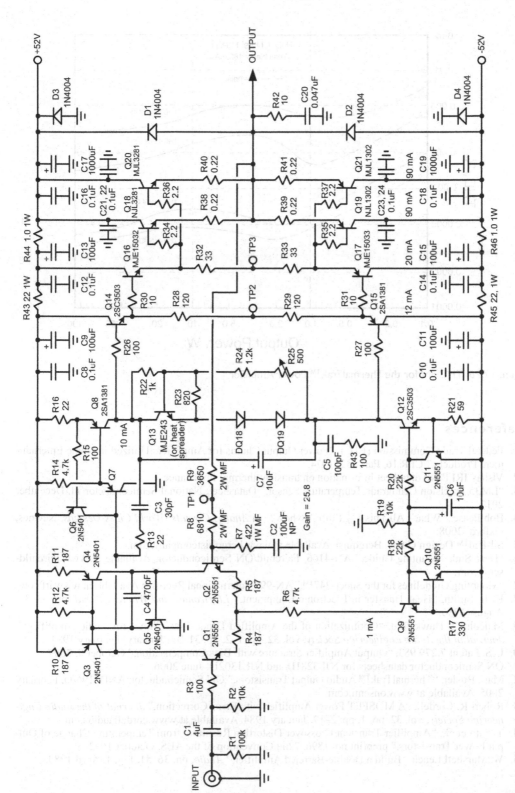

Figure 17.25 ThermalTrak™ Power Amplifier Example

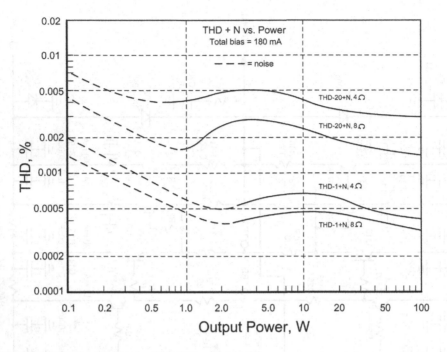

Figure 17.26 THD+N for the ThermalTrak™ Power Amplifier

References

1. Federal Trade Commission (FTC), "Power Output Claims for Amplifiers Utilized in Home Entertainment Products," CFR 16, Part 432, 1974.
2. Vishay IRFP240 datasheet information on transient thermal impedance.
3. "LM35 Precision Centigrade Temperature Sensor," Datasheet, National Semiconductor/TI, December 2017.
4. Bob Pease, "What's All This V_{BE} Stuff Anyhow?" in *Analog Circuits: World Class Designs*, Newnes, Oxford, 2008.
5. Sil-Pad™ Design Guide, Bergquist. Available at www.bergquistcompany.com.
6. "Heat Sink Mounting Guide," AN-4166, Fairchild/ON Semiconductor. Available at www.fairchild-semi.com.
7. "Mounting Guidelines for the Super-247™," AN-997, International Rectifier. Available at www.irf.com.
8. K. L. Smith, "Heat Transfer in Electronic Equipment," *Electronics and Wireless World*, pp. 33–37, August 1986.
9. Malcolm J. Hawksford, "Optimization of the Amplified Diode Bias Circuit for Audio Amplifiers," *Journal of the Audio Engineering Society*, vol. 32, no. 1–2, pp. 31–33, January–February 1984.
10. U.S. Patent 7,279,983, "Output Amplifier Structure with Bias Compensation," October 9, 2007.
11. ON Semiconductor datasheets for NJL3281D and NJL1302D, June 2006.
12. Mark Busier, "ThermalTrak™ Audio Output Transistors," ON Semiconductor, AND8196/D, February 2005. Available at www.onsemi.com.
13. Robert R. Cordell, "A MOSFET Power Amplifier with Error Correction," *Journal of the Audio Engineering Society*, vol. 32, no. 1, pp. 2–17, January 1984. Available at www.cordellaudio.com.
14. T. Sato et al., "Amplifier Transient Crossover Distortion Resulting from Temperature Change of Output Power Transistors," preprint no. 1896, 72nd Convention of the AES, October 1982.
15. W. Marshall Leach, "Build a Double-Barreled Amplifier," *Audio*, pp. 36–51, Fig. 1, April 1980.

Chapter 18

Safe Area and Short-Circuit Protection

18. INTRODUCTION

Amplifier output stages are subject to abuse from the outside world that lies beyond the speaker connectors. An obvious source of abuse is a short circuit. A less obvious one is a loudspeaker load that creates dangerous combinations of voltage and current in the output stage.

Output transistors can be destroyed by overheating or by sudden high currents when more than a certain amount of voltage is across them. There are three things to be concerned about: 1) long-term power dissipation and average junction temperature, 2) transient power dissipation and peak junction temperature and 3) secondary breakdown.

It is the job of an amplifier's protection circuits to prevent any kind of abuse from destroying the power transistors or any other part of the output stage, such as the driver transistors. Just as importantly, the protection circuit must prevent the expensive loudspeakers from being damaged by amplifier misbehavior. Often when an output stage fails, it will drive the output to a high DC voltage because the failure mechanism of output transistors usually results in a short circuit from collector to emitter.

This chapter provides an overview of protection issues, but not an in-depth treatment, especially with respect to protection circuit design details. Many other texts do a good job of treating protection circuits in greater depth [1, 2]. The BC-1 example amplifier of Chapter 4 includes a complete set of protection circuits that are described in detail.

18.1 Power Transistor Safe Operating Area

The safe operating area for a power transistor is one of the most important specifications for protection circuit design and sizing of the output stage. Figure 18.1 shows a typical BJT power transistor safe operating area diagram. The device here is the Fairchild 2SC5200 (AKA FJL4315) [3]. This is a 15-A, 230-V, 150-W power transistor in a TO-264 plastic package. The rated maximum junction temperature for this transistor is 150 °C. The horizontal voltage and vertical current axes of the SOA diagram are logarithmic, so a constant power boundary is a straight line. The DC SOA for the device is shown as the innermost boundary. Other boundaries are shown for 100 ms and 10 ms pulses.

The DC SOA line is the boundary of voltage and current conditions that the device can tolerate indefinitely. At low DC voltages the safe area is limited by the maximum rated current of 15 A. At V_{ce} above 10 V, the safe area is bounded with a sloped line that corresponds to 150-W power dissipation. If the case of the transistor is held at 25 °C, this line defines the voltage and current combinations that will keep the junction below its maximum rated temperature of 150 °C. At 60 V there is a break in the slope of the line where allowed operating current begins to fall off more steeply. In this region the maximum power dissipation starts to become smaller with increases

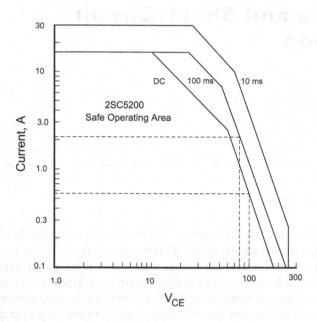

Figure 18.1 Safe Operating Area of the Fairchild 2SC5200

in voltage. This is where device capability is limited by secondary breakdown. The sustainable power dissipation is only 9 W ($I_c = 40$ mA) at the maximum voltage rating of the device (off the graph). More importantly, the device can dissipate about 55 W at $V_{ce} = 100$ V (550 mA). However, for 100 ms, the device can dissipate 2.1 A at 80 V, or 168 W. This shows how important the duration of an overload is.

The SOA boundary can also be plotted on linear coordinates, as shown in Figure 18.2. The linear presentation is more useful when plotting load lines to see if the SOA boundary is violated.

Yet another way to illustrate the SOA boundary is to plot allowable power dissipation as a function of V_{ce}. This will be shown in Figure 18.8. This illustrates the view of secondary breakdown corresponding to reduced power dissipation at high V_{ce}.

Secondary Breakdown Mechanism

The allowable power dissipation for a transistor decreases at higher V_{ce} because of hot spots that develop in the transistor structure. At high voltages the base thins because of the increased size of the base-collector depletion region. The thinner base has higher resistivity, allowing voltage drops across the base to influence the distribution of emitter current. This causes *emitter crowding*, in which most of the current conduction moves to the edges of the emitter, where there is less base voltage drop due to base resistivity. *Current hogging* is the result, causing those areas to have higher local power dissipation. Transistors draw more current as they get hotter, so these hot spots dissipate still more power in what leads to a regenerative process that can progress very quickly and lead to destruction of the device.

Temperature De-Rating of SOA

The curves in Figure 18.1 define the SOA when the transistor case is held at 25 °C. This is almost never the case, so these curves must be de-rated in consideration of the maximum anticipated operating case temperature. The de-rating of the constant-power portion of the SOA is straight-forward. As explained in Chapter 17, a simple thermal analysis to keep the junction below its maximum rated temperature is all that is needed. Given the rated junction operating temperature of 150 °C, the power dissipation is de-rated to about 64% (96 W) when the case is at 70 °C. It is notable that the 150 °C maximum junction temperature is due to a plastic packaging and reli-ability limitation. The MJ15024 in a metal TO-3 package, for example, sports a 200 °C maximum operating junction temperature.

De-rating of the second breakdown region is not as straightforward. Because the secondary breakdown failure mechanism is different from that for power dissipation, the way in which it is de-rated for higher case temperature is not the same. Fortunately, the appropriate de-rating is usually less severe. In some cases, it is de-rated as if the peak allowable junction temperature is much higher, like 250 °C. This makes a big difference for plastic devices where the maximum specified junction temperature may only be 150 °C. Using 250 °C, if the case temperature rose to 70 °C, then the second-breakdown de-rating factor would be $(250 - 70)/(250 - 25) = 80\%$ instead of $(150 - 70)/(150 - 25) = 64\%$.

Transient SOA

SOA curves like those in Figure 18.1 illustrate that the device can exhibit greater SOA for brief intervals. This conforms to the model of keeping peak junction temperature below a certain point, taking into account the thermal inertia of the device die. Sometimes this effect is described by what is called the *transient thermal impedance* (TTI) discussed in Chapter 17. For some transis-tors power dissipation is doubled for a 100 ms pulse and doubled again for a 10 ms pulse. This means that a 150-W device could withstand 600 W for 10 ms.

Similarly, allowed second-breakdown SOA may be greater by factors of 1.7 at 100 ms and 3–5 at 10 ms. At $V_{ce} = 100$ V, the Fairchild 2SC5200 can withstand 0.55 A at DC, 0.9 A for 100 ms and 3.1 A for 10 ms [3].

The increased SOA capabilities for short pulses like 100 ms and 10 ms are important for audio amplifier applications because the instantaneous peak power caused by a 20-Hz sinusoid, for example, will last for much less than the 25-ms half-cycle time when one polarity of output tran-sistor is conducting. This means that brief excursions outside the conservative DC SOA boundary can be tolerated.

Long-Term Reliability and Destruct Point

DC SOA is about die temperature and long-term reliability, not the device destruct point. Ignor-ing secondary breakdown for the moment, the typical 150 °C junction temperature limit for a plastic-packaged power device is set by long-term reliability criteria, perhaps something like 1% failure in 10 years of continuous duty at the limit. Indeed, this is often a package consideration, not a device limit. Heat is the enemy of reliability. For this reason, it is clear that margin must exist between the rated junction temperature and the destruct junction temperature—quite a bit of it, in fact. Bear in mind that a linear voltage regulator could be operating continuously on any part of the SOA boundary with some amount of reliability for years (assuming that it is properly de-rated for temperature).

Managing Risk

The discussions above show that designing an output stage and protection to assure that the load conditions never cause the de-rated DC SOA boundary to be violated is quite conservative in respect to the possibility of device destruction. Nevertheless, it is better to be safe than sorry, so it is recommended to err on the safe side when designing the protection circuits. The price paid for this will be a slightly higher probability that protection will be triggered. For those designers who opt for little or no protection, the discussions above can help lend insight to the nature of the risk involved.

18.2 Output Stage Safe Operating Area

In this section we discuss how different loads may cause the operating conditions of the output transistors to approach or even violate the SOA boundaries. This sets the stage for proper sizing of the output stage and the design of appropriate protection circuits.

Resistive Loads

The most fundamental and optimistic calculation of required safe area is for a resistive load. A resistive load line is plotted as a straight line on linear V-I coordinates. A typical plot for a 4-Ω load plotted against the linear version of the SOA curve is shown in Figure 18.2. This represents an ideal 100-W/8-Ω amplifier with 40-V rails driving a 4-Ω load with a single 2SC5200 on the high side of the output stage. The dotted portion of the curve at higher V_{ce} is an extension of the constant-power curve. It helps to illustrate the allowable operating region lost due to secondary breakdown.

In a class AB output stage with a resistive load the transistor current is greatest at maximum signal swing where the output voltage approaches the power supply rail. This is where the voltage across the transistor is conveniently lowest. By the same token, when the output is at its zero

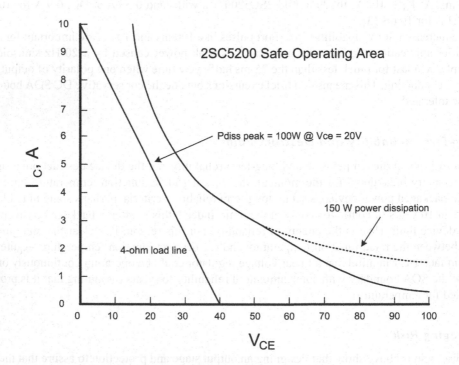

Figure 18.2 A 4-Ω Load Line Plotted with a Safe Area Curve

voltage crossing and more negative, current through the upper transistor is very low or zero. This is why resistive loads are fairly benign in terms of stressing the SOA boundaries of the transistor.

The instantaneous output transistor power dissipation in an ideal amplifier when feeding a resistive load is at its maximum when driving the load to half the rail voltage. That maximum is one-half the full average power capability into the load. A 100-W/8-Ω amplifier driving 40 V peak into a 4-Ω load (200 W full power) will dissipate a peak instantaneous power of 100 W when the load voltage is 20 V.

Reactive Loads

Loudspeakers present reactive loads, and these can be far more taxing of the SOA than resistive loads [4, 5]. Consider a pure 8-Ω capacitive load driven by a 40-V sinusoid, as shown in Figure 18.3. Such reactive load impedance is designated as j8 Ω. The current leads the voltage by 90°. The load line is a circle. In this case, the maximum load current occurs when the output voltage is zero and the output transistor has the full rail voltage across it. Instantaneous dissipation at this point is 200 W. This severely stresses the SOA boundary of the device. Notice that there is significant positive current flow even when the output is negative and the transistor has more than the rail voltage across it. The same holds true for a purely inductive load. The current lags the voltage by 90°, again forming a circular load line. A resistive 8-Ω load line has also been plotted as a straight line for reference.

Most real loudspeaker loads are not purely reactive, and virtually always have some effective resistance in series with the reactance. Often, this resistance will be the DC voice coil resistance of the woofer, referred to as R_e. This resistance is often on the order of 75% of the rated impedance of a loudspeaker (if it is rated honestly). In these cases an elliptical load line results. Such an elliptical load line with an impedance of 5.7 Ω plus j5.7 Ω is also shown in Figure 18.3. This load has a phase angle of 45° and a modulus of 8 Ω.

If only the positive-current portion of the load line is plotted as a function of $V_{ce} = V_{rail} - V_{out}$, the V-I combination can be plotted on linear coordinates along with the SOA curve, as shown in Figure 18.4. Here V_{rail} has been set to 45 V, corresponding to a 100-W/8-Ω amplifier. The 45° elliptical load line with $|Z| = 9$ Ω has been plotted with the SOA boundary of the 2SC5200. This reactive load line corresponds to a loudspeaker with $R_e = 4$ Ω. The resistive 8-Ω load line is also plotted for reference.

Impedance and Conductance as a Function of Phase Angle

Figure 18.5 shows a simple electrical model of a woofer in a closed box. A key observation is that there is a fixed value of resistance R_e in series with the reactive elements. This resistance represents the voice coil resistance. Even multi-way loudspeaker systems with much more complex models usually have a minimum DC resistance effectively in series with the reactive elements. To first order, when the loudspeaker impedance is at its minimum, it is this resistance with a phase angle ϕ of 0° that is presented to the amplifier. A good example of this is a vented loudspeaker system, where the resistance drops no lower than R_e at the tuning frequency of the port. What this means is that when the impedance of the loudspeaker becomes reactive and has a non-zero phase angle, it becomes that way by the addition of reactive impedance, increasing the magnitude of the impedance of the load.

The magnitude of the impedance is referred to as its modulus, designated by $|Z|$. Similarly its inverse, the magnitude of its admittance, is designated by $|Y|$. As the phase angle ϕ increases, $|Z|$ must necessarily increase and $|Y|$ must decrease [4]. This reality is easily apparent by looking at the impedance curve for a woofer. Such a curve is shown later in Figure 22.2. A woofer with $R_e = 6.4$ Ω cannot

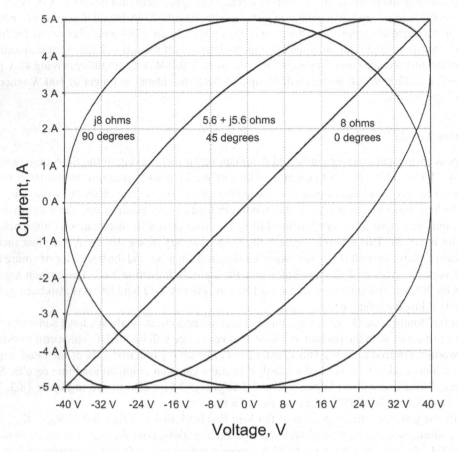

Figure 18.3 Circular Load Line of a Purely Reactive 8-Ω Load and Elliptical Load Line of a Partially Reactive 8-Ω Load

have |Z| = 6.4 Ω with a non-zero phase angle. As |Z| increases from its minimum, |φ| increases toward 90°, but the current flow decreases, mitigating the effect of the non-zero |φ|. An amplifier designed for a minimum load impedance of 4 Ω need not be designed to handle a load with a modulus of 4 Ω and a phase angle of 60°, for example.

As a simple example, consider a right triangle where the X axis is the real component of impedance and the Y axis is the imaginary component. The length of the hypotenuse represents |Z|. If the real component is 1 Ω and the imaginary component is 2 Ω, |Z| = √5 = 2.2 Ω and |φ| = 63°. A 4-Ω loudspeaker with R_e = 3.2 Ω and a 60° angle will have |Z| = 6.4 Ω.

Figure 18.6 shows the impedance plot for |φ| = 0°, 60° and 85° when R_e is held constant at 6.4 Ω. This is analogous to Figure 18.3, but here R_e rather than |Z| has been held constant. Notice how the size of the curve becomes smaller and tilts away from the 45° resistive line as |φ| becomes larger.

Overlapped Elliptical Load Lines

An audio power amplifier must be able to safely drive many combinations of load impedance and phase angle. It turns out, however, that if a large number of elliptical load lines with different

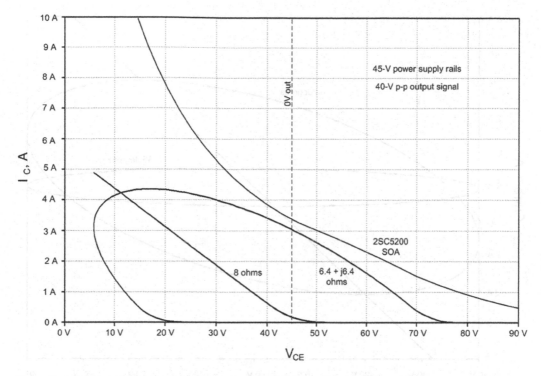

Figure 18.4 Positive-Current Portion of the Reactive Load Line Plotted Against V_{ce}

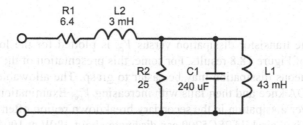

Figure 18.5 Electrical Model of a Woofer in a Sealed Enclosure

phase angles but the same minimum impedance R_e are overlapped, the maximum SOA boundary of the collection of load lines is a straight line with an equivalent resistive load line value of $2R_e$, [1] as shown in Figure 18.7. In contrast to the resistive load line, this virtual load line extends to $2V_{rail}$ (which is why its equivalent resistance is twice R_e). We can thus observe that the composite *V-I* locus for a reactive load with series resistance R_e is bounded by a resistive load line that extends from 0 V to twice the rail voltage and whose resistance is twice R_e. This is a valuable observation because it allows the evaluation of the general case without resort to every combination of reactive elements.

The above data can also be displayed on a power versus V_{ce} plot. Recall that the SOA for a transistor can be plotted this way. The constant-power portion of the SOA boundary will be a horizontal straight line. That line will then fall toward 0 for V_{ce} above the secondary breakdown

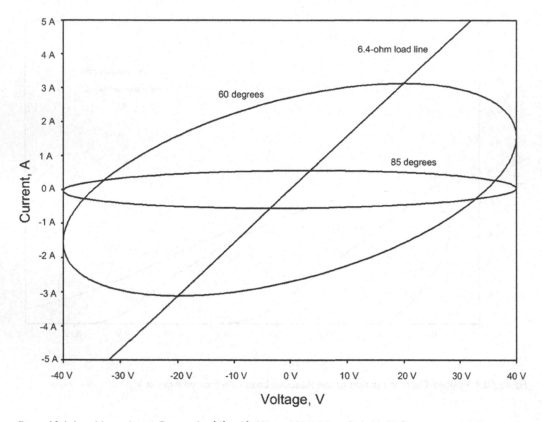

Figure 18.6 Load Impedance Curves for $|\phi|$ = 0°, 60° and 85° When R_e Is Held Constant at 6.4 Ω

knee voltage. If the transistor dissipation versus V_{ce} is plotted for the loads used in Figure 18.7, the diagram of Figure 18.8 results. For some, this presentation of the SOA boundary and transistor instantaneous dissipation may be easier to grasp. The allowable power dissipation is flat out to the SOA knee and then falls with increasing V_{ce}. Examination of the plot reveals that allowable power dissipation in the secondary breakdown region often falls off roughly as the square of V_{ce}. The Fairchild 2SC5200 can dissipate about 48 W at 100 V, but only 12 W at 200 V. The specified rate of falloff varies somewhat by manufacturer, but this behavior is not unusual.

18.3 Short-Circuit Protection

Short circuits happen. Without some sort of protection, enormous output currents can flow into a short circuit. The low output impedance of the amplifier will cause large currents to flow even when there is no signal present, since the feedback will be removed by the short, and the open-loop gain of the amplifier will magnify even a small offset in the amplifier's forward path. If a short circuit occurs, each output transistor will have the full rail voltage across it, assuming that the output is at 0 V because of a short to ground. The simultaneous high current and high voltage will surely violate the SOA of the output transistors.

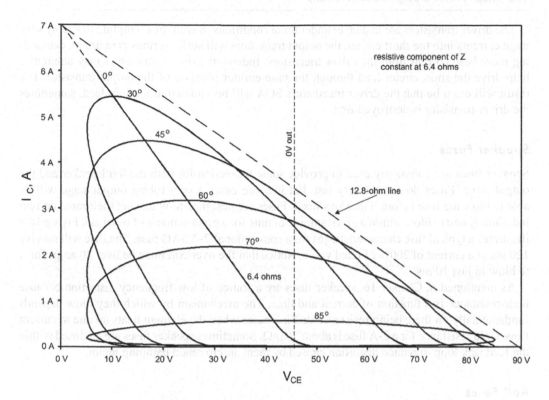

Figure 18.7 Overlapped Elliptical Load Lines with ϕ = 0°, 30°, 45°, 60°, 70° and 85° When R_e Is Always 6.4 Ω

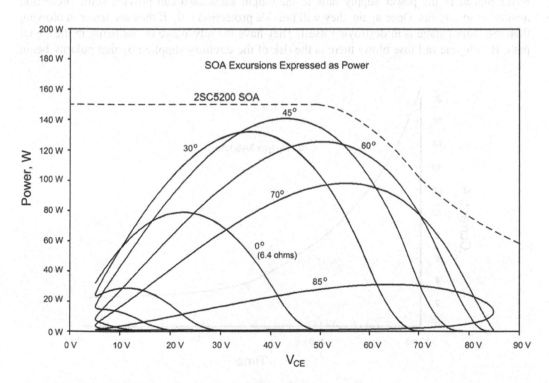

Figure 18.8 SOA and Transistor Stress Plotted in Terms of Power

The driver transistors are in danger under these conditions as well. In attempting to supply very large currents into the short circuit, the output transistors will suffer serious beta droop, demanding more base current from the driver transistors. Indeed, the driver transistors may attempt to help drive the short circuit load through the base-emitter junction of the output transistor. The result will often be that the driver transistor's SOA will be violated as well. Indeed, sometimes the driver transistor is destroyed first.

Speaker Fuses

Speaker fuses are commonly used to provide some protection for both the loudspeaker and the output stage. Fuses do not act very fast, but in some cases a very robust output stage will be able to blow the fuse before it destroys itself. Fuses are designed to withstand their rated current indefinitely and to blow within a given amount of time for a given amount of overload. Figure 18.9 illustrates a typical fuse characteristic [6] for a conventional 2-A 3AG fuse. The fuse will survive 120 ms at a current of 200% of rated value. Notice that the overload must be over 20 amps for it to blow in just 10 ms.

As mentioned in Chapter 16, speaker fuses are a source of low-frequency distortion because their resistance is a function of current and time. The mechanism by which they blow depends fundamentally on there being resistance that increases when the element heats up due to current flow. Cold resistance for a 2-A fuse is about 70 mΩ. Sometimes speaker fuses are enclosed within the feedback loop to reduce distortion caused by them and to retain damping factor.

Rail Fuses

Fuses placed in the power supply rails to the output stage also can provide some protection against short circuits. Once again, they will provide protection only if they are faster in blowing than the output stage is in destroying itself. They have the advantage of not being in the signal path. If only one rail fuse blows there is the risk of the circuitry supplied by that polarity being

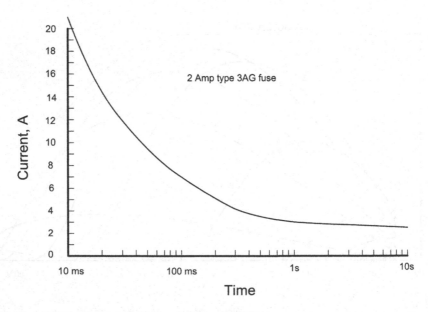

Figure 18.9 Typical Fuse Characteristic for a 2-A Type 3AG Fuse

reverse-biased and possibly suffering damage. For that reason, each rail should have a reverse protection diode to ground.

Current Limiting

A common approach to short-circuit protection is current limiting of the output stage. If the peak current can be limited to a reasonable value, this may allow time for a fuse to blow or for a protection relay to be opened. A 100-W/8-Ω amplifier must be able to deliver 10 A peak into a 4-Ω load. On the other hand, 40-V rails across the output stage with 10 A in the event of a short circuit will create 400 W of power dissipation. This underlines the fact that simple current limiting by itself is not enough.

In some cases the action of current limiters can be delayed by the incorporation of an R-C time constant to prevent the activation of the circuit on very short high-current bursts, recognizing the ability of the power transistors to survive brief overloads. The selection of the amount of delay can be a delicate balancing act. The reaction time must be long enough to provide the output current needed by legitimate loads at high power levels at low frequencies. This should include continuous sine wave testing at 20 Hz into the lowest-rated load impedance.

A Simple Current Limiter

Active current limiting employs a transistor that turns on to rob base drive current from the output transistor when the output current exceeds the threshold. An example of this is shown in Figure 18.10, where the top half of a Darlington output stage is shown with a current-limiting circuit set to a limit of about 5 A by the voltage divider formed by R1 and R2. If the voltage across the emitter resistor R_E exceeds 1.1 V, Q1 turns on and robs base drive from the output transistor by shunting the node from the VAS to the output node. This action turns the output stage into a current source, eliminating its damping influence on the loudspeaker and possibly allowing the energy stored in the loudspeaker to create a flyback pulse. This makes a terrible sound and can sometimes damage a tweeter.

The current-limiting circuit design presumes that the VAS is current limited in some way. D1 protects Q1 from negative voltages that can be present when the bottom half of the output stage is conducting. C1 slows the circuit down by introducing a 50-ms time constant to permit higher current for brief intervals. The current limit and time constant chosen here are just illustrative.

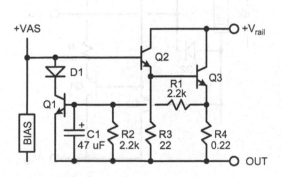

Figure 18.10 Simple Current-Limiting Circuit

Current-limiting circuits like these must also be designed with recognition that when active, these circuits create a negative feedback loop whose stability must be considered. In practice, this will occur only during the beginning of the turn-on of the current limiter and the turn-off of the current limiter. The presence of C1 helps reduce the likelihood of oscillations.

Natural Current Limiting

Sometimes the driver circuit can be designed so that clamping diodes limit the amount of base or gate voltage drive to a specific amount. Doing so essentially limits the amount of voltage that can be dropped across the emitter resistor, and thus the maximum amount of current that can be sourced. The *flying catch* diodes discussed in Chapter 14 accomplish this. Such an arrangement is shown in Figure 18.11.

The Darlington output stage shown requires a bias spreading voltage of about 2.6 V, of which about 1.3 V is needed to turn on Q1 and Q3 when the top half of the output stage is conducting. Current-limiting action occurs when the bottom end of the bias spreader rises to one diode drop above the output node. This will happen when high current is passing through R_{E1}. In the circuit shown, current limiting will begin when about 1.9 V appears across R_{E1}. This corresponds to a current limit of about 8.5 A, which is rather high. This current limiter is difficult to delay, but is simple in its operation and does not involve feedback. It is suitable for fast, simple, high-limit short-circuit protection. Different driver and bias spreader arrangements can be made to yield higher or lower limiting thresholds. In some cases this circuit will have a very high current limit threshold and its main function will be to allow the use of the shutdown circuit below and to limit catastrophic damage to the output stage in certain failure modes.

Shutdown Circuits

Some short-circuit protection circuits act by electronically shutting down the output stage when the output current exceeds a certain value. This value of current can be made to be a function of

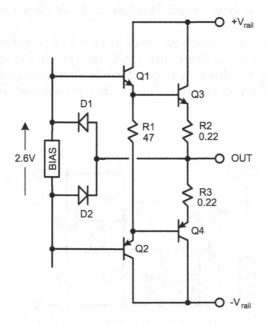

Figure 18.11 Natural Current Limiting with Flying Catch Diodes

the output voltage, and the action of this circuit can be delayed by the use of an R-C time constant. Such a shutdown circuit is usually configured to latch and stay in the shutdown mode until the mains power is cycled. This protection behavior is usually undesirable in professional sound reinforcement applications, however.

Figure 18.12 shows such a circuit that operates in conjunction with the natural current-limiting circuit shown above. It operates by collapsing the bias spreader. When a short circuit is detected, a TRIAC is fired that shorts out the bias spreader. With flying catch diodes in place, this effectively shuts down the output stage. Current from the VAS will flow harmlessly into the load through D1 or D2. A simple circuit using an optocoupler is shown for sensing high current and firing the TRIAC. If the voltage drop across either emitter resistor exceeds the forward voltage of the optocoupler LED, the circuit will fire the TRIAC. VAS bias current will keep the TRIAC on until power is cycled. This circuit can be made very fast and is a good candidate for protecting MOSFET output stages.

Other similar approaches to electronic shutdown can be employed as well. For example, a transistor which can short out the bias spreader and which can be driven by an optocoupler can be used in place of the TRIAC. Such a shunting transistor can also be employed as part of a *retry* circuit to turn off the TRIAC after a delay period or when an output impedance sensor has indicated that the short has cleared. A simple relay can also be used to collapse the bias spreader.

Speaker Relays

A relay in series with the speaker line can be used to provide short-circuit protection if it is driven by a circuit that senses an over-current condition. Relays are not always adequately fast, but they are often faster than fuses. Speaker relays are often incorporated into the overall protection scheme for other reasons as well, including protecting the loudspeaker from DC offset at the output of a failed amplifier and for power-on/power-off muting. These functions will be discussed in Section 18.8.

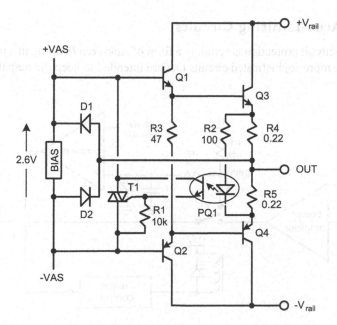

Figure 18.12 A TRIAC Disables the Output Stage by Collapsing the Bias Spreader

Load-Sensing Circuits

In more sophisticated amplifiers, circuitry is sometimes employed to electronically sense the condition of the load circuit for a short circuit during intervals when the speaker relay is open or when the output stage has been electronically shut down. Such circuitry can permit the protection circuits to reset once the short circuit in the load has been removed. The circuit can also be used to prevent the closing of the speaker relay at turn-on unless the load is deemed safe. Such circuits can be as simple as monitoring the DC output voltage across the load when a small direct current is passed through the load. Such a simple circuit obviously protects only against a DC short circuit or a load that has unreasonably low DC resistance (such as an excessive number of speakers connected in parallel).

Figure 18.13 shows a simple load-sensing circuit of this kind implemented in combination with a speaker relay. A current of 10 mA is applied by R1 to the loudspeaker through the normally closed contacts when the relay is open. The DC voltage will be 10 mV for every ohm of loudspeaker DC resistance (DCR). If the amplifier is designed to require no less than 2.5 Ω of DCR, the threshold for comparator U1 is set to 25 mV. The sensed voltage is filtered by R2 and C1 to reduce the effects of noise. D1 prevents the sense voltage from going above one junction drop when the relay is closed during normal amplifier operation. When the output of the comparator is positive it allows the relay control circuit to complete its speaker relay closure sequence. If the amplifier goes into a fault mode that causes K1 to be opened, the output of the load-sensing circuit can be used as one of a number of criteria to allow the amplifier to come back into operation and connect the loudspeaker.

One shortcoming of this circuit as drawn is that the normally closed terminal of the relay is not connected to ground. Such a connection to ground is desirable when the relay is opened for a fault. It can protect the speaker from flyback voltages and can reduce the amount of potential damage from arcing. For this reason, back-to-back diodes should be connected to ground from the normally closed terminal of the relay. This will not interfere with the small DC voltages present for load sensing.

18.4 Safe-Area-Limiting Circuits

Although short-circuit protection is certainly a form of *safe-area limiting*, this term is more often used to describe more sophisticated circuits that are intended to keep the output stage transistors

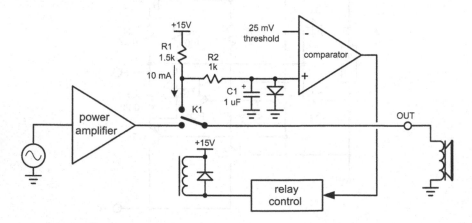

Figure 18.13 A Simple Load-Sensing Circuit

within their safe operating area during signal excursions into reactive loads with dangerous load lines. These circuits are often called *V-I* limiters because they usually limit the current to an amount that is governed by the voltage across the transistor at any given moment. These circuits often incorporate some delay so that the larger transient safe area capability of the output transistors can be utilized.

The challenge for *V-I* limiters is to follow the curved SOA boundary of the transistors as accurately as possible. This maximizes the use of the available SOA and minimizes the unnecessary activation of these circuits. *V-I* limiters have gained a reputation as being sonically intrusive. When properly designed and not being called on to act, they should not degrade the sound. However, many years ago they were designed to protect relatively undersized output stages, making them much more intrusive. With more robust output transistors and larger output stages common today, *V-I* limiters can be designed with high triggering thresholds. Of course, the newer output stages are less likely to need *V-I* limiters at all.

Single-Slope V-I Limiters

The simplest form of *V-I* limiter merely makes the current limiting threshold of Figure 18.10 smaller when V_{ce} is larger. This means that the threshold is larger for larger output voltage. When the output signal is near the power rail, this is when the highest output current is likely to be needed, and this is also when the output transistor can handle the highest amount of current and still be within its SOA. This is good for resistive loads, but such a limiter will be more likely to act in the case of a reactive load, where high currents may flow at smaller output voltages where V_{ce} is higher. Modulating the current-limiting threshold by the output voltage creates a single slope of current limit as a function of output voltage. The steepness of this slope is easily adjusted.

Figure 18.14 shows a simple single-slope *V-I* limiter circuit. It is largely the same as the current limiter circuit in Figure 18.10, with the mere addition of two resistors R5 and R6. These resistors simply inject current into the protection transistor's base circuit that is proportional to V_{ce} of the power transistor. This makes the current-limiting transistor turn on at a lower output current when there is a larger voltage across the output transistor.

While the circuit of Figure 18.10 tends to make the output stage look like a current source when current limiting occurs, the modulation of the current limit by output voltage in the *V-I* limiter of Figure 18.14 can actually create behavior more like that of a negative resistance. If an external current, as from loudspeaker EMF, pulls down on the output until the current limit is reached, the output will go negative. That will increase V_{ce} on the upper output transistor and further reduce its current limit, putting it more deeply into protection and allowing the output to get pulled down even harder by the external current. This is a form of positive feedback, and can cause the output to snap hard to the negative rail.

Figure 18.15 shows how the action of a single-slope *V-I* limiter maps onto the SOA boundary of a power transistor. It can be seen that its action is only a coarse approximation to the actual available SOA and that a lot of the available SOA of the transistor is wasted. This means that the amplifier may go into protection more often than it must for certain combinations of load impedance and phase angle.

Multi-Slope V-I Limiters

The *V-I*-limiting fit to the actual SOA boundary can be made more accurate if more than one slope is incorporated into the circuit [7]. In practice, there is a law of diminishing returns as more slope segments are added. On the other hand, the resistors, diodes and Zeners required to do so are quite

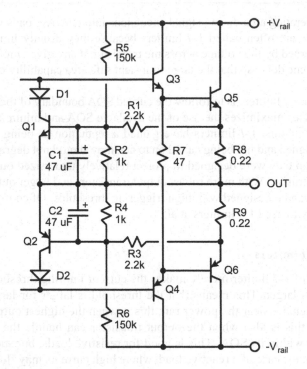

Figure 18.14 A Single-Slope *V-I* Limiter

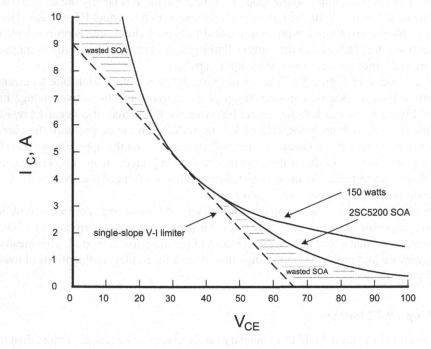

Figure 18.15 Single-Slope *V-I* Limiter Action Plotted with SOA Curve

inexpensive. The reactive load lines plotted against the SOA curve in Figure 18.8 provide guidance as to where more protection slopes can provide the most benefit. A number of other texts do a fine job of describing such V-I limiters in detail [1, 2].

Figure 18.16 illustrates a simple three-slope V-I limiter. Referring to the top half, R5 sets the nominal slope in the central region of operation, as in the single-slope design. For output signals more positive than about +15 V (where V_{ce} of Q5 is smaller), D3 turns on and R7 pulls current from the base node of Q1, increasing the current-limiting threshold. For output signals less than about −15 V, D4 turns on and R8 sources current to the base of Q1, reducing the current-limiting threshold. The breakpoint voltages shown in Figure 18.16 are conveniently derived from ±15-V supplies assumed to be available. If other breakpoint voltages are desired, they can be supplied by replacing R7 and R8 with resistive voltage dividers. Analogous circuitry is used on the bottom half of the output stage, which is not shown.

Alternatively, a pair of 15-V Zener diodes wired in series back-to-back to ground can be connected to the bases of Q1 and Q2 through 100-kΩ resistors to conduct current to/from the bases when the output voltage exceeds ±15 V.

Drawbacks of V-I Limiters

V-I limiters would be fine if they never had to act. Unfortunately, when they do act, they create terrible sounds and can cause damage to tweeters. Depending on their sophistication and the nature of the loudspeaker load, they sometimes act unnecessarily.

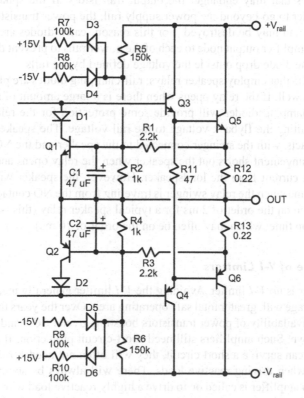

Figure 18.16 Three-Slope V-I Limiter

If *V-I* limiters only acted to clip the signal amplitude, as with ordinary clipping of an amplifier, they would not be so bad. Unfortunately, in most cases the *V-I* limiter causes the output stage to change from a voltage source to a current source (or even a negative impedance) when the *V-I* limiter engages. When this happens, there is almost surely a lot of stored energy in the loudspeaker and crossover network. This stored energy wants to cause current to flow somewhere and be dissipated. With the output stage in a current source mode of operation, this may not be possible. As a result, a large inductive spike or kick may result, often transitioning the output voltage to that of the opposite rail (i.e., in a direction opposite to that which the output stage was changing the signal).

This spike will be very audible, and its large amplitude may cause damage to the loudspeaker's tweeter. The action of a *V-I* limiter can turn an amplifier into a *tweeter eater*. The stored energy in the loudspeaker drivers and crossover will find its way to a place where it can be dissipated. The stored energy in the woofer and crossover coil(s) may be transferred to the tweeter.

As mentioned above, a *V-I* limiter can create a negative resistance effect. This happens because the current limit is a function of the voltage across the output transistor. Once the output transistor reaches its current limit, the speaker current will cause the voltage across the transistor to increase, further reducing its current limit. This results in a fast regenerative turn-off of that transistor.

Flyback Protection Diodes

The inductive flyback caused by the action of a *V-I* limiter with an inductive load can produce large voltage spikes that may endanger the output transistors. If the spike causes the output node of the amplifier to go beyond the power supply rail, the power transistor will be subjected to reverse polarity and may be destroyed. For this reason, catch diodes are almost universally installed from the amplifier output node to each of the power rails to prevent the output node from going more than one diode drop outside the voltage defined by the rails.

In some amplifiers that employ speaker relays, rail-clamping diodes are placed on the speaker side of the relay as well. If the relay opens when there is a large amount of stored energy in the loudspeaker, the clamping diodes will provide some protection for the relay contacts and the loudspeaker by limiting the flyback voltage to the rail voltage. The speaker relay should have double-throw contacts, with the swinger connected to the speaker and the NC contact connected to ground. This arrangement shorts out the speaker when the relay opens and provides a path to ground for flyback current from the loudspeaker. However, the speaker will still see an open-circuit during the time when the relay swinger is traveling from the NO contact to the NC contact. This interval is often on the order of 2 ms for a typical speaker relay (this is not the same as the total relay activation time, which may often be on the order of 10 ms).

Avoiding the Use of V-I Limiters

The best *V-I* limiter is no *V-I* limiter. Avoiding the *V-I* limiter generally means building a more expensive output stage with greater total safe operating area. Over the years this has become more practical with the availability of power transistors boasting greater SOA and a reduced tendency to second breakdown. Such amplifiers still need short-circuit protection. If these amplifiers are well designed and can survive a short circuit, they will often not need *V-I* limiters for purposes of adequate SOA when driving reactive loads. There will always be some risk, however. For example, if such an amplifier is called on to drive a highly reactive load with a very low effective series resistance of, say, 1 Ω, it might be in danger. A compromise condition is to design such an amplifier with a *V-I* limiter that is set for such high thresholds that it will only activate under conditions of extreme abuse. If executed well, such a *V-I* limiter will never activate in normal use and will be sonically transparent.

18.5 Testing Safe-Area-Limiting Circuits

Safe-area-limiting circuits can be tested by the use of large inductors and capacitors connected to the amplifier output to emulate a complex loudspeaker load, but this is not easy, especially in regard to obtaining an appropriate inductor that will not saturate under such conditions. Figure 18.5 showed a simplified electrical equivalent of an 8-Ω woofer in a closed box speaker system with a resonant frequency of 50 Hz. A 43-mH inductor and a 240-μF capacitor are required. Unlike the situation with a crossover inductor, some resistance in the inductor is not a problem and can be taken out of the 6.4-Ω series power resistor without degrading the behavior of the model. A motor starter capacitor can be used, as audio quality is not a big issue here. It is straightforward to extend this SOA load to emulate a two-way or three-way loudspeaker system.

Another possibility for SOA testing is to back-drive the amplifier under test from another larger, laboratory *load amplifier* through a load resistance. Such an arrangement is shown in Figure 18.17 [8]. If the amplifier under test is not fed with an input, its output will normally try to stay at zero, where the full rail voltage will be across each output transistor. This is often close to the most vulnerable situation encountered with realistic loads. The load amplifier can be driven with a sine wave at low frequencies that will exercise the output current at the fixed output voltage. One can then observe at what amplitude the amplifier under test goes into protection by seeing its output suddenly deviate from 0 V. Driving the load amplifier with a tone burst can further reveal the dynamics of the *V-I* limiter.

Another similar test can be carried out with two low-frequency sine waves at different frequencies. The second signal is applied to the input of the amplifier under test. As the sine wave signals beat against each other, virtually all combinations of voltage and current conditions for the amplifier will be exercised. The difference frequency of these two signals can be made large or small to exercise the SOA circuit's time constants in different ways. Finally, the load amplifier can be configured with active filter circuitry to emulate a loudspeaker with chosen reactive load characteristics [8].

Simulation of Protection Circuits

SPICE simulation is directly applicable to the analysis of *V-I*-limiting circuits and can provide a great deal of insight. Simulation allows the examination of *V-I* limiter behavior with nearly arbitrarily complex loudspeaker loads that would be difficult to emulate with real passive components

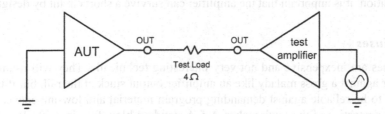

Figure 18.17 Back Driving an Amplifier to Test SOA Protection

in the laboratory. It goes without saying that the use of real loudspeakers should be avoided when evaluating V-I limiter circuits at high power levels. The testing of amplifier protection circuits in the real world can be dangerous to the amplifier, so it is wise to carry out protection simulations before doing laboratory tests.

18.6 Protection Circuits for MOSFETs

The absence of secondary breakdown in MOSFETs means that most of the time they will not need V-I limiting for purposes of the usual safe area protection. This does not mean that they do not need protection. While lateral MOSFET amplifiers tend to be forgiving, vertical MOSFET amplifiers are not very forgiving of short circuits. This is because the vertical MOSFETs will gleefully try to put huge amounts of current into a short circuit load, unhindered by BJT realities like beta droop or lateral MOSFET realities like significant $R_{DS(on)}$ with a positive temperature coefficient. For this reason, fast-acting short-circuit protection is desirable for vertical MOSFET power amplifiers. This was discussed in Chapter 14, Section 14.7. The fast electronic shutdown circuit for the output stage as described in Section 18.3 is especially useful here. If flying catch diodes are used as described previously, the shutdown circuit can be as simple as a small TRIAC or transistor that collapses the bias spreader, as was shown in Figure 18.12.

18.7 Protecting the Driver Transistors

If an output transistor is unable to drive the load (especially under short circuit conditions), the driver transistor may try to drive the load through the base-emitter junction of the BJT output transistor. This condition can also occur when the output is trying to drive a very low-impedance load to high current where the output transistors are suffering from serious high-current beta droop. The smaller driver transistor may fail due to secondary breakdown. Safe operating area of the driver transistors must be considered. These concerns were discussed in Chapters 12 and 14. The protection of the output stage can be made more conservative to take into account the need to protect the driver transistors.

18.8 Loudspeaker Protection Circuits

It is very important to protect the loudspeaker from the results of an amplifier failure. Output transistors often fail shorted, meaning that the full power supply rail voltage may be applied to the output terminal of the amplifier. It is important that such a dangerous voltage be removed quickly, before it can do damage to the loudspeaker. Sometimes a speaker fuse or a rail fuse will blow quickly enough to protect the loudspeaker. In other cases a circuit that senses DC on the output of the amplifier will open a speaker relay, fire a crowbar TRIAC or close a relay that shorts the amplifier output to ground. In the case of arrangements that deliberately short the output to ground under a fault condition, it is important that the amplifier can survive a short circuit by design.

Speaker Fuses

Speaker fuses are inexpensive and not very interesting technically. They will usually protect a loudspeaker against a gross malady like an amplifier output stuck at the rail, but if the fuses are big enough to be reliable against demanding program material and low-impedance loads, they will be less protective of the loudspeaker. A 5-A standard-blow fuse in a 100-W/8-Ω amplifier stuck at its 50-V rail and delivering 7.8 A DC into an 8-Ω loudspeaker with 6.4-Ω DCR will survive for fully 5 seconds [6]. Do your expensive loudspeakers want that? Fuses also introduce thermal distortion and can suffer degradation of their contacts that will also lead to distortion.

The Speaker Relay and Its Control

The ubiquitous speaker relay is the most common first line of defense for loudspeakers. Key to its operation is its control circuit. Space does not permit a detailed discussion of such circuits, and there is good coverage elsewhere [1, 2]. Moreover, their design is usually straightforward given a good understanding of the functions that should be performed and the pitfalls to be avoided. A good speaker relay control circuit provides a comprehensive protection system and a well-behaved amplifier. Functions to be performed by the speaker relay system may include:

* Protection against excessive DC at the speaker terminals
* Muting and thump elimination at turn-on and turn-off
* Power supply monitoring
* Speaker impedance monitoring
* Amplifier short-circuit protection
* Retry after a delay following a fault

Protection of the loudspeaker from excessive DC at the output of the amplifier is important because most electronic protection circuits will not prevent application of the rail voltage to the loudspeaker if an output transistor fails shorted. The key issue in activating DC protection is the time constant. It should be long enough so that it does not trigger on legitimate high-amplitude low-frequency signals, including during amplifier testing down to at least 10 Hz.

Many amplifiers produce a thump when power is applied or removed. This is not necessarily a sign of bad amplifier design, but it is annoying and potentially damaging to loudspeakers. The speaker relay should be closed only after a delay following turn-on and should be opened immediately on turn-off before the rails collapse. The latter usually requires a power supply circuit that quickly detects loss of AC mains power. When turn-off is detected, it is also important to reset the turn-on delay timing circuit so that a full turn-on mute delay will occur if power is suddenly restored.

The DC power supplies should be monitored and verified to be within proper limits before the turn-on delay circuit is allowed to begin its sequence. It is especially important that turn-on not be allowed if one rail is down. Verification of the rail voltages should occur downstream from any rail fuses.

Simple circuits that test the speaker load resistance before the speaker relay is closed can be used to prevent the turn-on mute timer from proceeding. In some designs it is desirable for the relay control circuit to execute a retry following a fault. Such a retry may often incorporate a delay similar to that of the power-on mute delay.

Many of these functions can be conveniently implemented with an LM339 quad comparator. A small power MOSFET makes a good relay driver in such circuits. It requires virtually no gate drive current, its gate drive can be the full voltage of a +15-V supply, and a relay catch diode is not required to protect the transistor, as it is built into the MOSFET as the body diode. The body diode will permit a much higher relay coil flyback voltage than the usual silicon diode that is forward-biased as soon as the coil voltage reverses. That slows down release of the relay. The higher reverse coil voltage allowed by the MOSFET allows faster release.

The TA7317 Loudspeaker Protection IC

No discussion of loudspeaker protection circuits would be complete without mention of the popular TA7317 integrated circuit expressly designed by Toshiba for the protection of loudspeakers [9]. This device is a 9-pin SIP for controlling a speaker relay. It performs the following functions:

* Over-current protection
* DC protection for the loudspeakers for two channels

- Delayed turn-on muting
- Fast turn-off muting
- Retry after about 3 seconds

Figure 18.18 shows a typical application circuit employing the TA7317. The IC is a convenient collection of transistors that provides the above functions. Most of its action is controlled by about 15 external passive components. The output at pin 6 controls the speaker relay with an open-collector Darlington driver. When the relay is energized, the loudspeaker is connected to the amplifier. Pin 6 can tolerate up to +60 V and can sink up to 130 mA through the relay. Although the relay is powered here from a +50-V main rail, it can be powered from a low-voltage supply instead. The relay driver is controlled by a Schmitt trigger that provides hysteresis to prevent relay chatter.

Positive supply current is sourced to pin 9, where the voltage is shunt regulated internally to +3.1 V. R4 limits the current to the shunt regulator to about 2.5 mA. The supply current can be sourced from a lower-voltage supply like +15 V if desired. R5 provides about 2.6 mA of negative bias current from a negative power supply.

The voltage applied to pin 8 controls muting. When power is applied, the voltage at pin 8 rises toward the 3.1-V supply at a rate controlled by R1 and C1. When the voltage on pin 8 reaches about +1.3 V, the speaker relay will be closed. If C1 = 22 µF and R1 is 100 kΩ, initial mute time is 2 seconds.

Pin 1 controls fast-off muting and over-current protection. It is connected to the base of an internal transistor whose emitter is connected to ground. When turned on, that transistor will discharge the pin 8 node and open the relay. If for any reason current is allowed to flow into pin 1, the relay will thus be opened. R6 sources a current of 50–170 µA to pin 1, depending on the voltage at pin 1. R7 sinks about 250 µA from pin 1, overcoming the current from R6. R7 is powered

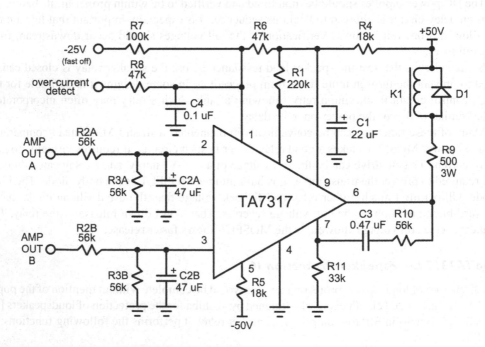

Figure 18.18 Typical Protection Circuit Using the TA7317

from a −25-V supply. This supply comprises a rectifier, small filter capacitor and bleeder resistor (typically 1 μF and 15 kΩ). Its output will fall very quickly when power is removed. On shutoff, the current through R6 will drive pin 1 positive and quickly open the relay.

Over-current shutdown is controlled by current-limiting resistor R8. It sources current into the pin 1 node to open the relay. R8 is fed from an over-current detection circuit within the amplifier that supplies positive rail voltage when the current limit has been exceeded. When the over-current detect voltage goes to the positive rail, pin 1 goes to +1 V_{be} and the relay is opened. A circuit similar to the opto-isolator circuit in Figure 18.12 can be used as the over-current sensor in the amplifier's output stage if it is connected to activate a PNP transistor whose emitter is connected to the positive rail.

Pins 2 and 3 are for DC detection for each of two channels. If the voltage at either pin goes high or low by 1 V_{be}, the relay will be opened. External components R2 and C2 filter the signal out and provide a delay for DC detection. R2 and R3 form an attenuator from the amplifier output signal to pin 2 (or 3). C2 must be large enough to prevent legitimate low-frequency signals from activating the protection circuit. If R2 = R3 = 56 kΩ and C2 = 47 μF, a DC level of 30 V will trigger shutdown in about 70 ms. A DC level of about 3 V will trigger the circuit in about 1 second.

R10, R11 and C3 act as a speedup circuit for opening of the relay by providing some initial positive feedback. When the relay is opened for any reason C1 of the turn-on mute circuit is discharged, forcing a new turn-on delay sequence. This means that the device will execute a retry after a few seconds. If the cause of the problem has been removed, normal operation will resume.

Another similar protection IC is the Unisonic Technologies/NEC μPC1237 [10]. The NTE 7100 is an equivalent IC.

Protecting the Speaker Relay Contacts

When a speaker relay does its job by breaking the circuit to the loudspeaker, it may suffer damage from arcing. Pitted contacts may result. That will lead to increased relay distortion (see Section 11 of Chapter 16). Opening a circuit in the presence of significant DC current flow can be particularly harmful (this can happen when the amplifier output goes to the rail). Relays are not good at breaking DC at high currents due to continuous arcing. Indeed, relays are usually rated for less current and voltage when used in a DC circuit. For these reasons it is important to use speaker relays with adequate contact area and generous current ratings; too many amplifiers employ flimsy speaker relays.

Figure 18.19 shows some circuitry that can be used to reduce the likelihood of contact damage when relay K1 opens. Always remember that the output stage Zobel network (R1-C1) must be placed upstream from the relay, since amplifier stability must be preserved when the relay is open. An R-C snubber (R2-C2) placed across the relay contacts can reduce the chance of arcing when the relay opens. The capacitor slows the rate of change of the voltage across the contacts on opening while the series resistor prevents excessive current flow through the capacitor if a voltage difference exists when the contacts close. This network will cause some audio leakage during the muting time. If the amplifier breaks into a continuous HF parasitic oscillation, and the protection circuit detects it and opens K1, R2 and C2 will provide a sneak path for this potentially harmful signal to get to the loudspeaker. This is a potential disadvantage. The NC contacts of K1 provide a path for speaker current flow when K1 opens. In this case, catch diodes D3 and D4 function only during the time that K1 is in transition from NO to NC (on the order of 2 ms). These diodes are more important if only an SPST relay without NC contacts is used.

An optional second Zobel network (R4-C3) at the speaker terminals can help with EMI ingress, but does not do much for relay contact protection. Relay K2 can be used to kill the audio input to the amplifier during muting, preventing any significant voltage differential across the

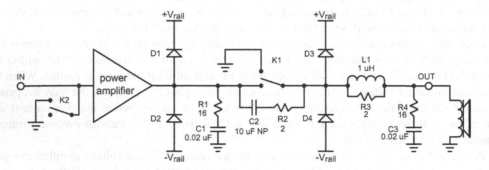

Figure 18.19 Some Relay Contact Protection Approaches

K1 contacts when they close. K2 should open, shorting the input, before K1 opens, even if by only a few ms.

It is also important to limit the maximum voltage that can appear across the relay contacts as a result of flyback from an inductive speaker load. This is no different than the reason for the traditional rail catch diodes (D1-D2) that are placed from emitter to collector of BJT output transistors. A second set of such catch diodes (D3-D4) should be placed on the downstream side of the relay contacts.

Closing the Feedback Loop Around a Protection Device

Fuses and relays in the signal path can introduce distortion, so it is tempting to enclose them in the global negative feedback loop so that their distortion will be reduced by feedback. This is fine, but with one major caveat. When the protection device opens, the feedback loop will be opened and the amplifier will operate open loop, possibly resulting in damage and undesired effects. If the loop is closed around a relay and that relay is used for turn-on mute delay, nasty signals could be initially presented to the loudspeaker while the feedback loop stabilizes. This could be especially problematic if the DC output of the amplifier has drifted to a high offset value. For this reason it is wise to keep the feedback loop closed in some way when the protection device is open. Figure 18.20 shows one approach to enclosing a speaker relay or a fuse in the feedback loop.

R1 and R2 make up the usual feedback network, but R3 is also put in the path to the amplifier output terminal after speaker relay K1. R3, at 100 Ω, allows the injection of negative feedback by diodes D1 and D2 even when the output is a short circuit. In normal operation the signal present at the junction of R3 and R2 is slightly less than the full amplifier swing. R4 and R5 attenuate slightly the output of the amplifier before application to diodes D1 and D2 so that in normal operation there is no signal across the diodes to cause distortion.

A second approach that uses a second set of relay contacts is shown in Figure 18.21. Resistor R3 provides negative feedback directly from the amplifier output node when the K1A speaker relay contacts are open. When the relay is closed, the K1B contacts short the feedback takeoff point to the downstream side of the K1A contacts, enclosing K1A in the feedback loop.

Notice that this amplifier uses a DC servo and that the servo takes its feedback directly from the amplifier output so that it can drive the DC output level to zero even when K1 is open. There are many variations and improvements that can be made to this approach. Notice that in Figure 18.21 a small amount of signal and possible thump will be allowed to sneak through R3 to the loudspeaker during the muting periods. In such amplifier designs it is also wise to short out the input signal during the muting interval.

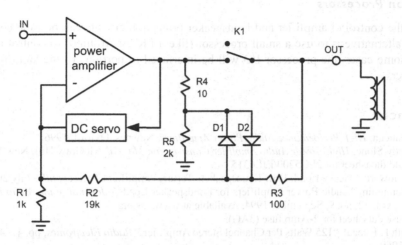

Figure 18.20 Enclosing the Speaker Relay in the Global Feedback Loop

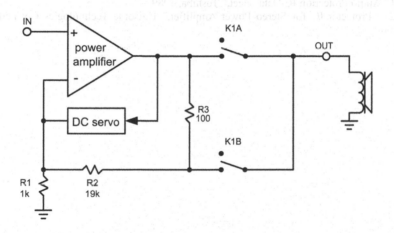

Figure 18.21 Enclosing the Speaker Relay in the Global Feedback Loop

Crowbar Circuits

Crowbar circuits act by deliberately placing a dead short across the output of the amplifier to protect the loudspeaker. These circuits will usually be triggered by the presence of an excessive DC offset at the output of the amplifier. A TRIAC or relay is usually employed for this function. When crowbar protection is used, the amplifier should obviously be designed to withstand an output short circuit. Nevertheless, the priority for the crowbar circuit is to protect the loudspeaker.

Avoiding Speaker Relays

If electronic shutdown circuits and an output crowbar are used, the speaker relay can be avoided entirely. The muting function can be implemented with the electronic shutdown circuit if some aspects of its operation are designed to be non-latching.

Protection Processors

Together the control of amplifier and loudspeaker protection circuits can become complex. An attractive alternative is to use a small processor (like a PIC or Arduino) to control these functions. In some cases, the processor I/O will be interfaced to portions of the amplifier through optocouplers.

References

1. Ben Duncan, *High Performance Audio Power Amplifiers*, Newnes, Oxford, 1996.
2. G. Randy Slone, *High-Power Audio Amplifier Construction Manual*, McGraw-Hill, New York, 1999.
3. Fairchild datasheet for 2SC5200/FJL4315.
4. Keith Howard, "Heavy Load: How Loudspeakers Torture Amplifiers," *Stereophile*, July 2007.
5. Eric Benjamin, "Audio Power Amplifiers for Loudspeaker Loads," *Journal of the Audio Engineering Society*, vol. 42, no. 9, September 1994. Available at www.aes.org.
6. Littelfuse datasheet for 2-Amp fuse (3AG).
7. Kenneth F. Buegal, "125 Watts Per Channel Stereo Amplifier," *Radio Electronics*, pp. 41–44 and p. 85, April 1969.
8. Harry C. P. Dymond and Phil Mellor, "An Active Load and Test Method for Evaluating the Efficiency of Audio Power Amplifiers," *Journal of the Audio Engineering Society*, vol. 58, no. 5, pp. 394–408, May 2010.
9. "TA7317 Audio Protection IC Data Sheet," Toshiba, 1989.
10. "μPC1237 Protector IC for Stereo Power Amplifier," Unisonic Technologies Co. and NEC. Also NTE7100.

Chapter 19

Power Supplies and Grounding

19. INTRODUCTION

Power supply design and grounding arrangements are too often the stepchild of the creative process when it comes to audio amplifier design. Poor power supply and grounding techniques can ruin an otherwise outstanding design by allowing grunge into the signal path in many different ways. Poor power supply regulation and inadequate power supply current capability can impair amplifier performance and limit the ability of amplifiers to drive difficult loads, especially at low frequencies.

19.1 The Design of the Power Supply

The power supply must convert the AC mains voltage into reliable and stable negative and positive DC rails. It must provide adequately small ripple and sufficient reserve current and regulation. Mains noise must not be communicated to the circuit side of the power supply.

Figure 19.1 shows a typical power supply for an audio power amplifier. In its simplest form it consists of a power transformer with a center-tapped secondary feeding a high-current bridge rectifier that produces positive and negative rails that are filtered by large reservoir capacitors. The design shown employs a 240 volt-ampere (*VA*) toroid with a pair of secondary windings each rated at 42 V RMS. The arrangement shown produces about 59 V DC on each rail with no load and about 52 V DC with a 2 A load on each rail.

Bleeder resistors are connected across the reservoir capacitors. This assures that the rail voltages fall to zero in a reasonable time after power is removed. So-called *X* capacitors are connected across the primary and the secondary windings to reduce noise. *X* capacitors (not shown) used on the mains side must be of the type designed for continuous connection across the mains. They should be mounted on the mains side of the power switch close to the mains wiring entrance and should be shunted by a bleeder resistor to discharge any residual voltage that may be present after power is disconnected. IEC power inlet connectors with integral EMI filters usually include *X* capacitors.

The large connection dot denotes a star ground, often implemented in power amplifiers with a large copper or aluminum plate joining C1 and C2, in the middle of which all of the other conductors are connected with a single terminal. The idea is to reduce voltage drops among the different ground lines by terminating them all at the same physical point. In this approach, however, the very large and impulsive charging currents of C1 and C2 flow through the star ground. A better arrangement is to sum and resolve these rectifier currents at a separate node prior to the star node, such as where R1 and R2 connect to the ground line.

Alternative Supply Arrangements

An alternative arrangement employs two secondary windings and two bridge rectifiers as shown in Figure 19.2. One possible advantage to this design is that it avoids the circulation of direct

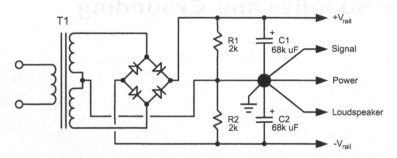

Figure 19.1 A Typical Amplifier Power Supply

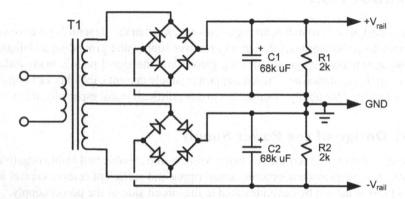

Figure 19.2 An Alternative Rectifier Arrangement

current through the transformer windings when the positive and negative rail load currents are different. The flow of direct current through transformer windings (primary or secondary) should be avoided, as it can degrade transformer performance and sometimes create transformer buzzing. Toroid transformers are sometimes more sensitive to this effect than transformers of conventional construction.

The rectifier current pulses for each half of the supply do not circulate through ground, but rather circulate locally. This may improve management of rectifier noise. Indeed, there is no direct connection of the secondary winding(s) to ground. This also means that the transformer is only connected to the amplifier during the brief rectifier conduction intervals. This may reduce the opportunity for mains EMI to enter the amplifier circuits. There does not appear to be universal agreement on the potential benefits of this connection.

This arrangement comes at a price. It suffers twice the voltage loss through the bridge rectifiers and costs more because of the doubling of the number of rectifiers. The added cost is not insignificant if expensive fast-recovery rectifiers are employed. On a fixed budget, I would opt for better rectifiers in the conventional arrangement.

Boosted Supply Rails

A number of amplifier designs employ boosted rails for the IPS and VAS circuits [1]. The availability of slightly higher rail voltages for these circuits provides extra headroom to reduce distortion or allow the introduction of VAS cascodes without wasting output stage headroom on the main high-current rails. The boosted rails also make the introduction of Baker clamps more convenient. Boosted supplies will often add 5–10 V to the main rails. Although boosted supplies represent an increase in complexity of the power supply, it is important to recognize that the current demands on them are very small compared to those on the main rails.

The most common way to realize boosted supplies is to add two low-current windings to the main windings of the power transformer, connected to the "ends" of the main winding of the transformer in Figure 19.1. The higher resulting AC voltages are then rectified with a single full-wave bridge equipped with its own small reservoir capacitors. In such an arrangement, when the output stage drags down the main rails, the boosted rails will not be dragged down with them because they have their own reservoir capacitors. Similarly, the boosted rails will tend to have smaller ripple voltages on them. There are many other ways to realize boosted supplies, including the use of a small dual-secondary auxiliary transformer if a main transformer with custom boost windings is not available. Even if extra transformer windings are not used, providing separate rails with their own rectifier and reservoir capacitors will result in some voltage boost and some isolation from the main rails being pulled down for short intervals. This can also help PSRR.

Power Supply Stiffness and Regulation

The ideal power supply has infinite *stiffness*, meaning that its output voltage will not change as a result of differing output current demands. This is referred to as *load regulation*. An amplifier built with an extremely stiff power supply will come close to putting out twice as much power into a 4-Ω load as into an 8-Ω load, and in some cases 4 times as much power into a 2-Ω load (assuming no output stage limitation).

An important trade-off in stiffness of the power supply is the amplifier's *dynamic headroom*, which is the ratio of maximum short-term burst power to maximum continuous power. This reflects the fact that the amplifier can produce more power when the power supply is lightly loaded and can maintain high rail voltage for a brief period of time due to the charge stored in its reservoir capacitors [2]. Eventually the rails will sag under the continuous power conditions and maximum output power will be smaller. A power supply with less stiffness will produce an amplifier with more dynamic headroom. Having significant dynamic headroom is good because it provides reserve power for signal peaks without costly increases in the power supply. However, sound quality may suffer because of the increased rail voltage fluctuations. PSRR will suffer. For a given continuous output power capability, the power supply rails will be at a higher voltage under no-signal conditions, putting greater demands on component voltage ratings and transistor SOA requirements. The short-term burst power available used to be called "music power" before the FTC published rules for amplifier power output claims in 1974 [3].

Figure 19.3 shows the output voltage of the power supply of Figure 19.1 as a function of load current drawn from each of the rails. This power supply employs a pair of 68,000-μF reservoir capacitors. The ripple voltage is shown as a function of load current on the right-hand scale.

Effective Power Supply Resistance

The curve shown in Figure 19.3 illustrates that it is not a bad approximation to assign an effective power supply resistance R_{eff} to each rail of the power supply for estimating power supply sag in typical power amplifiers. The effective resistance here is about 2.6 Ω for each rail at load currents greater than 1/10 the full amount shown.

The transformer's primary and secondary winding resistances influence power supply regulation. The primary resistance for this 240-VA transformer is 1.7 Ω and the secondary resistance (end-to-end) is 1.1 Ω. The mains voltage for these tests was 119 V and the mains impedance was 0.3 Ω.

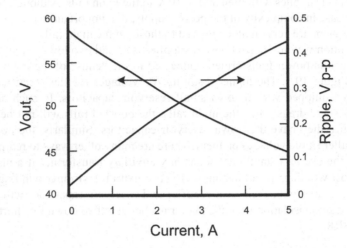

Figure 19.3 Output Voltage as a Function of Load Current; R_{eff} = 2.6 Ω

19.2 Sizing the Transformer

Picking the power transformer can be tricky because the VA and current ratings for power transformers are often given for AC loads, even though the relevant application here is for rectified DC loads. The same goes for load regulation. The VA rating is the product of the AC output voltage and the rated AC output current [4, 5]. The rated output current will typically be that current where the core temperature reaches 60 °C and is often the point where the AC output voltage sags between 5% and 10% from the no-load value. Using the 240 VA transformer of Figure 19.1, the AC secondary voltage sags from 90 V RMS at no load to 84 V RMS at its rated load of 2.8 A AC, for a sag of 7%.

Many manufacturers rate their transformers in volt-amperes assuming a resistive AC load. This is far from reality for a DC power supply for an audio amplifier, with very high peak rectifier currents and small conduction angles. Figure 19.4 shows the rectifier current waveform when the power supply of Figure 19.1 is supplying 2 A to each rail. The scale is 2 A per division.

Figure 19.5 shows the peak rectifier current as a function of load current for the power supply. The peak rectifier current does not increase linearly with load current because the rectifier *conduction angle* increases with load current as well, increasing the area under the conduction pulse. The second curve shows the peak currents as seen at the mains side of the power transformer.

There is no right or wrong choice of transformer because different transformers will largely affect the stiffness of the power supply as described above. To first order, one chooses a hefty enough transformer to achieve the degree of stiffness (load regulation) that is desired. Bear in mind that a transformer may be rated for its maximum VA on the basis of core temperature for continuous duty at that load, but in the case of an audio power amplifier the average load is much smaller. For consumer audio, sag under peak load is more important than core temperature.

VA Rating Rules of Thumb

A good rule of thumb is to employ a transformer with a VA rating that is twice the maximum rated power to be delivered by the amplifier. The 240-VA transformer in the power supply of Figure 19.1 sags by 5 V on each rail when the amplifier output is increased from 5 W to 125 W with a load impedance of 8 Ω. The rail voltages sag an additional 4 V down to 48 V when the amplifier is delivering 200 W into a 4-Ω load.

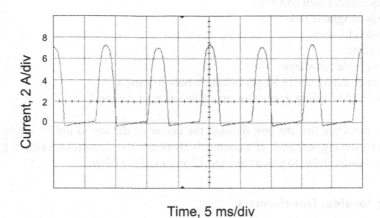

Figure 19.4 Rectifier Current Waveform When Load Current Is 2 A

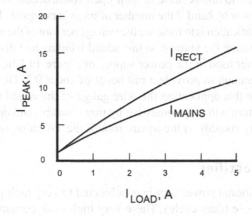

Figure 19.5 Peak Rectifier Current and Mains Current Versus Load Current

VA Versus Weight

The 240-VA toroidal transformer weighs 2.3 kg. This suggests that a reasonable approximation is 104 VA per kg. This is useful in estimating the VA of an unknown surplus power transformer. Other toroidal transformers appear to range between 100 and 140 VA/kg. One rule-of-thumb asserts that if VA is doubled, the weight of the transformer will increase by the factor $2^{3/4} = 1.68$ [5].

Toroid Versus Conventional Transformers

Toroidal power transformers tend to be more efficient than transformers of conventional construction. They also better confine the noisy magnetic fields that result from large rectifier spikes. There is little cost premium for toroidal transformers. Some claim that transformers with C-core or EI construction sound better, but this is unsubstantiated. The advantages of toroidal transformers over conventional laminated transformers include [6, 7]:

- High efficiency
- Small size

- Low mechanical hum and buzz
- Low stray magnetic field
- Low no-load losses
- Light weight
- More available turns/layer
- Less copper length due to longer effective bobbin length
- Magnetic field running in grain-oriented direction of the steel

It is very important that the bolt through the center of the toroid not be allowed to create a shorted turn by having both ends of it conductively tied to the chassis. One end of the bolt should float. It is also undesirable to mount one toroid on top of the other.

Modifying Toroidal Transformers

Sometimes a different voltage or an additional winding is needed on the power transformer. A good example is a pair of extra windings to implement a boosted rail supply. Toroidal power transformers lend themselves ideally to this because of their open construction. Additional turns can easily be wound on the transformer by hand if the number of turns is not great. The first thing to do when contemplating such a modification is to measure the voltage per turn of the transformer. Wind 10 turns of wire on the toroid, measure the voltage on this added winding and divide by 10.

The 240-VA transformer used in the power supply of Figure 19.1 has 0.36 V/T. Twenty turns will yield 7.2 V RMS, enough to provide a rail boost of about 9 V DC depending on the boost rectifier architecture. For this application the wire gauge of the added turns can be quite small. The number of volts per turn will vary somewhat by manufacturer and details of construction, but it appears to increase very roughly as the square root of the VA rating of the transformer.

19.3 Sizing the Rectifier

The rectifier in a conventional power supply is subjected to very high peak currents that have a fairly low conduction angle (duty cycle). These very high peak currents cause substantial voltage drops across the rectifier junctions, resulting in some reduction of the output voltage and in heating of the rectifier.

A common mistake is to undersize the rectifier by thinking in terms of the maximum delivered DC current. In such a case excessive rectifier forward voltage drops will occur and reliability of the rectifier will be degraded. Bridge rectifiers rated at 35–50 A are recommended for audio power amplifiers. The rectifier should be bolted to the chassis to keep it cool. The typical voltage drop for each diode in a 35-A full-wave bridge is about 1.0–1.2 V at a peak current of 30 A.

19.4 Sizing the Reservoir Capacitors

The primary job of the reservoir capacitors is to maintain the output voltage during the time between charging pulses from the rectifier; this will reduce ripple. Calculating the ripple for a given load current and reservoir capacitor is fairly straightforward. The peak-to-peak 120-Hz ripple will be a sawtooth waveform at 120 Hz. The peak-to-peak ripple voltage can be approximated by recognizing that the voltage developed across a capacitor is $V = I * T/C$, where T will be 8.3 ms for a full-wave rectifier operating at 60 Hz. For the circuit of Figure 19.1 operating at a load of 2 A, with 68,000-μF reservoir capacitance (0.068 F), the ripple on each rail is estimated to be $2.0 * 8.3e{-}3/6.8e{-}2 = 0.24$ V.

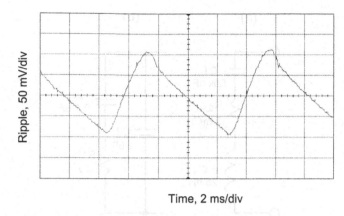

Figure 19.6 Ripple Waveform for the Circuit of Figure 19.1 with a 2-A Load

The measured ripple for that circuit is 0.2 V p-p, as shown in Figure 19.6. The slope of the rising edge is indicative of peak rectifier current, while the slope of the falling edge is indicative of the load current. Bear in mind that capacitor tolerances are wide. The 68,000-μF capacitor used here actually measured about 80,000 μF.

While the reservoir capacitor is important for minimizing ripple, a large reservoir capacitor is also desirable for providing very high output current on a transient basis for brief bursts of program material.

Equivalent Series Resistance (ESR) and Inductance (ESL)

Figure 19.7a shows a simplified equivalent circuit for a high-quality 50,000-μF reservoir capacitor. It is important to recognize that the reservoir capacitors have some *equivalent series resistance* (ESR) and that the extremely high peak currents of the rectifiers will create some voltage drop across the ESR independent of how large the capacitance is. This resistance varies considerably with the type and construction of the capacitor, but a value of 12 mΩ is not unusual. This means that a 7-A rectifier pulse may create a drop of 84 mV across the capacitor. Some evidence of this effect can be seen at the positive peaks of the ripple waveform in Figure 19.6.

Most capacitors also have a modest amount of *equivalent series inductance* (ESL). This is often the result of the way they are constructed with windings. This means that at high frequencies the impedance of the capacitors actually starts to rise instead of continuing to decrease. The 25-nH value for the capacitor of Figure 19.7 is quite low.

Reservoir capacitors are sometimes also characterized by a resonant frequency, where their impedance dips almost to the value of the ESR when the capacitance and inductance series resonate. This is often in the range of kHz to hundreds of kHz. The model in Figure 19.7a resonates at about 4.5 kHz. ESR and ESL can both be reduced significantly by employing several smaller capacitors in parallel instead of a single large capacitor.

Bypasses and Zobel Networks for Reservoir Capacitors

Because large reservoir capacitors are far from perfect at high frequencies, performance can be improved by the addition of extra filtering components that have better high-frequency characteristics. One example is shown in Figure 19.7b. The most common approach is to bypass the reservoir

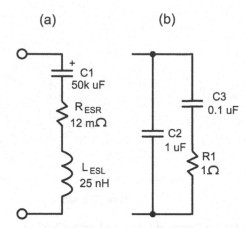

Figure 19.7 (a) Simplified Equivalent Circuit for a 50,000-μF Reservoir Capacitor (b) A Bypass/Snubber Arrangement

capacitors with a quality film capacitor of 1 μF or more as done with C2 in the figure. This will generally assure that the combined impedance will continue to fall at high frequencies above the resonant frequency of the reservoir capacitor. Even with a bypass capacitor in place there often remains the possibility of impedance peaks due to wiring and component inductances that can form tank circuits at high frequencies. The addition of the 1-μF capacitor in Figure 19.7b actually created a resonance at about 1 MHz where the net impedance more than doubled to about 0.3 Ω.

For this reason a high-frequency Zobel network can be added as well. The Zobel in Figure 19.7b consists of R1 and C3. At high frequencies it transitions to a damping circuit with a resistive behavior. Notice that 0.1 μF and 1 Ω are used here. The transition from capacitive to resistive for this network occurs at about 1.6 MHz. It is important that the capacitor has low ESR and ESL. This Zobel will not do much for the 1-MHz resonance, but will help at higher frequencies. Finally, connection of a low-ESR/ESL 100-μF electrolytic capacitor across the reservoir capacitor can be very helpful. Its ESR can actually help damp resonances like the one at 1 MHz mentioned here.

Split Reservoir Capacitors

Often two smaller reservoir capacitors will have an advantage over a single reservoir capacitor in terms of both ESR and ESL. If two reservoir capacitors are used an additional advantage can be had by placing a small resistor in series between them, as shown in Figure 19.8. A 0.22-Ω resistor is much smaller than the typical power supply effective resistance, and yet it can have a remarkable effect on reducing ripple and high-frequency noise content. The impedance of the 25,000-μF capacitors C1 and C2 is only 0.05 Ω at 120 Hz, less than 1/4 that of the 0.22-Ω resistor. In a typical arrangement with 1 A of load current, output ripple was 125 mV p-p with a single 50,000-μF capacitor. When it was split into two 25,000-μF capacitors by a 0.22-Ω resistor, output ripple fell to 43 mV p-p, better by a factor of almost 3. The spectral components of the ripple at 120, 240 and 360 Hz were decreased by 7 dB, 11 dB and 13 dB, respectively, when the split capacitance arrangement was used. The added 0.22 W of rail resistance is fairly small compared with the effective power supply rail resistance of 2.6 Ω, so that any added voltage loss is of little concern. Notice that the grounds for the two sets of capacitors should be physically separate in the grounding topology as shown in Figure 19.8. This will be discussed further in Section 19.9 on grounding.

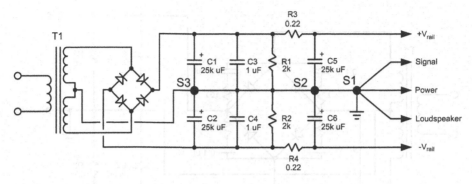

Figure 19.8 Split Reservoir Capacitor Arrangement

19.5 Rectifier Speed

During the forward conduction pulse the peak current through the rectifier diode can be very high. As with any *p-n* junction, there exists a large stored charge in the form of minority carriers during this time. When the driving AC waveform decreases below the DC level stored in the reservoir capacitor on the other side of the diode, the diode must turn off, but it cannot do that instantaneously as a result of the stored charge. This charge must be sucked out of the diode by means of reverse current flow through the diode. The amount of time it takes to pull out this charge is a measure of the diode's speed and is referred to as its *reverse recovery time* (t_{rr}). The pulse of reverse current flow is especially undesirable because it can lead to RF emissions.

The power supply of Figure 19.1 uses an ordinary 35-A bridge rectifier. The reverse recovery current manifested itself as a 500-mA p-p negative current pulse lasting 30 μs when the rectifier in the power supply turned off. The pulse had very sharp edges with rise times on the order of 1 μs. The load current for this test was 2 A. The reverse recovery pulse is not visible in Figure 19.4 due to the limited bandwidth of that measurement.

Soft Recovery and Fast Recovery

Diodes that are designed to have a fast recovery will create less noise because the amount of energy that must be sucked out of them during turn-off is smaller, leading to much shorter duration of the reverse current interval and consequent turn-off spikes. One type of fast recovery diode is the *fast recovery epitaxial diode* (FRED) [8]. While a conventional bridge rectifier may have a rated reverse recovery time of 350 ns, a FRED may have a t_{rr} of 35 ns. The trade-off is that FREDs cost more and sometimes have a larger forward voltage drop at a given forward peak current.

At the end of the reverse current interval, the diode finally turns off and its current returns to zero (where it should be in the off state). Once the excess carriers are swept out of the junction, some diodes will have their reverse current go to zero very quickly, almost snapping back. This high rate of change of reverse current can create more noise at high frequencies. Diodes that are designed to have a *soft recovery* have a more gradual return to zero of their reverse current. The smaller rate of change of falling reverse current causes less production of high-frequency noise. Some of the better fast recovery diodes are also characterized by soft recovery. The Vishay HEXFRED® devices are a good example [9].

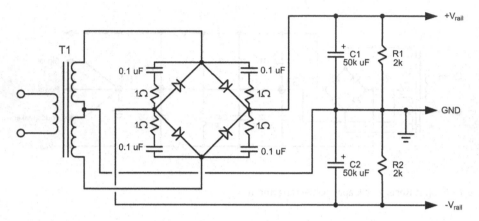

Figure 19.9 A Bridge Rectifier with Snubber Networks

Rectifier Noise and Snubbers

Rectifier noise due to turn-on and turn-off can be reduced by using snubber networks across each diode in a bridge, as shown in Figure 19.9 [10]. The resistors shown are often not used, but can be helpful in damping resonance effects. The snubber networks should be mounted right on the rectifier terminals. The use of damping resistors makes this more difficult, so there is a trade-off.

Measuring Rectifier Performance

Measuring the voltage across a rectifier diode with an oscilloscope under actual high-current DC load conditions can be very revealing. The two things of greatest interest are the peak forward voltage across the diode and the turn-off characteristic of the diode.

This measurement can be difficult to do in a conventional arrangement due to the common-mode and reverse voltages present. However, by reconfiguring a bridge rectifier arrangement, one can do this. The key is to arrange the circuit so that one side of the diode, the cathode, for example, is connected to ground. The remainder of the power supply is then allowed to float. Such an arrangement is shown in Figure 19.10.

The cathodes of the positive output diodes of the bridge rectifier are connected to ground while the rest of the secondary DC rectifier circuit, reservoir capacitors and dummy load resistors float. A 0.1-Ω resistor between the + output of the bridge rectifier and the positive reservoir capacitor allows the measurement of rectifier current with an oscilloscope. If the scope probe is moved to one of the AC input sides of the bridge, the voltage drop across the bridge diode can be observed. However, the reverse voltage will be large, making it difficult to read the forward voltage. For this reason, a resistor-diode combination is employed to clamp the reverse voltage to one diode drop, allowing a more sensitive oscilloscope setting.

19.6 Regulation and Active Smoothing of the Supply

One might think that an ideal audio power amplifier would employ voltage-regulated power supplies. While this would often make a fine power amplifier, it would add greatly to the cost and power dissipation. An amplifier with regulated power supplies does not have any *dynamic*

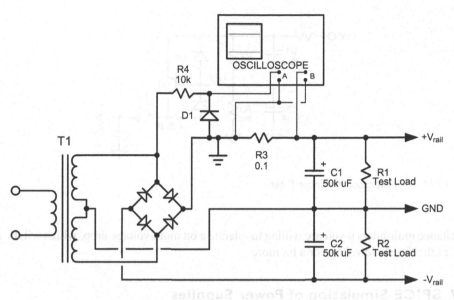

Figure 19.10 Measuring Rectifier Performance

headroom because the power supply does not rise during low-power periods to be able to provide larger bursts of power on transients.

The job of a power supply regulated for constant output voltage is made even more difficult when you consider the mains voltage fluctuations. If you design the supply to deliver a fixed voltage under conditions of worst-case low mains voltage (without falling out of regulation), then the voltage regulator (assuming it is a linear regulator) will be dissipating a large amount of heat as it throws away the extra rail voltage under conditions of worst-case high mains voltage.

Switching power supplies, on the other hand, usually produces a regulated output without the penalty of wasted power. These are discussed in Section 19.13.

Regulation of Input and VAS Power Supplies

It makes much more sense to regulate the power supplies feeding the input circuits and VAS because these circuits consume much less current. This is an alternative to the usual R-C filtering of the main supply that is used to power these circuits. Once again, regulation to a fixed voltage deprives the amplifier of operation at higher output power levels that are possible when the mains voltage is above the minimum value. For this reason, capacitance multiplier filters are often used for these circuits. Such a circuit is shown in Figure 19.11. The headroom loss can be designed to be less than 2 V if ripple on the main power supply is not excessive. A further advantage is that the filtered output of the capacitance multiplier can be referenced to the quiet ground without dumping a lot of ripple current into that ground.

Q1 is the pass transistor while R2 and R3 drop the input voltage by 1 V so that Q1 operates with 1-V collector to base. An additional 0.7 V is dropped through the base-emitter junction to arrive at the output voltage. R1 and C1 pre-filter the input to the capacitance multiplier. C2 is the multiplier capacitor and it forms a pole at 5 Hz for filtering purposes. Diodes D1 and D2 protect Q1 from voltage reversals. Vertical power MOSFETs also make good pass transistors for

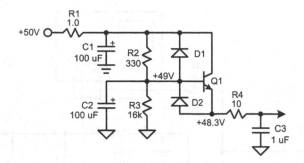

Figure 19.11 A Capacitance Multiplier Filter

capacitance multipliers if you are willing to tolerate a bit more voltage drop, which will be on the order of the threshold voltage or a bit more.

19.7 SPICE Simulation of Power Supplies

Simulating a power supply with SPICE is not always the first thing that comes to mind, but it can be valuable and provide a lot of insight. In particular, it can help in understanding peak recti- fier currents, rectifier turn-off issues, the effect of snubbers and the behavior of different kinds of reservoir capacitors and arrangements. The accuracy of the results can be greatly improved by having good models for the components used in a power supply, but valuable insight can be gained even with very approximate models. For the more adventurous, some of the power supply components can be modeled by straightforward measurements. These include transformer wind- ing resistances and inductances, rectifier *V-I* characteristics and reservoir capacitor ESR and ESL. It is also wise to include estimates of parasitic wiring inductances in these simulations.

19.8 Soft-Start Circuits

All amplifiers are characterized by *inrush current* when they are turned on. At minimum, this burst of mains current is required to quickly charge up the reservoir capacitors. Some trans- formers, toroids in particular, also can draw an initial large magnetizing current depending on where in the AC cycle the power is applied. The inrush current for large amplifiers employing efficient toroid power transformers and large amounts of reservoir capacitance can be very high. Figure 19.12 shows the AC inrush current for the power supply of Figure 19.1. This supply employs a 240-VA toroidal power transformer with 68,000-μF reservoir capacitors and produces 52-V rails with a 2-A load on each rail. Peaks in excess of 40 A can be seen. In some extreme cases this could damage power switch contacts or even blow circuit breakers. In other cases the inrush current can degrade reservoir capacitors. Bear in mind that the power supply of Figure 19.1 is not for a big amplifier, although it does include rather large reservoir capacitors.

Passive Soft-Start Circuits

Passive soft-start circuits rely on a temperature-dependent resistance placed in series with the mains line. Such devices are called *power thermistors* or *negative temperature coefficient resis- tors* (NTC). At room temperature these devices typically have a resistance on the order of 1 Ω to

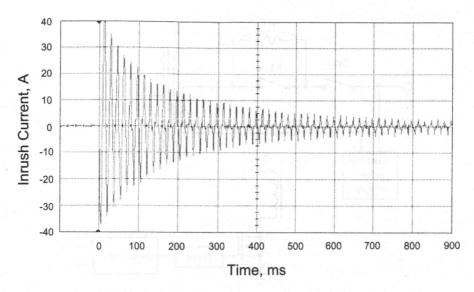

Figure 19.12 AC Inrush Current for the Power Supply of Figure 19.1

5 Ω. This puts a limit on the inrush current. Under load, these devices heat up and their resistance falls, often to tenths of an ohm or less. These devices get hot, sometimes reaching temperatures of 180 °C. The NTC device also takes time to cool down and increase its resistance after power is shut off. This means that if power is restored quickly after turn-off, an unexpectedly high surge current may result.

The high peak currents encountered at the transformer primary under load due to rectifier spikes also raise concerns about the resistance introduced by a passive inrush controller. The size of these peaks is shown in Figure 19.5. When the supply is loaded with 2 A on each rail, these peaks are about 6 A. A passive inrush controller might be rated at 0.1 Ω under full load and 0.2 Ω under a 50% load. These are RMS current values expressed as a percentage of the rated load current of the inrush controller.

When a power amplifier is operating at low sound levels most of the time, the resistance of the inrush controller may be quite a bit higher. This may deprive the amplifier of some headroom when a loud passage suddenly occurs. The popular CL30 inrush current limiter exhibits a resistance of 0.4 Ω in the power supply of Figure 19.1 when the power supply is delivering 1.5 A. At a load of only 0.5 A, the resistance of the inrush limiter rises to 0.8 Ω. A resistance of 0.4 Ω may not seem like much in the AC line until one looks at the peak rectifier current pulses. This device drops the mains peaks by almost 3 V under these conditions.

Active Soft-Start Circuits

Active soft-start circuits usually employ a relay or TRIAC in series with the line and in parallel with a soft-start resistance. The controlled switch shorts out a conventional resistor or a passive soft-start device after a fixed period of time or after the power supply rail voltages have come up to a certain point. Such a circuit is illustrated in Figure 19.13. If a conventional resistance is used for the soft-start, safety measures must be implemented to ensure that the resistor does not overheat in the event that the control circuitry does not activate the shunting device within a few seconds. It is much preferred that an NTC device be used for this reason. The use of a shunting

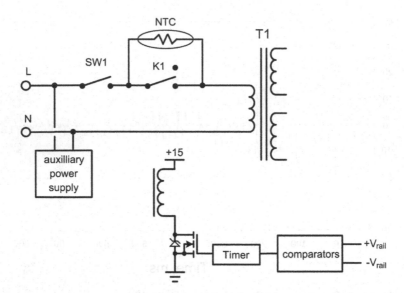

Figure 19.13 An Active Soft-Start Circuit

switch across an NTC device also allows it to cool down during normal amplifier operation, increasing its life and avoiding the problem of large inrush currents if the power is restored shortly after shutoff.

The control circuitry activates a vertical MOSFET relay driver a short time after both rail voltages are detected to be above a chosen minimum value. The circuitry opens the relay immediately on detection of a power-off condition. The control circuitry can be powered by a small auxiliary supply that does not require inrush control or the control circuit can be designed in such a way that it becomes properly operational during the time when the reservoir capacitors are charging. Many variations of these circuits are possible. An always-on auxiliary supply powered from a class B transformer may be employed in any case for auto-on or remote control circuitry. The soft-start relay can also be controlled by the same circuitry that controls the speaker relay.

19.9 Grounding Architectures

Ground is the reference for all signals in an amplifier, but not all grounds are the same. Currents flowing through grounds create voltage drops, even across seemingly heavy grounds. All ground paths have resistance and inductance, no matter how small. Magnetic fields can also induce voltages across ground conductors. The most obvious symptom of poor or inadequate grounding is hum and noise. However, a much more insidious result of poor grounding is distortion. Any ground "noise" that is correlated to the program signal in any way is considered distortion [11, 12, 13, 14].

Noisy and Quiet Grounds

In simple terms, there are two kinds of grounds—quiet signal grounds and noisy power grounds. All signal voltages that are single-ended should be referenced to quiet grounds. Quiet grounds are distinguished as grounds that have little or no current flowing through them.

When Ground Is Not Ground

Most ground points in an amplifier are interconnected with wires (never use the chassis). These wires have very finite impedance due to resistance and *self-inductance*. Heavy currents flowing through these wires will cause significant voltage drops. This is especially so for high frequencies or currents with sharp waveform edges (like rectifier spikes). Self-inductance of a 10-inch piece of 14-gauge wire is about 0.3 μH. Self-inductance of a straight piece of wire is about 30 nH per inch, nearly independent of wire gauge. The impedance of 0.3 μH at 20 kHz is about 36 mΩ. The DCR of the 10-inch length of 14-gauge wire is about 2.5 mΩ. A loudspeaker return line carrying 5 A RMS at 20 kHz through such a length of wire will cause a voltage drop of 180 mV.

Star Grounding

If all grounds are returned to a single point, then that point can be considered as the ground reference for the entire amplifier [13]. This is the idea behind a *star ground* arrangement as illustrated in Figure 19.1. Here the two reservoir capacitors are connected by a short heavy bus bar or piece of heavy-gauge wire. The center of this heavy conductor is defined as the star, and all other ground connections are returned to this point, including the center tap of the power transformer. In the arrangement shown, separate ground lines are shown for the quiet ground, the power ground and the loudspeaker return for each channel of a stereo amplifier.

If no current flows in a quiet ground connected to the center of the star, the potential on the quiet ground will remain unchanged; the same ground reference will exist at all places in that node. This will be the *true* ground for the amplifier. The exception to this is if there is a voltage that is magnetically induced in the connection of the quiet ground to the star ground point. Heavy noise, ripple and distorted audio currents can pass to the star ground without causing a disturbance to the quiet ground. Note that the "ground" potential at the far end of lines carrying such heavy currents will be moving with respect to the true ground reference at the star point.

There are some caveats, however. With heavy current flowing through it, the star ground node itself can develop some small voltage potentials among the several wires connected to it. In other words, it cannot really be a perfect equipotential point. Never use steel hardware to fasten together the elements of a star ground.

Star-on-Star Grounding

A star-on-star grounding architecture like that shown in Figure 19.8 is a better approach. All of the high-current rectifier spikes are resolved at the auxiliary star-3 ground (S3) independently of the main star ground. Split reservoir capacitor returns are merged at star 2. In addition to providing R-C filtering, resistors R3 and R4 prevent the formation of an AC ground loop among the reservoir capacitors. This arrangement keeps those dirty high currents away from the main star ground and keeps them from corrupting it.

An important concept is to have dirty currents circulate locally and be resolved before they are passed to a ground node involving other circuitry. If large electrolytic capacitors are used on the circuit board local to the output transistors, their grounds should be tied together before connection to another ground, such as the main star ground. This is because these capacitors are bypassing highly nonlinear class AB half-wave signal currents that should be resolved locally. The common ground of these capacitors can be thought of as another auxiliary star ground connected to the main star ground through a spoke.

The star-on-star approach is typical of the general approach to good grounding practice, namely, to follow the currents and beware of the voltage drops they can cause. More complex star-on-star grounding arrangements can be imagined when a thorough analysis of current flows is carried out. In many cases real amplifiers end up having some sort of a star-on-star grounding architecture without it having been intentionally designed that way.

Ground Corruption

One source of trouble that is often overlooked is the corruption of a quiet analog ground by currents injected into the ground by power supply bypass capacitors. This can happen when the main rail is routed to the input and VAS circuits through a small resistor or diode and then bypassed. Bypass capacitors can ruin an otherwise clean star grounding architecture by creating an AC ground loop and providing another path for alternating current flow.

Dual-Mono Designs

Even the best grounding architecture in a stereo amplifier cannot always prevent the formation of ground loops between the amplifier and the source (e.g., preamplifier). This can happen when the stereo pair of interconnects share the same ground at both ends of the interconnect cable. One way to avoid such ground loops is to completely isolate the power supplies of the two channels. The circuits of the two channels are then essentially floating with respect to each other. This is one advantage of dual-mono amplifier designs.

19.10 Radiated Magnetic Fields

The power transformer is usually the single biggest source of radiated magnetic fields. These fields can include both hum and rectifier spikes. However, any wiring that is carrying high currents can radiate a magnetic field that will induce voltages in nearby wiring. Toroidal power transformers can sometimes be rotated to minimize the impact of their radiated fields.

Antenna Loop Area

The best way to pick up an induced signal from a magnetic field is to build a loop with a fairly large area. Pickup from radiated magnetic fields is reduced if the loop area is small. This means that signal paths and their returns should be close to each other and not form a loop larger than absolutely necessary.

Circuit Path Crossing Angle

The possibility of magnetic coupling is maximized when wires pass parallel to each other. One wire carrying an aggressor current will create a magnetic field and induce a voltage into the victim wire. This transformer effect can be minimized if wires are crossed at a 90° angle.

19.11 Safety Circuits

Regardless of how good an amplifier sounds, safety must not be ignored. The two main aspects of safety are fire and the hazard of electrocution. The primary defense against fire is the line fuse or circuit breaker. A slow-blow characteristic is desirable, especially if an inrush limiting circuit is not used.

Safety Ground

Three-wire mains power cords include a safety ground as the third (green) wire. This wire runs right back to the house ground. This minimizes the danger if the neutral return path becomes

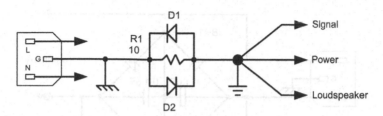

Figure 19.14 Breaking a Safety Ground Loop

open and is energized through the resistance of the load device (in this case the amplifier). The safety ground is connected to the chassis so that if there is a fault to the chassis it will be shunted by the safety ground. A typical fault that this guards against is a primary-secondary short in the power transformer. If the power transformer incorporates an inter-winding screen, it should be connected to the safety ground.

Interestingly, if the AC voltage between the safety ground and neutral is measured, a rough indication of the quality of the AC line can be inferred. Ideally, there will be no voltage measured when there is no load on the line. This measurement will give an indication of the voltage drop in the neutral line due to loading. If the only significant loading on the line is at the location of the amplifier, the total AC line drop from the breaker box to the amplifier can be inferred to be twice this measured voltage.

Breaking Safety Ground Loops

Several pieces of interconnected equipment, each with a safety ground connected to its circuit ground, may form a ground loop. This is undesirable. One approach, similar to that used in Reference 15 is illustrated in Figure 19.14. The safety ground is connected directly to the chassis ground. However, small-valued resistor R1 connects the safety ground to the circuit star ground. This is the only connection in normal operation, and the resistance breaks the ground loop. Back-to-back diodes D1 and D2 are connected in parallel with R1. In the event of a fault, the diodes conduct the fault current to the safety ground and prevent the circuit ground from ever deviating from the safety ground by more than one diode drop. In practice D1 and D2 are often replaced with a specially connected 35-A bridge rectifier wired in parallel with R1. The AC terminals are connected across R1 and the positive and negative "output" terminals of the bridge rectifier are shorted together. In this case the maximum allowed voltage drop between safety ground and circuit ground is about 1.4 V.

19.12 DC on the Mains

Although one would never think it to be there, DC voltages can exist on the mains power supply under some conditions. DC on the mains can cause the flow of direct current through the very low DC resistance of the primary of the power transformer, magnetizing the core. This can degrade transformer performance and sometimes cause audible buzzing. The cores of toroid transformers are more susceptible to the effects of DC on the mains. DC can be developed on the mains by the many loads shared among many utility customers. Some of those loads are highly nonlinear and may even employ half-wave rectification. Sometimes the presence of second harmonic distortion on the mains is accompanied with DC.

Some designers have taken measures to keep mains DC away from the transformer primary [15]. The obvious choice is to AC couple the mains through a large non-polarized capacitor, but this is

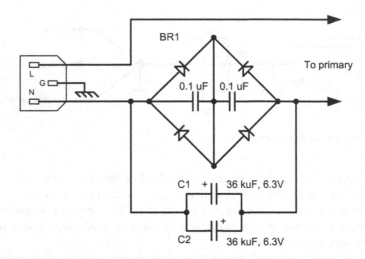

Figure 19.15 Keeping Mains DC off the Power Transformer Primary

impractical and unreliable. Another approach is to place the AC terminals of a bridge rectifier in series with the mains circuit as shown in Figure 19.15. The positive and negative "outputs" of the bridge are then shorted together. This arrangement will create two junction drops before conduction can begin, so that any DC on the mains less than about 1.4 V will not be passed.

This approach can be combined with AC coupling by placing large electrolytic coupling capacitors C1 and C2 in parallel with the bridge. The capacitors can be rated at very low voltage and will always be protected by the bridge. The bridge rectifier diodes do not conduct under normal conditions so snubber capacitors are not necessary. C1 and C2 should have very low ESR and should be rated for high ripple current. This arrangement is not recommended for amplifiers that do not incorporate a soft-start circuit. This approach does not necessarily guard the transformer against the AC effects of second harmonics on the mains, however. DC mains blocking circuits are rarely seen on consumer power amplifiers.

19.13 Switching Power Supplies (SMPS)

Switching power supplies (SMPS) are beginning to make their way into more power amplifiers. They offer high efficiency while being compact and lightweight. Their use is becoming widespread in pro audio amplifiers and in home theater receivers where these qualities are especially important. They are also moving into some audiophile power amplifiers. Their advantages include regulated outputs at no penalty in power dissipation. Of course, this means that amplifiers using these supplies will often have little or no dynamic headroom.

Switching supplies typically operate by rectifying the mains voltage on the line side and storing the DC on a reservoir capacitor. This DC is then switched at a high frequency (several hundred kHz) to become AC to drive an isolating transformer. The secondary voltage is then rectified and stored on secondary reservoir capacitors [16, 17, 18, 19]. High-frequency transformers are smaller, lighter and less expensive than transformers that operate at the mains frequency. The design of switching power supplies is highly specialized and is covered in Chapter 20.

There are some caveats when switching supplies are used for audio power amplifiers. For one, they can be a prolific source of EMI. Second, because they operate and rectify at high frequencies, the traditional ripple problem is not as great. This can lead designers to employ smaller reservoir capacitors. This is not always a good idea because it may deprive the amplifier of the large amount of energy storage necessary for reproduction of low-frequency transients.

There is also another phenomenon to watch out for. Switching power supplies are characterized by a constant power input for a given load current. If the mains voltage goes down, the current demanded from the mains will go up. Similarly, if the output voltage is constant and the load demands more current, the switcher will require more power from the mains and the front-end rectifier. If this causes the voltage at the input side to sag, still more current will be demanded by the switcher, since it is a constant-power device. The input will sag further and still more current will be demanded by the switcher. This is a potentially vicious circle. This behavior is not unlike that of a negative resistance, and can lead to instability on the line side. Such a power amplifier at the end of a long run of 14-AWG house wiring will be more prone to such instability. This also argues for a large reservoir capacitor on the mains input rectifier to better handle amplifier transients.

References

1. Robert R. Cordell, "A MOSFET Power Amplifier with Error Correction," *Journal of the Audio Engineering Society*, vol. 32, no. 1, pp. 2–19, January 1984. Available at www.cordellaudio.com.
2. Robert Carver, "A 700-watt Amplifier Design," *Audio*, pp. 24–34, February 1972.
3. Federal Trade Commission (FTC), "Power Output Claims for Amplifiers Utilized in Home Entertainment Products," CFR 16, Part 432, 1974.
4. D. Baert, "Designing Small Transformers," *Electronic and Wireless World*, pp. 17–19, August 1985.
5. K. L. Smith, "D.C. Supplies from A.C. Sources," *Electronic and Wireless World*, pp. 63–69, October 1984.
6. Toroid Corporation of Maryland, Technical Bulletins, No. 1, 3, 4. Available at www.toroid.com.
7. K. L. Smith, "D.C. Supplies from A.C. Sources—3," *Electronic and Wireless World*, pp. 24–27, February 1985.
8. IXYS VBE 26–12N07 datasheet for FRED 32A bridge rectifier.
9. Vishay datasheet for HFA25PB60 HEXFRED® Ultrafast Soft Recovery 25A diode.
10. Morgan Jones, "Rectifier Snubbing—Background and Best Practices," *Linear Audio*, vol. 5, pp. 7–26, April 2013.
11. Greg Ball, "Distorting Power Supplies," *Electronic and Wireless World*, pp. 1084–1087, December 1990.
12. Edward M. Cherry, "A New Distortion Mechanism in Class B Amplifiers," *Journal of the Audio Engineering Society*, vol. 29, no. 5, pp. 327–328, May 1981.
13. W. Marshall Leach, "Build a Double Barreled Amplifier," *Audio*, p. 52, May 1980.
14. National Semiconductor, AN-1849, "An Audio Amplifier Power Supply Design."
15. Bryston 4B-SST amplifier.
16. Keith Billings, *Switchmode Power Supply Handbook*, McGraw-Hill, New York, 2010.
17. Abraham I. Pressman, Keith Billings and Taylor Morey, *Switching Power Supply Design*, 3rd ed., McGraw-Hill, New York, 2009.
18. Robert Erickson, *Fundamentals of Power Electronics*, Springer, Boulder, CO, 1997.
19. Rudy Severns and G. E. Bloom, *Modern DC-to-DC Switch Mode Power Converter Circuits*, Van Nostrand Reinhold, Scarborough, Ontario, Canada, 1985.

Chapter 20

Switching Power Supplies

20. INTRODUCTION

Conventional linear power supplies are heavy and bulky due to the low line frequency of 50 Hz or 60 Hz. The power transformer is heavy and expensive. Large reservoir capacitors are required because of the low ripple frequency. Conventional supplies have poor power factor due to their highly nonlinear impulsive input current. The rectifier pulses of many tens of amperes under load cause voltage drops in the resistive mains supply that reduce the amount of power that can effectively be extracted from the mains.

Switch-mode power supplies (SMPS) are lighter and more efficient than conventional power supplies. They can also reduce the presence of 60-Hz or 120-Hz electrical and magnetic fields (and related harmonics) within the amplifier. Because the operating frequency is much higher than that of the mains, power supply ripple on the amplifier power rails is typically much smaller for a given amount of reservoir capacitance.

Switching power supplies are making their way into more power amplifiers. They offer high efficiency while being compact and lightweight. Their use is becoming widespread in pro audio amplifiers and in home theater receivers where these qualities are especially important. They are also moving into some audiophile power amplifiers. Their advantages include regulated outputs at no penalty in power dissipation.

Switching supplies operate by rectifying the mains voltage on the line side and storing the DC energy in a reservoir capacitor [1, 2, 3, 4]. Figure 20.1 shows a simplified arrangement. The intermediate DC bus voltage, here about 170 V, is switched at a high frequency (several tens to several hundreds of kHz) to become a pulse width modulated (PWM) AC waveform to drive an isolating transformer. The secondary voltage is then rectified and stored on secondary reservoir capacitors. High-frequency transformers are smaller, lighter and less expensive than transformers that operate at the mains frequency.

Switching is accomplished with a full-bridge switch in this example, usually consisting of four MOSFETs. Diagonally opposite switches are turned *on* at the same time. This forces +170 V across the primary on one half-cycle and −170 V across the primary for the other half-cycle, resulting in a 340-V p-p square wave. A controller implements a dead time when all four switches are *off* and no energy is being supplied to the circuit. The amount of dead time during each switching cycle controls the output voltage.

As with a conventional linear power supply, multiple secondary windings can be included on the transformer to create different secondary voltages, each with its own rectifiers, inductors and filter capacitors. Alternatively, a single rectified voltage can be created at the secondary side and if lower voltages are required they can be created by non-isolated SMPS converters.

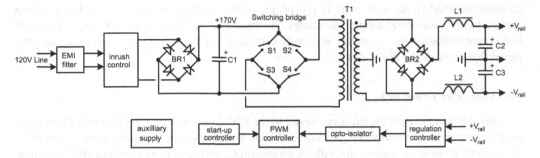

Figure 20.1 Switch-Mode Power Supply

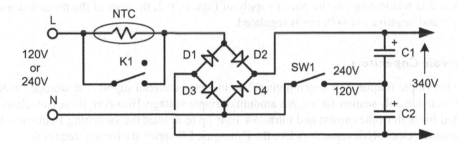

Figure 20.2 Line Supply

20.1 Line DC Supply

The line supply part of an SMPS often converts the mains voltage into a DC voltage of about 340 V for use by the subsequent isolated DC-DC converter. This allows operation from 120-V or 240-V line voltages. It is easy to see that for a 240-V input, the line supply shown in Figure 20.2 is merely a full-wave bridge rectifier feeding a filter capacitor consisting of C1 and C2 in series. These each may be on the order of hundreds of µF. Notice that SW1 is open for 240-V input operation. For 120-V operation, SW1 is closed, converting the arrangement into a voltage doubler [2]. Under these conditions, D3 and D4 are reverse-biased and not used. Inrush control is implemented with an NTC resistor. It is shunted by relay contacts that are closed shortly after power is applied.

20.2 Isolated DC-DC Converter

So-called *off-line* switching power supplies operate off the mains voltage and almost always provide isolation. Isolated switching converters are quite simple in concept. A high-frequency transformer is used instead of the line-frequency transformer in a conventional supply. The mains voltage is first rectified and filtered to produce a DC bus voltage of about 340 V. As shown in Figure 20.3, that voltage is then chopped into a square wave by a full-bridge switch to provide the AC signal necessary for feeding the high-frequency transformer. The conversion of the DC bus voltage to the AC square wave voltage is done with MOSFET switches. The diodes shown in the switching bridge are the body diodes integral to the MOSFETs. The secondary of the transformer then feeds a rectifier and reservoir capacitor.

The off-line DC-DC converter includes voltage regulation that is implemented via pulse width modulation (PWM). As with class D power amplifiers, voltage pulses with smaller *on* times applied to the transformer supply less energy and thus result in a lower output voltage. The rectified output voltage at the secondary is usually compared to a reference voltage to create an error signal that controls the PWM circuit.

Optocoupler Feedback

In many designs the error signal is conveyed to the PWM controller on the line side by an optocoupler to provide the necessary galvanic isolation. Because the optocoupler is only conveying error information in a closed-loop feedback arrangement, variation in its coupling efficiency matters little. If multiple secondary windings are used to implement different voltage supplies, all but the one connected in the error loop will have reduced load regulation. The term *cross-regulation* refers to this relationship. In the power supply of Figure 20.3, the sum of the magnitudes of the positive and negative rail voltages is regulated.

Reservoir Capacitors

Since the ripple frequency is much higher than in a conventional supply, the storage capacitors can be considerably smaller for a given amount of ripple voltage. However, these capacitors must be rated for high ripple current and ultra-low ESR up to at least the switching frequency. High-performance electrolytic capacitors like the Panasonic FR series are usually required.

Fairly large reservoir capacitors may still be necessary if the very large brief peak currents demanded by signal transients in the power amplifier cannot be quickly supplied by a fast-regulating SMPS that is capable of supplying the worst-case maximum current demanded when the amplifier is driving a low-impedance load. An SMPS for a 100-watt/8-Ω amplifier must be able to supply peak current of 20 A into a 2-Ω load if it is not to fall out of regulation. This number should be doubled for a stereo version of the amplifier. In a linear supply such large brief current demands are supplied by large reservoir capacitors. An SMPS can be designed to operate in the same way if equipped with large reservoir capacitors, but there may be a cost and space trade-off. In either case, a fairly large intermediate bus reservoir capacitor is usually required.

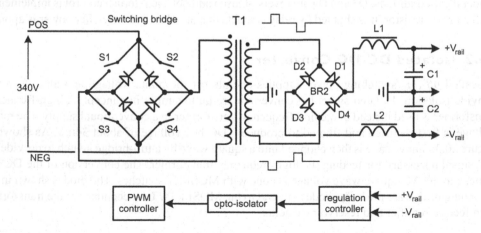

Figure 20.3 Isolated DC-DC Converter

Negative Input Impedance

In contrast to a linear supply, a typical SMPS will do whatever it takes to keep its output at the fully regulated value when the mains voltage falls for any reason. This means that the SMPS will demand more current from the mains, since its output power remains the same. This creates a negative input impedance characteristic. This is usually not a problem with a low-impedance mains supply. However, if the line voltage is supplied by a 50-foot run of 14-AWG wire from the other end of a house, the round-trip resistance is about 0.5 Ω.

Consider a 200-wpc/8-Ω stereo amplifier in the worst case where it delivers its full peak voltage capability of 56 V into 2 Ω at 20 Hz, requiring 28 A per channel for a total peak current of 56 A at a likely rail voltage of 60 V. This corresponds to peak power of 3360 watts. At the nominal mains voltage of 120 V RMS, this corresponds to about 30 A at the mains if the SMPS efficiency is 90% and the negative resistance effect is temporarily ignored. This peak load will drop the peak mains voltage by 15 V in the case of a 0.5-Ω mains wire resistance. The reduced mains voltage of 105 V will in turn cause the mains current drawn by the SMPS to increase to about 34 A in order to continue to supply the same wattage. This will result in an additional drop of 2 V. These rough numbers assume that the SMPS utilizes power factor correction so that the input current to the SMPS is reasonably sinusoidal.

This extra drop of about 2 V due to the negative resistance input characteristic of the SMPS exacerbates the mains sag problem, but does not come close to causing mains voltage instability. Nevertheless, the negative input resistance exhibited by an SMPS is a characteristic that should be kept in mind. Resistance in the EMI filter and line fuse can also play a role. A large intermediate bus reservoir capacitor will mitigate this effect in transient situations.

20.3 Buck Converters

The buck converter takes a DC input voltage and steps it down to a lower DC voltage. Understanding the buck converter is central to understanding most of the concepts underlying SMPS design. The buck converter uses a transistor switch, a diode, an inductor and a filter capacitor, as shown in Figure 20.4. It is basically a PWM arrangement that relies on the fact that an inductor tries to keep current flowing as its magnetic field collapses. It operates with high efficiency because the switching device is only *on* or *off*. The relative *on* time of the switch controls the amount of energy that is delivered to the load and thus the output voltage. In PWM terms, the ratio of switch *on* time to the switching period is referred to as *duty cycle*, or just *D*. In other words,

$$D = T_{on}/T_s \qquad (20.1)$$

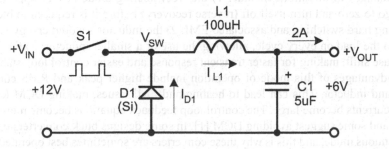

Figure 20.4 Buck Converter

The inductor is placed between the source and the load. If V_{in}-V_{out} is ΔV, then the current flowing in L1 when the switch is closed will rise with a slope of $\Delta V/L1$, reaching a maximum at the end of the time that the switch is closed. When the switch opens, the inductor tries to keep the current flowing, causing the voltage at the input side of the inductor to fly negative to minus one diode drop (V_d). The current is then sourced through catch diode D1, sometimes called the *freewheeling diode*. Inductor current continues to flow into the load as the magnetic field in the inductor collapses, decreasing with time at a rate ($V_{out} + V_d$)/L.

If the switch turns back on before the current reaches zero, the buck converter is said to be in the *continuous conduction mode* (CCM). However, if the current reaches zero before the switch turns back on, the converter is said to be in the *discontinuous conduction mode* (DCM).

Continuous Conduction Mode (CCM)

In the steady state, current is usually flowing continuously in the inductor during both the closed and open states of the switch. In other words, S1 turns back on before the magnetic field fully collapses. A triangular ripple current thus flows in the inductor. When the current in the inductor does not go to zero at any point in the cycle, the converter is said to be operating in the *continuous conduction mode* (CCM) of operation and the output voltage is a linear function of the duty cycle [1, 2].

Switch S1 is inevitably implemented with an N-channel MOSFET. In principle the gate drive circuitry would be simpler if a P-channel device were used, but the *on* resistance of P-channel devices is significantly higher for a given die size, and this is a serious disadvantage. For a given *on* resistance the larger P-channel die size increases cost and leads to significantly higher gate charge that results in slower speed.

A level translation and energy storage circuit is needed within the buck IC controller to drive the N-channel MOSFET gate high with respect to its source (which is at V_{sw}). Modern IC controllers are now available at low cost incorporating such level translation up to 600 V. Alternatively, pulse transformers are sometimes used to provide isolated gate-source drive for the floating switch, but this may add cost and space.

EMI is created in the CCM mode when the switch turns on and D1 goes through a hard-switched *reverse recovery*. High dI/dt in the MOSFET and diode can result, creating EMI in the 70–300 MHz range that can radiate off the PCB tracks. Using diodes with fast and soft recovery can mitigate this. Schottky diodes are even more preferable.

Discontinuous Conduction Mode (DCM)

If the load current is smaller than half the peak-to-peak value of the inductor ripple current, the converter will enter the so-called *discontinuous conduction mode* (DCM), where inductor current stops flowing before the end of the *off* time of the switch [2].

The advantages of this mode include: 1) the freewheeling diode has an opportunity for its current to go to zero and turn itself off (reverse recovery) before it is required to block voltage, thus avoiding hard switching and associated EMI; 2) the inductor's stored energy is completely delivered to the load on every cycle, allowing the use of a smaller inductor; 3) the control loop has less phase shift, making for faster transient response and easier control loop stabilization.

The disadvantages of this mode of operation include higher peak and RMS currents in the MOSFET and inductor. This can lead to heating and EMI issues, making DCM less desirable when load currents become large. The control loop feedback equations become more complex in this mode, and some suggest avoiding DCM [1]. In some designs buck converters operate better in the continuous mode, and this is why these converters are sometimes best operated with a load current greater than some minimum value.

If the inductance is too small, or if the switching frequency is too low, the current in the inductor will fall to zero before the end of the second half-cycle where energy is being transferred to the load. In this case, it can be shown that the relationship of output voltage to duty cycle is no longer linear [1]. The time slope of the current in the inductor is determined by the difference in the input and output voltage and the value of the inductance. A larger inductance will result in a shallower slope and less likelihood that current in the inductor will fall to zero. In the design shown, the boundary between DCM and CCM is I_{crit} = 150 mA for D = 0.5.

I_{crit} is a function of D for a given converter design. Excellent discussions of the location of the boundary between CCM and DCM can be found in these references [3, 4].

The minimum load current for CCM, I_{crit}, can be shown to be:

$$I_{crit} = V_{in} \, (T_s/2L) \, K(D) \tag{20.2}$$

where

$$K(D) = D(1 - D) \tag{20.3}$$

The term K is at its maximum value of 0.25 when D is 0.5. This means that I_{crit} is at its maximum at a 50% duty cycle. For V_{in} = 12 V, D = 0.5, T_s = 10 μs and L = 100 μH, I_{crit} is 150 mA. K has a broad maximum, and is down to 0.16 at D = 0.2 and D = 0.8.

All PWM converters require a minimum load current to operate in CCM. In the discontinuous portion of DCM, where current is not flowing in L1, S1 is open and D1 is not conducting. The V_{sw} node is thus not shunted and is free to oscillate with large amplitude at a frequency determined by L1 and the parasitic capacitance at the V_{sw} node. The discontinuous mode of operation can create increased EMI due to large busts of under-damped oscillations during the discontinuous interval. These oscillations can occur in the 2–20 MHz frequency range. A series R-C snubber network across D1 in Figure 20.4 can suppress this oscillation while dissipating little power. Here values of 200 Ω and 1000 pF work well. Further discussion of DCM is beyond the scope of this chapter, and the converters described are assumed to be operating in CCM.

The minimum load current in a class AB power amplifier can be much smaller than the peak or even average current under signal conditions. This means that the possibility of operation in the discontinuous mode must be considered. The buck converter will usually still work, but its control characteristics change and become nonlinear. This can affect control loop stability and recovery from the discontinuous mode to continuous mode operation.

Critical Conduction Mode (CrM) aka Boundary Conduction Mode (BCM)

Some converters are operated right at the border of CCM and DCM. This is called *critical conduction mode* (CrM), also known as *boundary conduction mode* (BCM). This switching behavior is accomplished by monitoring the inductor current and turning on the switch just when the inductor current goes to zero. For a given PWM duty cycle D, this means that the *off* time must be a function of the *on* time that is established when the inductor current falls to zero. As a result, the switching frequency changes. At light load, the inductor current will fall to zero more quickly after the switch turns *off*. This in turn means that the switch *on* time for a given D will be smaller, resulting in a higher switching frequency. This approach preserves the advantages of CCM at light loading.

In CrM the switch will turn *on* with zero current flowing through it. This is referred to as *zero-current switching* (ZCS). This *soft switching* event reduces switch power dissipation during turn-on.

CrM is used to strive for the best of CCM and DCM. The MOSFET is turned on as soon as the diode current reaches zero, minimizing RMS currents and resulting MOSFET and inductor heating. By letting the inductor current fall to zero each cycle, the entire stored energy of the inductor is utilized each cycle. This can allow the use of a smaller inductor than in CCM.

Buck Converter Waveforms

Operation of the buck converter is further explained by the waveforms in Figure 20.5, where the converter is delivering 2 A to the load in continuous conduction mode. During the first half-cycle when S1 is closed, the input voltage is applied to the inductor and the current will rise linearly with time in accordance with the voltage across the inductor and the inductance. It will reach a value I_{max}.

When S1 opens, the current in L1 will continue to flow due to its collapsing magnetic field. This current is usually referred to as the *flyback current* or the *commutation current*. During the second half-cycle the only path the current can flow is through D1 from ground. The switched output voltage V_{sw} thus snaps from +12 V down to about −0.7 V, the forward drop of D1. It is important to recognize that the current continues to flow through the inductor during this second half-cycle. During this time the current in the inductor will fall linearly to I_{min}. Switching frequencies and inductor values are usually chosen so that I_{min} is greater than zero at the end of the second half-cycle. The average of I_{max} and I_{min} is the output current into the load. The difference between I_{max} and I_{min} is called the ripple current.

As shown in Figure 20.5, D1 is in a conducting state at the end of the second half-cycle when S1 again closes and raises the switched output from −0.7 V to +12 V. Diodes in a conducting state cannot instantly stop conducting when the current is reversed. Silicon diodes that have been conducting contain stored charge in the form of *minority carriers* that must be swept out before the diode can allow reverse voltage across its terminals. This process is called *reverse recovery*. As a result, a brief large current spike will flow backwards through D1 when S1 closes, followed

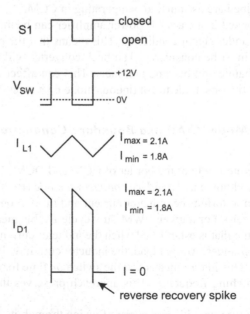

Figure 20.5 Buck Converter Waveforms (CCM)

by a snappy collapse to zero. This reverse current is undesirable and represents lost energy. The rapid current reversal and subsequent rapid collapse to zero both represent large dI/dt events that are sources of EMI. Fast diodes with small reverse recovery times are thus desirable to minimize energy loss. Schottky diodes are often employed for this reason. They do not involve minority carriers and are largely free from the reverse recovery effect altogether. They are also preferred because of their smaller forward voltage drop. However, they have larger junction capacitance and cost more.

Tips for Steady-State Converter Analysis

There are five "rules" that are very handy for analysis of converters in the steady state [3]. They may seem obvious, but they are nevertheless very helpful.

* Inductor volt-second balance
* Capacitor ampere-second balance
* Power in equals power out
* Inductor current never changes instantly
* Capacitor voltage never changes instantly

Over time, the number of volt-seconds applied to an inductor must equal the number of volt-seconds removed from the inductor. In the steady state, "over time" here generally means over one switching cycle. This is simply a re-statement that no net DC voltage can exist across an ideal inductor without resistance. The dual of this rule for a capacitor is that ampere-seconds applied to a capacitor must equal ampere-seconds removed from the capacitor. This is just a re-statement that DC current cannot flow through a capacitor. An ideal converter operates with 100% efficiency, so power in must equal power out. Combining these "rules" in studying converter behavior can be remarkably helpful. Because inductors play such a central role in most of the converters discussed here, the volt-second balance for inductors is probably the most important concept. For example, if one applies 10 V for 10 μs to an inductor, then 100 V will result if the inductor is allowed to collapse its magnetic field over a time interval of 1 μs.

Simulation

SPICE simulation for switching converters is extremely valuable, and not difficult for the simple PWM converters illustrated in this chapter. For example, operation in DCM is easily spotted in a plot of the inductor current waveform. Bear in mind that many cycles may have to be simulated to arrive at steady-state results. Pre-charging the output filter capacitor to the expected output voltage can sometimes reduce the number of cycles needed to reach steady state. If the simulation runs extremely slow, change something to get it out of the rut it is in. For example, change the output filter pre-charge voltage, even if it is not to the expected steady-state value. The converters here have been designed for a conservative 100-kHz switching frequency.

It is recommended that a load resistance be used, rather than a load current source. The latter will allow under-damped LF resonance between the switching inductor and the filter capacitor that can increase required simulation time to reach steady state. The use of SPICE switch elements in place of MOSFETs is fine, but a reasonable value for the switch resistance should be included and a silicon drain-source diode should be added to account for the MOSFET body diode and its capacitance. An RF2001NS2D soft-recovery diode was used in the simulations here.

Conversion Ratio and Duty Cycle

The conversion ratio M is the ratio of output voltage to input voltage, and it is a function of the PWM duty cycle D, where D is the ratio of *on* time to the switching period. For the buck converter in continuous conduction mode, $M(D)$ is linear, and $M = D$. For other converters, the relationship is not necessarily linear, and this must be taken into account in the feedback control design, since control loop gain will be dependent on an incremental value of M that can change with D.

$$M = V_{out}/V_{in} \tag{20.4}$$

For the ideal buck converter in CCM, the conversion ratio is simply:

$$M = D \tag{20.5}$$

It is worth noting that buck converters often work best when D is a larger value, since smaller peak current through the switch is required to transfer a given amount of power. This leads to smaller conduction losses. In applications where the nominal converter operating point can be chosen, it should be chosen to be at a larger value of D.

Open-Loop Load Regulation

Although most PWM converters are operated in a closed feedback loop with controllers, it is useful to understand the open-loop performance of the regulator. This pertains mostly to load regulation and output impedance. One might ask, if the load suddenly changes and the duty cycle is not changed, how much will the output voltage change? This is especially relevant if the feedback loop in the converter is relatively slow. The open-loop output impedance is quite low for the PWM regulators simulated in CCM. Load regulation was correspondingly good. Open-loop output impedance for the buck converter shown, at $D = 0.5$, is on the order of 40 mΩ when switch and inductor resistances of 20 mΩ are used. In contrast, open-loop load regulation is quite poor in DCM.

Switching Losses and EMI

Whenever a MOSFET turns on with voltage across it, power will be dissipated during the transition as its current increases and the voltage across it decreases. It is important to recognize that such transitions always take a finite amount of time. Available gate drive current in combination with Miller effect from C_{gd} can be one of the limiting factors, for example. Similarly, when a MOSFET that has been conducting turns off, power will be dissipated during the off transition as I_d decreases and V_{ds} increases. These are stressful intervals for the MOSFET and very high power can be dissipated during the brief transition interval.

Consider the buck converter of Figure 20.4 operating in continuous mode. Prior to the MOSFET (S1) turning on, D1 is turned on with the full output current flowing through it. The switching node is at about −0.7 V. When the MOSFET turns on, it has slightly more than the full input voltage across it. As the MOSFET's current increases over a finite amount of time, D1 will remain *on* until the current in the MOSFET rises to equal the full inductor current, finally allowing D1 to turn off. During this turn-on interval, the power dissipation in the MOSFET has been increasing to a high value. After D1 turns off, the voltage across the MOSFET will decrease toward zero, still dissipating power during the second half of the transition as V_{sw} rises and current through the MOSFET is high.

The V_{sw} node in Figure 20.4 has significant parasitic capacitance, C_p. This includes the drain-source capacitance of the MOSFET, which can be considerable. The capacitance of D1 and the winding capacitance of L1 are additional contributors. Total parasitic capacitance at V_{sw} can be several hundred pF. The sharp edges on the V_{sw} node in the idealized waveforms of Figure 20.5 will cause significant current to flow in charging and discharging this capacitance. In particular, note that the MOSFET switch has the full power supply voltage across it when it turns on. This means that it must quickly discharge that capacitance. The simultaneous presence of voltage across the switch and current through the switch means that power is being dissipated in the switch.

When the switch is off, the full supply voltage is across C_{ds} of the MOSFET, and significant energy is stored in C_p. When the switch closes, C_p undergoes a lossy voltage transition of magnitude V_{in}, and this energy is dissipated in the switch. If the switch turns on once per period T_s, then the average power dissipated in the switch is:

$$P_{diss} = 0.5 C_p (V_{in})^2 / T_s \qquad (20.6)$$

If C_{ds} is 500 pF, V_{in} is 100 V and the switching frequency is 100 kHz, this corresponds to 0.25 W in switching loss caused by capacitance alone. This switching loss goes up directly with switching frequency and with the square of input voltage. An off-line converter operating from 340 V and operating at 500 kHz could dissipate 29 watts. This gives a sense of why such high frequency operation is impractical in a hard-switched off-line converter.

The voltage at V_{sw} has a high edge rate (dV/dt), as does the current through the switch. These conditions create significant EMI over a wide spectrum of frequencies. The input current to the converter is pulsating because there is no inductor between the switch and the input node. This injects noise into the input port.

20.4 Synchronous Buck Converter

Figure 20.6 illustrates a synchronous buck converter. Here D1 is replaced with a switch S2 implemented with a MOSFET. This arrangement is more efficient because there is no junction drop when the inductor current is flowing from ground on the second half-cycle. That voltage drop represents lost power. S2 is sometimes referred to as a *synchronous rectifier*. It is a switch that is turned on when it is supposed to be conducting like a rectifier. In this arrangement, S2 is off when S1 is on and vice versa. Operation is largely identical to that of Figure 20.5 with the exception that V_{sw} falls close to zero during the second half-cycle instead of to −0.7 V.

There is one concern with the synchronous buck converter. If both switches are on even briefly the input power supply will be shorted to ground and a very large *shoot-through* current will flow.

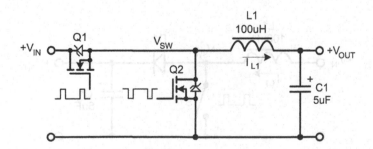

Figure 20.6 Synchronous Buck Converter

This must be avoided by adding *dead time* to the operation of the switches, when both switches are off for a brief time. If S1 opens and S2 has not yet closed, a negative flyback voltage will be created by the collapsing magnetic field of L1. The flyback voltage is clamped for this brief interval by adding a freewheeling diode across S2. Conveniently, such a diode is inherent within the structure of all MOSFETs and is called the *body diode*. The switched output node will then be prevented from going more negative than −0.7 V. This is why the negative portion of V_{sw} may exhibit negative −0.7 V "ears" at the beginning and end of the second half-cycle when both switches are open.

Because the dead time is usually kept very small compared to the period, most of the improved efficiency contributed by S2 is preserved. Importantly, the R_{DSon} of the MOSFET used for S2 must be small enough so that the voltage drop across it when it is *on* is considerably smaller than the *on* voltage of the diode.

Unfortunately, this arrangement does not prevent the reverse recovery current spike when S1 closes. This is because the freewheeling diode becomes conducting during the dead time just before the end of the second half-cycle. The reverse recovery shoot-through current can be reduced if a Schottky diode is connected externally in parallel with the MOSFET. It will prevent the slower silicon body diode from ever turning on. It will also reduce the size of the "ears" by about half.

20.5 Boost Converters

A buck converter can only create an output with a lower voltage than that of the DC source. In contrast, a boost converter can only create an output voltage greater than that of the DC source. A simple boost converter is shown in Figure 20.7. The DC source is connected to one end of inductor L1 and the other end of L1 is connected to ground through shunt switch S1 and to the load by series diode D1.

S1 is turned on during the first half-cycle of operation, causing current to flow from the source through L1 to ground. The current rises with time until S1 is turned off. During this time energy is stored in L1. When S1 is opened during the second half-cycle, the inductor seeks to keep the current flowing by creating a flyback voltage that forward biases D1. This allows the energy in L1 to be dumped into the load. The current in the inductor will then decrease with time as the energy is transferred to the load. Because there is no inductor between the switch and the load, pulsating current is delivered to the load and can create EMI on the load port.

As in the buck converter, there exists a *discontinuous mode* (DCM) that will be entered when the load current is smaller than a minimum amount. When current is not flowing in L1, S1 and D1 are open, leaving L1 to resonate at high amplitude with the parasitic capacitance at the switching node. In the design shown, the boundary between DCM and CCM is 150 mA for $D = 0.5$.

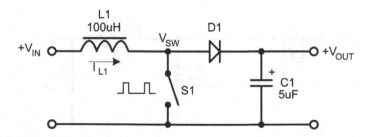

Figure 20.7 Boost Converter

In CCM, the control transfer function of the boost converter includes a *right-half-plane* (RHP) zero. This can make it more difficult to stabilize the control loop. For this reason, boost converters are sometimes operated in the discontinuous mode [1]. This mode of operation will not be discussed here.

The amount of energy transferred from the source to the load in CCM depends on the duty cycle D of the *on* time of S1. A PWM controller is used to control the *on* time and thus regulate the output voltage against input voltage and load current changes. During the second half-cycle the input source is also supplying the current that is flowing through the inductor into the load. This means that the energy flowing into the load is not just from the energy stored in the inductor, but also that flowing from the source into the inductor as well. The conversion ratio for the boost converter in CCM is:

$$M = 1/(1 - D) \tag{20.7}$$

As D approaches unity, M becomes very large and very high output voltages can be produced.

20.6 Buck-Boost Converters

The buck and boost converters are limited. The buck converter can only reduce the output voltage, while the boost converter can only increase the output voltage. The buck-boost converter shown in Figure 20.8 can provide increased or decreased output voltage. At $D = 0.5$, $V_{out} = V_{in}$.

A buck-boost converter can be obtained by a brute-force tandem connection of a buck converter and a boost converter. However, the resulting combination includes a shunt capacitor between two series inductors. It can be shown that this capacitor creates a third-order LPF, but that the capacitor is otherwise superfluous [3]. The capacitor can be removed and the inductors merged into one. With some further rearrangement, the output portion can be flipped and the second switch can be replaced with a commutating diode, arriving at the much-simplified arrangement of Figure 20.8.

As a result of this simplification, the output voltage is inverted and becomes negative. This can be advantageous in applications requiring a negative voltage. When the switch is closed, the input voltage is impressed across L1 and current rises linearly to I_{max}. When the switch is opened, the inductor seeks to keep the current flowing. Its output flies negative, sinking current from the load via D1. The current in L1 falls linearly as it discharges its energy into the load.

Like the other converters, this converter has continuous and discontinuous modes. If the current in the inductor falls to zero before the switch is turned on again, the converter will enter DCM, and the control transfer function will change. If the inductance is too small, or the load is too light, the converter enters the DCM. In order to avoid DCM, a minimum load current must

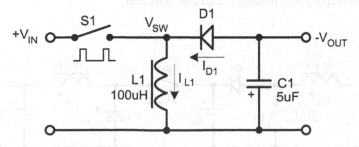

Figure 20.8 Buck-Boost Converter

be drawn. A larger value of inductance reduces the minimum load current required to remain in CCM. In the design shown, the boundary between DCM and CCM is 150 mA for $D = 0.5$.

The conversion ratio for the buck-boost converter is:

$$M = -D/(1 - D) \tag{20.8}$$

Because there is no inductor between the switch and the input or output ports of the buck-boost converter, it can be fairly noisy, injecting switching noise into both ports. Even when in CCM, there is pulsating current at both the input and output sides of the converter.

20.7 Boost-Buck Converters

The boost-buck converter family is topologically somewhat the reverse of the buck-boost family. A boost-buck converter is shown in Figure 20.9. It literally comprises boost and buck converters wired in tandem, and requires two each of every component. Both switches are opened and closed at the same time. It can provide boost or cut of the input voltage without inversion of the output voltage polarity. A further advantage is that there is no pulsating current present on either the input or output side of the converter as a result of the use of two inductors. This makes the converter relatively quiet.

In the design shown, the boundary between DCM and CCM is 290 mA for $D = 0.5$. The smallest boundary current occurs when L1 = L2, since CCM must be satisfied for both inductors. The conversion ratio for the boost-buck converter is:

$$M = D/(1 - D) \tag{20.9}$$

20.8 Ćuk Converters

The Ćuk converter (pronounced *Chook*) is an important member of the boost-buck family, and can be thought of as a reduced form of the boost-buck converter. It was named after Slobodan Ćuk, who presented it in 1976 at the IEEE Power Electronics Specialists Conference, co-authored with Robert Middlebrook [5]. It has sometimes been referred to as the "Optimum Topology Converter," but this assertion has understandably been controversial [6, 7, 8].

Figure 20.10 shows non-isolated and isolated versions of the Ćuk converter. It can produce a higher or lower output voltage than the input, but the non-isolated version can produce only a negative (inverting) output. It requires two inductors. It is somewhat like a boost converter followed by a buck converter with the capacitor of the boost converter coupling the energy to the buck converter. Because there are inductors on both sides of the switch, less EMI is transferred to the input and output ports, making it a quieter converter.

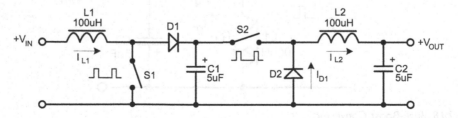

Figure 20.9 Boost-Buck Converter

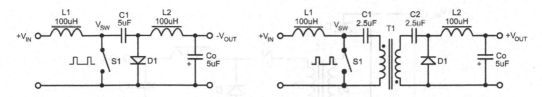

Figure 20.10 Non-Isolated and Isolated Ćuk Converters

In analyzing operation, first recognize that in the steady state C1 has $V_{in} - V_{out}$ across it, and that C1 is large enough that there is little ripple across it. When S1 is closed, the input voltage is impressed across L1. As current increases, more energy is stored in L1. When S1 is opened, L1 seeks to keep the current flowing, causing V_{sw} to fly positive, charging C1 through D1. When S1 again closes, V_{sw} is pulled negative, D1 is reverse-biased, and current is sunk through L2, storing charge in L2. All of the energy in L1 is thus transferred to L2 via coupling capacitor C1.

In the design shown, the boundary between DCM and CCM is 290 mA for $D = 0.5$. The smallest boundary current occurs when L1 = L2, since CCM must be satisfied for both inductors. The conversion ratio for the Ćuk converter is:

$$M = -D/(1 - D) \tag{20.10}$$

An isolated version of the Ćuk converter is easily implemented. A transformer is merely interposed between two halves of the coupling capacitor. Whatever impedance is seen on the secondary side is merely transformed back to the primary side by the square of the turns ratio. It is also straightforward to implement multiple secondary outputs, where the circuitry on each winding is merely replicated. The isolated Ćuk makes possible a non-inverted converter, as shown in the figure.

20.9 Forward Converters

A buck-derived *forward converter* is shown in Figure 20.11. It is essentially a buck converter wherein a transformer has been interposed in the forward path between the switch and the reactive components. The transformer merely transforms what is going on in the secondary back to the primary by its turns ratio (or the square of its turns ratio). The transformer does not just provide isolation. Its turns ratio can be chosen to provide the desired regulated output voltage at the optimum PWM duty cycle under nominal conditions.

When Q1 turns on, current flows in the primary, making the dotted ends of the transformer windings positive with respect to their opposite ends. D2 then sources secondary current into L1, just as did S1 in the buck converter of Figure 20.4. At the end of the *on* time, Q1 turns *off* and the secondary voltage snaps negative, reverse-biasing D2. Inductor L1 then seeks to maintain current flow by snapping its input end negative, forward biasing D3 and maintaining current flow into the load. L1 is central to the buck-derived operation of the forward converter. It is not just there for extra filtering. For multiple outputs, each additional secondary winding must be followed by the same diode-inductor-capacitor arrangement as for the first output circuit.

After Q1 and D2 turn off, the collapse of the magnetic field in the transformer would normally drive the drain of Q1 to a dangerously high voltage. This is prevented by catch diode D1 and the energy recovery winding P2, which normally has a 1:1 turns ratio with the primary winding P1. The current from the collapsing magnetic field of T1 flows through D1 and returns this energy to the input supply [1].

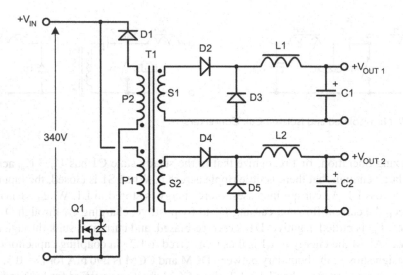

Figure 20.11 Forward Converter

20.10 Flyback Converters

If a second inductor L2 is wound on the same core as L1 in the boost converter, D1 and C1 can be fed from L2. This is a *flyback* converter, illustrated in Figure 20.12. The energy stored in the core by L1 during the first half-cycle is given up to the load by means of the L2 winding when Q1 is turned off, producing a positive output voltage via D1. This is much like the operation of a boost converter, but with isolation. Notice that no inductor is required in the secondary circuit. The figure also shows a second output voltage, this one negative, produced by L3, D2 and C2. Multiple outputs are economically supported by flyback converters, and their voltages are related by the turns ratio among the "secondary" inductors.

It is tempting to refer to the combination of L1 and L2 wound on the same core as a transformer, but it is more technically correct to view them as two inductors sharing the same core [2]. These are so-called *coupled* inductors. In transformer operation, current is usually flowing in both the primary and secondary at the same time. This is what happens in the *forward* converter described earlier. In the *coupled-inductor* flyback converter, current flows only in one inductor at a time. The flyback voltages appearing on L1 and L2 are still related by the turns ratio of L1 and L2. The turns ratio can be chosen to provide output voltages that are higher or lower than the source voltage. The term *flyback transformer* is often used for convenience. The Kettering ignition in automobiles is basically a flyback converter. Flyback converters are capable of producing very high voltages. The first DC-DC flyback converters were used in the horizontal deflection systems in televisions, where the high voltage for the CRT anode was generated. The fast horizontal *flyback* of the beam at the end of each line in the raster is the origin of the term *flyback*.

Energy Storage in the Core

In a transformer that is operating properly, little energy is stored in the magnetic core, since the load current flowing in the secondary opposes the creation of net magnetic flux by the current

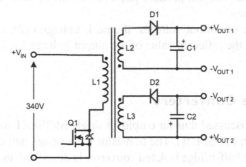

Figure 20.12 Flyback Converter

flowing in the primary. This is not the case for the coupled inductors in a flyback converter. Magnetic flux is created in the core in the first half-cycle and given up to the secondary in the second half-cycle. This means that all of the energy passing through the converter must be stored in the core at some point. The resulting large net magnetic flux can require that the core be larger in order to avoid saturation. This is a slight disadvantage compared to the forward converter whose primary and secondary windings operate in transformer mode.

The flyback converter is quite popular because a single component, the coupled inductor, provides the functions of energy storage, energy transfer and isolation [2]. The flyback converter is also economical in providing multiple outputs without the need for multiple inductors on the secondary side. However, flyback converters are usually limited to applications requiring less than 100 W. Because flyback converters source current to the load on only one half-cycle, and because they often are preferably operated in the discontinuous mode, RMS switch currents and output ripple voltage tend to be larger than that of the buck-derived forward converter. Overall, flyback converters tend to be noisier than others because of large dV/dt and dI/dt events at switch turn-on and turn-off, respectively.

Applications and Discontinuous Mode

Flyback converters are extremely popular for low-medium power applications because of their simplicity, requiring only one magnetic component and providing isolation. They are popular for audio amplifiers because multiple secondaries are readily provided and none of them require an output inductor. They do, however, require large filter capacitors because of the pulsating nature of the output current. They have been the converter of choice for televisions and many other consumer devices.

Flyback converters are often operated in the discontinuous mode, where the RHP zero is not as troublesome for compensation. There is substantial parasitic capacitance C_p at the drain node of the switch due to the sum of C_{ds}, the transformer winding capacitance and the output rectifier capacitance reflected through the transformer turns ratio. C_p will resonate with the inductance seen looking into the primary inductor. When Q1 is conducting and current is flowing in L1, there will be no resonance. When D2 is conducting and current is flowing in coupled inductor L2, C_{ds} will resonate with the small leakage inductance seen at L1. This resonance will be at a high frequency and will add significantly to the peak switch voltage and create EMI. An R-C snubber or R-C-D clamp network placed across the switch can reduce this peak voltage and suppress these oscillations. The impedance of the snubber network should be high enough to limit its dissipation to less than 1% of the operating power of the converter. During the discontinuous interval, the full primary inductance will be seen and the resonance will be at a much lower frequency.

Ignoring resonance or overshoot voltages, the peak voltage across the switch is the sum of the input voltage V_{in} and the reflected value of the output voltage V_R, as reflected by the flyback transformer turns ratio.

20.11 Half-Bridge Converters

Most of the converters discussed thus far employ a single MOSFET for switching (not counting synchronous rectification MOSFETs). The transistor is *on* or *off*, and control of the output voltage is by PWM. Here the half-bridge isolated converter is described, as shown in Figure 20.13. It employs two transistors in a push-pull half-bridge configuration and is suitable for higher-power applications.

A conventional half-bridge switch arrangement toggles its output from one input rail to the other by alternately turning on the top or bottom switch. The output is thus a square wave. It is critical in such half-bridge arrangements that some *dead time* be designed into the circuit. This means that one transistor begins its turn-on transition only after the other transistor completes its turn-off transition. Otherwise, wasteful and potentially destructive *shoot-through* current will result if both of the transistors are conducting any amount of current at the same time.

In Figure 20.13, output voltage is controlled by duty ratio modulation. One or the other of the half-bridge transistors Q1, Q2 is *on*, or both are *off*. Notice that this converter requires a "center-tapped" input voltage, here AC-coupled and provided by split intermediate bus reservoir capacitors. When Q1 is *on*, +170 V is applied to the primary. When Q2 is *on*, –170 V is applied to the primary. The percentage of *on* time as compared to the total time of the cycle controls the amount of energy injected into the circuit and thus the output voltage. The non-overlapping drive pulses for Q1 and Q2 are shown in the figure. These drive pulses can be derived from two 50% square waves that are phase-shifted with respect to one another. The pulses are then created with appropriate exclusive-OR circuitry. The advantage of this approach is that the switching frequency is held constant as the control is varied.

Staircase Core Saturation

Small differences in component parameters between the devices responsible for either half-cycle of operation can lead to a small build-up of DC magnetic flux on each cycle, potentially accumulating in staircase fashion over many cycles. This moves the core of the transformer toward

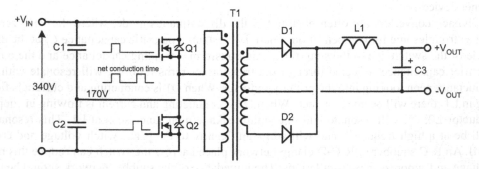

Figure 20.13 Half-Bridge Converter

saturation. Any tendency to core saturation is undesirable. AC coupling of the primary can mitigate this tendency on the primary side, but mismatches on the secondary side can still cause the
effect.

The tendency to saturation occurs in one polarity of the winding voltage, causing second harmonic distortion. Control ICs can employ active means to detect the resulting asymmetry and
compensate for it. Control can be accomplished by injecting an opposing DC current into a winding or by causing a deliberate switching time asymmetry so as to cause a DC magnetic flux that
cancels out the original DC flux. Negative feedback can be used to drive the net DC flux to zero.

20.12 Full-Bridge Converters

A simplified full-bridge converter is shown in Figure 20.14. Full-bridge converters are popular
for higher-power applications over 500 W. A full-bridge switching arrangement using four MOS-
FETs is employed to apply 340 V to the primary on one half-cycle (S2 and S3 closed) and −340
V to the primary on the next half-cycle (S1 and S4 closed). There can also be intervals when none
of the switches is closed, thus controlling the amount of energy applied to the transformer. When
none of the switches is closed, the primary of the transformer sees an AC open-circuit unless a
pair of the MOSFET drain-source diodes is conducting as a result of flyback voltage from the
transformer. As in the half-bridge converter shown above in Figure 20.13, the output voltage is
controlled by duty ratio modulation, as illustrated in Figure 20.14a.

The full bridge can also be viewed as two half bridges where each one closes to the positive or
negative rail and *off* time is only a small non-overlap interval. These half bridges can be driven
with 50% square waves. If the square waves are out of phase, the full supply is applied across
the transformer on alternate half-cycles and maximum energy is transferred. If the square waves
are in phase, both ends of the transformer winding receive the same signal and no net voltage is
applied across the primary. By applying phase-shifted versions of these square waves to the two
half bridges, a variable amount of energy can be delivered to the transformer, as shown in Figure
20.14b. In this arrangement the primary always sees an AC short circuit (with the exception of
the very small non-overlap interval). As with the half-bridge duty ratio arrangement, this technique allows the switching frequency to be constant, independent of duty ratio.

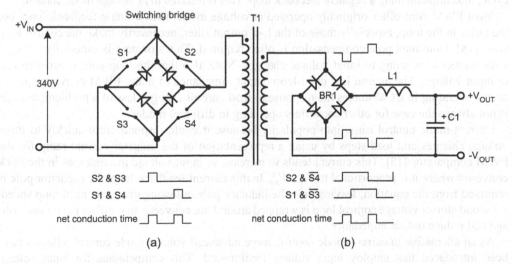

Figure 20.14 Full-Bridge Converter

20.13 Control ICs for PWM Converters

Numerous integrated circuits are available for implementing PWM converters from companies like ON Semiconductor [9, 10], Texas Instruments [11, 12], Linear Technology [13, 14, 15] and Maxim [16], to name only a few. Many of these devices can be configured to implement more than one type of converter. Moreover, they often include numerous housekeeping functions like over-voltage/under-voltage protection. Reviewing their datasheets and application notes can reveal a wealth of information on how these controllers implement PWM supplies—much more than can be covered here. Some controller examples include:

- Texas Instruments *UCC35701*
- Texas Instruments UC384x series
- ON Semiconductor *NCP1203*
- ON Semiconductor *NCP1271*
- Linear Technology LTC3721–1

Skip Mode

Under very light loading conditions it can be difficult for a converter to maintain regulation if it cannot shorten its minimum duty cycle as much as needed. In this case, many controllers will enter a *cycle-skipping* mode [10, 17]. In this mode, switching cycles are actually skipped under no-load or light-load conditions. The cycle proceeds normally, but the switch is simply not turned *on*. This maintains regulation under light load and reduces unnecessary switching of the MOS-FET. Cycle skipping allows efficiency to stay high even at very low operating power levels by reducing the amount of energy wasted by switching on every cycle. This approach is especially useful for *Green* devices such as mobile phone and laptop chargers that are energized at all times.

Voltage Mode Versus Current Mode Control

The control loop in a PWM converter simply adjusts the duty cycle to achieve the desired output voltage under a given set of input voltage and load current conditions. The simple approach is to compare the output voltage with a reference and create a feedback error signal to increase or decrease the duty cycle, thus implementing a negative feedback loop. This is referred to as voltage mode control.

Most PWM controllers originally operated in voltage mode, with a single feedback loop, but the poles in the loop, especially those of the L-C output filter, necessarily make the control loop slow [18]. Dominant pole compensation is often required. This slowness is especially the case with respect to reacting to input voltage changes. Note also that the loop gain is proportional to input voltage. The natural low open-loop output impedance of many PWM converter power stages operating in CCM makes the response to load current changes less of a problem, but this is not always the case for other converters operating in different modes.

Current-mode control came into popularity because it could respond more quickly to input voltage changes and load steps by using a representation of the inductor current ramp for the PWM comparator [18]. This current tends to increase as input voltage increases, as in the buck converter where it is proportional to $V_{in} - V_{out}$. In this current feedback loop, the capacitor pole is removed from the equation, leaving only the inductor pole, allowing greater control loop speed. A second slower voltage control loop is wrapped around the converter to regulate the output voltage and reduce output impedance.

As an alternative to current-mode control, more advanced voltage mode control schemes have been introduced that employ input voltage feedforward. This compensates for input voltage variations almost immediately, with little dependence on the voltage mode control loop. Furthermore, newer designs operating at higher switching frequencies naturally are able to react faster, since the output filter poles can be placed at higher frequencies. It is notable that converters that are preceded by PFC pre-regulators do not have a great need to have quick response to input voltage variations.

In audio power amplifiers, reaction to fast-moving input voltage changes is not nearly as important as reaction to load current changes. Higher speed in reacting to signal-induced load changes is important. It directly affects the frequency where suppression of load voltage changes transitions from the SMPS to the secondary reservoir capacitors. Low converter open-loop output impedance is desirable, since it reduces dependence on the control loop for regulation against load current changes, and minimizes the size required for the secondary reservoir capacitors.

20.14 Resonant Converters

The PWM converters described thus far are so-called *hard-switched* converters because voltages and/or currents with very fast edges are present. This is easily seen in the buck converter waveforms of Figure 20.5. The sharp-edged rectangular voltage and current waveforms in such conventional PWM converters cause switching losses and create a spray of EMI harmonics across a broad range of frequencies.

The switches often close or open with significant voltage across them or current flowing through them. This causes stress and power dissipation in the semiconductors, with corresponding losses. A MOSFET switch that closes when there is voltage across it shorts out its source-drain capacitance and dissipates all of the energy stored in that capacitance (and any other parasitic capacitance on the drain node). If the switch turns on once per period T, then the average power dissipated in the switch is $0.5C_p(V_{switch})^2/T$. Current flowing during the reverse recovery time of a commutating diode also causes loss.

$$P_{diss} = 0.5C_{ds}(V_{in})^2/T_s \qquad (20.11)$$

There is a strong incentive to operate an SMPS at higher frequencies, allowing the transformers, inductors and capacitors to be smaller. Unfortunately, if the switching interval takes a certain amount of time, it becomes a larger fraction of the switching cycle as frequency is increased. Moreover, if switching consumes a certain amount of energy, switching more frequently consumes more energy and reduces efficiency.

The general idea with the resonant class of converters is to introduce one or more reactive components to create a resonance, as with a tank circuit [1, 2, 3, 19]. The resulting sinusoidal waveforms provide an opportunity to operate the switch when there is zero voltage across it (ZVS) or zero current through it (ZCS). It is easy to see that if a switch is closed with no voltage across it, there will be no energy stored across its drain-source capacitance to be dissipated and no high current in discharging it quickly. Zero voltage switching (ZVS) is often preferred, especially for off-line applications, because it creates less EMI. Zero current switching converters are covered in these references [1–3].

In hard-switched converters, the switching node capacitance is the enemy. In resonant converters, it forms part of the tank, thus making it part of the solution rather than part of the problem.

A resonant tank circuit can be included in a switching converter that changes the square wave from a chopped DC input to a sinusoid for subsequent rectification or application to a transformer [19, 20, 21, 22]. A series-resonant tank can be placed in series with the primary of the transformer, for example. It will pass only the fundamental sine wave component of the applied square wave.

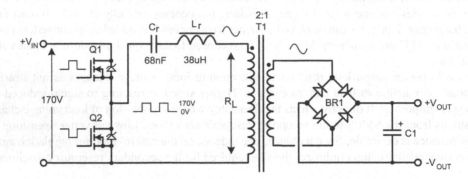

Figure 20.15 Half-Bridge SRC Resonant Converter

A simple half-bridge series resonant converter (SRC) is illustrated in Figure 20.15. Components C_r and L_r form a series resonant tank that resonates at frequency f_r, here 100 kHz. If excited with a square wave whose frequency equals that of the tank resonance, the tank passes the full amplitude of the fundamental of the switching waveform to the transformer, while filtering out the harmonics of the rectangular switching waveform. Current in the transformer is thus largely sinusoidal. Parasitic capacitance at the switching node is not shown. It participates in resonance action during the brief non-overlap intervals when both switches are *off*. The secondary uses a capacitor input filter so that the reflected impedance seen at the transformer primary is largely an AC short circuit at the switching frequency. This allows the series resonant circuit to operate with adequate Q.

$$f_r = 1/2\pi \left(\sqrt{L_r\, C_r}\right) \tag{20.12}$$

The key to achieving ZVS in a series resonant converter is to operate above the resonant frequency, where the series resonant circuit appears inductive. Current through the tank then lags the voltage applied to the tank. Assume that top transistor Q1 is initially *on* and conducting current. That same current will be flowing in L_r. As Q1 turns off, the current must continue to flow because of the inductive nature of the network. The voltage at the switching node therefore flies negative, discharging the switching node capacitance and then causing the substrate diode of Q2 to conduct the current when the voltage reaches about −0.7 V. If the current is 1 A and the node capacitance is 500 pF, the slew rate is 2000 V/µs. If the input voltage is 180 V, then this transition takes 90 ns. When Q2 subsequently turns on, it has only its inverse diode drop across it, so ZVS is essentially achieved. The half-bridge dead time must be greater than 90 ns in this case.

A half-cycle later, the inductive tank current is flowing in the reverse direction through Q2. When Q2 turns off, the inductor current must continue to flow, initially commutating the drain-source capacitances of Q1 and Q2, and then flowing through Q1's body diode, at which point Q1 can be turned on with ZVS.

Although ZVS switching is not entirely achieved as Q1 turns off, the presence of the switching node capacitance keeps the voltage across Q1 from rising extremely quickly as its current falls. If Q1 turns off much more quickly than the transition time of the switching node, then dissipation in Q1 during this time will be minimized. Extra capacitance is sometimes added from drain to source of the MOSFET to improve this characteristic, but care must be taken to make sure that the MOSFET is never hard-switched.

Because the MOSFET body diode is relied upon heavily in ZVS converters, it is desirable to use MOSFETs with fast or ultra-fast body diode recovery, small reverse recovery charge and a high reverse diode dV/dt rating.

In these converters, control of the output voltage is achieved by changing the switching frequency f_s, rather than by PWM. Zero-voltage switching is achieved by operating at a frequency above resonance, where the resonant circuit appears inductive. Moving the switching frequency further above the resonant frequency of the tank reduces the output voltage by increasing the series impedance of the tank. The impedance of the tank circuit forms a voltage divider with the effective load resistance R_L seen looking into the transformer primary, so that as the operating frequency is tuned away from resonance the output decreases.

The "gain" (*conversion ratio*) M of this voltage divider to the switching square wave fundamental can never be greater than unity. Note, however, that the peak fundamental amplitude of a square wave is $4/\pi$ greater than the peak value of the square wave. From this point of view, the SRC at resonance can effectively act as a modest boost converter with a gain of approximately $4/\pi$ (i.e., 1.27). The value of M as a function of f_s/f_r is illustrated in Figure 20.16. The family of curves represents different values of resistive loading R_L as seen looking into the primary.

The effective load resistance R_L seen looking into the primary is simply the output voltage divided by the output current, multiplied by the square of the transformer turns ratio N_p/N_s. The load resistance R_L can also be viewed as the power delivered to the primary divided by the square of the voltage across the primary. This load resistance determines the Q of the tank. Lighter loads result in a lower tank Q. Under high-Q conditions (Q = 2 – 5), the slope of impedance versus frequency is reasonably steep, so only modest frequency changes are needed to change M to accommodate mains voltage changes and load current variations. At high mains voltage or light load, M must be reduced by increasing the switching frequency further above f_r.

A significant problem occurs under light loading. The Q is low and the resonance is very broad, with a shallow impedance slope. In order to achieve a low value of M to reduce output voltage or accommodate high mains voltage, the switching frequency must go very high above the resonance frequency. One would normally like to operate the converter at no more than one octave above the resonant frequency. Note that the conversion ratio M is unity at f_r. Conversely,

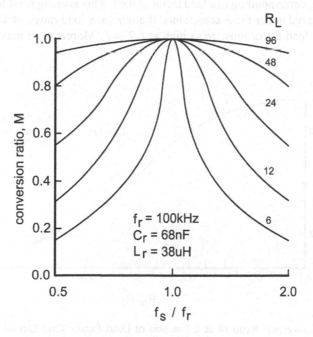

Figure 20.16 Conversion Ratio M as a Function of f_s/f_r for the SRC

under heavy loading, the switching frequency must be made lower and closer to resonance. If the frequency is allowed to venture below f_r, the ZVS operation is lost and the control relationship reverses, causing disastrous positive control feedback.

The problem occurs mainly with input line voltage variation. If the line has a ±15% tolerance, there must be at least a 30% range in M available to accommodate line variations. At low line, the SRC must operate at M close to unity. At high line, the SRC must operate at reduced M to compensate for the extra input voltage. In this case, it must operate at about 30% below unity, or down to about $M = 0.7$. This is easily achieved with medium to high loads, but is very difficult for no-load to light load conditions.

At 200 kHz ($2f_r$), the converter must be able to achieve $M < 0.7$ under the lightest load conditions that will be encountered. Otherwise, the output will go out of regulation and rise in proportion to the increase in line voltage if the switching frequency is constrained to be no more than 200 kHz. It can be seen in Figure 20.16 that the load resistance must be less than about 35 Ω for this to work.

The concept of a reference load and load factor is useful here. The reference load R_0 is defined as where the load resistance equals the reactance of C_r at the resonant frequency. When R_L is equal to R_0, the load factor is deemed unity. We can also define R_0 as the tank characteristic impedance:

$$R_0 = \sqrt{(L_r/C_r)} \tag{20.13}$$

$$\text{Load Factor} = R_0/R_L \tag{20.14}$$

Figure 20.17 illustrates conversion ratio as a function of load factor (R_0/R_L) when f_s is one octave above f_r. This shows how much loading is necessary to achieve adequately low M for a given SRC. As a point of reference, $M = 0.55$ at unity load factor.

If a ±15% line voltage range is to be accommodated, M must be no greater than about 0.7 at +15% line voltage, corresponding to a load factor of 0.67. This is the lightest load for which regulation can be achieved under these constraints. If a *min/max* load range of 10:1 is to be accommodated, then the load factor must go as high as 6.7 at f_r. Moreover, if maximum load is to be

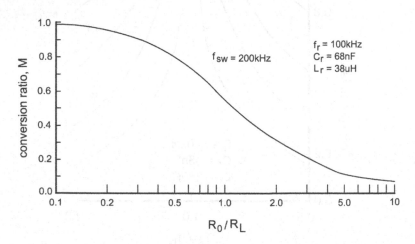

Figure 20.17 SRC Conversion Ratio M as a Function of Load Factor One Octave Above the Resonant Frequency

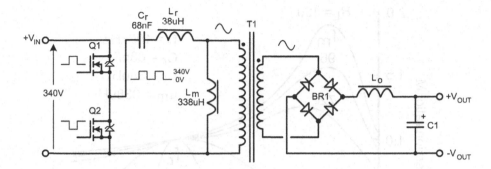

Figure 20.18 Half-Bridge LLC Resonant Converter

accommodated at −15% line, then the load factor must go as high as 6.7 + 30% = 8.6. Operating at high load factors near resonance can create high voltages across C_r and L_r.

To summarize, the SRC resonant converter excels at delivering high current, but its conversion ratio can never be greater than unity, so it should operate with M equal to unity at the lowest line voltage. At nominal line voltage, it will operate with M about 15% below unity if the line voltage range is ±15%. At high line voltage it must operate at M significantly less than unity, on the order of 0.7. This is difficult to achieve under light loading conditions. This converter thus does not perform well under light loading. In general, voltages and currents in resonant converters are larger than those in a PWM converter operating with the same input voltage and load current.

There is also a parallel-resonant configuration (PRC), where the tank resonant capacitor C_r is placed in parallel with the load to provide a predictable shunt impedance in order to maintain voltage regulation at light load when the load resistance approaches infinity. The PRC is discussed in more detail in these references [1–3].

LLC Resonant Converters

The *LLC resonant converter* was patented in 2002 [23]. LLC converters place a second inductor L_m across the load, as shown in Figure 20.18. This defines a second, lower, resonant frequency f_m, as can be seen where the value of M as a function of frequency is illustrated in Figure 20.19. The frequency f_m is defined by C_r and the sum of L_r and L_m. Under heavy load, the resonance behavior and M are still governed mainly by f_r. Under light load, L_m plays a greater role and M is governed more by f_m. Put another way, if the load is a short circuit, the resonance occurs at f_r. If the load is an open circuit, the resonance occurs at f_m.

$$f_m = 1/\left(2\pi\sqrt{C_r\left(L_r + L_m\right)}\right) \tag{20.15}$$

In this arrangement, the LLC tank circuit appears inductive at frequencies below f_r (but still above f_m), so ZVS is obtained in this region as well. Figure 20.19 also shows that M is greater than unity at frequencies between f_m and f_r for reasonable values of load current. The LLC arrangement thus generally provides gain as a function of switching frequency instead of loss as in the SRC resonant converter. Significantly, unity gain is provided at f_r, largely independent of the effective load resistance.

L_m denotes the magnetizing inductance of the transformer, but that of an un-gapped transformer by itself is usually too large for purposes here, so a discrete inductor L_m is introduced to

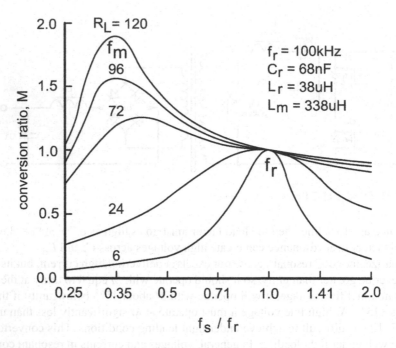

Figure 20.19 Conversion Ratio M as a Function of f_s/f_r for the LLC

provide a smaller (and better controlled) net value of inductance. Alternatively, a precise air gap is included in the transformer core to reduce its reluctance and provide the target value of L_m at no cost. The value of L_m is usually chosen to be on the order of 7–11 times the value of L_r [24]. If the ratio of L_m to L_r is too large, larger frequency variations are required for a given amount of change in M. If the ratio of L_m to L_r is too small, the circulating currents can be much higher and the efficiency can fall as a result of copper losses [24].

Note that L_r denotes the series-resonant inductance, which can be made of a combination of the transformer leakage inductance and a separate series inductor. Often the transformer windings are purposely separated to achieve a target leakage inductance, and in so doing minimize or eliminate the series inductor.

The LLC performs best when it is operated at frequencies closer to f_r, as this is the frequency where load regulation is naturally very good. This is especially advantageous for audio power amplifiers. Operation at frequencies above f_r is not very useful [24]. In this frequency range the LLC operates just like an SRC, and retains the problem of achieving conversion ratios much below unity with light loading. Thus the practical operating region lies between f_m and f_r, where the conversion ratio is equal to or greater than unity.

The biggest challenge in applying LLC converters is coping with mains voltage variations spanning light to heavy load. The converter must be operated close to f_r at worst-case maximum mains voltage, since the conversion ratio cannot reliably go below unity at light load. If worst-case minimum mains voltage is 30% less than the maximum, then the converter must be operated at a lower frequency, closer to f_m, where the conversion ratio is at least 1.3. Unfortunately, at these lower frequencies, load regulation is not as good, so this conversion ratio must be achieved under worst-case peak load conditions, requiring a still lower switching frequency. Thus, conditions of low mains voltage and heavy load tax the converter design and may require over-design of the converter to satisfy these worst-case conditions.

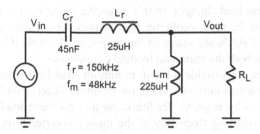

Figure 20.20 Simple LLC Network for Simulation

The simple LLC network of Figure 20.20 can be simulated to show the conversion ratio as a function of frequency for different loading for a given LLC design. There are just three main parameters that define the rough design of an LLC converter. These are f_r, L_r, and the ratio K of L_m to L_r. The value of C_r follows directly from f_r and L_r. Consider a simple LLC network with f_r = 150 kHz, L_r = 25 µH and K = 9. C_m will be 45 nF, L_m will be 225 µH and f_m will be 48 kHz. With a load of 75 Ω, the peak value of M = 1.28 and the switching frequency is 67 kHz.

It is imperative that the conversion ratio at frequencies below f_r and above f_m not go below its peak value as switching frequency falls to accommodate heavy loading or low line voltage. If this happens, the sign of the control feedback will change and the converter output will fold back and likely go into stressful hard-switching operation. Under light loading the peak value occurs very close to f_m, but as loading increases, the frequency of the peak increases somewhat.

The LLC converter is superior to the SRC converter because it does not have to operate at very high frequencies under light loading; its operating frequency range is reasonably well bounded. However, it still has limited tolerance to line variations and can be tricky to design. It performs best in off-line applications where it is preceded by a power factor correcting (PFC) preregulator (see Section 20.17).

To summarize, the LLC converter normally operates at M greater than unity (boost) and does not often operate at M less than unity. At low line voltage, it must operate at M significantly larger than unity and it must do this under maximum load. This can be difficult for the LLC converter to achieve without over-design. The LLC is thus less able to deliver high current over its full operating input voltage range.

Loaded Resonant Converters (LRC)

Were it not for its poor tolerance to high line voltage and light loading (with corresponding operation at frequencies far above resonance), the series resonant converter (SRC) would likely be the converter of choice for a great many applications. This is because of its relative simplicity and its ability to deliver high current over a wide range of operating conditions. What the SRC needs is a load that is heavy enough to keep it operating in a reasonable frequency range while being able to achieve small enough values of conversion ratio (M).

This *minimum load*, of course, must be *lossless*, or nearly so. The *loaded resonant converter* (LRC) achieves this goal by introducing a second, smaller *loading converter*. The output of the main SRC converter is connected to the input of the SMPS loading converter. The output of the loading converter (often a boost or flyback converter) is connected to the input of the main converter, returning the loading power to the input power source. The loading converter thus provides virtually *lossless loading* of the main converter. The loading converter is operated so that it injects current back into the power source, rather than trying to act as a voltage source that would fight with the power source.

Figure 20.21 shows the basic concept of the loaded resonant converter with illustrative power flow ranges as the system load changes. In this example, the system load changes over a 100:1 range, while the load on the main converter is constrained to change over only a 10:1 range. Converter efficiencies of 100% are assumed for simplicity. A single IC controller is shown that controls the switches for both the main and loading converters.

The loading converter can provide a fixed, minimum load for the main converter. It can be controlled to act as a constant current source, for example. Alternatively, the loading provided can be made adaptive, as by negative feedback, so that no more loading is provided than is necessary to keep the switching frequency of the main converter from exceeding some target frequency, such as an octave above the resonance frequency. The loading converter can be simple and small, and it enjoys a regulated input voltage. This *lossless loading* arrangement can also be applied to other converters that require a minimum load to perform optimally.

A simple LRC is shown in Figure 20.22. Its main converter is the half-bridge SRC converter of Figure 20.15. The loading converter is a simple PWM flyback converter powered from the DC output of the main converter. The loading converter can also be implemented with resonant techniques if desired. Diode D1 returns the loading power back to the DC input source via flyback pulses from the secondary side of T2.

Alternatively, the loading converter can be implemented as a PWM boost converter that is supplied DC from a tertiary winding on T1 whose output is rectified. The additional winding is often already present for provision of information to the controller IC and also for primary-side regulation if used. The boosted output of the loading converter is then connected to the DC input of the main converter. An isolated Ćuk converter could also be an attractive choice for the loading converter if reduced EMI from the loading converter is desirable. The LRC provides all of the benefits of the SRC while minimizing its shortcomings under light loading and high line voltage.

Application of LRC to LLC Converters

The *lossless* converter load topology can be applied to any converter that requires a minimum load to work as intended. In the case of the LLC converter, the *lossless* converter load arrangement can be used to substantially increase the useable control range of the LLC.

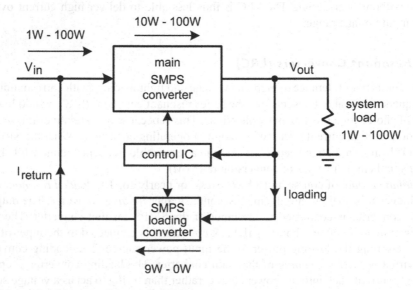

Figure 20.21 Loaded Resonant Converter (LRC)

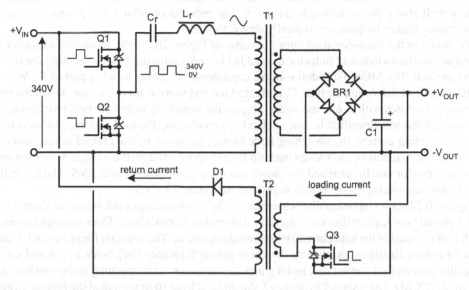

Figure 20.22 Half-Bridge Loaded Resonant Converter (LRC)

Recall that the LLC is often operated at a switching frequency between f_m and f_r, since under light loading there is not much to be gained by operating above f_r to achieve smaller M. However, if some *lossless* loading is added to the LLC, the frequency range above f_r becomes quite useful for achieving values of M below unity. This significantly increases the control range available to the LLC, resulting in a converter that can be efficiently used at frequencies above and below f_r and with conversion ratios above and below unity. By the same token, for a given required control range, the switching frequency can remain closer to f_r, where the converter is most efficient.

Control ICs for Resonant Converters

Numerous integrated circuits are available for implementing resonant converters [25–32]. They include:

- Fairchild Semiconductor FSFR1600
- STMicroelectronics L6599
- Unitrode/TI 1864
- ON Semiconductor MC34067

20.15 Quasi-Resonant Converters

The so-called *quasi-resonant* (Q-R) converter, patented in 1983 [33], is much different from the resonant converter. It represents an approach that seeks to combine the ZVS advantages of resonant conversion with the advantages of PWM converters. This is achieved by introducing a resonance into an otherwise-PWM arrangement [3, 33]. In fact, if the switching cell of virtually any PWM converter is replaced with a so-called *resonant switching cell*, a quasi-resonant converter results [3].

The resonant behavior is achieved by adding only one or two reactive components to the hard-switched converter circuit. The parasitic capacitances and inductances (like switching node capacitance and transformer leakage inductance) become part of the resulting resonant circuit.

The converter circuit works with these rather than against them. The resonant frequency f_r is usually well above the switching frequency f_s. The reduced switching loss permits operation at significantly higher frequencies, typically above 500 kHz.

As shown in the quasi-resonant buck converter of Figure 20.23, the resonance capacitor C_r is added across the switch, and inductor L_r is added in series with the switch, creating the resonant switching cell. The MOSFET's drain-source capacitance, C_{ds}, may be a big part of C_r. When the switch is *on* and delivering current, C_r is shorted out and there is no resonance. When the switch opens, C_r resonates with L_r and the voltage V_{sw} on the switching node falls negative, swinging in a sinusoidal fashion. Specifically, since L_r has been conducting the load current, and seeks to continue conducting current, the switching node swings negative, below ground in resonance with C_r. When the switching node swings back up to the supply voltage, the voltage across the switch becomes zero (or nearly zero) and the switch can once again turn on with ZVS. The controller is responsible for turning on the switch at just the right time for ZVS.

Figure 20.24 shows representative waveforms of the quasi-resonant buck converter. Output inductor L1 should be at least 10 times as large as L_r. This makes L_r look almost like a constant current sink (with a value equal to the load current) to the switching circuit. The resonant frequency of L1 and C1 must be at least 100 times less than the lowest switching frequency [34]. Node V_{D1} should see what looks like essentially a current sink looking into L1; the converter is operating in the continuous current mode (CCM). The resonant frequency f_r should be at least 10 times that of the highest switching

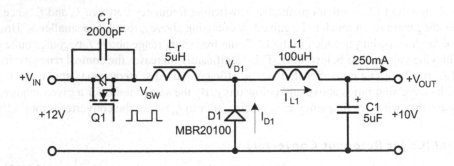

Figure 20.23 Quasi-Resonant Buck Converter

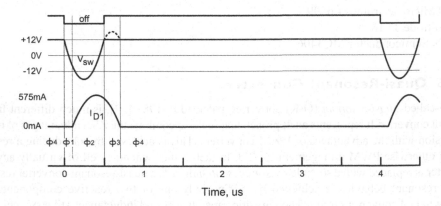

Figure 20.24 Waveforms of Quasi-Resonant Buck Converter

frequency. As depicted, the waveforms represent the converter operating at a fairly high duty cycle of 0.8, resulting in an output voltage of 10 V and sourcing about 250 mA to the load.

Four phases of operation can be identified [3]. The first phase begins when S1 turns *off*. The switch has previously been closed, $V_{cr} = 0$, $I_{Lr} = I_{load}$, and there was no resonance (C_r was shorted by S1). As S1 opens, L_r keeps the current flowing, driving the switching node voltage V_{sw} linearly in a negative direction and charging C_r. There is no resonance during this phase because D1 is not conducting to close the loop of the series-resonant tank. We are approximating L1 as an AC open circuit at this point. This is the time that the switching node would snap negative in a conventional PWM buck converter. In the Q-R converter, C_r prevents V_{sw} from immediately going negative. At the end of Phase 1, the voltage at the switching node has decreased to zero and the voltage across C_r has risen to equal V_{in}. At this point, the current through C_r and L_r equals the output current I_{load}.

Phase 2 begins when the voltage at the switching node goes through zero and slightly negative, causing catch diode D1 to turn on and allowing C_r and L_r to resonate. During Phase 2, L_r and C_r resonate, with I_{load} having been the stimulus. The resonance carries V_{sw} through a negative cosine swing, returning to, and passing through, zero. The peak voltage of the negative cosine swing is equal to the product of I_{load} and R_0 (the impedance of C_r at the resonant frequency f_r). At this point, the voltage across the switch V_{switch} is at its maximum, equal to $V_{in} + I_{load} * R_0$. After passing through zero, V_{sw} continues to rise to V_{in}. At this time, the voltage across the switch reaches zero and the switch can be turned on with ZVS.

$$V_{switch\_max} = V_{in} + I_{load} * R_0 \qquad (20.16)$$

Phase 3 begins when S1 closes, killing the resonance and beginning to supply current to L_r. The dashed overshoot portion of the V_{sw} waveform shows the trajectory that V_{sw} would follow if it were not clamped to the input supply line by S1 turning on. After S1 closes, D1 is still conducting I_{load} for a short interval. The voltage across L_r at this point equals V_{in}, and the current through L_r increases linearly, reducing the current in D1. When the current in L_r reaches I_{load}, the current in D1 reaches zero and D1 turns off. This marks the end of the third phase.

During Phase 4, S1 supplies the full input voltage to L1 through L_r, just as in the conventional buck converter. The longer S1 remains closed, the more energy is supplied to L1 of the buck converter. The control circuit adjusts the *on* time during this phase to provide regulation. If the output voltage needs to be increased or if the load current increases, the duration of Phase 4 will be increased, reducing the switching frequency.

During Phase 4, S1 is *on* and D1 is *off*. A series-resonant circuit is formed by L_r and the parasitic capacitance at the V_{D1} node. This can result in a high-amplitude under-damped oscillation at this node, often at a frequency in the tens of MHz. A series R-C snubber network can be placed across D1 to eliminate this ringing. In Figure 20.23, values of 500 Ω and 200 pF work well and dissipate virtually no power. To summarize operation:

Phase 1

- S1 turns *off*
- V_{sw} falls linearly to slightly below zero
- C_r charging current supplies the load

Phase 2

- D1 conducts, resonance begins
- D1 supplies half-cosine current to the load

- V_{sw} falls negative in cosine fashion
- V_{sw} rises back up to V_{in} as it resonates
- V_{sw} reaches zero

Phase 3

- S1 closes
- D1 is still supplying the load current
- Current through L_r increases
- Current through D1 decreases to zero
- D1 turns *off*

Phase 4

- S1 is *on* and D1 is *off*
- S1 supplies current to load as in normal buck operation

A controller must be used that senses the correct time for ZVS turn-on of S1. The switch is turned *on* at the valley of the voltage across the switch, and it is turned off at a time later when the necessary amount of energy has been transferred into L1. Thus, control of the output voltage is achieved by varying the *on* time of the switch by varying when it is turned *off*. The interval of time between turn-off and the next turn-on is governed by the resonant frequency of the switching node. It will correspond to very roughly half the period of the resonant frequency ($T_r/2$). For a given resonant frequency, this time is thus approximately fixed. As a result of these two considerations, it can be seen that the frequency changes as the output voltage is controlled. This is not a fixed-frequency PWM converter.

Moreover, while the maximum voltage across the MOSFET in a PWM converter is approximately equal to V_{in}, the maximum voltage across the MOSFET in a Q-R converter can be significantly greater than V_{in}, often more than twice as much. The peak voltage across the switch in a Q-R converter increases linearly with load current. Very high voltages can thus appear across the switch under heavy load current. The Q-R converter is thus less able to cope with heavy loads, and one must often use MOSFETs with a higher voltage rating and correspondingly higher *on* resistance [3]. This increases conduction loss in the switch. The substrate diode integral to the MOSFET will break down at the rated voltage V_{BR} of the MOSFET. If the peak voltage across the switch attempts to go greater than V_{BR}, the substrate diode will conduct and rob energy from the resonance and waste power.

The resonant swing must be sufficient in amplitude to reach back up to the input voltage so that ZVS may occur. Even with no damping, this means that the load current must be sufficient to cause the initial swing to take V_{sw} down to $-V_{in}$, making the minimum peak voltage for the switch equal to 2 V_{in} in the best case. If ZVS is not achieved, there will be a current spike when the switch closes. If a capacitor is added to C_{ds} to create the full value of C_r, the amplitude of the spike will be exacerbated. This is yet another example of a switching converter needing a minimum load to work as intended. Because the peak voltage across the switch is equal to $V_{in} + I_{load} * R_0$, and the second term must be at least unity, the minimum peak voltage is 2 V_{in}. If the load current range is 5:1, then the peak voltage across the switch will be 6 V_{in} [3].

The Q-R converter controllers often use a VCO and a one-shot. The one-shot interval sets the *off* time of the switch to a fairly fixed time range, on the order of one period of the resonant frequency. The zero crossing of the switch voltage initiates termination of the one-shot time interval.

To summarize, quasi-resonant converters act much like hard-switched PWM converters, but they achieve ZVS by introducing a resonance at the switch. Although they are a bit more complex to control insofar as timing for switch turn-on, controller ICs take care of this. Because Q-R converters operate on the PWM principle, but with a nearly fixed short *off* time, they produce more voltage by increasing the *on* time of the switch, thus causing the switching frequency to decrease with increased load current or load voltage demands. The magnetic components must be designed to be able to operate at the worst-case low switching frequency, potentially making them larger and more expensive.

Multiresonant Q-R Buck Converter

If a capacitor C_d is placed across D1 in Figure 20.23, a second resonance mode is introduced into the resonant switch cell. A ZVS multiresonant switch (ZVS-MRS) converter results [3]. In this arrangement, it is possible to have all semiconductors switch at zero voltage, resulting in further reduction of losses and EMI.

Q-R Flyback Discontinuous Mode Converters

The Q-R DCM flyback converter operates differently than the high-frequency Q-R class of converters described above [35, 36, 37]. This flyback converter must be operated in the discontinuous mode, and the quasi resonance occurs after the inductor current goes to zero. The resonance is created by the switch capacitance and the inductance of the coupled inductor (e.g., magnetizing inductance of the flyback transformer). This resonance was described in the earlier discussion of the PWM flyback converter operating in DCM. There is thus no added inductor required to create the quasi resonance. The Q-R DCM flyback converter is often used at lower switching frequencies, often below 150 kHz.

ZVS is not fully achieved. However, the switch voltage at turn-on is substantially reduced, resulting in a significant reduction of turn-on switching losses (which scale with V_{ds}^2) and EMI. Recall that switch turn-on power loss increases as the square of the switch voltage when the switch closes (Equation 20.6). When the inductor current goes to zero, the switch voltage rings down toward zero, reaching a valley. It is at this point that the switch is turned on with minimum loss. Under a sufficiently heavy load, the converter will enter CCM and near-ZVS operation will be lost. This Q-R converter does not suffer from the high peak switch voltages of the Q-R converter described above.

The amount by which the switch voltage swings down toward zero is equal to twice the value of the reflected load voltage V_R. We thus have:

$$V_{switch\_min} = V_{in} - V_R \qquad (20.17)$$

For a given output voltage, V_R can be manipulated by choosing the flyback transformer turns ratio, recognizing that this also influences the nominal switching duty cycle and frequency. Choosing a larger value of V_R reduces the switch voltage at the valley, getting closer to ZVS, reducing switching losses and EMI. Because V_R cannot exceed V_{in}, ZVS can only be approached.

Like the PWM flyback converter, this converter is very popular in low-to-medium power applications. Operation is very much like a PWM converter that operates with a fixed *off* time and a variable frequency to establish the switching duty cycle D. Because the switching frequency varies with M, ripple typically present in the DC input supply causes switching frequency modulation. This spreads the spectrum of the EMI and makes conformance to EMI standards easier. However, care must be taken when powering audio amplifiers, since the converter could slip into operation below 20 kHz at light load, causing unwanted noise pickup.

It is interesting to note that the PWM flyback converter incurs switching losses that vary with the particular time that the switch turns on. Turn-on is asynchronous with the ringing, and may occur when the switch voltage is in its valley or at its peak or anywhere in between.

Control ICs for Quasi-Resonant Converters

Numerous integrated circuits are available for implementing quasi-resonant converters [37, 38, 39, 40]. They include:

- Texas Instruments UC3861
- Texas Instruments UCC 28600, UCC 28610, TPS92070, TPS 92210, TPS 92010
- STMicroelectronics L6565
- Infineon ICE2QSO3G
- Infineon ICE2QSO2G

20.16 EMI Filtering and Suppression

Switching power supplies create a significant amount of high frequency electrical noise that can result in electromagnetic interference (EMI). This must be prevented from entering the AC power line or radiating into free space, where it can interfere with other electronic equipment. EMI emissions are governed by several agencies and standards, including FCC Part 15 and IEC BS 800 [1].

The switching frequencies usually fall in the range of 50 kHz to 500 kHz and the switching pulses are usually pulse width modulated (PWM). As a result, harmonics are generated in the radio-frequency range. This necessitates the incorporation of a sophisticated electromagnetic interference (EMI) filter to prevent the appearance of the switching frequency and its harmonics as differential-mode or common-mode signals on the AC supply lines. A typical EMI filter circuit is shown in Figure 20.25. Differential shunt capacitors C1, C4 and C5, in combination with the inductors, provide differential-mode attenuation in a straightforward and effective way.

The EMI filter must also be designed to prevent common-mode signals that are generated in the power supply from being conducted to the external supply mains while at the same time offering minimum series impedance to the differential 60-Hz AC of the supply mains. This is accomplished by having balanced inductors in each of the supply lines with positive mutual inductance between the inductors in the upper and lower lines. This arrangement maximizes the inductance for common-mode currents that flow in the same direction in both conductors.

For the oppositely directed differential currents of the 60-Hz mains, the mutual inductance is negative, thus forcing the overall series inductance to a small value. C2 and C3 shunt the common-mode signals to ground. The mutual inductors (called *common-mode chokes*) often have a small

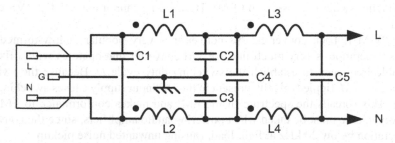

Figure 20.25 EMI Filter

amount of remaining series inductance that will provide some differential filtering at high frequencies above 150 kHz.

Radiated Emissions

Radiated emissions travel though the air as electromagnetic radio waves. Although the switching frequencies and their harmonics may appear to be low in frequency, the edges of the switched voltage and current waveforms are often very steep, especially in hard-switched converters, and generate very high-order harmonics of the switching frequency. These edges act like impulses, and will often excite natural resonant circuits in the PCB, transformer, components and system wiring in the 30–250 MHz region.

Anything in an SMPS that can act as a radio antenna can act as a source of radiated emissions if it carries a voltage or current at high frequencies. Radiated emissions are minimized by good PWB layout with small loop areas and minimal size of nodes with high dV/dt. The simple shielding provided by a steel enclosure can also be effective in suppressing radiated emissions, but care must be taken to dress the wires entering and exiting the enclosure with suitable filters. FCC Part 15 radiated emissions limits for class B (consumer) equipment are as follows [41]:

Radiated emissions limits

Frequency Range (MHz)	Amplitude dBμV/meter
30–88	+40
88–216	+43.5
216–960	+46
>960	+54

Conducted Emissions

Conducted emissions travel by conduction of currents or voltages through external connections to the amplifier, such as the power cord. These can be in the form of differential signals between line and neutral or common-mode signals between line/neutral and earth ground (safety ground). Differential emissions are fairly easy to suppress through the use of capacitors across line and neutral in combination with inductors, forming an effective low-pass filter. FCC Part 15 conducted emissions limits for class B (consumer) equipment are as shown below when measured with the signal conductor terminated to ground in 50 Ω [41].

Conducted emissions limits

Frequency Range (MHz)	Amplitude dBμV
0.15–0.5	56–46
0.5–5	46
5–30	50

Common-mode conducted emissions are much more difficult to suppress and are usually the cause for failure of emissions compliance. A big reason for this is the strict safety regulations limiting leakage current from line to ground. These include UL 1283 and IEC 335–1 [1]. EMI filter capacitors from line and neutral to ground are an important part of common-mode filtering, but they cause AC leakage current to flow into the safety ground. These are C2 and C3 in Figure 20.25. Depending on the regulating agency and the application, limits on earth leakage current lie typically between 0.5 mA and 3 mA in 120-V, 60-Hz systems [1]. This limits the total amount of common-mode EMI suppression capacitance that can be used to as little as 0.01 μF. This makes it difficult to achieve adequate suppression without resort to higher values of series inductance, which in turn can lead to power loss or larger components.

Most common-mode EMI results from parasitic capacitances between high edge rate primary-side switching voltages (as on a MOSFET drain or transformer primary winding) and chassis ground or secondary-side circuitry via transformer inter-winding capacitances. A good example is the capacitance between the case of the transistor and a chassis-mounted heat sink, across the thermally conductive insulator. A TO-220 MOSFET mounted to a heat sink with a 0.05 mm mica insulator will have about 40 pF between the drain and heat sink [42]. A 100-V transition in 100 ns has an edge rate of 1000 V/μs. This will create a current flow of 40 mA through a 40-pF parasitic capacitance. In an extreme example, if this 40 mA were to flow unattenuated into the load of the EMI test arrangement, 2 V would result, far in excess of the approximate 1–3 mV allowed.

The best solution is to eliminate EMI problems at the source, rather than trying to filter them away. Once EMI currents get into the ground plane they are difficult to control. Smaller PC-mounted heat sinks should have their pins routed to a single-point common node where the current can best be returned to the local circuit with small loop area. In some circuits, putting a small damping resistor or ferrite bead in series with this path can be helpful.

Common-mode problems are best handled by minimizing the parasitic capacitances that generate common-mode currents in the first place. Transformers should be wound with low-capacitance techniques. Where possible, primary-side transistors with high dV/dt on their mounting surfaces should be spaced away from heat sinks using thermal insulators made of thick material with low dielectric constant. Furthermore, using primary-side heat sinks that are not connected to the chassis can help to isolate this common-mode EMI.

Differential mode problems are best handled by using extensive planning in the PCB design process to contain current transition paths to tight (ideally overlapping) loop areas. The following measures can be helpful:

- Slow down dV/dt or dI/dt where possible
- Use high-frequency high-quality NPO or X7R surface-mount ceramic capacitors
- Use capacitors to steer or attenuate the RF currents that cannot be eliminated
- Add ferrite beads to attenuate RF currents
- Use upgraded parts with better RF characteristics
- Alter the size or geometry of an offending node to make it a less efficient radiator

The fully insulated TO-220 *FullPak* package can help significantly while reducing cost. This package incorporates the necessary insulation and requires only thermal grease between it and the heat sink. This comes at the expense of an increase in net junction-to-sink thermal resistance of about 0.5 °C/W in comparison with a conventional TO-220 package with a mica insulator. While a conventional TO-220 arrangement may have 40 pF of drain-to-sink capacitance, the capacitance for a TO-220 *FullPak* device may be as little as 12 pF [42]. Adding a 1.79-mm aluminum oxide pad to a TO-220 *FullPak* reduces this capacitance to about 4.3 pF.

Faraday Screens

A Faraday screen is essentially a conductive shield that is placed between the plates of the parasitic capacitance. This shield is usually connected to the common node of the circuitry creating the fast voltage edges, returning the current locally instead of through the earth ground [1]. Faraday screens are sometimes used in transformers to break the parasitic capacitance between the winding and the core. In some cases the use of a Faraday screen in a transformer can reduce efficiency slightly due to stray eddy currents.

Eddy-Current Magnetic Shielding

In suppressing EMI it is also important to keep fast-edge magnetic fields from inducing voltage into the ground plane of the power supply. These fields usually originate from inductors or transformers. The first line of defense, of course, is to design the magnetic circuit elements to radiate as little magnetic field as possible. The use of toroid cores, for example, helps in this regard. Nevertheless, suppression of radiated magnetic fields is desirable.

Magnetic shielding is generally more difficult and expensive than the electrostatic shielding that can be provided by a Faraday shield. Often, high-permeability materials like mu-metal are necessary. These approaches can add cost, size and weight. An alternative that can help significantly is use of the eddy-current effect to form an eddy-current magnetic shield [43, 44]. Such a shield is based on the principle that a time-varying magnetic field impinging on a conducting material will induce eddy current in that material that will produce a magnetic field that will oppose the radiated magnetic field. Eddy-current shielding depends on the time rate-of-change of the magnetic flux, and typically is effective at frequencies above 10 kHz.

A plate of such material interposed between the radiating device and the ground plane can largely prevent induction into the ground plane. One can think of the plate as acting like a large shorted turn. Indeed, at higher frequencies where skin effect comes into play, the circulating current will stay near the surface of the plate facing the source of the magnetic field. The resistance associated with the skin effect can actually dissipate the stray energy from the magnetic field. Because the conducting plate interferes with the magnetic flux lines of the radiating device, it can act to reduce inductance and Q, so care should be taken in its application.

The plate should be insulated from the ground plane and connected to it at a single point with one or multiple low-impedance straps to offer some electrostatic shielding as well. In a multi-layer PWB where the ground plane is a layer below the surface, the top copper layer can be used to create such a plate.

The eddy-current magnetic shield can also be used to reduce interaction between magnetic elements in the SMPS or amplifier with ferrous materials nearby, such as the steel cover of the amplifier. For example, it is desirable that the output coil used in a power amplifier should not interact with the steel chassis. Such interaction will cause distortion due to magnetic nonlinearity in the steel. The simple placement of a thin sheet of aluminum on the steel panel can interrupt any such interaction [45]. It is best if the aluminum sheet is insulated from the sheet metal and then connected to it at a single point so that it is not left floating.

Pot Core Transformers and Inductors

Pot core transformers and inductors comprise a ferrite core geometry in which the ferrite material completely encloses the windings [46]. The self-shielding properties of this design can significantly reduce EMI leakage. However, pot cores are more expensive and interfere with heat dissipation. The THX power supply design within the Benchmark AHB2 power amplifier uses self-shielding pot cores for all its transformers and inductors to contain their magnetic fields from the adjacent analog and power amplifier circuitry, which is co-planar and only an inch away.

Ferrite Suppressor Cores

It is common to see a cylindrical ferrite core on the power cord of modern equipment. This is basically a large ferrite bead that enhances suppression of common-mode conductive emissions. Ferrite beads create series impedance that is inductive at low frequencies and transitions to resistive at high frequencies. The series inductance impedes the flow of common-mode current at high frequencies. However, unlike a conventional inductor, the ferrite bead introduces loss into the path as well, due to its resistive nature at high frequencies. This behavior is just as important as the series inductance that is introduced by the ferrite bead. Bear in mind that the ferrite bead does not introduce significant impedance in the differential mode.

Depending on the size and type of ferrite material, a typical line cord suppressor core might introduce a rising impedance of 50 Ω at 5 MHz, corresponding to about 2 μH. By 15 MHz the reactive and resistive components of the impedance might be equal. By 100 MHz the impedance will have become almost entirely resistive, at perhaps 200 Ω [47, 48].

EMI Filter Interactions with Converters

Caution must be exercised in placing input filtering, such as an EMI filter, ahead of a switching converter [49]. Switching converters are often designed assuming that the input source has little or no impedance. Networks that are in the source path may invalidate this assumption. In the worst case, instability can result. To mitigate this issue, a suitable bypass capacitor should always be located at the input of any switching converter.

20.17 Power Factor Correction

Power factor describes the degree to which power is transferred for a given combination of delivered voltage and amperage. A resistive load has 100% of the apparent power (VA) delivered to it as real power (watts), and therefore has the ideal power factor of unity. The power factor is unity because the voltage and current wave-shapes are identical and in phase. Any departure from the wave-shapes being identical and in phase results in a power factor of less than unity. For example, in the extreme case of a purely inductive load, voltage and current are 90° out of phase and no power is delivered to the load, in spite of the fact that significant voltage is present and significant current flows (VA). Nonlinear loads, such as rectifiers with capacitor input filters, also represent loads with a poor power factor. Such inefficiency of power transfer is bad for the utility and sometimes bad for the user.

For linear loads, like inductive loads encountered in utility power distribution, power factor correction (PFC) is achieved passively with inductors and capacitors that shift the phase of the load current to be nearly the same as that of the voltage. Such passive power factor correction does not work for nonlinear loads. The highly distorted, high amplitude input current pulses of the front-end rectifier supply of a conventional switching power supply cause a great departure from an ideal power factor of unity. Once again, the definition of unity power factor is when the load acts like a pure resistance.

Active Power Factor Correction

The objective of active power factor correction is to alter the input current waveform to be one that has the same wave-shape and is in-phase with the mains voltage [1, 2, 3 ,50]. There are two purposes. The first is to reduce the RMS current in the mains and in the power supply for a given amount of real power delivered. The second is to reduce the harmonics in the current that is drawn from the mains.

At this time there is no PFC requirement for audio amplifiers and audio equipment sold into the North American market. However, all single-phase consumer electrical and electronic equipment sold into the European market is subject to the mains current harmonic limitations set forth in standard EN61000-3-2. For class A devices such as power amplifiers and audio equipment it should be noted that this standard specifies current limits at each 50-Hz harmonic rather than a power factor, and that these are absolute limits on current harmonics, and not percentages of the current drawn. For example, the third harmonic maximum permissible current draw is 2.3 amperes. A 2 X 50-W amplifier might meet this requirement without active PFC, whereas a 750-watt subwoofer amplifier might require active PFC to meet the requirement.

Active PFC circuits use a bridge rectifier on the mains, but no filter capacitor, as shown in Figure 20.26. Instead the full-wave-rectified voltage, called a *haversine*, feeds a dynamic boost converter. The boost converter creates a high-voltage DC intermediate bus supply of about 385 V DC that feeds the subsequent DC-DC converter. The filter capacitor, C1, is moved to after the boost converter. The boost converter can operate over a wide input voltage range while delivering the 385-V DC output. This allows the power supply to operate over the full universal input voltage range of 88–264 V AC without a mains voltage selector switch.

The 385 V DC bus will have some ripple at twice the mains frequency as a result of the full-wave rectification. The sizing of bulk capacitor C1 and its ESR determine the magnitude of this ripple, often around 10% p-p under maximum load. The DC-DC converter must deliver full output voltage and power across the span of this ripple. Ideally the DC-DC converter will also reject most of this ripple through feedback, so that it does not appear at the $\pm V_{out}$ rails in the power amplifier, where it could be injected into the amplifier output. If C1 is sized much larger to store lots of energy as a reservoir capacitor, ripple on the intermediate bus will be smaller and less of an issue.

The MOSFET switch in the boost converter is controlled in such a way that the average input current has the same waveform and phase as the input voltage. Average input current is adjusted by the controller in order to regulate the resulting boost voltage. This is accomplished with an IC specifically designed for the PFC function [4]. The controller IC uses feedback techniques to match the current waveform to the mains voltage waveform. The resulting boosted intermediate bus voltage must always be larger than the rectified mains voltage, since a boost converter cannot output a voltage lower than its input voltage. The regulated intermediate bus voltage permits the DC-DC converter to operate under its optimum conditions, even under substantial line voltage variations. This is especially advantageous for resonant converters.

The boost topology is not capable of limiting inrush current, so a separate inrush current-limiting circuit is still required for switch-on including a resistor or NTC device to limit the current. A standard-recovery pre-charge diode with more robust non-repetitive peak current rating can be used to bypass the boost rectifier and inductor, since the boost rectifier is usually optimized for speed rather than inrush current.

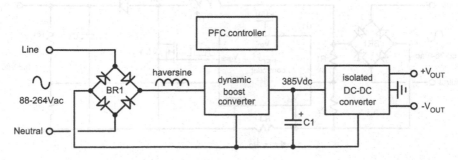

Figure 20.26 Active Power Factor Correction

The Dynamic Boost Converter

The boost converter can take virtually any input voltage and boost it to a higher voltage, where the energy is then dumped into a storage capacitor. For the PFC application, it must dynamically adjust its boost ratio as the instantaneous voltage of the haversine waveform goes from its valley to its peak. The switching frequency is much higher than the mains frequency, so in a relative sense the line voltage waveform at the input to the PFC is changing very slowly, and the switching converter has plenty of time to adapt its operating parameters to the "slowly changing" haversine input voltage waveform.

Near the bottom of the input voltage waveform, the boost ratio of the PFC boost converter must be large to achieve the higher regulated output voltage from the small input voltage of the haversine. At the same time, the input current should be small. We note that the power consumed by a real resistance from its source is proportional to the square of the voltage, so when the haversine input waveform is at 10% of its peak value, for example, the instantaneous power being drawn from the full-wave-rectified source by the PFC converter will be only 1% of the maximum it will draw at the peak of the waveform.

Figure 20.27 illustrates the boost converter input current with ripple as compared to the desired mean haversine current waveform. The boost converter should operate in the continuous conduction mode (CCM), otherwise substantial amounts of high-frequency ripple current will be introduced into the mains supply.

The PFC System

The PFC control circuit must monitor both the PFC output voltage and its input current so as to provide a regulated output DC voltage while drawing input current that is proportional to

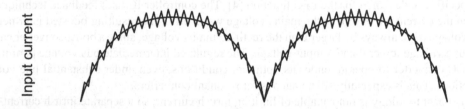

Figure 20.27 Power Factor Correction Waveform

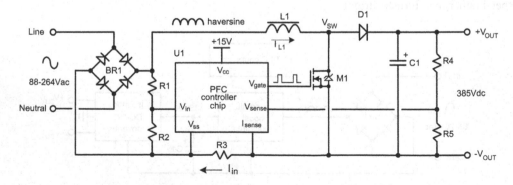

Figure 20.28 Power Factor Controller

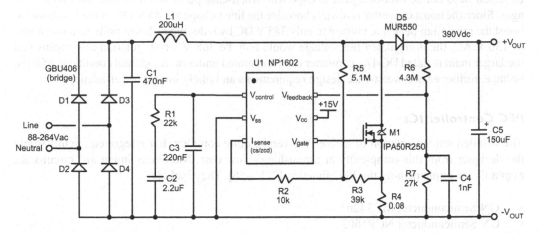

Figure 20.29 Power Factor Controller Circuit Using NP1602 IC Controller

the input haversine voltage waveform. To achieve the latter, it must also monitor its full-wave-rectified input voltage.

Figure 20.28 shows a very simplified PFC circuit, including the mains bridge rectifier, the boost converter and the PFC controller chip. R3 senses the input current and allows the PFC controller to match the input current waveform to the input voltage haversine waveform reference presented at the V_{in} pin. R4 and R5 provide a scaled version of the output voltage for output regulation by feedback. Note importantly that the controller chip has two feedback jobs to do, one for input current and one for output voltage. The controller chip is powered from a small auxiliary supply.

Once the initial inrush period is over, and capacitor C1 has been charged to nearly the peak value of the line voltage, the control IC often tailors its PWM to shape the input current waveform, as sensed at the current feedback point, for gradual charging of C1 the rest of the way to 385 V DC.

Bear in mind that the subsequent DC-DC converter will draw more input current when its input voltage is low, in order that it can deliver the same needed output power. This represents a negative resistance effect; as the input voltage falls, the current flow increases. Therefore, when the intermediate bus voltage is being brought up gradually, the switching converter's turn-on behavior must be controlled in some way. For example, the DC-DC converter should not be enabled until the bus voltage rises to its nominal regulated value.

Power factor correction is not employed merely to allow the amplifier to act as a "good neighbor" or to satisfy utility standards. It also enables the amplifier to extract the maximum amount of energy out of a mains supply with finite impedance and with a finite-amperage circuit breaker, since power is being extracted over the full mains voltage cycle, not just in high-current pulses of short duration. An actual PFC application circuit is shown in Figure 20.29.

Advantage of Pre-Regulation

Most active PFC circuits produce a regulated intermediate bus voltage. This immunizes the amplifier against large mains voltage variations at a very early stage. A large portion of the energy storage in an SMPS lies in the intermediate bus reservoir capacitor. Because of the high storage density of high voltage electrolytic capacitors, this is a less costly place to store energy than at, say, the ±50-V DC rails in the power amplifier. Regulation by the active PFC means that this expensive capacitor need not be over-designed to cope with worst-case permitted high and low mains voltages. Since the boost converter is always boosting the line voltage to 385 V DC, it has headroom to boost the maximum peak line voltage to only 385 V DC by reducing its boost ratio. At a maximum of 130 V AC, the unregulated bus voltage would only be 368 V. Pre-regulation also means that the larger main isolated DC-DC converter can be operated under more optimal conditions of input voltage, further easing worst-case design requirements and likely improving efficiency.

PFC Controller ICs

The detailed implementation of active PFC can be quite complex, but integrated circuits shield the designer from this complexity at astonishingly low cost. Numerous integrated circuits are available for implementing the PFC function [2, 51–56]. They include:

* ON Semiconductor MC34261
* ON Semiconductor NCP1602
* Texas Instruments UC3854/3855A
* Fairchild Semiconductor ML4826

20.18 Auxiliary Supplies

Many amplifiers require low-voltage auxiliary supplies on the secondary side for housekeeping functions and powering of operational amplifiers and protection circuits. These supplies typically are in the range of ±15 V and draw relatively little current. However, they must be very clean and well regulated. Traditionally they have been implemented with a small class-2 transformer, especially if some always-on functionality is needed.

In an amplifier with a switchmode main supply, one approach is to add extra windings to the switching transformer and implement separate rectifiers for each of these lower voltages. The extra windings take up space on the transformer bobbin and can lead to slightly lower efficiency. These auxiliary voltages may also be somewhat poorly regulated due to the wide range of current demands of the main power rails and limited cross-regulation performance.

An alternative approach is to employ small buck converters on the secondary side that are powered from the main rails of the amplifier. These can be simple non-isolated SMPS that require only a single small inductor. Although these supplies add another step of conversion with perhaps only 90% efficiency, the small power involved does not significantly affect the overall efficiency of the power amplifier as a whole. Simple low-current buck supplies with a known, regulated DC input source can be operated at 2 to 4 times the switching frequency of the larger main supply. This allows them to be quite small and inexpensive. Because auxiliary circuits often require little current, minimum load current requirements for such switching supplies must be kept in mind.

These separate non-isolated auxiliary converters provide excellent regulation for the low-level circuits and may obviate the need for linear regulators in these supplies. In perfectionist applications, the outputs can be post-regulated by a conventional linear regulator. With a known, regulated input having small ripple, linear regulators can be implemented with low headroom and consequently high efficiency.

Some SMPS designs need an initial source of power to operate the control circuits and allow start-up of the power supply. These auxiliary circuits are most often in an always-on state. These supplies can also supply isolated power to other functions, such as DC servo operational amplifiers, protection circuits, remote control receivers and signal-detect for auto power-on amplifier functions. In this case small low-cost stand-alone switchmode supplies can serve both line-side and secondary-side functions. Indeed, the same SMPS technology as used in USB chargers can be used to implement such supplies. Because these supplies are always consuming power, Green regulations require that such auxiliary supplies consume less than 0.5 watt, and these supplies easily meet that requirement [57].

Control ICs for Auxiliary Converters

Numerous integrated circuits are good candidates for implementing low-current auxiliary switching supplies [58, 59, 60]. All of these devices are available in small packages with 8 pins or less. They include:

*   Texas Instruments UCC28720
*   ON Semiconductor NCP1200
*   ON Semiconductor NCP1015

The UCC28720 is suitable for USB adapters and other low-cost, low-power devices. It features a simple circuit that uses an inexpensive BJT switch. Because it uses supply-side regulation it does not require an optocoupler for feedback. No-load power is less than 10 mW. The NCP1200 uses a single external MOSFET and also achieves very low standby power consumption. Although it employs an optocoupler, the entire circuit is extremely simple. The NCP1015 includes a built-in 700-V MOSFET switch, permitting even simpler implementations.

20.19 Switching Supplies for Power Amplifiers

Switching power supplies are obviously advantageous for audio power amplifiers, but the special needs and behavior of power amplifiers must be considered, especially for amplifiers of audiophile sound quality.

Special needs of power amplifiers include:

*   Energy storage
*   High peak current
*   Large load current range (peak-average ratio)
*   Multiple outputs
*   Auxiliary supplies
*   Low noise

Benefits of switching supplies for amplifiers include:

*   Line and load regulation
*   Less margin needs to be added for mains and load voltage regulation
*   Less waste heat in linear output stages because less voltage margin is needed
*   If PFC is used, more efficient use of mains power

The Advantage of Regulation

Conventional power amplifiers fitted with linear power supplies must be designed to accommodate a fairly wide range of worst-case power supply conditions. The rail voltages can vary over a large range when mains voltage variations and amplifier current load variations are considered. Mains voltage can easily vary over ±10%, corresponding to a range of 108 V to 132 V. Depending on transformer regulation, power supply rails can rise by 20% from full-load to no-load conditions. A 100-W/8-Ω amplifier with full-load rail voltages of ±45 V may see those rails rise to ±54 V under idle conditions. However, if the amplifier is required to produce its rated power at a low mains voltage of 108 V, the full-load rail voltages will nominally have to be ±50 V when the mains is at 120 V, and these may rise to ±60 V under idle conditions.

Under high mains voltage of 132 V, those no-load voltages will rise still further to ±66 V. This means that the output stage safe operating area and reservoir capacitors must be designed to handle ±66 V instead of what would have been ±45 V. Put another way, the amplifier must be designed so that it can produce bursts of 233 watts into 8 Ω. The substantial amount of over-design made necessary by unregulated supplies is eliminated if the rails are regulated, as in the case of an amplifier equipped with an SMPS power supply. This translates directly to cost, weight and heat.

The absence of 120 Hz ripple also buys another volt or more of useable rail voltage. The 100-W amplifier must source 5 A on low-frequency peaks that could span several cycles of 120 Hz. With 10,000-µF reservoir capacitors, ripple under these conditions can approach 3 V p-p.

Energy Storage and Peak Current Demand

An important design choice concerns how the large peak current demand will be met and where the corresponding energy storage will be located. The intermediate bus supply is one choice. In this case, the subsequent isolated supply must be able to supply the peak current and must be able to react quickly to load current changes. The high switching frequency enables this approach due to the much more frequent replenishment of the storage capacitors on the secondary side. The speed with which the SMPS regulating feedback loop can track the power required depends on the size of the secondary reservoir capacitors. At some audio frequency, the primary responsibility for maintaining the secondary voltage transitions from the SMPS feedback to the secondary reservoir capacitors. Smaller reservoir capacitors require a faster SMPS feedback loop.

A second choice is the use of large reservoir capacitors on the secondary side, as is done with power amplifiers using a linear power supply. This allows very high peak current to be supplied for brief periods. Just because the switching frequency is high and less filter capacitance is needed to keep ripple adequately small does not mean that reservoir capacitors should necessarily be smaller by a huge amount—like the ratio of 120 Hz to 120 kHz.

The circuit designer should keep in mind that a capacitor can only deliver energy when its voltage is allowed to drop, in accordance with the formula: $E = 1/2\, C\, (V_2{}^2 - V_1{}^2)$. This means that planning for only say 5% voltage drop on a secondary reservoir capacitor will access only 10% of the stored energy in the capacitor. A primary reservoir capacitor situated on the intermediate bus may be allowed to fall further, say 20%, thereby extracting 36% of the stored energy, and relying on the DC-DC regulation to prevent this input voltage sag from reaching the amplifier. Two other factors in favor of primary-side storage are: 1) a joule of energy is more economically stored on higher voltage capacitors, and 2) in a split-rail amplifier where it is unknown whether the positive or negative reservoir capacitor will need a large amount of low frequency energy, it can be more economical to store it on a single primary reservoir capacitor rather than on two oversized secondary reservoir capacitors.

Since the input of an amplifier looks either like a current sink (class AB) or a negative imped-ance (class D), the SMPS should never "clip"; i.e., reach the point where it cannot supply more current because its pulse width is at its maximum. Traditional linear supplies fall out of regula-tion gracefully, supplying more current, but at the expense of sagging output voltage. However, an SMPS such as a flyback converter, when loaded with an amplifier, can exhibit abrupt output voltage drop to a new equilibrium point if its current limit is hit. Bear in mind that most switching power supplies use negative feedback to maintain their output voltage where it should be. This can be a blessing and a curse.

Regulation of Multiple Secondary Supplies

Regulation of multiple secondary supplies is also an issue. Most switching supplies regulate the voltage at the secondary side, often using an optocoupler to feed an error signal back to the PWM controller on the primary side. This provides a very stiff supply, but only for the secondary voltage supply being sensed. The sensing can only be done for one secondary winding, so regulation of both the positive and negative rails may not be the same, since only one is usually sensed and tightly regulated.

At least one commercial amplifier deals with this in a rather elegant way. Comparator circuitry is used to effectively choose the rail that is sagging the most under load and regulate it. In other words, the rail needing the regulation the most is the one that is regulated.

Yet another approach is to regulate the difference between positive and negative rail voltages. This results in very tight control and limiting of peak voltage stress exerted on the output transistors, allowing a designer to push a set of semiconductors to higher output voltage.

Yet still another approach is to regulate on the primary side, and let all voltages on the secondary side be slightly less regulated. In this case a sense winding is wound to supply the sample voltage to be regulated. In this approach an opto-isolator is not needed. This arrangement essentially regulates the flux in the transformer core. The same number of volts per turn is generated by each winding, but for copper losses and secondary leakage inductance losses. The approach does regulate against copper losses and leakage inductance losses in the primary. All windings on the secondary are treated the same, as in a linear supply. Firm, but not perfectly stiff, regulation is achieved for each secondary voltage. The slightly softer regulation allows the reservoir capacitors to play a larger role in energy storage. This approach can also be implemented by monitoring the rectified output of a low-current low-voltage auxiliary supply on the secondary side. An error signal is then fed back to the primary side via an optocoupler.

Figure 20.30 illustrates a forward converter using primary-side control implemented with a small extra winding on the transformer. This supply provides two sets of output rails. The second set might be used for low-level circuits.

The current demands in a power amplifier can be highly asymmetrical. If there is a heavy load on one rail regulated by the PWM controller, the voltage on the other, more lightly loaded rail

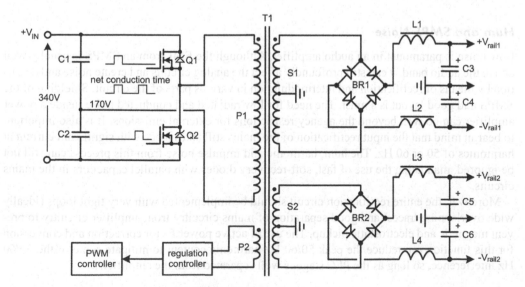

Figure 20.30 Switching Supply for a Power Amplifier

(at that time) may increase. This see-saw effect on the positive and negative rails can be a source of concern because the lightly loaded rail on any given half-cycle will be moving with the audio signal, possibly influencing the operation of the low-level stages and degrading audio quality in demanding audiophile applications. This is an example of the sort of thing that conventional continuous-power THD testing may not pick up. Of course, some amount of see-saw action occurs in amplifiers equipped with linear supplies as well.

This problem can be dealt with by using regulated non-isolated boost converters on the secondary side to supply the rails for the low-current IPS-VAS circuits preceding the output stage. This kills two birds with one stone by also providing these circuits with more headroom and permitting the implementation of IPS-VAS circuits that may need higher voltages. Providing such boost supplies in amplifiers with linear supplies can add significant expense, while this is much less the case with SMPS designs.

Light Loading Behavior

Idle current in a class AB amplifier can be very low under no-signal or small-signal conditions. In some cases it might be as low as 50 mA for a 100-watt/8-Ω amplifier. At the same time, if this amplifier is driving 40 V peak into a 2-Ω load the transient current demand will be 20 A. This amounts to a 400:1 range of current demand. Many SMPS designs require a minimum amount of load current to operate properly and stay in regulation. This can be a problem, since deliberately designing the amplifier to draw at least so much current at all times is very wasteful of energy, not to mention the need for additional heat sinking. The lossless loading approach described earlier in connection with the loaded resonant converter (LRC) can be helpful here.

Shared Inductor Cores

Virtually all secondary supplies in an SMPS, with the notable exception of flyback and LLC resonant types, require that an inductor be included in front of the rectifier. This implies that a power amplifier requiring positive and negative rail supplies will need two such inductors, adding to cost and space. However, both of these inductors can share the same core, as in a coupled inductor.

Hum and SMPS Noise

Low noise is paramount in an audio amplifier. Although the EMI from an SMPS is usually well above the audio band, it can still profoundly affect the analog circuits and create noise and distortion by means of rectification and intermodulation in various parts of the circuit. Shielding of the SMPS and good layout is a must. The need for low radiated and conducted noise inside a power amplifier can go well beyond the agency restrictions for external emissions. It is also important to bear in mind that the input rectification of the mains still takes place with significant current at harmonics of 50 or 60 Hz. The hum, harmonic and impulse noise from this process can still not be ignored, mandating the use of fast, soft-recovery diodes with parallel capacitors in the mains circuits.

Moreover, the entire rectification circuit should be implemented with very tight loops (ideally wide overlapping traces) and good separation of mains circuitry from amplifier circuitry to prevent magnetic and electrostatic pickup. The use of active power factor correction and conversion for this function can reduce the peak 50/60 Hz harmonic content to mitigate some of the 50/60 Hz interference, so long as the PFC stage is well separated from the amplifier.

Class G Amplifiers

Class G amplifiers require multiple high-current power rails at different voltages. The extra rails must be provided by additional windings on the main SMPS transformer. Several choices must be considered for implementing these multiple-output high current supplies, recognizing that not all rails can be regulated at the same time. The lower-voltage rails in a class G amplifier supply the output stage current 80–90% of the time. Consequently, the quality of the lower-voltage rails in terms of regulation and noise is audibly much more important. The detailed performance of the high voltage rails is less important during the brief loud transients that bring those supplies into play. For this reason, one approach is to regulate the low-voltage rails in the same way as is done in an amplifier requiring one set of rails. The high voltage rails, which incidentally are the ones usually requiring the most amount of peak current, can be largely unregulated and fitted with larger reservoir capacitors. Their output voltages will depend on the degree of cross-regulation with the regulated output.

Bear in mind that in a 2-rail class G amplifier (total of 4 rails when negative rails are included), only 1 of the 4 rails is demanding significant current on any given portion of the signal waveform. Flux regulation on the primary side can be especially advantageous for providing the same soft regulation of all of the secondary rail supplies. This is analogous to the way that linear power supplies work, but with a smaller transformer, far better load regulation and regulation against mains voltage variations.

Recommended SMPS Topologies

As seen earlier in this chapter, there exist a plethora of SMPS types and topologies that may be candidates for use in a power amplifier. However, the unique needs of an audio power amplifier tend to eliminate some of them while making others more preferable. Hard-switched PWM converters are straightforward and perform well, but can create increased EMI and are less capable of operating at higher frequencies. These shortcomings must be put into proper perspective and these converters should not be ruled out. Indeed, the relative importance of operating at higher frequencies and thus having smaller components may be different for their application in linear or class D power amplifiers.

The resonant converters are appealing, but they also have shortcomings, often including less tolerance to line voltage variations. Their ability to operate at much higher frequencies may have limited value-added, especially in linear power amplifiers. However, in power supply systems that employ PFC and pre-regulation, the smaller tolerance to line voltage variations of some of the resonant converters becomes nearly a moot issue. At the same time, one cannot ignore the fact that many active PFC circuits are themselves based on hard-switched PWM approaches.

An LLC converter that is preceded by active PFC and using primary-side regulation (flux regulation) appears to be an attractive choice. It is highly desirable that the isolated LLC be able to have quick response time and the ability to provide high peak current. The addition of a loading converter may enhance the ability of the system to deal with the very wide range of load current encountered in an audio power amplifier.

For lower-powered amplifiers a bridge rectifier with a flyback converter is a straightforward and very common way to generate positive and negative rails, and some housekeeping rails if necessary. To function across global nominal mains voltages of 100–240 V AC ±10%, there are several options: 1) an automatically switched doubler, 2) a factory-select doubler or 3) a user-select 120/240 V switch, 4) two factory BOMs (including choice of 120-V or 240-V transformer) or 5) a wide-range flyback design.

20.20 Switching Supplies for Class D Amplifiers

Power supplies for class D amplifiers must generally satisfy all of the needs discussed above, but also may include some additional challenges. Some audiophile amplifier designers are fond of saying that the output of a power amplifier is little more than a modulated version of the power supply voltage. Never was this more true than for a class D power amplifier.

As discussed later in Chapter 34, many class D amplifiers have poor power supply rejection ratio (PSRR). This is simply because these switching amplifiers switch the full positive power supply rail to the output of the amplifier for a controlled interval and then do the same with the negative power supply rail for a different interval. This is simply pulse width modulation (PWM). The amplifier's output filter reconstructs the resulting rectangular output waveform into an audio signal by low-pass filtering the signal. Any noise or ripple on the supply in the audio band will not be removed. Although some class D amplifier designs have means to increase PSRR, like high-order negative feedback loops and compensating arrangements, the requirements for low noise on the power supply rails are still quite stringent if the highest audio quality is to be achieved.

SMPS have an advantage over linear supplies for class D amplifiers because they are relatively free of low-frequency ripple at 120 Hz and its harmonics. Nevertheless, variations in the SMPS supply voltages can directly affect the gain of the class D amplifier in many designs, leading to intermodulation distortion. Also, if much smaller reservoir capacitors are used in the SMPS than would be used in a linear supply, then power supply noise in the audio band may be greater. There are also opportunities for noise modulation as the SMPS responds to the current demands that change at the audio frequency rate. There can also be opportunities for the switching frequency of the SMPS to interact with the switching frequency of the class D amplifier, leading to possible beat frequency and aliasing distortion. Many of these effects may not show up on traditional static THD measurements.

Bus Pumping

There is a phenomenon in some half-bridge class D amplifiers called *bus pumping*, as discussed in Chapter 34.6 and illustrated in Figure 34.6 [61–64]. It represents the transfer of energy from the loaded power supply rail to the unloaded power supply rail, tending to elevate the voltage of the unloaded rail. This is due to the switching power supply nature of class D amplifiers and energy storage in the output inductor.

With most linear supplies and many SMPS, the supply has no means to keep the power supply voltage from rising above its nominal value. This is because their main function is to deliver power, not to receive or absorb power. This is obvious by the unidirectional current flow enforced by the power supply rectifiers. If bus pumping causes one power supply rail to rise in a somewhat uncontrolled way, then the poor PSRR of the class D amplifier will allow the audio signal to be affected. This phenomenon is mitigated somewhat by the use of large reservoir capacitors, pushing the main effects of bus pumping to lower frequencies. However, smaller reservoir capacitors are often used with SMPS.

Some SMPS architectures are able to absorb the reverse pump current with very little loss and return it to the opposite supply or to a common supply source [62–64]. Figure 20.31 shows how a synchronous buck converter provides bi-directional regulation. If the output voltage rises above the set point as a result of current pumping, the current is fed back through the converter and transferred to the DC source. This reverse current flow through the buck converter is possible because the synchronous buck converter functions as a boost converter in the opposite direction.

It is notable that full-bridge (BTL) class D amplifiers do not suffer from the bus pumping effects because the two amplifiers comprising the bridged amplifier are drawing opposite currents from

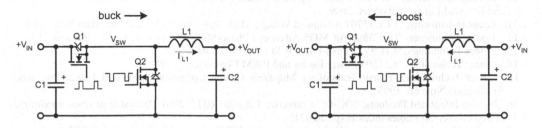

Figure 20.31 Bi-Directional Regulation of a Synchronous Buck Converter

the power supply rail. When the amplifier for the positive output signal is sourcing current to the load, it pumps some current into the negative rail. At the same time, the other amplifier for the negative output signal is sinking current from the opposite side of the load. It is drawing current from the negative rail and pumping some current back into the positive rail. The bus pumping effects of the two amplifiers thus cancel.

Indeed, the same switching phenomenon that causes supply pumping can be used in reverse. Supply pumping can be counteracted by use of a dedicated switch-mode circuit to return the pump current to the opposite supply. In one example, visualize bus pumping compensation by taking advantage of the bus pumping effect. A crude "dummy" half-bridge class D amplifier channel or similar switching arrangement can be used to provide controlled low-loss bi-directional power transfer from one rail to the other. In such an arrangement the "output" of the dummy class D amplifier would be connected to ground and its output capacitor removed.

The switches in this "amplifier" could be operated alternately at an equal duty cycle of 50% less some dead time. If an imbalance exists between the positive and negative rails it will appear momentarily as a net voltage across the inductor, causing its average current to rise in the desired direction and shuttle charge from the higher magnitude to lower magnitude rail, thereby restoring balance. The technique can be made faster still via a more aggressive closed-loop feedback technique that actively changes the duty cycle of the switches when faced with an imbalance. This technique amounts to summing the positive and negative rails to create an input signal for the dummy class D amplifier. That sum is then forced to zero by the action of the dummy amplifier.

References

1. Keith Billings, *Switchmode Power Supply Handbook*, McGraw-Hill, New York, 2010.
2. Abraham I. Pressman, Keith Billings and Taylor Morey, *Switching Power Supply Design*, 3rd ed., McGraw-Hill, New York, 2009.
3. Robert Erickson, *Fundamentals of Power Electronics*, Springer, Boulder, CO, 1997.
4. Rudy Severns and Gordon E. Bloom, *Modern DC-to-DC Switch Mode Power Converter Circuits*, Van Nostrand Reinhold, Scarborough, Ontario, Canada, 1985.
5. Slobodan Ćuk and Robert D. Middlebrook, "A New Optimum Topology Switching DC-to-DC Converter," proceedings of the IEEE Power Electronics Specialists Conference, June 1977.
6. Rudy Severns, "High Frequency Switching Regulator Techniques," proceedings of the IEEE Power Electronics Specialists Conference, June 1978.
7. Rudy Severns and Hal Wittlinger, "High Frequency Power Converters," Intersil Application Note AN9208, April 1994.
8. Emanuel E. Landsman, "A Unifying Derivation of Switching DC-DC Converter Topologies," IEEE Power Electronics Specialists Conference, San Diego, 1979.
9. ON Semiconductor, "UNCP1203 PWM Current-Mode Controller for Universal Off-line Supplies Featuring Standby and Short Circuit Protection," datasheet, 2007.

10. ON Semiconductor, "NCP1271 Soft-Skip Mode Standby PWM Controller with Adjustable Skip Level and External Latch," datasheet, 2009.

11. Texas Instruments, "UCC35701 Advanced Voltage Mode Pulse Width Modulator," datasheet, 2005.

12. Texas Instruments, "UCC3895 BiCMOS Advanced Phase-Shift PWM Controller," datasheet, 2013.

13. Linear Technology, "LTC3721 Push-Pull PWM Controller," datasheet.

14. Linear Technology, "LT1509 Power Factor and PWM Controller," 1995.

15. Linear Technology, "Linear Technology Magazine Circuit Collection Volume II Power Products," Application Note 66, 1996.

16. Maxim Integrated Products, "DC-DC Converter Tutorial 2031," 2001. Available at www.maximintegrated.com/en/app-notes/index.mvp/id/2031.

17. Sanjay Pithadia and Jeff Falin, "Understanding TPS61175's Pulse-Skipping Function," TI Application Report SLVA353, 2009.

18. Robert Mammano, "Switching Power Supply Toplogy Voltage Mode vs. Current Mode," Unitrode/TI Design Note DN-62, 1994.

19. Robert Mammano, "Resonant Mode Converter Topologies—Topic 1," Unitrode Corp.

20. Robert Mammano, "Resonant Mode Converter Topologies—Topic 6," Additional Topics, Unitrode Corp.

21. Robert L. Staigerwald, "A Comparison of Half-Bridge Resonant Converter Topologies," *IEEE Transactions on Power Electronics*, vol. 3, no. 2, April 1988.

22. Freescale MC56F80xx controller applications, Resonant Converter Topologies and Features.

23. Guisong Huang et al., "LLC Series Resonant DC-to-DC Converter," US Patent 6,344,979, Delta Electronics, 2002.

24. Hangseok Coi, "Design Considerations for an LLC Resonant Converter," Fairchild Power Seminar, 2007.

25. Fairchild Semiconductor, "Half-Bridge LLC Resonant Converter Design Using FSFR-Series Fairchild Power Switch (FPS)," AN-4151, 2007.

26. STMicroelectronics, "An Introduction to LLC Half-Bridge Converter," AN2644, 2008. Available at www.st.com.

27. Silvio De Simone, "LLC Resonant Half-Bridge Converter Design Guideline," AN2450, STMicroelectronics, 2014. Available at www.st.com.

28. Gyana Ranjan Sahu, Rohit Dash and Bimal Prasad Behera, "Design and Implementation of ZCS Buck Converter," Rourkela National Institute of Technology, August 2010.

29. Sanjaya Maniktala, "Understanding and Using LLC Converters to Great Advantage," Microsemi Corporation, 2013.

30. Hong Huang, "Designing an LLC Resonant Half-Bridge Power Converter," 2010 Texas Instruments Power Supply Design Seminar. Available at www.power.ti.com/seminars.

31. Unitrode Products from Texas Instruments, "Resonant-Mode Power Supply Controllers," UC1861–1868 family, September 2007.

32. ON Semiconductor, "NCP1399 High Performance Current Mode Resonant Controller With Integrated High-Voltage Drivers," datasheet, January 2016.

33. Patrizio Vinciarelli, "Forward Converter Switching at Zero Current," U.S. Patent 4,415,959, 1983.

34. Bill Andreycak, "Zero Voltage Switching Resonant Power Conversion," Unitrode Application Note U-138.

35. Lisa Dinwoodie, "Exposing the Inner Behavior of a Quasi-Resonant Flyback Converter," 2012 Texas Instruments Power Supply Design Seminar. Available at www.ti.com/psds.

36 . Fairchild Semiconductor, "Design Guidelines for Quasi-Resonant Converters Using FSQ-series Fair-child Power Switch (FPS)," Application Note AN4146.

37 . Larry Wofford, "A New Family of Integrated Circuits Controls Resonant Mode Power Converters," Unitrode (TI) Application Note U-122.

38 . STMicroelectronics, "L6565 Quasi Resonant SMPS Controller," datasheet, 2003.

39 . Infineon, "Converter Design Using Quasi-resonant PWM Controller ICE2QSO3G," Application Note ANPS005, 2010.

40 . Infineon, "Converter Design Using Quasi-resonant PWM Controller ICE2QSO2G," Application Note ANPS0027, 2008.

41 . J. Patrick Donohoe, Professor, Mississipi State University, "ECE4323 EMC Requirements." Available at my.ece.msstate.edu/faculty/donohoe/ece4323EMCreq.pdf.

42 . Peter Wood, et al., "Thermal and Mechanical Considerations for FullPak Applications," International Rectifier Application Note AN-972B.

43 . LearnEMC, "Practical EM Shielding," 2015. Available at www.learnemc.com .

44. Field Management Services, "Shielding AC Magnetic Fields." Available at www.fms-corp.com.
45. Bruce Hofer, "Designing for Ultra-Low THD+N in Analog Circuits," 139th Convention of the Audio Engineering Society, Session PD3, October 29, 2015.
46. Magnetics Division of Spang & Company, "Magnetic Cores for Switching Power Supplies." Available at www.mag-inc.com.
47. Fair-Rite Products Corp., "How to Choose Ferrite Components for EMI Suppression." Available at www.fair-rite.com.
48. Robert West, "Common Mode Inductors for EMI Filters Require Careful Attention to Core Material Selection," Magnetics Division of Spang & Butler. Available at www.mag-inc.com.
49. Robert D. Middlebrook, "Input Filter Considerations in Design and Application of Switching Regulators," IEEE Industry Applications Society Annual Meeting, October 1976.
50. ON Semiconductor, "Power Factor Correction (PFC) Handbook," 2011. Available at www.onsemi.com.
51. Fairchild Semiconductor, "ML4826 PFC Controller," datasheet, 2001.
52. Unitrode/TI UC3855A datasheet, "High Performance Power Factor Preregulator," 2005.
53. Jim Noon, TI Application Report SLUA146A, "UC3855A High Performance Power Factor Preregulator," 2004.
54. ON Semiconductor, "NCP1602 Enhanced High-Efficiency Power Factor Controller," datasheet.
55. ON Semiconductor, "5 Key Steps to Designing a Compact, High-Efficiency PFC Stage Using the NCP1602," Application Note AND9218/D, May 2015.
56. ON Semiconductor, "NCP1602 Evaluation Board User's Manual," NCP1602GEVB.
57. European Commission EC Regulation 1275/2008 (EU Ecodesign). See also the International Energy Agency (IEA) One Watt Initiative.
58. Texas Instruments, "UCC28720 Constant-Voltage, Constant-Current Controller with Primary-Side Regulation," datasheet, September 2015.
59. ON Semiconductor, "NCP1200 PWM Current-Mode Controller for Low-Power Universal Off-line Supplies," datasheet, April 2015.
60. ON Semiconductor, "NCP1015 Self-Supplied Monolithic Switcher for Low Standby-Power Offline SMPS," March 2011.
61. Jun Honda and Jonathan Adams, "Class D Audio Amplifier Basics," International Rectifier Application Note AN-1071, February 2005.
62. Vicent Sala et al., "Study of Hybrid Active Control Strategies for the Bus-Pumping Cancellation in the Half-Bridge Class-D Audio Power Amplifiers," 2011 IEEE International Symposium on Industrial Electronics (ISIE), Gdansk, Poland, June 2011.
63. Mehrzad Koohian, "Synchronous Bus Converter Supplying Class D Amplifier Virtually Eliminates Bus Pumping," *Electronic Design*, January 26, 2011.
64. Planet Analog, "Minimize Power Supply Pumping for Single-ended-output Class-D Audio Amplifiers," June 20, 2009. Available at www.planetanalog.com.

Chapter 21

Clipping Control and Civilized Amplifier Behavior

21. INTRODUCTION

The question is often asked why amplifiers with otherwise very good measured performance sound different. There are many possible reasons for this, but some of those reasons may lie in how the amplifier behaves under conditions for which it was not tested or optimized.

Overload conditions are a good example. Does the amplifier clip gracefully or does it create more than just a cleanly clipped waveform? Does the amplifier current limit prematurely under some conditions? Does the amplifier produce a burst of oscillation when it clips into certain kinds of loads?

21.1 The Incidence of Clipping

Amplifiers clip more frequently than some realize. This is especially the case on well-recorded music where the dynamic range has been well preserved. While low-power amplifiers are often thought to be more prone to clipping at realistic sound levels, even high-power amplifiers may clip when driving low-efficiency loudspeakers. Bear in mind that a loudspeaker having sensitivity of 84 dB requires 10 times as much power as one with 94 dB.

The dynamic range that must be reproduced is strongly influenced by the *crest factor* of the music being played. This is the ratio of the maximum power that occurs on peaks to the average power. Well-recorded music can have a crest factor exceeding 14 dB, while music from a typical FM station may have a crest factor of less than 3 dB. The latter is a result of processing that deliberately reduces dynamic range so that the station can sound louder while still remaining within its peak modulation limits. Music with a 14-dB crest factor will require 10 times the power of music with a 4-dB crest factor played at the same perceived sound level.

It is important to keep in mind that a 10-dB increase in sound level corresponds to a factor of 10 in power level. It is also important to recognize that the human perception of loudness corresponds more closely to the average power and that brief percussive transients are not bothersome. Consider program material with a 14-dB crest factor where loudspeakers with an efficiency of 83 dB SPL at 1 meter are being driven. Assume that realistic levels are being played such that the average SPL at 1 meter is 96 dB from each channel, on average (the level at the listening position will be considerably less). The average power must be +13 dB-W, or 20 W. The maximum power on the peaks will be 14 dB higher, corresponding to a factor of 25. Thus the amplifier must be rated at no less than 500 W per channel if it is not to clip.

Clipping Experiments

Several experiments were carried out to evaluate clipping with real program material [1, 2]. The objective was to measure average power and peak power so as to be able to compare them and determine the crest factor of the program material. A special peak/average meter was designed and built for this purpose [3]. The average-responding side of the meter uses a true RMS IC with a moderately long time constant. Its output is calibrated in average power into an 8-Ω load. The peak-responding side of the meter employs a two-stage peak-hold detector that can resolve peaks as short as 10 μs with good accuracy. This circuit holds the highest peak for 3 seconds to allow time for the reading. This circuit is calibrated so that its peak voltage reading corresponds to equivalent average power into 8 Ω. If both meters are fed with a sine wave, their readings will be identical.

A percussive CD track was played through loudspeakers with a sensitivity of 84 dB. A 250-W amplifier was used. The music was played at a realistic but not overly loud level in a 400-square-foot hotel room. Average power hovered between 1 W and 2 W. Power peaks exceeded 260 W, clipping the power amplifier.

21.2 Clipping and Sticking

Given the inevitability of clipping, it is important to consider the sonic behavior of the amplifier when it clips. Indeed, some believe that the more civilized way in which vacuum tube amplifiers clip is responsible for their perceived better sound (by some) in spite of their lower power.

The cleanest of solid-state power amplifiers will neatly clip the peaks off the signal. Many solid-state amplifiers will suffer what is called *sticking* or *overhang*. Such amplifiers go into clipping cleanly, but don't come out of clipping cleanly when the amplitude demanded by the program falls below the clipping point. Instead, such amplifiers *stick* to the higher clipped level for a brief instant and then fall more quickly down to the level required by the signal. Such a waveform is shown in Figure 21.1. Here we have also shown the waveform of the difference between the correct output and the actual output. Notice that the waveform includes spikes in the area where the sticking has occurred. These are noticeable and objectionable [4, 5, 6].

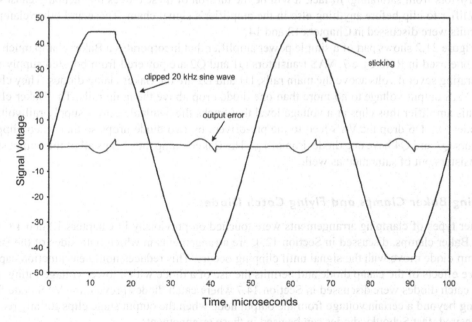

Figure 21.1 Output Waveform of an Amplifier That Is Clipping and Sticking

21.3 Negative Feedback and Clipping

Negative feedback can exacerbate undesirable clipping behavior. When clipping occurs, the error signal at the input of the amplifier becomes very large and may overload one or more stages of the amplifier. Those stages, having been overdriven, will take time to recover and get back to their proper signal voltages after the input is no longer being overdriven. The VAS is especially likely to become overloaded. Delays introduced by frequency compensation capacitors may lengthen the time required for recovery. It is especially important that output transistors not be allowed to saturate.

Negative feedback sharpens up the clipping edges. This happens largely because the gain from input to output at the onset of clipping attempts to go from the closed-loop gain to the open-loop gain. This happens because there is no global negative feedback once the amplifier clips.

21.4 Baker Clamps

If an amplifier must clip it is important that it clip cleanly to achieve the highest sound quality. Ideally, power amplifiers should clip softly, but even when they clip hard, as most solid-state amplifiers do, it is important that there be no sticking or other failure to come out of the clipped state quickly.

It is also desirable that the amplifiers not lose their power supply rejection during clipping intervals, as that would allow ripple and grunge from the power supply to enter the signal path during the clipping interval. Amplifiers that clip to the main rail will essentially be applying the dirty main rail voltage directly to the loudspeaker under clipping conditions.

By proper design negative feedback amplifiers can be made to clip quite cleanly. In order to accomplish this it is very important that no transistors go into saturation. One approach to this is to employ so-called Baker clamps on the output of the VAS [7, 8]. These are diodes that will turn on and prevent the VAS output signal from swinging toward the rail beyond a certain point. This clamping action prevents the VAS transistor from saturating and also prevents the output transistors from saturating. In fact, it will be the turn-on of these diodes that actually causes the amplifier to clip before anything else in the amplifier's signal chain. These and other clamping circuits were discussed in Chapters 12 and 14.

Figure 21.2 shows part of a simple power amplifier that incorporates a Baker clamp much like the one used in Reference 7. VAS transistors Q1 and Q2 are powered from boosted supply rails operating several volts above the main rails. D1 and D2 are the Baker clamp diodes. They clamp the VAS output voltage to no more than one diode drop above the main rails. The Baker clamp in this amplifier thus clips at a voltage level that tracks the available power supply rail voltage. Diodes D3–D6 drop the VAS feed to the pre-drivers by two diode drops so that when clipping occurs Q3 and Q4 have one diode drop of V_{cb}. The Baker clamp thus keeps all of the output stage transistors out of saturation as well.

Flying Baker Clamps and Flying Catch Diodes

Other types of clamping arrangements were touched on previously in Chapters 12 and 14. Flying Baker clamps, discussed in Section 12.4, are arrangements in which both sides of the Baker clamp diode move with the signal until clipping occurs. This reduces nonlinear junction capacitance effects of the clamp diode and permits the use of a diode with a lower voltage rating. Flying catch diodes were discussed in Section 14.7 where catch diodes prevent the VAS node from going beyond a certain voltage from the output node when the output stage clips for any reason. If desired, fast Schottky diodes can be used in these arrangements.

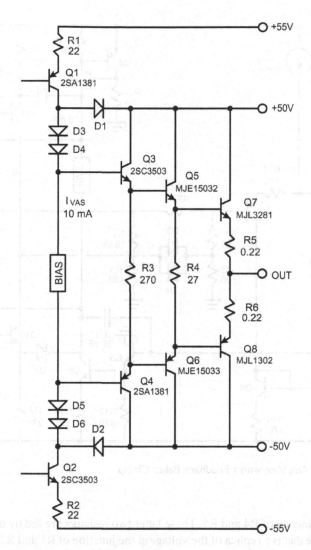

Figure 21.2 A Simple Power Amplifier with a Baker Clamp

Feedback Baker Clamps

Feedback Baker clamps keep the global feedback loop closed during clipping by diverting the clamping current back to the input of the amplifier, bypassing the output stage. As shown in Figure 21.3, the Baker clamp reference is provided from emitter followers Q9 and Q10 whose complementary collectors are joined. A signal is fed back from that junction to the input stage to complete the feedback connection around the amplifier when the Baker clamps are acting. This keeps the whole front-end linear and prevents sticking. It is important that this bypass feedback loop be stable.

R1, R2 and R3 implement the normal global feedback network that is fed by the output stage, which is not shown. When Baker clamp diodes D1 or D2 conduct, their current is conducted by

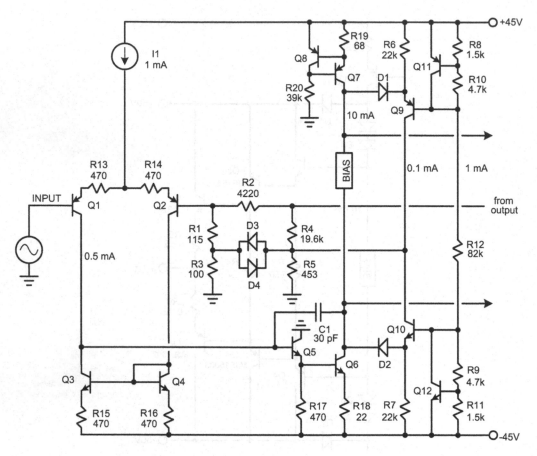

Figure 21.3 A Power Amplifier with a Feedback Baker Clamp

Q9 or Q10 to the junction of R4 and R5. These latter two resistors are fed by the amplifier output and create a voltage that is a replica of the voltage at the junction of R1 and R3. This arrangement keeps the signal voltage across isolating diodes D3 and D4 at zero during normal operation when the amplifier is not clipping. This eliminates the possibility of distortion currents being injected into the feedback node. When the amplifier clips, the diverted VAS current closes the feedback loop through D3 and D4. This allows the input stage to stay linear for input signals that far exceed the clipping point of the amplifier.

The use of the tapped shunt feedback resistor (R1, R3) illustrates the general case, but the clamp feedback can be connected directly to the feedback node if desired. The ratio of R1 and R3 allows some control of the behavior of the feedback Baker clamp under heavy-overload conditions. R4 and R5 must always be adjusted so that the normal signal voltage across D3 and D4 is zero for whatever ratio of R1 and R3 is chosen.

Q11 and Q12 are simply V_{be} multipliers that set how far from the rails the Baker clamp threshold is. As shown here, they produce a drop of 4 V_{be} so that Q6 and Q7 will have one V_{be} of collector-base voltage when clipping occurs. This design is also compatible with amplifiers using boosted rails for the IPS-VAS. Because the global feedback never disappears, the clipping afforded by the Baker clamp diodes here is a bit softer than without Baker clamp feedback.

21.5 Soft Clipping

The best amplifier is one that never clips. If clipping must occur, it is best accomplished with a passive circuit ahead of the amplifier input. Such a circuit can clip the signal quickly and cleanly, without creating sharp edges. Ordinary silicon diodes can fulfill this task. Of course, with soft clipping, there will be some gradual rise in THD prior to clipping. As a result, such amplifiers might not measure as well as amplifiers of the same rated power that are not preceded by a soft-clip circuit.

The Klever Klipper

While diodes clipping to fixed thresholds can provide the desired soft-clipping behavior, they will not allow for the dynamic headroom normally available for brief signal bursts in normal amplifiers. Thus there is the need for dynamic clipping thresholds that track the available peak output of the amplifier in real time. Figure 21.4 shows what I call the Klever Klipper circuit [9].

Diodes D1 and D2 provide the soft-clip function in combination with R1. The soft-clip threshold voltages are created at op amps U1A and U1B. These voltages track the short-term average power supply rail voltage. Adaptive soft clipping occurs just shy of output amplifier hard clipping at any given power supply rail voltage condition. As a result, little or no dynamic headroom is sacrificed.

The Klever Klipper control circuit takes the negative rail supply voltage and scales it down to approximately 1/20 of its value, to roughly match the gain of the amplifier. This moving reference voltage is filtered by C1 and C2. The U1A buffer then adds one diode drop (D3) to this voltage to account for the approximate conducting diode drop of the soft-clip diodes when they begin

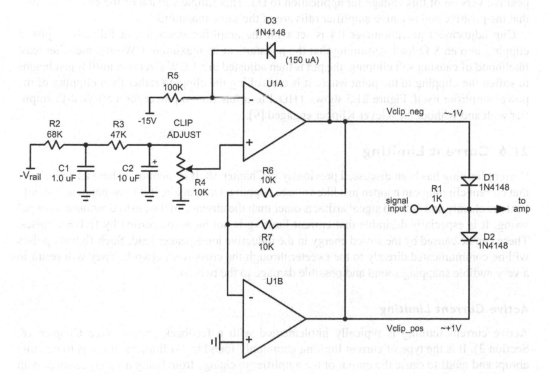

Figure 21.4 The Klever Klipper Soft-Clip Circuit

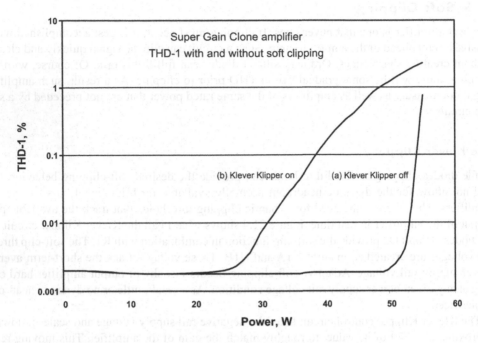

Figure 21.5 THD-1 Versus Power for an Amplifier Using the Klever Klipper

to clip. The output of U1A is the negative clipping voltage that is applied to D1. U1B creates a positive version of this voltage for application to D2. This simple version of the circuit assumes that the positive and negative amplifier rails are of the same magnitude.

Clip adjustment potentiometer R4 is set with the amplifier operating at full power just at clipping into an 8-Ω load. Assuming that the pot starts at its maximum CW end, meaning least likelihood of causing soft clipping, the pot is then adjusted in a CCW direction until it just begins to soften the clipping to the point where it is controlling the clipping rather than clipping of the power amplifier itself. Figure 21.5 shows THD-1 as a function of power for a 50-W/8-Ω amplifier with and without the Klever Klipper engaged [9].

21.6 Current Limiting

Current limiting has been discussed previously in Chapter 18. It is mentioned here as a reminder that current clipping can happen just like voltage clipping. Once again, if it must happen, it should be done cleanly and with no signal artifacts other than the absence of the desired amount of signal swing. It is especially desirable that current limiting is not be accompanied by flyback pulses. These can be caused by the stored energy in the inductive loudspeaker load. Such flyback pulses will be communicated directly to the tweeter through the crossover network. They will result in a very audible snapping sound and possible damage to the tweeter.

Active Current Limiting

Active current limiting is typically implemented with a feedback process (see Chapter 18, Section 3). It is the type of current limiting commonly found in *V-I* limiters. It tends to be quite abrupt and tends to cause the output of the amplifier to change from being a voltage source with low output impedance to a current source with high output impedance.

Natural Current Limiting

Natural current limiting is achieved by the use of flying catch diodes, as described in Section 3 of Chapter 18. This type of current limiting tends to be softer and tends to retain moderately low output impedance during current limiting.

21.7 Parasitic Oscillation Bursts

Another thing that can degrade the sound quality of an otherwise good-measuring amplifier is the occurrence of parasitic oscillation bursts. The operating points of many of the transistors in an amplifier signal path change quite a bit when there are large signal swings. Different operating points can cause altered transistor dynamic parameters like speed and capacitance. This can lead to a loss of feedback stability margin and possible parasitic oscillation bursts on signal peaks. This goes double for when the amplifier clips.

Amplifiers without negative feedback can also be prone to parasitic oscillations under certain signal swing and load conditions. These oscillations may be local in nature, such as in the case of an oscillating output emitter follower. Such parasitic oscillation bursts often may not show up on bench tests, especially when a resistive load is being used. A *parasitic oscillation sniffer* that can be used to detect bursts under a large variety of signal and load conditions is discussed in Chapter 27.

21.8 Selectable Output Impedance

In some cases it is preferable to have controlled output impedance of a couple of tenths of an ohm. Some loudspeakers seem to sound better when fed with such a source. In some ways this is like the output impedance of a vacuum tube amplifier. This results in a low damping factor, often less than 20. A damping factor of 20 can be obtained by simply adding a 0.4-Ω non-inductive power resistor in series with the output of the amplifier.

However, it is desirable to make the introduction of such a resistance into the output of the amplifier optional. This can be done with a relay across the added series resistor, but there is a more elegant way. The series resistor is placed between the output stage and the output coil. The negative feedback is then taken off either before or after the series resistor, as controlled by a small-signal relay. Figure 21.6 shows such an arrangement.

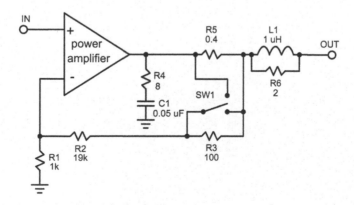

Figure 21.6 Selection of Amplifier Output Impedance

References

1. Rocky Mountain Audio Fest, Denver, CO, 2006.
2. Home Entertainment Show (HE2007), New York, 2007.
3. "A Peak/Average Power Level Meter." Available at www.cordellaudio.com.
4. Tomlinson Holman, "New Factors in Power Amplifier Design," *Journal of the Audio Engineering Society*, vol. 29, no. 7–8, pp. 517–522, July–August 1981.
5. Tomlinson Holman, "Amplifier Design & Sound Quality," *Audio*, pp. 26–31, November 1996.
6. "Apt Corp. Apt 1 Amplifier" (review), *Audio*, pp. 44–47, March 1981.
7. "The Apt1 Power Amplifier Owner's Manual," Apt Corporation, 1979.
8. Robert R. Cordell, "A MOSFET Power Amplifier with Error Correction," *Journal of the Audio Engineering Society*, vol. 32, no. 1, pp. 2–17, January 1984. Available at www.cordellaudio.com.
9. "The Super Gain Clone." Available at www.cordellaudio.com.

Chapter 22

Interfacing the Real World

22. INTRODUCTION

More than any other component in the audio chain, the power amplifier must deal with the realities of the real world. This is particularly so because it drives the loudspeaker with high currents over typically unshielded cables and because the loudspeaker itself is an electromechanical device that makes for a highly complex load. Other realities include *electromagnetic interference* (EMI) ingress from the input and mains ports, high-current rectifier noise and other sources of trouble. Some aspects of the interface have already been covered in the earlier chapters on protection circuits and power supplies.

22.1 The Amplifier-Loudspeaker Interface

The fact that the power amplifier drives an electromechanical device distinguishes it from most of the other elements in the audio signal chain. The loudspeaker load has complex impedance and can store substantial amounts of energy. It can also require high currents to drive it.

The Loudspeaker Is Not a Resistive Load

The loudspeaker consists of multiple electromechanical drivers connected together by a passive LCR crossover network. Together, the elements of this arrangement form a complex nonlinear load for the amplifier. In some cases highly capacitive speaker cables add to the complexity. At radio frequencies there can be transmission line effects from the speaker cables as well. Figure 22.1 illustrates a simple equivalent electric circuit for a loudspeaker woofer in a sealed enclosure. Figure 22.2 shows the impedance curve for the driver of Figure 22.1.

Peak Output Current Requirements

The loudspeaker can store energy in many different ways, but the woofer velocity and displacement against the restoring force of the suspension are usually the two greatest ways in which energy is stored. Figure 22.3 shows the results of a SPICE simulation that reveals the high peak load currents that are possible under conditions where the loudspeaker is driven by a particular contrived waveform [1]. Other authors have also studied phenomena at the amplifier-loudspeaker interface [2, 3, 4].

The driving waveform in Figure 22.3 was deliberately chosen to maximize the expected peak load current. The signal swings between large positive and negative values, rather than simply starting from zero. The waveform begins at −28 V and remains there for 16 ms to allow load

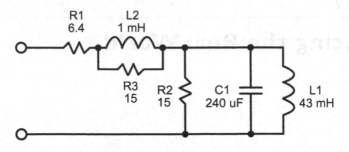

Figure 22.1 Electrical Model of a Loudspeaker Woofer

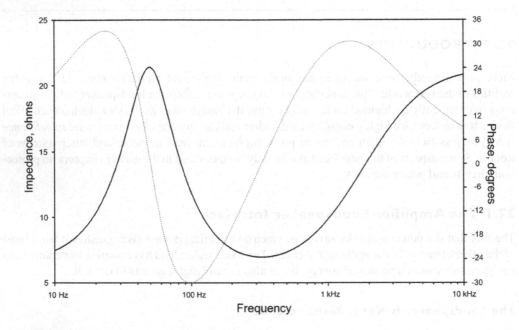

Figure 22.2 The Impedance Curve for the Driver of Figure 22.1

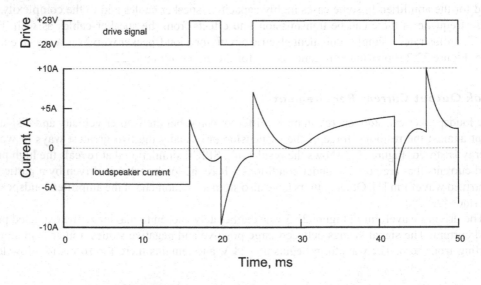

Figure 22.3 High Peak Load Currents When the Loudspeaker Is Driven with a Contrived Waveform

current to rise to at least 90% of its final value. The signal then goes to the positive extreme for 4 ms. At the end of this 4-ms interval the counter-emf of the loudspeaker is at its maximum (maximum cone velocity). The polarity is then reversed, going to the negative extreme. The counter-emf now opposes the new signal polarity and enhances the peak current flow. The very large negative current flow of nearly 10 A is the result. After 4 ms, the signal polarity goes back to the positive extreme and the sequence is repeated for the opposite polarity so that the average value of the signal is zero.

While an amplifier delivering this waveform to an 8-Ω resistive load would normally see a peak load current of about 3.5 A, we see from Figure 22.3 that the RLC load develops a peak load current of 10 A. This is where the somewhat arbitrary factor of 3 for desired reserve current capability of an amplifier mentioned in Chapter 1.5 came from. While the probability and extent of this kind of occurrence in the real world with musical program material may be questioned, the exercise does provide some food for thought. The lesson to be learned here is to be prepared to handle larger currents than are encountered with a simple resistive load.

Transmission Line Effects of Speaker Cables

The loudspeaker cable is a transmission line that is usually mis-terminated. As such, it is subject to reflections, where energy is reflected back from one end to the other. Even if the speaker cable had a characteristic impedance of 8 Ω, it would probably be mis-terminated at the high frequencies where it matters most. In reality most loudspeaker cables have *characteristic impedance* Z_o between 50 Ω and 150 Ω. Transmission line effects are unlikely to matter much at frequencies in the audio band, but they can cause load impedance variations that can provoke amplifier instability at high frequencies.

Consider a 10-foot speaker cable. The speed of light is on the order of 0.7 foot per nanosecond in cables with typical dielectrics. The propagation delay will be about 14 ns. The round-trip delay for a reflection will be about 28 ns. At a frequency whose period is 56 ns, this round-trip delay will represent 180°. This frequency is 18 MHz. An ideal transmission line with no far-end termination (at high frequencies) will exhibit very low impedance at this frequency. One with a shorted termination will exhibit very high impedance at this frequency. This frequency is well above the gain crossover frequency of virtually all audio power amplifiers, but it can be in the range where local output stage oscillations can be provoked in wideband output stages that are not isolated from the output by an L-R network. Amplifiers that misbehave under certain loading conditions will certainly sound different with different speaker cables.

Figure 22.4 shows the impedance versus frequency looking into the end of a 10-foot length of a very popular brand of speaker cable that employs a ZIP cord type of construction. The three plots are for the conditions where the far end is open, shorted and terminated in the characteristic impedance of 120 Ω. The wild impedance gyrations in the open and shorted cases may drive some amplifiers to instability. Bear in mind that the terminating impedance of a loudspeaker at frequencies above 1 MHz is anyone's guess. Figure 22.4c shows what a remarkable difference a proper termination makes even to frequencies approaching 100 MHz.

The data in Figure 22.4 suggests that a Zobel network located at the far end of the speaker cable is a good idea, even if it consists of little more than a 100-Ω resistor in series with a 0.01-μF capacitor. More complex networks that provide a controlled transition of the terminating impedance to 100 Ω as frequency increases can easily be imagined. Loudspeakers could also incorporate such a network internally.

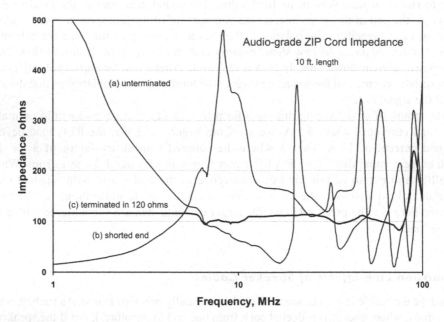

Figure 22.4 Impedance Versus Frequency Looking into a 10-Foot Length of Speaker Cable (a) Far End Open (b) Far End Shorted (c) Far End Terminated

22.2 EMI Ingress: Antennas Everywhere

Electromagnetic interference is all around us, across the complete frequency spectrum from the audio band up to several GHz and beyond. Operate a cell phone near your inexpensive powered computer loudspeakers, and you will likely hear the interference. Microwave ovens, light dimmers and electric drills are other common sources of EMI.

These sources of interference can all create noise in the audio band, even though their frequencies are often way above the audio band. The noise, if present, is created by demodulation that creates audible artifacts. Noise is annoying, but is not the same as distortion. There are, however, mechanisms whereby the in-band noises created by EMI are correlated to the audio signal, and this then *is* distortion.

Finally, one must ask the question: *If in the absence of the audio signal there is no audible noise from EMI, is it then safe to conclude that there must not be enough EMI to cause distortion of the audio signal?* It is tempting to say yes, but the answer is probably "no." If there is intermodulation, the EMI may produce artifacts only when an audio signal is present. The use of good RF design techniques is called for, even though these are audio amplifiers.

RFI and EMI

Radio frequency interference (RFI) and electromagnetic interference are treated as one and the same in this book. The term *EMI* is broader, so it will be used almost exclusively in these discussions.

EMI Ingress from the Amplifier Input

A monophonic amplifier driven by a well-designed shielded interconnect cable from the preamp should be subject to very little EMI from its input, assuming that the preamplifier is not generating rogue high-frequency signals. However, such rogue high-frequency signals can sometimes originate with lower-quality SACD players. One SACD player tested produced a 100-mV RMS ultrasonic output at the 96-kHz sampling rate. This may have been due to inadequate reconstruction filtering after the DAC. The power amplifier should be designed so that it can handle these rogue signals in stride without creating distortion. Heavy-handed attempts to filter them out in the power amplifier will likely cause phase and frequency response aberrations.

It is important to recognize that the interconnect cable is an un-terminated transmission line. As such, it can exhibit unusual behavior at frequencies that are related to its quarter-wave frequency. As with the speaker cable, the quarter-wave frequency for a 10-foot length of interconnect cable will be about 18 MHz. A transmission line that is terminated in its characteristic impedance at only one end will usually be quite benign in terms of this kind of behavior. For this reason it is wise to terminate the interconnect cable at approximately its characteristic impedance Z_o at high frequencies with a Zobel network at the amplifier input. Most interconnect cables will have a characteristic impedance Z_o between 50 Ω and 100 Ω, so a 75-Ω termination at high frequencies is a good compromise. The network should begin to look resistive at least one octave below the quarter-wave frequency, here 18 MHz. A Zobel network consisting of a 75-Ω resistor and a 220-pF capacitor will satisfy this requirement.

It is best to place the Zobel network right at the input connector or at the input to the circuit board with the input connected to the circuit board via a short piece of coaxial cable. Figure 22.5 illustrates an amplifier input circuit that incorporates the Zobel network. Here the network is connected with R3 in series and C1 in shunt to ground, the Zobel terminating effect is still preserved and in addition a low-pass filter at about 10 MHz is created. The added filter provides an additional defense against RFI at high frequencies.

Many preamplifiers source-terminate the interconnect cable by their output impedance, even though this may not be a good approximation to the characteristic impedance of the interconnect cable. Good preamps with a low-impedance emitter follower or op amp output include at least 50 Ω of resistance in series with their output. Those that directly connect a low-impedance active circuit output to the interconnect cable seriously mis-terminate the line and risk oscillation. Most audio interconnect is similar to RG-58 or RG-59 coaxial cable, with Z_o of 50 Ω and 75 Ω, respectively. These cables have capacitances of 100 pF and 67 pF per meter, respectively. For those worried about the effect of a source-terminating resistor on frequency response, bear in mind that the 3-dB frequency of 600 Ω with 2 meters of RG-58 is about 1.3 MHz.

Implications for Input Stage Design

The need for immunity from EMI at the input means that the input stage must have good dynamic range and linearity up to high frequencies. This provides tolerance to EMI that makes its way to

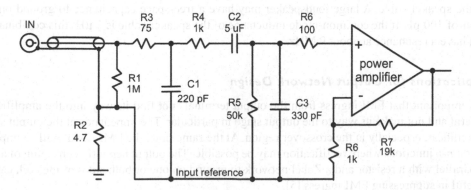

Figure 22.5 Amplifier Input Circuit with a Zobel Network

the input stage even after passing through whatever input low-pass filtering is in place. This also reduces the effects of any EMI that makes its way back to the input stage from the output of the amplifier (e.g., EMI picked up by the speaker cable antenna).

JFET input stages are especially good in this regard because the input of the stage does not include a forward-biased *p-n* junction. Instead, the input sees a reverse-biased *p-n* junction at the gate. The amount of voltage swing difference between the gate and source that would be required to cause rectification at the JFET input is typically greater than the threshold voltage of the device. The situation is different for a BJT input stage. The amount of voltage swing that is required between the base and emitter to disturb a conducting transistor is quite small; a differential as small as 60 mV will cause the transistor to conduct about 10 times less current, pushing it to near cutoff and essentially into rectification.

Things are not as bad for a BJT differential input stage if it is operated at a reasonable current and it incorporates significant emitter degeneration. In such a case, the potential at the emitter will largely follow the changes at the base caused by the EMI, and the *p-n* base-emitter junction will not be pushed into rectification. At very high frequencies, however, the emitter may not do as well following the base. In this situation the stage will be more vulnerable to rectification from incident EMI. Large RF signal voltage acting on a nonlinear capacitance can also cause demodulation and other effects (this is related to the operating principle of *parametric amplifiers*).

EMI Ingress from the Loudspeaker Cable

The loudspeaker cable is a big antenna. As mentioned above, the propagation delay of a 10-foot length of speaker cable is about 14 ns. This corresponds to a quarter-wave line at 18 MHz. The speaker cable can act as a quarter-wave antenna at this frequency even though both conductors of the speaker cable may be intimately twisted together. Any inductance in the ground return of the speaker output of the amplifier will increase vulnerability to pickup of radio signals and other EMI at these high frequencies where the speaker cable is an effective antenna. The speaker return lead inside the amplifier may have self-inductance between its point of entry and the point where it reaches the star ground reference. Straight wire typically has self-inductance of about 30 nH per inch. A 6-inch ground return run will have 180 nH of inductance (there are many caveats here beyond the scope of this book). Its impedance at 100 MHz will be about 110 Ω. The impedance of this inductance may actually tune the antenna to some extent. In any case, RF that is developed across this length of wire may be able to cause some EMI vulnerability in the amplifier.

The loudspeaker hanging off the end of the speaker cable is also part of the antenna. The wiring inside the loudspeaker often forms a loop that can pick up EMI. The free-space capacitance of the loudspeaker hanging on the end of the speaker cable can also act to "tune" the antenna formed by the speaker cable. A large loudspeaker may have a free-space capacitance to ground on the order of 100 pF. If the common-mode inductance of the speaker cable is 3 μH, this combination will have a resonance at about 9 MHz.

Implications for Output Network Design

It is important that EMI ingress from the output terminals not find its way into the amplifier in general and not make its way to the output stage in particular. The impedance at the output stage is nonlinear, especially in the crossover region. At the same time, the EMI signal will be impinging on *p-n* junctions where rectification may be possible. The output network, consisting of a coil in parallel with a resistor and a Zobel network to ground on one or both sides of the coil, can be helpful in suppressing EMI ingress [5].

Implications for Feedback Network Design

Some amplifiers incorporate a phase lead capacitor across the feedback resistor. This often improves stability if it is not overdone (too large a capacitor). However, this capacitor provides a direct path for EMI from the speaker cable to enter the sensitive input stage. I prefer to achieve good feedback loop stability without resort to such a capacitor.

EMI Ingress from the Mains

Often the mains supply contains a lot of garbage, both in differential mode and common mode. Unfortunately, the mains safety ground also often contains garbage that can ultimately be conveyed to the power amplifier chassis ground. So-called *X capacitors* between the hot and neutral leads help to suppress differential-mode EMI, while capacitors from each of the hot and neutral leads to the safety ground help suppress common-mode EMI to some extent. EMI input filters integrated with the IEC power connector help. Ideally, EMI input filters should be rated for the peak currents so that the coils do not saturate and degrade performance.

Unfortunately, the power transformer does not always do a very good job of isolating EMI on the primary from the secondary. The capacitance from primary to secondary for one 500-VA toroidal power transformer measured over 1300 pF. Toroidal transformers usually have greater primary-secondary coupling capacitance than transformers of conventional construction. A 500-VA laminated E-I core transformer measured just under 600 pF. If possible, a transformer with an inter-winding screen should be used connected to the safety ground. To the extent possible, connect line neutral to the end of the primary winding(s) closest to the secondary center tap.

EMI Distortion Mechanisms

EMI can cause audible noise to be generated even though the EMI itself may lie at very high frequencies. This happens as a result of nonlinear effects in the amplifier which either cause different elements of the EMI signal to intermodulate with each other or which cause rectification to occur so that AM detection of the EMI *carrier* occurs.

As long as the EMI causes only audible noise, this is not distortion per se. However, inevitably the voltage swing of the audio signal will affect in some way the behavior of the nonlinearity that brought the EMI down into the audio band. The EMI products are now correlated with the audio signal and bona fide distortion of the audio signal has occurred. Similarly, the presence of the EMI can affect the operating points of the circuitry within the amplifier, thereby causing distortion of the audio signal by impairing the operation of those circuits.

The big question is whether EMI can distort or otherwise impair the audio signal if the EMI is not audible by itself when there is no audio signal. In other words, if we can't hear noises produced by EMI, are we OK? Maybe not. For example, it could be the case that in the absence of an audio signal the amplifier has sufficient margins to handle the EMI without encountering enough nonlinearity to cause audible artifacts of the EMI to emerge. When the operating points are being moved around by an audio signal, those margins may no longer be adequate and EMI may rear its ugly head only in the presence of the audio signal. One would hope, however, that such behavior would show up on a distortion test if the offending EMI is present when the distortion test is being carried out.

22.3 Input Filtering

The best defense against EMI at the input of the amplifier is input filtering that very substantially reduces the amplitude of any high frequency signal before it reaches the first active devices. Such filtering must be effective to very high frequencies in the GHz range; this is not always as easy as it seems. Filtering of the input signal in both the differential mode and the common mode (even for single-ended inputs) can be important. Placing some very high frequency filtering immediately at the point where the signal is no longer in a coaxial environment is very helpful.

Achieving a Linear Phase Response

If multi-pole filtering is employed at the input of the amplifier, it is desirable that such filtering be a good approximation to a linear phase response to well beyond 20 kHz. This can be accomplished by placing multiple poles at successively higher frequencies. It then becomes possible to design the combined filter to have a better approximation to a flat time delay, where the phase lag introduced by the filter increases linearly with frequency. A simple rule of thumb is that each successive pole should be higher in frequency by one octave than the previous one in order to achieve a reasonable approximation to linear phase out to an octave beyond the highest pole frequency. While a Bessel filter provides a better approximation to linear phase, it requires complex poles, not something whose complexity is warranted or desirable in a power amplifier input filter. Because the input filter is in the signal path, NPO/COG capacitors should be used and ferrites should be avoided.

22.4 Input Ground Loops

Most stereo amplifiers are susceptible to input ground loops. This results from the two shielded interconnect cables carrying the left and right channels. In simple designs, the shields go to the same ground at both the source and amplifier ends. If the two interconnects do not follow exactly the same physical path, a loop will be formed and ground currents will flow in the loop. These ground currents, potentially representing hum and EMI, can induce garbage into the signal path [6, 7, 8, 9].

Ground Break Resistor

The first order of business in dealing with the input ground loop problem is to minimize the circulating currents. This can be accomplished by putting resistance into the loop. The induced voltages are usually small, so even a few ohms will greatly reduce the amount of circulating ground current. This is commonly done in many power amplifiers by inserting a 4.7-Ω resistor between the input ground and the analog circuit ground. The negative feedback shunt resistor is returned to the input ground node. This was included in Figure 22.5.

Balanced Inputs

Fully balanced amplifier inputs with true differential signals provide the best immunity to pickup of interference. Balanced inputs with high common-mode rejection to high frequencies are most immune to EMI.

Interconnect Alternatives

Some single-ended interconnect arrangements employ a shielded twisted pair. In this case both the signal and the return for the signal (ground) are carried on a twisted pair so that the signal and return lines both experience exactly the same electromagnetic environment. The shield is connected to ground at only one end, usually the source end. This prevents the circulation of ground loop currents in the shield. While it does not prevent the circulation of ground loop currents in the ground return conductor, it reduces the effect by causing the same voltage to be induced into the signal conductor, forcing it into the common-mode signal domain.

22.5 Mains Filtering

Just about everything electric in your home (and your neighbor's home) is connected to the mains. As such, the mains network is a cesspool of EMI.

Line Filters

Prefabricated line filters are convenient and often quite effective. They are professionally designed and often well shielded. A more common and less-expensive approach is the line filter that is often built into an IEC receptacle. This has the advantage of being physically located right at the entrance of the power cord into the equipment, making it possible to stop the incoming EMI at the earliest point.

Ferrites and Inductors

The use of ferrites in audio amplifiers has gotten a bad reputation. These will often be in the form of ferrite beads and inductor cores. They can be in the signal path or in power supply circuits on either the mains side or the circuit side. Inductors in particular must be carefully placed because they can radiate electromagnetic fields. Toroidal inductors will be less prone to this, but are less common.

A common-mode choke should be put right at the mains entrance, before the mains fuse and X capacitor. All three conductors should be run through the same ferrite [10].

22.6 EMI Egress

A power amplifier can also create EMI that can disturb other components. The best example of this is rectifier EMI radiated through the mains power cable. It may appear as EMI that is conducted into other equipment via their power cords, or it may be electromagnetically coupled to signal interconnects that pass near the power amplifier's power cord. This kind of coupling must be avoided when carrying out sensitive distortion tests on power amplifiers. Rectifier noise was discussed at length in Chapter 19.

22.7 EMI Susceptibility Testing

As applied to audio amplifiers, EMI testing is an inexact science at best and completely absent at worst. Here we suggest a few procedures that can bring some experimentation and measurement into the picture.

Cell Phones and Electric Drills

Although they are certainly not calibrated instruments, cell phones, hair dryers and electric drills operated in close proximity to a power amplifier can provide a helpful qualitative indication of the amplifier's susceptibility to EMI. For these tests, the amplifier should be connected to a real source or an emulated source through interconnects of typical length. It should likewise be connected to a loudspeaker load or emulated loudspeaker load through speaker cable of typical length. The premise here is that susceptibility to such relatively high levels of local EMI will reveal itself as audible output from the amplifier or output that is measurable on an oscilloscope or spectrum analyzer.

EMI Generators

A slightly more scientific approach to EMI evaluation involves the use of toroidal inductors or power transformers through which an interconnect cable or speaker cable is passed. This causes the cable to become one-half turn of what amounts to a transformer. The transformer or inductor can then be excited by EMI using flyback techniques where a current is set up in the winding and then interrupted, resulting in a large flyback voltage. This can be done with a relay hooked up as a buzzer. One can also envision exciting the coil with a power MOSFET whose gate is turned on and off with some kind of generated signal, perhaps even a pseudo-random word generator.

References

1. Robert R. Cordell, "Open-Loop Output Impedance and Interface Intermodulation Distortion in Audio Power Amplifiers," 64th Convention of the AES, preprint no. 1537, 1982. Available at www.cordellaudio.com.
2. Matti Otala and Jorma Lammasniemi, "Intermodulation Distortion in the Amplifier Loudspeaker Interface," 59th Convention of the Audio Engineering Society, preprint no. 1336, February 1978.
3. Edward M. Cherry and G. K. Cambrell, "Output Resistance and Intermodulation Distortion of Feedback Amplifiers," *Journal of the Audio Engineering Society*, vol. 30, pp. 178–191, April 1982.
4. Matti Otala and Pertti Huttunen, "Peak Current Requirements of Commercial Loudspeaker Systems," *Journal of the Audio Engineering Society*, vol. 35, no. 6, June 1987.
5. A. Neville Thiele, "Load Circuit Stabilizing Networks for Audio Amplifiers," *Journal of the Audio Engineering Society*, vol. 24, no. 1, January–February 1976, pp. 20–23.
6. Bill Whitlock and Jamie Fox, "Ground Loops: The Rest of the Story," presented 5 November 2010 at the 129th AES convention in San Francisco. Available from AES as preprint no. 8234.
7. Bill Whitlock, "An Overview of Audio System Grounding and Signal Interfacing," Tutorial T5, 135th Convention of the Audio Engineering Society, New York City, October 2013.
8. Bill Whitlock, "An Overview of Audio System Grounding and Shielding," Tutorial T2, Convention of the Audio Engineering Society, Star-Quad, p. 126.
9. Bill Whitlock, "Design of High Performance Audio Interfaces," Jensen Transformers, Inc.
10. IEC 61000–4–6, Electromagnetic Compatibility (EMC)—Part 4–6: Testing and Measurement Techniques—Immunity to Conducted Disturbances, Induced by Radio-Frequency Fields, 2013, ISBN 978-2-8322-1176-2.

Part 4

Simulation and Measurement

SPICE simulation can be extremely important to power amplifier design, and its use in this area is described in detail in Part 4. Even those designers with no SPICE experience will be able to employ this valuable tool. The excellent SPICE simulator LTspice®, made available free of charge from Linear Technology Corporation (now Analog Devices, Inc.), is the central focus. Accurate transistor models suitable for audio amplifier simulations can be difficult or impossible to obtain from manufacturers. This is especially so for BJT and MOSFET power output transistors. Chapter 24 is devoted to enabling you to create accurate models for transistors used in audio amplifier simulations, armed with only datasheet information and some simple measurements.

The many approaches to distortion measurement are also explained in Part 4. Much attention is paid to the techniques needed in order to achieve the high sensitivity required to measure the low-distortion designs discussed in the book. Less well-known distortion measurements, such as TIM, PIM and IIM are also covered here. In the quest for meaningful correspondence between listening and measurement results, other non-traditional amplifier tests are also discussed.

Chapter 23

SPICE Simulation

23. INTRODUCTION

The SPICE circuit simulator has been around for over 40 years. Developed at U.C. Berkeley, the acronym stands for *Simulator with Integrated Circuit Emphasis* [1]. Because integrated circuits are almost impossible to breadboard and probe, there was a great need to simulate designs before committing to the expensive process of laying out and fabricating an integrated circuit. Although that was its original mission, it is equally useful for discrete circuits at the circuit board level.

This chapter is a brief audio-centric description of SPICE and its application to audio power amplifier design. It is definitely not a comprehensive treatment of SPICE, but it does provide quick access to most of the features and nuances needed for design of audio circuits. It is presented in a tutorial style. This chapter will not make a SPICE guru out of you, but it will arm you with most of the techniques that are valuable in the design and analysis of audio power amplifiers.

The use of SPICE simulation can save hours in reaching the point where you can build a working amplifier. Intuition is not always right when it comes to circuit design, and SPICE helps here.

23.1 LTspice®

Although there are many SPICE packages available, one that stands out for audio amplifier design is LTspice® from Linear Technology Corporation, now Analog Devices, Inc [2, 3, 4]. The program is free, well supported, widely accepted and easy to use. It is also one of the best-performing SPICE simulators. It is easily downloaded from the Analog Devices site at www.analog.com. LTspice® runs on a PC.

All of the simulations done for this book were carried out with LTspice®, and this discussion will focus on LTspice®. Like most other software tools, LTspice® has many options and capabilities (and a very large user manual). Here we will just scratch the surface to get you started in an efficient way. The 90–10 rule applies: 90% of what you need to do can be accomplished with 10% of the features. LTspice® also comes with a good set of device libraries and a very helpful Educational directory where many example designs illustrate how LTspice® is used.

Installation

Download the LTspice® software from the Analog Devices website at www.analog.com. After LTspice® has been downloaded and installed, click on the icon that it places on the desktop. If you double-click on the icon, the program will come up with a toolbar with most of the features grayed out.

The Toolbars

Familiarize yourself with the toolbar by placing the cursor under each icon and reading the name of the function that appears. Most of these functions are obvious. In most cases a click on the icon will activate the function. In some cases a window will appear and selections will be available. Details for these can be found in the LTspice® manual or in Help. The functions listed below are used the most. The name of the item is followed by a description of the icon. A brief description of each function is also provided.

Save   **(diskette)**

Saves the current schematic

Run   **(runner)**

Runs the simulation

Cut   **(scissors)**

Deletes elements from the schematic

Copy   **(two pages)**

Copies windowed elements or groups of elements

Wire   **(pencil and line)**

Places a wire

Ground   **(ground symbol)**

Places a ground with its symbol

Label Net   **(tag with A)**

Gives a node a name

Resistor, capacitor, inductor, diode

Places one of these frequently used components

Component   **(AND gate)**

Places other components, like transistors, selected from a list

Move and Drag   **(big hand, little hand)**

Windows an element and moves or drags it

Rotate and Mirror   **(E with arrow)**

Rotates or mirrors an element when it is selected with *Move*

Text   **(Aa)**

Adds text to the schematic

SPICE Directive   **(.op)**

Defines aspects of the simulation to be run

Directory Organization

The LTspice® material of greatest interest is located in two directories:

C:\Program Files\LTC\LTspiceXVII\examples
C:\Program Files\LTC\LTspiceXVII\lib

The . . .*examples* directory contains an *Educational* subdirectory where a great many example circuits are illustrated as simulations.

The . . .*lib* directory contains the models for the devices used in LTspice® simulations. It contains three important subfolders. The first is the . . .*cmp* folder where files for components are stored. This directory contains both passive and active components. An example file in this directory is the *standard.bjt* file. This single text file contains the SPICE models of all of the bipolar transistors supplied with LTspice® and is the place where the user can append additional models. Adding models will be discussed later. If your copy of LTspice® is in the *C:\program files* directory, the full path for the *cmp* folder will be

C:\program files\LTC\LTspiceXVII\lib\cmp

The . . .*sub* folder contains subcircuits. These are circuits made up of components that can be used just like a component on a schematic. An op amp is a good example of a subcircuit. Every subcircuit requires a symbol. These are stored in the . . .*sym* folder. LTspice® allows you to create your own components and subcircuits and store them in these folders. You can thus create your own libraries.

Control Panel

Go to the *Control Panel* by clicking on the hammer icon on the toolbar. There are many important controls here, most of which are best described in the LTspice® documentation. A couple of controls will be explained as useful examples. Open the *Waveforms* tab, and notice that you can select thick lines for the plots and can change the color scheme for the plots. The color scheme for each of many traces can be selected. You can click on the background and set its color as well. In cases where a plot will be placed in a document, it may be desirable to put black traces on a white background. The color scheme selection also allows you to select the color scheme for the schematic and netlist presentations.

Under the *Operation* tab it is usually wise to select the option to automatically delete *raw files* to avoid saving unnecessarily large amounts of data. Under *Drafting Options* you may wish to have drawing grids visible and you may wish to draft with thick lines, especially if you will be placing the drawing in a document. You can also control the font.

Help

The Help files for LTspice® are well organized and comprehensive. The examples shown in this chapter will quickly get you started, but many useful capabilities are not covered. Review the Help files to get a feeling for what is there and how it is organized.

The LTspice® Users' Group

There is a very active, independent LTspice® users' group on Yahoo. The users' group is at groups. yahoo.com/LTspice [3]. Good discussions can be found there as well as many files for download and a good manual.

23.2 Schematic Capture

Schematic capture will be explained by actually creating a simple circuit comprising a differential pair amplifier stage. This demonstration circuit is shown below in Figure 23.1. The differential pair is operated with 1 mA in each transistor supplied by the 2-mA tail current source I1. Emitter resistors R3 and R4 provide 10:1 emitter degeneration; that is, they reduce the gain by a factor of 10 compared to what it would be without degeneration.

Collector load resistors R1 and R2 allow a nominal DC voltage drop of 2.5 V and provide for a differential gain of 10. The gain is 10 because each collector resistor is 10 times the value of the effective emitter resistance of its associated transistor. Single-ended gain to the output node V_{out} is 5, or about 14 dB. Capacitor C1 provides for some high-frequency roll-off by creating a pole at about 640 kHz.

Placing Components

Open LTspice® and click on *File → New Schematic*. Place the resistors first. Click on the resistor toolbar button, bring the cursor to the location for the resistor and click. R1 will be placed. Move the cursor to the position for R2 and click. R2 will be placed. Do the same for R3 and R4. Notice that the resistors are numbered in sequence as they are placed. Right-click to exit from the resistor placement mode. Right-click on R1, enter its value of 2500 in the dialog box that appears and click *OK*. Enter the values for the other three resistors. Place capacitor C1 in the same fashion and enter its value as 100p or 100pF. Do not leave a space in this entry.

Place the voltage sources by clicking on the *Component* button on the toolbar. A list of available components will come up. Click on the component labeled *voltage*. The voltage source

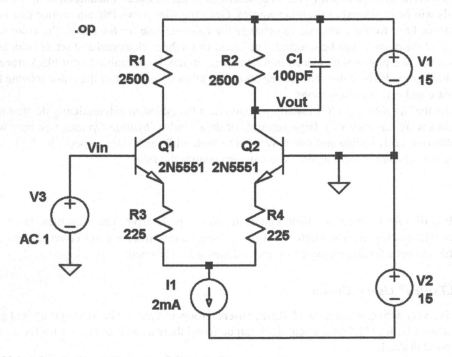

Figure 23.1 A Simple Circuit Illustrating Schematic Capture

symbol will appear in a dialog box. Click *OK* and then place the voltage source by clicking on the location where it is to go. Click again on the locations for V2 and V3. Right-click to exit the voltage source entry mode. Right-click on V1 and enter its DC value of 15. Do the same for V2. Right-click on V3 and then click *Advanced*. A dialog box will appear with many options describing the behavior of the voltage source. For now, simply enter *1* in the box for *AC Amplitude*. The button *(none)* under *Functions* should be turned on. This prepares the voltage source to act as an AC voltage source for a small-signal AC simulation.

Click on the component button on the toolbar and select *current* from the list. Place the current source, and then right-click on it to enter its value of 2 mA. Place the two ground symbols by clicking on the *Ground* button on the toolbar.

Picking and Placing Transistors

Place transistor Q1 by clicking on the *Component* button in the toolbar and selecting *npn* from the list that appears. Hit *OK* and place the transistor by clicking when it has been moved to the proper location. A second transistor will be attached to the cursor. Move the cursor up to the *Mirror* button on the toolbar and click it. This will flip the orientation of the transistor as needed for Q2. Now place the transistor. Right-click to exit transistor placement. Note that any component can be mirrored (*Ctrl* E) or rotated (*Ctrl* R) while it is attached to the cursor, including when using the *Move* command from the toolbar.

Right-click on Q1. A dialog box will appear, showing the type and some of the characteristics of Q1. At this point the information will be missing or generic. Click on *Pick New Transistor*. A list will appear showing all of the NPN transistors in the library. Select the 2N5551 and hit *OK*. Right-click on Q1 and notice now that the type and some of the characteristics are filled in. Define transistor Q2 in the same way.

Other Components and Subcircuit Libraries

When you clicked on the *Component* button in the toolbar, many different component selections appeared. Some of those were named in brackets. These are additional parts directories that can be accessed in the same way by double-clicking on them. Take note especially of what lies in the *Misc* directory.

Many of these components are constructed of so-called subcircuits, which are themselves circuits composed of numerous basic devices. A good example of a subcircuit is an operational amplifier. Subcircuits have symbols and can be placed on the schematic and used just like any other component.

Parameterized Elements

Sometimes it is desirable to parameterize the value of an element, like a resistor, rather than committing it to a fixed value. As we will see later, this can allow the value of the element to be stepped across a range of values, producing a simulation that has a family of curves. A component (like a resistor) can be parameterized by placing a variable inside curly brackets, where the component value would normally go. If the variable for a resistor is *Rseries*, the parameterization is completed by adding the SPICE directive:

param Rseries=1k

where 1k is the nominal value of *Rseries*.

Completing the Schematic

Move the components into their desired position if necessary. Do this by clicking on the *Move* button on the toolbar and then windowing the component you wish to move. Move the component as desired and then left-click. Right-click to exit the *Move* mode. Note that you can move the component designation or the value in the same way. You can change the component designation by right-clicking on the text of the designator.

Wire the circuit by clicking on the *Wire* button on the toolbar. Click on the beginning point of the wire and each place where it is to turn. Click on the final destination of the wire. Right-click to exit the wire mode. Do the same for all of the remaining wiring of the circuit. If any wire or component is put down that must be deleted, click on the *Cut* button on the toolbar, designated by a pair of scissors. Click on the element to be deleted, and it will be removed. Right-click to exit the *Cut* mode.

Add node names to selected nodes by clicking *Label Net* in the toolbar. A dialog box will appear. Enter the name for the node and click *OK*. A rectangle with a diamond at its bottom will be attached to the cursor. Place the diamond on the net in a convenient place and click. Right-click to exit the *Label Net* mode. Label the base net of Q1 as V_{in} and the collector net of Q2 as V_{out}. Nodes with the same label will be connected for the simulation even if not connected by a wire in the schematic. This can sometimes reduce clutter on the schematic.

Add text to the schematic by clicking on the *Text* option in the toolbar. Type the desired text into the dialog box, hit *OK* and place the text on the schematic. The schematic of the differential amplifier stage is now complete and should look like Figure 23.1.

Save the schematic to a directory of your choice. Call it *diffamp*. If you then look in that directory you will see a file named *diffamp.asc*. This is the schematic file.

The SPICE netlist can be obtained from the schematic by using the command *View → SPICE netlist*. Select all the text and copy it to the clipboard (*Ctrl C*) to paste it into a different editor, like Notepad.

The schematic can be copied to an application like *Word* or *Paint* by typing *Ctrl C* and putting it onto the clipboard. This is also true for plots. *Paint* can be used to convert the file to convenient formats like *TIF*, *GIF* and *JPG*. The complete LTspice® screen can be copied to the clipboard by clicking *Ctrl Print Screen*. This is useful for putting the material into a Microsoft Word™ document.

23.3 DC, AC and Transient Simulation

The differential amplifier stage will be simulated in this section. The three most important and basic types of simulation are DC, AC and transient. The DC simulation biases the circuit and shows all of the node voltages and operating points. The AC analysis shows small-signal frequency and phase response. The transient simulation shows the actual waveforms at different points in the circuit when the circuit is driven by an input signal. The result of a transient simulation is much like what one would see on an oscilloscope.

The DC Operating Point

Set up the DC simulation by clicking on *Simulate → Edit Simulation Cmd*. The simulation dialog box will appear with tabs for the following types of simulation:

- Transient
- AC Analysis
- DC sweep

- Noise
- DC Transfer
- DC operating point

Click on the tab for *DC operating point*. Note that *.op* will appear in the syntax box. This is the SPICE directive for a DC simulation. Hit *OK* and notice that a rectangle will be attached to the cursor. Place this rectangle on the schematic in a convenient place and click. The *.op* SPICE directive will now appear on the schematic.

The simulation is now ready to be run. Click on the *Run* button on the toolbar. The simulation will run quickly, and an information box will appear showing all of the node voltages in the circuit. Nodes you have named will appear with their names. Other nodes will be designated with sequential node numbers. Interpreting results is easier if you label all of the nets of interest. The box will also show the direct current flow in all of the transistors and passive elements. Note that the voltage at V_{out} is about 12.5 V, as expected, and that the collector current of each transistor is just shy of 1.0 mA.

Close the DC simulation information window and move the cursor over a component on the schematic. The lower left portion of the window will show the voltages, currents and power dissipation for each component. Move the cursor over a wire. The information window will display its node name and voltage. This approach to viewing DC simulation results is usually much more convenient than viewing the results in the initial DC simulation results window.

The SPICE Error Log

There is more to the SPICE error log than one would think. It is not just telling you what is wrong. In the case of a DC operating point simulation, it reveals the operating point and small-signal model parameters for every transistor in the circuit.

Click on *View → Spice Error Log* and the information will appear. Information like DC and AC beta and f_T of the transistors at the actual operating point in the circuit can be very valuable. Virtually all of the important hybrid pi parameters are shown. This information provides good insight into how the model for the transistor is behaving. One can actually set up a circuit to bias the transistor at a chosen operating point and compare these numbers to the datasheet values for the transistor in order to see how well the model compares to the datasheet.

Convergence

LTspice® is very good at converging to solutions, but there will be rare occurrences when the iteration limit is reached and the simulation fails. The message *Analysis Failed: Iteration limit reached* will be displayed.

This is often a result of the particular circuit topology being simulated and sometimes can be the result of a circuit design or schematic entry error. Apart from checking the circuit for errors, there are three things that can often solve the convergence problem. The first is to use the alternate solver available in LTspice®. One can switch to the alternate solver by going to *Control Panel → SPICE* and selecting the alternate solver in the *Engine* area in the dialog box.

The second thing to try is more specific to power amplifiers and similar topologies. Connect a large-value resistor from the high-impedance VAS output node to ground. This resistor can often be as large as 1 MΩ or even 10 MΩ and still fix the convergence problem while causing minimal interference to normal circuit operation. Once convergence is achieved this way, the operating points of all of the transistors can be checked to see if something is wrong with the circuit itself. There is also the occasional case where the presence of a very large resistance to ground on the high-impedance node (e.g., 100 MΩ) can actually impede convergence.

I have also found that certain transistor models can be more prone to causing convergence problems. This possibility can be tested by temporarily changing one or more of the transistors to a different type. The VAS and pre-driver transistors connected to the high-impedance node are prime candidates in this case.

AC Analysis

Click on *Simulate* → *Edit Simulation Cmd* and then hit the *AC Analysis* tab in the dialog box to set up the AC simulation. Fill in the information requested. Select a decade sweep with 100 points per decade. Enter start and stop frequencies of 10k and 100Meg, respectively. The use of M for MHz will not work; *MEG* will work. Click *OK* and place the SPICE directive on the schematic in a convenient place by clicking there. Notice that the .*op* SPICE directive for the DC operating point will remain on the schematic but will change to ;*op*, signifying that it is not the simulation to be run this time.

Right-click on V3 and click on the *Advanced* box if necessary. Make sure that there is a *1* in the *AC Amplitude* box. The other boxes can remain blank. The circuit is now ready for AC analysis.

Click on *Run* in the toolbar. The AC analysis will be performed and a second window will come up with frequency labeled across the X axis. Place the cursor on the V_{out} net. The cursor will change to a red probe symbol. Click on the net. The frequency and phase response will appear in the plot window. Gain will be shown in dB on the left axis and phase will be shown in degrees on the right axis. Notice that the frequency response indicates approximately 14 dB of gain at low frequencies, as expected. Move the cursor over the frequency response curve at some low frequency and notice that the frequency, gain and phase are shown numerically in the lower left corner of the main window. Move the cursor to the point on the curve where the gain is 11 dB, which is approximately 3 dB down. The frequency readout will indicate about 604 kHz, the 3-dB frequency of the circuit.

Move the mouse probe to the collector net of Q1 and click. The frequency response at that point will also be shown. Notice that it exhibits very little reduction in gain even at 10 MHz. This is because there is no external capacitor connected to that node to create a significant HF roll-off.

Click in the area of the plot. Now right-click. A list of options will appear. Select *Visible Traces*. A list of all possible traces to plot will appear. $V(V_{out})$ and $V(n001)$ will be highlighted because they have already been selected for plotting. Select an entry without deleting the others by control clicking on that entry. The status of an entry can be changed by control clicking on it as well. Do this for $I(C1)$ and hit *OK*. The plot of the current in C1 as a function of frequency will now appear.

Position the cursor over the V_{out} net. Push and hold down the left mouse button while the red probe is displayed. Hold down the button and move the probe cursor to the collector net of Q1. Notice that the probe color changes to black, signifying that it is now a negative probe. Release the left mouse button. A trace will appear showing the differential output voltage from the collector of Q1 to the collector of Q2. This curve will be up 6 dB compared to the single ended output curve, as expected.

Click in the area of the plot and see an entry in the toolbar called *Plot Options*. Select *Save Plot*. The same plot will automatically appear the next time the simulation is run.

PSRR analysis can be done by making the positive supply the AC input and running an AC analysis on the differential output.

Transient Simulation

Click on *Simulate* → *Edit Simulation Cmd* and then hit the *Transient* tab in the dialog box to set up the transient simulation. The transient simulation dialog box will appear. Enter 10 ms for *Stop*

Time. All of the remaining boxes can be left blank. Hit *OK* and place the transient SPICE directive on the schematic.

Create the time-varying input for the transient simulation by right-clicking on input voltage source V3. Because the *Advanced* dialog box was previously used to specify an AC source for AC simulation, that dialog box will again come up. In the function area, define the function by clicking the *Sine* button. Enter 0.1 in the amplitude box to generate a 0.1-V peak sine wave. Enter 1000 for the frequency and 10 for *Ncycles*.

Hit *Run*. The simulation will run and a waveform window will appear with time running from 0 to 10 ms along the *X* axis. Probe the V_{out} net, and the sine wave output of the amplifier stage will appear along with the voltage shown on the *Y* axis. Using the mouse, create a rectangle on the waveform window that extends left to right between a pair of cycle peaks and vertically from the bottom of the waveform to the top. Hold the mouse button. Notice the information box in the lower left corner of the screen. The period of the waveform is shown as *dx* equaling about 1000 µs and the peak-to-peak amplitude of the waveform indicated by *dy* as about 985 mV. Release the mouse button. This portion of the waveform will now appear zoomed in. Hit *Zoom Full Extents* in the toolbar to restore the waveform to its original presentation.

Probe the collector wire of Q1 and see the other output out of phase with V_{out}. Probe V_{in}, and see this smaller waveform presented swinging about zero volts and notice that the combined waveform presentation is auto-scaled by LTspice®.

Move the cursor over C1 and see that the cursor changes to a representation of a clip-on ammeter. Click on C1 and see its current plotted in the plot window. Notice that it is 90 degrees out of phase with respect to the voltage waveforms, as expected. The current in any of the wires in the circuit can be plotted by *Alt* left-clicking on the wire.

Click on the display, then right-click on it and select *Visible Traces*. Clear all of the visible traces. Go back to the schematic window and probe from V_{out} to the collector of Q1. The differential output voltage will be displayed. Now click on V_{in} and see it displayed as well, with both traces scaled properly without the presence of a DC component on the output, making them easier to see and compare.

Right-click on V3 and set the function to a pulse. Enter $V_{initial}$ = −0.1, V_{on} = 0.1, *Tdelay* = 0, *Trise* = 10 ns, *Tfall* = 10ns, *Ton* = 5 µs, *Tperiod* = 10 µs and *Ncycles* = 10. This will create a 100-kHz square wave at V_{out} with 1 V p-p amplitude. Go into *Simulate > Edit Simulation Cmd* and enter a stop time of 100 µs in the transient dialog box. Hit *Run*, and then probe V_{out}. Ten cycles of the 100-kHz square wave will be shown. Notice the rounded leading edges due to the circuit bandwidth of less than 1 MHz.

Notice that in the *Transient* dialog box there was a place to enter *maximum timestep*. LTspice® normally does a fairly good job of automatically adjusting the timestep in a transient simulation. The size of the timestep affects the accuracy of the simulation. In some critical simulations it is sometimes desirable to set a maximum timestep that LTspice® is allowed to take. For the simulations done so far, it has been unnecessary to set a maximum timestep.

Transient simulations can create fairly large *.raw* files that will take up unnecessary space in the simulation directory. I recommend you have LTspice® automatically delete them after a run is closed. Go to the Control Panel by clicking on the hammer on the toolbar and select the *Operation* tab. Answer *Yes* to *Automatically delete .raw files*.

If you *Alt*-click a device in the schematic, a thermometer will appear and power dissipation of the device will be plotted. If you *Ctrl*-click the label of a waveform (at the top of the plot pane), a display box will appear that shows the RMS and average values of the waveform. If you do this for a power waveform of a device, the average power dissipation of the device will be displayed in the box. This is especially useful in looking at output stage transistors.

23.4 Distortion Analysis

The Fast Fourier Transform (FFT) capabilities built into LTspice® make convenient and detailed distortion analysis possible. In this section distortion analysis will be illustrated by analyzing the distortion of the differential amplifier stage described above. Most conventional distortion analysis is carried out with a transient simulation employing a sinusoidal input. In preparation for the FFT analysis, attach the SPICE directive *.option plotwinsize=0* to the schematic.

This turns off data compression that LTspice® normally uses to save space. The accuracy and dynamic range of FFT analysis can be degraded when the data from a transient run is compressed. Data compression can also be disabled on the *Compression* tab of the Control Panel, but that setting will not be remembered the next time LTspice® is invoked.

Run a transient simulation of the differential amplifier stage with a 1-kHz input with 0.1 V peak amplitude for 16 cycles. When you click on the sine function, the new window that comes up will still have some of the boxes populated (left over from the square-wave run). Clear these boxes. Set the transient run duration for 16 ms and run the simulation. Click on V_{out} to obtain the output waveform. Its amplitude will be 1 V p-p.

FFT Spectral Plots

Click on the waveform window and then right-click on it. Select *View → FFT* at the bottom of the list of selections. In the *Time range to include* section, click the button *Specify a time range*. The *End Time* will already be listed as 16 ms or something very close to it. If it is not exactly 16 ms, enter 16 ms. Enter 8 ms in the *start time* box. This causes the FFT to be performed on the last eight cycles of the simulation, suppressing start-up artifacts. The default number of FFT points is 65,536, and this number is fine. Hit *OK*. A new window will appear with the FFT spectrum of the waveform shown in dBV as a function of frequency. Using the mouse, drag a window over the main area of interest in the spectral plot. This will provide a close-up view for measuring the harmonic amplitudes.

Place the cursor on the fundamental peak at 1 kHz and notice that the cursor value in the lower left corner of the window reads −9.5 dB. This is dB relative to 1 V RMS. Place the cursor on the second peak at 2 kHz, representing the second harmonic. It will read approximately −104 dB. The second harmonic is thus down about 94 dB, or at about 0.002%. Measure the third harmonic at 3 kHz in the same way. It will read about −79 dB, down about 69 dB from the fundamental, or at about 0.035%.

Notice that the fourth and fifth harmonics read −138 dB and −125 dB, respectively. The spectral noise floor, often called the *grass*, is at −142 dB. Notice also the large amount of spectra extending out to high frequencies. There are high-frequency artifacts at −81 dB, for example. This is evidence that the FFT has not been optimized.

Reading out the amplitudes of the spectral lines is made much easier by attaching a cursor to the spectral plot. Just left-click on the plot label and a crosshairs will appear. Right-click on the plot label, select one cursor and hit *OK*. A readout box will appear showing the frequency and amplitude of the point on the plot where the cursor crosshairs are located. The crosshairs are dragged to the desired location by clicking on and holding at or near the crosshairs where the number 1 (designating cursor number 1) appears. You can also move the crosshairs left and right by convenient increments with the ← and → keys to zero in on a particular peak. These keys will usually land the crosshairs on a harmonic frequency of the fundamental. The step size of the cursor movement is the frequency increment corresponding to the period processed by the FFT. For example, if the number of cycles processed corresponds to the last 8 ms of the simulation, the step size will be 125 Hz.

Optimizing FFT Simulations

There are seven things that are important in optimizing the FFT:

- Turn compression off
- Simulate an adequate number of cycles
- Simulate an integer number of cycles of the lowest frequency
- Ignore an adequate number of cycles before beginning the FFT
- Choose an appropriate number of FFT points
- Choose a small enough and optimal *maximum timestep*
- Beware of long time constants in the circuit

The first five items were satisfied above, but the *timestep* was left to SPICE to determine. Better results are achieved with the FFT when the *maximum timestep* size is limited to a small value that is related to the period of the sinusoid being simulated and to the length of time of the FFT analysis.

It has been recommended in the past that the *maximum timestep* be set to the FFT analysis time divided by 16,383 [5, 6]. This caused the timesteps of the transient simulation to coincide with the FFT sample points when the number of FFT points in the analysis time was 16,384. I have used this relationship over the years, even though the default number of FFT sample points in LTspice® has increased in later releases, and it has always worked well for me. For an FFT analysis time of 8 ms, this corresponds to about 0.488 μs. Enter the SPICE directive .*options* *maxstep=0.48831106u* on the schematic.

Run the simulation again, noting that it will now take longer because of the smaller *maximum timestep*. Perform an FFT analysis on the result. Notice the significant reduction in spectra at higher frequencies. The high-frequency artifacts are at about −153 dB as compared with −81 dB in the previous simulation. Use the cursor to measure the fourth harmonic at 4 kHz. It will read about −153 dB, as compared with −138 dB above where *maximum timestep* control was not implemented. The grass is at about −162 dB as compared with the previous −142-dB number. These very substantial improvements in FFT resolution are due to the enforcement of the appropriate *maximum timestep*.

My experience has been that the above rule for *maximum timestep* may be conservative and that the precision of the chosen number is not sacred. A doubled maximum timestep rounded up to 1 μs raised the high-frequency artifacts from −153 dB to −132 dB, but the other quoted numbers changed by an insignificant amount. My advice is to experiment with the *maxstep* number so that you get good results without resorting to an unnecessarily small number.

The FFT results can be degraded by the presence of circuitry with long settling times, such as AC-coupled inputs. Wherever possible, perform the FFT simulations on DC-coupled versions of the circuit. The number of cycles of the waveform that you employ for the FFT has a direct impact on the frequency resolution of the FFT. The width of the frequency bin for each spectral line will be $2/FFT_{time}$. A simulation with FFT processing of 8 ms of data will have spectral lines with a total width of 250 Hz.

The number of cycles that you simulate before starting the FFT analysis is a judgment call. The settling time is intended to ensure that all initial conditions and transients have had time to settle out so as not affect the FFT results. For high-quality FFT results I recommend that you simulate 16 cycles and perform the FFT on the last eight cycles. Too small a settling time or too small an FFT time may result in artifacts and loss of FFT dynamic range. However, I have often obtained satisfactory results with as little as one cycle of simulation ignored before the FFT interval.

Total Harmonic Distortion (THD)

Total harmonic distortion (THD) for a sinusoidal transient simulation can be displayed in the SPICE error log by adding the *.four* SPICE directive to the schematic. Total harmonic distortion and the level of each harmonic will be listed up to the requested number of harmonics. Add the SPICE directive *.four 1kHz 10 8 V(V_{out})* to the differential pair schematic.

This directive specifies that the analysis will be done on a 1 kHz sine wave and that readout of the first 10 harmonics be provided. The number 8 indicates that the FFT is based on the last eight cycles of the waveform. Run the simulation, and then go to *View → SPICE Error Log*. THD will read about 0.035%. Second and third harmonic levels relative to the fundamental will read about 2e-5 and 3.5e-4, respectively.

23.5 Noise Analysis

Perform a noise analysis by going to *Simulate → Edit Simulation Cmd → Noise*. Fill in the dialog box with the output node to be measured, $V(V_{out})$, the input source (V3), the type of frequency plot (decade), number of points per decade (20) and start (1k) and stop (100k) frequency. Hit *OK* and place the resulting SPICE directive *.noise V(V_{out}) V3 dec 20 1k 100k* on the schematic.

Run the simulation. A plot window will appear, much like the one for AC analysis. Probe the V_{out} node and the plot of noise density versus frequency will be displayed. In this case it is fairly flat at about 16 nV/$\sqrt{Hz}$. The input-referred noise voltage is often of greater interest. It is just the output noise density divided by the gain of the circuit. With a single-ended gain of 5, the input-referred noise voltage for this differential stage is 3.2 nV/$\sqrt{Hz}$.

To put things in perspective, consider how much noise voltage is produced in a 20-kHz bandwidth. There are about 141 $\sqrt{Hz}$ in a bandwidth of 20 kHz, so the total output noise in this bandwidth would be 2.3 µV. If the maximum useful output of the circuit were 5 V RMS, the SNR would be a factor of 2.17 million, corresponding to about 126 dB.

Control-click on the trace label to see up a pop-up box that will display the total RMS noise on the plotted node over the plot frequency interval. This can be very helpful in deriving a signal-to-noise ratio.

Noise of Individual Contributors

The noise contribution of an individual component can be plotted by clicking on the component. The noise contribution for a transistor will include the effects from both its input voltage noise and input current noise.

Weighted Noise Simulations

As mentioned in Chapter 1, weighted noise measurements are often performed on amplifiers to better reflect those parts of the noise spectrum to which the ear is most sensitive. A good example of this is the A-weighting curve. This type of weighting can also be applied to simulations by following the circuit being simulated with a simulation circuit that implements the desired filter function.

Figure 23.2 is a schematic of a filter that implements the A-weighting function. The VCVS at the input buffers the signal. The VCVS at the output implements the necessary mid-band gain and buffers the output. This is a handy function to implement as a subcircuit.

23.6 Controlled Voltage and Current Sources

LTspice® includes many types of voltage and current sources that can be controlled by voltages or currents. All of these sources can be assigned gain values. These provide a quick way of implementing certain functions within a simulation, such as buffering, chunks of gain and so on.

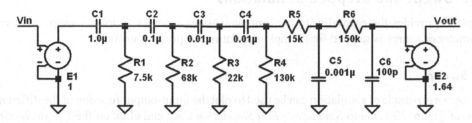

Figure 23.2 A-Weighting Noise Filter

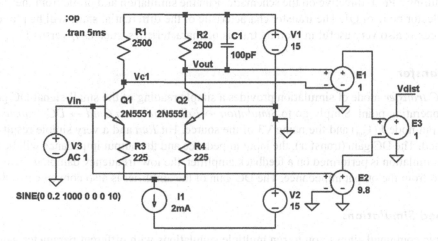

Figure 23.3 Simulation Circuit for Viewing Input-Output Error

The voltage-controlled voltage source (VCVS) can be used to implement voltage gain. It is designated by the letter *e* in the toolbar *Component* list. Setting its gain to unity implements an ideal buffer whose inputs and outputs are conveniently isolated. This allows one to pick off differential voltages, for example. Setting the gain to a very large number implements a near-ideal operational amplifier.

The voltage-dependent current source is designated by *g* (transconductance). LTspice® also includes current-controlled voltage sources (*h*) and current-controlled current sources (*f*), but these are less frequently used. Their control current is the current flowing through a designated voltage source.

The circuit in Figure 23.3 subtracts an ideally amplified version of the input from the output to allow the viewing of the input-output error of the circuit. Go to the toolbar component selector and choose a VCVS by selecting the *e* listing. Place this on the schematic and connect it as shown. Right-click on the VCVS to set its gain. In the resulting dialog box highlight *Value* and type a gain of 1 in the box above. Place, connect and set the gains of the other controlled sources *E2* and *E3*. *E1* merely derives the differential output voltage across the collectors of Q1 and Q2. *E2* multiplies the input signal by a value just slightly less than the differential gain of the stage. *E3* then takes the difference of the outputs of *E1* and *E2* and presents it as a distortion residual output.

Run the 1-kHz sinusoidal transient simulation for 5 ms with peak input amplitude of 0.1 V. Probe the V_{dist} node and see a small, slightly distorted replica of the sine wave. Rerun the simulation with a peak input voltage of 0.2 V and look at V_{dist}. Very visible distortion will now be evident as a result of the increased signal level.

23.7 Swept and Stepped Simulations

LTspice® provides the capability to perform simulations where a source is swept or where a parameter or source is stepped for multiple simulations to produce useful plots.

DC Sweep

The *DC sweep* mode of simulation can be used to plot the input-output function of the differential stage of Figure 23.1. Go to *Simulate > Edit Simulation Cmd* and click on the *DC sweep* tab. A dialog box will come up. Select V3 as the source and pick a linear sweep. Enter start and stop voltage values of −1.0 and 1.0. Enter 0.01 for a sweep in increments of 10 mV. Hit *OK* and place the resulting SPICE directive on the schematic. Run the simulation and probe from the V_{out} net to the collector node of Q1. The transfer characteristic of the differential stage will be plotted. The DC sweep is also very useful in plotting transistor characteristics, such as I_c versus V_{be}.

DC Transfer

The *DC transfer* mode of simulation provides a simple reading of the small-signal DC gain at a given operating point. Simply go to *Simulation → Edit Simulation Cmd → DC Transfer*. Enter the output node $V(V_{out})$ and the name V3 of the source. Hit *Run* and a very simple result will be presented. The DC gain (transfer), the input impedance and the output impedance will be shown. If this simulation is performed on a feedback amplifier, the low-frequency damping factor can be inferred from the output impedance. The DC gain of the amplifier is also conveniently shown.

Stepped Simulations

The *.step* command allows you to run multiple simulations with different parameter values and have the results plotted together as a family of curves. The stepped simulation can be illustrated by running the above DC sweep, but with the tail current source *I1* stepped from 1 mA to 3 mA in increments of 0.5 mA. Add the *.step* SPICE directive shown below to the schematic.

.step I1 1mA 3mA 0.5mA

The syntax is simple. Listed in order are the source to be stepped, the starting value, the ending value and the value of the increment. Run the simulation and probe the differential output by dragging the mouse from the V_{out} net to the Q1 collector net. The resulting plot will show the differential transfer function for all six values of tail current. Notice that the output swing range before clipping occurs becomes smaller as tail current becomes smaller.

The stepped simulation is especially useful for plotting the output characteristic of a transistor, where V_{ce} is swept and base current is stepped. An AC analysis can also be stepped. Step the tail current in the AC analysis above from 0.1 mA to 2 mA in steps of 0.1 mA and see the effect on gain of the differential stage.

Example: A Wingspread Simulation

An important example of the use of stepped swept simulations is the so-called *wingspread* simulation of a class AB output stage. This simulation shows the gain of the output stage as a function of output current for several different values of quiescent bias current I_q. This is valuable in evaluating crossover distortion.

The circuit of Figure 23.4 implements a wingspread simulation of a simple class AB output stage driving an 8-Ω load over a range of −10 V to +10 V. The two voltage sources $V_{spreadp}$ and $V_{spreadn}$ provide the bias spreading function and set the quiescent bias current. These voltage

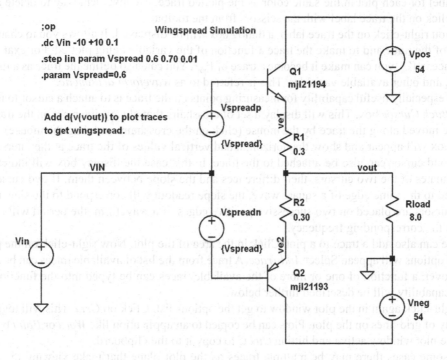

;op
.dc Vin -10 +10 0.1
.step lin param Vspread 0.6 0.70 0.01
.param Vspread=0.6

Wingspread Simulation

Q1
mjl21194

Vpos
54

Vspreadp
{Vspread}

Add d(v(vout)) to plot traces
to get wingspread.

R1
0.3

VIN

vout

R2
0.30

Vspreadn
{Vspread}

Rload
8.0

Vin

Q2
mjl21193

Vneg
54

Figure 23.4 Wingspread Simulation of a Class AB Output Stage

sources are parameterized by placing a variable inside curly brackets in place of the value of the voltage source. Here that variable is called V_{spread}. The parameterization is made complete by adding the SPICE directive

.param Vspread=0.6

where 0.6 V is the nominal value of V_{spread}. The value of V_{spread} is then stepped for multiple simulations by adding the SPICE directive

.step lin param Vspread 0.6 0.70 0.01

where the starting and ending points for V_{spread} are 0.6 V and 0.70 V, respectively. The stepping is done in increments of 10 mV. Each simulation is done in the *DC sweep* mode, where V_{in} is swept from −10 V to +10 V in increments of 0.1 V.

The gain of the output stage is plotted by adding a trace whose function is defined as $d(V(V_{out}))$. This is the derivative of V_{out} as a function of the input voltage. The resulting family of curves illustrates how crossover distortion is influenced by the quiescent bias of the output stage.

The nominal gain is about 0.95. With too little bias, the gain is seen to fall seriously at the crossover point to about 0.93. With the bias set too high, the gain is too high in the middle range of the crossover, at about 0.97, illustrating what is called *gm* doubling. Although an optimum compromise bias can be chosen, it is clear that there is no bias value that completely eliminates the crossover distortion.

23.8 Plotting Results

Many of the plotting capabilities of LTspice® have been used and explained above. Some additional useful plotting functions are described here. At the top of the plot area (the plot plane) there is a label for each plot in the same color as the plotted trace. A convenient way to delete a trace is to click on the trace label with the scissors from the toolbar.

If you right-click on the trace label, a dialog box will be displayed. It allows you to change the color of the trace and to make the trace a function of the variable being plotted. For example, if the trace is V_{out}, you can make it become a trace of V_{out}^2. You can also define the trace as a function of V_{out} and other available variables. This is referred to as *waveform arithmetic*.

An especially useful capability for measuring points on the trace is to attach a cursor to it using the *Attach Cursor* box. This will display a set of crosshairs that always intersect on the trace and can be moved along the trace by the mouse (click on the crosshairs and drag the mouse). A display box will appear and show the horizontal and vertical values of the trace at the intersection. A second cursor can also be attached to the trace. In this case the display box will indicate the coordinates of the two cursors, their differences and the slope between them. If two cursors are applied to the same edge of a square wave, the slope readout will correspond to the slew rate. If two cursors are placed on two successive positive edges of a waveform, the period will be read out as the corresponding frequency.

One can also add a trace to a plot. Click in the area of the plot. Now right-click on the plot. A list of options will appear. Select *Add Trace*. A trace from the list of available traces can be added. Moreover, a function of one or more of the available traces can be typed into the function box. This capability will be described further below.

Right-click again in the plot window to get the options list. Click on *Grid*. This will toggle the display of grid lines on the plot. Plots can be copied to an application like *Word* or *Paint* by making the plot window active and hitting *Ctrl C* to copy it to the clipboard.

In some cases there may be multiple traces on the plot plane that make viewing the results confusing. Traces in the plot plane can be displayed separately in an additional plot plane by right-clicking in the original plot plane and then selecting *Add Plot Plane*. A blank plot plane will appear. Click on the waveform label of a trace and drag it to the new plot plane. One can add as many plot planes as desired and place the traces in any of them as desired by dragging them into the desired plot plane.

The default is to present plots with a black background. Sometimes this is undesirable, especially for printing. The background can be changed. Go to *Tools* → *Color Preferences*. Select *Background*, and then adjust the color sliders to obtain the desired color. Moving all three sliders to the right will provide a white background.

You can obtain the average and RMS values of a trace by Control left-clicking on the trace label.

Gummel Plot

The *Gummel plot* shows log of I_c and log of I_b versus V_{be} for a BJT. The Gummel plot illustrates the use of a logarithmic Y axis. The circuit of Figure 23.5 can be used to simulate a transistor and obtain its Gummel plot.

The MJL21194 NPN power transistor is used for illustration. This circuit merely applies a swept base-emitter voltage to the transistor under test. The base and collector currents are then probed to create the plot. Finally, the Y axis displaying current is given logarithmic coordinates by clicking the mouse just to the left of the Y axis. A dialog box will appear. Check the *Logarithmic* box and enter the desired minimum value of current to display.

Beta Versus I_c

Plotting the current gain of a transistor is a good way to illustrate the ability to plot functions of trace variables. The circuit of Figure 23.6 is used for this example. The emitter current of the

Gummel Plot Test Circuit

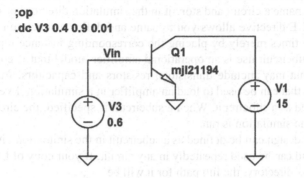

;op
.dc V3 0.4 0.9 0.01

Figure 23.5 Generation of a Gummel Plot for a Transistor

Beta vs. Ic Test Circuit

.dc dec I1 1mA 10 10

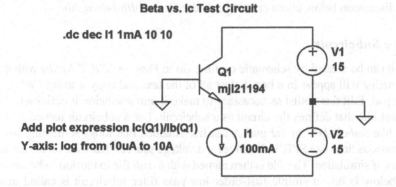

Add plot expression Ic(Q1)/Ib(Q1)
Y-axis: log from 10uA to 10A

Figure 23.6 Sweeping the Emitter Current of a Power Transistor

power transistor is swept from 1 mA to 10 A with a decade sweep. Sweeping the emitter current introduces a small error due to the finite transistor beta. After the simulation is run, right-click on the plot plane and select *Add Trace*. In the dialog box enter the function $I_c(Q1)/I_b(Q1)$ and hit *OK*. Transistor beta will then be plotted. Notice how the current gain falls at both low and high current, illustrating so-called beta droop.

The small error created by sweeping emitter current instead of collector current can be corrected in the plot by changing the X-axis function that is plotted from the default swept value of emitter current to the value of the collector current. Simply left-click in the area below the X-axis on the plot. A dialog box will appear. In the area labeled *Quantity Plotted* replace I1 with IC(Q1).

Transconductance Versus I_c

This example illustrates the use of another function, the derivative function. Sweep the emitter current from 1 μA to 10 A. Add a trace with the function $1/\{-d[V(V_{e1})]\}$ to the simulation of Figure 23.6. This shows the inverse of the derivative of the emitter voltage with respect to emitter current, which is transconductance. Set the Y axis to be logarithmic from 1e-5 to 1e2. The log-log plot of transconductance versus transistor current will be a straight line over several decades, indicating that transconductance is proportional to operating current. However, the line begins to bend at currents above about 1 A, indicating the effect of ohmic transistor components (parasitic elements) on the transconductance at high current.

23.9 Subcircuits

Sometimes it is desired to employ the same circuit numerous times in a simulation. Subcircuits make it possible to name a circuit and store it in the simulation directory or in a library.

The .*subckt* SPICE directive allows you to name and define a subcircuit that can be put in a schematic multiple times merely by placing the corresponding instance repeatedly. One of the best examples of subcircuit use is an operational amplifier model that is composed of a collection of elements that may include transistors, resistors and capacitors. Another example is a loudspeaker model that can be used to load an amplifier in a simulation. Even a simple low-pass filter can be realized as a subcircuit. When a subcircuit is specified, the circuit is expanded to a flat netlist before the simulation is run.

A circuit of your design can be defined as a subcircuit in the simulation directory or in the . . .\ *lib\sub* directory and can be used repeatedly in any circuit. If your copy of LTspice® is located in the *C:\program files* directory, the full path for it will be

C:\program files\LTC\LTspiceIV\lib\sub

See the discussion below about creating a folder like . . .*lib\sub\mylib*.

Creating a Subcircuit

A subcircuit can be created by schematic capture. Go to *View → SPICE Netlist* with the schematic open. The netlist will appear in a box. Select all of the text and copy it using *Ctrl+C*. Then paste it into Notepad. Edit that netlist as necessary to make it into a subcircuit definition. This includes adding a first line that defines the circuit as a subcircuit. For a subcircuit named *LPF* it will look something like .*subckt LPF in out gnd*. A last line is added that looks something like .*ends LPF*. Delete extraneous lines like SPICE directives and voltage sources that were on the circuit schematic for purposes of simulation. The file is then named with a .*sub* file extension to become *LPF.sub*.

Shown below is how a simple first-order low-pass filter subcircuit is called in a *netlist*. Its instance is called U1 on the schematic. The instance name of the subcircuit is preceded by an *X* in the SPICE netlist.

XU1 Vin Vout 0 LPF1

The subcircuit that defines the 100-kHz LPF is shown below. Its filename should be *LPF1.sub*.

.subckt LPF1 in out gnd
R1 in out 10k
C1 out gnd 160pF
* 100-kHz first-order low-pass filter
.ends LPF1

It is helpful if the terminals in the .*subckt* line (here *in*, *out* and *gnd*) are labeled in accordance with the pin names that will be put on the symbol. However, this is not entirely necessary, as the pin order in the symbol specification will correlate pin names between the symbol and the subcircuit definition.

The elements of the subcircuit can be parameterized and defined when the subcircuit is called in the SPICE netlist. The LPF below with parameterized *R* and *C* also includes a VCVS acting as a unity gain output buffer. The intermediate node at the input to the VCVS was labeled N001 by LTspice®.

.subckt LPF2 in out gnd
R1 in N001 {R}

C1 N001 gnd {C}
E1 out gnd N001 gnd 1
.ends LPF2

The parameterized subcircuit as it appears in the circuit netlist is

XU1 Vin Vout 0 LPF2 R=10k C=160pF

In the schematic of the circuit using the subcircuit, the *.param* directive is used for each of *R* and *C* in the usual way.

The Symbol Editor

Every model must have an associated symbol for entry onto the schematic. The symbols are defined in *.asy* files. LTspice® has its own symbol editor for creating *.asy* symbol files. It can be used to create a symbol from scratch or to modify an existing symbol. The symbol editor can be accessed by going to *File → New Symbol*.

The usual drawing tools are available in the symbol editor. For example, one can go to *Draw → (Line, Rectangle, Text)* to add those elements to the symbol. Scissors are available to delete elements.

After the symbol is drawn the connecting pins must be added. Go to *Edit → Add Pin/Port*. Fill in the *Label* with the pin name, hit the radio button for its desired location on the symbol perimeter, hit *OK* and place the pin on the symbol.

The *Pin Table* contains the assigned (and visible) names of the pins and the order in which they appear on the SPICE line. Go to *View → Pin Table*. Double-click on a pin in the left column to change its name. Double-click on the numbers in the right column to rearrange their *Spice Order*. The order of the pins must correspond to the order of the pins listed in the *.sub* file.

The *Attribute Window* is used to select what attributes will be visibly attached to the symbol. These will often be the *InstName* and *Value*, where *Value* is the name of the part, i.e., *LPF1*. Go to *Edit → Attributes → Attribute Window*. Select a desired attribute to appear on the symbol and hit OK. The attribute will appear on the symbol attached to the cursor. Move the cursor to the desired location and click. Right-click on it to enter the desired character string. Attributes can be removed with the scissors. Visible attributes can be changed by right-clicking on them. This is not true for the instance name, which will be altered on the part when it is placed on the schematic.

The *Attribute Editor* allows you to set or modify both visible and hidden attributes of the symbol. Go to *Edit → Attributes → Attribute Editor*. Attributes that were added in the Attribute Window can be modified here as well. The hidden attributes (including those not selected to be visible in the Attribute Window) are also set here. These include *Prefix*, *SpiceModel* and *ModelFile*. The prefix for a subcircuit should be *X*. Leave *Spice Model* blank and enter the name of the *.sub* file for *ModelFile*. The *.asy* symbol file should be saved with the same name as the *.sub* file but with the *.asy* file extension.

Modifying an Existing Symbol

In many cases a suitable symbol for your subcircuit is already a standard one that is available in the simulator library. Sometimes, when a new device is created, one of the standard symbols must be modified. The symbol editor can be used to do this. Go into the . . .\lib\sym directory and its subdirectories to find symbols that may work for your subcircuit. Then just choose one, copy it and edit it to suit your needs. Rename it to the name of the device you are installing. Double-click on it and it will open in the symbol editor. Change its name and model attributes to reflect the name of the model file and what you want the symbol to say on its text on the schematic. The model name should be the same as the subcircuit name and subcircuit filename. Bear in mind that you can right-click on an existing symbol to see its visible attributes. You can *Ctrl* right-click on a symbol to access the attribute editor for it.

Summary for Creating the LPF1 Symbol

Here is a summary for creating the symbol for the subcircuit LPF1. Go into the *Symbol Editor* from the subcircuit schematic by hitting *File* → *New Symbol*. Go to *Draw* → *Rect* and draw a good-sized rectangle. It is helpful if the lines of the rectangle are on grid lines. Go to *Edit* → *Add Pin/Port*. Fill in the Label as *in*, select the *left* radio button to put it on the left edge, hit *OK* and place the pin on the left side of the rectangle. Place the *out* and *gnd* pins accordingly. Go to View → *Pin Table* and verify that the pin names are as desired and that the pin order corresponds to that in the *.subckt* definition. Notice that the pin order will be in the order that they were placed. Change the pin names or order if needed.

Go to *Edit* → *Attributes* → *Attribute Window*. Click on *inst name* and place it somewhere outside the rectangle. Go back into the Attribute Window and click on *Value* and place it somewhere inside the rectangle. Right-click on it and enter *LPF1* as the *string*. Go into *Edit* → *Attributes* → *Edit Attributes*. Select *Prefix* and enter *X*. Go back into *Edit Attributes*, select *ModelFile* and enter *LPF1.sub*. Save the file in the simulation directory as *LPF1.asy*. If desired, copy the symbol file to a library directory, such as . . .\lib\sym\mylib (assuming that the *LPF1.sub* file was placed in . . .\lib\sub\mylib).

Using the Subcircuit in a Schematic

With the subcircuit and corresponding symbol file in the simulation directory, the subcircuit can be used in the schematic just like a component. Click on the *Component* button. Scroll to the simulation directory and click on it. All of the subcircuits in the simulation directory will be listed. Click on the desired subcircuit, and its symbol will appear in the window. Hit *OK* and place it.

Installing Subcircuit Models in a Library

Subcircuit models can be accessed from a library that you can assemble. Create the folder . . .\lib\ sub\mylib. This is where the subcircuit file(s) will be stored. This folder will be available via the *Component* button. Place the *.sub* file in this folder. Similarly, create the folder . . .\lib\sym\mylib. This is where the symbol file for the device should be stored.

To place the subcircuit on a schematic, click on the component button. Then double-click on *[mylib]* within the group of selections. A list will come up with all of the subcircuits that you have placed in . . .\lib\sub\mylib. Click on the one you want. Its symbol will appear in a window. Hit *OK* to place it on the schematic in the usual way. If you wish to go up one level back to the full component selection, double-click on *[..]*.

23.10 SPICE Models

Good models are at the heart of SPICE simulations. Most manufacturers provide models for their transistors and integrated circuits, and there are a great number of models available for other components, like transformers and loudspeakers. A tremendous amount of model information is available on the Web, both from manufacturer's sites and private sites.

Go to the Internet and search device manufacturers' websites. You can create your own model by modifying an existing one and renaming it. Software for creating models is also available at www.intusoft.com/spicemod.htm. Chapter 24 will cover device models and the creation or tweaking of them in much more detail. Here we will focus on using the models in the LTspice® library and adding acquired models.

Bipolar Junction Transistors

BJTs are supported with the symbol names NPN and PNP. BJT models are located in the *standard.bjt* file in the . . .\lib\cmp *folder*. Many models are included in this one file. LTspice® comes with this file populated with a large number of models, but you can add more models simply by pasting the transistor model file into the *standard.bjt* file. Those models will then be available via the component button just like any other BJT models. A simple SPICE model for a BJT is

.model 2N3904 NPN (IS=1E-14 VAF=100
+ Bf=300 IKF=0.4 XTB=1.5 BR=4
+ CJC=4E-12 CJE=8E-12 RB=20 RC=0.1 RE=0.1
+ TR=300E-9 TF=400E-12 ITF=1 VTF=2 XTF=3 Vceo=40)

This is the kind of model that can be pasted into the *standard.bjt* file. The meaning of each of the parameters in this file and how to tweak or derive them will be discussed in the next chapter.

To install a BJT model, start with the *.model* file in Windows Notepad. Set the file type to *All Files*. Append the file to . . .\lib\cmp\standard.bjt with a paste. It will then show up in the list of available transistors when using *pick new transistor*.

Junction Field Effect Transistors

The JFET transistor is chosen with a symbol name of NJF or PJF from the library. These correspond to N-channel and P-channel devices, respectively. A simple N-channel JFET model looks like

.model J310 NJF (BETA=0.004 VTO=-4 LAMBDA=0.06
+ CGS=16E-12 CGD=12E-12 PB=0.5 M=0.5 FC=0.5 N=1
+ RD=5 RS=30 IS=4E-14 KF=6E-18 AF=0.6)

JFET models can be added to the *standard.jft* file in the . . .\lib\cmp folder with a simple paste operation.

Power MOSFETs

LTspice® supports two different kinds of models for power MOSFETs. The first is the proprietary LTspice® VDMOS model for vertical double-diffused power MOSFETs. It is supported in the library in much the same way as a BJT, with a *standard.mos* file in the . . .\lib\cmp folder. It is placed using the component button by selecting *nmos* or *pmos*, placing the transistor, then right-clicking on it to select the particular transistor. A simple VDMOS model looks like

.model IRFP240 VDMOS (nchan Vto=4 Kp=0.1 Lambda=0.003
+ Rs=0.02 Rd=0.1 Cgdmax=30p Cgdmin=6p a=0.4
+ Cgs=40p Cjo=7p m=0.7 VJ=2.5 IS=4E-06 N=2.5)

Additional VDMOS models can be installed by pasting them into the *standard.mos* file. Such VDMOS files are not generally available from other than LTspice®, but the next chapter shows how they can be created.

The second kind of power MOSFET model supported is based on a subcircuit that includes at its core a SPICE Level 1 MOSFET model like that used in an integrated circuit. The remainder of the subcircuit adds some of the peculiarities of power MOSFETs, such as the body diode and nonlinear gate-drain capacitance. This is the way much of the industry models power MOSFETs. Models in this form are widely available from vendors. A very simplified version of such a model (where gate-drain capacitance is constant) is shown below. Many vendor models have elaborate circuits within the subcircuit to attempt to model the nonlinear gate-drain capacitance, but many are not very accurate, especially for linear applications.

```
.subckt IRFP240 1 2 3
* pins: Drain Gate Source
M1 9 7 8 8 MM
D1 3 1 MD
RS 8 3 0.05
RD 9 1 0.1
RG 2 7 5
CGS 2 3 1e-9
CGD 1 2 0.5e-9
.MODEL MM NMOS LEVEL=1 IS=1e-32
+ VTO=4 LAMBDA=0.002 KP=3
.MODEL MD D IS=1e-12 N=1
+ CJO=1e-09 VJ=4 M=1 FC=0.5
.ENDS
```

The power MOSFET subcircuit is treated largely the same as any other subcircuit. However, its *.subckt* file should be saved with a *.mod* file extension instead of a *.sub* file extension. Similarly, its *ModelFile* definition in the symbol attributes should have a *.mod* file extension instead of a *.sub* file extension.

Include Statements

The *.include* spice directive causes a file to be included in the simulation file as if it was written there. The *.include* directive can be used to put models or libraries of models or subcircuits in the simulation.

.include 2N3904.mod

The above directive will include the model of the 2N3904 BJT for use by the simulation. LTspice® will first look for the file in the *. . .\lib\sub directory*. If it is not found there, LTspice® will look in the directory where the simulation is being run. Alternatively, a complete path name can be specified for the file. This could be a directory in your top-level simulation folder where you place your own models.

The *.include* directive applies to a single file, but that file can contain many models, just like the LTspice® *standard.bjt* library file contains many BJT models. Your file of BJT models might be called *mybjt.bjt*. The *.include* directive can also be used to include a text file with a large number of models on the schematic as shown below.

.include Cordell Models.txt

The above directive will include a model file containing numerous models created by me. They are available at cordellaudio.com. Virtually all of the models used in this book and many additional similar models are available in this file. These models were created because manufacturers' models for these devices were not available or were judged to be less accurate than desired. These models were usually created by measuring actual sample devices.

You can place your own library of BJT models in the . . .\lib\sub directory by placing the file *mybjt.bjt* there. The directive *.include mybjt.bjt* should then be placed on the schematic. Notice that the full path name does not need to be included when this approach is used. The transistors in *mybjt.bjt* will not show up in *pick new transistor* if you right-click on a transistor, so you must edit the name of the transistor to be that of the one in *mybjt.bjt* that you want. You can also place the BJT library file in . . .\lib\sub\mylib. In that case the directive *.include mylib\mybjt.bjt* would be used.

Libraries

The *.lib* directive is similar to the *.include* directive. It includes the model and subcircuit definitions of the specified file in the simulation file. LTspice® looks first in the . . .\lib\cmp directory when the *.lib* directive is used. It will then search the . . .\lib\sub directory and finally the simulation directory. A complete path name can also be specified. In contrast, LTspice® does not look for the file in the . . .\lib\cmp directory when the *.include* directive is employed.

As an example, you can place your own library of BJT models in the . . .\lib\cmp directory by placing the file *mybjt.bjt* in the . . .\lib\cmp directory. The directive *.lib mybjt.bjt* should then be placed on the schematic. Notice that the full path name does not need to be included when this approach is used. The transistors in *mybjt.bjt* will not show up in *pick new transistor* if you right-click on a transistor, so you must edit the name of a transistor on the schematic to be that of the one in *mybjt.bjt* that you want.

Indeed, the directive *.lib . . .\lib\cmp\standard.bjt* is automatically included by LTspice® in the netlist whenever a BJT is used from the library in the usual way.

23.11 Simulating a Power Amplifier

Here I'll describe the simulation of the simple power amplifier first introduced in Chapter 1. This will provide a good start so that a *put it all together* understanding can be achieved. This template simulation can be modified extensively for use with virtually any amplifier design. Indeed, I usually begin with a previous amplifier schematic and copy it to a new name and just modify it to the new design, reusing many of the components and SPICE directives. The simulation circuit is shown in Figure 23.7. Notice that it is in DC-coupled form to optimize FFT analysis.

DC Analysis

Select *Simulate → Edit Simulation Cmd → DC operating point*. Hit *OK*, and then hit *Run*. Check V_{out} in the pop-up simulation results window for acceptably low output offset. If necessary, adjust V_{offset} to obtain offset voltage of less than 1 mV at the output. Check the operating currents of the transistors. Click out of the simulation results box and go to *View → SPICE error log*. Look at the operating points and model parameters of the transistors in the circuit. Are they all operating at expected current levels? Are any transistors in saturation?

Close the SPICE error log and mouse-over various circuit nodes to see their voltage levels in the lower left display window. The node number will also conveniently be displayed. Mouse-over the *vt* and *vb* nodes at the power transistor emitters in the output stage to infer the output stage operating bias. Under optimum conditions, these nodes would be at or a bit less than ±26 mV. However, this amplifier has been slightly over-biased. Place the mouse over the resistors to see their operating currents and power dissipations in the same display. Place the mouse over a transistor to see its power dissipations.

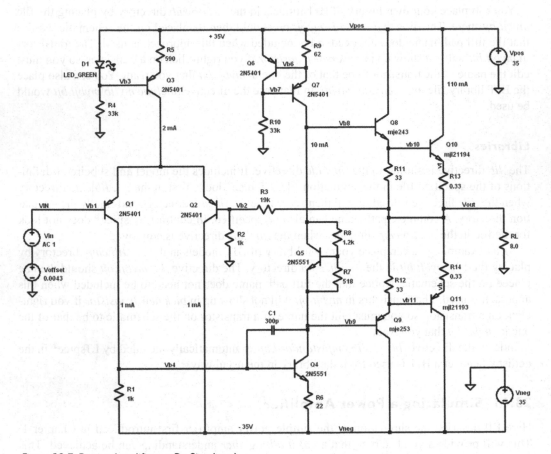

Figure 23.7 Power Amplifier to Be Simulated

Frequency Response

Select *Simulate → Edit Simulation Cmd → AC Analysis*. Select a decade sweep with 100 points per decade, a start frequency of 100 and a stop frequency of 1000k or 1Meg (note that M by itself is interpreted as *milli*).

Hit *OK* and hit *Run*. Probe V_{out} with the mouse. The frequency and phase response will appear in the plot window. Left-click on the trace label. A crosshairs cursor will appear. Mouse-over it and the number 1 will appear, designating cursor number 1. This crosshairs cursor will be attached to the gain trace. Move it to a low frequency and note that the gain is indicated as 26 dB in the pop-up display box. Move the cursor up in frequency until the gain is down by 3 dB. Note that the frequency is indicated as about 540 kHz.

Click on *vb8*, the VAS output node. An additional trace showing frequency response to this new point will be added to the plot window. Drag the mouse in the plot window to zoom in on the pair of traces at 1 kHz. Notice from the spacing of the traces that the output stage has a small-signal gain loss of about 0.35 dB.

1-kHz Transient Analysis

In this section a 1-kHz sine wave will be simulated by a transient analysis that will show voltages and currents in the circuit as a function of time. Select *Simulate → Edit Simulation Cmd → Transient*. Set a stop time of 4 ms to simulate four cycles of the waveform.

Specify the signal source V_{in} to be a 1-kHz sine wave with amplitude of 1 V peak and comprising four cycles. This will produce an output of 20 V peak, corresponding to 25 W into the 8-Ω load. Run the simulation and click on the V_{out} node to see the expected output waveform. *Alt*-click on Q10's collector wire and see the class AB half-wave rectified current waveform for Q10.

Increase the input signal amplitude to 1.6 V peak, rerun the simulation and probe V_{out}. Notice that there is no obvious clipping. *Alt*-click on Q1's collector wire to show the waveform for Q1's collector current. Notice the onset of clipping as revealed by the sharp peaks in the collector current of Q1. The amplifier just clips at 64 W.

Increase the input signal to 2 V peak to see the amplifier operating in fairly strong clipping. Here the amplifier is attempting to produce an output of 100 W. Probe V_{out} and see the obviously clipped output waveform. *Alt*-click on Q1's collector wire to show Q1's collector current waveform. Probe from the base of Q1 to the base of Q2 to see the negative feedback error signal. Notice how the error signal suddenly pops up to a large value on the peaks where clipping is occurring. Probe various nodes to see voltages and currents. Look especially for signs of transistor saturation and unusually large currents and evidence of sticking.

20-kHz Transient Analysis

In this section a 20-kHz sine wave will be simulated. Select *Simulate → Edit Simulation Cmd → Transient*. Set a stop time of 400 μs to capture eight cycles of the waveform.

Specify the signal source V_{in} to be a 20-kHz sine wave with amplitude of 1 V peak and comprising 16 cycles (even though only eight cycles will be simulated at this time). This will produce an output of 20 V peak, corresponding to 25 W into the 8-Ω load. Run the simulation and click on the V_{out} node to see the expected output waveform.

Alt-click on Q1's collector wire to show Q1's collector current. Notice that it falls dangerously close to zero, indicating that the amplifier's input differential pair is operating at close to its maximum capability. This means that the amplifier is close to its slew rate limiting point. Mouse-over Miller compensation capacitor C1 and click on it to see its current. *Alt*-click on Q4's base wire to plot Q4's base current. Notice that most of the signal current supplied by the LTP is flowing into the compensation capacitor.

Increase the input signal amplitude to 1.3 V peak and run the simulation. Again plot the current in Q1's collector wire, seeing that it is clipped; this indicates that slew rate limiting is occurring. Plot the input differential error signal by dragging the mouse probe from the base of Q1 to the base of Q2, noting the large error voltage present when slew rate limiting is occurring.

Square-Wave Response

The 50-kHz square-wave response will now be simulated. Select *Simulate → Edit Simulation Cmd → Transient*. Choose a stop time of 80 μs to cover four cycles. Set the input source to provide a 50-kHz square wave with amplitude of 0.1 V peak. Set the initial voltage to −0.1 V and the pulse voltage to 0.1 V. Set 10 ns rise and fall times. Set a pulse time of 10 μs and a period of 20 ms.

Run the simulation and probe V_{out}, noting the rounded leading edges that result from the finite bandwidth of the amplifier. Probe the collector current in Q1's collector wire. The current waveform exhibits large pulses at the times of the square-wave edges. These current pulses are charging and discharging the compensation capacitor in order to produce the required output.

Increase the input amplitude to 0.2 V peak, rerun the simulation and probe V_{out}. The amplifier is now exhibiting slew rate limiting. Window in on the straight-sloped portion of a waveform edge. Right-click on the V_{out} label and select two cursors in the pop-up window. Manipulate the cursors to be 1 μs apart on the straight slope. Read the indicated slew rate in the bottom right portion of the information box. The slew rate is approximately 3 V/μs.

Close the information box and un-zoom the plot. Probe the current in the collector wire of Q1. Notice the clipped spikes, indicative of slew rate limiting.

1-kHz Total Harmonic Distortion

Change the input voltage source to be a 1-kHz sine wave with amplitude of 1 V peak. This corresponds to 25 W into the 8-Ω load. Select *Simulate → Edit Simulation Cmd → Transient*. Set a stop time of 8 ms to simulate eight cycles of the waveform. Place the following two SPICE directives on the schematic:

.option plotwinsize=0
.maxstep=0.48831106u
.four 1kHz 10 4 v(vout)

As described earlier, the first directive turns off compression for greater transient analysis accuracy. The second sets an optimum *timestep* for Fourier spectral analysis of the 1-kHz sine wave. The third causes a THD summary of the Fourier analysis to be printed in the SPICE error log result. The amplitudes of the first ten harmonics, normalized to that of the fundamental, will be printed. The Fourier analysis will be carried out during the last four periods of the 1-kHz sine wave.

Run the simulation and look at the information window at the bottom of the screen to view simulation progress. When the simulation is completed, click on V_{out} to plot its trace. Left-click and right-click in the plot window; then select *View → FFT* from the pop-up window. The trace selection $V(V_{out})$ will already be highlighted in the list of available traces. Select *Specify a time range*. The *End Time* should already be shown as 8 ms. If it is not exact, enter 8 ms. Enter a *Start Time* of 4 ms. The first 4 ms allow the simulation to stabilize and the last 4 ms are analyzed by the FFT.

Hit *OK* and a new window will appear with the FFT results. Window-in on the range of interest, extending from about 1 kHz to about 20 kHz, and extending down to −140 dB. Mouse-over the peaks of the spectra and observe the corresponding amplitude in the bottom display window. The amplitudes are reported in dB with respect to 1 V (dBV). The fundamental at 1 kHz lies at about +23 dB, corresponding to 14.14 V RMS.

The second and third harmonics are at −55 dB and −59 dB, respectively. They are 78 dB and 82 dB below the fundamental, suggesting THD on the order of 0.013%, not including all of the higher harmonics. The fairly rich spectrum of higher harmonics suggests the presence of output stage crossover distortion.

Close the FFT window and click on the schematic window. Then hit *View → SPICE error log*. A list of the amplitudes of the first ten harmonics, relative to the fundamental, will be shown. At the bottom of this list the THD will be shown as 0.016%. This display is often more convenient than looking at the raw spectral trace.

20-kHz THD

Total harmonic distortion at 20 kHz is one of the more difficult and stressful tests for an amplifier. Set the input source to a 20-kHz sine wave with amplitude of 1 V peak. This will produce 25 W into the 8-Ω load. Set a transient simulation stop time of 400 μs to simulate 8 cycles of the waveform. Alter the SPICE directives on the schematic as follows:

.maxstep=0.02441555u
.four 20kHz 10 4 v(vout)

Here four cycles, or 200 µs, will be processed by the FFT. Run the transient simulation until it is completed. Click on V_{out} to display the output waveform. Left-click and right-click in the plot window. Select *View → FFT* from the pop-up box. The end time should be listed as 400 µs. Enter 200 µs for the start time. Hit *OK*, and the FFT spectrum will come up in a new window. Note the substantially increased levels of distortion as compared to the 1-kHz simulation. With the fundamental at +23 dBV and the third harmonic at −23 dBV, the worst harmonic is only 46 dB down, suggesting THD of at least 0.5%. Notice that significant spectral lines are present out to fairly high frequencies. Bear in mind that the power level here is only half the rated 50 W.

Window-in on the plot region extending from about 20 kHz to about 500 kHz, and from +30 dBV down to −100 dBV. Sometimes this will require more than one windowing operation to obtain the properly zoomed area. This is the region of greatest interest, as harmonics more than 120 dB below the fundamental are of relatively little significance and may even be simulation artifacts.

A big advantage of 20-kHz THD simulations is the ability to view the higher harmonics. A major limitation of THD-20 lab measurements is that the THD analyzer sometimes limits the bandwidth of the residual to 80 kHz. This barely allows room for the second through fourth harmonics.

Close the FFT window and click on the schematic window. Go to *View → SPICE error log* and review the THD data. Note that this data is available even if the plot of output voltage and its FFT are not made. The total harmonic distortion reads 0.52%, with significant harmonic content out to the tenth harmonic.

This is surely not a good amplifier design and is one that would do poorly on the DIM-30 test for Transient Intermodulation distortion (TIM). This illustrates that such an amplifier will do poorly on a THD-20 test as well, especially when a spectral analysis of the harmonics out to at least 200 kHz is performed.

CCIF Intermodulation Distortion

The 19 + 20-kHz CCIF intermodulation test is implemented by feeding the amplifier a mixed combination of 19-kHz and 20-kHz sinusoids in equal proportions. The test is best conducted at or near a voltage swing corresponding to peak voltage at full power. For this amplifier the test will be conducted at a smaller total peak input level of 1 V, the same peak level that was used for the 20-kHz THD test. This corresponds to 25 W into 8 Ω.

The CCIF IM test produces intermodulation products at integer sum and difference combinations of the two test frequencies. These IM products are best viewed on a spectrum analyzer or with an FFT plot as will be done here. The order of the products is equal to the number of instances of the two test tone frequencies in the sum and difference equation that creates a given IM frequency product. Suppose frequency A is 20 kHz and frequency B is 19 kHz. The second-order product will be produced at A-B = 1 kHz. The fourth-order product will be produced at 2A-2B = 2 kHz.

The third-order product will be produced at 2B-A = 18 kHz. The fifth-order product will be produced at 3B-2A = 17 kHz. The seventh-order product will be produced at 4B-3A = 16 kHz. And so on. Notice that all of these components are in-band. Of course, there are many other components created with other sum/difference combinations of the test frequencies, many lying out of band. Of particular interest is the fact that the odd-order distortion products lie at intervals 1 kHz apart going down from 18 kHz and jumping two orders at every 1-kHz grid point. This produces a picket-fence appearance in the spectrum. The CCIF IM test is thus very capable of displaying high-order nonlinearities without resort to testing frequencies beyond the audio band.

Simulate the CCIF IM test by adding a second sinusoidal voltage source set for 19 kHz and a peak level of 0.5 V. This source can be simply placed in series with the existing 20-kHz source. Set the level of the 20-kHz source for 0.5 V peak.

The lowest frequency of interest in the simulation and FFT analysis is 1 kHz. Moreover, all of the IM products of interest are on a 1-kHz grid. For this reason, the FFT time should be an integer multiple of the 1-kHz period. Here the FFT will be performed on four cycles of the 1-kHz beat frequency.

Alter the SPICE directives on the schematic as follows:

> .maxstep=0.12207776u
> remove or disable the .four 20kHz 10 4 v(vout) directive.

Here four cycles of the 1-kHz difference signal, or 4 ms, will be processed by the FFT. The *max step* is set to 0.5 times 4 ms/16,383. The FFT will be run with 65,536 points. The above selection of *maximum timestep* will provide an FFT analysis with grass that is below −120 dBV. Some experimentation with the FFT interval and the *maximum timestep* is often needed in simulations involving multiple tones in order to obtain a result with adequately low spurious grass.

Set the transient simulation for a stop time of 5.5 ms. The first 1.5 ms of the simulation is to be ignored by the FFT. The use of 1.5 ms causes the beginning and end of the FFT analysis to occur at a point where the 1-kHz beat signal goes approximately through zero. This helps slightly to reduce FFT windowing artifacts.

Run the transient simulation and click on V_{out}. Notice the beat-frequency action between the 19-kHz and 20-kHz sinusoids as they produce a signal envelope that goes from essentially zero amplitude to the full amplitude of 20 V peak.

Select the FFT. The stop time should be listed as 5.5 ms. Enter a start time of 1.5 ms and hit *OK*. The FFT will appear in a new window. Select the region of interest by windowing the plot from 1 kHz to about 50 kHz and from 30 dBV down to −160 dBV. Notice the picket fence of odd-order distortion spectra extending both above and below the pair of input sinusoids at 19 kHz and 20 kHz. There is also a prominent picket fence of intermodulation products centered at 39 kHz. The even-order products can be seen beginning at 1 kHz.

In a very good amplifier, all of the CCIF IM products will be 100 dB or more below the level of the two test tones. In the simulation of this amplifier, the test tones are at +17 dBV, while the third-order product at 18 kHz is at −34 dBV, down only 51 dB. The second-order product at 1 kHz is at −41 dBV, down by only 58 dB from the test tones. These are not very good results and are consistent with the poor high-frequency distortion results obtained from the THD-20 simulations.

Signal-to-Noise Ratio

LTspice® can be used to perform a simulation that will predict the *signal-to-noise ratio* (SNR) of an amplifier in a number of different ways. One caveat is that environmental sources of noise, like power supply ripple and EMI, will not be taken into account. For this reason, noise predictions based on simulation may be a bit optimistic.

The bandwidth over which the noise is being characterized is of paramount importance. In the specification of power amplifiers there are usually three ways to characterize noise. The first is an unweighted wideband measurement where the bandwidth is substantially greater than the audio band, usually limited by the closed-loop bandwidth of the amplifier under test or by a filter in the test equipment. This yields the poorest signal-to-noise ratio, and often the actual measurement bandwidth is not specified. The second is the noise in a 20-kHz bandwidth. The effective noise bandwidth can be set to 20 kHz by using a first-order low-pass filter at 12.7 kHz (the noise bandwidth of a 20-kHz first-order filter is 31.4 kHz because it does not have a brick-wall characteristic). In general, the ENBW of a first-order filter is 1.57 times its 3-dB bandwidth.

The last is the A-weighted noise measurement (whose effective noise bandwidth is approximately 13.5 kHz). All of these were described briefly in Chapter 1. Simulation of A-weighted SNR can be conveniently carried out using the A-weighting filter subcircuit described earlier in Figure 23.2.

In a well-designed amplifier with very little hum, the SNR will be governed mainly by the high-frequency measurement bandwidth, since thermal noise is proportional to the square root of bandwidth.

Select *Simulate → Edit Simulation Cmd → Noise*. Set the output to $V(V_{out})$ and set the input to V_{in}. Select a decade sweep from 20 Hz to 20 kHz with 20 points per decade. Run the simulation and click on V_{out}. A plot of voltage noise will appear. The noise density will be flat at 86 nV/$\sqrt{Hz}$. Because the amplifier gain is 20, this corresponds to an input-referred noise of 4.3 nV/$\sqrt{Hz}$. *Ctrl*-click on the plot label. The RMS noise within the 20-kHz bandwidth is 12.2 μV RMS. This is 124 dB below the maximum 20 V RMS output of the amplifier at 50 W. The unweighted SNR with respect to 1 W is about 107 dB.

Click on R2, the feedback network shunt resistor. It contributes about 77 nV/$\sqrt{Hz}$ to the total output noise of 86 nV/$\sqrt{Hz}$. Click on Q1. It contributes about 12 nV/$\sqrt{Hz}$. Click on Q2. It contributes about 30 nV/$\sqrt{Hz}$. Why the difference? Q2 contributes more noise because of its input current noise. The input current noise of Q1 is not a contributor because the impedance at its base is zero in this simulation.

Damping Factor and Output Impedance

The damping factor of a power amplifier is the factor by which the output impedance is smaller than 8 W. An amplifier with an output impedance of 0.08 Ω will have a damping factor of 100. The output impedance of an amplifier is usually a function of frequency.

The best way to measure the output impedance is to apply a known alternating current to the output of the amplifier and measure the resulting voltage at the output node. This is done while the input to the amplifier is connected to ground. This can best be accomplished by connecting a voltage source to the output through a resistor whose value is high compared with the expected output impedance of the amplifier. This is how it will normally be done in the laboratory. The voltage gain from this voltage source to the output node is then indicative of the output impedance. If 10 V is applied through an 800-Ω series resistor and 1 mV is measured at the amplifier output terminals, the DF is 100.

The way in which the output impedance changes with frequency can also be an indicator of stability. If the impedance rises sharply in a particular high-frequency region, this may indicate instability. If this frequency region is near the expected global feedback gain crossover frequency, then stability of the overall negative feedback loop may be inadequate. If instead the output impedance peak occurs at a much higher frequency, the possibility of a local instability, such as in the output stage, is suggested.

Stability

Select the *AC Analysis* tab in the simulation options and select a frequency range by decade from 10 kHz to 10 MHz. Probe the base node of Q2. This node should normally act like a unity-gain voltage follower to the input signal. You should see unity gain at this point with little or no gain peak. If this is done with a parameterized Miller compensation capacitor that is stepped in value, a family of stability curves will be produced.

Inferring Loop Gain

The open-loop gain (OLG) characteristic can be inferred by setting the closed-loop gain (CLG) very high. This *exposes* the open-loop gain. This will be reasonably accurate for OLG readings that are less than the high CLG for which the amplifier is configured. Change R3 in the simulated amplifier to 1 MΩ and carefully adjust input offset for small output offset, recognizing that the CLG has been set to about 1000. Run the AC simulation and see the open-loop gain. Above 10 kHz the gain drops below 60 dB. Above this frequency the plot is a reasonable depiction of OLG. The gain crossover is at 400 kHz, where the OLG equals 26 dB, the normal CLG for the amplifier. Phase at this frequency is lagging by 101°, suggesting a phase margin of 79°.

Measuring Loop Gain

The negative feedback loop gain can be measured directly by breaking the loop. Replace R3 with a 1000-GH inductor. This will keep the feedback path closed at DC. Connect the base of Q1 to ground. Connect the AC source to the base of Q2 through the 19-kΩ resistor that was R3. Loop gain will be plotted directly. The plot now will go to 0 dB at 400 kHz.

Output Stage Power Dissipation

The instantaneous power dissipation of any component can be plotted. Move the cursor to that component and press the *Alt* key. A thermometer will appear on the cursor. Left-click and the power waveform will be displayed. If you do this for a transistor, its instantaneous power dissipation will be revealed. You can also do this for a voltage source.

The average of the waveform (here the average power dissipation) can be displayed in an information box. Simply *Control*-left-click on the label of the instantaneous power dissipation waveform. If you do this for the amplifier load resistor and subtract the result from the sum of the dissipations for the rail voltage sources, the power dissipation of the amplifier will be revealed. These procedures also prove very convenient for determining the power dissipation of output stage emitter resistors and Zobel networks.

Output Current Limiting

The maximum available output current of the power amplifier can be checked by back-driving current into the output of the amplifier from a voltage source connected to the output through a low-value resistance, typically 8 Ω or less. This is not unlike the technique for measuring output impedance, but is carried out in a large-signal sense. The input of the amplifier is grounded and the voltage at the output node of the amplifier is viewed in a transient simulation or a DC sweep. The point at which the voltage at the output of the amplifier suddenly becomes large is the point at which current limiting is occurring.

Notice that this technique measures output current limiting when the output voltage is near zero. Alternatively, one can measure output current limiting by driving large signals through the amplifier into progressively smaller output load resistances.

You can also adopt this technique to measure output current limiting at voltages other than zero by applying a DC offset to the amplifier, assuming the amplifier is DC coupled. This can be useful in evaluating the behavior of *V-I* limiting circuits used for safe area protection, as the current limit for these circuits is often made a function of the output voltage.

Safe Operating Area

Plot the voltage across the power transistor against its current, using the *XY* plot capability. Load the amplifier with a series *R-C* combination of 8 Ω and 30 μF. Run a transient simulation with a 1.4-V peak sinusoidal input at 1 kHz. Run the simulation with a start time of 1.5 ms and a stop time of 4 ms. Plot I_c(Q10). You will see the expected half-wave current plot as a function of time.

With the plot plane active, move the cursor just below the *X* axis of the plot plane. A ruler will appear. Click on it. A dialog box will appear for the *X* axis. Normally it will show time as what is being plotted (*Quantity Plotted*) in a transient simulation, but you can type in an expression to plot in the window. Type the expression for the voltage across the output power transistor, $V(V_{pos})$–$V(V_{out})$, into the *Quantity Plotted* box. You will now see a plot of collector current as a function of V_{ce} for the transistor. If you wish, you can make both axes logarithmic to portray the usual safe area plot. You can also do this for a loudspeaker load, plotting current as a function of voltage and getting an ellipse.

23.12 Middlebrook and Tian Probes

Previous methods discussed here for measuring loop gain have involved breaking the loop for AC signals by using an extremely large inductor to keep the loop closed at DC to maintain the proper operating point. The test signal is then injected into the input side of the loop with a large capacitor and a small-signal AC analysis is run. This works well for many cases where the source impedance is low compared to the input impedance of the feedback network; else loading of the source would not be properly taken into account in the measurement of loop gain. Breaking the loop for AC destroys relevant interactions of impedances on either side of the break, and this can lead to large inaccuracies in some cases.

It is often desirable to evaluate the loop gain around a section of circuitry without breaking the loop or otherwise disturbing the circuit. This can be especially important where the circuit is enclosed by multiple feedback loops, such as global and local feedback loops that can each contribute to the total loop gain. A simple example is a 2-EF VAS, where local feedback comes from the Miller compensating capacitor and additional feedback comes from the amplifier's global feedback loop. Proper evaluation of the *in situ* stability of the VAS depends on taking both of these contributors into account. Another example is finding the total loop gain surrounding the output stage in an amplifier that employs transitional Miller compensation (TMC), as described in Chapter 11.

A Simple Probe for Loop Gain Evaluation

Figure 23.8a shows a power amplifier with a simple voltage loop gain probe that does not require breaking the loop. More accurately, one breaks the loop at the point of interest and then re-closes the loop with V_{probe}, a floating AC voltage source whose DC value is zero. Because the voltage source looks like a short circuit, the DC and AC operation of the circuit remain the same. Impedances on both sides of the probe are un-disturbed and their interactions are preserved. The probe is put in series with the feedback path at a location through which all of the feedback around the portion of the circuit of interest passes. Its nodes are labeled *x* and *y* for purposes here. This point should be located at a position where it would break all return loops in the circuit [7]. The voltage source probe creates a forward test signal in the direction that the feedback signal will travel. The other side of the voltage source receives the return signal that has been amplified by the loop gain. The loop voltage gain G_v is then simply $V(x)/V(y)$.

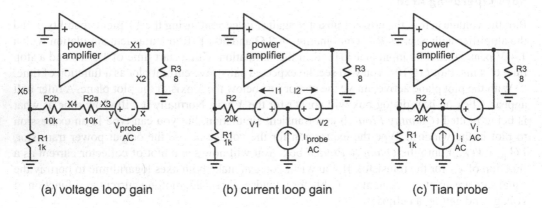

Figure 23.8 Voltage (a) Current (b) and Tian Probes (c)

This simple voltage probe works with adequate accuracy in the location shown (labeled X2) because the impedance of the return source node x (the output of the amplifier) is much lower than the input impedance of the forward path at node y. Simple math will show this. However, the figure shows four other places where the probe can be inserted (X1, X3, X4, X5), but results from the simple voltage probe may not be accurate if it is placed in these locations. The feedback resistor has been split for purposes of illustration to create a probe point X4 that will not produce accurate results. The reader can apply some analysis to show that these four other probe points may not produce accurate results. However, a full-featured probe should produce the same loop gain results when located at any of these five points.

Figure 23.8b shows a probe that can measure the current gain of the loop. The probe I_{probe} operates by injecting a current into the loop at the probe point. The current flow into each side of the injection point is then measured by 0-V floating voltage sources (resistors of very small value could be used as well). The ratio of the return current to the forward current is then the current gain G_i of the loop. Taking account of the current loop gain is key to enabling a loop gain probe to give the same results at any of the five probe points illustrated in the figure.

Middlebrook Probe

Middlebrook showed that the voltage loop gain G_v and the current loop gain G_i derived from the measurements above can be combined mathematically to produce the loop gain G_L equation and plot [7–10]. The governing formula is:

$$G_L + 1 = (G_v + 1) \| (G_i + 1) \tag{23.1}$$

This can be rearranged to give:

$$G_L = (G_i * G_v - 1)/(G_i + G_v + 2) \tag{23.2}$$

The simulation is run with two copies of the circuit, one configured to measure G_v and the other configured to measure G_i. The results are then combined using the equation above to formulate the plot command. Specifically, add a trace to the plot using the formula below.

$$((I(V3)/I(V4)) * (-V(x)/V(y))-1)/((I(V3)/I(V4)) + (-V(x)/V(y)) + 2) \tag{23.3}$$

A working LTspice® example of the Middlebrook probe called "LoopGain" can be found in the LTspice® examples/Educational folder [11]. By taking into account G_i, the Middlebrook technique avoids the need in the simple voltage probe approach above for the source impedance of the return signal to be much lower than the input impedance of the circuitry receiving the forward signal.

The Middlebrook probe requires that the feedback in the path be essentially unidirectional. Notice that the smaller of the voltage and current gains dominates the total loop gain in Equation 23.1. If the current gain is far higher than the voltage gain, then the simple voltage-only approach discussed above will give adequately accurate results. Because the output impedance of a power amplifier is usually quite low in comparison with the input impedance of the feedback network, loop current gain is very high, so this condition is usually satisfied if the probe is placed at location X2 between the output of the amplifier and the input of the feedback network.

Tian Probe

Figure 23.8c shows the Tian probe, developed by Michael Tian et al. in 2001 [12, 13]. It is a significant improvement over the Middlebrook probe. It does not require two copies of the circuit and properly takes account of bi-directional signals in the feedback path.

Once again, it works by placing a voltage source probe V_i in series with the loop signal path and a current source probe I_i in parallel with the point being probed. It also depends on performing two simulations, but it can be set up with only one copy of the circuit. Two simulations are run automatically in sequence using the *step param* function to disable the parameterized current probe in the first simulation and disable the parameterized voltage probe in the second simulation. The results of the sequential simulations are then combined arithmetically in a single plot command to produce the loop gain and phase plot. The ability of LTspice® and some other simulators to run two different simulations in sequence and combine the results arithmetically is key to implementing the Tian probe. The magic is in the plotting formula [7, 13]. Add a trace to a blank plot using the formula below.

$$-1/(1-1/(2 * (I(Vi)@1 * V(x)@2 - V(x)@1 * I(Vi)@2) + V(x)@1 + I(Vi)@2)) \qquad (23.4)$$

The @ symbol designates that the value is from simulation #1 or simulation #2. A working LTspice® example of the Tian probe called "LoopGain2" can be found in the LTspice® examples/ Educational folder [14].

References

1. Lawrence W. Nagel and Donald O. Pederson, "Simulation Program with Integrated Circuit Emphasis," Proceedings of Sixteenth Midwest Symposium on Circuit Theory, Waterloo, Canada, April 12, 1973; available as Memorandum No. ERL-M382, Electronics Research Laboratory, University of California, Berkeley.
2. LTspice® User's Manual. Available at www.analog.com.
3. LTspice® Users' Group. Available at www.groups.yahoo.com/group/LTspice.
4. Gilles Brocard, *The LTSPICE® IV Simulator*, 1st ed., Wurth Electronik, Waldenburg, Germany, 2013, ISBN 978-3-89929-258-9.
5. DIYaudio Forum. Available at www.diyaudio.com.
6. Andy Connors, private communication.
7. Frank Wiedmann, "Loop Gain Simulation," November 2, 2014. Available at https://sites.google.com/site/frankwiedmann/loopgain.
8. David Middlebrook, "Measurement of Loop Gain in Feedback Systems," *International Journal of Electronics*, vol. 38, no. 4, pp. 485–512, April 1975.

9. "Simulating Loop Gain—Spring 1997," Spectrum Software. Available at www.spectrum-soft.com/news/spring97/loopgain.shtm.

10. Sergio Franco, "Loop Gain Measurements," *EDN*, September 13, 2014.

11. Program Files/LTC/LTspiceXVII/examples/Educational/LoopGain, LoopGain2.

12. Michael Tian, V. Visvanathan, Jeffrey Hantgan and Kenneth Kundert, "Striving for Small-Signal Stability," *Circuits & Devices*, January 2001.

13. "Plotting Loop Gain Using the Tian Method—Spring 2011," Spectrum Software. Available at www.spectrum-soft.com/news/spring2011/loopgain.shtm.

14. "LTspice® Getting Started Guide," Linear Technology/Analog Devices. Available at www.analog.com/media/en/simulation-models/spice-models/LTspiceGettingStartedGuide.pdf?modelType=spice-models.

Chapter 24

SPICE Models and Libraries

24. INTRODUCTION

This chapter covers the adjustment and creation of SPICE models for BJT, JFET and power MOSFET devices. SPICE is only as good as the models used with it. In some cases, manufacturer-supplied models are quite inaccurate. One would think that those models would almost perfectly fit the data supplied in the transistor specification sheet. However, in many cases they don't even get the basics accurate, like V_{be} versus I_c.

Often, manufacturers subcontract the generation of SPICE models, especially for older devices where SPICE models were not developed for the part when the part was new. Often, they will just give a paper specification sheet to the subcontractor and have him or her glean data from the sheet. That data will then be entered into an automated curve-fitting software tool that quickly spits out a model. This is why we often see SPICE parameters with a ridiculous number of decimal places when they are in reality not even within 10% of where they should be.

You can do a much better job creating a SPICE model armed only with the same datasheet information. SPICE models can also be created or improved with relatively simple laboratory measurements that you can make using standard test equipment.

We all know that many transistor parameters vary quite a bit. Transistor β, for example, may easily vary by a factor of 2 from part to part. Base-emitter voltage can also vary, especially from one manufacturer to another. We therefore know instinctively that high precision in SPICE parameters is not important. Yet, the *behavior* of the transistor that they model is important.

SPICE is extremely valuable even with the use of inaccurate off-the-shelf models. It provides insight and exposes design behavior. However, poor models may give misleading results, especially under certain conditions. Poor models will often lead to inaccurate results in distortion simulations. This can be especially important in BJT output stages where beta droop and f_T droop at high current cause distortion. If these effects are not modeled reasonably well, distortion results will be significantly in error under high-swing conditions.

This chapter addresses the quality and accuracy of manufacturers' SPICE models with a strong emphasis on those for power BJTs. The chapter also provides understanding of what each SPICE parameter means. It also illustrates possibilities for simplifying the models by showing which parameters are more important. We also show how to create or fine-tune a model from information on a datasheet. SPICE models can be extracted from laboratory measurements; you don't have to have expensive measuring equipment for this.

The emphasis is on BJT power transistors, but the techniques and procedures are applicable to small signal and driver transistors as well. JFET and power MOSFET models are also covered. A substantial amount of material on device behavior, device operating equations and SPICE parameter meanings is provided in this chapter that has been gleaned from many sources [1–4].

24.1 Verifying SPICE Models

Manufacturer-supplied SPICE models are helpful and convenient. Some manufacturers do a good job on their SPICE models, but others do not. For this reason, it is wise for the serious designer to verify manufacturers' SPICE models against the supplied datasheets for the transistor. Surprisingly, sometimes there will be significant errors, even in simple things.

SPICE models can be checked by simulating the transistor at a specified operating point. After running a DC operating point simulation in LTspice®, you can go to View -> SPICE Error Log. There you will see the transistor parameters in effect at the chosen operating point. These include V_{be}, AC and DC beta, f_T and many others.

For example, you might check the parameters of a power transistor at $V_{ce} = 20$ V and at collector currents of 0.1, 1.0 and 10 A. This will give a quick indication of how close the SPICE parameters are.

Figure 24.1 shows a simple simulation circuit for verifying the operating parameters of an NPN transistor, in this case the MJL21194 power transistor. The circuit does little more than cause the transistor to operate at a particular collector current with a specified V_{ce}. The simulation includes the option of sweeping the emitter current so that plots can be viewed.

The Hybrid Pi Model

The hybrid pi model was introduced in Chapter 2. It models the small-signal behavior of a transistor. A very simplified version of the model is shown below in Figure 24.2. The fundamental active

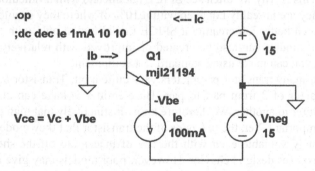

Figure 24.1 A Simple SPICE Simulation Circuit for Model Verification

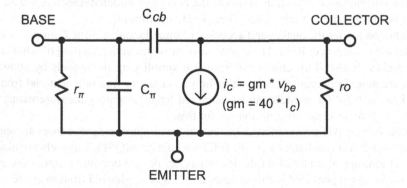

Figure 24.2 Hybrid Pi Model of a BJT

element of the transistor is a voltage-controlled current source; namely a transconductance gm. The remainder of the model comprises passive parasitic components. AC current gain is taken into account by the base-emitter resistance r_π. Early effect is taken into account by r_o. Collector-base capacitance is shown as C_{cb}. Current gain roll-off with frequency (f_T) is modeled by C_π. Because this is a small-signal model, element values will change with the operating point of the transistor. This behavior is what makes SPICE modeling more complex.

24.2 Tweaking SPICE Models

In some cases it may only be necessary to tweak a couple of key SPICE parameters in the model. In this section, we cover the most basic of tweaks, such as those for V_{be} and β as a function of I_c. This will set the stage for what follows later where a complete SPICE model is actually created.

Perhaps the best example is that of getting V_{be} right at a low operating current. This is done by adjustment of the parameter IS (saturation current). The transistor is simulated at a chosen V_{ce} and I_c. V_{be} is then checked and compared to the datasheet value for the same operating point, and the error in millivolts is noted. A factor of 2 increase in the parameter IS will decrease V_{be} by 18 mV; a factor of 10 increase in IS will decrease V_{be} by 60 mV. This information guides the adjustment of IS to eliminate the V_{be} error.

A Typical SPICE Model File

A typical BJT model file is shown below. Throughout this chapter SPICE parameters like TF will be shown in capital letters, while a corresponding transistor parameter, typically at a given operating point, will be shown with a different designation, like T_f. Similarly, the value of C_{jc} (C_{cb}) for a transistor will be governed largely by the SPICE parameter CJC.

.MODEL mjl21194 npn

+	IS-4e-12	BF=65	VAF=500	
+	IKF=14	ISE=1.2e-9	NE=2.0	NF=1.01
+	RB=3.4	RBM=0.1	IRB=1.0	RC=0.06
+	CJE=8e-9	MJE=0.35	VJE=0.5	RE=0.01
+	CJC=1e-9	MJC=0.5	VJC=0.6	FC=0.5
+	TF=21e-9	XTF=90	VTF=10	ITF=100
+	TR=100e-9	BR=5	VAR=100	NR=1.1
+	EG=1.1	XCJC=0.96	XTB=0.1	XTI=1.0
+	NC=4	ISC=0.3e-12	mfg=OnSemi060708	

The model may appear complex, but it is important to understand that it consists of primary and secondary parameters. The primary parameters include IS (V_{be}), BF (current gain), TF (f_T), CJE (base-emitter capacitance) and CJC (base-collector capacitance).

Many of the secondary parameters govern the detailed way in which the primary parameter elements are modulated by the transistor operating point. These secondary parameters are often less important and often can be left at their default values or their values as listed in the manufacturer's SPICE model. For example, MJC and VJC govern the way in which the parameter CJC acts when the voltage across the base collector junction changes. Similarly, IKF, NE and ISE control beta droop at high and low collector current.

Base-Emitter Voltage

Adjustment of V_{be} is very straightforward. We just tweak IS to get V_{be} right at a single operating point. Figure 24.3 shows a plot of V_{be} versus log I_c. We note the approximate relationship

$$I_c = IS(e^{V_{be}/V_t} - 1) \tag{24.1}$$

As mentioned above, V_{be} changes by 60 mV per decade of change in collector current. This also means that V_{be} changes by 18 mV for a doubling of collector current. If simulated V_{be} is high by $60 + 18 = 78$ mV, IS should be increased by a factor of 20. If V_{be} is low by $60 - 18$ mV, IS should be decreased by a factor of 5.

Notice, however, that V_{be} does not follow the logarithmic relationship dictated by Equation 24.1 at high collector currents in Figure 24.3. This is largely due to the presence of base resistance.

Current Gain

A plot of β versus log I_c is shown in Figure 24.4. The parameter BF does not always represent peak beta, but it controls it. BF can be tweaked to get beta correct at a single operating point. The current gain of a transistor can vary over a large range from one device to another, so getting β exact is a foolish pursuit. We will, however, see later that properly modeling the behavior of β as a function of collector current and voltage is important for power transistors.

Speed

The variation of f_T with collector current for a typical power transistor is illustrated in Figure 24.5. The SPICE parameter TF (transit time, τ_f) should be tweaked to get the transistor f_T correct at the collector current for which f_T is maximum. We note the idealized relationship

$$f_T \approx 1/(2\pi * \tau_f) \tag{24.2}$$

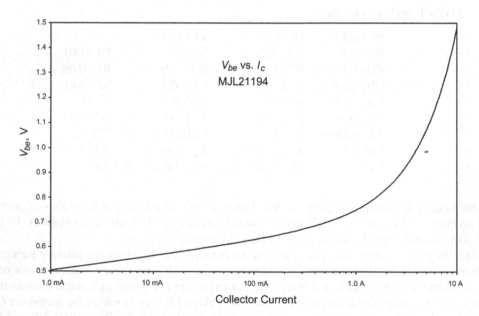

Figure 24.3 V_{be} Versus Log I_c for a Typical Power Transistor

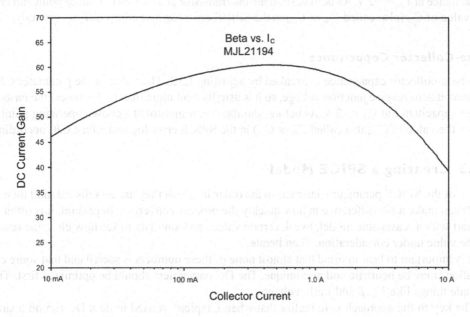

Figure 24.4 Current Gain Beta Versus Collector Current

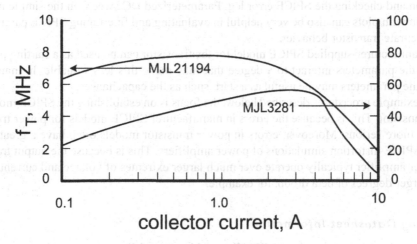

Figure 24.5 Transistor f_T as a Function of Collector Current

As will be discussed later, there are other factors that cause f_T at a given operating point to be smaller than the value given in Equation 24.2. A simple simulation like that in Figure 24.1 can be run at any given DC operating point, and then the SPICE error log can be checked for the value of f_T at that operating point. If maximum f_T is too low, TF should be decreased.

Base-Emitter Capacitance

The base-emitter capacitance is tweaked by adjusting CJE. The value of the parameter CJE is defined at zero reverse junction voltage, so it is usually a bit more than (~ 2 times) the base-emitter capacitance at $V_{be} = -2$ V. As before, simulate the transistor at a chosen operating point and check the value of C_{je} (also called C_{be} or C_{ib}) in the SPICE error log and adjust CJE accordingly.

Base-Collector Capacitance

The base-collector capacitance is tweaked by adjusting CJC. The value of the parameter CJC is defined at zero reverse junction voltage, so it is usually a bit more than (~ 1.3 times) the base-collector capacitance at $V_{cb} = -2$ V. As before, simulate the transistor at a chosen operating point and check the value of C_{jc} (also called C_{cb} or C_{ob}) in the SPICE error log and adjust CJC accordingly.

24.3 Creating a SPICE Model

Many of the SPICE parameters interact, so the order in which they are established (and then iterated) can make a big difference in how quickly the process converges. In general, it is often best to start with a reasonable model, tweak certain values and simulate to see how close the result is to the value under consideration. Then iterate.

It is important to bear in mind that almost none of these numbers is sacred and that some combinations may be heuristic and non-unique. The DC parameters should be optimized first. These include things like V_{be}, β and Early voltage.

The key to the approach is to realize that when LTspice® is used to do a DC run on a simple one-transistor circuit, the operating parameters of the transistor are listed in the resulting SPICE error log. Thus, changes can be made to the model and then quickly evaluated by doing a DC simulation and checking the SPICE error log. Parameterized DC sweeps in the simple transistor simulation with plots can also be very helpful in evaluating and fine-tuning SPICE parameters to obtain accurate transistor behavior.

The manufacturer-supplied SPICE model for the transistor can be used as a starting point, but many of the parameters interact to a degree that may make this less valuable. In many cases, some of the parameters may be useful as a start, such as the capacitances.

In the example procedures described below, the focus is on establishing the SPICE model for a power transistor. This is because the errors in manufacturer SPICE models for power transistors are often more serious. Moreover, errors in power transistor models often have a greater influence on SPICE distortion simulations of power amplifiers. This is because the output transistors in a power amplifier typically operate over much larger extremes of voltage and current, experiencing larger degrees of beta droop, for example.

Gathering Datasheet Information

The first step in creating or revising a SPICE model is to carefully review the datasheet curves for the device and list some key characteristic values as a function of operating point. This data can be taken right off the graphs that are supplied with the datasheet. You can also use the program *Engauge Digitizer* (available at download.cnet.com) to capture data from a .PNG or .JPG file of the datasheet. The data needed is described below.

Record V_{be} versus I_c from a minimum value to a reasonable maximum value of I_c. For a power transistor, this should typically include a minimum of four points at 100 mA, 1 A, 5 A and 10 A.

Record β versus I_c from a minimum to a maximum value of I_c, like those mentioned above. Record these data at both the higher and lower values of V_{ce} for which data are usually supplied (typically 5 V and 10 V).

Record f_T versus I_c. Do this for a minimum to a maximum value of I_c. Record these data at both the higher and lower values of V_{ce}.

Record C_{be} and C_{bc} versus reverse V_{be} and reverse V_{bc} at a low reverse voltage and a medium reverse voltage. For a power transistor this should typically include reverse collector-base junction voltages of 2 V and 10 V. For most transistors the reverse base-emitter breakdown is less than 7 V, often more like 4–6 V. For this reason, it is best to measure C_{be} at 0 V and 2 V.

Measuring Device Data

In some cases, the best models will only result if some lab measurements are done on actual sample devices. This is because the range of current over which the data are shown in datasheets is sometimes not great enough or the precision of the information is not sufficient. This may especially be true for V_{be}-versus-I_c and β-versus-I_c measurements in the range of I_c from 1 mA to 1 A. These measurements can be done without specialized pulsed measurements in many cases, but should be done quickly so that junction temperature does not rise excessively and shift the parameters.

It is especially important to get low-current V_{be} versus I_c data in order to determine the parameter NF. The value of NF will usually be near unity. If supporting data cannot be gathered, NF should be set to unity. Defaulting NF to unity may just make it a bit more difficult to get good model matches.

Figure 24.6 illustrates a Gummel plot for a transistor. It is simply a plot of base current and collector current as a function of V_{be}. The base and collector currents are plotted on a log scale. The distance from the base current to the collector current is the current gain of the transistor at that point.

Figure 24.7 shows a measurement setup that can be used to gather data for the Gummel plot. The transistor is connected with a grounded emitter and a high resistance supplying current to the base. The collector current is measured by measuring the voltage drop across a small collector resistor. The base current is calculated from the applied base voltage source voltage less the measured V_{be}, divided by the base resistance. The measured data can be entered into a spreadsheet.

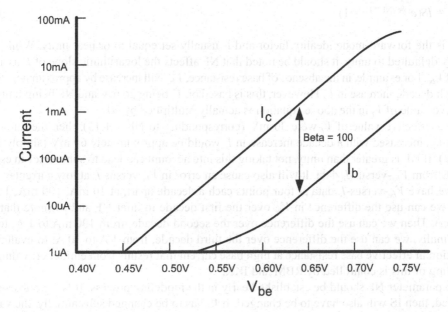

Figure 24.6 Transistor Gummel Plot

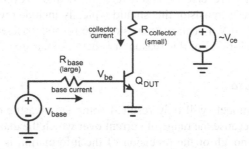

Figure 24.7 A Measurement Setup for Gathering Gummel Plot Data

Saturation Current and Nominal V_{be}

The V_{be} of a transistor is determined by the saturation current I_s. The saturation current parameter *IS* should be determined first, as it controls V_{be} at lower values of I_c. V_{be} at higher current is influenced by *RB* and base current, and thus by β. For this reason, estimation of RB will be done after BF is determined.

Beta is also influenced by V_{ce} via the Early effect. So before we adjust the BF parameter for β, we should make an estimate of the Early effect parameter VAF.

A good starting point for IS can be had by realizing that a value of IS = 200e-12 will result in a V_{be} of about 600 mV at a collector current of 100 mA. Keep in mind that a factor of 10 increase in *IS* will result in V_{be} being reduced by about 60 mV, while a factor of 2 change in IS will change V_{be} by about 18 mV. If, for example, V_{be} is 540 mV at 200 mA, then one might want to try increasing IS by a factor of 10 to account for the 60 mV shortfall from 600 mV, and another factor of 2 to account for the bias point being at 200 mA instead of 100 mA. This would result in an *IS* = 4000e-12 = 4e-9. The equation below governs I_c.

$$I_c = IS(e^{Vbe/(\text{NF} * Vt)} - 1) \tag{24.3}$$

NF is the forward-mode ideality factor and is usually set equal to or near unity. While NF is usually defaulted to unity, it should be noted that NF affects the logarithmic slope of I_c as a function of V_{be}. For example, in the absence of base resistance, V_{be} will increase by approximately 60 mV for each decade increase in I_c. However, this is based on V_t being 26 mV and NF being unity. The effective value of V_t in the above equation is actually multiplied by NF.

If the effective value of V_t were 30 mV (corresponding to NF = 1.15), then the amount by which V_{be} increased with a decade increase in I_c would be approximately 69 mV (simply 1.15 * 60 mV). If NF is greater than unity, not taking this into account can lead to some error in estimating RB from V_{be}-versus-I_c data. It will also cause an error in V_{be} versus I_c at low currents.

If we have V_{be}-versus-I_c data at four points each a decade apart, at 10 mA, 100 mA, 1 A and 10 A, we can use the difference in V_{be} over the first decade to infer NF and IS (note that these interact). Then we can use the difference over the second decade, from 100 mA to 1 A, to infer RB. Finally, we can use the difference over the third decade, from 1 A to 10 A, to evaluate the reduction in effective base resistance at high base current that results from emitter crowding. The modeling of this is controlled by RBM and IKR.

The parameter NF should be established early in the modeling process. If NF is subsequently changed, then IS will also have to be changed. If IS has to be changed substantially, the value of ISE, which determines low-current beta droop, will have to be changed.

In the procedure below the values of IS and NF are determined. If the proper value of NF is not known, it should be set to unity. NF usually falls between 0.9 and 1.1. Setting BF to the transistor's peak beta value and RB to 5 Ω provides a better starting approximation in this procedure.

PROCEDURE 1   (determine IS and NF)

- Measure V_{be} versus I_c at collector currents from 1 mA to 100 mA
- Record the ratio of observed decade changes in V_{be} to 60 mV
- Record octave changes in V_{be} compared to 18 mV
- Set NF as the average ratio to the ideal values of 60 mV and 18 mV
- Typical NF will be between 0.9 and 1.1
- Set initial BF = peak beta datasheet value
- Set RB = 5, RBM = 0.1, IRB = 10
- Set IS so that simulated V_{be} is correct at $I_c = 1$ mA
- Typical IS might be 1e-12 to 1000e-12

Early Voltage

The Early voltage *VA* determines the output resistance of the transistor. In other words, if the base current is held constant and the collector-emitter voltage is increased, by how much will the collector current increase? The collector current will increase because the current gain of the transistor increases slightly as collector-base voltage is increased. The current gain, in turn, increases because the increased collector voltage causes the depletion region of the collector-base junction to enlarge, thinning the base.

Figure 24.8 illustrates a typical transistor output characteristic that would be provided on a datasheet.

Estimating *VA* can be done from the transistor's output curves. It is essentially governed by the way in which β changes with V_{ce}. One way to find *VA* is to determine the slope of the I_c-versus-V_{ce}

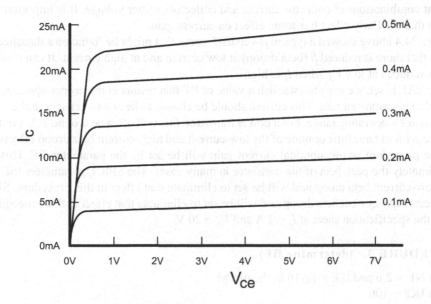

Figure 24.8 Transistor Collector Current Characteristic

curve on the output characteristic at a medium value of I_c. This is output resistance r_o. We then know that $r_o = VA/I_c$, so

$$VA = r_o * I_c \tag{24.4}$$

Of course, the value of I_c that is used for this calculation matters, but it does not change a lot over the range of V_{ce} that is being used because a reasonable Early voltage result is a relatively shallow slope. Once again, it is important to estimate VA first, because the presence of finite VA means that actual operating beta will be higher at the usual 20 V operating point for determining β than at low voltages. The presence of VA can sometimes result in a β that is higher than BF for this reason.

The value of VA for a high-voltage power transistor will usually be greater than its V_{ceo} voltage rating. Bear in mind that when V_{ce} is equal to VA, the beta has doubled with respect to its very low V_{ce} value.

In the procedure below, the Early voltage parameter VAF will be estimated from the slope of the curves in the transistor output characteristic.

PROCEDURE 2   (determine VAF)

- Measure the I_c versus V_{ce} slope at a nominal collector current
- This slope is r_o in ohms
- Set VAF = $r_o * I_c$
- Typical VAF might be twice rated maximum collector voltage

Note that VA can be inferred from $\Delta\beta$ over a range ΔV_{ce} as

$$VAF = \Delta V_{ce} * \beta/\Delta\beta$$

Nominal Beta

The current gain β of a transistor is readily found on the datasheet. Beta will often be given for different combinations of collector current and collector-emitter voltage. It is important to keep in mind that the Early effect has some effect on current gain.

Figure 24.4 above showed a typical β-versus-I_c curve that might be found on a datasheet. It can be seen that there is reduced β (beta droop) at low current and at high current. It can also be seen that β is reduced at low V_{ce} when I_c is high.

Once VAF is set, we wish to establish a value of BF that results in the proper operating beta at a desirable operating current. This current should be chosen to lie at a sweet spot in the middle of the transistor's operating range. For a power transistor, this will often be around 1 A. For this purpose, we wish to have little or none of the low-current and high-current beta droop effects in play.

In the procedure below, nominal current gain will be set by the parameter BF. This will be approximately the peak beta of the transistor in many cases. The SPICE parameters ISE and NE affect low-current beta droop and will be set to eliminate that effect in this procedure. Similarly, IKF affects high-current beta droop and will be set to eliminate that effect in this procedure. Find β from the specification sheet at $I_c = 1$ A and $V_{ce} = 20$ V.

PROCEDURE 3   (determine BF)

- Set NE = 2.0 and ISE = 1e-16 in the model
- Set IKF = 100

- Set BF to the value of β at $I_c = 1$ A and $V_{ce} = 20$ V
- Typical BF might be 50–200
- Verify β with a DC simulation at this operating point
- Adjust BF to obtain correct β

Beta Droop at High and Low Current

The current gain of a transistor is often fairly constant over a middle range of collector current. However, β falls off at very low current and fairly high current. This is often referred to as beta droop. High-current beta droop is especially important in power transistors for amplifier output stages because of the high current they may be called on to handle. The procedure below will facilitate determination of the SPICE parameters that control the modeling of beta droop at low and high currents.

Beta droop at high current is first approximated by finding a value for the parameter IKF. Set IKF to a large value like 100. Check β at 1 A and at 10 A; these two values should be about the same. Now introduce a lower value of IKF of about 10 and observe how it has affected β at 10 A and 1 A. Iterate until a reasonable fit is obtained. Some adjustment of BF may be necessary as well.

Now try to establish beta droop at low currents. Low-current beta droop is controlled by ISE and NE. The parameter NE is the base-emitter leakage emission coefficient. It controls the slope of the I_b line on the Gummel plot in the low-current beta droop region. Keep NE at 2.0 and increase ISE until β at 100 mA begins to fall toward the datasheet value of β at 100 mA. The SPICE default value of NE is 1.5, but a larger value often results in a better model in power transistors.

Figure 24.9 shows a Gummel plot that has been modified to show the effects that contribute to low-current beta droop. The key thing to recognize is that there are two contributors to base current that result in finite transistor current gain. The first component of base current is determined by BF. It creates a base current that is numerically equal to I_c/BF. If this were the only contributor, β would equal BF and there would be no beta droop at low currents. The second component is created by the combination of ISE and NE. These create a current that is of the same exponential form as a base or collector current, but with a different slope that is controlled by NE. This latter component of base current tends to dominate at lower collector currents, causing transistor beta to be reduced.

$$I_b = I_c/BF + ISE(e^{Vbe/(NE * Vt)} - 1) \tag{24.5}$$

while

$$I_c = IS(e^{Vbe/(NF * Vt)} - 1) \tag{24.6}$$

The parameter NE will virtually always be greater than NF. While NF is often close to unity, NE is often 1.5–2.0. This difference is what creates the difference in slope of the two contributors to base current on the Gummel plot. If NE = 1.5, for example, the slope of the ISE contributor will be 90 mV/decade. Because the difference in slope created by NF and NE is what determines low-current beta droop, it will often be the case that a transistor with a larger value of NF will need a correspondingly larger value of NE for a given degree of low-current beta droop behavior.

There is significant interaction between the IKF control of beta droop at high current and the ISE control of beta droop at low currents. While IKF attacks β at 10 A, it also has a depressing

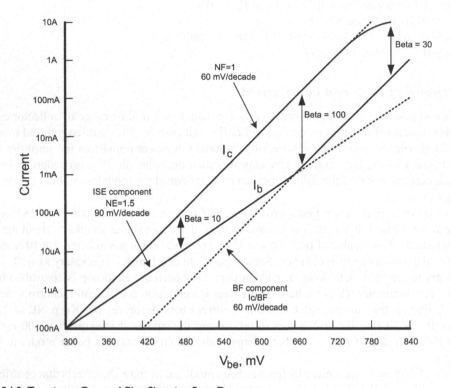

Figure 24.9 Transistor Gummel Plot Showing Beta Droop

effect on peak β at 1 A. Similarly, while ISE attacks β at 0.1 A, it also substantially reduces β at 1 A. In other words, both IKF and ISE reduce β at the nominal point of 1 A.

Indeed, because IKF reduces β with increasing current even beginning at 1 A, it will have the effect of making β at 0.1 A larger (in a relative sense) than that at 1 A before the introduction of ISE. This happens because IKF depresses β more at 1 A than at 0.1 A. This interaction makes the needed effect of ISE to be even more aggressive in attempting to make β lower at 0.1 A than at 1A if low-current beta droop is being experienced at 0.1 A.

Similarly, the effect of ISE lingers so as to tend to reduce the high-current beta droop. As a result, when both IKF and ISE are introduced, BF will have to be increased to move β back toward peak β. If IKF is set first and β from 1 A to 10 A looks good, and then ISE is adjusted to bring β at 0.1 A down to below that at 1 A, then it will be found that not only does BF need to be increased, but IKF may need to be made more aggressive (a smaller value). As mentioned below, if BF becomes more than about twice the value of peak beta, the model may be questionable.

A significant part of the problem is that the influence of ISE is often not very strong as a function of I_c. Thus, achieving a significant effect at 0.1 A results in an undesired amount of effect at 1 A. Also, the effect at even 0.01 A is surprisingly less than one would expect. The Gummel slope NE affects this. A larger value tends to sharpen the effect of ISE as a function of I_c.

With many power transistors, beta droop at 0.1 A compared to 1 A is relatively insignificant, often being only 10%. Indeed, in some cases β may be the same or greater at 0.1 A. For this reason, it may be unwise to try too hard to increase ISE aggressively to obtain the desired beta droop at 0.1 A. Such action may jeopardize the high-current beta droop match, which is much

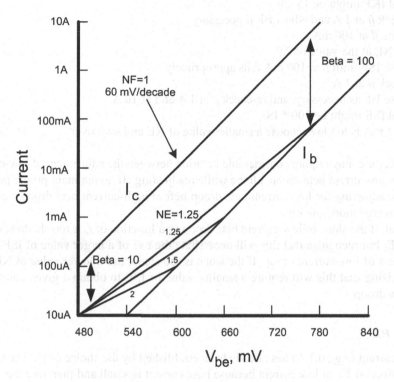

Figure 24.10 Effect of NE on β Versus I_c

more important. A compromise is to introduce just enough ISE to bring beta at 0.1 A down to the value of beta at 1.0 A. A transistor with significant high-current AND low-current beta droop will be more difficult to fit.

Figure 24.10 illustrates the effect of different values of NE on low-current beta droop by showing a Gummel plot with a family of curves having three different values of NE. As seen in the figure, changing ISE moves the more steeply sloped curve up and down, while changing NE adjusts its slope; together, they establish the point of intersection where low-current beta droop begins.

With some model parameter combinations, β at nominal operating current can become hypersensitive to NE. This sensitivity occurs when the value of NE is chosen to be inadequately larger than NF, while seeking some moderate low-current beta droop. This results in the effects of the ISE contribution to base current still being large (and maybe dominant) even at the peak beta point. This can result in a very large value of BF being required to achieve correct peak beta. If BF is greater than twice the value of peak beta, this may be an indication of an unhealthy model.

In the procedure below, the high-current and low-current beta droop parameters will be set. Be aware that this procedure can decrease simulated peak β somewhat, so that some revision of BF may be needed.

PROCEDURE 4 (determine *IKF, NE* and *ISE*)

- Set NE = 2.0 and ISE = 1e-16 in the model
- Evaluate β at 10 A
- Reduce IKF from 100 until β at 10 A is a good fit

- Typical IKF might be 15–20
- Re-check β at 1 A and adjust BF if necessary
- Evaluate β at 100 mA
- Leave NE at the value of 2.0
- Increase ISE until β at 100 mA falls appropriately
- Re-check β at 1 A
- Increase BF as necessary, and re-check β at 1 A and at 10 A
- Typical ISE might be 300 * IS
- If β at 1 mA is too low, choose a smaller value of NE and start over

This procedure may require considerable iteration between the adjustment of low-current beta droop and high-current beta droop, all the while readjusting BF to maintain proper peak beta. In some cases, adjusting for low-current beta droop before high-current beta droop may make the process converge more quickly.

In general, if the slope of low-current beta droop as a function of I_c is too shallow, use a larger value of NE, but recognize that this will necessitate the use of a larger value of ISE to obtain a given amount of low-current droop. If the slope is too steep, use a lower value of NE, closer to 1.5, recognizing that this will require a smaller value of ISE to obtain a given amount of low-current beta droop.

Establish RB

V_{be} at low current (e.g., 0.1 A) has already been established by the choice of IS. The value of RB has little effect on V_{be} at low current because base current is small and therefore the drop across RB is small.

Base resistance has an influence on V_{be} at higher collector currents. With reasonable values of β as a function of I_c, we can now establish base resistance RB. The parameter RB corresponds to the value of base resistance r_{bb} at low and moderate currents. Actual base resistance effectively decreases at high current due to emitter crowding. Another term is introduced to allow for modulation of r_{bb} at high base current. This term is RBM, the minimum value of resistance to which base resistance falls at high base current.

It is important to realize that the excess V_{be} caused by voltage drop across r_{bb} is due to base current and thus is a function of β. That is why it was important to establish β prior to establishing RB. Moreover, β droops at high currents, further increasing excess V_{be}, so it was important to establish high-current beta droop prior to this exercise as well.

RB can be established by choosing a value of RB that makes V_{be} correct at a medium current like 1 A. The influence of base resistance on V_{be} at a collector current of 1 A can be easily illustrated. Assume $\beta = 50$. Base current I_b will then be 20 mA. If RB = 5, then an extra 100 mV will be added to V_{be} by RB.

Bear in mind the relation

$$I_c = IS(e^{Vbe/(NF * Vt)} - 1) \tag{24.7}$$

See the comments above about the influence of non-unity values of forward ideality factor NF on the determination of RB. If NF is wrong, it will cause the change in V_{be} from 0.1 A to 1.0 A to be wrong. This "error" in V_{be} could then be wrongly attributed to RB.

In the procedure below, RB is determined by adjusting it to achieve the correct V_{be} at a medium value of collector current. Setting RBM = 0.1 and IRB = 10 in the model prevents modulation of RB by collector current.

PROCEDURE 5 (determine RB)

- Set IRB in the model to 10
- Set RBM in the model to 0.1
- Measure the value of V_{be} at $I_c = 1$ A
- Adjust RB so that simulated V_{be} is correct at 1 A
- Check simulated V_{be} at 5 A; increase RB if necessary
- Check simulated V_{be} at 10 A; it should be too high
- If not, increase RB to make it so
- Typical RB might be 3–5 Ω

If IS and NF have been chosen correctly, the non-zero value of RB should cause an increase in V_{be} at high current above that which would be expected from the basic equation for V_{be} as a function of I_c. We refer to this ohmic drop across r_{bb} as excess V_{be}.

We should point out that the Gummel plot in Figure 24.6 is an idealized one that assumes no base resistance. With base resistance included, the I_b-versus-V_{be} line would tend to bend down at high currents along with the I_c-versus-V_{be} line.

Establish RB at High Base Current

As mentioned above, the effective value of base resistance decreases at high collector current. The parameter IRB is the base current (not collector current) at which r_{bb} has fallen halfway from RB to RBM. The parameters RBM and IRB may be omitted in simpler models, but may result in excessive V_{be} at high I_c (e.g., 10 A) where a reasonable value of RB has been chosen to properly reflect increased V_{be} at moderate current (e.g., 1 A).

Suppose β has dropped to 33 at 10 A. Base current will now be 300 mA. If $r_{bb} = 5$, then the excess V_{be} voltage drop across r_{bb} will be 1500 mV. If V_{be} was 700 mV at 1 A, then V_{be} will now be 2.2 V. This is way too high. A more typical value would be about 1.6 V. This is why RBM and IRB are needed to model the effective decrease in base resistance at high collector current.

To get the maximum slope of effective base resistance as a function of collector current (it will often be needed), set RBM to a very small value, on the order of 0.1 Ω or less. At the RBM point, excess V_{be} will then be 30 mV if base current at that point is 300 mA. The value of IRB can then be chosen to get V_{be} down to about what it should be at 10 A.

This adjustment of IRB will have some effect on the value of V_{be} at 1 A (V_{be} will have been reduced a bit). Some iteration is then necessary. Often, V_{be} at middle currents like 1 A will be a bit less than the datasheet sheet value due to the limited slope of base resistance afforded by the model. This is a better compromise than allowing V_{be} to become unrealistically high at high currents.

In the procedure below, the model parameters RBM and IRB are determined. We will usually set the minimum base resistance RBM to a very small value to achieve the highest slope of change in r_{bb} at high currents. We will then set the center of the transition point from RB to RBM. This is controlled by the parameter IRB, which is the base current (not collector current) at which r_{bb} has fallen halfway from RB to RBM. IRB will be set to yield the best fit for V_{be} at high collector current.

PROCEDURE 6 (determine RBM and IRB)

- Set RBM = 0.1 Ω
- Set IRB to a high value like 10 A
- Simulate V_{be} at $I_c = 10$ A; V_{be} will be too high
- Reduce IRB until V_{be} at $I_c = 10$ A falls to the correct value

- Simulate V_{be} at $I_c = 1$ A, 2 A, 5 A
- Iterate RB and IRB if necessary for a reasonable compromise fit
- A typical value of IRB might be 1 (1 A base current)

This procedure is carried out with a DC sweep simulation of V_{be} versus $\log I_c$ from 100 mA to 10 A.

In evaluating the effect of excess V_{be} resulting from r_{bb}, it is helpful to bear in mind the following. If you know V_{be} at 100 mA, then ideally with no r_{bb} V_{be} at 1 A will be larger by NF times 60 mV. Similarly, ideal V_{be} at 10 A will be greater than that at 1 A by NF times 60 mV. Were it not for decreasing r_{bb} at high collector currents, excess V_{be} would be far too much at very high currents.

Bear in mind that excess V_{be} at high currents and the choice of IRB depends on the amount of high-current beta droop and the choice of RBM. If the downward slope of change of r_{bb} is too high (as indicated by excess V_{be} not sufficient at high current), increase the value of RBM. Too high a slope in r_{bb} will result in inadequate excess V_{be} at high current.

Establish Nominal Transit Time and f_T

The speed of a transistor is mainly characterized by its f_T, the frequency at which its current gain has fallen to unity. The value of f_T, in turn, is primarily determined by the so-called transit time τ_f of the transistor. In this section the nominal peak f_T of the transistor will be established by setting the transit time τ_f. We start using the approximation that

$$f_T \approx 1/(2\pi * \tau_f) \tag{24.8}$$

at its highest point (if we are lucky). It is useful to note that $\tau_f = 10$ ns $= > f_T = 16$ MHz.

In the procedure below, the model parameter TF is set by finding the peak f_T of the transistor and setting TF so that the simulation yields this f_T at the operating point where f_T is at its peak. The parameter XTF controls f_T droop at high current. It is set to zero for this exercise so that there is no effect from high-current f_T droop. The SPICE parameter CJE will be set to a relatively small value so that low-current f_T droop effects largely disappear at the nominal peak f_T operating conditions.

PROCEDURE 7   (determine TF)

- Find the peak f_T from the datasheet
- Set XTF = 0 in the model
- Set CJE to 1/10 the value given in the model
- Set TF = $1/(2\pi * f_T)$
- Simulate f_T at the peak f_T operating point
- Tweak TF if necessary
- A typical value of TF would be $20e$-9, corresponding to 8 MHz

Establish f_T Droop at High Current

The f_T of a transistor declines at high operating current as a result of the Kirk effect [5]. It further declines at high operating current when V_{ce} is low. The f_T droop at high current (and low voltage) is established by

		Default
XTF	Coefficient for bias dependence of τ_f	0
VTF	Voltage for V_{bc} dependence of τ_f	∞
ITF	High-current dependence of τ_f	0

The f_T droop behavior at high current is modeled in SPICE by modulating the transition time by the variable *ATF* as

$$\tau_f = ATF * TF \tag{24.9}$$

The variable ATF is defined as

$$ATF = 1 + XTF * e^{(Vbc/(1.44*VTF))} * [I_c/(I_c + ITF)]^2 \tag{24.10}$$

It is important to recognize that the value of V_{bc} in the above equation is negative, reflecting the reverse-biased condition of the base-collector junction. This means that the exponent is negative. The value of τ_f is multiplied by the factor ATF, where ATF = 1 + XTF at very high current while V_{bc} is still reasonably high (e.g., 10–20 V).

ITF controls the introduction of the factor XTF as I_c increases relative to ITF. The variable VTF will be discussed later in connection with f_T droop at low collector voltages.

The slope of the transition from nominal f_T to drooped f_T is controlled by the relationship of ITF to the current level at the high-current data point. If there is substantial f_T droop at 10 A and ITF is set to 10 A, then a medium slope of the droop transition results. This all assumes that the value of XTF has been set to achieve the proper level of f_T reduction at the high current data point.

If instead ITF is made about 10 times the value of the current at the high current data point, then nearly the maximum achievable slope between the medium current region and the high current region will be obtained. This assumes that a different value of XTF has been chosen to once again achieve the proper level of f_T reduction at the high-current data point.

Adequacy of the slope can be checked with an evaluation of f_T at 1 A and 10 A. Starting with ITF = 10 A is not unreasonable, but may often result in too shallow a slope in the transition region. A larger value of ITF may be necessary. Bear in mind that the choice of ITF directly affects the value of XTF required to achieve the desired amount of high-current f_T droop.

The f_T droop at high current for power BJTs often has a fairly steep slope. For this reason, it may often be that as much slope as possible is desired. This would lead to choosing a large value of ITF, like 100. In this case, the factor $I_c/(I_c + ITF)$ will be 10/110 = 0.091 at I_c = 10 A, and the squared value will be 0.008, leading to an effective needed multiplier on XTF of 1/0.008 = 125. Of course, if one chooses ITF = 9 times the high-current data point, then the factor becomes 0.1 and the squared value becomes simply 0.01.

In the procedure below, the values of ITF and XTF in the model are determined so that f_T will fall at high collector currents in accordance with the datasheet information. The value of ITF is set to a large value of 100 so as to maximize the slope of beta droop in the high-current region, as this is usually necessary to obtain the best fit to the data. VTF here is set to a nominal value that will usually give a reasonable degree of f_T droop at low collector voltages in the high-current operating region.

PROCEDURE 8   (determine XTF and ITF)

- Set ITF = 100
- Set VTF = 10
- Set XTF = 0
- Simulate f_T at 10 A
- Increase XTF to the point where f_T at 10 A falls to the desired value
- Re-check f_T at 1 A; iterate TF and XTF as necessary
- Also check f_T match at 3 A
- Typical XTF may be 10–100

Establish f_T Droop at Low Voltage and High Current

The droop of f_T at high current tends to become more pronounced at low collector-emitter voltages. This behavior is controlled by the parameter *VTF*. Once the f_T droop is established at $V_{ce} = 10$ V, one can then set the further amount of f_T droop experienced when V_{ce} is small, for example, 5 V. In actuality, it is more proper to refer to base-collector voltage V_{bc} as the controlling variable. Differences between V_{bc} and V_{be} on the specification sheet in the way that f_T droop is displayed should be noted.

In the procedure below, the *VTF* parameter will be set so that f_T at 10 A and 5 V is appropriately less than f_T at 10 V and 10 A. A smaller value of *VTF* causes greater f_T droop at low voltages. A value of *VTF* = 10 will often give satisfactory results. It is important to note that a different value of *VTF* will usually require a change in *XTF* in order to re-establish the proper amount of f_T droop at high current and nominal V_{ce}.

PROCEDURE 9   (determine VTF)

- Set VTF = 10, as used in establishing f_T droop at higher voltages
- Simulate f_T at $V_{ce} = 5$V and $I_c = 10$ A
- See if adequate f_T reduction due to the lower voltage has occurred
- If not, choose a smaller value of VTF
- Adjust XTF for correct f_T droop at 10 V
- Simulate f_T at 5 V and 10 A
- A typical value of VTF is 10

Establish f_T Droop at Low Current

The f_T of a transistor droops at low collector current because the base-emitter junction capacitance remains relatively fixed while both diffusion capacitance and transconductance decrease. This decreases f_T. Thus, low-current f_T droop is set by CJE.

Recall from the hybrid pi model that

$$f_T = gm/(2\pi * C_\pi)$$

or that

$$C_\pi = gm/(2\pi * f_T)$$

However, in reality,

$$C_\pi = C_{je} + gm * \tau_f$$

At high current, C_π is dominated by the product $gm * \tau_f$; this is the diffusion capacitance component of C_p. In the limit, $f_T \sim 1/(2\pi * \tau_f)$. However, at low current the somewhat fixed value of junction capacitance C_{je} comes into play and decreases f_T by limiting how small C_π can go. At very low current, f_T devolves to $f_T = gm/(2\pi * C_{je})$. Because gm is proportional to collector current, one can see that under these conditions, f_T decreases directly as collector current decreases.

Consider a transistor operating at 100 mA with f_T of 8 MHz. Its gm will be 4 S. If diffusion capacitance alone is considered, its C_π will be

$$C_\pi = 4/(2\pi * 8 \text{ MHz}) = 79{,}577 \text{ pF}$$

If, instead, its f_T has actually sagged to 6.5 MHz due the additional presence of the junction capacitance, then $C_\pi = 97{,}941$ pF, an increase of 18,364 pF. This increased amount of capacitance relates approximately to the junction capacitance C_{je}.

The junction capacitance at work here is a bit larger than CJE due to the forward bias of the base-emitter junction. Bear in mind that CJE is the capacitance of the base-emitter junction at zero bias.

The junction capacitance C_{je} is governed by the SPICE parameter CJE. We set CJE in the SPICE model by increasing it from a low value until the f_T at a low current (e.g., 100 mA) has drooped by the appropriate amount.

There is a compromise here. The resulting CJE may conflict with the datasheet value of C_{je} at low reverse base-emitter voltages like –2 V. This is because C_{je} is a function of reverse junction voltage. Adjusting the MJE and VJE may help this situation.

In the procedure below, the model parameters CJE, MJE and VJE will be established to obtain the datasheet amount of f_T droop at low collector current. The most important parameter is CJE, while MJE and VJE can usually be set to typical values or those provided in the manufacturer's SPICE model, if available.

PROCEDURE 10   (determine CJE, MJE and VJE)

- Note datasheet C_{je} at $V_{be} = -2$ V and $V_{be} = -10$ V
- Set CJE = 2 * C_{je} @ –2 V
- Set MJE = 0.5
- Set VJE = 0.5
- Simulate f_T at 1 A and at 100 mA
- See if low-current f_T droop is appropriate
- If f_T droop is too great, decrease CJE
- Tweak TF if peak f_T has fallen too much due to CJE
- Simulate C_{je} at 0 V, 2 V and 10 V, check match

The parameter CJE will often be about double the value of C_{je} at 2 V reverse bias. Moreover, MJE = 0.5 and VJE = 0.5 are reasonable for the base-emitter junction. If the manufacturer's SPICE model is available, use their values for MJE and CJE. Otherwise, use 0.5 for both parameters. The error will not be serious.

For those interested, SPICE computes C_{je} as

$$C_{je} = \text{CJE}[1 - (V_{be}/\text{VJE})]^{-\text{MJE}}$$

where *VJE* is the built-in potential and MJE is the capacitance exponent. Junction capacitances are usually of this form.

Determine Base-Collector Capacitance

In the procedure below, the base-collector capacitance C_{jc} (also called C_{bc} or C_{ob}) is established by setting CJC, MJC and VJC. The parameters MJC and VJC are less important and will usually be set to typical values or those values found in the manufacturer's SPICE model, if available.

PROCEDURE 11   (determine CJC, MJC and VJC)

- Set VJC = 0.6
- Set MJC = 0.5

- Read datasheet value of C_{jc} at –2 V
- Set CJC = 1.3 times C_{jc} @ –2 V
- Check behavior of C_{jc} against datasheet
- Adjust CJC as needed

Check the Model

In the procedure below, the SPICE model that has been created is spot-checked for accuracy. This helps assure that no serious problems have crept into the model as it was being created, either by error or by unanticipated interactions.

PROCEDURE 12

- Check V_{be} at low, medium and high current
- Check β at low, medium and high current
- Check β at low and high voltages
- Check f_T at low, medium and high current at medium V_{ce}
- Check f_T at low, medium and high current at low V_{ce}
- Check C_{je} at low and medium reverse bias
- Check C_{jc} at low and medium reverse bias

All of these items can be checked by viewing the SPICE error log after performing a simple simulation that biases the transistor at the operating point of interest.

BJT Model Example

Below is a typical BJT power transistor model that was created using these procedures. The model is of the MJL21194 NPN transistor that has been used in several of the power amplifier examples in other chapters.

```
.MODEL mjl21194 npn
+   IS=4e-12      BF=65        VAF=500
+   IKF=14        ISE=1.2e-9   NE=2.0      NF=1.01
+   RB=3.4        RBM=0.1      IRB=1.0     RC=0.06
+   CJE=8e-9      MJE=0.35     VJE=0.5     RE=0.01
+   CJC=1e-9      MJC=0.5      VJC=0.6     FC=0.5
+   TF=21e-9      XTF=90       VTF=10      ITF=100
+   TR=100e-9     BR=5         VAR=100     NR=1.1
+   EG=1.1        XCJC=0.96    XTB=0.1     XTI=1.0
+   NC=4          ISC=0.3e-12  mfg=OnSemi060708
```

24.4 JFET Models

JFETs operate on a different principle than BJTs. Think of a bar of N-type doped silicon connected from source to drain. This bar will have a resistance and act like a resistor. Now form a *p-n* junction somewhere along the length of this bar by adding a region with p-type doping. As the P-type gate is reverse-biased, a larger depletion region will be formed and this will begin to pinch off the region of conductivity in the N-type bar. This reduces current flow. This is called a *depletion device*. It is normally on (at zero gate-source voltage) and is turned off by controlling the amount of depletion by applying a negative voltage to its gate (for N-channel devices). The JFET will be nominally *on* and its degree of conductance will decrease as reverse bias on its gate is increased until the channel is completely pinched off.

The amount of reverse gate-source voltage that causes complete pinch-off of the channel is called the pinch-off voltage V_p or more commonly the threshold voltage V_t. Put differently, V_t is the gate-source voltage at which current flow begins as reverse bias is decreased from the threshold voltage value. The threshold voltage is often on the order of 0.5 V to 4 V for most small-signal JFETs.

For small values of V_{ds}, the JFET acts as a resistance whose value is controlled by the degree of channel pinch-off caused by the reverse bias of the gate junction. This is called the *linear* or *triode region*. JFETs are often used as voltage-controlled resistors in this region. As V_{ds} increases, it also tends to pinch off the channel by creating a reverse bias with the gate junction. This opposes increased I_d caused by V_{ds} and causes the JFET to enter into what is called the *saturation region*, where I_d is largely independent of V_{ds}. The demarcation between the linear region of operation and the saturation region occurs at the point where $V_{ds} = V_t$ if $V_{gs} = 0$. The discussion here will focus on JFET behavior in the saturation region.

DC Behavior of JFETs

The JFET *I-V* characteristic (I_d vs. V_{gs}) obeys a square law, rather than the exponential law applicable to BJTs. The simple model below is valid for $V_{ds} > V_t$ and does not take into account the influence of V_{ds} that is responsible for output resistance of the device.

$$I_d = \beta(V_{gs} - V_t)^2 \qquad (24.11)$$

The equation is valid only for positive values of ($V_{gs} - V_t$). The factor β (not to be confused with BJT current gain) governs the transconductance of the device. The variable name K is sometimes used instead of b. When $V_{gs} = V_t$, the $V_{gs} - V_t$ term is zero and no current flows. When $V_{gs} = 0$ V, the term is equal to V_t^2 and maximum current flows.

The maximum current that flows when $V_{gs} = 0$ V and $V_{ds} >> V_t$ is referred to as I_{DSS}, a key JFET parameter usually specified on datasheets. Under these conditions the channel is at the edge of pinch-off and the current is largely self-limiting. The parameter β is the transconductance coefficient and is defined in terms of I_{DSS} and V_t. It has units of amperes per volt squared.

$$\beta = I_{DSS}/V_t^2 \qquad (24.12)$$

In real devices the drain-source voltage also has some influence on the drain current, even in the saturation region. This is not unlike the Early effect in BJT devices. The influence of V_{ds} on I_d is a result of what is called *channel length modulation*, described by the SPICE parameter LAMBDA (l). The relationship below includes the effect of l.

$$I_d = \beta(V_{gs} - V_t)^2 * (1 + \lambda V_{ds}) \qquad (24.13)$$

LAMBDA controls the output resistance of the JFET. Smaller values of l correspond to higher output resistance r_0. LAMBDA is referred to as the *channel-length modulation parameter* and has the units of inverse volts (1/V).

Transconductance for a JFET at a given operating current is smaller than that of a BJT by a factor of about 10 in many cases. Its turn-on characteristic (once the threshold voltage has been overcome) is much less abrupt than that of a BJT.

The JFET SPICE Model

A typical model for a JFET is shown below. As with BJTs, there are primary and secondary model parameters, where the former are fairly fundamental and the latter are less important in many applications. The first line includes the primary parameters that govern DC behavior. BETA governs the transconductance above threshold, VTO is the threshold voltage and LAMBDA controls the output resistance of the device.

> .MODEL J310 NJF
> + BETA = 0.004 VTO = -3.75 LAMBDA = 58E-3
> + CGS=16E-12 CGD=12E-12
> + PB=0.52 M=0.54 FC=0.5 RD = 5 RS= 33 N=1
> + IS=41E-15 KF =5.6E-18 AF = 0.56

The second line includes the primary AC parameters, which are merely gate-source and gate-drain capacitance. The third line includes secondary parameters influencing the capacitance behavior and the drain and source parasitic resistances. The last line includes IS, which characterizes the saturation current of the gate junction, and parameters that govern the $1/f$ noise (flicker noise) of the device. KF is the $1/f$ noise coefficient and is the most important of the two noise parameters. For a given JFET, it essentially sets the frequency where noise begins to increase at low frequencies. AF is the $1/f$ noise exponent, and its default value of 1 is usually satisfactory to use. Noise will typically be up 3 dB in the range of 50–100 Hz. Well below this corner frequency it will rise at about 3 dB per octave.

Creating and Tweaking the JFET Model

JFET behavior is mainly modeled by the three parameters BETA, VTO and LAMBDA, and these are the only ones that will be discussed here. Numbers from manufacturers' models, if available, should generally suffice for the others. Manufacturers' JFET SPICE models are usually decent, but the large range of threshold voltage and I_{DSS} sometimes make creation of a model from measurements of real devices quite useful.

Measure I_{DSS} as the current that flows with $V_{gs} = 0$ V and $V_{ds} = 10$ V. Measure V_T as the reverse bias needed to reduce drain current to 1% of I_{DSS}. This is not necessarily the same as what the manufacturer specifies as V_{gs_off}, which may be specified at extremely low current values. Calculate BETA as $\beta = I_{DSS}/V_t^2$. Set VTO as the measured threshold voltage (VTO is always a negative value, regardless of device polarity). Set LAMBDA = 0.002 as a starting value.

Simulate the family of I-V characteristics for the device and iterate the three parameters to obtain a good match. If the slope of the I_d-versus-V_{ds} curve at $V_{gs} = 0$ V is too shallow, increase LAMBDA. If the spacing of the curves as a function of V_{gs} is too small, increase BETA.

24.5 Vertical Power MOSFET Models

Unlike the JFET, the MOSFET power transistor is an enhancement-mode device; this means it is normally off at $V_{gs} = 0$ and must be turned on by increasing V_{gs} to a voltage greater than the threshold voltage. The MOSFET gate is insulated from the underlying source-drain structure by a thin oxide, forming capacitances to the source and drain nodes.

The DC characteristics of power MOSFETs are usually modeled in essentially the same way as those for small-signal MOSFETs like those used in integrated circuits. An example of the basic DC SPICE Level 1 model for a power MOSFET is shown below.

.MODEL IRFP244 NMOS LEVEL=1
+KP=2.9 VTO=4.2 LAMBDA=0.003
+CGSO=0 CGDO=0 IS=1e-32

The threshold voltage VTO is positive for enhancement mode N-channel devices and negative for P-channel enhancement devices. The parameter KP governs device transconductance. The drain current obeys a square-law relationship.

$$I_d = \tfrac{1}{2}KP(V_{gs} - V_t)^2 \tag{24.14}$$

Notice the similarity to the JFET model. The threshold voltage V_t (VTO) is a positive number, reflecting the enhancement nature of the device. In comparison to the JFET model, the parameter KP serves the same transconductance function as β.

The influence of V_{ds} on I_d is modeled in the MOSFET by the parameter LAMBDA as shown below. LAMBDA controls the output resistance of the MOSFET. The effect controlled by LAMBDA is not unlike the Early effect in BJT devices. Smaller values of l correspond to higher output resistance r_0.

$$I_d = \tfrac{1}{2}KP(V_{gs} - V_t)^2 * (1 + \lambda V_{ds}) \tag{24.15}$$

Establishing the DC Parameters

It is important to recognize that the threshold voltage cannot be accurately estimated just by looking at the datasheet plot of I_d versus V_{gs} where I_d appears to go to zero. This is because of subthreshold conduction that will be discussed later. It causes I_d to be larger than predicted by the square law at V_{gs} near threshold.

The parameters VTO and KP can easily be estimated by checking V_{gs} at two values of drain current that differ by a factor of 4. Recall the square-law relationship.

$$I_d = \tfrac{1}{2}KP(V_{gs} - V_t)^2$$

Consider the datasheet values of V_{gs} at 1 A and 4 A (V_{gs1} and V_{gs2}, respectively). We know that $(V_{gs} - V_t)^2$ at 4 A will be 4 times that at 1 A. Similarly, we know that $(V_{gs} - V_t)$ will be twice that at 4 A as at 1 A. We can use these relationships to yield estimates for VTO and KP.

$$VTO = 2V_{gs1} - V_{gs2}$$

$$KP = 2I_{d1}/(V_{gs1} - VTO)^2$$

These may be useful starting values, but simulation of I_d versus V_{gs} and iteration to refine the fit will be necessary. As long as iteration is necessary, a useful alternative is to estimate VTO as the voltage where I_d for a power MOSFET is down to about 50 mA, then adjust KP as necessary to achieve the right V_{gs} at some medium current like 4 A. VTO and KP will have to be juggled to optimize the fit of I_d versus V_{gs}. The I_d-versus-V_{gs} curve may bow above or below the target curve at values of I_d that are smaller than the I_d value chosen for the initial setting of KP. If V_{gs} for $I_d =$ 2 A is low, increase VTO and revise KP to obtain correct V_{gs} at 4 A.

The drain current of a MOSFET is a mild function of V_{ds}, resulting in output conductance. This is not unlike the Early effect in a BJT. This behavior is modeled by LAMBDA in the MOSFET model. Simulate I_d versus V_{ds} at a gate voltage that produces a medium value of drain current, like 3 A. Adjust LAMBDA so that the difference in I_d from $V_{ds} = 10$ V to $V_{ds} = 50$ V matches that on the device datasheet. In other words, adjust the shallow upward slope of the curve to be correct. LAMBDA will often be in the range of 0.001 to 0.1. Because LAMBDA works by influencing the effective value of KP, the value of KP used in the model may have to be trimmed a bit after LAMBDA is set. If the approximate value of LAMBDA is known beforehand, it should be put in the model before VTO and KP are adjusted.

The recommended procedure for creating a typical Level 1 power MOSFET model is

- Review the I_d versus V_{gs} datasheet curve
- Estimate VTO as V_{gs} at I_d of about 50 mA
- Set KP to a reasonable starting value, like 3.0
- Set LAMBDA to a reasonable starting value, like 0.003
- Simulate I_d versus V_{gs} of the model at the same V_{ds} as used for the datasheet curves
- Adjust KP to get desired V_{gs} at 4 A
- Juggle VTO and KP to get desired curve shape
- If V_{gs} for $I_d = 2A$ is low, increase VTO and revise KP to obtain correct V_{gs} at 4 A
- Simulate I_d versus V_{ds} at 2 A
- Adjust LAMBDA to obtain the same slope as found on the device output curves
- Iterate

Gate-Source Capacitance

The gate-source capacitance C_{gs} for a power MOSFET is fairly constant with changes in V_{gs}. Some engineers have been fooled into thinking that this capacitance increases dramatically by looking at the steep incline that begins at turn-on in the usual gate charge plot provided in power MOSFET datasheets. In reality, this steep incline is caused by the transistor turning on and allowing the Miller effect on the gate-drain capacitance to take effect.

The input capacitance of a power MOSFET is referred to as C_{iss}, and this is usually what is plotted in datasheets. It is the sum of C_{gs} and C_{gd}. The rise in C_{gd} is what is responsible for the rise in C_{iss} at low V_{ds}. C_{gs} is about 1200 pF for an IRFP240. This is a fairly large capacitance, but it is bootstrapped to a much smaller effective value in a source follower output stage.

Gate-Drain Capacitance

The gate-drain capacitance C_{gd} becomes very large at low values of V_{ds}, as measured when $V_{gs} = 0$ V. The nonlinear nature and large maximum values of C_{gd} are a concern for audio amplifier design at high frequencies and high levels. If high slew rate is to be supported at output voltages near the rails, the driver must be able to source and sink the current required to charge and discharge this increased gate-drain capacitance.

The value of gate-drain capacitance for an IRFP240 MOSFET is plotted below in Figure 24.11 as a function of V_{ds} when $V_{gs} = 0$ V. At high V_{ds}, C_{gd} is a modest 50 pF. However, C_{ds} grows rapidly at reduced V_{ds}. At $V_{ds} = 10$ V it has already climbed to about 700 pF. At $V_{ds} = 1$ V, it is a very large 1250 pF. Note that C_{gd} is also referred to as C_{rss}.

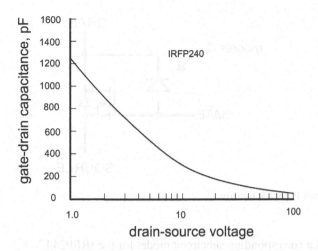

Figure 24.11 Power MOSFET C_{gd} Versus V_{ds} for IRFP240

C_{gd} Test Circuit

The individual inter-electrode capacitances in a MOSFET can be difficult to isolate; when two terminals are shorted for a capacitance measurement, two of the three capacitances are being measured in parallel. It is possible to measure C_{gd} by driving an AC signal into the drain and then measuring the signal voltage that appears from gate to source while a shunting resistance is tied from gate to source. This resistor should have a resistance that is at least 10 times smaller than the reactance of the estimated C_{gs} capacitance at the test frequency. The attenuation of the applied AC signal from drain to gate can then be used to infer C_{gd}.

The Subcircuit Model

The common SPICE model used for power MOSFETs comprises a basic core SPICE MOSFET model that is encapsulated in a subcircuit that includes additional components to model the drain-source body diode and the capacitance effects.

The power MOSFET C_{gd} nonlinearity is difficult to model. Some manufacturers have rather elaborate subcircuit models with numerous diodes and passive elements to model the capacitances. In those cases, it may be best to use the manufacturer's model and just tweak the DC core part of the model to obtain desired DC behavior. However, bear in mind that manufacturer's MOSFET SPICE models are created primarily for correct behavior in switching applications, not linear applications. Some of the subcircuits used to model the gate-drain capacitance can introduce nonlinearities that do not exist in the real device.

Figure 24.12 shows a schematic of a simple subcircuit model for a power MOSFET. This model employs only a diode for simulation of the nonlinear gate-drain capacitance. This is a very oversimplified approach for purposes of illustration only. Clearly, this model does not behave properly if the gate becomes forward-biased relative to the drain. Such a condition can occur in amplifiers that employ boosted supplies for the driver circuitry. MOSFET amplifier designs should avoid this condition because of the very high C_{gd} that results and the consequences for output stage bandwidth shrinkage and increased dynamic driver current requirements.

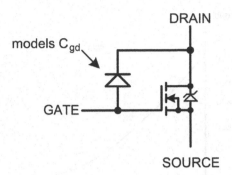

Figure 24.12 Subcircuit Model Schematic

Shown below is a corresponding subcircuit model for the IRFP244.

```
* Drain Gate Source
.SUBCKT IRFP244 1 2 3
M1 10 20 30 MOSFET L=100u W=100u
RD 1 10 0.05
RG 2 20 3
RS 3 30 0.005
CGS 20 30 1000p
DGD 2 1 GDDIODE
DDS 3 1 DSDIODE
*
.MODEL MOSFET NMOS LEVEL=1
+KP=2.9 VTO=4.2 LAMBDA=0.003
+CGSO=10p CGDO=10p IS=1e-32
*
.MODEL DGDIODE D
+IS=1E-32 N=100 RS=10
+CJO=1200p M=0.5 VJ=0.5
*
.MODEL DSDIODE D
+IS=4E-6 N=1.0 RS=0.02
+CJO=1400p M=0.7 VJ=2.5
*
.ENDS
```

The parameters $L = 100m$ and $W = 100m$ are sometimes seen in the first line of the call to the MOSFET model. These parameters define what is called the *W/L ratio*. The appropriate *W/L* ratio when using the Level 1 SPICE model for power MOSFETs is usually unity. These parameters can be left off, as the default is unity. However, these parameters should be explicitly specified when using the EKV model to be discussed later.

Capacitor CGS models the relatively constant gate-source capacitance of the vertical MOS-FET. For this reason, CGSO is set to a fairly small value in the core MOSFET model. Diode DGDIODE models the nonlinear gate-drain capacitance, so CGDO is set to a small value in the core MOSFET model. Diode DSDIODE models the body diode.

Although not shown, bond wire and package inductances can be incorporated into the model for improved modeling of behavior that might influence parasitic oscillations in a real circuit. The inductances for the bond wires can be increased somewhat to account for local board-level trace inductances as well. Typical values for bond wire inductances are 10 nH for gate and source and 5 nH for the drain.

Subthreshold Conduction

The DC behavior of the basic SPICE model is inaccurate at low current for V_{gs} in the vicinity of the threshold voltage and below. This is because weak inversion is not modeled in the Level 1 model. The power MOSFET does not behave as a square-law device in this region of operation. This is a problem for simulation of MOSFET amplifier biasing and crossover distortion.

The simple square-law equation for drain current goes to zero at the threshold voltage; this causes a discontinuity in transconductance. This is simply not accurate for MOSFETs. In fact, at low current, the MOSFET characteristic transitions to an exponential law that is much like that followed by BJTs, but with far different coefficients. Because the simulation of crossover distortion involves behavior of the devices at low current (150 mA is considered to be in the transition region of the model), the normal SPICE models will give misleading results. For example, at the threshold voltage, where the simple model would have $I_d = 0$, I_d for an IRFP240 is between 50 and 100 mA. There are better models for MOSFETs, one of which is called the EKV model. However, EKV model parameters for power MOSFETs are quite rare.

Applicability

The conventional subcircuit power MOSFET model is adequate for many power amplifier simulations with the major exception of crossover distortion. The subthreshold region is encountered as the signal passes through the crossover region in a class AB amplifier. The typical bias point of 150 mA often is in the transition region between the subthreshold and square-law parts of the model, so that the actual bias voltage required to obtain the desired bias current may be off a bit.

Power Amplifier Design Concerns

Datasheet information on MOSFET C_{gd} is incomplete for the operating region relevant to audio. The datasheet usually plots C_{gd} for $V_{gs} = 0$ V. This is a condition that never occurs in audio amplifiers at low values of V_{ds} where C_{gd} becomes large. V_{ds} is small when the output is near the rails and the amplifier is sourcing considerable current, requiring V_{gs} to be quite positive, that is, nowhere near 0 V. The region of greatest concern relevant to audio is near clipping, where V_{ds} is 5 V or less and V_{gs} is > 5 V, *and* where the source is largely still following the signal. This may have important implications for the behavior of C_{gs} and C_{ds} and how they influence effective C_{gd}.

The effect of the increased C_{gd} in source follower amplifier output stages largely manifests itself as a substantially increased need for dynamic gate drive current at output amplitudes near clipping. Output stage bandwidth also decreases as C_{gd} increases, especially if gate stopper resistors are employed. Consider the case where C_{gd} has risen to 1000 pF and a 50 Ω gate stopper resistor is in place. This combination will introduce a pole at 3.2 MHz. Increased C_{gd} is likely much more serious in MOSFET CFP output stages because of the increased influence of Miller effect in that arrangement.

24.6 LTspice® VDMOS Models

The LTspice® VDMOS model was created specifically for vertical double-diffused power MOSFETs. It eliminates the need for the subcircuit approach and incorporates modeling of the source-drain body diode and the nonlinear capacitances right into the model itself. As a result, the model runs faster and more accurately models the nonlinear capacitances. The VDMOS model uses the basic Level 1 SPICE core model for DC behavior of the power MOSFET with two exceptions. Subthreshold behavior is modeled using a parameter named KSUBTHRES. Improved modeling of the transition to the triode region is accomplished with a parameter named MTRIODE. A typical VDMOS model is shown below.

> .MODEL IRFP244 VDMOS NCHAN
> +KP=2.9 VTO=4.2 LAMBDA=0.003
> +CGDMAX=2.3e-9 CGDMIN=6.3e-12 a=0.34
> +CGS=1340p CJO=1300p M=0.68 VJ=2.5
> +RS=0.05 RD=0.1 RDS=1e7 IS=4.0e-6 N=2.4
> +MTRIODE=2.0 KSUBTHRES=197m

The second line defines the DC behavior in the same way as the conventional Level 1 SPICE model. The third line defines the behavior of the nonlinear gate-drain capacitance in a way unique to the VDMOS model. The fourth line defines the gate-source capacitance C_{gs} and the body diode capacitance C_{jo} (a junction capacitance). The fifth line defines the internal resistances and the body diode saturation current. The sixth line defines the triode region and the subthreshold conduction region.

Establishing the VDMOS Capacitance Parameters

The gate-drain capacitance is modeled as a nonlinear function of V_{gd} as illustrated in Figure 24.13 for an N-channel device. For negative values of V_{gd} (e.g., high drain-source voltage), C_{gd} becomes smaller, approaching C_{gdmin} at large values of reverse bias. For positive V_{gd}, C_{gd} increases to C_{gdmax}. The C_{gd} behavior is modeled by the parameters CGDMIN, CGDMAX and a. The parameter a controls how abrupt the change of C_{gd} is as a function of V_{gd}. The default value of a is 1.0 and typical values lie between 0.3 and 1.0. The computation of C_{gd} by CGDMAX, CGDMIN and a is described in the LTspice® User Manual [6].

The behavior of C_{gd} is usually described on a datasheet with a plot of C_{gd} (C_{rss}) as a function of V_{ds}, with V_{gs} = 0. This means that V_{ds} is the same as the reverse bias amount of V_{gd}. The plot often

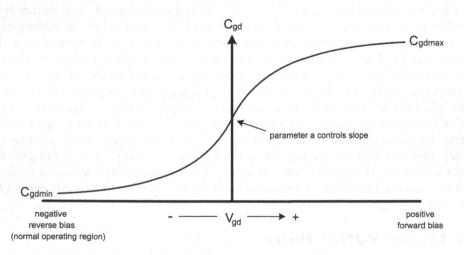

Figure 24.13 C_{gd} *as a Function of* V_{gd} *in the VDMOS Model, for* V_{gd} *Large Negative to Large Positive*

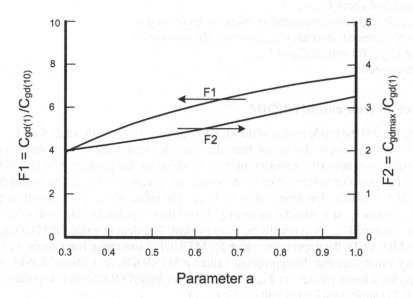

Figure 24.14 $F1 = C_{gd(1)}/C_{gd(10)}$ and $F2 = C_{gdmax}/C_{gd(1)}$ as a Function of a

extends from $V_{ds} = 1$ V to $V_{ds} = V_{dsmax}$, where V_{dsmax} may be 50 V or more. C_{gd} is at its minimum value at V_{dsmax}. This can be seen in Figure 24.11. The VDMOS parameters are chosen to fit this curve.

The parameter C_{gdmin} can be chosen to equal the minimum value of C_{gd} on the plot, at V_{dsmax}, although often it will be chosen to be smaller. The parameter C_{gdmax} is not the same as the maximum value of C_{gd} on the plot at V_{ds} of 1 V or less. It can be seen from Figure 24.12 that C_{gdmax} is a larger value than this. Bear in mind that the parameter a controls the sharpness of the transition from C_{gdmin} to C_{gdmax} as V_{gd} changes from negative values to positive values.

To help establish the VDMOS parameters we define two factors that will be a function of the parameter a. The factor $F1 = C_{gd(1)}/C_{gd(10)}$, where $C_{gd(1)}$ and $C_{gd(10)}$ are the values of C_{gd} at V_{ds} equal to 1 V and 10 V. This factor captures the slope of the transition from the datasheet plot. Similarly, the factor $F2 = C_{gdmax}/C_{gd(1)}$ helps establish the parameter C_{gdmax} from the value of C_{gd} at $V_{ds} = 1$ V.

Figure 24.14 is a plot showing typical values of $F1$ and $F2$ as a function of the parameter a. In this plot the values of C_{gdmin} and C_{gdmax} were held constant. The plot provides insight into the influence of a. It also provides a means of estimating starting values of the parameters when fitting the datasheet plot. Increasing a increases the slope of the transition of C_{gd}, as can be seen by how it affects $F1$. Increasing a while C_{gdmax} is held constant decreases the value of $C_{gd(1)}$. The procedure below should be iterated as needed. The value of C_{gd} for the device model is found from simulation with $V_{gs} = 0$ and various values of V_{ds}. The simulated value of C_{gd} can be found by looking in the SPICE error log.

Below is a procedure for establishing C_{gdmin}, C_{gdmax} and a.

- Find $F1 = C_{gd(1)}/C_{gd(10)}$ from the datasheet
- Set a based on $F1$ and Figure 24.14
- Find $F2 = C_{gdmax}/C_{gd(1)}$ from the datasheet
- Set C_{gdmax} based on $F2$ and Figure 24.14
- Set C_{gdmin} to 1/10 the value of C_{gd} at V_{dsmax}
- Simulate and adjust C_{gdmax} for correct $C_{gd(1)}$

- Simulate and check $C_{gd(10)}$
- If $C_{gd(10)}$ is high, increase transition slope by increasing a
- If a was increased, increase C_{gdmax} to re-establish correct $C_{gd(1)}$
- Adjust C_{gdmin} for correct C_{gd} at V_{dsmax}
- Iterate as needed

The Triode Region and MTRIODE

MTRIODE models the triode region of the MOSFET, where I_d is very dependent on V_{ds}. In power MOSFETs, the transition to this region from the saturation region as V_{ds} decreases is not related to the threshold voltage. The transition point is modeled by the parameter MTRIODE. Here I refer to the V_{ds} transition voltage from triode to saturation region as V_{triode}. The appropriate value for MTRIODE is smaller for larger values of V_{triode}. The value of V_{triode} can be estimated from a graph of I_{ds} versus V_{ds} at a medium value of I_d. If that data is plotted in the form of $d(I_{ds})$ versus log V_{ds}, the value of V_{triode} will usually be more evident. The default value for MTRIODE is 1.0. For power MOSFETs, the appropriate value for MTRIODE can range from about 0.25 to 5.0. It has been my experience that the appropriate value for MTRIODE for vertical MOSFETs is 1.0 or more (implying a lower voltage for V_{triode}), while that for lateral MOSFETs is less than 1, perhaps as low as 0.25 (implying a higher voltage for V_{triode}).

The relationship between MTRIODE and V_{triode} in a VDMOS simulation appears to be dependent on the model and type of MOSFET, so knowing V_{triode} does not help much for establishing MTRIODE. Moreover, measurements or looking at a datasheet can only yield a rather approximate value for V_{triode}. If you plot $d(I_{ds})$ versus log V_{ds} in a VDMOS simulation, you will see output conductance, and it will have a sharp change at V_{triode} from the moderate and changing value in the triode region to the very small and hardly changing value in the saturation region. This is a limitation of the VDMOS model. A reasonably reliable way to establish MTRIODE is to start with a VDMOS model with MTRIODE = 1.0, and which yields the proper I_{ds} at V_{ds} well into the saturation region. At an intermediate value of I_{ds} (perhaps 1–2 A) adjust MTRIODE to achieve the proper value of I_{ds} at a V_{ds} voltage that is well into the triode region. For an IRFP240, MTRIODE is about 2.0 and V_{triode} varies very approximately as the square root of MTRIODE. For a lateral 10N20, MTRIODE is about 0.25, and V_{triode} varies very approximately as 2/MTRIODE.

Subthreshold Conduction and KSUBTHRES

The power MOSFET is not purely a square-law device governed by the usual equation where I_d goes to zero at $V_{gs} = V_t$. The MOSFET has substantial subthreshold conduction in the so-called weak inversion region. Indeed, at very low current, the behavior of the device transitions from a square-law characteristic to an exponential characteristic that is much like that of the BJT. The basic SPICE model does not handle this behavior. It has I_d go to zero at the threshold voltage.

Because the simulation of crossover distortion involves behavior of the devices at low current (150 mA is considered to be in the transition region between the subthreshold and square-law parts of the model), the normal SPICE models will give misleading results. For example, at the threshold voltage, where the simple model would have $I_d = 0$, I_d for an IRFP240 is between 50 and 100 mA.

The VDMOS model includes the parameter KSUBTHRES to take account of subthreshold conduction. It provides a smooth transition in the I_d versus V_{gs} characteristic from the square-law behavior to exponential behavior as I_d passes through the subthreshold region. The KSUBTHRES parameter establishes the behavior of the model in the exponential subthreshold region, and controls the slope of log I_{ds} versus V_{gs}, in mV/decade. Note that this number also applies to the slope of log I_c versus V_{be} in a bipolar transistor, which is on the order of 60 mV/decade. The slope in the subthreshold region for MOSFETs is usually on the order of 120–500 mV/decade. In this region, transconductance is proportional to drain current, and is significantly smaller than gm for a bipolar transistor. Without the KSUBTHRES parameter, the drain current and transconductance will go to zero when V_{gs} equals the threshold voltage, creating a serious discontinuity in the transconductance characteristic.

The appropriate value for KSUBTHRES is almost equal to the subthreshold slope, but usually a bit smaller. If subthreshold slope is 250 mV per decade, the value for KSUBTHRES will be on the order of 197 milli (KSUBTHRES = 197m). When KSUBTHRES is incorporated into a VDMOS model that was created without KSUBTHRES, the value of VT may have to be tweaked to obtain the right value of I_{ds} as a function of V_{ds} in the transition region. If VT is tweaked, KP may also need to be tweaked, since VT affects I_d for a given V_{gs}.

The subthreshold slope for a typical power MOSFET can be obtained by measuring V_{gs} for values of I_d at 100 µA and 1 mA, that is, over one decade of I_d in the subthreshold region. The difference in V_{gs} at those two points is the subthreshold slope. Divide that number by 1.27 and you have a good value for KSUBTHRES. The accuracy of the model in the subthreshold region is not critical; what is most important is that the subthreshold conduction is there, so that there is no discontinuity in gm. A very good tutorial on VDMOS parameter extraction that presents another technique can be found in this reference [7].

Applicability

The VDMOS model is suitable for most amplifier simulations. The VDMOS model is much more convenient to use because it does not require a subcircuit and tends to model the nonlinear capacitances more accurately.

24.7 The EKV Model

The EKV model [8] is a more sophisticated model that accurately accounts for subthreshold behavior in power MOSFETs. The EKV model can be used to replace the DC core model in a subcircuit-based power MOSFET model. It does not contain the built-in body diode and the nonlinear gate-drain capacitance modeling. The EKV model is assigned as SPICE Level 12 in LTspice®. A simple version of an EKV core model for the IRFP240 is shown below.

```
.MODEL MOSFET nmos level=12
+VTO=4.1 PHI=0.7 GAMMA=5.0
+KP=6.0 LAMBDA=100
```

The parameters in the second line establish the behavior in the weak inversion region where drain current is small. These parameters should be adjusted first. The parameters VTO, GAMMA and PHI are adjusted to optimize the fit of I_d versus V_{gs} at low current. VTO is the threshold voltage parameter. Transconductance in the subthreshold region is controlled by drain current via an exponential that is dependent on GAMMA and PHI. I have found that GAMMA can be set to 5.0 for power MOSFETs without compromising the achievable fit to device data. Throughout the remainder of this section GAMMA is assumed to be 5.0.

The parameters KP and LAMBDA in the third line are associated with the DC characteristics in the strong-inversion region. In combination with VT0, they are analogous to the parameters in the conventional Level 1 square-law model. However, it is important to recognize that the value of KP may be quite different in the EKV model compared to its value in the Level 1 model for the same transistor. These parameters should be adjusted next. KP is adjusted to fit the I_d-versus-V_{gs}

curve at moderate current well above the threshold voltage but at current levels that are not significantly limited by device resistances (primarily source resistance). KP is the transconductance parameter and has units of A/V^2.

Shown below is a corresponding subcircuit model for the IRFP240.

```
                    * Drain Gate Source
                    .SUBCKT IRFP240 1 2 3
                    M1 10 20 30 MOSFET L=100u W=500u
                    RD 1 10 0.05
                    RG 2 20 3
                    RS 3 30 0.005
                    CGS 20 30 1000p
                    DGD 2 1 GDDIODE
                    DDS 3 1 DSDIODE
                    *
                    .MODEL MOSFET nmos level=12
                    +VTO=4.1 PHI=0.7 GAMMA=5.0
                    +KP=6.0 LAMBDA=100
                    +CGSO=10p CGDO=10p IS=1e-32
                    *
                    .MODEL DGDIODE D
                    +IS=1E-32 N=100 RS=10
                    +CJO=1200p M=0.5 VJ=0.5
                    *
                    .MODEL DSDIODE D
                    +IS=4E-12 N=1.0 RS=0.02
                    +CJO=1400p M=0.9 VJ=4.4
                    *
                    .ENDS
```

It is very important to recognize that KP in the EKV model can be numerically quite different from KP in the SPICE Level 1 model for the same transistor. KP in the EKV model will generally need to be larger to obtain the same I_d for a transistor with the same W/L ratio. KP in the EKV model will often be about 3–5 times larger than KP in the SPICE Level 1 model for the same transistor. The required value of KP in the EKV model is strongly influenced by PHI.

I have chosen to use the W/L ratio to bring KP in the EKV model into line with KP in the Level 1 model. The values $L = 100\mu$ and $W = 500m$ in the first line above create a W/L ratio of 5. This ratio is by default unity in the Level 1 model and was not shown earlier. Specific values for L and W should always be used in the EKV model for power MOSFETs, with L in the range of 100μ being a good number. Undesired behavior not relevant to power MOSFETs will occur, for example, if instead $L = 1\mu$ and $W = 5m$ are used. The value of W/L required to bring KP into conformance with the Level 1 value of KP is largely dependent on the choice of PHI and the resulting subthreshold slope created. I have found empirically that the desired value of W/L will be very roughly inversely proportional to 40 mS/mA divided by subthreshold slope in mS/mA.

Figure 24.15 shows how the choice of PHI affects subthreshold slope and required W/L ratio to achieve the same I_d in the EKV model as in the Level 1 model with the same value of KP. If subthreshold slope is known, required PHI can be found from the left-hand Y axis. Once PHI is determined, the recommended starting value of W/L can be found from the right-hand Y axis. We note that a subthreshold slope of 10 mS/mA corresponds to V_{gs} changing with I_d at a rate of 240 mV/

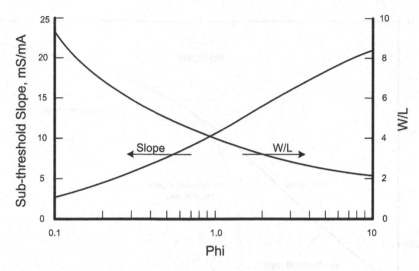

Figure 24.15 Subthreshold Slope and Desired *W/L* as a Function of PHI

decade. This is analogous to the familiar rule for BJTs where V_{be} changes by 60 mV/decade of increase in I_c. The relationships illustrated in Figure 24.15 are approximate, and are valid only for GAMMA = 5.0.

The parameter LAMBDA is responsible for controlling I_d behavior as a function of V_{ds}. Adjust LAMBDA for the best fit to the I_d-versus-V_{ds} data (slope) at medium current like 2 A. As an example, the change in I_d from $V_{ds} = 10$ V to $V_{ds} = 50$ V at a nominal drain current of 2 A is about 100 mA for an IRFP240. This is the change that should be matched by adjusting LAMBDA when the model is simulated. While LAMBDA in the EKV model acts in the same way as LAMBDA in the Level 1 model, the numerical value of LAMBDA required for the same transistor behavior is quite different in the EKV model. While a value of LAMBDA of 0.003 is about right for the IRFP240 in the Level 1 model, the required value of LAMBDA is about 100 for the same device in the EKV model.

Those familiar with the EKV model know that there are many more parameters that have not been mentioned here. I have found that these can be left out of the model and remain at their default values while still allowing a very good model fit for power MOSFETs. In particular, THETA and *UCRIT* are left at their defaults in the model creation procedures here.

Figure 24.16 shows plots of log I_d versus V_{gs} for the conventional SPICE model and the EKV model. Notice the transition region around 100 mA where the EKV model correctly shows much more drain current than the standard SPICE model. The incorrect behavior of the simple square-law model will cause a large discontinuity in *gm* in the region of low current where V_{gs} will be close to V_t.

Subthreshold MOSFET Measurements

Few MOSFET datasheets show subthreshold I_d-versus-V_{gs} data, so the proper creation of an EKV model should include measurements of the subthreshold behavior. There are two pieces of data that are important. The first is the subthreshold slope in mS/mV. Alternatively this can be measured in millivolts per decade and converted to mS/mV by dividing 2400 by the former number. A relationship of 240 mV/decade thus corresponds to 10 mS/mA. The second piece of data is the amplitude of I_{ds} at a value of V_{gs} that is well into the subthreshold region. Put simply, slope

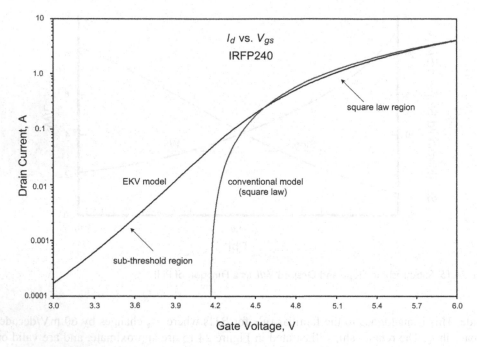

Figure 24.16 MOSFET I_d Versus V_{gs} for Standard and EKV SPICE Models

and amplitude describe the subthreshold region. The necessary data can be obtained for a typical power MOSFET by measuring V_{gs} for values of I_d from 10 μA to 1 mA, that is, over two decades of I_d in the subthreshold region.

Model Creation Procedure

The procedure below summarizes the approach that can be taken to arrive at the parameters for the DC EKV power MOSFET model.

- Review the I_d versus V_{gs} datasheet curve
- Estimate VTO as V_{gs} at I_d of about 50 mA
- Set KP to a reasonable starting value, like 3.0
- Set LAMBDA to a reasonable starting value, like 100
- Measure subthreshold slope (typically 5–20 mS/mA)
- Set PHI based on Figure 24.15 (typically 0.5–3.0)
- Set W/L from Figure 24.15 by setting $L = 100μ$ and W as needed
- W will typically be 200μ to 600μ
- Set KP to Level 1 SPICE model value if available
- Simulate I_d versus V_{gs} of the model at the same V_{ds} as used for the datasheet curves
- Adjust KP to get desired V_{gs} at 4 A
- Juggle VTO and KP to get desired curve shape
- If V_{gs} for $I_d = 2$ A is low, increase VTO and revise KP to obtain correct V_{gs} at 4 A
- Simulate I_d versus V_{ds} at 2 A
- Adjust LAMBDA to obtain the same slope as found on the device output curves
- Iterate

I have found that the required value of LAMBDA in the EKV model (as created here) is much larger for a given target I_d-versus-V_{ds} slope (output conductance) than the corresponding value of LAMBDA for the Level 1 model.

Applicability

The EKV model provides a reasonably accurate simulation of crossover distortion effects because it properly handles the subthreshold and transition region where crossover distortion issues are most prominent. The DC accuracy of the EKV model is somewhat better than that of the VDMOS model, but its modeling of the drain-gate capacitance leaves a lot to be desired.

24.8 Lateral Power MOSFETs

The discussion on power MOSFET modeling thus far has focused on vertical double-diffused power MOSFETs. The lateral power MOSFET has many of the same electrical characteristics and is modeled in the same way, but there are some distinctions worth noting. The lateral power MOSFET acts somewhat more like a conventional small-signal MOSFET that would be found in an IC.

Most importantly, C_{gd} for lateral MOSFETs is much smaller and exhibits relatively little non-linearity. The 2SK1058/2SJ162 complementary pair of lateral MOSFETs typically have C_{gd} on the order of 50 pF or less [9]. Gate-source capacitance is generally between 500 pF and 1000 pF. The lateral MOSFET includes a body diode and its capacitance is typically on the order of 400 pF.

The lateral MOSFET can be modeled with the subcircuit model described in Section 24.5 with a simple fixed capacitor for C_{gd} as shown below.

```
* Drain Gate Source
.SUBCKT 2SK1058 1 2 3
M1 10 20 30 MOSFET
RD 1 10 0.5
RG 2 20 75
RS 3 30 0.5
CGS 20 30 400p
DDS 3 1 DSDIODE
*
.MODEL MOSFET NMOS LEVEL=1
+KP=1.5 VTO=0.5 LAMBDA=0.1
+CGSO=0 CGDO=50E-12 IS=1e-32
*
.MODEL DSDIODE D
+IS=2E-12 N=50 RS=0.05
+CJO=1000E-12 M=0.9 VJ=0.3
*
.ENDS
```

The lateral power MOSFET devices have internal back-to-back Zener diodes from gate to source to limit V_{gs} to safe values. These diodes are not modeled above and usually need not be modeled for amplifier simulations where very large gate-source voltages are not applied. Typical breakdown voltage for these diodes is 15 V. Like any MOSFET, the lateral MOSFET exhibits sub-threshold conduction, and that can be modeled using the KSUBTHRES parameter in the VDMOS model. Values for subthreshold slope in lateral MOSFETs tend to be smaller than those for vertical MOSFETs. A typical value for KSUBTHRES for a lateral MOSFET might be about 85m.

Lateral MOSFETs suffer from a fairly large value of internal gate resistance, on the order of 75 Ω as shown in the example model above. This creates a pole in the AC transconductance of the device. For C_{gs} = 400 pF, this pole is at 5.3 MHz. Unfortunately, it is often necessary to employ gate stopper resistors of 200–500 Ω with lateral MOSFETs for HF stability. This will bring the *gm* pole down to 1.4 MHz or below.

24.9 Installing Models

Place the subcircuit model file in the simulation directory and give it a *.mod* file extension. Take a symbol for the transistor and edit its attributes to include the name of the model file.

- Open the symbol
- Go to *Edit > Attributes > Edit Attributes*
- Enter the transistor name for *Value*
- Enter the name of the model file on the *ModelFile* line

To place a transistor modeled with a subcircuit on the schematic, click on the *Component* tab on the toolbar. A directory path to the library will appear in the top window. Click the down arrow to select another option. Select the simulation directory path. A list of available models in the local simulation directory will appear. Click on the desired device name. The symbol will come up in the large window. Hit *OK* and place the device.

References

1. Adele Sedra and Kenneth Smith, *Microelectronic Circuits*, 6th ed., Oxford University Press, New York, 2010.
2. Gordan W. Roberts and Adele S. Sedra, *SPICE*, 2nd ed., Oxford University Press, New York, 1997.
3. Thomas M. Frederiksen, *Intuitive IC Electronics: A Sophisticated Primer for Engineers and Technicians*, McGraw-Hill, New York, 1982.
4. Paul R. Gray and Robert G. Meyer, *Analysis and Design of Analog Integrated Circuits*, 2nd ed., Wiley, New York, 1984.
5. C. T. Kirk, "A Theory of Transistor Cut-off Frequency, f_T, Falloff at High Current Density," *IEEE Transactions on Electron Devices*, ED-9, March 1964.
6. LTspice® User's Manual. Available at www.analog.com.
7. Ian Hegglun, "VDMOS Parameter Extraction," June 2018. Available at www.paklaunchsite.jimdo.com/spice-models/vdmos.
8. Christian C. Enz and Eric A. Vittoz, *Charge-based MOS Transistor Modeling*, Wiley, New York, 2006.
9. Renesas Tehnology 2SK1058 and 2SJ162 datasheets. Available at www.renesas.com.

Chapter 25

Audio Instrumentation

25. INTRODUCTION

Designing and building an audio amplifier without test instrumentation is like flying blind. For the better part of my many years in audio, test equipment was a limiting factor. It was expensive. I had access to great test equipment by Tektronix and HP at work, but not at home. Often I relied on kits from Heathkit and EICO, among others. Forget about a decent THD analyzer, much less a spectrum analyzer! As a result, I resorted to designing and building my own test equipment. That was a very satisfying endeavor, and even today in some situations it is really the way to go, especially if you need a custom or specialized function.

Fortunately, much has changed with the introduction of the modern PC. Although there is a lot of test equipment that is PC-based that uses a proprietary card that goes into a PC slot, there is more software-based electronic test equipment that is based on the ubiquitous sound card. Measurement capabilities are limited only by the performance of the sound card. That, coupled with the availability of very high-performance sound cards with sample rates as high as 192 kHz and word sizes as large as 24 bits, and you have the makings for a very valuable test bench at reasonable cost. Much of the software is available as shareware or freeware.

The industry evolution from analog test equipment to digital test equipment has put a lot of very good analog test equipment on the surplus block, including HP and Tektronix equipment that was once far out of the reach of the individual designer. Some of the traditional hardware-based instrumentation can be obtained at affordable prices from surplus outlets and on eBay. In some cases the equipment may need repair, but in many cases the service manuals are available from various sources.

In this chapter we will describe audio testing instruments and methodology, both traditional and PC-based.

25.1 Basic Audio Test Instruments

The audio oscillator, the AC voltmeter and the oscilloscope are the most fundamental building blocks for testing and measuring power amplifiers.

Audio Oscillator

Choose an audio oscillator that will go as low as 10 Hz and as high as 1 MHz or more. It should have a constant output with frequency and preferably a decade attenuator to provide output levels over a wide range. The oscillator will be used for stability testing as well as frequency response measurements, thus the need for the high-frequency capability. It should also have a square-wave output unless you also have a function generator. While it is nice to have a low-distortion oscillator, in most cases the low-distortion signal source will come from the THD analyzer if one is available. Good choices include the HP 651, 652 and 654 test oscillators. HP 200 series oscillators are also very good, but do not include decade attenuators. The Tektronix SG502 and SG505 are also excellent choices. Function generators are readily available, and you can get by with one as the sine wave and square-wave test source if resources are very limited. Function generator levels of distortion are not a problem for frequency response and stability measurements.

AC Voltmeter

A good AC voltmeter is a must. It should be very flat over the audio band and should go down to 1 mV full scale. For amplifier bandwidth and stability testing, it should have a bandwidth extending to at least 1 MHz. It need not be *true RMS* responding, but that feature helps provide more accuracy in noise and distortion measurements. The meters in the HP 400 series are widely available on the surplus market and are an excellent choice. One of my favorites is the HP 400EL. It is flat to 10 MHz and has a scale that is linear in dB. The Tektronix DM502 is also a good candidate. The HP 3400A is a very good *true RMS* meter.

Analog Oscilloscope

Used analog oscilloscopes are widely available at low cost. The unit should have a bandwidth of at least 100 MHz to allow the viewing of parasitic oscillations. Tektronix portable oscilloscopes are probably the best choice and most widely available. Good choices include the Tektronix 465, 475 and 485 series.

Digital Storage Oscilloscope

Digital storage oscilloscopes (DSO) generally pass the input signal through an analog-to-digital converter (ADC) and then store it in a fairly deep memory. The digitized signal in the memory is processed by a microcontroller or computer and then displayed. Advances in technology, especially in the area of high-speed ADCs, have made such DSOs available at reasonable prices.

25.2 Dummy Loads

It is very important that amplifiers be properly loaded when they are tested. In fact, many vacuum tube amplifiers can be damaged if they are operated without a load. The load can be as simple as an 8-Ω power resistor for low-power investigations. Metal-cased 50-W power resistors mounted on heat sinks can support high-power measurements. In some cases a fan-cooled heat sink may be necessary.

Choose Load Resistors Wisely

Some wire-wound power resistors are poor performers and will induce distortion into the measurement. Applying a nonlinear load to an amplifier with finite output impedance will cause distortion. This kind of load resistor distortion may be caused by the way in which the wire-wound resistive element is affixed to the solder terminals. This behavior can be evaluated by placing a known-good small resistor in the return leg of the resistor under test. The distortion across the small resistor is then measured. The distortion measured in this way should be no larger than the distortion measured at the output of the amplifier under the same conditions.

Inductive Versus Non-Inductive

It is usually recommended that dummy load resistors be non-inductive. However, this is not absolutely necessary in most cases. The inductance of even fairly large wire-wound power resistors is typically less than 10 µH. For such an 8-Ω load resistor, the inductive reactance reaches 8 Ω at about 100 kHz. If desired, an appropriate Zobel network can be connected across such a resistor to make it essentially non-inductive. Non-inductive load resistors should be used for high-frequency stability testing, but for such testing high power dissipation is not as important; a small array of metal oxide resistors will usually suffice.

A really high-quality non-inductive load resistor can be made from a large array of series-parallel connected 3-W metal oxide film resistors. An array of sixty-four 512-Ω resistors can be configured to provide a 192-Ω load. A second array can be connected in parallel to provide a 4-Ω load that can dissipate 384 W.

Power Dissipation and Cooling

The most straightforward approach to a high-power dummy load is to use several 50-W chassis-mountable 8-Ω power resistors and mount them on a large heat sink. One can also fan-cool the heat sink for the higher-power measurements if it is not sufficiently large. Fan cooling also works well if the dummy load is to be mounted in a ventilated box. A more sophisticated approach would be to monitor the temperature of the heat sink and energize the fan when the temperature exceeds 50 °C.

Connecting to the Dummy Load

Where high currents and low impedances are involved, it is wise to make Kelvin-like connections to the load. This can be done right at the output of the amplifier. This reduces the chance of high current flowing through load connectors introducing distortion. It can also be beneficial to pick off the amplifier output signal differentially to avoid forming ground loops. This makes the cable to the distortion analyzer less vulnerable to pickup of noise and rectifier spikes that may be in the amplifier power cord. It is wise to place a 50-Ω resistor in series with the hot amplifier output terminal to avoid damage to the amplifier in the event of an accidental short circuit in the connection with test equipment.

25.3 Simulated Loudspeaker Loads

More sophisticated amplifier testing can be carried out with simulated loudspeaker loads. This is done in some amplifier reviews to show the influence of load impedance on amplifier frequency response. This reveals behavior due to amplifier output impedance and damping factor. The results can be a real eye-opener and can sometimes explain differences heard among amplifiers that otherwise test the same. The frequency-dependent impedance of a loudspeaker load can also influence amplifier distortion. For small-signal frequency response tests, actual loudspeakers can be used as the load. The disadvantage here is that it is not a standardized load.

Figure 25.1 is a schematic for a simulated loudspeaker load. Even though it is not a real loudspeaker, power levels must be kept to reasonable levels unless the load is constructed from high-power components. The 6.4-Ω resistor represents the DC voice coil resistance while the 240-µF capacitance and 43-mH inductance model the mechanical resonance. The 15-Ω resistor models the losses in the speaker and largely defines the maximum impedance at resonance. The 1-mH inductor corresponds to the voice coil inductance. This model represents a single driver

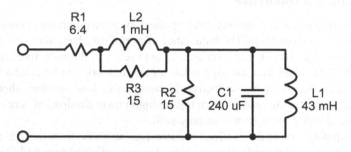

Figure 25.1 A Passive Simulated Loudspeaker Load

in a closed box with a resonance at 50 Hz. More complex models can easily be constructed to represent multi-way loudspeaker systems with crossovers.

If the model of Figure 25.1 is implemented, it is important that the capacitor not be implemented as an electrolytic, but rather as a large Mylar or polypropylene metal film capacitor. A parallel combination of several polypropylene AC motor run capacitors can suffice. The inductor must be rated for high current so that its core does not saturate under the testing power levels. One or more inductors used for high-power woofer crossovers are usually suitable. The inductor used in the load circuit here need not have very low resistance, as some resistance in L2 can be absorbed as a reduction in the value of R1.

For the more adventurous an active loudspeaker load can be built using a second power amplifier to back-drive the amplifier under test (AUT) through a load resistance equal to the DC voice coil resistance of the load to be simulated [1]. A second signal from the signal source is sent to the back-drive amplifier through an appropriate active filter to create at its output a signal representing the back-emf of the simulated loudspeaker. An alternative arrangement is to directly synthesize the desired impedance characteristic with the back-drive amplifier using appropriate feedback around it. The back-drive amplifier must have adequate power handling and SOA capability for this approach to be safe and reliable.

Protection Circuit Testing

Protection circuits like *V-I* limiters are more likely to act in the presence of a reactive load. For this reason, a simulated loudspeaker load can be used to test them. The use of such loads can get the bench testing a step closer to the real world where some amplifier anomalies may show up.

25.4 THD Analyzer

The most common distortion measurement is of total harmonic distortion. In this test a low-distortion sine wave is applied to the amplifier and the output is fed through a deep notch filter that eliminates the fundamental signal. Everything that remains is considered distortion. What remains will also include amplifier noise, so the test is actually referred to as *THD+N*. For detailed information on how THD analyzers work, see this reference [2].

Figure 25.2 shows a block diagram of a THD analyzer. The output signal from the power amplifier is scaled down by an input attenuator and sent through a voltage-controlled notch filter. This filter removes the fundamental without adding significant attenuation at any of the harmonic frequencies. The residual contains the distortion and noise from the amplifier. It also contains a very small amount of the fundamental that has not been removed if the notch is not

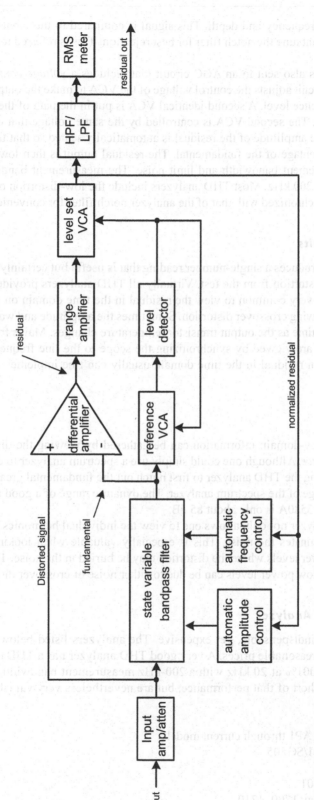

Figure 25.2 Block Diagram of a Typical THD Analyzer

perfectly tuned for frequency and depth. This signal is compared to the input signal to create DC control signals that tune the notch filter for best rejection. This is referred to as the *auto-null function*.

The input signal is also sent to an AGC circuit that includes a *voltage-controlled amplifier* (VCA). The AGC circuit adjusts the control voltage of the VCA to make the output signal amplitude equal to a reference level. A second identical VCA is put in the path of the residual output from the notch filter. The second VCA is controlled by the same voltage that controls the first VCA. As a result, the amplitude of the residual is automatically scaled so that the output is representative of a percentage of the fundamental. The residual output is then low-pass filtered to establish the measurement bandwidth and limit noise. The measurement bandwidth is usually between 80 kHz and 200 kHz. Most THD analyzers include the low-distortion oscillator. Its frequency setting is synchronized with that of the analyzer notch filter for convenience.

Interpreting Results

The THD analyzer produces a single-number reading that is useful but certainly not complete in characterizing the distortion from the test. Virtually all THD analyzers provide the residual as a signal output. It is very common to view the residual in the time domain on an oscilloscope. This is useful for viewing crossover distortion. Sometimes the amplitude and wave-shape can be seen to change with time as the output transistor temperature changes. Mains frequency-related distortion and noise are viewed by synchronizing the scope to the line frequency. The ability to view the distortion residual in the time domain usually can't be implemented by PC-based solutions.

Spectral Analysis

Much more frequency-domain information can be gathered by viewing the distortion residual on a *spectrum analyzer*. Although one could simply use a spectrum analyzer to directly view the amplifier output, using the THD analyzer to first notch out the fundamental greatly improves the useable dynamic range of the spectrum analyzer. The dynamic range of a good analog spectrum analyzer like the HP 3580A is only about 85 dB.

The spectrum analyzer not only allows one to view the individual harmonics in the spectrum, but also largely eliminates the noise. This is especially valuable when looking for crossover distortion at low power levels where the distortion may be buried in the noise. The typical rising curve of THD+N at low power levels can be due to either noise or crossover distortion.

Obtaining a THD Analyzer

THD analyzers are indispensable, yet expensive. The analyzers listed below can sometimes be obtained used at reasonable prices. A very good THD analyzer has a THD+N *measurement floor* of less than 0.001% at 20 kHz with a 200-kHz measurement bandwidth. Some of these analyzers fall a bit short of that performance, but are nevertheless very valuable for amplifier evaluation.

* Audio Precision AP1 through current models
* Tektronix AA501/SG505
* HP 334A, 339A
* Amber 5500, 3501
* Sound Technology 1700, 1710

25.5 PC-Based Instruments

Good audio test equipment used to be expensive and often out of the reach of the average enthusiast. The PC has largely changed that, with the combination of high-performance sound cards and appropriate software. While PC-based instrumentation is not a complete substitute for traditional laboratory test equipment, it can provide many very important functions with surprisingly good performance and sophisticated functionality.

PC-based solutions include digital storage oscilloscopes (DSOs), signal generators, AC voltmeters and spectrum analyzers. Some include THD analyzer functions. Much of the software is freeware or shareware. The level of performance that can be obtained depends mainly on the performance of the sound card. Useful functionality can even be had from the computer's standard onboard sound card.

Sound Card Software

Given a good sound card to get the input and output signals from the analog domain to the digital domain, the computing power of the PC can be used to carry out sophisticated analysis. The PC also provides a very user-friendly virtual front panel.

Most of the available software packages support the basic capabilities needed, but ease of use and quality of documentation vary widely. Many packages support sound card sampling rates up to 192 kHz and data widths of 24 bits (192/24), but some may not. Compatibility with the many different sound cards available also varies. Listed below is a sample of the numerous test/measurement software packages available. ASIO driver compatibility is important.

- ARTA—www.fesb.hr/~mateljan/arta
- Virtins Multi-Instrument 3.1—www.virtins.com
- Visual Analyzer—www.sillanumsoft.org
- RMAA: RightMark Audio Analyzer—http://audio.rightmark.org
- SpectraPlus—www.spectraplus.com

Sound Cards

Sound cards of very high quality can be had for roughly $100–$300. I recommend choosing one with 192/24 ADC to maximize useable frequency range and dynamic range. Note that some sound cards advertised as 192 kHz may only be 192 kHz on the output and only 96 kHz on the input. Having a 192-kHz sampling rate on the input is what is important for measurement applications. The input bandwidth should be 85 kHz or more. The majority of available sound cards are only capable of 96-kHz input sampling. If available, sound cards with balanced inputs and outputs are desirable. Excellent sound cards are available in both internal and external USB formats. Below is a brief list of some sound cards to consider. Some may no longer be available. Some good sound cards can often be found on eBay at low cost. In any case, carefully check manufacturer's specifications to see if it meets your needs.

- Focusrite Scarlett 2i2 2nd Generation (USB)—us.focusrite.com
- Steinberg UR22 mk II (USB)—www.steinberg.net
- Creative Sound Blaster ZxR (PCIe)—www.creative.com
- Asus Xonar Essence STX II (PCIe)—www.asus.com
- ESI Juli@ (PCI)—www.esi-audio.com
- E-MU Tracker Pre (USB)—www.creative.com

- E-MU 1616M (PCIe)—www.creative.com
- Sound Blaster X-Fi (USB)—www.creative.com

A key limitation for use of sound cards in audio instrumentation is how well they perform in measuring THD-20, since even its fifth harmonic extends to 100 kHz. A 192-kHz sound card can only "see" up to about 90 kHz because of the Nyquist sampling theorem. A 96-kHz card can only see up to about 45 kHz, corresponding to only the second harmonic of a 20-kHz fundamental. ASIO (Audio Stream Input/Output) driver compatibility is important. Use of sound cards for audio measurement was covered extensively by Stuart Yaniger in *AudioXpress* magazine [3].

PC-Based Audio Analyzer

As an attractive all-in-one alternative to a sound card and PC software, the nifty and convenient QuantAsylum QA401 provides excellent performance and functionality for its price [4]. Think of the QA401 as a proprietary high-performance stereo USB sound card mated to proprietary audio analysis software all optimized for audio measurement applications in a convenient turn-key system that avoids the hassles of selecting and mating sound cards and software from different manufacturers. It is in a compact box and is powered by the USB cable by which it connects to the PC.

The 24-bit/192-kHz hardware enables its performance to meet or exceed that of the best sound cards, while the software features compare well to software offerings like ARTA. The QA401 includes balanced I/O and a selectable 20-dB input attenuator to make it suitable for measuring power amplifiers. Robust input protection circuits make the unit less likely to be damaged by large input signal voltages. Visit the QuantAsylum website to learn more about this unit's performance and measurement capabilities [4].

Supported measurements include:

- THD and THD+N
- Spectrum analysis to over 75 kHz
- Frequency response
- RMS voltage
- Noise and SNR
- SMPTE IM using 60 Hz and 7000 Hz
- CCIF IM using 19 kHz and 20 kHz

Bandwidth of the unit is about 80 kHz, which is typical of the measurement bandwidth often used for THD measurements. This allows measurement of THD up to 20 kHz with the prominent second and third harmonics included. Its THD+N measurement floor is on the order of −120 dB at 1 kHz.

PC-Based Oscilloscopes

Although several of the sound-card-based PC instruments above support an oscilloscope function, the bandwidth is strictly limited to the audio band (or just slightly beyond). There are PC-based digital storage oscilloscopes (DSO) designed as instruments employing special-purpose hardware that supports sample rates at 10 MHz to beyond 100 MHz. These solutions provide the higher bandwidth necessary for general-purpose oscilloscope use. They are also much more flexible in regard to input signal levels they can handle and often include probes.

The wideband DSOs connect to the PC via USB. Many of these instruments have data paths that are only 8- or 10-bits wide, but that is entirely satisfactory for the oscilloscope function.

The software that comes with these devices is often capable of many other functions, such as FFT spectrum analysis. These functions may be useful for audio analysis if the dynamic range is extended by use of the Distortion Magnifier or fundamental notch filters described below. Some examples of PC-based DSOs in the $250 to $400 range are listed below.

- PicoScope 2206B 50 MHz, 8-bit $349—www.picotech.com
- Virtins VT DSO 2810 40 MHz, 2-channel (USB) $260—www.virtins.com
- QuantAsylum QA101B 70 MHz, 10-bit $379—www.quantasylum.com

25.6 Purpose-Built Test Gear

Off-the-shelf test equipment often does not exist for the specialized needs of audio measurements. To fill this gap, many purpose-built *gadgets* can be very helpful. Many can be constructed with little difficulty. The dummy loads described above fall into this category, but many other functions are important. A very good example is an interface box for the sound card in a PC-based test setup.

Sound Card Interface Boxes

Sound cards are not designed for use as test instruments. They are designed for line-level sound processing. As such, it is important for both the safety and utility of the card to have an interface device between it and the equipment to be measured. Figure 25.3 is a block diagram of such an interface box.

The most important function is input attenuation and protection. The output level of power amplifiers can be well above the maximum input that can be handled by the sound card and can easily damage the device. For this reason a stepped attenuator followed by protection diodes should be implemented in the input interface. It is also desirable to implement a preamplifier in this part of the interface so that very small signals can be analyzed without sacrificing available dynamic range of the sound card. The combination of attenuation and preamplification can enable the use of the sound card near its "sweet spot" input level. Differential inputs are another example of desirable functionality on this side of the interface. The 1200 series of balanced input line receivers made by THAT Corporation are a convenient high-performance solution [5]. The box might also incorporate A-weighting input filters to support SNR measurements if that function is not available in the software.

The output of the sound card should be buffered and amplified or attenuated. The buffering and amplification protects the sound card outputs and makes available signal levels up to 10 V RMS. The attenuator allows signals in the millivolt range to be produced without sacrificing sound card dynamic range. A solid-state floating differential output stage would be convenient

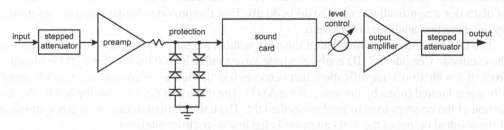

Figure 25.3 A Sound Card Interface Box

for driving differential amplifier inputs and for breaking ground loops. The model 1646 balanced driver IC made by THAT Corporation is a convenient high-performance solution [6].

The interface box might also take advantage of the two-channel capability of stereo sound cards, so that two-tone test signals can be generated, such as CCIF IM using 19 kHz and 20 kHz and SMPTE IM using 60 Hz and 7000 Hz. For this application, the box should include a very high-quality output summing operational amplifier. In such an arrangement the IM distortion of the composite test signal will not be dependent on the sound card electronics.

All of these functions can be easily implemented at low cost with high-quality operational amplifiers. The TI/National LM4562 dual operational amplifier is an excellent choice [7].

Millett Sound Card Interface

Pete Millett has developed a Sound Card Interface/AC RMS Voltmeter [8]. Pete sells the PCB and front/rear panels for about $50 on his eBay store. The sound card interface can accept a wide range of inputs with a four-position decade attenuator and input protection circuits. The Interface also includes output protection that limits the output to the sound card to about 2 V RMS. The unit also includes a digital true-RMS meter so that the input level can be conveniently and accurately read. More details can be found on Millett's web page [8].

Autoranger

It is usually important to scale the input voltage to a sound card, especially in the case of a power amplifier where the voltages can be large and could damage the sound card. Scaling can also put the amplitude of the test signal into the "sweet spot" of the sound card where its dynamic range is optimized. Jan Didden has developed an Autoranger for this purpose [9]. It will automatically scale the output level to about 1 V RMS, often in the optimum range for sound cards (the output level typically lies between about 0.9 and 1.4 V as the autoranging takes place when the input level is changing). The unit is conveniently powered by a USB power source. It includes single-ended and balanced inputs and outputs, digital readout indications of input voltage, output voltage and the gain that it is providing and other helpful features.

The Distortion Magnifier

The Distortion Magnifier (DM) [10] increases the sensitivity of a THD analyzer or spectrum analyzer by subtracting most of the original test signal from the output of the *amplifier under test* (AUT) before it is applied to the input of the analyzer. If 90% of the test signal is subtracted, the distortion percentage in the resulting signal will be magnified by a factor of 10.

A block diagram of the DM is shown in Figure 25.4. The DM subtracts a version of the input signal from a scaled version of the signal from the AUT to form a deep (> 60 dB) null at the fundamental. A controlled amount of the source signal is then added back to achieve a known amount of distortion magnification, either 20 dB or 40 dB. This also provides fundamental energy needed by the distortion analyzer to lock onto.

The DM magnifies the distortion of the AUT without magnifying the distortion (and noise) of the oscillator. Consider a THD analyzer whose lowest range is 0.003% FS. If the DM is placed in front of it with 40-dB magnification, that becomes 0.00003% FS. Of course, this result is generally noise limited (often by the noise of the AUT). The distortion floor is also limited by the distortion of the op amps used to implement the DM. The noise limitation can be largely eliminated if the residual output of the THD analyzer is fed into a spectrum analyzer.

The DM includes coarse and fine amplitude and phase adjustments to optimize the null. The phase adjustment is simply a first-order LPF that constitutes a very simplified model of the AUT.

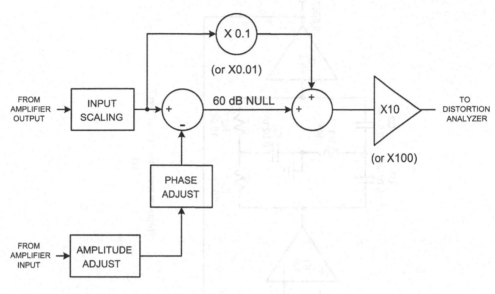

Figure 25.4 The Distortion Magnifier

Amplitude and phase must be tweaked a bit if the fundamental frequency is changed. Because of the simple model, there will be some amplitude and phase error at the higher harmonic frequencies that will detract a bit from the cancellation process, allowing a bit of oscillator noise and distortion to come through. A more complex model of the AUT high-frequency response can be used, but I have not found that necessary for my use. Similarly, the DM can incorporate low-frequency phase matching for low-frequency THD measurements on AC-coupled power amplifiers, but that has not been implemented. More detail on the Distortion Magnifier can be found in Reference 10.

The DM is also useful in extending the dynamic range of analog spectrum analyzers and PC instruments using sound cards. With the DM set to 40 dB of magnification, the HP 3580A, with its 85-dB dynamic range, can resolve spectral components down to about −125 dB.

Twin-T Notch Filters

A sharp notch filter can be used to attenuate the fundamental to measure distortion components. It can also be used to increase the sensitivity of a THD analyzer or spectrum analyzer by attenuating most of the original test signal from the output of the *amplifier under test* (AUT) before it is applied to the input of the analyzer. If 10% of the test signal is allowed through by attenuated bypass of the notch filter, the distortion percentage in the resulting signal will be magnified by a factor of 10.

Such filters can usually be made so that they introduce very little distortion of their own (especially passive notch filters). The twin-T filter, shown in Figure 25.5a, is the most widely used in this application. It can be implemented entirely as a passive circuit that creates a very deep notch at the fundamental frequency [11, 12, 13]. However, it does introduce loss at the second and third harmonic frequencies of about 9.4 dB and 5 dB, respectively [11]. Its center frequency is simply $f_{notch} = 1/(2\pi RC)$. An active version of the twin-T, shown in Figure 25.5b, includes two op amps that introduce positive feedback that greatly reduces the attenuation at the harmonic frequencies at the expense of some reduction in the depth of the notch. K1 buffers the output of the twin-T

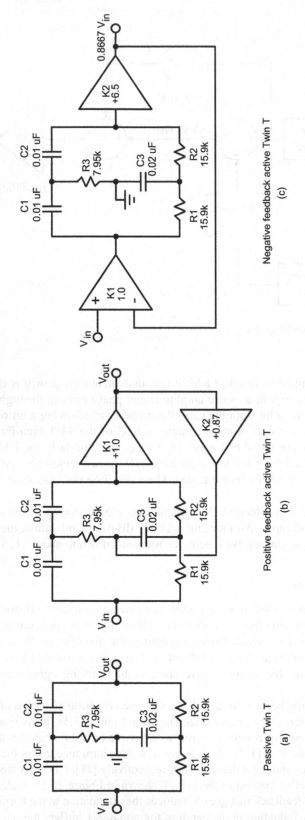

Figure 25.5 Passive and Active 1 kHz Twin-T Filters

network while K2 provides a slightly attenuated positive feedback signal to the twin-T network with low impedance. The value of K1 sets the Q of the filter. As shown, K2 = 0.87 allows 0.5 dB of attenuation at the second harmonic frequency.

Figure 25.5c shows an active twin-T filter that uses negative feedback to reduce attenuation at the harmonic frequencies. K2 provides the loop gain, while K1 implements the feedback summing node. As shown, K2 = 6.5 sets the Q so that 0.5 dB of attenuation is allowed at the second harmonic. As shown, gain of this filter is slightly less than unity at 0.8667 because the NFB loop gain is only 6.5. The positive and negative feedback twin-T filters provide equivalent notch depth, Q and tolerance to imperfect component matching. The op amps used for K1 and K2 should have high unity-gain bandwidth to maintain good filter performance at higher frequencies like 20 kHz. The 50-MHz GBW LM4562 op amp is a good choice here.

The twin-T filter must be implemented with precision components to maximize the depth of the notch. Moreover, because of this and the fact that there are six tuning elements, it is difficult to implement continuous tuning with the twin-T. Small trimming of the shunt resistor can be used to fine-tune the notch depth if the other components are matched to 1% or better. An un-trimmed passive twin-T network with one component off by 1% will be limited to a notch depth of about 64 dB. If one component is off by only 0.1%, notch depth will be limited to about 84 dB. If C1 is 0.1% high, reducing R3 by 4 Ω will restore the full notch depth of over 120 dB.

Bainter Notch Filter

Another active notch filter is the Bainter filter, shown in Figure 25.6 [14, 15]. Requiring only two resistors and two capacitors for tuning, it can provide a very deep, high-Q notch without resort to precision-matched resistors and capacitors. It can be built with just three op amps without resort to op amp stages that have fundamental signal on their positive input, reducing common-mode distortion introduction. The notch frequency $f_{notch} = 1/(2\pi R_1 R_2 C_1 C_2)$ can be tuned with either R1 or R2. R3 should equal R1 to obtain a symmetrical filter characteristic. R4 independently determines filter Q, and maximum Q is obtained if R4 is infinity. In this case, the 3-dB points are at 952 and 1052 Hz, notch depth is 82 dB and loss at 2 kHz is 0.02 dB. If R4 is set to 40 kΩ, notch depth is 94 dB and loss at 2 kHz is 0.5 dB. A 50-MHz GBW op amp (like the LM4562) is

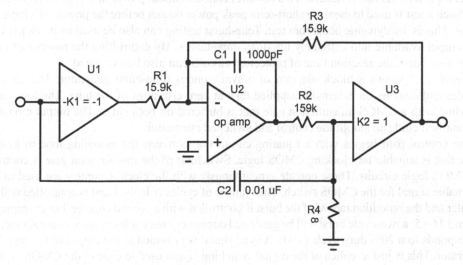

Figure 25.6 1-kHz Bainter Notch Filter

recommended for this application, especially for higher-notch frequencies like 20 kHz. A small resistor placed in series with C1 can mitigate op amp GBW limitations.

Balanced Interfaces

Balanced input and output interfaces make low-noise, low-distortion measurements easier and help break ground loops among instrumentation and devices under test. Even if the target signals are single-ended, as often is the case with power amplifiers, the use of balanced interfaces can help with sensitive measurements. Test signals from the AUT are fed to a differential amplifier [5]. Similarly, balanced versions of the test signals can be delivered to the AUT by implementing differential output buffers. The simplest balanced output buffer merely provides two polarities of the test signal, each referenced to ground. This does not help break ground loops. A more sophisticated arrangement implements an active floating differential output buffer—the electronic equivalent of an output transformer (mentioned above in connection with the sound card interface box) [6].

IM Test Signal Combiner

This device is simply a very low-distortion mixer with two inputs. It is used to generate the test signals for the SMPTE IM (60 + 7000 Hz, 4:1) CCIF IM (19 + 20 kHz, 1:1) and DIM (3.18 kHz square wave & 15-kHz sine wave, 4:1) distortion tests. The outputs of two oscillators are combined in the mixer by a very low-distortion op amp, so that IM products in the resulting test signal depend only on the low-distortion performance of the mixer and not the individual generators. The mixing should be done by applying both signals to the inverting virtual ground input of an op amp so that there is no common-mode signal at the input to cause distortion. If the combiner is to be used for DIM tests, it should include selectable first-order 30-kHz and 100-kHz low-pass filters in the output path.

Synchronous Tone-Burst Generator

Tone-burst testing can be valuable for power amplifiers when performance at very high power levels for brief intervals is needed. This obviates concerns about power dissipation and overheating. Such a test is used to measure short-term peak power output before the power rails have time to sag. This is the dynamic headroom test. Tone-burst testing can also be used to check peak current output available into extremely low-load impedances. By controlling the number of cycles in the tone burst, the reaction time of protection circuits can also be evaluated.

Figure 25.7 shows a block diagram of a synchronous tone-burst generator. The generator enables and disables an externally supplied tone at zero crossings of the tone. The tone can be switched with a CMOS transmission gate that is buffered on both sides. The output circuit can optionally include an amplitude control and switched attenuator.

The control path begins with a squaring circuit that converts the incoming tone to a square wave that is suitable for clocking CMOS logic. Switching of the transmission gate is controlled by CMOS logic circuits. These operate synchronously with the clock. Counters are used to form the enable signal for the CMOS switch. The number of cycles N in the burst is controlled with one counter and the repetition rate M of the burst is controlled with a second counter. For example, if N = 2 and M = 5, a two-cycle burst will be generated once every ten cycles of the incoming tone. This corresponds to a 20% duty cycle ($1/M$). A sync signal is provided at the output of the tone-burst generator. This is just a replica of the digital switching signal used to control the CMOS switch.

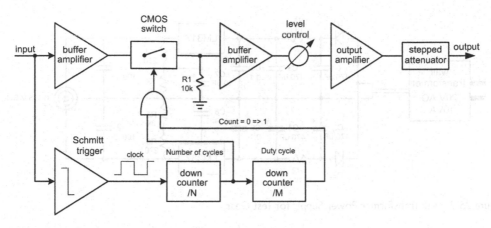

Figure 25.7 A Synchronous Tone-Burst Generator

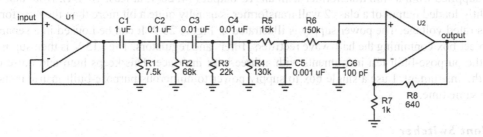

Figure 25.8 A-Weighting Filter

Signal-to-Noise Measurement Preamp with a Weighting

This device enables accurate SNR measurements to be made if another piece of test equipment with this capability is not available. It is a high-gain preamp with controlled bandwidth and optional A-weighted frequency shaping. The unit should have a gain of 100 or 1000 in order to provide a sensitivity of 10 μV full scale when used with an external AC voltmeter (preferably true RMS, like the HP 3400A). The preamp should incorporate input protection. A power amplifier with a very good input-referred noise level of 5 nV/√Hz and a voltage gain of 20 will produce about 14 μV RMS at its output when the measurement bandwidth is limited to 20 kHz. A first-order LPF at 12.7 kHz has a noise bandwidth of 20 kHz. A third-order Butterworth LPF at 19.1 kHz also has a noise bandwidth of 20 kHz. The amplifier will produce a smaller noise reading if the measurement is A-weighted. The equivalent noise bandwidth of the A-weighting function is 13.5 kHz.

Figure 25.8 is a schematic of an A-weighting filter that can be used for SNR measurements. The filter is an entirely passive R-C design that just needs to be buffered at its input and output and which needs a gain of 1.64 in one of the buffers to make up for loss and provide the +1.2 dB of gain at the 2.5-kHz pass-band peak.

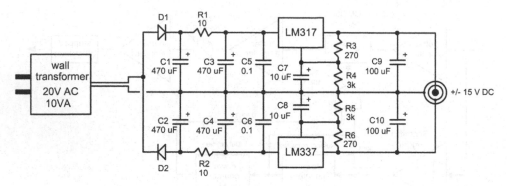

Figure 25.9 Wall Transformer Power Supply for Test Gear

Powering Purpose-Built Test Equipment

Most purpose-built audio test equipment requires very little power at typically ±15 V. Power can be supplied from a wall transformer with an AC output (14–20 V, at least 10 VA). Note that the lightly loaded output of a class 2 wall transformer is usually quite a bit more than the transformer's rated voltage. The power supply is illustrated in Figure 25.9. It can be housed in a separate project box containing the half-wave rectifiers, filters and regulators. The ±15 V is then supplied to the purpose-built test instruments with a three-wire interface. This keeps hum and noise out of the instrument. I use a single box to supply power to numerous purpose-built instruments at the same time.

Silent Switcher

An attractive alternative for powering purpose-built test equipment is the "Silent Switcher," developed by Jan Didden [16]. It is based on a small IC switching power supply that takes in 5 V DC and outputs ±15 V DC up to 150 mA. The actual output voltage is provided from low-noise linear regulators powered with a higher voltage from the switcher. It also has a 5-V output capable of up to 0.5 A. The unit is powered from a conventional USB 5-V, 2.1-A USB charger. The Silent Switcher is used to power Jan's Autoranger and is built into the unit.

References

1. Harry C. P. Dymond and Phil Mellor, "An Active Load and Test Method for Evaluating the Efficiency of Audio Power Amplifiers," *Journal of the Audio Engineering Society*, vol. 58, no. 5, pp. 394–408, May 2010.
2. Robert R. Cordell, "Build a High Performance THD Analyzer," *Audio*, vol. 65, July–September 1981. Available at www.cordellaudio.com.
3. Stuart Yaniger, "Practical Test & Measurement: Sound Cards for Data Acquisition in Audio Measurements," 7 part series in *AudioXpress* magazine, beginning April 13, 2017. Available at www.audioxpress.com/article/ptactical-test-measurement-sound-cards-for-data-acquision-in-audio-measurements.
4. QuantAsylum USA, "QuantAsylum QA401 Audio Analyzer." Available at www.quantasylum.com.
5. THAT Corporation 1200-series Ingenious® Differential Line Receiver Data Sheet. Available at www.thatcorp.com.
6. THAT Corporation 1646-series OutSmarts® Balanced Output Driver Data Sheet. Available at www.thatcorp.com.
7. National Semiconductor/Texas Instruments LM4562 Dual Operational Amplifier Data Sheet. Available at www.ti.com.

8. Pete Millett, "Sound Card Interface/AC RMS Voltmeter." Available at www.pmillett.com/ATEST.htm.
9. Jan Didden, "The L|A Autoranger." Available at www.linearaudio.nl/la-autoranger.
10. Bob Cordell, "The Distortion Magnifier," *Linear Audio*, September 2010. Available from www.linear audio.net.
11. Dick Moore, "Active Twin-T Notch Filter—A Path to High-resolution Distortion Analysis." Available at www.tronola.com/moorepage/Twin-T.html.
12. "Twin T Notch Filter Mini Tutorial," *Analog Devices*, MT-225. Available at www.analog.com/media/en/training-seminars/tutorials/MT-225.pdf.
13. "The Twin-T Notch (band-stop) Filter." Available at http://fourier.eng.hmc.edu/e84/lectures/ActiveFil ters/node4.html.
14. James R. Bainter, "Active Filter has Stable Notch, and Response Can be Regulated," *Electronics*, pp. 115–117, October 2, 1975.
15. "Bainter Notch Filters, MT-203," *Analog Devices*. Available at www.analog.com.
16. Jan Didden, "The Silent Switcher." Available at www.linearaudio.nl/silentswitcher.

Chapter 26

Distortion and Its Measurement

26. INTRODUCTION

In this chapter we'll look at some distortion theory and some approaches to measuring distortion. Throughout the discussion it is important to distinguish between distortion mechanisms and distortion measurement techniques. Many of the mechanisms were discussed in Chapter 16.

26.1 Nonlinearity and Its Consequences

Nonlinearity is the underlying mechanism of distortion. When a circuit parameter changes as a function of signal, nonlinearity exists. Stimulation of the nonlinearity is also necessary for creation of distortion from that nonlinearity. Signal voltage or current is usually the stimulus at the location of the nonlinearity. The resulting distortion will usually be in proportion to the amplitude of the stimulus (or to a power of it). When the value of a capacitance changes as a function of signal voltage, distortion is created that is stimulated by the voltage. The consequences of that capacitance change are what a distortion test measures. If the capacitor is across a very low-impedance source, its signal-dependent capacitance change may not make much difference in the signal and measured distortion will be low.

More often, the capacitance will be associated with a resistance, causing a pole in the circuit to move up and down in frequency. One distortion test may measure the time-varying frequency response that results, while another test might measure the time-varying signal phase that results. Yet another test might measure the harmonic frequencies that are created. The same underlying nonlinearity will cause distortion to be seen by many different types of measurement. This is a very important point and is sometimes misunderstood.

This is why it is virtually impossible to have one type of measured distortion without having another type of measured distortion. Having said that, it is important to realize that different distortion measurements can have vastly different sensitivities to the same nonlinearity, depending on how effective they are in stimulating (exercising) that nonlinearity and how sensitive they are in measuring the resulting distortion products. This is why there are numerous different types of distortion tests. A given nonlinearity creates THD, TIM and CCIF IM; these are just different ways of stimulating the nonlinearity and measuring its consequences.

High-frequency distortion is a very good example. It is largely a function of the rate of change of the stimulus that is exercising the amplifier. THD-20 and TIM are measuring the same nonlinearity by stimulating the amplifier with a high rate of change.

Indeed, a given type of distortion measurement is really the observation of the symptoms of the nonlinearity. An analogy here would be that in medicine the presence of many diseases is inferred from symptoms or from the presence of antibodies, not necessarily the disease itself.

The Order of a Nonlinearity

As signal level is increased, the increase in distortion percentage is a function of the order of the distortion being considered. For example, with second-order distortion, a 1 dB increase in signal level will cause the magnitude of the second-order distortion product to go up by 2 dB. This means that the distortion expressed as a percentage will go up by 1 dB. For example, if second-harmonic distortion is 1% at a fundamental level of 1V RMS, then the second harmonic will rise to 2% at a fundamental level of 2 V RMS.

For third-order distortion, a 1 dB increase in fundamental level will result in a 3-dB increase in the product magnitude and a 2 dB increase in the distortion percentage. If the third-order distortion is −80 dB relative to the fundamental at a fundamental level of 0 dBV, it will rise to −78 dB relative to the fundamental at a fundamental level of +1 dBV. Distortion of order n will go up by $(n − 1)$ dB relative to the fundamental when the signal level is increased by 1 dB.

This known behavior can be helpful in inferring whether the distortion component being observed (as on a spectrum analyzer) is from the source or the *amplifier under test (AUT)*. Increase the level to the AUT by 1 dB. The magnitude of the third harmonic (for example) should go up by 3 dB. If it goes up by only 1 dB, it has probably originated in the source.

26.2 Total Harmonic Distortion

Total harmonic distortion (THD) is one of the most common measures of distortion. It is based on the fact that when a sine wave encounters a nonlinearity, harmonics will be created at integer multiples of the fundamental frequency of the sinusoid.

As explained in Chapter 25, the amplifier under test is fed a low-distortion sine wave. The fundamental frequency of the sine wave is removed with a very narrow and deep notch filter. What remains after the notch filter is the residual, consisting of noise and harmonics [1]. This is why the measurement is usually referred to as *THD+N*. The residual is passed through a low-pass filter that limits the measurement bandwidth to typically 80 kHz or 200 kHz. This improves the SNR of the measurement. The amplitude of the residual is displayed on the instrument meter as a distortion percentage and the residual signal is made available at an output jack for viewing with an oscilloscope or a spectrum analyzer [1].

Interpretation of THD

Single-number THD specifications are of limited use in characterizing the sound quality of an amplifier because they do not convey the nature of the distortion. For example, they do not distinguish between low-order and high-order distortions. Worse, single-number THD will often be quoted at 1 kHz, where it is easy to achieve very small numbers. This is a big part of the reason why it is common in many circles to dismiss THD as having little or no relationship to perceived sound quality. Such a view paints THD with too broad a brush. THD is a much better indicator of amplifier performance when it is measured at high frequencies (THD-20) and a full spectral analysis is presented of the amplitudes of the individual harmonics.

Advantages of THD Measurements

While not an airtight guarantee of good sound, a very low THD number for an amplifier leaves little room for most other distortions to be present. By very low THD we mean THD well under 0.01% under all conditions with thorough testing and observation of the amplitudes of the harmonics. This is not the only path to good sound, however. Benign distortions that elevate

THD readings may not audibly degrade sound quality. Second-harmonic distortion would be an example.

Very low THD assures exceptional overall circuit linearity under static conditions. Low THD at all frequencies means that most other distortions will be very small as well. These include CCIF IM, SMPTE IM, TIM and PIM. It is very difficult for these other measured distortions to exist without there also being at least small amounts of THD. This is especially so if the THD residual has been evaluated on a spectrum analyzer and if the THD has been measured in a bandwidth that is at least 10 times that of the fundamental.

Very low THD virtually guarantees the absence of audible crossover distortion. This is especially so if THD without noise is shown to be very small at lower power levels by the use of spectral analysis of the residual. There is also a stronger assurance if the measurements include loading with 2 Ω.

Very low THD-20 indicates that magnetic coupling distortions from power supply rails are absent. Magnetic coupling of the highly nonlinear half-wave-rectified signal currents in class AB output stages readily shows up as high-frequency THD. Very low THD also assures that power supply coupling distortions due to limited PSRR are very small. Low THD also indicates that ground-induced distortions due to imperfect grounding are very small.

Very low THD-20 suggests that parasitic oscillations are absent when driving the load used in the test setup. In thorough testing, these THD tests should be done with capacitive loads and simulated speaker loads. The presence of parasitic oscillations usually causes subtle increases in THD. These subtle increases will go unnoticed in amplifiers with higher THD.

Very low 20-Hz THD strongly suggests that many low-frequency thermal distortions are absent. These include fuse distortion, feedback resistor thermal distortion and transistor junction thermal distortion. It also suggests that some measurable capacitor distortions, such as from electrolytic capacitors at low frequencies, are absent.

Very low THD at 20 Hz or 50 Hz virtually guarantees that power supply ripple and its harmonics are not entering the signal path. These will show up in the distortion residual even though they are not harmonics of the fundamental signal stimulating the amplifier.

Finally, low THD when driving a 2-Ω load assures that the amplifier has very good high-current capabilities. Effects of beta droop in the output stage will often be unmasked in this THD test.

In general, the attention to design detail and implementation necessary to achieve very low THD will tend to result in a better amplifier (as long as something stupid is not done to achieve low THD at the expense of something else).

Limitations of THD Measurements

THD-20 is one of the tougher and more revealing distortion measurements that can be done on an amplifier. However, the higher harmonics lie well above the audio band and many audio spectrum analyzers cannot display spectra above 50–100 kHz. If the THD analyzer has an 80-kHz filter engaged to improve its sensitivity, these harmonics will be attenuated. Single-number THD measurements do not tell the whole story because they do not distinguish between the low-order nonlinearities and the more troublesome high-order nonlinearities.

THD+N measurements are of limited value at low power levels because there is no way of knowing whether the reported THD+N level is primarily noise or crossover distortion. If THD+N is very low, it means that both THD and noise are both low.

Low THD does not assure the absence of many sonic shortcomings. For example, it does not reveal distortion resulting from thermal bias instability, as when an output stage becomes temporarily under-biased following a high-power interval. It also does not assure that there is no flabby low-end performance due to a sloppy power supply. Nor does it assure the absence of sonic degradation due to poor transient performance and ringing.

Low THD does not guarantee civilized amplifier behavior under clipping conditions or when protection circuits are activated. It does not indicate the absence of frequency response coloration due to the effects of frequency-dependent variations in load impedance (damping factor). It also does not assure the absence of poor sound quality resulting from unforeseen interactions with the loudspeaker load under dynamic conditions. Nor does it prove the absence of instability under all possible cable and loudspeaker loads.

Good THD readings do not assure that the amplifier is reproducing music faithfully in the presence of EMI ingress. It does not address some linear and nonlinear distortions that are less well understood, such as the influence on sound quality of passive component quality. In fairness to THD, many other distortion tests also do not reveal these sonic shortcomings.

Single-number mid-band THD, like THD-1, can give a misleading impression of good amplifier performance. This may be the single biggest reason why some eschew THD measurements, claiming that it has little correlation with sound quality. The fact that some amplifiers with relatively high THD sound very good further contributes to this impression.

26.3 SMPTE IM

SMPTE intermodulation distortion (SMPTE IM) is another distortion measure that has long been in use. It is based on the observation that nonlinearity can be represented as a change in the incremental gain of a circuit as a function of instantaneous signal amplitude. This measurement is also referred to as *amplitude intermodulation distortion* (AIM). The dynamic changes in incremental gain are observed by creating a test signal with a small-amplitude high-frequency *carrier* on top of a large-amplitude lower-frequency signal.

After the test signal passes through the amplifier under test, the low-frequency signal is filtered out and the carrier signal is AM-detected. The SMPTE IM test employs test signals at 60 and 7000 Hz mixed in a 4:1 ratio. A typical SMPTE IM measuring arrangement is shown in Figure 26.1. The sensitivity of the analyzer is largely determined by the rejection characteristics of the various filters. Most SMPTE IM analyzers use a conventional rectifying AM demodulator, but better ones employ *synchronous detection* where phase-locked loops are required to recover

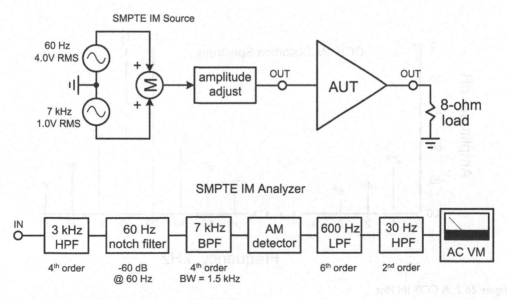

Figure 26.1 Measuring SMPTE Intermodulation Distortion

the 7-kHz carrier. Synchronous detection does a far better job of keeping AUT noise out of the measurement.

26.4 CCIF IM

The CCIF IM distortion test takes advantage of the fact that if two tones are passed through a nonlinear circuit, spectral components will be created at frequencies at $mf_1 \pm nf_2$ where f_1 and f_2 are the frequencies of the two tones. This test, often referred to as the twin-tone test, is usually conducted with equal-level sine waves at 19 kHz (f_1) and 20 kHz (f_2). A second-order nonlinearity will produce a distortion component at 1 kHz, the difference of the two frequencies. Notice here that two frequencies are involved in the calculation of the distortion product frequency and that $m + n = 2$. This characterizes the nonlinearity as second order. The sum of m and n always designates the order of the nonlinearity that the spectral component represents. The second-order nonlinearity will also produce a spectral *line* at 39 kHz, representing the $f_1 + f_2$ component (19 kHz + 20 kHz).

A third-order nonlinearity will produce components at $2f_1 - f_2 = 18$ kHz and $2f_2 - f_1 = 21$ kHz. A fifth-order nonlinearity will produce components at $3f_1 - 2f_2 = 17$ kHz and $3f_2 - 2f_1 = 22$ kHz. It is easy to see the progression as the order of the nonlinearity increases. Figure 26.2 shows a typical CCIF IM plot that illustrates the result when nonlinearities at second through seventh order are present.

Notice that the fourth-order nonlinearity shows up at $2f_2 - 2f_1 = 2$ kHz and that the sixth-order nonlinearity shows up at $3f_2 - 3f_1 = 3$ kHz. The CCIF IM test reflects even-order nonlinearities down to low frequencies. Early uses of the test simply employed a low-pass filter to attenuate the higher test frequencies so that the lower products could be measured with an AC voltmeter, with particular emphasis on the second-order product at 1 kHz. Unfortunately this was a very incomplete test. Proper use of the CCIF IM test requires the use of a spectrum analyzer. This enables the odd-order distortion products to be viewed.

The great advantage of the CCIF IM test is that representatives of all of the distortion orders are present in-band, allowing the use of spectrum analyzers of modest high-frequency capability. The individual oscillators do not have to have very low distortion, and this is a major advantage

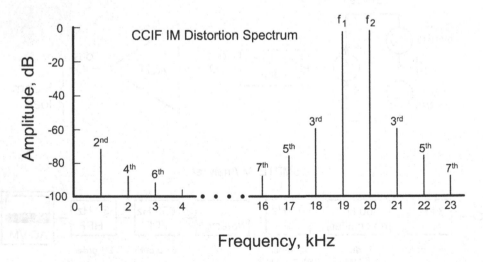

Figure 26.2 A CCIF IM Plot

of this test. However, the summing circuit where their outputs are combined must have very low distortion.

26.5 Transient Intermodulation Distortion (TIM) and SID

Transient intermodulation distortion (TIM) received a great deal of attention in the 1970s and early 1980s [2–10]. It is a distortion mechanism that is often described in time-domain terms. It has wrongly been blamed on the use of large amounts of negative feedback and small open-loop bandwidth. If an input signal to a feedback amplifier changes very quickly—too fast for the output of the "slow" amplifier to respond—the input stage may be overloaded and clip. The stage will be overloaded by the large error signal that arises before the feedback from the output catches up to the input. The overload occurs because the input stage of a feedback amplifier is not usually designed to be able to handle the full amplitude of the input signal, since under normal conditions it need only handle a much smaller error signal.

Slew Rate Limiting and Input Stage Stress

The *slew rate limiting* distortion mechanism was known many years before the term *TIM* was coined. TIM has in fact been described as *slewing-induced distortion* (SID) [7, 8]. *Hard TIM* occurs when the input stage clips and the amplifier is in slew rate limiting. *Soft TIM* occurs when the stress on the input stage increases as the slew rate limit is approached, resulting in input stage nonlinearity. It is important to recognize that TIM results from signal stress on the input stage.

Amplifiers with large amounts of negative feedback and small open-loop bandwidth can achieve very high slew rates. This is why large amounts of negative feedback and small open-loop bandwidth are not a root cause of TIM. Poor amplifier design without adequate slew rate and input stage dynamic range is what is responsible for TIM [10].

The Dynamic Intermodulation Distortion (DIM) Test

TIM is a dynamic distortion that results from fast changes in the signal. For this reason it is also a distortion that is more prominent at high frequencies. The original test developed for this distortion is referred to as *dynamic intermodulation distortion* (DIM) [11–13].

As shown in Figure 26.3, the DIM test signal consists of a 3.18-kHz square wave and a 15-kHz sine wave mixed in a 4:1 ratio. The combined signal is then low-pass filtered with a first-order network at 30 kHz (DIM-30) or 100 kHz (DIM-100). The fast edges of the square wave stress the amplifier at high frequencies with high-voltage rates of change while the 15-kHz carrier signal is modulated as a result. The result of the test must be viewed on a spectrum analyzer, and the

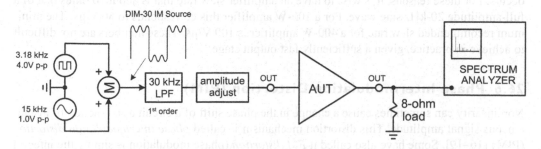

Figure 26.3 The DIM Test for Transient Intermodulation Distortion (TIM)

amplitudes of all of the relevant spectral lines must be added on an RMS basis and then referred to the amplitude of the 15-kHz carrier. Notice that the square-wave portion of the input signal contains components at the fundamental and its odd harmonics. Each of those harmonics can interact with the 15-kHz fundamental on an $m \pm n$ basis. This means that nonlinearities of low order will still produce a very rich spectrum.

The dynamic range of the spectrum analyzer limits the measurement floor for the DIM-30 test. The popular HP 3580A analog audio spectrum analyzer has a dynamic range of 85 dB on a good day. The 15-kHz component of the DIM-30 test signal is down 14 dB from the full peak-to-peak amplitude of the composite DIM-30 test signal. This means that the measurement floor is at about −71 dB, or about 0.03%. Some newer spectrum analyzers based on high-performance sound cards, or ones like the QuantAsylum QA-401, can do much better [14].

THD-20 Will Always Accompany TIM

With some caveats, I have never seen an amplifier that had measurable TIM that did not also have measurable THD-20. The essential difference between these tests is that TIM exercises the amplifier with a high peak rate of change that has a small duty cycle, while THD-20 exercises the amplifier with a smaller rate of change but with a much larger duty cycle. The peak slew rate for a 20-kHz sine wave is 0.125 $V/\mu s$ per volt of peak signal amplitude ($V/\mu s/V_{pk}$). The peak slew rate for the DIM-30 test signal is 0.319 $V/\mu s/V_{pk}$. A 20-kHz sine wave at 100 W into 8 Ω has a slew rate of 5 $V/\mu s$.

THD-20 will tend to track DIM fairly well as long as the amplifier is not pushed into slew rate limiting (hard TIM) by the higher peak slew rate of the DIM test signal. THD-20 will often be about 4–7 dB below the DIM-30 number [15]. However, the measurement floor with a good THD analyzer (about 0.001% or −100 dB) lies well below that of the DIM-30 test. Because of the large rate of change applied to the amplifier in order to stimulate the high-frequency nonlinearity, it is possible for a DIM test to cause an amplifier to go into slew rate limiting when a THD-20 test would not. In this case it is possible to have large amounts of DIM with fairly small amounts of THD-20. More often, low THD-20 virtually guarantees the absence of TIM, especially if the amplifier is known to have a healthy slew rate of 50 $V/\mu s$ or more depending on power rating.

Recommended Amplifier Slew Rate

The maximum slew rate from a CD source is limited by the very steep anti-alias filtering required by the *Red Book* standard for audio CDs. A square wave recorded on a CD will have a slew rate of about twice that of a 20 kHz sine wave of the same peak amplitude, or about 0.25 $V/\mu s/V_{pk}$. Newer recoding standards, like SACD and high-rate PCM, increase this maximum, at least in principle. Many amplifier input stages begin to exhibit nonlinearity well before slew rate limiting occurs. For these reasons, it is wise to have an amplifier slew rate that is about 10 times that of a full-amplitude 20-kHz sine wave. For a 100-W amplifier this corresponds to 50 $V/\mu s$. The minimum recommended slew rate for a 400-W amplifier is 100 $V/\mu s$. These numbers are not difficult to achieve in practice, given a sufficiently fast output stage.

26.6 Phase Intermodulation Distortion (PIM)

Nonlinearity can sometimes cause a change in the phase shift of a circuit as a function of instantaneous signal amplitude. This distortion mechanism is called *phase intermodulation distortion* (PIM) [16–19]. Some have also called it *FM distortion* (phase modulation is simply the integral of frequency modulation).

The PIM measurement is analogous to SMPTE IM (AIM), but involves phase modulation instead of amplitude modulation. Two test signals at 60 Hz and 7 kHz are mixed in a 4:1 ratio and applied to the AUT. A simple example of a circuit that creates PIM is a nonlinear capacitance in an R-C low-pass filter arrangement. If the capacitance changes as a function of signal amplitude, the corner frequency of the low-pass filter will change and, correspondingly, the phase shift of the filter will change.

PIM is also created by AIM in a feedback amplifier. If the signal amplitude modulates the gain of the input stage, the changing open-loop gain of the amplifier will result in a changing closed-loop bandwidth, as illustrated in Figure 26.4. Just as in the passive case above without feedback, the movement of the closed-loop 3-dB frequency will cause a change in phase shift, which corresponds to PIM. This is an example of amplitude-to-phase conversion caused in this case by the action of the negative feedback loop. Only a portion of the AIM is converted to PIM, leaving AIM as well as PIM. For this reason, an amplifier of conventional design cannot exhibit PIM without also exhibiting AIM.

Differential Gain and Phase

The measurement of AIM and PIM is not new to the analog video world. There it is referred to as *differential gain and phase*, referring simply to gain and phase shift that are a function of signal.

Measuring PIM

PIM is measured in much the same way as AIM, but with a phase detector substituted for the AM detector. The test signal for PIM and AIM is the same [16–18] A large low-frequency signal, typically 60 Hz, is used to cause the operating points in the amplifier to traverse a large-signal range. A smaller signal, typically 4 times smaller and at a high frequency like 7 kHz, then has its

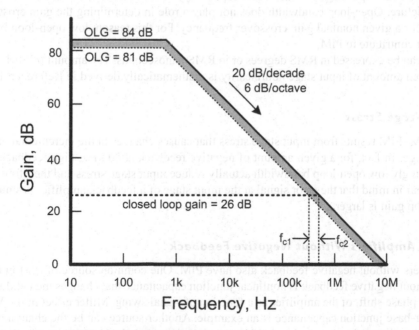

Figure 26.4 Creation of PIM by Modulation of the Open-Loop Gain

amplitude and phase modulations measured. Modulation on the 7-kHz small signal at 60 Hz or its harmonics is either AIM or PIM depending on whether amplitude or phase detection is used.

This is a more challenging measurement that usually requires a phase-locked-loop and synchronous detection, as described in Reference 19. The arrangement is referred to as a coherent IM analyzer. The instrument uses in-phase and quadrature coherent detection to extract both conventional SMPTE IM (AIM) and PIM, respectively.

The 19 + 20-kHz CCIF test is also sensitive to PIM. Phase and amplitude distortions both create spectral lines at frequencies that have an $mf_1 \pm nf_2$ relationship. For this reason, an amplifier with PIM will exhibit spectral lines using the 19 + 20-kHz CCIF test. However, one will not be able to distinguish PIM from AIM with this test.

Negative Feedback and PIM

If the open-loop gain of a feedback amplifier is a function of signal swing (this is AIM), the gain crossover frequency of the global negative feedback around the amplifier will change. If the open-loop gain decreases, the gain crossover frequency and the closed-loop bandwidth of the amplifier will become smaller. When this happens, the amplifier will exhibit slightly higher in-band phase lag. This is PIM. The negative feedback has effectively caused some of the AIM to be converted to PIM. Put simply, movement of the closed-loop pole of a feedback amplifier is the source of feedback-generated PIM. Even though the closed-loop pole is well above the audio band (typically above 50 kHz), it causes a small amount of phase lag in-band, whose change creates PIM. For this reason, PIM has been blamed on negative feedback.

Moreover, PIM has wrongly been blamed on low open-loop bandwidth [16]. It seems intuitive that if movement of the closed-loop pole causes PIM, then the lower the frequency of the open-loop pole, the worse the PIM. This is not true. Consider a conventional Miller-compensated amplifier. Input stage *gm* and the size of the Miller capacitor set the gain crossover frequency for a given closed-loop gain. If you simply look at the calculation for gain crossover frequency as a function of input stage *gm* and Miller capacitor, you see that the open-loop bandwidth is not in the picture. Open-loop bandwidth does not play a role in determining the gain crossover frequency for a given nominal gain crossover frequency. For this reason, low open-loop bandwidth does not contribute to PIM.

PIM can be expressed in RMS degrees or in RMS nanoseconds. The amount of PIM generated by a given amount of input stage nonlinearity is mathematically derived in Reference 19.

Input Stage Stress

Like TIM, PIM results from input stage stress that causes changes in the incremental gain of the input stage. In fact, for a given amount of negative feedback at 20 kHz, high feedback and correspondingly low open-loop bandwidth actually reduce input stage stress and therefore AIM and PIM. Bear in mind that the error signal at the input stage of a feedback amplifier is smaller if the open-loop gain is larger.

PIM in Amplifiers Without Negative Feedback

Amplifiers without negative feedback also have PIM. One common source of PIM in an amplifier without negative feedback is nonlinear junction capacitance that changes the bandwidth, and thus the phase shift, of the amplifier as a function of signal swing. Miller effect in the VAS from collector-base junction capacitance is an example. Another source can be the changing f_T of the output transistors affecting the bandwidth as their f_T droops at signal extremes.

Feedback amplifiers thus have PIM in the open loop even before negative feedback is applied. Interestingly, the application of negative feedback reduces this component of PIM in the same way that it reduces distortion from any other open-loop nonlinearity.

The results in Reference 18 show that PIM is not a problem in contemporary amplifiers and that negative feedback reduces total PIM in most cases. The instrument built in Reference 19 has a PIM measurement floor of 1 ns. An amplifier of ordinary design tested in Reference 19 had PIM of less than 3 ns. The amplifier of this reference [20] was also measured for PIM. It has very high negative feedback and very low open-loop bandwidth and measures less than 0.001% THD-20. It had measured PIM of only 0.04 ns (40 ps). That measurement required the use of a spectrum analyzer connected to the residual output of the coherent IM analyzer [19].

26.7 Interface Intermodulation Distortion (IIM)

Interface intermodulation distortion (IIM) is caused by interaction of the load with the nonlinearity of the output impedance of a power amplifier [21, 22]. The nonlinearity can be stimulated by *back-emf* that originates in the loudspeaker. One example of this mechanism is the change in open-loop output impedance of the output stage as the output current goes through a zero crossing.

The output impedance of a feedback amplifier is approximately equal to its open-loop output impedance divided by the feedback factor. If the open-loop output impedance is large, the low closed-loop output impedance of the amplifier will depend on negative feedback for its reduction. IIM can thus be influenced by negative feedback. If the input stage experiences greater stress when the amplifier must deliver high currents, its incremental gain may decrease. This will decrease the amount of open-loop gain and reduce the amount of negative feedback. The change in feedback factor as a function of output current thus causes a change in closed-loop output impedance, and thus IIM.

Loudspeaker EMF and Peak Current Requirements

The typical loudspeaker is anything but a benign resistive load. It is a reactive load that comprises mechanical elements that can store energy. This is especially the case with woofers, where significant electromotive force can be generated by cone motion. This can cause large excursions in output current that can stimulate the production of IIM.

The loudspeaker presents a fairly complex load to the amplifier, often with several significant resonances. The impedance can sometimes rise to over 10 times its rated value and fall to much less than 80% of its rated value (sometimes to less than half). The electromechanical system of the speaker (particularly the woofer) also represents an energy source and generation capability. Any movement of the cone will cause an emf to be generated by the voice coil or magnet system. This movement is often due to cone momentum developed by earlier excitation. The capability thus exists for the speaker to inject a signal back into the output of the amplifier. This can cause unexpectedly large currents to flow. These currents, in turn, can excite nonlinearity of the open-loop output impedance.

Figure 26.5 shows a simple RLC electrical model of a loudspeaker woofer in a closed box. The loudspeaker resonance is at 50 Hz. In this model, R represents the DC voice coil resistance, C accounts for the mass of the cone and L accounts for the suspension compliance. The model ignores the effects of crossovers, other drivers and so on. In all cases we make the conventional assumption that the minimum speaker impedance at low frequencies is equal to 80% of rated impedance.

Figure 26.6 shows the speaker current as a function of time when driven by the 28-V peak waveform shown. The driving waveform was deliberately chosen to maximize the expected peak load current [22]. The signal swings between large negative and positive values, rather than

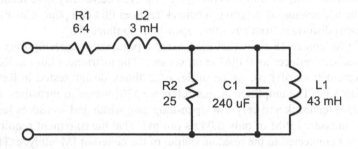

Figure 26.5 Loudspeaker Model for Simulation of Peak Current Flow

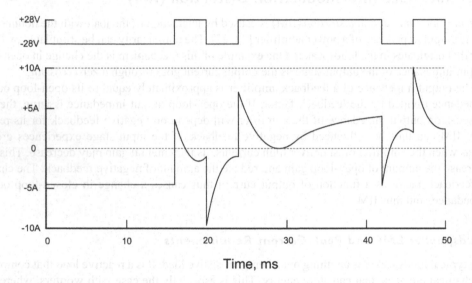

Time, ms

Figure 26.6 Peak Loudspeaker Currents with a Special Test Signal

simply starting from zero. It stays at one extreme for 16 ms to allow load current to rise to at least 90% of its final value. A 4-ms pulse then follows, with the trailing edge of this pulse reversing the applied polarity just when the counter-emf of the speaker is at its maximum. This causes a large current to flow because the counter-emf is now enhancing current flow rather than opposing it. The situation is then repeated for the opposite polarity sequence so that the average value of the signal is zero.

While an amplifier delivering this waveform to an 8-Ω resistive load would normally see a peak load current of about 3.5 A, we see here that the *RLC* load develops a peak load current of 10 A. While the probability and extent of this kind of occurrence in the real world with musical program material may be questioned, the exercise does provide some food for thought. The lesson to be learned here is to be prepared to handle larger currents than are encountered with a simple resistive load.

High-Current Amplifier Design

The message here is that the power amplifier should be designed for high-current output capability. Use an output stage with lots of current gain, employ a good number of output pairs, avoid current limiting and current limiters to the extent possible and run the VAS at a healthy bias current (at least 10 mA). Note that a good number of MOSFET output pairs will provide very low open-loop output impedance because of their nearly infinite current gain at audio frequencies.

Measuring IIM

IIM is measured with a very interesting variation of the SMPTE IM arrangement. The IIM test is illustrated in Figure 26.7 [21]. Tones at 60 Hz and 1000 Hz are used in a 1:1 amplitude ratio. The 1-kHz carrier signal is fed forward through the amplifier. The amplifier is connected to an 8-Ω load resistor, the other end of which is tied to another amplifier instead of to ground. The second amplifier is fed the 60-Hz component of the IM test. The amplitude of the 60-Hz signal at its output is the same as the 1-kHz carrier signal at the output of the amplifier under test (AUT).

This is a back-feeding distortion test, where current is forced to flow in the output stage of the AUT. The current created by the 60-Hz back-feed signal will cause modulation of the 1-kHz carrier via the IIM distortion mechanism. The 60-Hz modulation present on the 1-kHz carrier at the output of the AUT is then measured and referenced to the 1-kHz signal level to arrive at a distortion number.

Open-Loop Output Impedance

Open-loop output impedance plays a role in the creation of IIM. If it is nonlinear, it will contribute significantly to the IIM nonlinearity. Unfortunately, open-loop output impedance is sometimes misunderstood.

The high impedance of the VAS output node, combined with finite output stage current gain, makes it intuitive that open-loop output impedance should be high. This is only partially true. The shunt feedback on the VAS output node created by the Miller compensation capacitor greatly reduces the impedance at that node, especially at high frequencies. For this reason the open-loop output impedance of the amplifier is a function of frequency. It is naturally lower at high frequencies. This tends to compensate for the fact that there is less feedback at high frequencies to reduce output impedance when the loop is closed. Indeed, over a good part of the upper frequency range, the practice (intuitive to some) of adding VAS load resistors to reduce open-loop output impedance is both ineffective and harmful. It is ineffective because at upper frequencies the Miller shunt feedback has already reduced the impedance at that node much lower than the value of any reasonable VAS shunt resistors. It is harmful because the shunt resistors make the VAS work harder and create more distortion.

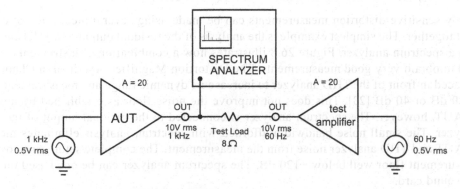

Figure 26.7 Measurement Setup for Interface Intermodulation Distortion

The amount by which Miller compensation reduces VAS output impedance is a strong function of the circuit design details. High impedance at the input node of the VAS tends to elevate the amount of shunt feedback provided by the compensation capacitor. The amplifier of Figure 3.10 can serve as an example to put the matter in perspective. The VAS output impedance at 1 kHz for that design is less than 5 kΩ. When that impedance is divided by the current gain of the Triple output stage (over 100,000), the VAS output impedance contribution to open-loop output impedance becomes a mere 50 mΩ. The VAS output impedance contribution will be much lower than this at higher frequencies.

The argument in Reference 21 asserts that when open-loop output impedance is higher, negative feedback must play a greater role in keeping the output voltage from changing in the presence of counter-emf from the loudspeaker. A larger error must circulate through the amplifier, as it were. In reality, the error signal amplitude is largely independent of the open-loop output impedance [22].

26.8 Multitone Intermodulation Distortion (MIM)

The *multitone intermodulation* (MIM) distortion test employs three tones instead of the two tones used in the CCIF test [15]. This allows odd-order distortion products to be reflected to a low frequency in the audio band so that they can be measured without the need for a spectrum analyzer. This is also referred to as a *triple-beat* test because three tones create beat frequencies among them when they interact with nonlinearities. Three equal-level tones at 20.00 kHz, 10.05 kHz and 9.00 kHz are fed to the AUT. If these three frequencies are designated f_a, f_b and f_c, then an even-order nonlinearity will produce a beat frequency distortion product at $f_b - f_c = 1050$ Hz and an odd-order nonlinearity will produce a distortion product at $f_a - f_b - f_c = 950$ Hz.

The MIM distortion products can be measured with simple equipment. The output of the AUT is passed through a sixth-order 2-kHz LPF to discard the three high-frequency test tones. The signal is then passed through a fourth-order 1-kHz band-pass filter with a bandwidth of 150 Hz, which is wide enough to pass the distortion product frequencies at 950 Hz and 1050 Hz. This arrangement provides very good sensitivity because the noise bandwidth of the analyzer is quite small. Details of the test setup and how well it compares to THD-20 and DIM-30 tests can be found in Reference 15. MIM produces smaller distortion percentage numbers, but it has a correspondingly lower measurement floor. MIM is largely sensitive to distortion products of only second and third order.

26.9 Highly Sensitive Distortion Measurement

Highly sensitive distortion measurements can be made using several pieces of test equipment together. The simplest example is the analysis of the residual output of a THD analyzer with a spectrum analyzer. Figure 26.8 illustrates how a combination of instruments can be used to obtain very good measurements. The Distortion Magnifier described in Chapter 25 is placed in front of the THD analyzer to increase its dynamic range and measurement floor by 20 dB or 40 dB [23]. This does not improve the noise floor as established by noise in the AUT, however. The spectrum analyzer is connected to the residual output of the THD analyzer. The small noise bandwidth afforded with spectrum analysis eliminates most of the AUT and THD analyzer noise from the measurement. The complete setup can provide a measurement floor well below −120 dB. The spectrum analyzer can be one based on a PC and sound card.

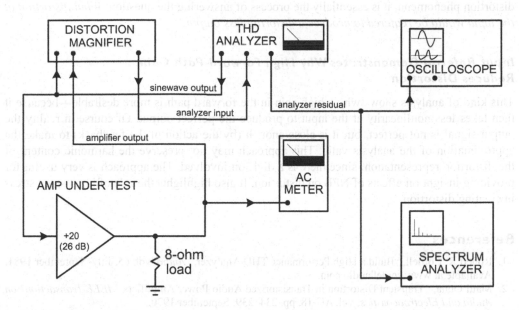

Figure 26.8 Highly Sensitive Distortion Measurement

26.10 Input-Referred Distortion Analysis

The effect of negative feedback on distortion is most easily understood by working backward from the output. We assume a perfect output and evaluate the input-referred distortion required to generate that perfect output, just as we do in calculating input-referred noise. Because the feedback signal under these conditions is perfect, the level of the input-referred distortion is the same for either open-loop or closed-loop conditions. Distortion percentage is reduced by feedback simply as a result of the larger pure component of the input signal required under closed-loop conditions. This technique is quite accurate when the referred distortion products are small compared to the total closed-loop input, that is, when closed-loop distortion is small. It is important to remember that the gain involved in referring a distortion product back to the input may be a strong function of frequency.

Consider one of the major contributors to IM distortion: output stage beta variations with current. The output stage requires a nonlinear driver current from the VAS to produce a perfect output. This results in an input-referred distortion voltage when the transconductance of the IPS-VAS is considered. Notice that the value of VAS load resistors (if present) will have virtually no effect on the level of this particular product because they have little effect on the transconductance.

Input Referral Breaks the Feedback Loop

While distortion analysis of the open-loop forward path of an amplifier can be difficult enough, a good understanding of distortion behavior can be even more difficult under closed-loop negative feedback conditions. For this reason, sometimes the technique of examining input-referred distortion can be helpful. This is not unlike input-referred noise techniques. By referring these phenomena back to the input, the feedback loop is effectively broken. When used in looking at distortion phenomena, it is essentially the process of answering the question: *What distortion at the input would be required to produce a distortionless output?*

Input Referral Demonstrates Why High Forward-Path Gain Reduces Distortion

This kind of analysis shows why a high gain in the forward path is more desirable—because it then takes less nonlinearity at the input to produce the perfect output. Of course in reality the output signal is not perfect, but it is close enough (by the action of the feedback) to make the approximation of the analysis valid. This approach may not preserve the harmonic content of the distortion representation, since there is a division involved. The approach is very useful for providing insight on effects of NFB on distortion. It also highlights the role of input stage stress in creating distortion.

References

1. Robert R. Cordell, "Build a High Performance THD Analyzer," *Audio*, vol. 65, July–September 1981. Available at www.cordellaudio.com.
2. Matti Otala, "Transient Distortion in Transistorized Audio Power Amplifiers," *IEEE Transactions on Audio and Electroacoustics*, vol. AU-18, pp. 234–239, September 1970.
3. Matti Otala and E. Leinonen, "The Theory of Transient Intermodulation Distortion," *IEEE Transactions on Acoustics, Speech and Signal Processing*, vol. ASSP-25, no. 1, pp. 2–8, February 1977.
4. W. Marshall Leach, "Transient IM Distortion in Power Amplifiers," *Audio*, pp. 34–41, February 1975.
5. W. Marshall Leach, "Suppression of Slew-Rate and Transient Intermodulation Distortions in Audio Power Amplifiers," *Journal of the Audio Engineering Society*, vol. 25, no. 7–8, pp. 466–473, July–August 1977.
6. Richard A. Greiner, "Amp Design and Overload," *Audio*, pp. 50–62, November 1977.
7. Walter G. Jung, Mark L. Stephens and Craig C. Todd, "Slewing Induced Distortion and its Effect on Audio Amplifier Performance—With Correlated Measurement Listening Results," AES preprint no. 1252 presented at the 57th AES Convention, Los Angeles, May 1977.
8. Walter G. Jung, Mark L. Stephens and Craig C. Todd, "An Overview of SID and TIM," *Audio*, vol. 63, no. 6–8, June–August 1979.
9. Peter Garde, "Transient Distortion in Feedback Amplifiers," *Journal of the Audio Engineering Society*, vol. 26, no. 5, pp. 314–321, May 1978.
10. Robert R. Cordell, "Another View of TIM," *Audio*, February–March 1980. Available at www.cordellaudio.com.
11. Eero Leinonen, Matti Otala and John Curl, "A Method for Measuring Transient Intermodulation Distortion (TIM)," *Journal of the Audio Engineering Society*, vol. 25, no. 4, pp. 170–177, April 1977.
12. Susumo Takahashi and Susumo Tanaka, "A Method of Measuring Transient Intermodulation Distortion," 63rd Convention of the Audio Eng. Soc., preprint no. 1478, May 1979.
13. Eero Leinonen and Matti Otala, "Correlation Audio Distortion Measurements," *Journal of the Audio Engineering Society*, vol. 26, no. 1–2, pp. 12–19, January–February 1978.
14. QuantAsylum USA, QuantAsylum QA401 Audio Analyzer. Available at www.quantasylum.com.
15. Robert R. Cordell, "A Fully In-band Multitone Test for Transient Intermodulation Distortion," *Journal of the Audio Engineering Society*, vol. 29, September 1981. Available at www.cordellaudio.com.
16. Matti Otala, "Feedback-Generated Phase Modulation in Audio Amplifiers," 65th Convention of the Audio Engineering Society, London, 1980; preprint no. 1576.
17. Matti Otala, "Conversion of Amplitude Nonlinearities to Phase Nonlinearities in Feedback Audio Amplifiers," pp. 498–499, Proceedings of IEEE International Conference on Acoustics, Speech and Signal Processing, Denver, CO, 1980.
18. Matti Otala, "Phase Modulation and Intermodulation in Feedback Audio Amplifiers," 69th Convention of the Audio Engineering Society, Hamburg, 1981; preprint no. 1751.
19. Robert R. Cordell, "Phase Intermodulation Distortion: Instrumentation and Measurements," *Journal of the Audio Engineering Society*, vol. 31, March 1983. Available at www.cordellaudio.com.

20. Robert R. Cordell, "A MOSFET Power Amplifier with Error Correction," *Journal of the Audio Engineering Society*, vol. 32, January 1984. Available at www.cordellaudio.com.
21. Matti Otala and Jorma Lammasniemi, "Intermodulation Distortion in the Amplifier Loudspeaker Interface," 59th Convention of the Audio Engineering Society, preprint no. 1336, February 1978.
22. Robert R. Cordell, "Open-loop Output Impedance and Interface Intermodulation Distortion in Audio Power Amplifiers," preprint no. 1537, 64th Convention of the AES, 1982. Available at www.cordellaudio.com.
23. Bob Cordell, "The Distortion Magnifier," *Linear Audio*, pp. 142–150, September 2010.

Chapter 27

Other Amplifier Tests

27. INTRODUCTION

Even with the best conventional testing of amplifiers there remains a good deal of discrepancy between measurement results and perceived quality of sound. Some of this can be attributed to the many poorly controlled variables in subjective listening tests, but it is also likely that the relevant differences are measurable but they are not being measured. For example, the often relatively static conditions of conventional tests may not stimulate that which is responsible for two amplifiers having a different sound.

In this chapter some other tests not mentioned elsewhere in the book will be discussed. Most of these tests fall into the unconventional category. Amplifiers often sound different because they misbehave differently. One of the objectives of some of these tests is to get the amplifiers to misbehave.

27.1 Measuring Damping Factor

The importance of damping factor is sometimes underestimated, particularly in regard to its frequency-dependence and its effect on frequency response. Every amplifier review should start with a frequency response measurement as seen at the terminals of the speaker system being used for the listening test. This would be very revealing in many cases. The effect of frequency response differences on perceived sound quality differences must never be underestimated.

Damping factor is defined as the ratio of 8 Ω to the output impedance of the amplifier. An amplifier whose output impedance is 0.16 Ω will have a damping factor of 50. The output impedance of the amplifier forms a voltage divider with the speaker load impedance, creating attenuation between the idealized output and the actual output. The damping factor can be inferred by measuring the amplifier frequency response under no load and with a known load. This is a fairly crude approach.

A better approach is to inject a signal current into the output of the amplifier and measure the resulting voltage. Such an arrangement is shown in Figure 27.1. This can be done by *back-feeding* from another amplifier through a 100-Ω, 2-W resistor. The back-feeding amplifier is set for an output level of 10 V RMS. This will create a "probe" signal current of approximately 100 mA RMS. The voltage across the output terminals of the amplifier under test (AUT) is then measured with an AC voltmeter. A 10-mV reading will correspond to an output impedance of 100 mΩ, which in turn corresponds to a DF of 80. The test frequency should be swept across the audio band to obtain a plot of output impedance versus frequency.

It is wise and instructive to view the signal at the output terminals on an oscilloscope as well. First, make sure that wideband noise from the amplifier is not dominating the reading. If it is,

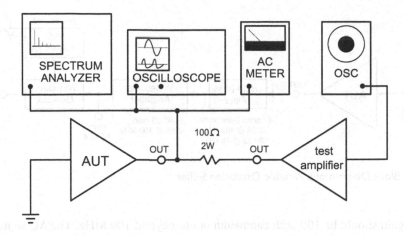

Figure 27.1 Measurement of Damping Factor

insert a 100-kHz low-pass filter ahead of the AC voltmeter. Second, make sure that hum from the AUT is not dominating the reading. If it is, insert a high-pass filter ahead of the AC voltmeter and measure DF only above 1 kHz.

Finally, the oscilloscope waveform may unmask some crossover distortion whose percentage has been magnified by the DF. The signal at the output terminals of the AUT can also be viewed with a spectrum analyzer. This is especially useful when the damping factor is very high and the fundamental of the probing signal is very small.

Do not underestimate the influence of damping factor across the full audio frequency band. Some claim to be able to hear a 0.1 dB (1%) frequency response deviation. Some loudspeakers dip below 2 Ω somewhere in the frequency band. An output impedance of about 0.02 Ω is required to meet this combination of requirements, corresponding to a DF of about 400. This DF is virtually unattainable at frequencies above about 5 kHz. Expecting a DF of 400 out to high frequencies is unrealistic, but this exercise puts the importance of DF into perspective. A DF of 40 could lead to a 1-dB frequency response deviation when a speaker's impedance dips to 2 Ω. Loudspeaker impedances go through major changes in the vicinity of woofer resonances, but they often also go through significant gyrations in the vicinity of crossover frequencies, which can lead to coloration.

27.2 Sniffing Parasitic Oscillations

Some amplifiers are designed with inadequate stability margins and may break into bursts of parasitic oscillations under some signal and loading conditions. This can lead to audible differences that may not show up in conventional bench testing. Sometimes these oscillations will be visible as small bursts on a sine wave as viewed on a wideband oscilloscope. A purpose-built test instrument can be made to sniff such oscillations, even in the presence of music played through a loudspeaker and speaker cables. Such a device is shown in block diagram form in Figure 27.2.

The instrument comprises a sharp high-pass filter with a high-order cutoff at about 600 kHz. The filter can be implemented with a passive first-order HPF at the input followed by a pair of second-order active emitter follower (Sallen-Key) filters. Rejection at 20 kHz is in excess of 80

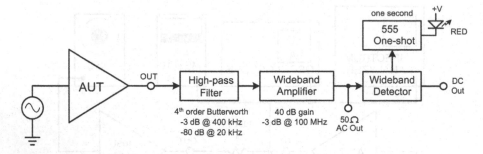

Figure 27.2 Block Diagram of a Parasitic Oscillation Sniffer

dB. Total gain should be 100 with bandwidth out to beyond 100 MHz. The AC signal is available as a 50-Ω output that can be viewed on a wideband oscilloscope. The signal is also fed to a wideband peak detector for display of detected parasitic oscillation activity as a DC signal. If the signal strength is above a certain threshold, it will also trigger a 555 timer that will illuminate an LED for 1 second.

27.3 EMI Ingress Susceptibility

EMI ingress was discussed in Chapter 22. Here we discuss a possible method to test for EMI ingress susceptibility that is more objective than operating a cell phone or hair dryer near the amplifier under test. The idea is to generate an RF signal that can be applied to any of the three conductive EMI ingress ports of an amplifier (input, output and mains). The test uses an RF multi-tone signal in the low MHz range with three tones whose intermodulation products lie in the audio band. These three tones are located at 990 kHz, 1001 kHz and 2000 kHz. The intermodulation products of interest lie at 11 kHz (1001–990) and 9 kHz (2000–1001–990).

These signals can be generated by three audio generators and combined passively for injection into the amplifier port. An optional passive high-pass filter can be placed in the injection path if there is any concern about IM distortion created by interaction among the three oscillators (such interaction should be very small due to the isolating effect of the attenuation inherent in the summing process). The distortion products at 9 kHz and 11 kHz can be viewed at the output of the amplifier with a spectrum analyzer. They can also be viewed using a 10-kHz band-pass filter and an oscilloscope. Each frequency should be accurate within 0.1% to land the IM products to within 1 kHz of the target frequency. A frequency counter is a must in setting up the frequencies. Figure 27.3 illustrates how the EMI test signals might be injected into the three ports of the amplifier.

The test signal will be especially effective when injected at the input port because of the high impedance there and the ease of passive combining of the three tones. Caution must be used in selecting the input signal amplitude because some power amplifiers may still have significant gain at 1 MHz. Each tone should be applied with amplitude of 10 mV RMS. Three 1-V RMS tones can be fed to the amplifier input through 10-kΩ resistors. The input is shunted with a 100-Ω resistor to complete the input attenuator/combiner.

The combining at the output port can take place directly at the output node of the amplifier with 10 mA RMS injected by each of the oscillators. This is simply accomplished by routing each signal through a 600-Ω resistor with the oscillator set to generate 6 V RMS into a 600-Ω load.

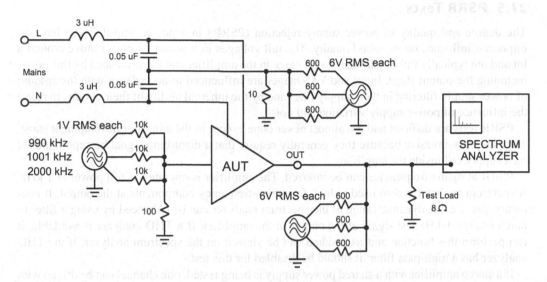

Figure 27.3 Injection of the RF Multitone EMI Test Signals into the Three Ports of the Amplifier

This test should be done with a standard 8-Ω load connected across the output terminals. Notice that this test produces peak currents of 42 mA into the output port with a peak current rate of change of 56 mA/μs.

Signal combining and injection at the mains port requires a bit more thought. Caution must be observed in light of the high mains voltages involved. The test signal is most easily applied in the common mode to the mains input. A 3-μH air-core inductor can be connected in series with each of the hot and neutral lines (the impedance of each inductor equals about 19 Ω at 1 MHz). Each of those lines is connected to a summing point through a 0.05-μF capacitor rated at 600 V. Notice that this summing junction will have half the mains voltage on it in the absence of any loading at the summing point. The impedance of each 0.05-μF capacitor at 60 Hz is about 60 kΩ (it is about 3 W at 1 MHz). The summing node is connected to ground through a 10-Ω resistor and each of the tones is connected to the summing node through a 600-Ω resistor. With each tone set to 6 V RMS, 10 mA RMS is injected into the summing node by each tone. This will result in about 70 mV RMS for each tone in the common mode on the mains lines when the impedance of the series inductors is considered. Other arrangements are also possible. For example, a small air-core transformer can be used to inject the signal to one side or the other of the mains, or to both sides if the transformer has three windings.

27.4 Burst Power and Peak Current

Conventional tests do not adequately probe an amplifier's ability to deliver large bursts of power or output current on a momentary basis. The closest that they have come was the old dynamic headroom test. Tone-burst testing of amplifiers can help evaluate these short-term performance capabilities without damaging the amplifier. A two-cycle tone burst at 50 Hz with load resistances of 8 Ω, 4 Ω, 2 Ω and 1 Ω can probe both an amplifier's short-term power capability (dynamic headroom) and its maximum current output capability. The duty cycle of the tone burst should be 10% or less. The tone-burst generator was discussed in Section 25.6.

27.5 PSRR Tests

The degree and quality of power supply rejection (PSRR) in a power amplifier can have an important influence on its sound quality. The rail voltages in a power amplifier move around a lot and are typically full of garbage. Every stage in the amplifier can be influenced by this noise, including the output stage. Input and VAS stages are influenced in accordance with the amount of power supply filtering in their supply lines and by the inherent ability of their circuits to reject the influence of power supply variation and noise.

PSRR tests are difficult and are almost never done except in the amplifier development stage. They are very invasive because they generally require that a disturbing signal be superimposed on a supply rail inside the amplifier.

PSRR at mains frequencies can be inferred. The amplifier is operated at full power at 1 kHz. A spectrum analyzer is then used to look for mains-frequency components at the output. If necessary, the useable dynamic range of the spectrum analyzer can be enhanced by using a filter to notch out the 1-kHz test signal at the output of the amplifier. If a THD analyzer is available, it can perform this function and its residual can be viewed on the spectrum analyzer. If the THD analyzer has a high-pass filter, it should be disabled for this test.

If a stereo amplifier with a shared power supply is being tested, one channel can be driven with a sine wave at medium to high power into a load while the output of the other channel, with no input, can be measured or viewed. The loading sine wave could be at any frequency, but a low or even subsonic frequency might be of greatest interest. One could also run a THD test on the channel under test. In another variation on this test, the loading channel can be fed two tones that create a beat-frequency power envelope to dynamically load the power supply.

27.6 Low-Frequency Tests

It is often mistakenly thought that achieving low THD at low frequencies in a solid-state power amplifier is easy, especially when large amounts of negative feedback are available at low frequencies. This is not always so. This can lead to low-frequency performance being taken for granted and low-frequency tests being overlooked. THD at 50 Hz (or lower) should be measured, and the residual subjected to spectral analysis. Some surprises are often in store. Such tests stress power supply rail rigidity and invite intermodulation with the 120-Hz ripple present on the power supply rails.

Beat Frequency Tests

Tests with a very low frequency component of output power variation can also stress an amplifier. Consider what may be called the low-frequency CCIF test. In this test sinusoidal signals of equal amplitude at 19 Hz and 20 Hz are applied to the amplifier. The output power of the amplifier then rises and falls in accordance with the 1-Hz beat frequency of the applied low-frequency tones.

This test can be particularly brutal and revealing, as the output stage is called on to change from no power output to full power output at a 1-Hz rate as the 19-Hz and 20-Hz signals beat against each other. This could reveal some thermal distortion mechanisms. No energy should be present in the output at 1 Hz, 2 Hz, 17 Hz, 18 Hz or 39 Hz. If an analog spectrum analyzer is used for this test, the sweep can take quite some time because the analysis bandwidth must be set small (1 Hz or lower).

27.7 Back-Feeding Tests

The IIM test proposed by Otala [1, 2] was an early example of employing a back-feed test to an amplifier. That test was similar to the SMPTE IM test, but using 60-Hz and 1000-Hz signals in a 1:1 ratio. However, the high-frequency signal was fed forward through the amplifier under test while the low-frequency signal was fed backward through a dummy load resistor from a laboratory amplifier. This test exercised the output current range of the output stage while evaluating any resultant modulation of the forward-propagated high-frequency signal. The back-feed testing concept can be generalized to many other useful testing procedures.

Back-Fed Beat Frequency Test

A variant of the low-frequency 19 + 20-Hz beat frequency test is to apply one of the tones in the forward direction and the other one via back-feeding. Such a test exercises a large portion of the SOA of the output stage without resort to use of a reactive load. This test will really exercise the protection circuits. Such a test arrangement is shown in Figure 27.4.

Both amplifiers will be subjected to various combinations of high voltage swing and high current swing while the phase relationship between voltage and current will change, in some ways emulating what may be seen when driving a reactive load. This test is not for the faint of heart. The reference amplifier providing the back-feed signal should be a large hefty one with substantial SOA. The output of the AUT should be observed on an oscilloscope.

THD-20 in the Presence of Low-Frequency Back-Feed

This test feeds a 20-kHz test signal through the amplifier at the power level where crossover distortion is highest and THD-20 is measured. A 20-Hz back-drive signal is applied through an 8-Ω load resistor to produce a substantial back-feed current. THD-20 is measured and the residual is then viewed at the output of the THD analyzer. The change in the residual as a function of the low-frequency back-drive signal is noted. This allows you to see the effect of the larger output stage current swings on the THD distortion products. The effect of crossover distortion on the THD-20 will vary as the 20-kHz signal moves in and out of the crossover region. To perform this test you need a THD analyzer setup that has good rejection of the 20-Hz back-drive signal.

Current-Induced Distortion (CID) Tests

Two of the more villainous distortions in power amplifiers are output stage crossover distortion and pickup from class-AB half-wave-rectified signal currents. Both of these distortions are caused by signal current rather than signal voltage. These are *current-induced distortions* (CID). Both of these distortions have potentially broadband distortion spectra that are not at all benign.

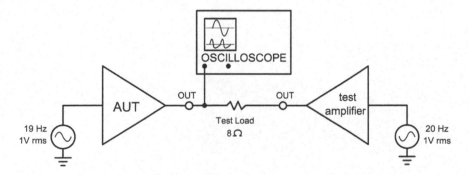

Figure 27.4 Two-Tone Back-Feeding Test Using 19-Hz and 20-Hz Test Signals

To isolate the current-induced distortions the amplifier is forced to produce a significant output signal current without making it produce any significant voltage output. This can be done by back-feeding current into the output of the amplifier through a load resistor from another amplifier. The amplitude of the distortion spectra present at the output of the amplifier under test is then observed with a spectrum analyzer.

One version of this test is implemented by feeding a 20-kHz sine wave at different power levels from one channel of a stereo amplifier through an 8-Ω load resistor into the output of the channel of the amplifier under test, whose input is grounded. The signal at the output terminals of the channel under test is then fed to a THD analyzer, oscilloscope or spectrum analyzer. Caution is advised with a test like this because it exercises a significant portion of the safe area of the output stage. With the output held essentially at ground, the peak current corresponding to the power output of the driving amplifier is drawn from the output stage while the full rail voltage of the amplifier under test is impressed across its output stage. In addition to unmasking crossover distortion, this test can also reveal distortion from poor PSRR, and distortion from nonlinear ground currents.

References

1. Matti Otala and Jorma Lammasniemi, "Intermodulation Distortion in the Amplifier Loudspeaker Interface," 59th Convention of the Audio Engineering Society, preprint no. 1336, February 1978.
2. Robert R. Cordell, "Open-loop Output Impedance and Interface Intermodulation Distortion in Audio Power Amplifiers," 64th Convention of the AES, preprint no. 1537, 1982. Available at www.cordellaudio.com.

Part 5

Topics in Amplifier Design

Part 5 covers many of those other important topics that do not fit neatly into the other parts. Advanced designers as well as audiophiles will find many interesting discussions in this part. Some of the controversies in audio, such as the use of negative feedback, are addressed here. Chapter 29 discusses the challenges faced in the design of amplifiers that do not use negative feedback or which maintain wide open-loop bandwidth for those who believe these features to be important. Part 5 also covers other amplifier designs, such as fully balanced designs and power amplifiers that can be implemented all or in part with integrated circuits. Finally, the unique features and challenges of professional power amplifiers are discussed in Chapter 32.

Chapter 28

The Negative Feedback Controversy

28. INTRODUCTION

The use of negative feedback has been controversial for many years in the hi-end audio community. Some argue that none should be used, while others argue that only small amounts should be used. Some argue that wide open-loop bandwidth is required to achieve the best sound quality. Some argue that global negative feedback is bad but that local negative feedback in IPS and VAS stages is OK.

28.1 How Negative Feedback Got Its Bad Rap

Negative feedback has gotten a mostly undeserved bad rap. Much of this is because poorly designed solid-state amplifiers of the 1970s happened to use large amounts of negative feedback. These were poor-performing designs in the first place, but negative feedback got the blame. Some designers carelessly believed that negative feedback could be used to linearize a design that was inherently not very linear to begin with.

Amplifier Limitations of the 1970s

Audio power amplifiers of the 1970s, being part of the early era of solid-state power amplifier technology, suffered many problems. These are some of the problems that contributed to poor sound quality.

- Slow power transistors with inadequate SOA
- Too few output devices
- Frequent use of quasi-complementary output stages
- Aggressive protection circuits that misbehaved
- Excessive crossover distortion
- Inadequate current gain
- Input stages with little dynamic range, leading to inadequate slew rate
- Poor stability margins and occasional parasitic oscillations
- Output coils wrapped around aluminum electrolytic power supply capacitors
- Poor capacitor choices

Guilt by Association

Most of these early designs achieved decent distortion measurements at 1 kHz by using negative feedback, but nevertheless did not sound good. In a sense, the negative feedback allowed designers to make bad choices or cut corners. A very good example of such poor choices was the use of un-degenerated differential input stages in the misguided belief that the resulting higher gain would provide more negative feedback and lower distortion, at least in the mid-band. Those designers did not realize that they were crippling the amplifier's ability to deliver high slew rate. All of this was compounded by the fact that many designers were struggling to harness the new solid-state technology. Power transistors were slow and had poor SOA, and so required intrusive protection circuits.

It was not feedback itself that was responsible for the poor sound, but its inability to perform miracles on fundamentally poor circuit designs; that was the problem. Nevertheless, feedback got the blame for the poor sound. In a sense, a form of architectural profiling emerged, in which some treated any amplifier using negative feedback with suspicion.

TIM, PIM and IIM

During the 1970s and early 1980s several researchers sought to identify logical and measurable phenomena that were correlated with poor sound [1–9]. This effort was noble, but often the wrong conclusions were drawn. In many cases a distortion mechanism would be identified and negative feedback would be given the blame. A measurement technique for the identified distortion would then be defined [8–11]. Those very measurements ultimately disproved the assertion that negative feedback was the villain [12–14].

The distortions that were identified do indeed exist and can be measured. That is not the controversy. The point is that negative feedback itself, when properly applied, does not exacerbate these distortions. All of these distortions are in fact quite measurable in amplifiers that do not even have any global negative feedback.

28.2 Negative Feedback and Open-Loop Bandwidth

It may seem intuitive to many that the open-loop bandwidth of a feedback amplifier should extend to the highest audio frequencies. This allows NFB to act equally on all frequencies. It was also demonstrated by Otala that error overshoot would occur in the input stage of a feedback amplifier when the open-loop bandwidth was substantially less than the low-pass filtered bandwidth of a square wave.

The amount of NFB applied at the highest audio frequencies (e.g., 20 kHz) is necessarily limited by feedback stability considerations. The open-loop gain must usually fall at 6 dB per octave so that the NFB loop gain reaches 0 dB at a sufficiently low unity gain frequency, often on the order of 1 MHz or less. In such a case the NFB at 20 kHz will be about 34 dB. If the open-loop bandwidth is 20 kHz, the amount of NFB at low frequencies will also be about 34 dB. If the unity gain frequency is kept the same and the open-loop bandwidth is allowed to decrease, the amount of negative feedback at lower frequencies will increase. For example, if the open-loop bandwidth is 1 kHz, the NFB at 1 kHz and below will be about 60 dB. This gives rise to the association of high feedback with bad sound, since high feedback occurs coincidentally with low open-loop bandwidth when the unity gain frequency is held constant.

The Input Stage Error Signal

Figure 28.1 shows the input stage error signal for two amplifiers when driven with a 5-kHz square wave that is rolled off with a first-order filter at 30 kHz (as in the DIM-30 test). Both amplifiers have the same 500-kHz unity gain frequency and the same amount of negative feedback at 20 kHz. The amplifier in Figure 28.1a has a 20-kHz open-loop bandwidth and open-loop gain of

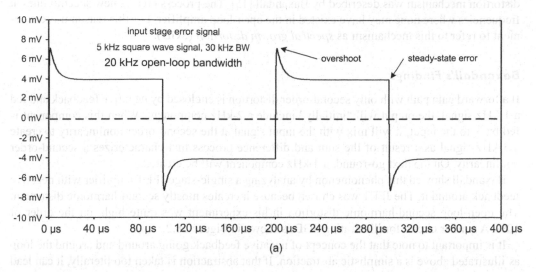

(a)

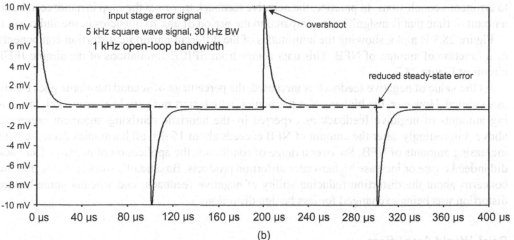

(b)

Figure 28.1 Input Stage Error Signal (a) With High Open-Loop Bandwidth (b) With Low Open-Loop Bandwidth

about 54 dB. The amplifier in Figure 28.1b has a 1-kHz open-loop bandwidth and open-loop gain of about 80 dB. Both amplifiers are driven to a peak output voltage of 2 V.

The early papers on TIM made much of the concern about the overshoot in amplifiers with low open-loop bandwidth. However, it is clear that when an apples-apples comparison is done, the "overshoot" is really created by a major *reduction* in error as time progresses. The peak stress on the input stage is similar in both cases (7 mV and 10 mV, respectively), but the average stress is much larger in the case where open-loop bandwidth has been made large.

28.3 Spectral Growth Distortion

If distortion products are created in the forward path of an amplifier, negative feedback will feed these new frequencies back to the input stage, where they will have another opportunity to mix with the input signal where the nonlinearities in the forward path are encountered. This *reentrant* distortion mechanism was described by Baxandall [15]. The process creates new spectral lines at frequencies where none may have existed in the open-loop amplifier. For this reason it is convenient to refer to this mechanism as *spectral growth distortion* (SGD).

Baxandall's Findings

If a forward gain path with only second-order distortion is enclosed by negative feedback and fed a 1-kHz signal, the output will "initially" include a 2-kHz component. When this component is fed back to the input, it will mix with the input signal at the second-order nonlinearity to create a 3-kHz signal as a result of the sum and difference process that characterizes a second-order nonlinearity. On the next go-round, a 4-kHz component will be created.

Baxandall showed this phenomenon by analyzing a single-stage JFET amplifier with negative feedback around it. The JFET was chosen because it creates mostly second-harmonic distortion. The open-loop second-harmonic distortion in his experiment was quite high, on the order of 10%. A similar circuit for illustrating SGD is shown in Figure 28.2.

It is important to note that the concept of negative feedback going around and around the loop as illustrated above is a simplistic abstraction. If that abstraction is taken too literally, it can lead to erroneous conclusions. In practice, the negative feedback traverses the loop in nanoseconds, an amount of time that is insignificant compared to the period of any frequencies in the audio band.

Figure 28.3 is a plot showing the amplitudes of the different harmonic distortion components as a function of amount of NFB. This data comes from SPICE simulations of the simple JFET circuit of Figure 28.2.

As the value of negative feedback is increased, the percentage of second harmonic goes down, as expected. However, the higher-order harmonics, which start out very low, grow with increasing amounts of negative feedback as expected by the heuristic remixing argument presented above. Interestingly, after the amount of NFB exceeds about 15 dB, all harmonics decrease with increasing amounts of NFB. So, over a range of conditions, the application of negative feedback did indeed create or increase higher-order distortion products. Baxandall's work raised legitimate concerns about the distortion-reducing ability of negative feedback, and whether some benign distortion was being exchanged for less benign distortion.

Real-World Amplifiers

Baxandall's work was incomplete in that it only dealt with a single amplifier stage. Multistage amplifiers have a higher-order distortion characteristic due to the multiplication of individual characteristics that occurs as the stages are cascaded.

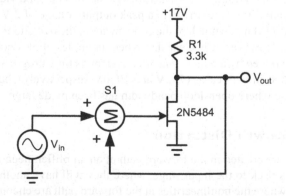

Figure 28.2 One-Stage Circuit for Evaluating Spectral Growth Distortion

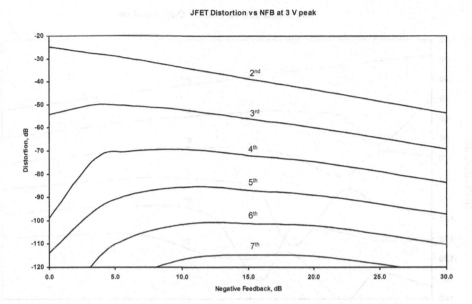

Figure 28.3 Harmonic Distortion Components Versus Amount of Negative Feedback

Moreover, real circuits are not pure second order, but more often exponential. The BJT has an exponential characteristic that is naturally rich in high-order distortion components. Even real circuits implemented with square-law devices do not have a square-law characteristic. A JFET differential pair does not have a square-law characteristic. A class AB output stage certainly does not have a square-law characteristic. The bottom line is that real amplifiers have complex nonlinearities in their open loop to begin with. In many cases, any spectral growth that occurs may be smaller than these initial high-order products.

Baxandall was operating the JFET amplifier stage at a very high distortion level in order to demonstrate the point. Open-loop distortion was on the order of 10% in the single common-source stage he demonstrated. The effects described become much smaller at more reasonable operating levels, especially when one recognizes that the higher-order distortion products increase much faster with increases in signal level than the low-order products (e.g., fifth-order distortion products go up 5 dB for every dB increase in operating level).

Degeneration and SGD

SGD is not limited to global negative feedback. Indeed, Baxandall's experiment involved rather local feedback. It turns out that even emitter degeneration can be shown to exhibit the SGD Baxandall effect. A single-ended BJT amplifier stage exhibits spectral growth if its gain is reduced by applying NFB in the form of emitter degeneration, as shown in Figure 28.4.

Once the amount of total negative feedback around the stage exceeds about 20 dB the spectral growth stops and all orders of distortion decrease as feedback is increased. It is tempting to generalize that this 20 dB number includes emitter degeneration or other local feedback. The existence of the SGD effect seems to make a good case for designing the amplifier for good open-loop linearity. The first 20 dB of feedback can be applied locally, typically as emitter degeneration, to get beyond the starting region of the SGD effect. The application of global negative feedback will then not be expected to cause any SGD.

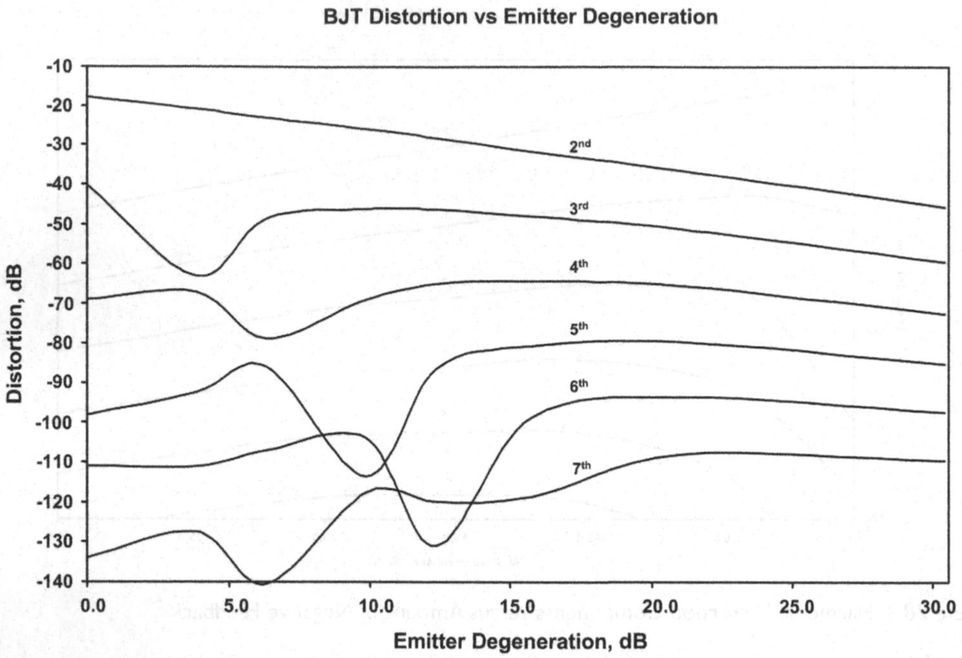

BJT Distortion vs Emitter Degeneration

Figure 28.4 Harmonic Distortion Versus Amount of Emitter Degeneration

SGD and Crossover Distortion

Crossover distortion is usually the biggest and most audible distortion in a properly designed amplifier. It starts out being rich in high-order products. Figure 28.5 shows that the application of NFB to an amplifier reduces all orders of crossover distortion, right from the beginning.

Although the Baxandall effect was an intriguing eye-opener, the audio community read too much into it, wrongly generalizing the results and asserting that feedback did not really reduce the net imperfection of the signal.

28.4 Global Versus Local Feedback

Issues of open-loop bandwidth and frequency compensation largely pertain to global feedback loops that usually enclose virtually all of the amplifier stages. In contrast, local negative feedback rarely needs compensation and typically has very wide bandwidth. Emitter degeneration is a form of local negative feedback. Shunt feedback around a single stage is also local negative feedback.

Some who are opposed to the use of negative feedback are also opposed to the use of local feedback, but do not consider emitter degeneration to carry with it the supposed ills of negative feedback. Analysis shows that even emitter degeneration causes spectral growth distortion just like any other form of negative feedback.

28.5 Timeliness of Correction

Some critics of negative feedback argue that NFB represents an electronic attempt to correct an error after it has happened and that the finite time delay and sequence of events make the correction faulty. The electronic time-of-flight delay and phase delay due to feedback

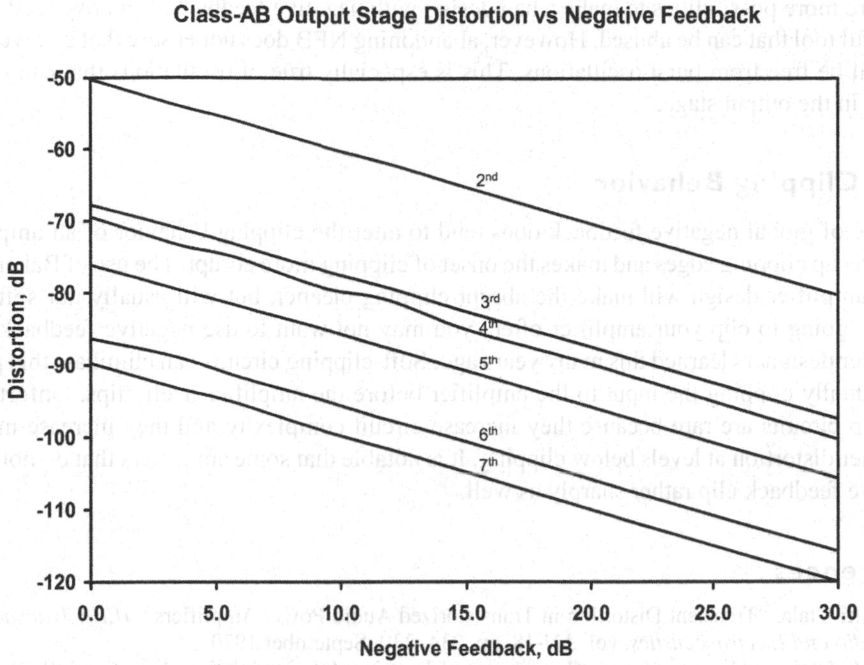

Class-AB Output Stage Distortion vs Negative Feedback

Figure 28.5 Harmonic Distortion Versus Feedback for a Class AB Output Stage

compensation do indeed exist, but these delays must be small in order for the circuit to be stable. In an amplifier with a 1-MHz unity gain frequency the delay must certainly be less than 0.5 μs. This is 100 times smaller than the period of a 20-kHz sinusoid. This delay is merely a different way of recognizing that the distortion-reducing properties of negative feedback are less effective at very high frequencies. It does not suggest that negative feedback is failing to correct an error at 20 kHz.

28.6 EMI from the Speaker Cable

The speaker cable is a big antenna. The concern about negative feedback here is that EMI from the loudspeaker cable will get back to the input via the feedback path [16]. This concern is not completely unfounded. In fact, the use of a phase lead capacitor across the feedback resistor can make the input stage unnecessarily vulnerable to EMI that makes its way into the amplifier via the speaker cable. Such EMI will be attenuated by the shunting impedance of the output stage and by the feedback network before it arrives at the input stage. Nevertheless, this is a good reason to employ an input stage that has good signal-handling capability to high frequencies, such as a JFET stage or a well-degenerated BJT stage operated at a healthy bias current. Such an input stage is more resistant to EMI effects from the input port as well.

28.7 Stability and Burst Oscillations

An amplifier that does not employ global negative feedback does not need to be properly compensated (it does not need to be compensated at all) for global loop stability. Such amplifiers will tend to be less prone to burst oscillations due to global feedback loop instability. It is true that there are more possibilities to make a bad design with negative feedback. Negative feedback is a powerful tool that can be abused. However, abandoning NFB does not ensure that a power amplifier will be free from burst oscillations. This is especially true of oscillations that can originate locally in the output stage.

28.8 Clipping Behavior

The use of global negative feedback does tend to alter the clipping behavior of an amplifier. It sharpens up clipping edges and makes the onset of clipping more abrupt. The use of Baker clamps in the amplifier design will make the abrupt clipping cleaner, but will usually not soften it. If you are going to clip your amplifier often, you may not want to use negative feedback. Guitar amplifier designers learned this many years ago. Soft-clipping circuits can eliminate this problem by gradually clipping the input to the amplifier before the amplifier itself clips. Unfortunately, soft-clip circuits are rare because they increase circuit complexity and they increase measured amplifier distortion at levels below clipping. It is notable that some amplifiers that do not employ negative feedback clip rather sharply as well.

References

1. Matti Otala, "Transient Distortion in Transistorized Audio Power Amplifiers," *IEEE Transactions on Audio and Electroacoustics*, vol. AU-18, pp. 234–239, September 1970.
2. Matti Otala and Eero Leinonen, "The Theory of Transient Intermodulation Distortion," *IEEE Transactions on Acoustics, Speech and Signal Processing*, vol. ASSP-25, no. 1, pp. 2–8, February 1977.
3. W. Marshall Leach, "Transient IM Distortion in Power Amplifiers," *Audio*, pp. 34–41, February 1975.
4. W. Marshall Leach, "Suppression of Slew-Rate and Transient Intermodulation Distortions in Audio Power Amplifiers," *Journal of the Audio Engineering Society*, vol. 25, no. 7–8, pp. 466–473, July–August 1977.
5. Richard A. Greiner, "Amp Design and Overload," *Audio*, pp. 50–62, November 1977.
6. Matti Otala, "Feedback-Generated Phase Modulation in Audio Amplifiers," 65th Convention of the Audio Engineering Society, preprint no. 1576, London, 1980.
7. Matti Otala, "Conversion of Amplitude Nonlinearities to Phase Nonlinearities in Feedback Audio Amplifiers," Proceedings of IEEE International Conference on Acoustics, Speech and Signal Processing, Denver, CO, 1980, pp. 498–499.
8. Matti Otala, "Phase Modulation and Intermodulation in Feedback Audio Amplifiers," 69th Convention of the Audio Engineering Society, preprint no. 1751, Hamburg, 1981.
9. Matti Otala and Jorma Lammasniemi, "Intermodulation Distortion in the Amplifier Loudspeaker Interface," 59th Convention of the Audio Engineering Society, preprint no. 1336, February 1978.
10. Eero Leinonen, Matti Otala, and John Curl, "A Method for Measuring Transient Intermodulation Distortion (TIM)," *Journal of the Audio Engineering Society*, vol. 25, no. 4, pp. 170–177, April 1977.
11. Eero Leinonen and Matti Otala, "Correlation Audio Distortion Measurements," *Journal of the Audio Engineering Society*, vol. 26, no. 1–2, pp. 12–19, January–February 1978.
12. Robert R. Cordell, "Another View of TIM," *Audio*, February–March 1980. Available at www.cordellaudio.com.
13. Robert R. Cordell, "Phase Intermodulation Distortion—Instrumentation and Measurements," *Journal of the Audio Engineering Society*, vol. 31, March 1983. Available at www.cordellaudio.com.
14. Robert R. Cordell, "Open-Loop Output Impedance and Interface Intermodulation Distortion in Audio Power Amplifiers," 64th Convention of the AES, preprint no. 1537, 1982. Available at www.cordellaudio.com.
15. Peter J. Baxandall, "Audio Power Amplifier Design—5," *Wireless World*, December 1978.
16. A. Neville Thiele, "Load Stabilizing Networks for Audio Amplifiers," *Journal of the Audio Engineering Society*, vol. 24, no. 1, pp. 20–23, January–February 1976.

Chapter 29

Amplifiers Without Negative Feedback

29. INTRODUCTION

Some designers eschew the use of negative feedback, while many others embrace it. There are two sides to this controversy, and both deserve exploration. In this chapter we'll look at the design trade-offs in amplifiers that use little or no negative feedback and discuss some approaches to high-quality implementations of these amplifiers.

The greatest focus here will be on amplifiers with no negative feedback of any kind, since this is the most difficult challenge. Amplifiers with no global negative feedback but with liberal use of local negative feedback (not enclosing the output stage) will also be discussed. Finally, although a bit off topic, amplifiers with wide open-loop bandwidth will be discussed.

29.1 Design Trade-Offs and Challenges

The basic assumption here is that no-feedback amplifiers do not employ global negative feedback from the output of the amplifier. They may employ local negative feedback within the IPS, VAS and driver circuits. However, some designers prefer a more strict definition of no-feedback and philosophically limit themselves to emitter degeneration. The challenges described here assume the more strict view of having no negative feedback.

The challenges in a *no-negative-feedback* (NNFB) amplifier design begin with the input stage. It must handle the full line-level signal swing with very low distortion. It will typically require a large amount of emitter degeneration to do this and noise will be increased as a result.

The VAS must be able to produce the full output swing of the amplifier with equally low distortion and with gain that is well defined by a load resistance. The substantial current swing required to drive the load resistance will create distortion. The large voltage swing at the output of the VAS may allow the Early effect to cause distortion as well.

The gain of most audio power amplifiers lies in the range of 20–30. The VAS gain must therefore be held to about 10–30 to allow at least unity gain for the input stage. This means that the VAS must be heavily degenerated; this may also increase VAS noise. Because the input stage gain will be typically small, much of the VAS noise will be referred back to the input, further compromising amplifier noise performance. Power supply noise making its way into the input stage and VAS will not be mitigated by negative feedback, so inherent PSRR of these circuits must be high and the power rails must be very quiet.

The output stage is one of the most significant contributors to distortion in an amplifier, so any amplifier that does not include global negative feedback faces a big challenge here. Without feedback, it is especially important to minimize crossover distortion. The most straightforward way to do this is to use more pairs of output transistors optimally biased with fairly small emitter resistors. The amplifier will run hotter, but it will benefit from a larger class A region without *gm* doubling.

Beta droop in the output stage will cause signal-dependent loading on the VAS. This will cause distortion at the moderately high impedance output node of the VAS. For this reason, the current gain of the output stage must be made very high. An output Triple will provide adequate current gain, but four levels of output emitter followers (an output Quad) may be better able to minimize this distortion. The Early effect in the output transistors can also contribute distortion.

The damping factor in some NNFB amplifiers can be poor because there is no NFB to reduce output impedance. Fortunately, if more output pairs are used in combination with an output Quad, the damping factor will be good.

DC offset can be a problem for NNFB amplifiers because the amplifier gain will usually be maintained at its full value down to DC. In the absence of NFB, there is no feedback network shunt capacitor (often an electrolytic) to reduce gain down to unity at DC. Input stage DC offset is thus multiplied by the full gain of the NNFB amplifier. Moreover, because the input stage gain may be as low as unity in some designs, DC offsets in the VAS will add to offset problems. Strict adherence to the no-feedback philosophy rules out the use of a DC servo.

Input Stage Dynamic Range and Distortion

The input stage must handle the full input signal swing with generous margin so as not to distort. Large amounts of emitter degeneration are required and input noise characteristics may be compromised. It is very important to use balanced differential architectures to maintain good PSRR.

BJTs will usually perform better in this situation, but their input bias current and input noise current can exacerbate the DC offset problem and compromise SNR in some arrangements. Many prefer the sound of JFET inputs. In some cases JFET inputs are more resistant to EMI effects. However, the linearity of a JFET input stage is a bigger challenge. In some cases it is better to employ the JFETs as source follower buffers ahead of a BJT LTP.

Figure 29.1 shows a BJT and a JFET input pair. Each BJT emitter has a 650-Ω degeneration resistor to set the gain. This corresponds to a generous degeneration factor of 52:1. The degeneration resistors for the JFET pair are smaller, at 270 Ω because of the lower JFET transconductance.

Both stages have small-signal gain of 1.5 (to each side of the differential output) and tail current of 4 mA. Each stage has 2-kΩ load resistors to permit a theoretical maximum output of 8 V p-p on each side of the output. The nominal voltage drop across the collector resistors is 4 V. This eats into power supply headroom, and boosted supplies for the IPS and VAS are desirable

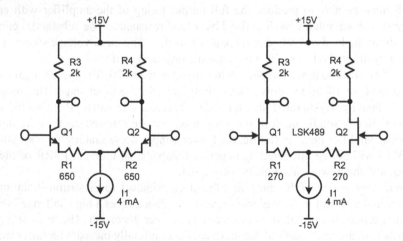

Figure 29.1 BJT and JFET Input Pairs Each with a Gain of 1.5 to One Output

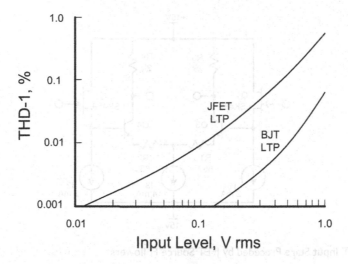

Figure 29.2 THD Versus Input Level for BJT and JFET Input Pairs

in amplifiers like this. Figure 29.2 shows THD as a function of input level for each pair. Notice the substantially larger THD from the JFET pair. The JFET has smaller transconductance than the BJT for a given operating current, and so signal-dependent changes in transconductance of the JFET have a greater influence on incremental voltage gain of the stage. This results in higher distortion.

Whenever a large amount of emitter degeneration is used in an input stage, resistor noise is a concern. From Chapter 5 we know that resistor noise is:

$$E_n = 4.2 \text{ nV}/\sqrt{\text{Hz}} \text{ per } \sqrt{\text{k}\Omega}$$

The total emitter-to-emitter resistance of the BJT pair is 1300 Ω, so the resistor noise contribution is 4.8 nV/√Hz. Simulation shows total input-referred noise of 5.8 nV/√Hz, reflecting additional smaller contributions from the transistors and load resistors. This is very good for a power amplifier, but input-referred VAS noise has not yet been considered. For comparison, the simulated input noise of the JFET pair, using LSK489 devices, is 5 nV/√Hz. Interestingly, the amount of resistance needed to degenerate the BJT LTP to the same transconductance as the JFET LTP makes the BJT circuit slightly noisier. Bear in mind that these simulations were done with zero-impedance input sources. Finite input source impedance will allow additional noise to be generated in the BJT case because of BJT input noise current.

JFET Input Buffers

The BJT IPS can be used while retaining its distortion advantage and eliminating its input current disadvantage by preceding both inputs with matched JFET source followers. The JFETs will add little noise in comparison to the noise of the degenerated BJT LTP and will eliminate DC input bias current and input current noise. This can actually mitigate DC offset issues. Total input noise of this arrangement, shown in Figure 29.3, is 7.0 nV/√Hz. Current sources I1 and I2 can be replaced with 330-Ω base-emitter resistors if I3 is increased to 8 mA.

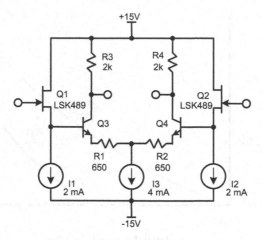

Figure 29.3 A BJT Input Stage Preceded by JFET Source Followers

JFET CFP Input Stage

Even better input stage distortion can be had while retaining the benefits of a JFET input stage by employing JFET-BJT CFP transistors in the input pair, as discussed in Chapter 9. In such a stage, each of the four transistors could be run at only 1 mA, and distortion would still be much less. Smaller degeneration resistors could also be used while retaining very low distortion as well.

Cascoding the Input Stage

It is especially desirable to employ a cascoded input stage to achieve better PSRR. Moreover, further reduction in common-mode input stage distortion can be had if the cascode bases are driven with a replica of the input common-mode signal so as to minimize the effects of common-mode nonlinearities.

Gain Allocation

Another issue that makes for difficulty in designing NNFB amplifiers is what to do with all the VAS gain. It is actually difficult to get that gain down. Shunting resistive loads are often used on the VAS output node to control the gain. This makes the VAS work harder and increases distortion. Large amounts of emitter degeneration are also often used in the VAS to control gain and reduce distortion. If local negative feedback is permitted, it can be used quite effectively to rid the VAS of much excess gain without compromising its distortion characteristics.

The forward gain in an amplifier without global negative feedback must be carefully controlled and allocated. Overall, best performance is achieved if the gain of the amplifier is made to be on the high side. A gain of about 30 is a good number. A simple allocation sets a differential gain of three in the input stage (1.5 to each output side) and a gain of 10 in the VAS.

Controlling VAS Gain

Achieving a gain of only 3 in the input stage is not difficult, but achieving a gain of only 10 in the VAS requires large emitter degeneration resistors if the VAS collector load resistor is to be of a reasonable value. The key challenge with the VAS is to achieve the large voltage swings with controlled gain and low distortion. Resistively loading the VAS output node to ground is necessary in designs that use no negative feedback of any kind. The changes in transistor current needed to create the large signal across the VAS load resistor tend to increase VAS distortion.

Consider the simple VAS of Figure 29.4a. It consists of a differential pair loaded by a current mirror. The differential-to-single-ended gain of the VAS is set to 10. The net load resistance on the single-ended output of the VAS is 10 kΩ. The quiescent bias of the top and bottom VAS transistors is 10 mA. This means that 20 mA can be driven into the 10-kΩ load resistance, producing a theoretical output voltage of 400 V p-p were it not for rail voltage limiting. A 100-W amplifier requires 80 V p-p. This means that the swing margin is 5:1. Only 1/5 of the theoretical swing is actually used; this means that the operating currents change by only 20% to generate the actual needed output swing. This reduces distortion due to current variations in the transistors.

To achieve a gain of only 10 to the single-ended output, a differential emitter degeneration resistance of 2 kΩ is required. The use of the twin emitter current source arrangement makes this possible without incurring very large DC drops across the emitter degeneration resistors that would be employed in a conventional circuit.

An improved practical version of the VAS is shown in Figure 29.4b. Input emitter followers have been added to isolate the VAS from the input stage and make loading on the input stage very light. The differential pair and the current mirror have been cascoded to reduce Early effect distortion. A second cascode has been added on the side driving the current mirror to equalize

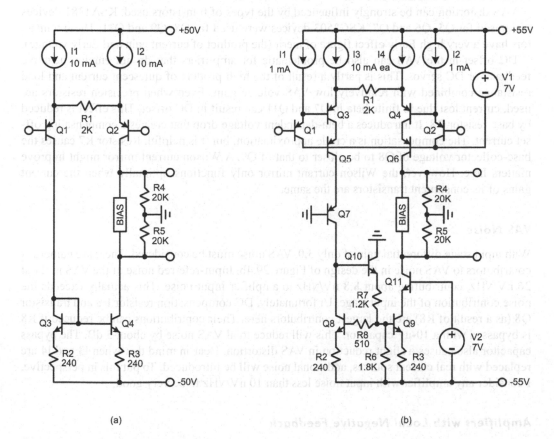

(a)   (b)

Figure 29.4 (a) A Suitable VAS with a Gain of 10 (b) An Improved Version

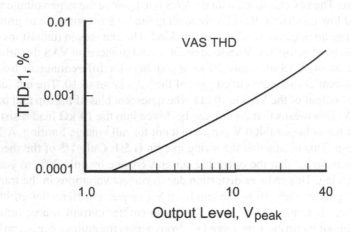

Figure 29.5 THD of the VAS of Figure 29.4b

nominal quiescent collector voltages and power dissipations of the upper stage. Distortion of this VAS as a function of peak output signal level is shown in Figure 29.5. Simulated THD-1 at a peak output of 40 V is less than 0.004%.

VAS distortion can be strongly influenced by the types of transistors used. KSA1381 devices were used for Q5, Q6 and Q7. KSC3503 devices were used for Q8, Q9 and Q11. These transistors have a very high Early effect figure of merit (the product of current gain and Early voltage).

DC offset in the VAS can also be problematic for amplifiers that do not employ negative feedback or DC servos. This is partly a result of the high product of quiescent current and load resistance combined with relatively low VAS voltage gain. Even when precision resistors are used, current lost due to finite beta in Q7 and Q11 can result in DC offset. This effect is reduced by base resistor R8. It introduces a beta-dependent voltage drop that creates a compensating offset current. The compensation is a crude approximation, but it is helpful. Resistor R7 causes the base-collector voltage of Q8 to be closer to that of Q9. A Wilson current mirror might improve matters here. However, the Wilson current mirror only functions optimally when the current gains of its constituent transistors are the same.

VAS Noise

With input stage differential gain of only 3.0, VAS noise must be considered. There are numerous contributors to VAS noise in the design of Figure 29.4b. Input-referred noise of the VAS is about 26 nV/$\sqrt{Hz}$, contributing about 8.8 nV/$\sqrt{Hz}$ to amplifier input noise. This actually exceeds the noise contribution of the input stage. Unfortunately, DC compensation resistor R8 and transistor Q8 (as a result of R8) are the largest contributors here. Their contributions can be reduced if R8 is bypassed with a 10-μF capacitor. This will reduce total VAS noise by about 3 dB. The bypass capacitor also causes a slight reduction in VAS distortion. Bear in mind that when I3 and I4 are replaced with real current sources, additional noise will be introduced. To put this in perspective, I consider any amplifier with input noise less than 10 nV/$\sqrt{Hz}$ to be very good.

Amplifiers with Local Negative Feedback

If local negative feedback is permitted, one can control the VAS gain with feedback and many of the design challenges are mitigated. The local feedback also makes the VAS stage more linear and reduces its output impedance. Local shunt feedback in the VAS would be an example of this. If local feedback is also extended to the input stage, dynamic range problems there will also be reduced, although then opportunities for balanced inputs may be diminished. Such *local* feedback encompassing more than one stage may have to incorporate frequency compensation. Such frequency compensation can be much lighter, however, since a higher closed-loop bandwidth is allowable when the slower output stage is not enclosed.

Output Stage Distortion

The class AB output stage is often the single largest source of distortion in any amplifier, whether global negative feedback is used or not. In amplifiers with no negative feedback, it is desirable to use an output stage with a large number of output pairs. This will provide a fairly large class A region and will reduce crossover distortion by reducing the nonlinear output impedance of the stage. Good thermal stability can be achieved by employing ThermalTrak™ output transistors. This may permit the use of smaller emitter resistors and correspondingly higher quiescent current, further reducing crossover distortion.

Four output pairs employing 0.15-Ω emitter resistors will have a total quiescent bias current on the order of 650 mA when optimally biased. The amplifier will be able to supply 1.3 A peak before exiting the class A region, corresponding to almost 7 W into 8 Ω. A 200-W version of such an amplifier with 65-V rails will dissipate about 85 W at idle. Figure 29.6 shows simulated THD-20 for the BJT output stage described here when driving 8-Ω and 4-Ω loads. 2.2-Ω base stopper resistors were used. Open-loop output stage distortion at 20 kHz is less than 0.06% when delivering 200 W into an 8-Ω load and 0.1% when operating at 400 W into a 4-Ω load.

The output stage should be at minimum a Triple so as to present an extremely light load to the VAS. A Quadruple including a diamond driver may be helpful here. High current gain in the output drivers is also important in minimizing the output impedance of the amplifier. In essence, we want the output impedance to be entirely governed by the output transistor circuit, as if it were being driven by a voltage source instead of a VAS. This also means that the use of large base stopper resistors on the output and driver transistor bases should be avoided to the extent possible.

An output stage Quad using a diamond buffer to drive four output pairs can be used as illustrated in Figure 12.9b.

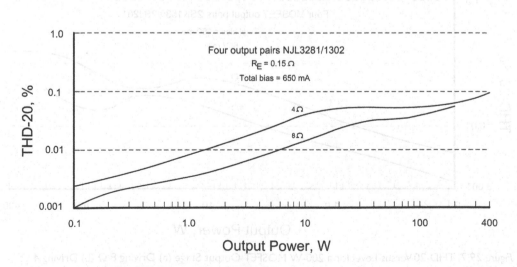

Figure 29.6 THD-20 for a 200 W BJT Output Stage (a) Driving 8 Ω (b) Driving 4 Ω

Early effect in the emitter follower output stage should be minimized to reduce nonlinearity and provide good rejection from the main power supply rails. For this reason, it is advantageous to cascode the output stage. This will cost headroom. It is also desirable to maintain a decent value of V_{ce} for the output transistors. This translates to a further price paid in amplifier power dissipation and need for higher power supply voltages.

MOSFET Output Stages

The use of MOSFET output transistors will result in somewhat more distortion in the NNFB amplifier because it is difficult to achieve adequate linearity in view of the transconductance droop problem. Nevertheless, four vertical MOSFET pairs with each pair biased at 165 mA will be able to remain in the class A region up to a peak output current of 1.3 A, corresponding to about 7 W into 8 Ω. The transconductance of each MOSFET will average about 0.7 S at 165 mA; eight devices will provide total *gm* of about 5.6 S at the crossover point, corresponding to an output impedance of about 0.18 Ω.

Figure 29.7 shows simulated THD-20 for the MOSFET output stage described here when driving 8-Ω and 4-Ω loads. EKV models were used for the 2SJ201 and 2SK1530 vertical power MOSFETs. Rail voltages of ±64 V are assumed. The MOSFET output stage includes asymmetrical source resistors for distortion reduction. The source resistors for the N-channel devices were 0.33 Ω, while those for the P-channel devices were 0.15 Ω. Open-loop output stage distortion at 20 kHz is about 0.16% when operating at 400 W into a 4-Ω load. This is only 1.6 times that of the BJT stage described earlier, with both designs operating at the same bias current.

If bias current is increased to 250 mA per MOSFET and source resistances are reduced to 0.22 Ω and 0.1 Ω, THD-20 falls below 0.09%. The output stage will then dissipate 130 W under quiescent conditions.

Damping Factor

An amplifier without global negative feedback will tend to have higher output impedance because there is no impedance-reducing feedback involved. Achieving low output impedance

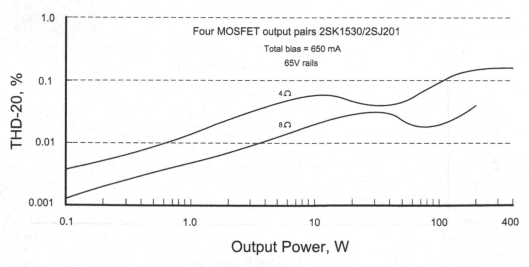

Figure 29.7 THD-20 Versus Level for a 200-W MOSFET Output Stage (a) Driving 8 Ω (b) Driving 4 Ω

first requires that there be large enough output stage current gain to make sure that VAS output impedance is completely inconsequential in determining output impedance. Consider a VAS with 10-kΩ output impedance paired with a total output stage current gain of 125,000 (a Triple with current gain of 50 in all transistors). The output impedance resulting from the VAS contribution will be 10 kΩ/125,000 = 0.08 Ω, corresponding to DF = 100. This is just on the border of acceptability, especially recognizing the likelihood of beta droop effects that will introduce nonlinearity into this impedance. A Quad, with four levels of emitter followers, can be used to provide enough additional current gain to take VAS output impedance completely out of the picture. The first two stages of the Quad can be realized with a diamond buffer as shown in Figure 12.9.

Once VAS output impedance effects are dealt with, the remainder is largely the output impedance from the output emitter resistors and output transistor intrinsic output impedance. The best approach here is to use a large number of output pairs in parallel, each with a fairly low-value emitter resistor of 0.15 Ω. Four output pairs, each with R_E = 0.15 Ω and biased at 165 mA, will yield an output impedance contribution of about 0.038 Ω, corresponding to a healthy damping factor of 210.

The effect of intrinsic base resistance in the output transistors should not be neglected. A 4-Ω base resistance combined with a beta of 50 will result in 0.08 Ω of additional emitter resistance. An additional 4 Ω of base resistance in the form of a base stopper resistor will double this contribution. At high output current, where re' can be neglected, the actual output impedance could approach 0.3 Ω per pair when the 0.15-Ω emitter resistance is added to the effective emitter resistance. This underscores the need for output transistors with high current gain and minimal beta droop. Notice that the use of a larger number of pairs to reduce output impedance has the secondary effect of reducing beta droop effects. If the NPN and PNP output transistors have different amounts of intrinsic base resistance, it may be helpful to compensate for this by using slightly different values of base stopper resistance for the NPN and PNP output transistors.

Power Supply Rejection and Power Supply Design

Without negative feedback the amplifier is more vulnerable to power supply ripple and noise. For this reason, the circuits must be designed to have inherently high PSRR. This can be achieved by the use of a fully differential design and liberal use of cascode stages. It is also important to have a very quiet power supply. The input and VAS stages should be fed with supply rails that have been filtered by capacitance multipliers. The extra headroom needed by these circuits dictates that boosted rails should be used to supply the IPS-VAS for best performance.

Output stage PSRR is often ignored in designs that use negative feedback. It cannot be ignored when global negative feedback is absent. Although it is intuitive that a simple emitter follower stage would have good PSRR, the Early effect can degrade it. Fluctuations on the rail will modulate transistor beta and V_{be}, causing noise ingress. The consequences of the Early effect can be reduced (but not eliminated) by feeding the output transistors from a very low-impedance driver.

Beware that the Early effect will act on all three transistor stages in a Triple, causing a β^3 effect. At minimum, the collector rails of the pre-driver and driver transistors should receive extra R-C filtering.

A very high-quality no-feedback amplifier should incorporate a capacitance multiplier filter into the main rails for the output stage if a cascoded stage is not used. This involves a cost in power supply headroom and amplifier power dissipation. It is also important to minimize the amount of magnetic induction noise and nonlinearity from the output stage half-wave rectified currents. Having a large class A region helps to reduce the production of such noise at small signal levels where it will not be masked by program material.

DC Offset

Without negative feedback there will be little correction for DC offset. It will be amplified by the full amount of the amplifier gain. We assume here that the use of a DC servo violates the mandate of no negative feedback. A fundamentally DC-stable design must be used, and some amount of offset trimming may be required. Although JFETs have a higher DC voltage offset than BJTs, they likely to be better in regard to DC offset overall because they do not require input bias current. It is also possible to choose well-matched JFET input pairs. As mentioned earlier, DC offset and DC offset drift contributed by the VAS can also be problematic.

Balanced Inputs

Balanced inputs are naturally available by the balanced input of the degenerated LTP because it has no negative feedback connection to the opposite side. When balanced inputs are used, there is no common-mode signal at the input LTP.

29.2 Additional Design Techniques

Here we discuss some additional circuit techniques that can further improve the performance or versatility of an NNFB amplifier.

A Complementary IPS-VAS

The complementary IPS with unipolar JFETs described in Figure 9.13 can form the basis of a fully complementary IPS-VAS for an NNFB amplifier. As shown in Figure 29.8, the IPS employs BJT input transistors instead of JFETs. This keeps the distortion low. JFET source followers at the inputs provide very high input impedance and freedom from input bias current. The differential input stage operates at an equivalent tail current of 4 mA. The top and bottom differential outputs are resistively loaded and applied to LTP VAS stages through emitter follower buffers. The differential VAS stages employ twin 20-mA current sources with generous degeneration as described earlier. Output cascodes provide freedom from Early effect distortion.

This VAS provides full-swing outputs of opposite polarity for free. These can be used to drive a pair of output stages to provide a balanced output. This configuration can have advantages in the way power supply current is used. It also provides double the output power for given power supply rail voltages. Balanced output arrangements like this are especially easy to implement in no-feedback amplifiers.

The Cascomp Input Stage

The cascomp stage shown in Figure 29.9 is a low-distortion configuration often used with bipolar transistors. The circuit was designed for use in oscilloscopes where wide bandwidth and high linearity is required [1, 2]. The cascomp linearizes the LTP by taking account of the nonlinear difference signals across the LTP emitters and injecting a compensating signal to cancel those differences.

A cascode stage placed above the LTP creates a replica of these nonlinear differences. Across its emitters appear the same base-emitter signals, with their nonlinearities, as appear across the LTP emitters. A second differential pair amplifies the difference of these nonlinear emitter signals and injects a correction signal at the collectors of the replica cascode in the opposite phase, using crossed collector connections.

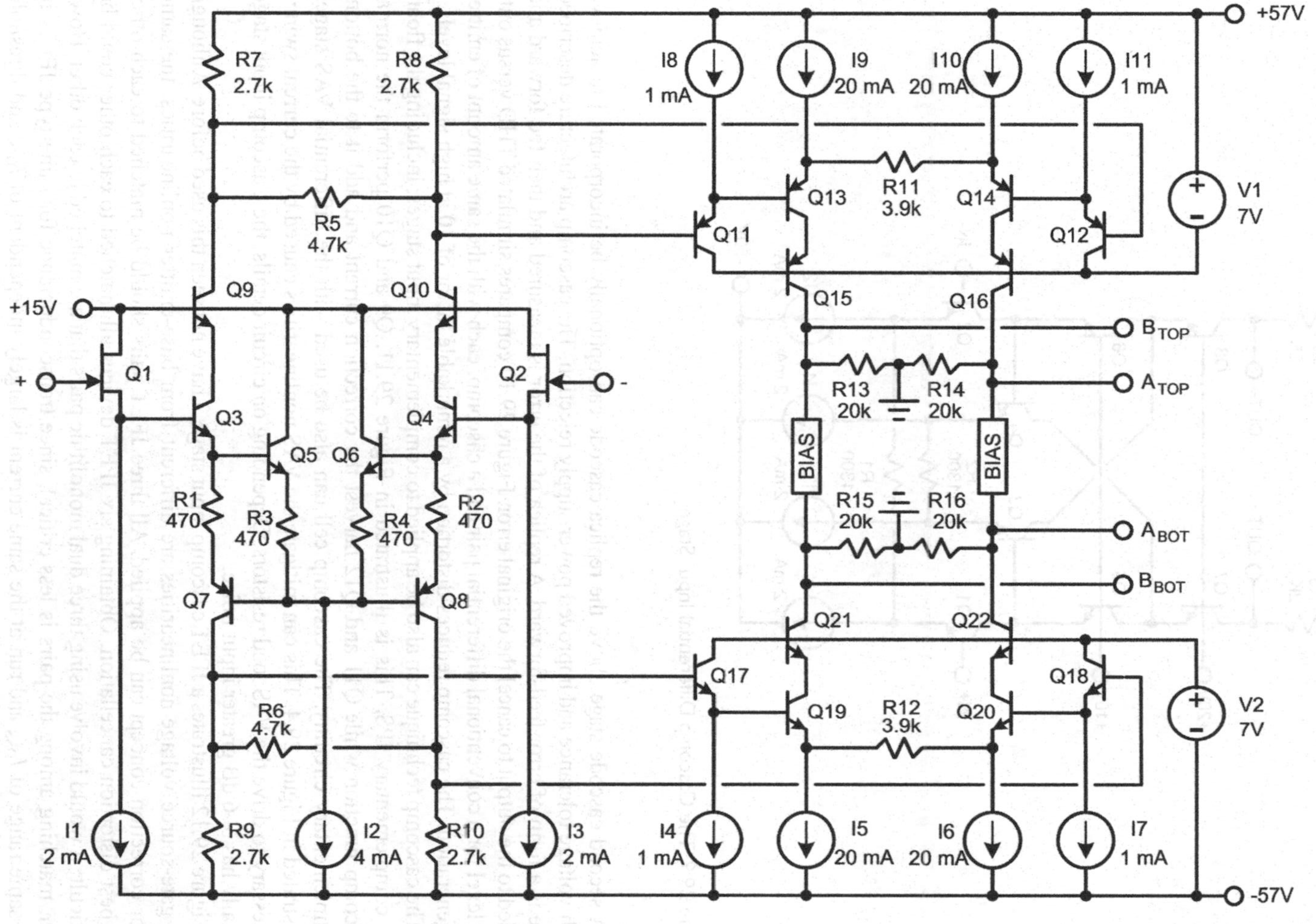

Figure 29.8 A Complementary IPS-VAS for an NNFB Amplifier

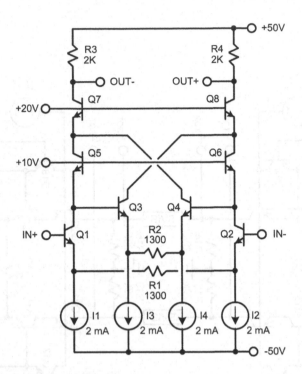

Figure 29.9 The Cascomp Differential Input Stage

A second cascode stage above the replica cascode can optionally be incorporated to achieve high voltage tolerance and improved power supply rejection. The cascomp architecture described here is a form of error feed-forward. A replica of the error is measured and then fed forward and added to the output to cancel the original error. Figure 29.10 compares simulated THD versus output level for a conventional differential pair and a cascomp, each with the same amount of emitter degeneration. The cascomp reduces distortion by a remarkable factor of 10 at high signal levels.

The cascomp technique can also be applied to complementary input stages, including the floating complementary IPS. This is illustrated in Figure 29.11. Q9 and Q10 perform the normal cascomp function while Q11 and Q12 harvest the correction current and add it to the bottom complementary circuitry. The cascomp cell can also be used with the differential VAS stages illustrated in Figure 29.4. This can mitigate the VAS nonlinearities caused by the current swings necessary to drive the VAS load resistors. Depending on circuit details, the Cascomp input stage usually has 3–6 dB greater input noise.

Figure 29.12 illustrates a JFET cascomp input stage. I have not seen this used before. Although the gate-source voltage nonlinearities are different from base-emitter nonlinearities, the same error correction concept can be applied. All three JFET pairs should be matched to each other for best distortion cancellation. Obtaining six JFET devices all matched to each other could be difficult; it would involve using three dual monolithic pairs that are matched to each other. However, matching among the pairs is less critical, since transconductance for same-type JFETs in the same range of I_{DSS} and run at the same current is largely independent of I_{DSS} and threshold voltage. The JFET cascomp reduces distortion by a factor of over 6:1 at high signal levels. It may provide sufficient linearity to allow the use of JFETs instead of BJTs in input differential pairs where distortion must be kept very low.

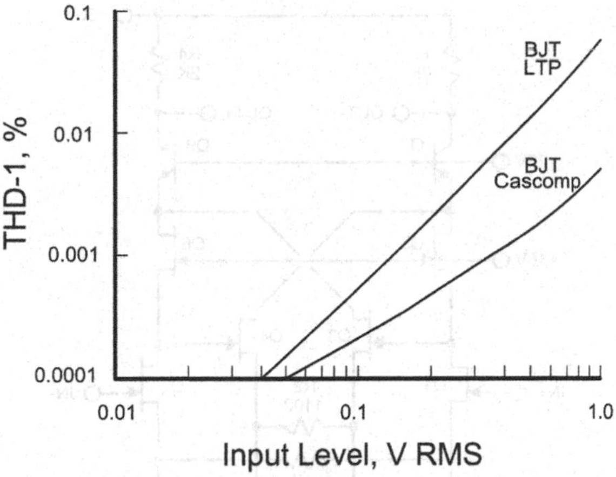

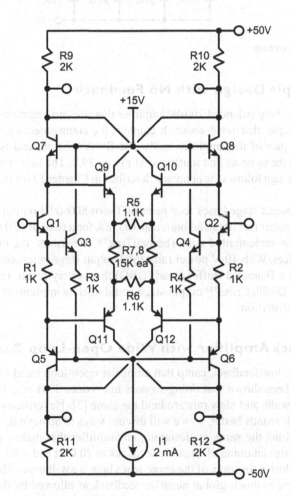

Figure 29.10 Cascomp THD Versus Differential Pair THD

Figure 29.11 Complementary Cascomp Input Stage

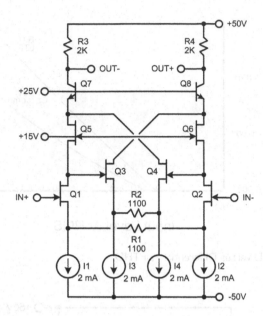

Figure 29.12 A JFET Cascomp

29.3 An Example Design with No Feedback

Figure 29.13 shows a fully balanced 200-W amplifier that uses no negative feedback. This design is based on the principles discussed above. It employs the complementary IPS-VAS arrangement of Figure 29.8. The gain of the amplifier is 30. The IPS-VAS front-end is shown as a block in Figure 29.13, as it is the same as that illustrated in Figure 29.8. The bias spreaders are omitted for simplicity, but their design follows the approach described in Chapter 17 for use with ThermalTrak™ transistors.

Each of the two output stages uses four pairs of ThermalTrak™ output transistors, each pair employing 0.15-Ω emitter resistors and biased at 163 mA for a total of 650 mA. Only one pair on each side needs to be implemented with ThermalTrak™ transistors; the other three pair can be MJL3281/1302 devices. With 40-V power rails, total output stage quiescent dissipation is 104 W. Each output stage is a Diamond Buffer Quad (DBQ) that is very linear and provides extremely high current gain. A DoubleCross™ output stage could also be implemented in this design for a further reduction in distortion.

29.4 A Feedback Amplifier with Wide Open-Loop Bandwidth

It is often said in the low-feedback camp that amplifier open-loop bandwidth should be greater than 20 kHz. It has been shown that doing so does not reduce TIM or other distortions as long as closed-loop bandwidth and slew rate are held the same [3]. Nevertheless, some say that wide open-loop bandwidth sounds better, so we will discuss ways to achieve it.

Notice that this is not the same as designing an amplifier that makes minimal or no use of negative feedback. High amounts of negative feedback at 20 kHz and wide open-loop bandwidth are not mutually exclusive. In some of the approaches here we will make liberal use of local feedback and will also use as much global negative feedback as allowed by the chosen closed-loop bandwidth and stability criteria.

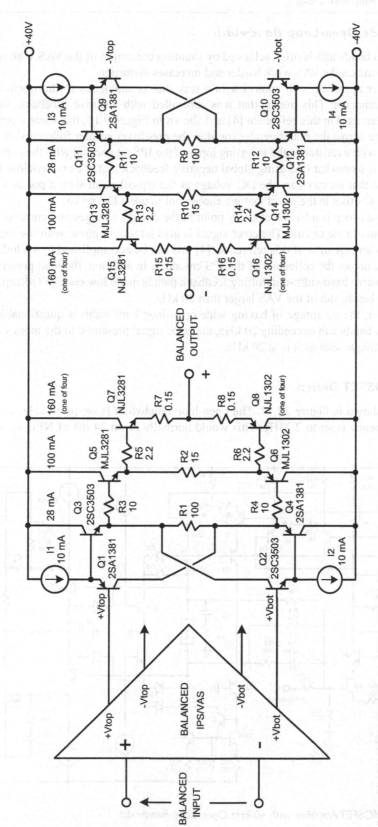

Figure 29.13 A 200 W Balanced Amplifier Without Negative Feedback

Achieving Wide Open-Loop Bandwidth

Wide open-loop bandwidth is often achieved by shunting the output of the VAS with resistances to ground. This makes the VAS work harder and increases distortion.

A much better approach is to control it in a way that is analogous to the operation of the compensation capacitor. This means that it is controlled with negative feedback. In the IPS-VAS arrangement used in this reference [4] and shown in Figure 9.16, this means tapping off a midpoint voltage across the bias spreader (or after the pre-driver emitter followers) and feeding back the signal with a resistor to the inverting input of the IPS, in parallel with the compensating capacitor. This is somewhat like using global negative feedback that does not enclose the output stage. Note also that we can vary the DC voltage at the tap-off point with a pot as a trimming adjustment of DC offset in the event that we choose not to use a DC servo.

With wide open-loop bandwidth at every point in the circuit, voltages and currents remain in phase at all points in the circuit. The error signal is also largely in phase with the input signal. In the IPS-VAS arrangement of this reference, [4] this can be accomplished by including a light shunt resistance across the collectors of the IPS cascodes. In addition, the load presented to the VAS by the resistive bandwidth-controlling feedback path is made low enough to keep the inherent stand-alone bandwidth of the VAS larger than 20 kHz.

As a reminder, the advantage of having wide open-loop bandwidth is questionable, at best. With open-loop bandwidth exceeding 20 kHz, the error signal presented to the input stage is just as large at low frequencies as it is at 20 kHz.

A 200-W MOSFET Design

The design is shown in Figure 29.14. The open-loop bandwidth is set to 40 kHz, and the gain crossover frequency is set to 2 MHz. This would normally mean 34 dB of NFB up to 40 kHz.

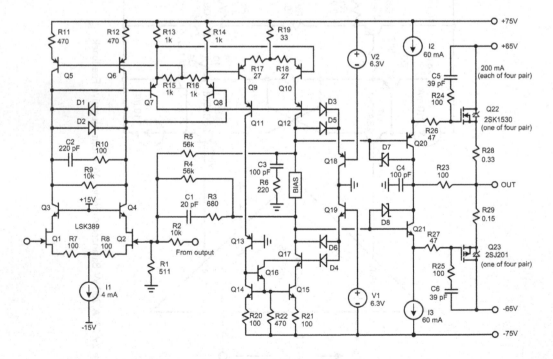

Figure 29.14 A MOSFET Amplifier with 40-kHz Open-Loop Bandwidth

However, some mild increase in the slope of the open-loop frequency roll-off is used to make the open-loop response a closer approximation to linear phase in the upper part of the audio band. This slightly steeper roll-off provides a bit more feedback at 40 kHz without introducing the overshoot that is usually associated with two-pole compensation. The output stage employs four pairs of vertical MOSFETs (2SJ201/2SK1530). Each pair is biased at 200 mA and optimized asymmetric source resistors are employed.

References

1. Patrick A. Quinn, "Feed-Forward Amplifier," U.S. Patent 4,146,844, March 27, 1979.
2. Malcolm J. Hawksford, "Distortion Correction Circuits for Audio Amplifiers," *Journal of the Audio Engineering Society*, vol. 29, no. 7–8, July–August 1981, pp. 503–510.
3. Robert R. Cordell, "Another View of TIM," *Audio*, February–March 1980. Available at www.cordellaudio.com.
4. Robert R. Cordell, "A MOSFET Power Amplifier with Error Correction," *Journal of the Audio Engineering Society*, vol. 32, January 1984. Available at www.cordellaudio.com.

Chapter 30

Balanced and Bridged Amplifiers

30. INTRODUCTION

Balanced is beautiful—that's what a mentor of mine told me many years ago. In this chapter three aspects of differential operation of power amplifiers will be discussed. Single-ended power amplifiers that have balanced inputs will first be considered. These are the most common variant of differential operation in consumer audio. Most of these can also accommodate a single-ended input as well.

Bridged amplifiers will next be discussed. These amplifiers are actually best referred to as bridge-connected amplifiers, as they are usually composed of a stereo pair of amplifiers that are fed the same signal but fed in opposite phase. Twice the voltage is available across their output terminals, so in this configuration they can supply 4 times the power into a given load impedance. Some of these amplifiers can be configured to accept balanced inputs as well.

Balanced amplifiers will finally be discussed. These are much like bridged amplifiers, and differences will be highlighted. In some cases, amplifiers that are called balanced amplifiers are merely what used to be called bridged amplifiers. The term *balanced* is much more appealing than the term *bridged*.

30.1 Balanced Input Amplifiers

Most professional, and a number of hi-end audio power amplifiers, incorporate balanced inputs using XLR connectors with pin 2 hot. This provides for improved rejection of external common-mode noise.

Achieving balanced inputs for a single-ended (SE) power amplifier is not always straightforward, especially if the highest sound quality is to be maintained and if the amplifier must allow the user to operate the amplifier with either XLR balanced inputs or RCA single-ended inputs (with or without the need for a mode switch). The purpose of balanced inputs is to achieve good CMRR, independent of the nature of the source, as long as the output impedance of the source is the same to ground on both sides.

Gain and Input Impedance Considerations

Other important considerations include the gain of the amplifier in the two modes and the input impedance of the amplifier in the two modes. In particular, in the balanced input mode it is especially desirable that both the positive and negative inputs of the amplifier have the same input impedance so as to maintain the best possible common-mode rejection in the presence of finite driving impedances from the source.

Often, the incorporation of a balanced input capability will involve the introduction of additional circuitry and possible compromise of the signal-to-noise ratio of the amplifier. In high-end applications, it is sometimes frowned on to use an op-amp, which is the obvious and easy choice in implementing a balanced input capability.

Single and Triple Op-Amp Solutions

An op-amp can be used to implement a differential-to-single-ended converter, as shown in Figure 30.1a [1–6]. Notice that this circuit presents different load impedance to single-ended sources on the positive and negative inputs. A single-ended source on the positive input sees an input impedance of 20 kΩ. A single-ended source on the negative input sees an input impedance of 10 kΩ. This is undesirable. For common-mode signals, the situation is different. A common-mode source sees 20 kΩ on each side for a net of 10 kΩ. If the source impedance on both sides of the source is the same, common-mode rejection will be acceptable if resistor tolerances are tight.

In Figure 30.1b the resistor values are chosen so that the positive and negative inputs exhibit the same input impedance to a single-ended source (the resistors on the inverting side have been doubled to 20 kΩ). However, when a balanced signal is applied (with each of its signals referenced to ground at the source), the signal current flowing in the plus and minus inputs is not the same. Similarly, if a common-mode signal is applied to both inputs, the current flowing into the positive input is larger than the current flowing into the negative input. This happens because the virtual short at the op-amp inputs forces 1/2 of the voltage at the positive input to appear at the negative terminal of the op-amp. This seriously compromises CMRR in proportion to the output impedance of the source. If the positive and negative source impedances are not zero, common-mode rejection (CMRR) will suffer.

The single op-amp solution also forces a compromise between input impedance and noise. The circuit as shown has single-ended input impedances of only 20 kΩ and yet puts 20-kΩ resistors in the signal path, generating noise. The 20-kΩ input impedance also forces the use of large-value input-blocking capacitors if they are to be used.

The three op-amp instrumentation amplifier shown in Figure 30.2 does not have these limitations. It can be used to achieve high common-mode rejection and symmetrical input impedances under all drive conditions. Input impedances on each side are the same for both single-ended and common-mode signals. It also provides high input impedance while allowing the use of relatively

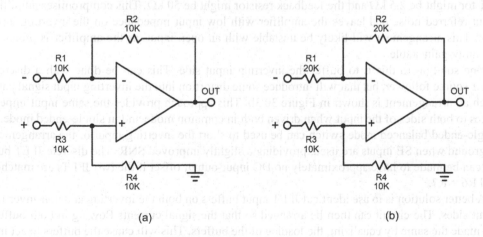

(a)   (b)

Figure 30.1 Single Op-Amp Differential-to-Single-Ended Input Circuits

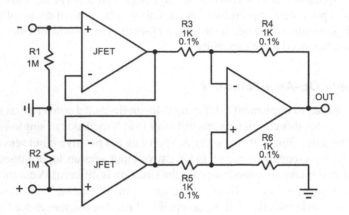

Figure 30.2 Triple Op-Amp Instrumentation Amplifier

low-value resistors in its implementation. However, the circuit is more complex, and concerns about passing the signal through multiple op-amps are magnified.

The circuit is composed of unity-gain input buffers followed by a differential-SE converter with a gain of 1. Very high input impedances can be obtained by using low-noise JFET op-amps for the input buffers. The resistors surrounding the third op-amp should have 0.1% tolerance in order to maintain high CMRR. The third op amp can actually be the power amplifier, configured as Figure 30.1a.

Configuring the Power Amplifier as a Differential Amplifier

Another approach is to configure the main amplifier itself as a differential input amplifier in a way analogous to that of Figure 30.1a. This is illustrated in Figure 30.3a. Unfortunately, this results in relatively low input impedance on the inverting side because the power amplifier has gain on the order of 26 dB. It also forces up the choice in value of the amplifier's feedback resistor. For example, if one wants a single-ended-to-single-ended gain of 20, then the negative input resistor might be 2.5 kΩ and the feedback resistor might be 50 kΩ. This compromises amplifier input-referred noise and leaves the amplifier with low input impedance on the inverting input side. This arrangement will likely be unstable with an open input, as the amplifier is probably not unity-gain stable.

One solution to this is to buffer the inverting input side. This can be done with a discrete JFET source follower, but that will introduce some distortion into the inverting input signal path. Such an arrangement is shown in Figure 30.3b. This approach provides the same input impedances to both sides of the input when driven both in common mode and in single-ended mode. A single-ended/balanced mode switch can be used to short the inverting input of the arrangement to ground when SE inputs are used, providing a slightly improved SNR. The discrete JFET buffer can be made to have approximately no DC input-output offset if the two JFETs are matched and R6 = R7.

A better solution is to use identical JFET input buffers on both the inverting and non-inverting input sides. The circuit can then be arranged so that the signal currents flowing in both buffers are made the same by equalizing the loading of the buffers. This will cause the buffers to act in a balanced, differential fashion, canceling out second harmonic distortion. Both of the differential

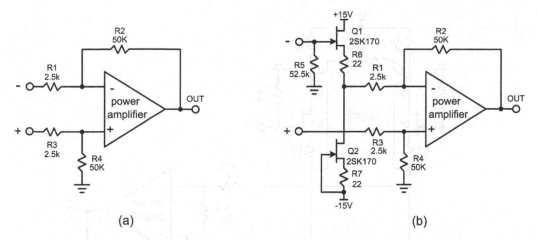

Figure 30.3 (a) Power Amplifier Configured as a Differential Amplifier (b) with JFET Input Buffer on the Inverting Side

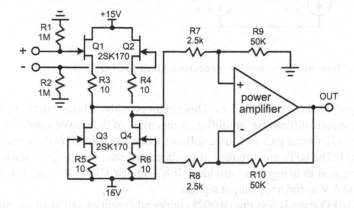

Figure 30.4 Configuring the Power Amplifier as a Differential Amplifier with Dual Input Buffers

inputs can have identical, high values of input impedance and both positive and negative signals encounter the identical signal path for symmetry.

The discrete JFET input buffers must be run at fairly high currents to keep the distortion low, given the subsequent low impedances (on the order of 2.5 kΩ) that they must drive in the power amplifier's differential-to-single-ended conversion input stage. This arrangement is illustrated in Figure 30.4. This makes the power amplifier part of an instrumentation amplifier arrangement. The non-zero output impedance of the discrete JFET buffers can detract from common-mode rejection, however. Q1 and Q2 can be replaced with a dual monolithic matched JFET pair like the LSK389. Q3 and Q4 can also be replaced with an LSK389 with similar I_{DSS}.

The Differential Complementary Feedback Quad (DCFQ)

The differential buffer shown in Figure 30.5 is referred to here as a *differential complementary feedback Quad* (DCFQ). It employs two JFETs and two BJTs in a differential arrangement

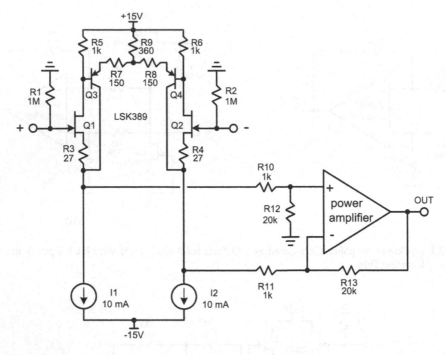

Figure 30.5 The Differential Complementary Feedback Quad (DCFQ) Buffer

analogous to a pair of JFET-bipolar CFPs. This circuit provides a low-distortion, low-impedance drive to the subsequent differential amplifier arrangement of the power amplifier.

JFETs Q1 and Q2 form a pair of source-followers that are *helped* by their respective complements Q3 and Q4. The BJTs are arranged as a differential pair, resulting in much-improved bias stability as compared to using two individual JFET-bipolar CFPs. R3 and R4 drop the source voltages down to 0 V to form the output signals.

THD of the DCFQ stage is less than 0.002% under all conditions of positive, negative and differential drive. This arrangement takes advantage of the good offset matching of the dual JFET, with pair-to-pair variations in V_{gs} resulting in only a common-mode offset that is ignored by the subsequent differential amplifier of the power amplifier. If desired, the common-mode level presented to the subsequent differential amplifier can be trimmed to zero with adjustment of R9. The low output impedance of the DCFQ buffer allows for the possibility of using lower-value resistors for R10–R13 to achieve lower noise. In principle, one could servo the currents in I1 and I2 to make the common-mode output DC level equal to zero.

Line Receiver ICs for Balanced Inputs

A single-chip audio-quality solution exists for achieving balanced inputs with exceptional CMRR performance for virtually all types of audio sources and conditions. The 1206 Ingenious® Balanced Input Line Receiver made by THAT Corporation achieves these goals with very low distortion [6]. Widely used in professional audio applications, this device is also ideal for consumer audiophile applications as well. On-chip laser-trimmed precision resistors combined with a very clever circuit make its exceptional performance possible. The datasheet for the THAT 1206 is also an outstanding tutorial on balanced input circuits and challenges.

THAT Corporation also provides a companion single-chip balanced output solution [7]. The THAT 1646 OutSmarts® Balanced Line Driver provides a transformer-like floating balanced output.

30.2 Bridged Amplifiers

Bridged amplifiers are used widely where high power is required. This is especially so in the pro-sound arena. A simple bridged amplifier arrangement is illustrated in Figure 30.6. Two channels of a stereo amplifier are driven out of phase and the loudspeaker is connected across the hot outputs of the two amplifiers. A bridged power amplifier can theoretically produce 4 times the power into a given load compared to its non-bridged counterpart (one channel of the stereo amplifier) using the same rail voltages. This is because the voltage across the loudspeaker is doubled and power goes as the square of voltage. Under these conditions, each of the two amplifiers "sees" an effective load resistance equal to half that of the loudspeaker impedance. For this reason, the bridged amplifier may produce somewhat less than 4 times the power.

Sound Quality

Bridged amplifiers have not always enjoyed a reputation for the highest-quality sound. This is partly because they are seeing half the impedance of the loudspeaker and the distortion of power amplifiers is virtually always higher when driving lower impedance loads. Moreover, the peak output current requirements are doubled, and some amplifiers may not be up to the task. Indeed, their protection circuits may be activating. Finally, bridged amplifiers are often abused. Each channel may not be rated to drive a 2-Ω load, but the amplifier may often be asked to drive a 4-Ω load in bridged mode.

Bridged amplifiers provide only half the damping factor into a given load because there are essentially two amplifier output impedances in series with the load. This can also detract from sound quality.

Power Supply Advantages

Amplifiers operating in bridged mode often enjoy some advantage in regard to the power supply. Single-ended amplifiers draw power from either the positive rail or the negative rail on each half-cycle

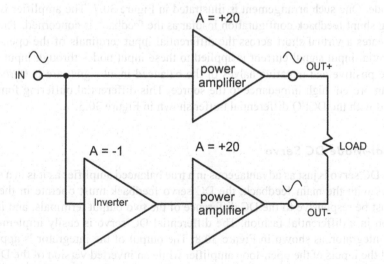

Figure 30.6 A Bridged Power Amplifier

of the signal. This current flows through ground. The duty cycle of current flow from each rail is only 50%. In bridged mode current is being drawn from both rails simultaneously and little or none of the output current flows through ground. Each rail is being used 100% of the time. The waveform of the current being sourced by each rail is full wave rather than half wave. There is no DC in the secondary of the power supply, which is especially advantageous for toroid power transformers.

30.3 Balanced Amplifiers

Any amplifier that can accept balanced inputs and produce balanced outputs can be called a *balanced amplifier*. However, there are degrees of balance. A bridged amplifier with a balanced input can be called a balanced amplifier, and many so-called balanced amplifiers are made that way. For the audiophile, balanced is beautiful and bridged is brawn. That is why the term *balanced* is preferred when what would otherwise be a bridged amplifier is marketed to the audiophile community.

True *Balanced Amplifiers*

True balanced amplifiers comprise a single amplifier whose circuitry is balanced from input to output, not two separate amplifiers wired together in bridged mode. One version of an amplifier without negative feedback described in Chapter 29 was a truly balanced design. Indeed, in some ways it is easier to design a true balanced amplifier in the absence of negative feedback.

Building the differential open-loop amplifier is not difficult. It is fairly straightforward to adopt a differential VAS arrangement and drive two copies of the output stage. Applying the negative feedback is where the added challenge lies.

Differential-Mode Feedback

The output of a balanced amplifier is a differential (balanced) signal. The information applied to the loudspeaker is in the differential mode. That means that the negative feedback must be taken as the difference of the voltages existing across the two *hot* output terminals. The negative feedback must therefore be in the differential mode and be applied to the input circuits in the differential mode. One such arrangement is illustrated in Figure 30.7. The amplifier is operated in the inverting shunt feedback configuration insofar as the feedback is concerned. The differential feedback creates a virtual short across the differential input terminals of the open-loop amplifier. Differential input signal current is applied to these input nodes through input resistors R1 and R2. The positive and negative halves of the balanced input signal are buffered so that the amplifier can present high impedance to the source. This differential buffering function can be implemented with the DCFQ differential buffer shown in Figure 30.5.

Differential-Mode DC Servo

The use of a DC servo is just as advantageous in a true balanced amplifier as it is in a single-ended amplifier. As with the main feedback, the DC servo feedback must operate in the differential mode. It must be responsive to the DC difference of the two output terminals, and it must apply its correction in a differential fashion. The differential DC servo is easily implemented with a differential integrator, as shown in Figure 30.8. The output of the integrator is applied through R5 to one of the inputs of the open-loop amplifier while an inverted version of the DC correction signal is applied to the input of opposite polarity by R6.

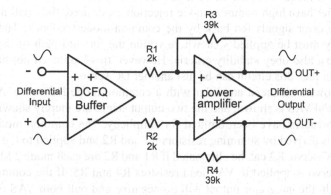

Figure 30.7 A True Balanced Amplifier with Differential-Mode Feedback

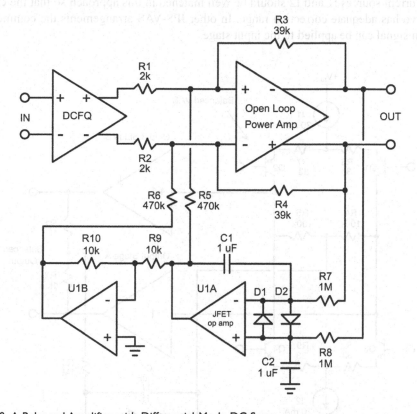

Figure 30.8 A Balanced Amplifier with Differential-Mode DC Servo

Common-Mode DC Servo

In single-ended amplifiers the negative feedback establishes the output DC level and drives DC offset to zero. Differential-mode feedback is unaware of the common-mode output level of the amplifier and therefore cannot stabilize it to a value near zero. For this reason a second feedback loop called a *common-mode feedback loop* is needed in most true balanced amplifiers to sense the output common-mode level and drive it to zero. However, if the differential inputs of the open-loop amplifier have high common-mode rejection, as desired, they will not be responsive to common-mode error signals fed back by the common-mode feedback. The common-mode feedback typically must be applied somewhere within the forward path of the amplifier. This feedback loop must also obey stability criteria. However, this loop need not have high closed-loop bandwidth. In fact, this circuit can be just another DC servo.

Figure 30.9 shows a balanced amplifier with a common-mode DC servo. A simplified version of a differential VAS is shown, and the two output stages are merely shown as gain blocks. The usual bias spreaders have been omitted for simplicity. The common-mode component of the output signal is derived by summing resistors R1 and R2 and applied to the integrator of the common-mode DC servo. R3 can be eliminated if R1 and R2 are each made 2 MΩ. The DC error signal from the servo is applied to VAS load resistors R4 and R5. If the common-mode output voltage is too high, the integrator output will go negative and pull both VAS output nodes in a negative direction to correct the error. Notice that the VAS load resistors are not unlike the load resistors employed in low-feedback amplifiers, but in this application they can have much higher values. Current sources I1 and I2 should be well matched in this approach so that the common-mode servo has adequate correction range. In other IPS-VAS arrangements the common-mode correction signal can be applied to the input stage.

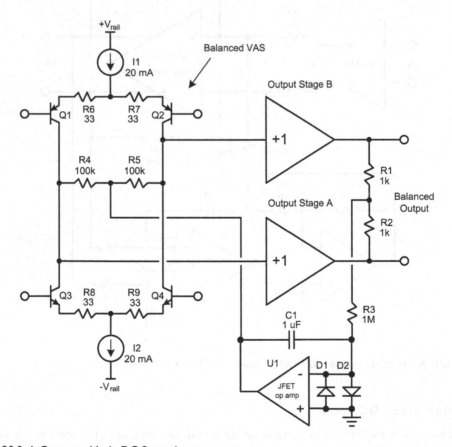

Figure 30.9 A Common-Mode DC Servo Arrangement

References

1. Bruno Putzeys, "The G word, or How to Get Your Audio off the Ground," *Linear Audio*, vol. 5, pp. 105–126, April 2013.
2. Bill Whitlock, "An Overview of Audio System Grounding and Signal Interfacing," Tutorial T5, 135th Convention of the Audio Engineering Society, New York City, October 2013.
3. Bill Whitlock, "An Overview of Audio System Grounding and Shielding," Tutorial T2, Convention of the Audio Engineering Society, Star-Quad, p. 126.
4. Bill Whitlock, "Design of High Performance Audio Interfaces," Jensen Transformers, Inc.
5. Walter G. Jung, *Op Amp Applications Handbook*, Elsevier/Newnes, Burlington, MA, 2005, pp. 121–148, ISBN 0-7506-7844-5.
6. Ingenious® Balanced Input Line Receiver ICs, THAT 1206 datasheet, THAT Corporation. Available at thatcorp.com.
7. OutSmarts® Balanced Line Driver ICs, THAT 1646 datasheet, THAT Corporation. Available at thatcorp. com.

Chapter 31

Integrated Circuit Power Amplifiers and Drivers

31. INTRODUCTION

In this chapter we'll discuss power amplifiers that can be built from integrated circuits that are designed to function as complete power amplifiers on a chip. Power amplifier driver chips that can implement most of the complexity of a power amplifier on a single chip will also be discussed. Such chips typically need only a BJT or MOSFET output stage attached to them to form a complete power amplifier.

31.1 IC Power Amplifiers

Integrated circuit power amplifiers have been around for quite some time and have been used extensively in high-volume consumer gear. Often, these have been of relatively low power capability and without a strong focus on true high-fidelity performance. Bob Widlar designed what was probably the first serious monolithic power amplifier [1].

Several companies have introduced IC power amplifiers that are capable of substantial power and high-fidelity performance. Power capabilities in the 25-W to 80-W range with very good distortion specifications are readily available. Although such amplifiers are not the primary focus of this book, an overview is presented here for completeness. Numerous companies produce such amplifiers, but perhaps the best-known is National Semiconductor (now Texas Instruments). These products are the ones with which the author has had direct experience.

Integrated circuit power amplifiers have made tremendous strides in the last 20 years. The National Semiconductor LM3886 and its cousins have even been popularized for use in some audiophile power amplifiers [2, 3]. These applications are most often referred to as *Gain Clones*. The term arises from an early audiophile amplifier based on the LM3875 chip called the *Gaincard* [4, 5]. It was rated at 25 W per channel into 8 Ω. The design approach was then copied and modified by many, thus the name *Gain Clone*. In addition to its very good performance, the allure of the device is that it allows many more DIY audiophiles to get involved with building power amplifiers without getting deeply involved in the complexities and risks of discrete power amplifier design.

31.2 The Gain Clones

Quite some years ago a Japanese designer named Kimura-San of 47 Laboratory introduced an IC-based power amplifier to the serious audiophile community and called it the *Gaincard* [4]. It was based on the National Semiconductor LM3875 [5]. The amplifier was greeted with surprising enthusiasm and many variants on the basic approach emerged, especially in the DIY community. Such IC high-fidelity power amplifier variants, usually employing one of several different National chips, have become known as *Gain Clones*. Most of the information needed to build a Gain Clone can be found in the National Semiconductor datasheets and application notes for these devices.

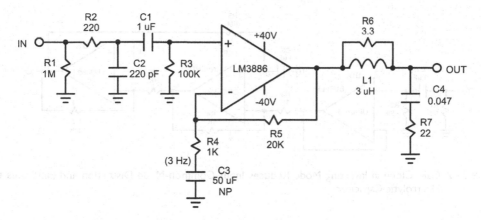

Figure 31.1 A 50-W Gain Clone Amplifier Based on the LM3886

A Basic Gain Clone Design

The LM3886 is not unlike a power op amp and can be used in a multitude of op-amp topologies. Figure 31.1 shows a typical LM3886-based Gain Clone similar to the design recommended by National in its datasheet [2]. This is a simplified diagram and does not show compensation details.

As with many power amplifiers, a large electrolytic capacitor is included in the return leg of the feedback network to reduce the closed loop gain to unity at DC and reduce offset voltage at the output. Unfortunately, the presence of such electrolytic capacitors in the signal path can create distortion and otherwise impair sound quality. Shortly we'll see a couple of ways to eliminate this electrolytic capacitor.

A Gain Clone Using the Inverting Mode

Most operational amplifiers produce somewhat lower distortion when operated in the inverting mode, and the LM3886 is no exception. Operation in the inverting mode improves performance by eliminating common-mode distortion in the input stage of the LM3886. A disadvantage of the inverting mode is its relatively low input impedance. For this reason the amplifier should be preceded by an inverting buffer. This arrangement is shown in Figure 31.2. The inversion in U1B also restores the overall polarity of the signal path to non-inverting. The inverter should be implemented with a high-quality audio-grade op amp. Here the OPA2604 dual JFET op amp is used.

The inverter also has a moderate input impedance of only 10 kΩ. This would require a fairly large input coupling capacitor. For this reason, it is preceded by unity gain JFET input buffer U1A employing the other half of the JFET op amp. This provides very high input impedance, allowing the use of a high-quality film input capacitor of reasonable size and cost.

Avoiding Electrolytic Capacitors While Controlling Offset

The inverting stage preceding the LM3886 provides an opportunity to eliminate the electrolytic capacitor in the feedback network of the LM3886 in the form of its unused non-inverting input. This input provides a non-inverting gain of 2 to the output of the stage. As shown in Figure 31.2, DC feedback from the output of the LM3886 can be separately fed back to this input through a large resistance and then filtered with a film capacitor of modest value.

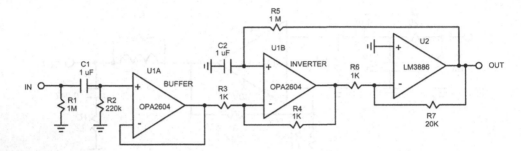

Figure 31.2 Gain Clone in Inverting Mode Reduces Input Common-Mode Distortion and Eliminates the Electrolytic Capacitor

31.3 The Super Gain Clone

The *Super Gain Clone* is an inexpensive quality power amp block based on the popular National LM3886 IC power amplifier chip [6]. This circuit encapsulates the LM3886 IC and maximizes its achievable quality.

A DC servo is employed for offset control instead of the passive DC feedback circuit illustrated above. The availability of the extra inversion in the signal path makes it possible to employ a simple inverting integrator for the servo. The schematic of the Super Gain Clone amplifier is shown in Figure 31.3. It includes the compensation and filtering circuitry recommended by National Semiconductor. The Super Gain Clone includes an input buffer stage, an inverter, the LM3886 power amplifier operating in inverting mode and an integrating DC servo for offset control.

Input Circuits

The amplifier begins with the usual 4.7-Ω resistor that isolates analog ground from circuit ground so as to prevent ground loops. A passive LPF input network provides two poles of roll-off and minimizes RFI ingress. U1A is a conventional non-inverting unity gain buffer employing half of an OPA2604 dual FET op amp. This is a good audio op amp, especially for the price.

U1B just acts as the unity gain inverter. It cancels the inversion that results from using the LM3886 in its inverting mode. This inverter is also the point at which the offset control signal from the DC servo is injected via R17.

Power Amplifier

The inverter feeds the LM3886 power amplifier through R9 (1 kΩ) in the inverting mode. In combination with 20-kΩ feedback resistor R11, the power amplifier produces a gain of −20. Capacitors C4 and C5, in combination with R12, provide feedback compensation for the LM3886, as recommended in its datasheet [2].

Output Network

The output network is a pi-section arrangement with Zobel networks on both sides of the L-R network. This provides enhanced resistance to RFI ingress via the speaker cables. The inductor is optionally constructed as an air-core toroid to keep the magnetic field circulating largely inside the toroid, reducing possible magnetic coupling to other circuits. The first Zobel, comprising C7 and R13, is located close to the output of the LM3886. The L-R network and second Zobel are

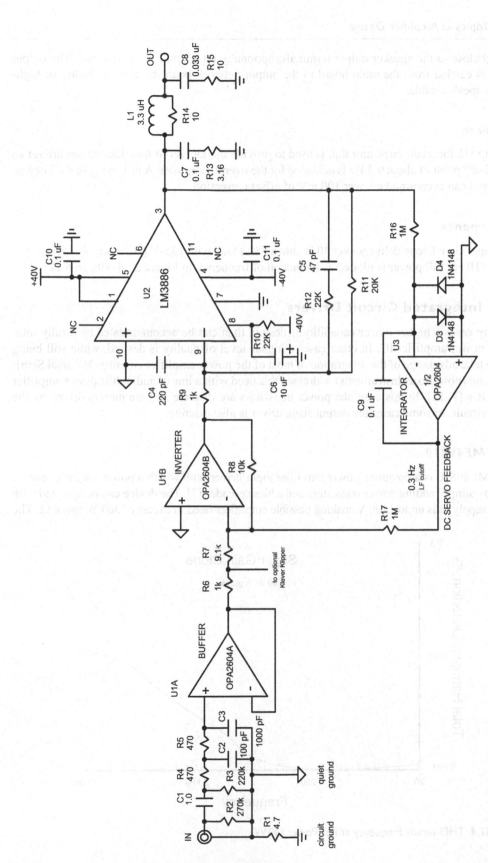

Figure 31.3 The Super Gain Clone Employing a DC Servo

located close to the speaker output terminals, optionally on a small separate board. The output signal is carried from the main board to the output network board by about 6 inches of high-quality speaker cable.

DC Servo

Op-amp U3 forms an integrator that is used to provide the DC servo function. Values are set so that a 3-dB point of about 0.3 Hz is achieved for the overall amplifier. A full swing of the integrator output can accommodate over 100 mV of offset correction.

Performance

The Super Gain Clone delivers over 40 W into an 8-Ω load with $\pm$35-V power supply rails (at full load). THD at full power is plotted as a function of frequency in Figure 31.4 with an 8-Ω load.

31.4 Integrated Circuit Drivers

In many cases a higher power capability is desired than can be accommodated by a fully integrated power amplifier IC. In other cases a higher level of quality is desired, while still being able to take advantage of the integration of most of the power amplifier circuitry. National Semiconductor (now Texas Instruments) addressed this need with a line of audiophile power amplifier driver ICs [7–12]. In this case, the power transistors are discrete and are merely driven by the driver circuit. In some cases the output stage driver is also discrete.

The LME49810

The LME49810 is a complete power amplifier input/driver with which a power amplifier can be made by simply adding power transistors and a bias spreader [7]. The device can be operated with power supply rails up to $\pm$100 V, making possible amplifiers rated in excess of 300 W into 8 Ω. The

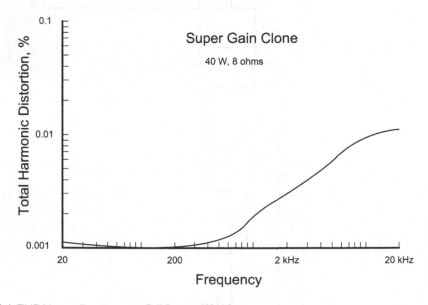

Figure 31.4 THD Versus Frequency at Full Power (40 W)

device includes emitter follower output driver transistors that can supply 50 mA to the external output stage. Separate pins are provided for connection of a bias spreader, such as a V_{be} multiplier. Figure 31.5 is a simplified schematic of a power amplifier built with the LME49810 and one pair of BJT output transistors. Input offset voltage is less than 3 mV, and input bias current is less than 200 nA. Input-referred noise is about 10 nV/$\sqrt{Hz}$. A bias current of 2.8 mA flows through the V_{be} multiplier.

Better performance and higher amplifier output current can be achieved by employing an output Triple as shown in Figure 31.6. The output emitter followers in the LME49810 serve as the pre-drivers while external discrete drivers are used to drive the output transistors. They have their own bias spreading diodes inside the chip. The V_{be} multiplier implemented with Q1 provides 4 V_{be} of temperature-compensated bias spread for Q2–Q5. Drivers Q2 and Q3 are mounted on the heat sink.

The maximum bias-spreading voltage available from the LME49810 is only 10 V. For this reason the device may not be capable of properly driving some vertical power MOSFETs [9] in source-follower mode. Vertical power MOSFETs with smaller turn-on voltages, such as the Toshiba 2SJ201/2SK1530 pair, are compatible with the LME49810. Lateral MOSFETs, with their still smaller turn-on voltages, can also be driven by the LME49810.

A more complete amplifier is shown in Figure 31.6. This design also includes a bias spreader that employs the tracking diodes in the ThermalTrak™ output transistors. Depending on the detailed thermal arrangement, other variations of the bias spreader may provide better thermal tracking. For example, the bias spreader of Figure 17.19a, where a resistor connected from base to emitter of Q1 is added, may be preferable. This amplifier employs two output pairs and ±55-V power supply rails (no load).

The amplifier delivers 125 W into an 8-Ω load when the rails have sagged to ±49 V. It will deliver 180 W into a 4-Ω load when the power supply rails have sagged to ±44 V.

All of the techniques employed in the Super Gain Clone can be applied to the amplifier of Figure 31.6. These include the DC servo and the approaches that allow the use of small film capacitors in place of electrolytic capacitors.

Sadly, the LME49810 has been discontinued by Texas Instruments, who bought National Semiconductor in 2011. This material does, however, illustrate what can be achieved with integrated circuit drivers.

The LME49830

The LME49830 is similar to the LME49810, but it is optimized for use with power MOSFETs instead of bipolar output transistors [11, 12]. It is capable of supporting bias spread voltages as high as 16 V; this makes it fully capable of properly driving vertical power MOSFETs like the IRFP240/9240. The LME49830 lacks the Baker clamps and clipping indicator outputs of the LME49810. Its input circuits have similar characteristics to those of the LME49810. The device feeds 1.6 to 2.7 mA through the bias spreader. It would be nice if this number were a bit bigger. The outputs that drive the MOSFET gates are push-pull, so they can source or sink current. Figure 31.7 illustrates a power amplifier employing two pairs of the popular IRFP240/9240. Total bias current for the two pairs is set at 300 mA.

The bias spreader employs a transistor mounted on the circuit board and a diode-connected transistor mounted on the heat sink to achieve the proper degree of temperature compensation for the IRFP240/9240 MOSFET devices used.

Sadly, the LME49830 has been discontinued by Texas Instruments, who bought National Semiconductor in 2011. This material does, however, illustrate what can be achieved with integrated circuit drivers.

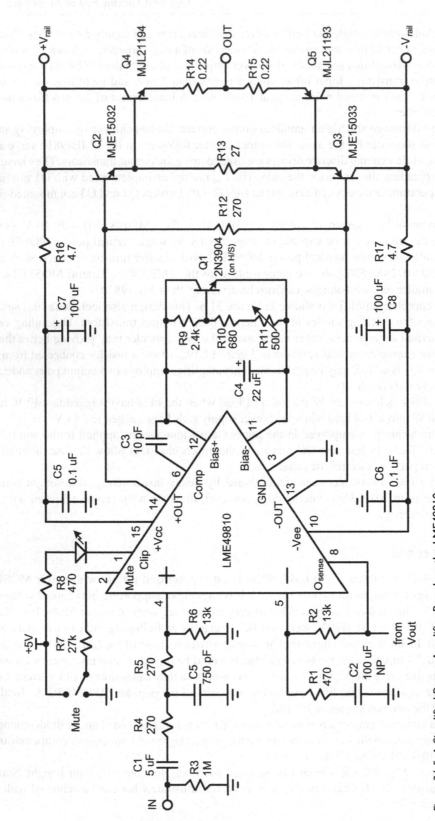

Figure 31.5 A Simplified Power Amplifier Based on the LME49810

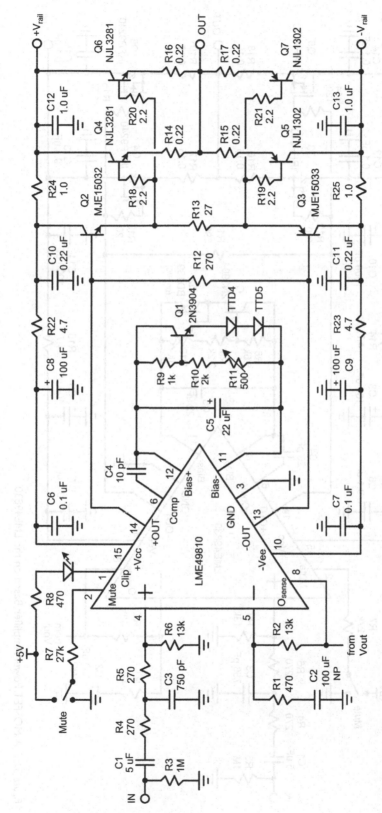

Figure 31.6 An Amplifier Employing the LME49810 with a Triple Output Stage. The Pre-Driver Is Inside the LME49810

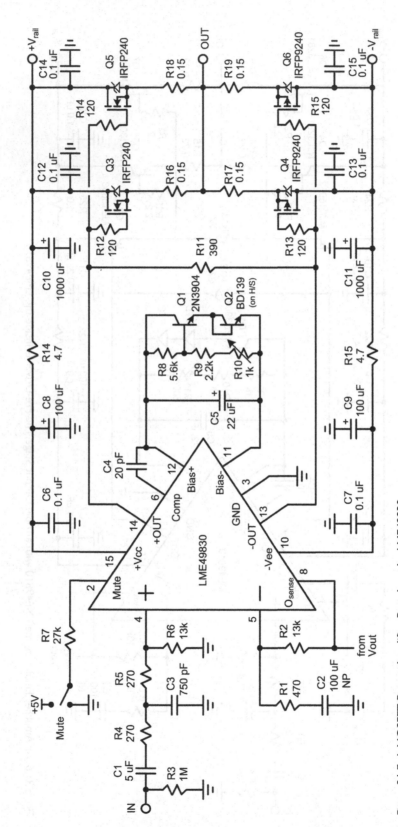

Figure 31.7 A MOSFET Power Amplifier Based on the LME49830

31.5 Summary

While integrated circuit audio power amplifiers and drivers may not meet the needs of all serious audiophiles, they do present a very convenient and valuable option for achieving surprisingly high performance at greatly reduced cost and complexity.

References

1. Bob Widlar, "A Monolithic Power Amp," *IEEE Journal of Solid State Circuits*, vol. 23, no. 2, pp. 527–535, April 1989.
2. National Semiconductor/TI datasheet, "LM3886 Overture Audio Power Amplifier Series High-Performance 68W Audio Power Amplifier w/Mute." Available at www.ti.com.
3. National Semiconductor/TI AN-1192, "Overture Series High Power Solutions," Application Note. Available at www.ti.com.
4. "47 Laboratory Model 4706 Gaincard," Sakura Systems. Available at www.sakurasystems.com.
5. National Semiconductor/TI datasheet, "LM3875 Overture Audio Power Amplifier Series High-Performance 56W Audio Power Amplifier." Available at www.ti.com.
6. "The Super Gain Clone." Available at www.cordellaudio.com.
7. National Semiconductor/TI datasheet, "LME49810 200V Audio Power Amplifier Driver with Baker Clamp." Available at www.ti.com.
8. National Semiconductor/TI AN-1490, "LM4702 Power Amplifier." Available at www.ti.com.
9. National Semiconductor/TI AN-1645, "LM4702 Driving a MOSFET Output Stage." Available at www.ti.com.
10. National Semiconductor/TI datasheet, "LME49811 Audio Power Amplifier Series High Fidelity 200 Volt Power Amplifier Input Stage with Shutdown." Available at www.ti.com.
11. National Semiconductor/TI datasheet, "LME49830 Mono High Fidelity 200 Volt MOSFET Power Amplifier Input Stage with Mute." Available at www.ti.com.
12. National Semiconductor/TI AN-1850, "LME49830TB Ultra-High Fidelity High Power Amplifier Reference Design." Available at www.ti.com.

Chapter 32

Professional Power Amplifiers

32. INTRODUCTION

Professional power amplifiers are used to drive the loudspeakers in professional sound reinforcement systems. In professional audio amplifiers, it's all about power, reliability, size, weight, convenience of setup and reduction of system wiring [1].

In the 1970s professional amplifiers were not unlike higher-powered consumer audio amplifiers that were made more rugged. They often included balanced inputs, power monitoring displays and clipping indicators. They were frequently operated in bridged mode to achieve higher power with given supply voltages. The march to higher power levels in the 1980s saw the introduction of class G and class H amplifiers. These designs reduced heat generation and allowed more power to be made available in a given amount of precious rack space. The legendary Crest 8001 was a good example of such high-power amplifiers of the time. It provided over 1200 W per channel into a 4-Ω load or over 2400 W into an 8-Ω load in bridged mode in a 3 rack-space design [2].

Figure 32.1 illustrates a traditional pro amplifier that was common in the 1980s. These amplifiers were built with a linear power supply and a class AB, G or H power stage. Features often included the following:

- Balanced inputs
- Clipping indicators and output level metering
- Clipping control
- Fan

Second-generation professional power amplifiers began to appear in the late 1980s and early 1990s. Figure 32.2 illustrates some of their features. In more recent years there has been a strong migration to the use of class D amplifiers in these applications. The need for higher power and less heat generation, weight and rack space for a given power has driven this trend. These amplifiers usually employ switch-mode power supplies (SMPS) that greatly reduce size and weight. Many newer amplifiers incorporate power factor correction (PFC) to make the most efficient use of the mains power and comply with governmental directives. Modern SMPS also enable great agility in powering for many different mains standards, often over a range of 88 to 264 V AC. In many cases this is achieved without the need for a voltage selector switch. Amplifiers with SMPS and PFC can usually deliver their full power when the mains voltage sags and do not need to be over-designed to handle worst-case high mains voltages.

Balanced and single-ended inputs have always been a must for professional power amplifiers, but many have also included AES3 (AES/EBU) digital audio transport standard inputs for quite some time [3]. These digital inputs are available with electrical and optical interfaces.

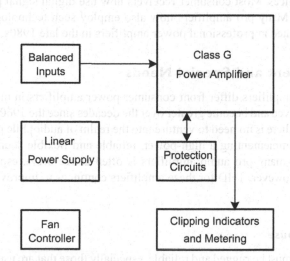

Figure 32.1 Traditional Professional Audio Amplifier

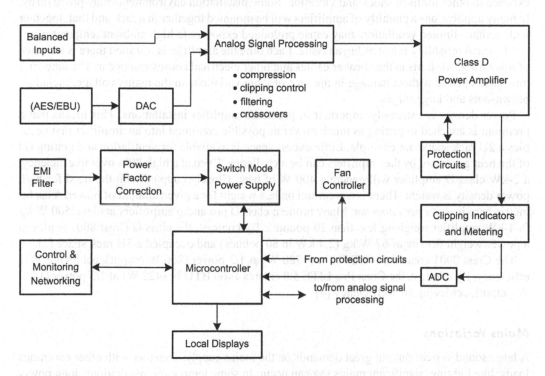

Figure 32.2 Second-Generation Professional Audio Amplifier

Beginning around 1996, digital audio inputs were also made available over proprietary Ethernet connections.

The introduction of microprocessors and DSP changed the landscape. Think of the modern consumer AV receiver and all of the control processing and DSP in it. Such approaches were made possible by the availability of economical microprocessor, DSP, A/D, D/A and non-volatile

FLASH memory devices. Most consumer receivers now use digital signal processing and class D power amplifiers. Many pro amplifiers now also employ such technology. Microprocessors began to be incorporated in professional power amplifiers in the late 1980s.

32.1 Environment and Special Needs

Professional power amplifiers differ from consumer power amplifiers in many ways, and these differences have evolved and become greater over the decades since the 1960s. While sound quality is still important, there is no need to venture into the realm of audiophile performance. Rather, the practicalities of implementing a high-power, reliable and flexible sound system dominate. Worst-case THD for many pro audio amplifiers is often above 0.1%, especially those implemented in class D. However, THD for these amplifiers continues to improve, and overall sound quality is quite good.

Reliability and Abuse

Pro audio amplifiers must be rugged and reliable, especially those that are used in touring applications where they may be moved frequently. There is always the risk that they may be dropped or exposed to other forms of shock and vibration. Some installation environments may prove dusty. In many applications a plurality of amplifiers will be mounted together in a rack and that, together with perhaps limited ventilation, may cause prolonged exposure to high ambient temperatures.

Electrical reliability is just as big an issue. Each time the amplifier is installed there is a chance of miswiring, so shorts in the speaker cables and other electrical abuses can occur. The amplifier must also operate without damage in the face of large variations in the mains voltage, including brown-outs and large surges.

Power density is extremely important in pro audio amplifier installations. This means that a premium is attached to getting as much power as possible crammed into an amplifier that occupies a 2U rack space, for example. Little excess space is available for ventilation and getting rid of the heat generated by the amplifiers can be a challenge. Even at a high 80% overall efficiency, a 2-kW class D amplifier will generate 400 W of heat. Closely coupled with the need for high power density is weight. There is a premium on less weight for a given amount of power. A useful metric here is *watts per kilogram*. Many modern class D pro audio amplifiers achieve 500 W/kg (a 4-kW amplifier weighing less than 20 pounds). By contrast, the class G Crest 8001 achieved a power/weight density of 67 W/kg (2.4 kW in 80 pounds) and occupied a 3U rack space [2].

The Crest 8001 created 6140 BTU/hr (1780 W) at 1/3 power (800 W output), achieving 31% efficiency. By contrast, the Crest Pro-LITE 5.0 creates 1460 BTU/hr (422 W) at 1/3 power (1730 W output), achieving 80% efficiency [4].

Mains Variations

A large sound system can put great demands on the mains supply. Together with other enormous loads, like lighting, significant mains sag can occur. In some temporary installations, long power cable runs can add to the problem. It is important that drops and peaks in the mains voltage not damage the amplifier. The mains may drop from a nominal of 120 V down to 90–100 V, or spike to perhaps 150 V. It is also desirable that the output power capability not be reduced under mains sag. Amplifiers with well-designed switching power supplies can often do just this.

Finally, in an international setting, the amplifier must be able to be operated with a variety of combinations of mains voltage and frequency [5]. Amplifiers that require manual switching or altered wiring to accommodate different mains voltage may be subjected to damage from accidental over-voltage if connected with the wrong mains voltage setting. Some pro amplifiers can automatically adapt to different mains voltages, and this is a huge plus for tours that occur on both sides of the ocean.

Integration of Functions

Space, weight and installation labor are at a premium in pro audio applications. Whenever it is possible to integrate multiple functions into a single box there is a big advantage. As a result, many modern pro audio amplifiers are more than just an amplifier and may incorporate other signal processing functions as well, such as compression, limiting and equalization [4, 6, 7, 8]

Remote Control and Monitoring

In a large venue there may be amplifiers in many different places, certainly not all conveniently accessible in a single control room. Often, for reasons of reduced amounts of heavy speaker cabling, amplifiers are installed closer to the loudspeakers than to the control room or mixing desk. This is especially the case in high-power systems where 2-Ω loads are often used. Note that 5 kW into a 2-Ω load corresponds to 100 V RMS and 50 A RMS. The ultimate example of close proximity of the amplifier to the loudspeaker is in the growing use of self-powered loudspeakers, where the amplifier is located inside the loudspeaker enclosure. Thus the need for remote control and monitoring becomes important.

32.2 Output Stages and Output Power

Efficiency is very important because power levels can be very high and heat removal requires rack space and weight. For many years pro amplifiers used the class AB output stage commonly found in consumer power amplifiers. This stage is simple and reliable. It provides good sound, but it has fairly low efficiency. This becomes a special concern in high-power professional power amplifiers.

Class AB Output Stages

The venerable class AB output stage, so thoroughly covered in this book, is seeing less and less use in pro audio amplifiers, simply because it is relatively inefficient. It remains in some lower-power pro amplifiers, say below 200 W, but is rapidly being made obsolete even in those applications by small and inexpensive class D amplifiers that sound just as good if properly designed.

Class G and Class H Output Stages

The quest for higher power and higher efficiency led to the development of class G and class H amplifier designs beginning in the 1980s [5]. The basic concept is to provide the class AB output stage with rail voltages that are no larger than is necessary for the output voltage demanded by the signal at any given time. This greatly reduces wasted power. High rail voltages then need only be used on signal peaks. In the discussion below, note that the terms *class G* and *class H* are sometimes used in reverse [5] See Chapter 7 for a more detailed discussion of classes G and H.

A class G design supplies the class AB output stage with a fixed smaller rail voltage for low power situations. The positive rail voltage supplied to the upper class AB output transistor increases and becomes variable with signal when the signal amplitude exceeds that which can be provided by the smaller rail voltage. The variable rail voltage is maintained several volts above the peak signal amplitude by a power transistor connected to a higher rail voltage. That transistor takes over and supplies the variable rail voltage to the class AB stage so as to provide headroom for the class AB part of the output stage. This is illustrated in Figure 32.3a, where the changing rail voltages for the class AB output stage are illustrated as a function of the amplitude of a sine wave.

In class H amplifiers, two or three different fixed rail voltages are applied to the class AB output stage depending on signal amplitude conditions. The rails are hard-switched in real time in accordance with the needs of the signal. In Figure 32.3b, the switches apply the higher rail voltage to the class AB output stage when the peak audio amplitude exceeds about 25 V.

A highly simplified class G output stage is illustrated in Figure 32.4. Power transistors Q3 and Q4 have a level-shifted version of the signal applied to their bases. The level shift establishes the headroom for the class AB portion of the output stage when signal peaks exceed the voltage of the lower supply rail. Diodes D1 and D2 prevent reverse bias of the bases of Q3 and Q4 when signal amplitude is smaller than the 40-V rails. R1 and R2 help turn off Q3 and Q4 (in this simplified example the VAS would have to supply current for R1 and R2). Diodes D3 and D4 provide the 40-V rail voltages when Q3 and Q4 are off.

The Crest 8001, introduced in 1987, is a good example of a large pro amplifier that operated in class G. It employed two rail voltages. The 8001 can deliver 750 W per channel into 8 Ω, and used rail voltages of 67 V and 135 V. The amplifier could provide 1200 W per channel into a 4-Ω load. It was capable of producing a peak output voltage of 129 V [2].

Class D Output Stages

The class D switching power amplifier has always been the Holy Grail of high efficiency in power amplifiers, but for a very long time was unable to provide satisfactory sound quality. Beginning in the late 1990s, technology and circuit approaches improved to the point where class D provided good sound quality and was capable of very high power. Some of the technology improvements were driven by the widespread use and improvement of switch-mode power supplies. A simplified class D amplifier is illustrated in Figure 32.5. The output transistors are always alternately *on* or *off*, so in theory they dissipate no power. The duty cycle of the *on* and *off* periods establishes the average value of the output signal, which is then low-pass filtered to become the analog output signal. This process is referred to as pulse width modulation (PWM). Class D amplifiers are discussed in more detail in Chapters 33–36.

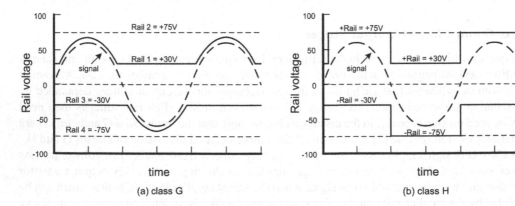

Figure 32.3 Class G and Class H Rail Voltage

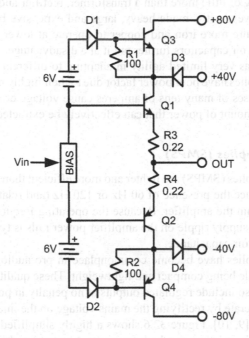

Figure 32.4 Simplified Class G Output Stage

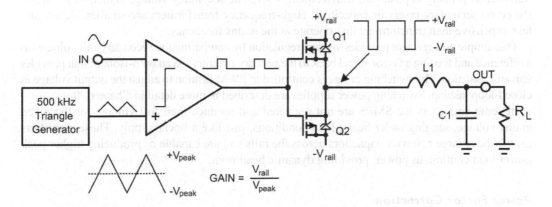

$$GAIN = \frac{V_{rail}}{V_{peak}}$$

Figure 32.5 Simplified Class D Output Stage

32.3 Power Supplies

Much of the weight and bulk in pro amplifiers lies in the power supply. For this reason, many designers have turned to switching power supplies (SMPS) [9, 10]. They are smaller, lighter, more efficient and less expensive.

Linear Power Supplies

The so-called linear power supply (only as to distinguish it from a switching power supply—it is not very linear in any regard by traditional descriptions) has been around since the beginning of electronics. Consisting of little more than a transformer, rectifier and reservoir capacitor, it is simple and reliable. However, it is made heavy, large and expensive by its power transformer. Power transformers require more iron and copper to operate at lower frequencies, like 60 Hz. The need for large reservoir capacitors further put it at a disadvantage. The linear supply is usually not regulated and has very limited agility in adapting to different mains voltages. Finally, conventional power supplies have poor power factor due to their highly nonlinear impulsive input current. The rectifier pulses of many tens of amperes cause voltage drops in the resistive mains supply that reduce the amount of power that can effectively be extracted from the mains.

Switching Power Supplies (SMPS)

Switch-mode power supplies (SMPS) are lighter and more efficient than conventional power supplies. They can also reduce the presence of 60 Hz or 120 Hz (and related harmonics) electrical and magnetic fields within the amplifier. Because the operating frequency is much higher than that of the mains, power supply ripple on the amplifier power rails is typically much smaller for a given amount of reservoir capacitance.

Switching power supplies have become commonplace in pro audio power amplifiers. They offer high efficiency while being compact and lightweight. These qualities are especially important. Their advantages also include regulated outputs at no penalty in power dissipation.

Switching supplies operate by rectifying the mains voltage on the line side and storing the DC on a reservoir capacitor [9, 10]. Figure 32.6 shows a highly simplified arrangement. The intermediate DC bus voltage is switched at a high frequency (several hundred kHz) to become AC to drive an isolating transformer. In this simple illustration, the switching bridge (S1–S4) drives the transformer primary in push-pull with a square wave. The secondary voltage is then rectified and stored on secondary reservoir capacitors. High-frequency transformers are smaller, lighter and less expensive than transformers that operate at the mains frequency.

This simple arrangement provides voltage regulation by comparing the secondary rail voltages to a reference and feeding an error signal back to the primary circuitry via an opto-isolator that provides galvanic isolation. The switching bridge is controlled in PWM fashion to adjust the output voltage in closed-loop fashion. Switching power supplies are described in more detail in Chapter 20.

In some amplifiers the SMPS are not regulated and produce their full voltage for the given mains voltage, sagging under heavy load conditions, just like a linear supply. These amplifiers usually have large reservoir capacitors across the rails and are capable of producing higher peak power than continuous power, providing dynamic headroom.

Power Factor Correction

Power factor describes the degree to which power is transferred for a given combination of delivered voltage and amperage. A resistive load has 100% of the apparent power (VA) delivered to it as real power, and therefore has the ideal power factor of unity. The power factor is unity because the voltage and current wave-shapes are identical and in phase. Any departure from the wave-shapes being identical and in phase results in a power factor of less than unity. In the extreme case of a purely inductive load, voltage and current are 90° out of phase and no power is delivered to the load, in spite of the fact that significant voltage is present and significant current flows (VA). Nonlinear loads, such as rectifiers with capacitor input filters, also represent a load with a poor power factor. Such inefficiency of power transfer is bad for the utility and sometimes bad for the user.

For linear loads, like inductive loads encountered in utility power distribution, power factor correction (PFC) is achieved passively with inductors and capacitors that shift the phase of the load current to be nearly the same as that of the voltage. Such passive power factor correction

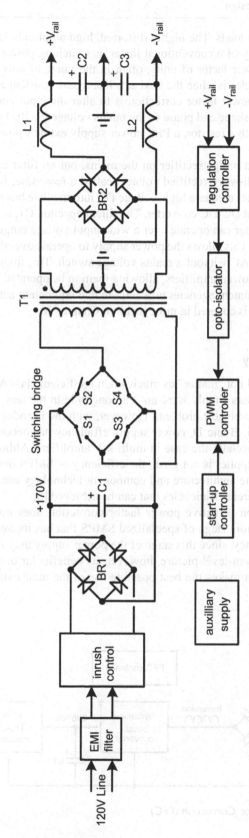

Figure 32.6 A Simple Isolated Switching Power Supply

does not work for nonlinear loads. The highly distorted, high amplitude input current pulses of the front-end rectifier supply of a conventional linear or switching power supply cause a great departure from an ideal power factor of unity, often in the range of only 0.6. Once again, the definition of unity power factor is when the load acts like a pure resistance.

The objective of active power factor correction is to alter the input current waveform to be one that has the same wave-shape and phase as the mains voltage [9, 10, 11, 12]. This makes the load act resistive. Just as with a resistor, a PFC power supply extracts power from the mains on all parts of the cycle.

Active PFC circuits use a bridge rectifier on the mains, but no filter capacitor, as shown in Figure 32.7. Instead, the full-wave-rectified voltage, called a *haversine*, feeds a dynamic boost converter. The boost converter creates a high-voltage DC intermediate bus supply of about 385 V DC that feeds the subsequent DC-DC converter. The filter capacitor, C1, is placed after the boost converter. The boost converter can operate over a wide input voltage range while delivering the regulated 385 V DC output. This allows the power supply to operate over the full universal input voltage range of 88–264 V AC without a mains voltage switch. This mains powering agility is important for professional power amplifiers, allowing them to be operated virtually anywhere in the world. Numerous government agencies now require that equipment achieve high power factor. Power factor correction is covered in more detail in Chapter 20.

Power Supply Efficiency

Power supply efficiency did not "matter" as much when inefficient class AB output stages were used. The power supply losses simply were not the long pole in the tent. The weight and bulk of the power supply was the biggest problem. However, with the introduction of more efficient output stages like classes G, H and D, power supply efficiency has become more important in a relative sense. This is especially the case in multi-kW amplifiers. Although the efficiency of unregulated linear power supplies is not great, the efficiency of SMPS designs can vary quite a bit as well, depending on the architecture and component technology used. Duncan provides a very good overview of different efficiencies that can be achieved [5].

Interestingly, the addition of active power factor correction does not increase efficiency *per se*. It amounts to one more stage of specialized SMPS that has its own efficiency. In local terms, it costs some efficiency, since this stage of the power supply may itself be only 90–95% efficient. In the bigger system-level picture, however, its benefits far out-weigh its slight cost in efficiency. Once again, it makes the best possible use of the mains supply that is available in a venue.

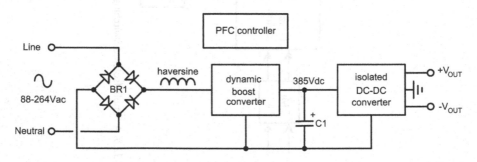

Figure 32.7 Active Power Factor Correction (PFC)

32.4 Cooling and Heat Removal

A primary objective in pro power amplifiers is to get the highest power into the smallest rack space and with the least amount of weight. High power amplifiers create a lot of heat, even the ones using switching power supplies and class D output stages (they produce higher power in the given space). Heat is the enemy of electronics reliability. The lifetime of a typical electronic component is reduced by 50% for every 10 °C rise in temperature.

Heat sinks relying only on natural convection are used widely in consumer amplifier applications, but they are heavy, bulky and expensive for the amount of heat removal they provide. Fans are rarely seen in consumer applications because of the noise they make.

Fans are a must in pro amplifiers where a lot of heat is generated in a small space. Fan noise is a bit less of a concern in these applications as well. However, a large number of amplifiers in a rack may create quite a bit of noise, so many amplifiers incorporate a variable-speed fan or a multi-speed fan. The fans are often microprocessor-controlled and usually include tachometer feedback to verify that the fan is operating. It is also very desirable that the fan continues to run after an over-temperature amplifier shut-down. The amplifier should always be designed to recover automatically in the event of such a shut-down.

A common heat removal approach incorporates a heat sink tunnel through which the fan blows air. The inlet end of the tunnel will be cooler than the outlet end, so a plurality of paralleled devices on blown tunnel heat sinks often operate at slightly different temperatures [5]. The use of fans brings with it the issue of increased exposure of the interior of the amplifier to the outside environment and the need for air filters. This is especially the case in dusty areas or where the amplifiers may be in proximity to "pollutants," like smoke machines [5]. Filters need to be changed, and this can be a maintenance headache.

32.5 Microcomputers

Microcomputers in many amplifiers have permitted the implementation of advanced protection and monitoring schemes [6]. The use of FLASH memory now allows amplifiers to retain settings and implement different software modes. In many cases the microcomputer (μC) implements power-on self-test (POST) and diagnostic aids. Later models of amplifiers can employ a USB port for local access to diagnostics and even for software upgrades in the field. A laptop or tablet computer can easily be connected to the amplifier, especially if a USB port is located on the front panel. The modern professional power amplifier is truly a system with many advanced and intelligent features, as illustrated by the block diagram in Figure 32.8.

The heart of the microcomputer system is a microcomputer with a non-volatile FLASH memory. Here the μC program and amplifier parameters are stored. At power-on, the program and some of the other information will be transferred into the RAM in the μC. Many modern microcomputer chips include a good amount of on-chip RAM, but additional RAM can be added externally if necessary. Some include on-chip FLASH as well, so the FLASH block may be internal to the μC as well if it provides adequate non-volatile storage.

The μC may also be implemented on a field programmable gate array (FPGA) where other logic functions for the amplifier may be implemented. Some FPGAs include considerable amounts of RAM and FLASH and can be programmed and configured to form a fairly complete subsystem. Some FPGAs include a "hard-core" (fixed circuitry) processor on-board, while others can implement a "soft-core" processor, wherein the processor is configured from the logic cells by a firmware module supplied by the manufacturer. Leading manufacturers of FPGAs include Altera and Xilinx.

Serial, Ethernet and USB ports are shown as means for communication with the external world. Some system-on-chip microcomputers include these interfaces on-chip. These ports make

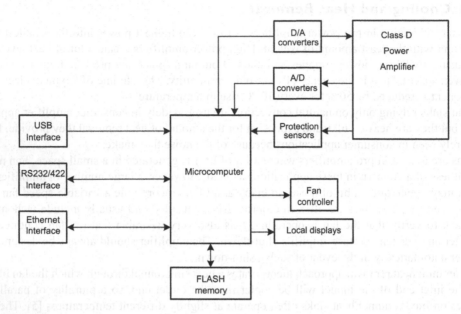

Figure 32.8 Microcomputer System for Professional Power Amplifier

possible the external monitoring and control of the amplifier. They also permit programming of the μC and amplifier as a whole.

Logic inputs and outputs on the μC allow it to monitor and control a host of functions that include relays, LEDs and other displays, power sequencing at turn-on and turn-off, fans and many other functions. Many fans include a tachometer function that permits even simple microcomputers to verify fan operation and control speed.

Most amplifiers equipped with a μC include a small auxiliary power supply to enable the mC to function at some level all of the time. This is especially important if the μC is to participate in power sequencing, but also allows the μC to perform some portions of a power-on self-test (POST) before the main power is applied. If the small auxiliary supply is always on, this also allows a networked system to "see" the amplifier. The auxiliary supply is also often needed for start-up and control of the power supply if an SMPS is used.

Several general purpose ADCs, or one ADC with multiple selectable input channels, are used to monitor various DC levels that represent operating conditions like temperature, rail voltages and mains voltage. An ADC and a DAC may also be incorporated into intelligent amplifier and loudspeaker protection systems.

Even without a DSP, some real-time signal control functions can be managed by the μC, sometimes including functions like clip detection and clip prevention. As microcomputers become faster, the distinction between the μC and a DSP becomes blurred for some functions.

32.6 Networked Control and Monitoring

Remote control and monitoring has long been important for professional audio amplifiers, especially in large venues. In the 1980s Crown introduced the IQ system, where RS232/422 serial interfaces made it possible to connect many amplifiers in a daisy chain that was connected to a host computer for control and monitoring of the amplifiers [13].

In 1993 a major step forward was made by Crest Audio (owned by Peavy since 1999) with the introduction of the MediaMatrix® networking system based on Ethernet connectivity. This was far superior to interconnecting power amplifiers with serial ports [14].

32.7 Digital Signal Processing

The availability of economical digital signal processing (DSP), A/D conversion and D/A conversion in the early 1990s made it inevitable that DSP would be introduced into professional power amplifiers and play a major role in reducing system costs while greatly increasing system flexibility and capability. In many cases, DSP has allowed the power amplifier to absorb many functions that were formerly performed by external boxes. One of the most important areas in which DSP allows system integration in one place is the group of functions called loudspeaker management [4, 6, 7, 8]. These functions include:

- Filtering and equalization
- Delay for time alignment
- Compression and limiting
- Electronic crossovers

In conjunction with the microcomputer system, different DSP functionality, processing programs and settings can be conveniently selected. While the DSP plays a major role in the signal path, it also makes possible other monitoring and control functions.

- Amplifier protection
- Loudspeaker protection
- Load sensing
- Test signal creation
- Amplifier measurement and diagnosis

Latency

The use of digital signal processing can introduce time delay, also referred to as latency. It is important that the DSP in power amplifiers introduce only a very small amount of latency, preferably less than 2 μs.

System Installation and Setup

Labor for system installation and setup is a significant expense in any venue and is even more so in a touring situation. The resulting sound quality is also very dependent on the amount of labor put into setup. DSP makes it practical to include system testing features in the amplifier, such as the generation of sine wave tones, pink noise, white noise, MLS and so on. Such test signals can be used during installation and setup to verify loudspeaker operation, balance levels and even tailor frequency response to the acoustic environment.

32.8 DSP-Based Protection and Monitoring

DSP also makes it possible to implement more sophisticated protection and monitoring systems for the amplifier and the connected loudspeaker [6]. The introduction of the microcomputer and general-purpose ADCs and DACs made it possible to monitor all kinds of functions and power amplifier operating parameters, such as operating temperature, rail voltages, mains voltages. The addition of the DSP greatly expanded the number of things that could be monitored readily. This capability goes beyond the audio-path signal processing capabilities already added by the DSP.

Protection Circuits

Protection circuits are crucial to maintaining reliability of the amplifier and the connected loudspeakers. Replacement of either is costly and an amplifier blowing up and failing during a performance is unacceptable. The protection circuits described in earlier chapters are crude in comparison to what can be done with a microcomputer or DSP. The more sophisticated protection circuits can allow the amplifier to operate closer to its maximum capability, reduce the failure rate and reduce the incidence of brief protection events that are unnecessary. Real-time calculation of SOA margins and limits based on current operating temperature is possible, for example.

Power Levels and Load Impedance

Output power is readily calculated in real time with a DSP simply by multiplying the output voltage by the output current if the output current is somehow being measured (there are numerous means for doing so). Figure 32.9 illustrates such an approach. ADCs digitize measures of output signal voltage and current. That information is then fed back to the DSP. These ADCs need not be of high audio quality, but should capture the actual signal waveforms. 8-bit converters sampling at 48 kHz with crude anti-alias filters are adequate for coarse measurements.

Average power and peak power are both of great interest, the former for anticipating voice coil heating in the loudspeaker and the latter for seeing how close one is to clipping. Notably, the DSP can calculate the phase relationships of the output voltage and current signals so as to determine real power and complex impedance. Knowledge of these phase relationships can also play a role in more sophisticated protection circuits.

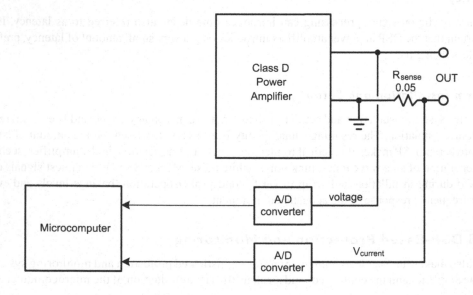

Figure 32.9 Output Voltage and Current Measurement

Such calculations also permit the evaluation of the crest factor and dynamic range of the program material. In some cases it can also be valuable to determine the spectrum of the power, recognizing that multi-way loudspeakers may have vastly different average power handling capabilities in different frequency ranges.

Pro amplifiers may be connected to a very wide variety of loudspeakers, depending on the function of that channel and the venue for the system. The performance of an amplifier can be quite dependent on the load impedance. Most fundamentally, it is important to know whether the loudspeaker connection is open or shorted. Moreover, it is very helpful to know approximately what the impedance is in order to optimize the performance of the amplifier and anticipate output power levels. DSP makes it possible to report effective load impedance in real time while the amplifier is passing program material [6]. In some cases, a power amplifier capable of changing its rail voltages will select lower rail voltages when driving a load having lower impedance.

Testing and Diagnostics

Put together a sound card and a PC with today's technology and you can do a great deal of sophisticated audio testing at low cost. This is how a lot of audio testing instrumentation is now being implemented. With this combination, you literally have a test bench. Such arrangements are used in audio labs for testing and characterizing both amplifiers and loudspeakers.

Analogously this same capability already exists in a pro amplifier that incorporates a microcomputer, DSP, DAC and ADC. The presence of a DSP in the signal path further enhances this capability. It enables detailed amplifier self-test and diagnostics. Fairly thorough testing can be done at turn-on even before the output of the amplifier is un-muted. The concept of the power-on self-test (POST) is readily implemented. Testing of the overall system setup is also made possible.

32.9 The DSP to Class D Interface

Amplifiers employing DSP with class D output stages must incorporate an interface between the PCM digital signal coming from the DSP and the class D amplifier. This depends heavily on the type of class D amplifier employed. It seems a shame in an otherwise mostly digital environment to have to go back to analog to implement this interface, but a great many class D amplifiers expect an analog input (class D amplifiers are not "digital amplifiers" as the D in class D might lead some to believe). Conventional PWM class D amplifiers are an example. This is explained in Chapter 33. For these amplifiers, there is little choice but to pass the output of the DSP through a digital-to-analog converter (DAC). The resulting analog signal is then applied to the class D amplifier.

Alternatively, there are direct-digital approaches to class D amplifier designs wherein a PCM input is received and converted by specialized digital signal processing into a class D format for driving the class D output switching stage. In principle, it is possible to incorporate the direct-digital interface into the main DSP of the amplifier. Unfortunately, if negative feedback is to be used to reduce distortion and otherwise improve performance of the class D amplifier, an analog-to-digital converter (ADC) must be used to convert the analog output of the class D amplifier into digital form to be incorporated as negative feedback in the digital domain in the direct-digital conversion step.

32.10 Programming

With so much functionality controlled by software and firmware running in the microcomputer and DSP, the opportunities presented by the ability to change programs in the field are enormous. Firmware and software upgrades are commonplace today in consumer electronics, where even your TV or CD player may be upgraded to improve performance or functionality (or address a

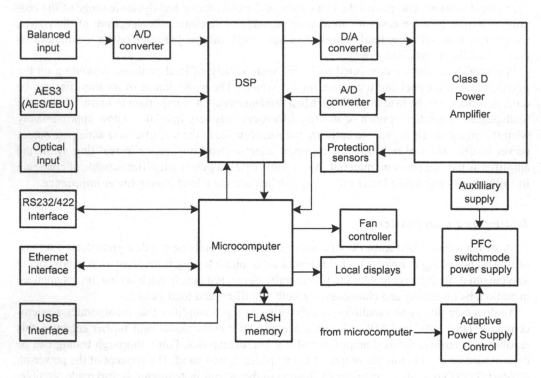

Figure 32.10 Modern Professional Power Amplifier Including Programming

bug in the original software). This capability has been incorporated in many modern professional audio power amplifiers [4, 6, 7, 8, 15].

Figure 32.10 illustrates a modern pro amplifier that incorporates a programming arrangement. As discussed earlier, the use of modern non-volatile FLASH memory is key to the programmability of these amplifiers. Both the microcomputer and DSP can be programmed. The DSP's RAM is usually loaded with its program by the microcomputer. Programming means may include RS232/422, Ethernet, USB or even Bluetooth.

The ability to carry out firmware upgrades in the field is only a very small piece of this picture. The most significant feature is the ability to program the signal processing functionality of the DSP. This includes setting gain, equalization frequency and phase response, signal delay, compression and limiting behavior, active crossover functions and many other functions. A given amplifier can thus be configured to perform a particular task in a venue where many others of the same amplifier can be tasked to perform other functions with different setup parameters. If the amplifiers are networked, this programming can be done over the network.

Portable programmers can also be used to load information into the amplifier and even retrieve setup information from the amplifier [15]. If desired, a plurality of amplifiers can be cloned with the same setup information. An RS232 or USB connector can be put right on the front panel of the amplifier. As with hearing aids, even Bluetooth can be used to program the amplifier.

32.11 Audio Networking

The combination of digital audio and Ethernet has made audio networking possible. For power amplifiers, this means the digital delivery of the audio signal and control over the Ethernet infrastructure in an integrated way. The present-day audio networking approach is referred to as *Audio over Internet Protocol*, or simply AoIP. It uses existing Ethernet infrastructure for professional audio transport in virtually any venue, providing for far less cabling. It also makes possible more complex and sophisticated applications. Easy reconfiguration and expansion is another benefit brought by AoIP.

The audio networking approach began in earnest in 1996 when CobraNet® was introduced by Peak Audio [16, 17]. This Audio over Ethernet system (AoE) has enjoyed considerable success, but does not allow one to use much off-shelf Ethernet hardware and software. It also introduces signal latency (delay) on the order of as much as 5 ms [18, 19, 20] While such latency is acceptable for many pro sound applications, it is a significant concern for live sound applications. Live sound is the most stringent networked audio application because of its need for very low latency.

There have been numerous other system approaches to networked audio that were based on AoE or Audio over Internet Protocol (AoIP), such as RAVENNA [21, 22, 23] and Livewire [24]. Another noteworthy approach is Audinate's Dante® system [18, 19, 20]. This system has largely solved the latency problem while enabling the full use of Ethernet off-shelf hardware and routing capabilities.

To address networking compatibility and facilitate seamless communication among different manufacturers' networking products, the Audio Engineering Society issued interoperability standards for AoIP systems and technology in 2013 [25, 26, 27]. The AES67–2013 standard was published in September 2013 and updated in 2015. The AES67 standard is the result of an industry collaborative effort and does not seek to obsolete existing AoIP approaches or hardware. Much of the compliance with AES67 can be achieved with software changes, and it has been largely adopted throughout the industry.

Of course, AoIP pertains to the networked interconnection of all professional audio equipment, not just power amplifiers. AoIP is a rapidly moving technology that has already revolutionized the professional sound industry [28, 29]

While AES67 pertains to *transport* of audio, AES70 was defined by the AES in 2015 as the Open Control Architecture (OCA) standard for *control* of media networks consisting of audio and video devices. It is a standardized protocol that allows control and monitoring of devices on a network. It also provides for management of that network. It permits compatible support of devices from diverse manufacturers. AES70 does not provide transport of program material [30].

References

1. Glen Ballou, *Handbook for Sound Engineers*, Focal Press, Burlington, MA, 2015.
2. Crest 8001 Power Amplifier. Available at www.peaveycommercialaudio.com.
3. AES3–1992 (ANSI S4.40–1992), "AES Recommended Practice for Digital Audio Engineering—Serial Transmission Format for Two Channel Linearly Represented Digital Audio Data." See also IEC 60958.
4. Crest Audio (Peavey Commercial Audio), "Pro-LITE Series Amplifiers." Available at www.peaveycommercialaudio.com.
5. Ben Duncan, *High Performance Audio Amplifiers*, Newnes, Oxford, 1996.
6. Dennis Fink, "Digital Signal Processor for Amplifier," U.S. Patent 5,652,542, assigned to Crest Audio, July 29, 1997.
7. QSC Audio Products, "CXD Series Amplifiers." Available at www.qscaudio.com.
8. Crown International "*Cdi* and *I-Tech* Series Amplifiers." Available at www.crownaudio.com.
9. Abraham I. Pressman, Keith Billings and Taylor Morey, *Switching Power Supply Design*, 3rd ed., McGraw-Hill, New York, 2009.
10. Keith Billings, *Switchmode Power Supply Handbook*, McGraw-Hill, New York, 2010.
11. ON Semiconductor, "Power Factor Correction (PFC) Handbook," 2011. Available at www.onsemi.com.
12. Dennis Fink, "Audio Amplifier Having Power Factor Correction," U.S. Patent 6,023,153, assigned to Crest Audio, February 8, 2000.

13. Crown IQ system, 1988, IQ2000 system.
14. Don Davis, Eugene Patronis, Jr. and Pat Brown, *Sound System Engineering*, 4th ed., Focal Press, Burlington, MA, 2013.
15. Dennis Fink, "Apparatus and Method for Programming an Amplifier," U.S. Patent 5,652,544, assigned to Crest Audio, July 29, 1997.
16. Yamaha System Solutions, "An Introduction to Networked Audio," White Paper. Available at www.cobranet.info.
17. Crown by Harman, "CobraNet Primer," White Paper. Available at www.crownaudio.com.
18. João Martins, "Audinate Dante (Part 1)—Making Digital Audio Networking Easy," *AudioXpress*, February 2014. Available at www.audioxpress.com.
19. João Martins, "Audinate Dante (Part 2)—Audio Networks Fast to Market," *AudioXpress*, March 2014. Available at www.audioxpress.com.
20. Audinate, "Audio Networks Past, Present and Future (CobraNet and Dante)," White Paper. Available at www.audinate.com.
21. João Martins, "Ravenna (Part 1)—The Scalable Audio Network Solution," *AudioXpress*, May 2014. Available at www.audioxpress.com.
22. João Martins, "Ravenna (Part 2)—A New Framework for Audio-Over-IP," *AudioXpress*, June 2014. Available at www.audioxpress.com.
23. ALC NetworX, "RAVENNA," White Paper/Brochure, Ravenna@alcnetworx.de.
24. Axia Audio, "A Closer Look at Livewire," Informational web page. Available at www.axiaaudio.com/livewire.
25. João Martins, "AES67–2013—The New Networked Audio-Over-IP (AOIP) Interoperability Standard," *AudioXpress*, January 2014. Available at www.audioxpress.com.
26. Audio Engineering Society, "AES Standard for Audio Applications of Networks—High Performance Streaming Audio-over-IP (AoIP) Interoperability," AES67–2013, 2015.
27. João Martins, "AES67 2015 Revision Confirms Stability of the Audio Network Interoperability Standard," *AudioXpress*, October 2015.
28. RAVENNA, "Your Practical Guide to AES67," Part 1, The Broadcast Audio Bridge, August 23, 2017. Available at www.thebroadcastbridge.com.
29. João Martins, "Audio Network Development," Parts 1–5, AudioXpress, January 2016–May 2016. Available at www.audioxpress.com.
30. Arie van den Broek, "AES70—The Future of Audio Control," Part 1, The Broadcast Audio Bridge, December 11, 2017. Available at www.thebroadcastbridge.com.

Part 6

Class D Amplifiers

Traditional audio power amplifiers have served us well for many decades, and will continue to do so. However, class D amplifiers are the wave of the future. They have already found their way into many consumer and professional audio products. The proliferation of multi-channel receivers in the home theater market has created an increased need for amplifiers that require little space and generate a minimal amount of heat. Powered loudspeakers and mobile electronics are also very important applications of class D amplifiers. At the same time, the quality achievable with class D amplification has improved dramatically. Although the *D* in class D does not stand for *digital*, digital signal processing techniques are playing an increased role in some class D amplifier implementations.

Part 6 covers the class D amplifier technology. The design of class D amplifiers requires a somewhat different skill set that may be unfamiliar to designers of conventional power amplifiers. I have found that there is a great deal of information on class D amplifier design available, but it is scattered across many sources. I have tried to bring much of that together in Part 6 of this book in a way that is easily assimilated by those new to this exciting field of power amplifier design.

Chapter 33

Class D Audio Amplifiers

33. INTRODUCTION

Class D amplifiers operate on an entirely different principle from those discussed elsewhere in this book. The output stage in a class D amplifier comprises switches that are either *on* or *off*. The switches apply the positive supply to the output for one brief period and then connect the negative supply rail to the output for the next brief period. The process is then repeated indefinitely. This results in a square wave at the output. If these two intervals are the same, the net output is zero. If the first is longer than the second, the output has a net positive value. A low-pass filter extracts the average value to drive the loudspeaker. The cutoff of the LPF often lies in the 30-kHz to 60-kHz range.

This process is referred to as *pulse width modulation* (PWM). These switching intervals alternate at a high frequency, often in the range of 500 kHz. Thus, the average value of the square wave drives the load. Because the switches are either on or off, they dissipate no power. Virtually all of the input power from the power rails is transferred to the load, so efficiency is very high and power dissipation is very low. Efficiency of 85% to 95% is not uncommon. Small amplifiers that run cool are the result. Because the output transistors act as switches, and are either *on* or *off*, class D amplifiers are largely free of SOA issues. A big challenge in class D amplifiers is the proper driving of the output stage switches so that the on and off timing intervals accurately reflect the input signal.

Class D amplifiers have long suffered from poor distortion performance. Matters have improved dramatically since the late 1990s. The need to squeeze more power capability into a smaller space while generating less heat has driven their development. This has been especially the case in portable battery-operated consumer devices, but also includes the following important areas in audio:

- Home theater receivers
- Subwoofers
- Pro sound
- Self-powered loudspeakers
- Professional sound reinforcement

As we'll see, the implementation of class D amplifiers is far more involved than the simple description above. Although the *D* in class D does not stand for digital, it is true that the implementations of class D amplifiers are moving more toward digital, and in fact there are approaches that involve direct digital conversion from PCM input streams to class D audio outputs.

There are numerous ways to build a class D power amplifier, but those based on PWM are the oldest and still very popular. This chapter will discuss PWM in depth; other approaches will be covered in Chapter 35.

The emphasis in these chapters will be on class D amplifiers designed for high sound quality, as opposed to smaller or more efficient amplifiers designed for more pedestrian applications like handheld consumer electronics. Entire books can be devoted to the subject of class D amplifiers. The limited space here permits only the scratching of the surface as a primer on class D amplification.

33.1 How Class D Amplifiers Work

Figure 33.1 shows a simple arrangement that converts an analog input to a digital PWM signal [1–3]. This circuit is referred to as a PWM modulator. A triangle waveform at several hundred kilohertz is applied to one side of a comparator while the input signal is applied to the other side. Whenever the input signal is more positive than the reference triangle wave, a positive pulse is produced that lasts as long as the input signal is above the threshold set by the triangle wave. With a perfect triangle wave, it is easy to see that the pulse width is linearly proportional to the input amplitude. Conversely, when the input signal is below the time-varying threshold set by the triangle wave, the output of the comparator is negative.

The output of the comparator is thus a square wave whose duty cycle corresponds to the amplitude of the input signal voltage. The frequency of the square wave is referred to as the carrier frequency. It is easy to see that the average value of the square wave is an accurate representation of the input signal. If the comparator output is used to drive power MOSFETs on and off, the average output will reflect the input signal value multiplied by the power supply rail voltage. The average output is extracted from the high-power pulse stream by a low-pass output filter, as shown in Figure 33.2. The technical term for this process of retrieving the analog signal from a pulse train is *reconstruction*; a discrete-time switched signal is converted to a continuous-time signal. The output filter also suppresses high-frequency EMI that is a part of the square wave.

The gain of this amplifier is equal to the ratio of the power supply rail voltage to the peak voltage of the triangle wave. If the peak value of the triangle wave is 2 V and the peak value of the input signal is just equal to 2 V, the output of the comparator will be high all of the time and the positive rail will be connected to the load 100% of the time. If the rail voltage is 40 V, the peak output will be 40 V and the gain of the amplifier will be seen to be 20. The positive duty cycle of

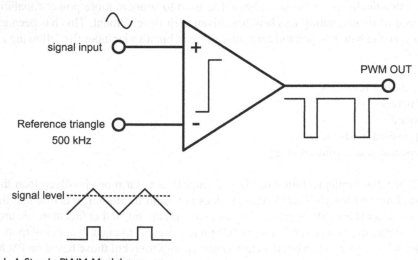

Figure 33.1 A Simple PWM Modulator

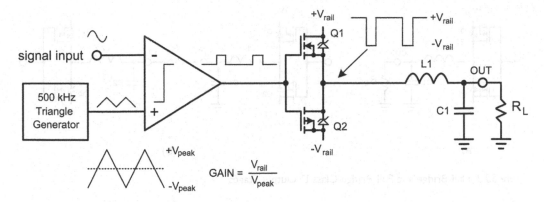

Figure 33.2 A Simple PWM Class D Amplifier

the square wave is reflective of modulation depth, which is the relative degree of departure from a 50% duty cycle in either direction.

Analog Class D and Digital Class D

The arrangement described above is described as analog class D because the pulse width modulator is implemented in the analog domain. There also exist various forms of digital class D in which digital techniques are employed to construct a PWM signal whose duty cycle is a faithful representation of the signal amplitude. Digital techniques can also be used to implement *pulse density modulation* (PDM) wherein the pulse density represents signal amplitude. The digitally generated signal is then used to drive the output switching devices. Finally, there is direct digital class D. In this case a PCM digital input signal is converted in the digital domain to the string of pulses required to drive the output MOSFET switches.

Synchronous and Asynchronous Class D

The fixed-frequency triangle wave PWM modulator described in Figure 33.1 is a synchronous modulator because the period of the output frequency is always the same; only the duty cycle changes. There also exist asynchronous modulators like the so-called self-oscillating modulators where the switching frequency is not determined by a fixed-frequency oscillator and may actually change as a function of the signal. There are also sigma-delta modulators where a fixed high-frequency clock is employed, but where the resulting pulse density stream does not reflect a particular carrier frequency.

33.2 Class D Output Stages

The output stage is where many of the challenges lie in class D amplifiers [2, 4]. This is where the power supply rail voltages are alternately switched onto the output bus at very high frequencies with very fast rise and fall times. Typical carrier frequencies are in the 500-kHz range while typical rise and fall times are in the 20 ns range. In an amplifier with ±50-V rails where the output transitions through 100 V in 20 ns, the voltage rate-of-change at the output switching node is about 5000 V/μs. To put this in perspective, 2.5 A is required to drive a 500-pF capacitor at this voltage rate of change (slew rate).

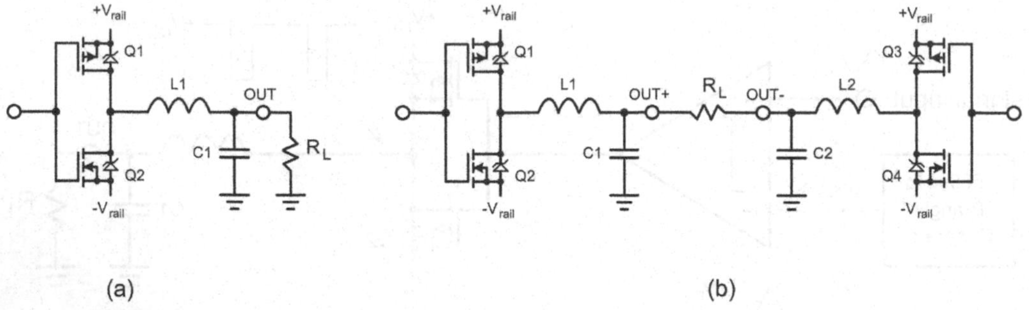

(a)   (b)

Figure 33.3 Half-Bridge and Full-Bridge Class D Output Stages

Single-Ended and H-Bridge Output Stages

Figure 33.3 illustrates two common class D output stages [2]. The first is a single-ended version employing complementary MOSFETs. An N-channel MOSFET is used for the low side switch, and a P-channel MOSFET is used for the high side switch. Both of these devices are in a common-source configuration. A typical second-order output filter is shown. The second arrangement is what is called an *H-bridge*. This circuit is driven so that when one side is high, the other side is low, doubling the output voltage available to the load. Interestingly, the H-bridge conveniently has an off position where no current flows in the load if both sides of the bridge are high or low. The single-ended output arrangement is commonly referred to as a *half bridge*, while the H-bridge arrangement is referred to as a *full bridge*.

The full bridge requires twice as many components, but only half the power supply voltage to realize a given output power capability. The full-bridge arrangement is often referred to as a *bridge tied load* (BTL) amplifier. This is analogous to linear amplifiers that are bridged. The full bridge also has some technical advantages in regard to the class D amplification process that will be discussed later. MOSFETs with desirable switching characteristics are more readily available with lower voltage ratings (below about 150 V), so high-power class D amplifiers will often employ a full bridge.

N-Channel Output Stages

N-channel MOSFETs inevitably have better switching characteristics than P-channel devices. Their FOM is usually less than half that of corresponding P-channel devices. For this reason, output stages often employ N-channel devices for both the low side and high side switches, as shown in Figure 33.4. These designs require more complex drivers that include level shifters and boost supplies [1, 4]. The gate drive for the high side N-channel MOSFET must float on top of the output signal, since it is referenced to the source of the device. The boost supply is required because the high side N-channel switch requires gate drive voltages above the positive rail in order to turn on. Fortunately, integrated circuit drivers are available that take care of most of the complexity. An example is the International Rectifier IR2011 [4].

The boost voltage is usually obtained by a bootstrap circuit consisting of a diode and a capacitor pumped from the square wave on the output node. The bootstrap supply can suffer performance problems when the PWM duty cycle approaches 0% or 100% as a result of the very narrow pulses that it has to work with under these conditions. For the highest sound quality and maximum achievable depth of modulation, a separate linear or switching supply can be used to

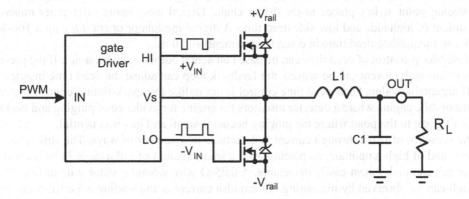

Figure 33.4 Half-Bridge Output Stage with N-Channel High Side and Low Side MOSFETs

provide the gate drive boost voltage. Such a supply will then not depend on the nature of the audio signal.

Gate Drive Control

The input impedance of a MOSFET gate is very high at low frequencies, but becomes quite low at high frequencies as a result of gate-source (C_{gs}) and gate-drain (C_{gd}) capacitances, which can be quite high. These gate capacitances must be charged and discharged at a very high speed in class D output stages. Further adding to the gate-drive burden is the Miller effect involving the gate-drain capacitance. Peak gate drive currents can lie in the ampere range even in a 100-W class D amplifier. Further complicating this is the need to provide highly precise gate timing to minimize distortion.

Dead Time Control

When switches are connected from both the positive and negative rails to a single output node, there is always the possibility that both switches will briefly be on at the same time. This will cause shoot-through current to flow directly from the positive supply to the negative supply. At minimum, this will result in wasted power. In some cases it will result in the destruction of the output stage. For this reason there is a very small dead zone incorporated into the MOSFET drive circuit. This ensures that there is always a very small time when both devices are turned off. With slight variations in timing, there should never be a situation where both devices are on simultaneously. Unfortunately, the presence of this dead time is a cause of distortion [4]. This will be discussed in Chapter 34.

Adaptive Dead Time Control

Proper dead time is critical for most class D solutions to prevent shoot-through current and minimize distortion. The amount of dead time is measured in nanoseconds, yet it may vary from part to part and unit to unit and with temperature and operating conditions. If the presence and amount of shoot-through current can be measured, dead time for the high side and low side switches can be independently controlled. This can be done by a set-and-forget approach or by a control loop.

Dead time controllability can be implemented in analog PWM circuits by changing the bias and slicing point at key places in the driver chain. Digital modulators offer other means for adjustment of high side and low side dead time. A digital modulator operating with a 100-MHz clock can manipulate dead time in discrete increments of 10 ns.

Closed-loop control of dead time can be based on actual output stage behavior. If the presence of shoot-through current can be sensed, the feedback loop can adjust the dead time inserted to a small acceptable value. Such dead time control is not unlike the spark-timing control system in an automobile engine where a detector monitors the engine for evidence of pinging and backs off the timing just to the point where the pinging becomes small and inconsequential.

The presence of shoot-through current is detectable in a number of ways. The spikes are very narrow, and of high amplitude, so placing even a small amount of inductance in series with the power rail can make them easily detectable. A 0.05-Ω wire wound resistor will suffice. Shoot-through can be observed by measuring the impulse current at the leading edge (turn-on) of the driver with the smaller pulse width. The leading edge is not affected by body diode reverse recovery. If shoot-through is detected, turn-on of that device should be retarded. Dead time for the low side device is evaluated on positive half-cycles, while dead time for the high side device is evaluated on negative half-cycles.

33.3 Bridge Tied Load Designs (BTL)

Class D bridged amplifier arrangements are referred to as *bridge tied load* (BTL) designs. They are analogous to bridging a linear amplifier. They are very popular because they can produce almost 4 times as much power into a given load impedance with a given rail-to-rail supply voltage. This can be especially important because integrated circuit modulator/drivers are limited in how much voltage can be placed across them. BTL amplifiers can be operated with a single supply and yet can be DC coupled to the load. Often, the two output terminals will float at an average DC potential of half the single-supply voltage. Single-supply operation inevitably costs less to implement. A single-supply half-bridge amplifier will need to be AC coupled to the load through a large electrolytic capacitor.

A BTL arrangement can either be made by using a full bridge (H bridge) output stage, as discussed earlier, or by connecting two complete half-bridge amplifiers, one of whose inputs is inverted, just as is done with a bridged class AB amplifier. In fact, BTL class D amplifiers built with ICs and using two identical half-bridge channels in the IC for each stereo channel are more popular, since they can be configured in a stereo mode or paralleled in a mono mode with almost twice the output power. The Texas Instruments TPA3255 is a good example of such a class D amplifier chip [5]. Indeed, it actually contains four half-bridge amplifiers, and can be configured as a four-channel amplifier with each channel operating in single-ended (SE) mode.

33.4 Negative Feedback

While traditional class AB power amplifiers profit from negative feedback, closing a feedback loop around a class D power amplifier can be challenging [3, 6]. The feedback can be taken from before or after the output filter. The use of negative feedback is especially important for improving the very poor PSRR of many class D amplifier architectures. Many class D amplifiers that run open loop also suffer high distortion.

If the negative feedback is taken from before the filter, the signal has not yet been reconstructed. Even at this point the sampling nature of the amplifier fundamentally limits the bandwidth over which the feedback loop can operate. Some low-pass filtering or integration must be placed somewhere in the loop for purposes of reconstructing the feedback signal. The signal must be converted from a discrete-time representation to a continuous-time representation (analog). Feedback taken from the near side of the filter (pre-filter) will not reduce output filter distortion, improve frequency response or improve the high-frequency damping factor. It will also not reduce distortion that may be created by nonlinearity in the output inductor.

If the feedback signal is taken from the output of the amplifier at the far side of the filter (often second-order L-C), quite a bit of lagging phase shift will be introduced into the feedback loop by the output filter. This will require somewhat more complex feedback compensation, such as PID control. If the closed-loop bandwidth of the feedback loop is restricted to achieve adequate stability, then there may not be enough loop gain at the higher audio frequencies to make much of an improvement. The ability to operate at higher PWM carrier frequencies pays many dividends when applying negative feedback. However, other issues may then come into play, such as increased losses.

Closing the Loop Before the Output Filter

One approach to implementing negative feedback from the near side of the filter is shown in Figure 33.5 [6]. An inverting Miller integrator is placed in front of the class D amplifier. The analog input is applied through R1 while the switched PWM feedback is applied through R2. The forward path integrator serves the need for reconstruction of the raw switched output that is fed back. The closed loop gain is simply the ratio of R2 to R1. The feedback gain crossover frequency is at $\omega_0 = k/(R2 * C_M)$ where k is the forward gain of the class D amplifier. This is a fairly conventional dominant pole approach to compensation of the feedback.

There is inevitable sampling delay between the input signal and the output signal in a PWM amplifier, even without considering the effects of a reconstruction filter. The delay is a result of the sampling process that is fundamental to analog-to-PWM conversion. In a very simplified sense, this is not unlike an added pole in the feedback loop that decreases phase margin. The analog input signal is sampled twice per period of the PWM carrier, once when it slices the positive slope of the triangle reference and once when it slices the negative slope of the triangle. A change in the analog signal will not result in a change in the PWM pulse ratio until an average of 1/4 of a sample period has elapsed. In this case, the sample period is the period of the PWM carrier frequency. If the carrier frequency is 500 kHz, the period is 2 µs and the effective signal delay is 0.5 µs.

If the phase lag created by sampling delay is approximated as a single pole, then the pole frequency is at 250 kHz. This happens to be the Nyquist frequency for a 500-kHz sample rate. This sampling pole places a serious constraint on the unity gain frequency of negative feedback placed

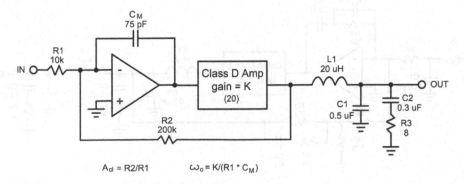

$$A_{cl} = R2/R1 \qquad \omega_0 = K/(R1 * C_M)$$

Figure 33.5 PWM Amplifier with Negative Feedback Taken Before the Filter

around a PWM amplifier. This suggests that the gain crossover frequency should be less than 250 kHz to ensure reasonable phase margin. This will provide about 22 dB of negative feedback at 20 kHz in the simple arrangement shown in Figure 33.5.

Many more sophisticated approaches to pre-filter feedback and its compensation are applied in practice [7]. The big disadvantage of pre-filter feedback pickoff is that it does nothing about filter distortion or filter degradation of damping factor.

Closing the Loop Around the Output Filter

Output filters introduce distortion and reduce damping factor. For this reason, it is desirable to close the negative feedback loop around the output filter. This is referred to as *post-filter feedback* (PFFB). However, PFFB can raise potentially serious stability challenges [4].

Taking the negative feedback from the far side of the filter incurs the phase lag of the filter and the consequent design challenge for stability and flat closed loop frequency response. The allowable gain crossover frequency in such an arrangement may be fairly low if only simple loop compensation is used. Then the feedback may be of limited help in reducing high-frequency distortion and in maintaining an adequate damping factor at 20 kHz. Phase lag introduced by the output filter is influenced by load impedance, and it is helpful in this case to employ a Zobel network that will keep the net load impedance low at the higher frequencies where the filter is active.

The output filter response falls at 12 dB/octave beyond its cutoff frequency if it is of second order. At its cutoff frequency it will typically introduce 90° of phase lag that will climb to 135° about an octave above the cutoff frequency. This would normally be the highest feedback gain crossover frequency allowed, providing about 45° of phase margin (if phase lag due to sampling delay is ignored). If a zero is introduced into the loop at this high frequency, it will contribute 45° of leading phase shift, bringing the phase margin back to 90°. This will extend the permissible gain crossover frequency. Better performance is achievable with a PID control approach to the PFFB.

Figure 33.6 shows one possible arrangement for an amplifier with PFFB. The forward path integrator includes a resistor in series with C3 for insertion of a zero. The feedback path also includes capacitor C4 that can insert another zero. The insertion of zeros in the feedback path must normally be done with caution because it allows the introduction of EMI from the speaker cables to the input circuitry.

The loop gain in this arrangement can be rolled off at an average rate of about 9 dB per octave. This is analogous to the increased slopes attainable with two-pole compensation in linear amplifiers. This will cause some overshoot, but the output filter will likely suppress that. If a gain crossover frequency of 200 kHz can be achieved, about 30 dB of feedback will be available at 20

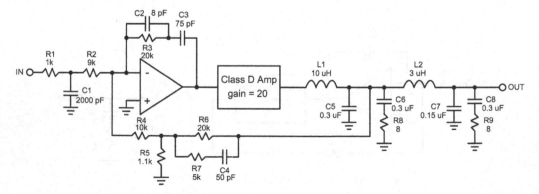

Figure 33.6 Negative Feedback Taken After the Output Filter

kHz. Other feedback arrangements are possible where a combination of pre-filter and post-filter feedback is employed [3, 7]. A good example of the use of PFFB is with the TI TPA3245 Class-D amplifier IC [8, 9].

An innovative approach that allows the use of post-filter feedback and allows the loop to oscillate will be discussed in the next chapter [10, 11].

33.5 Noise Shaping in PWM Modulators with Feedback

Noise shaping is often associated with oversampled D/A converters and Delta-Sigma modulators [12]. It shifts noise that would otherwise be in the audio band to higher frequencies. Noise shaping will be discussed in more detail in Chapter 35. Many sampled systems can employ noise shaping if properly designed, including PWM modulators. If an integrator is placed in front of the comparator in a PWM modulator that incorporates negative feedback, first-order noise shaping will take place [13]. Some PWM modulators can be designed with higher-order noise shaping.

33.6 Summary

While the operating principle for class D amplification is fairly simple, the devil is truly in the details. As we have already begun to see, there are many potential sources of imperfection and many design challenges associated with class D. Chapter 34 will delve more deeply into the challenges posed by simple PWM class D amplifiers. Chapter 35 will cover alternative modulators, such as the self-oscillating PWM design by Bruno Putzeys, which have contributed to significantly improved performance [10, 11].

References

1. Jun Honda and Jonathan Adams, "Class D Audio Amplifier Basics," International Rectifier Application Note AN-1071, February 2005.
2. Karsten Neilsen, "A Review and Comparison of Pulse-Width Modulation (PWM) Methods for Analog and Digital Input Switching Power Amplifiers," paper no. 4446, 102d AES Convention, March 1997.
3. Eric Gaalaas, "Class D Audio Amplifiers: What, Why, and How," *Analog Dialog*, vol. 40, pp. 1–7, June 2006.
4. "High Power Class D Audio Power Amplifier Using IR2011S," International Rectifier Application Note IRAUDAMP1, 2005. Available at www.irf.com.
5. Datasheet, "TPA3255 315-W Stereo, 600-W Mono PurePath Ultra-HD Analog Input," Texas Instruments, 2016.
6. W. Marshall Leach, Jr., *Introduction to Electroacoustics and Audio Amplifier Design*, 2nd ed., Kendall/Hunt, Dubuque, IA, 2001.
7. Stephen Cox and Bruce Candy, "Class-D Audio Amplifiers with Negative Feedback," 117th Audio Engineering Society Convention, San Francisco, October 2004.
8. Dan Kisling and Matthew Beardsworth, "TPA324x and TPA325x Post-Filter Feedback," Application Report SLAA788A, Texas Instruments, March 2018.
9. Datasheet, "TPA3245 115-W Stereo, 230-W Mono PurePath Ultra-HD Analog-Input Class-D Amplifier," Texas Instruments, October 2016.
10. Bruno Putzeys, "Power Amplifier," U.S. Patent #7,113,038, September 26, 2006.
11. Bruno Putzeys, "Simple Self-Oscillating Class D Amplifier with Full Output Filter Control," presented at the 118th AES Convention, May 2005.
12. Richard Schreier and Gabor Temes, *Understanding Delta-Sigma Data Converters*, Wiley-IEEE Press, Piscataway, NJ, 2004.
13. "Class D Amplifiers: Fundamentals of Operation and Recent Developments" Application Note 3977, Maxim Integrated Products, 2007.

Chapter 34

Class D Design Issues

34. INTRODUCTION

This chapter really gets into the nuts and bolts of class D amplifier design, focusing on the many challenges that must be overcome to achieve high sound quality from what is essentially a switching process that alternately applies the positive and negative power supply rails to the output to drive the loudspeaker. Imperfections in the high-speed switching process are a major contributor to distortion in class D amplifiers, and these are discussed in detail. This is an area that is somewhat foreign to many designers of conventional linear amplifiers.

Negative feedback must be applied to most class D amplifiers so that acceptable performance can be achieved, and yet the application of negative feedback to these amplifiers is not always straightforward. The necessary output low-pass filter in class D amplifiers usually causes a significant increase in the output impedance of the amplifier, which leads to poor damping factor, especially at high frequencies. The output filter can also cause frequency response aberrations and tonal coloration that depend on the frequency-dependent load impedance of the particular loudspeaker being driven. Some class D amplifiers enclose all or part of the output filter in the feedback loop, but this can make achieving stability quite difficult.

The high-speed switching in class D output stages can be a powerful source of EMI. This can result in unacceptable EMI emissions that may interfere with radio transmissions and associated audio equipment. Without adequate precautions, the loudspeaker cable can become an effective antenna for spreading these emissions.

34.1 The Output Filter and EMI

The raw output of the class D output stage is a sharp-edged rectangular waveform whose pulse width varies. Rise times of the switched output pulses are often between 5 ns and 40 ns. The output filter serves two very important purposes. First, the filter extracts the low-frequency average from the high-frequency rectangular waveform to provide the audio output signal. In this regard, it is a PWM-to-analog converter. Second, it must filter out the very high-frequency carrier and its harmonics to prevent EMI radiation from the amplifier. The filter must be designed to operate at high current with low distortion. The radiation of the filter itself within the amplifier must also be considered carefully as this could cause distortion in the analog circuits.

The simplest output filter is the second-order filter with a single inductor and capacitor. Its response will fall off at 12 dB per octave (40 dB/decade) above its cutoff frequency. It is not unusual to employ an output filter with a cutoff frequency that is about a decade below the PWM carrier frequency. This will provide about 40 dB of attenuation at the carrier frequency and about 59 dB of attenuation at the third harmonic of the carrier frequency. The PWM carrier frequency should be kept below the AM radio band unless very strong output filtering is employed.

It is important that the filter capacitor has low ESR and low ESL so that it maintains low impedance at high frequencies. For this reason it is unwise to employ only a film capacitor. A film capacitor can be bypassed with a large-capacitance ceramic NPO capacitor, however caution is advised in doing this because two high Q capacitors in parallel may form a resonant circuit with sufficient Q to be problematic. Axial-leaded capacitors should be avoided because of their lead inductance. Radial or surface-mount capacitors should be used.

The stray capacitance in the filter inductor can cause EMI leak-through at high frequencies; above their self-resonant frequency, inductors look like capacitors. For this reason, inductor geometry can be important. Output inductors with very high self-resonant frequencies should be chosen, especially when output filters are of only second order.

Higher-order filters employing a multiplicity of inductors and capacitors can provide improved attenuation at the higher harmonic frequencies that are often more troublesome for EMI conformance. This does not necessarily mean that the filter has to grow a lot in size or total series inductance. In many cases it means a redistribution of the existing amount of induc-tance and capacitance into a multiplicity of smaller elements. It is worth noting that smaller shunt capacitors are usually better at suppressing high frequencies because of their lower ESL and ESR.

Example second-order and fourth-order output filters are shown in Figure 34.1. The fourth-order filter is an example of a higher-order filter as discussed above. Half of the original 20-μH inductance of the second-order filter is placed in the first inductor and only 3 mH is placed in the second inductor, for a net reduction in total inductance. Most of the original capacitance is split to lie on either side of the second inductor. The second inductor can be made physically quite small. The third inductor is only 100 nH and comes from the wiring to the amplifier output terminals. A few inches will create this much inductance. Care should be taken to prevent this length of wiring from radiating EMI within the chassis of the amplifier. The third capacitor, only 0.02 μF, is placed right across the output terminals. These last two components technically make the filter sixth order, but the last pair of poles is at a much higher frequency. This arrangement does not change the attenuation at 500 kHz (both 40 dB), but it greatly improves the attenuation at higher frequencies. It also greatly reduces high-frequency EMI *sneak-through* that may result from the real-world imperfections of simpler filter implementations. The output amplitude from this filter at the seventh harmonic of the carrier frequency (3.5 MHz) is over 35 dB lower than from the second-order filter.

Compared to the second-order filter, the higher-order filter has less variation in response at 20 kHz when the load impedance varies from 4 Ω to 8 Ω, and also has a higher 3-dB bandwidth (60 kHz with a 4-Ω load and 90 kHz with an 8-Ω load). The phase characteristic of the higher-order filter is also more linear than that of the second-order filter. A Bessel characteristic will do the best job of preserving phase linearity. These output filters work well with a 50-kHz single-pole filter at the input of the amplifier.

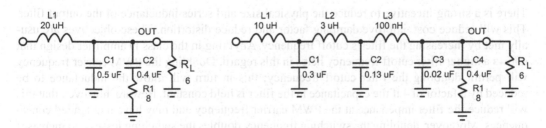

Figure 34.1 Example Second- and Fourth-Order Output Filters with Zobel Networks

The Zobel Network

Proper operation of the output filter depends on it being loaded with the correct impedance. Unfortunately, the impedance of loudspeakers is all over the map. This is of particular concern at high frequencies where the filter action is taking place. Loudspeakers can become inductive at high frequencies, as can the speaker cable driving them. For this reason a Zobel network is often placed at the output of the filter so that at high frequencies there appears a resistive load. [1] This network will usually comprise an 8-Ω resistor in series with a capacitor, much like the Zobel network used in linear amplifiers. The corner frequency of the Zobel network impedance will often be set to be in the vicinity of the cutoff frequency of the filter. For a Zobel impedance corner of 50 kHz with 8 Ω the capacitance will be about 0.4 µF. Such a Zobel network is illustrated in the filter of Figure 34.1. As with linear amplifiers, some dissipation will occur in the Zobel network and the resistor must be sized accordingly. It is especially desirable that the Zobel resistor in a class D amplifier be non-inductive.

Differing Loudspeaker Impedance

While the Zobel network can at least keep the load impedance fairly resistive at high frequencies, the need for a class D amplifier to drive loudspeakers of differing nominal impedances can create even more problems for filter design. Indeed, even resistive loads ranging from 2 Ω to 8 Ω can play havoc with filter frequency response and force serious compromises. Worse yet, the filter characteristic driving a very high impedance or no-load condition can develop serious peaking. This is not something we are used to with conventional amplifiers.

Linear Phase Approximation

Because the filter cutoff frequency is less than a decade above the audio band it becomes important to do the best job possible with its phase response and group delay. This means attempting to approximate a linear phase response that corresponds to a constant time delay. This means that maximally flat filters like the Butterworth may not always be the best choice. The overall linear phase characteristic of the amplifier can also be achieved in part by designing the amplifier input filter to work with the output filter. A fourth-order output filter that has a shallow initial roll-off will provide a better approximation to a linear phase characteristic than a second-order filter. The objective is to have a phase characteristic that approaches a straight line when phase is plotted with an X axis that is linear in frequency, like that of a Bessel filter. The need to drive speaker loads of differing nominal impedances makes the goal of achieving a linear phase filter characteristic for all conditions elusive.

Reducing Output Filter Size

There is a strong incentive to reduce the physical size and series inductance of the output filter. This will reduce cost, improve damping factor and reduce distortion. These objectives are usually met by increasing the filter's cutoff frequency. Anything in the class D amplifier design that allows a higher filter cutoff frequency helps in this regard. Doubling the PWM carrier frequency will permit doubling the filter cutoff frequency; this in turn will allow the inductance to be reduced by a factor of 4 if the capacitance in the filter is held constant. Beware, however, that this will reduce the filter impedance at the PWM carrier frequency and may have unintended consequences. Moreover, doubling the switching frequency doubles the switching losses and increases distortion, so this approach must be used with great caution.

The PWM carrier frequency is the highest-amplitude EMI frequency and lies at the point of lowest output filter EMI attenuation. If carrier frequency EMI is a problem, sometimes a filter with a notch at the carrier frequency (or an elliptic filter) will be used to enhance carrier rejection.

Input Filter and Aliasing

While on the subject of filters it is worth mentioning the importance of the amplifier input filter. All class D amplifiers involve a sampling process of one kind or another applied to the analog input signal. This means that aliasing is always possible. For this reason the input filter that is commonly found on linear power amplifiers takes on even greater importance. The input filter need not necessarily have a low cutoff frequency, but it must have good attenuation at frequencies approaching half the PWM carrier frequency and above.

The different frequencies that can be produced when the triangle wave and the audio signal are together passed through the odd-order nonlinearity of the comparator are discussed in Reference 2. This process can produce lower frequencies that must be kept above the audio band. If f_c is the PWM carrier frequency and f_s is the signal frequency, then aliases will be produced at numerous frequencies including the following:

$$f_c, \quad f_c \pm 2f_s, \quad 2f_c \pm f_s, \quad 2f_c \pm 3f_s, \quad \ldots$$

The lowest of these cited by Leach is $f_c - 2f_s$. If f_c is 500 kHz and f_s is 20 kHz, then the alias frequency will be 460 kHz [2]. However, there is not a brick wall filter at 20 kHz. The output of an SACD player often has significant high-frequency content, sometimes including some sampling tones at ultrasonic frequencies. One SACD player that was measured produced about 50 mV RMS at its output at 80 kHz. A tone at 80 kHz will produce an alias at 340 kHz. Fortunately, this is still well above the audio band. It is also important to recognize that noise from the signal source in the upper ultrasonic band can be aliased down into the audio band. Of course, aliasing products that lie in the audio band are reduced by negative feedback, just like other distortions, reducing somewhat the need for pre-filtering.

Other EMI Issues

There are many sources of EMI in a switching amplifier by the very nature of the fast, sharp-edged signals transitioning at high frequencies within the circuit. In many cases these exist as high currents that can create high-frequency magnetic fields, especially if the loops through which these currents circulate are not kept very small. If the body diode of the output MOSFET is allowed to conduct, its reverse recovery time will lead to brief high currents when the diode is quickly changed from a forward-biased condition to a reverse-biased one. The brief high-current flow required to sweep out the minority carrier charge in the diode will be a source of EMI. The use of Shottky commutating diodes from drain to source on output transistors can help here.

In general, one should use many of the same EMI management and suppression techniques as were discussed in Chapter 20 on switch-mode power supplies.

Output Filter Distortion

The components employed in the output filter must pass high currents. This is a problem for the output inductor, which almost always includes a magnetic core of some kind to keep the size reasonable. The core is subject to nonlinearity and ultimately saturation. [3] Indeed, the value of the inductance is current-dependent, so that the tuning of the output filter may actually change as a function of load current. This can cause decreased effectiveness of tuned notch filters and can certainly cause PIM in addition to increased THD. Any degree of core saturation will also introduce switching losses, which will decrease the efficiency of the amplifier. Powder core toroidal inductors are often employed for the output coils, but gapped ferrite cores are also quite linear below their saturation point.

In applications for highest sound quality, air-core toroids can be considered, especially if space permits and the value of inductance can be kept below about 4 μH. A non-ferrous plastic toroidal former can be employed.

34.2 Spread Spectrum Class D

EMI from PWM class D amplifiers can sometimes be a problem, especially with filterless designs. This is because the PWM carrier (and its harmonics) is at a nearly fixed frequency or varies over a small range of frequencies most of the time. To mitigate this, sometimes spread spectrum techniques are applied to PWM. [4] In this case, the triangle reference frequency is dithered or changed in a random manner over a range of about ±10% of the carrier frequency. The spread spectrum approach does not reduce the total EMI energy, but spreads it over a wider range of frequencies.

34.3 Filterless Class D Amplifiers

Output filters add cost and bulk to class D amplifiers. This is of special concern in small portable devices where power is low and the loudspeaker is close to the amplifier. Under these conditions a BTL design can often achieve adequately low EMI emissions without the use of output filters as long as the PWM carrier can be presented to the loudspeaker as a common-mode signal, while the audio signal is presented to the loudspeaker in a differential fashion. [4, 5, 6] Conventional full-bridge amplifiers present two PWM outputs that are of opposite phase, causing the PWM carrier to be presented in differential mode. This results in a large square wave appearing across the load in the absence of the usual output filters. This can damage the load, create a large amount of EMI and result in losses.

In a filterless amplifier, a BTL design is implemented as two identical amplifiers, one of whose inputs is inverted. In this case the two class D amplifiers employ the same triangle reference generator. The PWM carriers on both the positive and negative outputs are thus in phase, and appear in the common mode. With no signal, the same square wave carrier appears on both sides of the load and virtually no carrier current flows in the load. As the signal departs from zero, the pulse duty cycles on the two sides depart from 50% in opposite directions, resulting in a differential pulse that appears across the load whose width and polarity correspond to the signal. The width of this pulse is only as large as required to deliver the signal energy to the load. When the signal amplitude is small, the differential pulse is very narrow. In some cases, where EMI must be made still smaller, an output filter merely consisting of a small ferrite bead and capacitor on the order of 1000 pF can be included to further reduce EMI.

34.4 Buck Converters and Class D Amplifiers

Those familiar with switch-mode power supplies (SMPS) will recognize some similarities with class D amplification. [7, 8] One good example is the synchronous buck converter, which takes a DC input voltage and steps it down to a lower DC voltage. It was discussed in detail in Chapter 20. The buck converter operates with high efficiency because its output devices switch on and off. The relative *on* time of these switches controls the amount of energy that is delivered to the load and thus the output voltage. It is basically a PWM arrangement. Having a basic

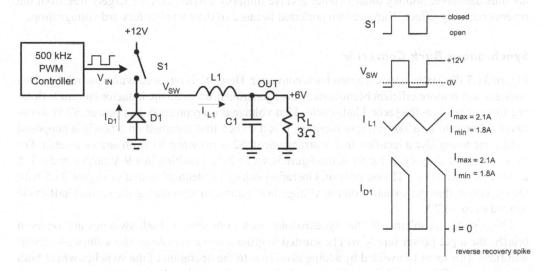

Figure 34.2 A Buck Converter Switching Supply

understanding of the buck converter is a very big step toward understanding a class D amplifier and some of its issues. For that reason we'll take a quick detour here and discuss some of the issues and challenges common to class D amplifiers and buck converters.

Figure 34.2 shows a very simple diagram of a buck converter. It consists of a single switch that connects the supply rail to the output inductor for a variable amount of time. In this case the duty cycle is 50% and the output voltage will be half the supply voltage of 12 V. The converter also includes diode D1 and an output filter capacitor C1. A 3-Ω load is shown which will draw 2 A from the 6-V output of the converter. In a real circuit S1 will be implemented with a MOSFET. During the first half-cycle when S1 is closed, the rail voltage is applied to the inductor and the current will rise linearly with time in accordance with the voltage across the inductor and the inductance. It will reach a value I_{max}. This is also illustrated in Figure 34.2.

When S1 opens, the current in L1 will continue to flow due to its collapsing magnetic field. Inductors resist current change and try to keep current flowing. This current is usually referred to as the *flyback current* or the *commutation current*. During the second half-cycle the only place the current can flow from is through D1 from ground. The switched output voltage thus snaps from +12 V down to about −0.7 V, the forward drop of D1. It is very important in switching circuit design to recognize that the current continues to flow through the inductor during this second half-cycle. During the second half-cycle, the current in the inductor will fall linearly to I_{min}. Switching frequencies and inductor values are such that I_{min} is greater than zero at the end of the second half-cycle. The average of I_{max} and I_{min} is the output current into the load. The difference between I_{max} and I_{min} is called the *ripple current*.

Diode D1 is often called a *freewheeling, flyback* or *commutating diode*. As shown in Figure 34.2, D1 is in a conducting state at the end of the second half-cycle when S1 again closes and raises the switched output from −0.7 V to +12 V. Diodes in a conducting state cannot instantly stop conducting and become an open circuit when the current is reversed. Silicon diodes that have been conducting contain stored charge in the form of *minority carriers* that must be swept out before the diode can allow reverse voltage across its terminals. This process is called *reverse recovery*. As a result, a brief large current spike will occur when S1 closes. This is undesirable and represents lost energy and a source of EMI. Fast diodes with small reverse recovery times are thus desirable. Shottky diodes do not involve minority carriers and are largely free from the reverse recovery effect. They are also preferred because of their smaller forward voltage drop.

Synchronous Buck Converter

Figure 34.3 illustrates a synchronous buck converter. Here D1 is replaced with a switch S2. This arrangement is more efficient because there is no junction drop when the inductor current is flowing from ground on the second half-cycle. That voltage drop represents lost power. S2 is sometimes referred to as a *synchronous rectifier*. It is a switch that is turned on when it is supposed to be conducting like a rectifier. In this arrangement, S2 is off when S1 is on and vice versa. For illustration, the duty cycle for S1 in the figure is set to 75%, resulting in a 9-V output and a 3-A current flow into the 3-Ω load resistor. Operation is largely identical to that of Figure 33.3 with the exception that the switched output voltage falls almost to zero during the second half-cycle instead of to −0.7 V.

There is one problem with the synchronous buck converter. If both switches are on even briefly, the input power supply will be shorted to ground and a very large *shoot-through* current will flow. This must be avoided by adding *dead time* to the operation of the switches where both switches are off for a brief time. This is much the same way a class D output stage works.

If S1 opens and S2 has not yet closed, a very large negative flyback voltage will be created by the collapsing magnetic field of L1. The flyback voltage is clamped by adding D2. The switched output node will then be prevented from going more negative than −0.7 V. For this reason the negative portion of the switched output voltage will exhibit negative "ears" at the beginning and end of the second half-cycle when both switches are open.

Because the dead time is usually kept very small compared to the period, most of the improved efficiency contributed by S2 is preserved. Importantly, the R_{DSON} of the MOSFET used for S2 must be small enough so that the voltage drop across it when it is on is considerably smaller than the *on* voltage of D2. All MOSFETs designed for switching applications have D2 built into them in the form of the source-drain silicon body diode.

Unfortunately, this arrangement does not prevent the reverse recovery current spike when S1 closes. This is because D2 becomes conducting during the dead time just before the end of the second half-cycle. The reverse recovery shoot-through current can sometimes be reduced if a Shottky diode is connected in parallel with the MOSFET. In principle, it will prevent the slower silicon diode (D2) from turning on. However, the round-trip lead inductance of the MOSFET and

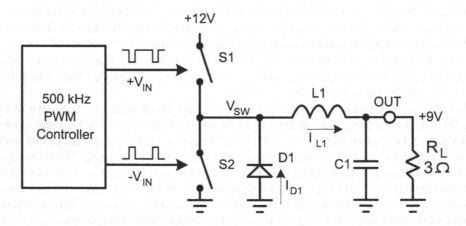

Figure 34.3 A Synchronous Buck Converter

Shottky diode will slow down the transfer of the current from the MOSFET to the Shottky diode, in some cases allowing D2 to go into conduction anyway. MOSFETS that incorporate an internal monolithic Shottky diode are available to mitigate the problem caused by lead inductance [9].

The switched output waveform depends only on the edge times of the high side switch S1. The *on* time of the low side switch S2 has no influence on the timing of the output edges as long as there is finite dead time. This is because a significant average current is being sourced to the load by the high side switch. The output goes low as soon as the high side switch turns off, before the low side switch turns on. This is due to the flyback behavior.

The output goes high only when the high side switch turns on, after the low side switch has turned off. The flyback current keeps the output low even after the low side switch has turned off. As long as current is being sourced to the load through the inductor, flyback current will be flowing through D2 until the end of the second half-cycle. This means that the timing of the low side switch will have no influence on the output edge times. This observation will be important when the influence of dead time on distortion in a class D amplifier is discussed. This process is called *commutation*, and it explains the integral role that the output inductor plays in optimum performance of a class D amplifier, beyond just providing output filtering.

Gate Charge

Gate charge is important in any switching application because it governs the amount of current required to charge the effective gate capacitance to turn the MOSFET on. This takes into account the change that is occurring in the drain voltage V_{ds} during turn-on. Gate charge is the effective charge contained in both the gate-source capacitance C_{gs} and the gate-drain capacitance C_{gd} when the device has turned on and V_{ds} is near zero. Gate charge also takes into account the strong nonlinearity of C_{gd} as a function of V_{ds}. Gate charge is defined in units of *coulombs*, where one *nanocoulomb* (nC) is the charge on a 1 nF capacitor with 1 V across it. The coulomb is also defined as a current flow over a period of time. One mA flowing for 1 μs corresponds to 1 nC.

MOSFET Figure of Merit

The *figure of merit* (FOM) for MOSFETs in switching applications is the product of *on* resistance R_{DSON} and total gate charge Q_g. [10] Smaller FOM is better. R_{DSON} is important because it determines power losses when the device is turned on. Typical values lie in the range of 0.02 Ω to 0.2 Ω. The total gate charge is important because it determines how much power is expended in turning the transistor on and off. The power required to turn the gate on and off at a given frequency is simply $Q_g * V_g * f$. This is called *switching loss*. Typical FOM for 200-V MOSFETs ranges between 5 Ω-nC and 25 Ω-nC. FOM tends to be larger for higher-voltage devices because they have higher R_{DSON} for a given amount of gate capacitance.

Conduction Loss

Consider a MOSFET with R_{DSON} = 0.1 Ω that must operate at 10 A with a 50% duty cycle. Its average conduction power loss will be 5 W. This is plainly not insignificant. It represents a 2.5% efficiency loss in a 200-W amplifier, and that is for only one half of the output stage. A good design will strike a proper balance between conduction loss and switching loss.

Switching Loss

Whenever a capacitance is charged and discharged at a certain frequency, power is expended. This is where gate charge comes into play. Consider a MOSFET with Q_g = 40 nC and operating at a switching frequency f_s of 500 kHz. Assume that the gate drive voltage is 10 V p-p. The power required to drive it will be $Q_g V_{gs} f_s = 40$ nC * 10 V * 500 kHz = 0.2 W.

Reverse Recovery Loss

There is a second significant source of switching loss. This one results from the reverse recovery time t_{rr} of the MOSFET body diode. If the low side body diode is in conduction when the high side switch turns on, a very large current will flow from the positive rail into the diode to sweep out its minority carrier charge. This is a form of shoot-through current. Normally the low side switch will have been on prior to this event, shunting the body diode and preventing it from being in conduction. However, if inductor flyback current is flowing during the necessary dead time prior to the turn-on of the high side switch, the inductor current will flow through the body diode during the dead time, forward biasing it and allowing it to accumulate a minority carrier charge.

Body diode conduction is another issue that underlines the importance of low R_{DSON} for the switching MOSFETs. If R_{DSON} is so large that the flyback current causes a voltage drop across the MOSFET equal to a one junction drop, the body diode will be on during a large portion of the half-cycle when the MOSFET should be shunting it and keeping it off. This will add to distortion by allowing the peak value of V_s to be larger by one junction drop than the rail voltage. It may also exacerbate body diode reverse recovery current spikes, especially relative to what they might have been if dead time is kept very small.

34.5 Sources of Distortion

As suggested in Chapter 33, there are numerous sources of distortion in class D amplifiers. A few of these are listed below and each is discussed briefly [11].

Triangle Reference Linearity and Bandwidth

Any departure from linearity in the reference triangle waveform in a class D PWM modulator will translate directly to distortion in the audio output [1, 11]. While the production of a very linear ramp is not difficult at low frequencies, it becomes progressively more difficult at higher frequencies. Limited bandwidth in the generation or processing of the triangle wave will result in rounded edges rather than straight and pointed edges. This corresponds to nonlinearity as well, and therefore causes distortion.

Pulse Width Quantization

Simple digital approaches to the production of a PWM signal rely on functions like counters and timers to determine the width of the modulator output pulses. This leads to quantization in the time domain. Consider a simple digital modulator that employs a 100-MHz clock. This will result in pulses that are quantized to 10 ns. Now assume that the PWM switching frequency is 500 kHz, with alternating 1 μs positive and negative pulses at idle. This means that the width of each pulse has a resolution of only 1%. This corresponds to a factor of 100, which in turn corresponds to a granularity of only about 7 bits. Fortunately, applying negative feedback will tend to noise shape the quantization error. Alternatively, one can use direct digital synthesis to minimize quantization effects. It accumulates and carries forward the quantization error to the next quantization.

Dead Time

The necessary dead time in the PWM output stage causes distortion [11]. For this reason dead time must be minimized. However, if nominal dead time is made too small, there may be insufficient time for lossless commutation of the output stage at low output currents. In the worst case, there may be a greater risk of shoot-through current unless precise timing in the output stage can be maintained. There is thus a delicate trade-off between distortion, losses and design margin. Dead time subtracts from effective pulse width and causes the relationship between modulator pulse width and output pulse width to become nonlinear and so results in distortion.

Low distortion in the output stage depends on the areas of the positive and negative pulses at the switching output being in the same proportion as the corresponding pulse areas in the waveform driving the output stage from the PWM modulator. If the positive and negative supply voltages are the same, this translates to faithful reproduction of the duty cycle ratio from the PWM modulator output to the switching output. Within the output stage, the positive and negative pulses are each shortened on their leading edges to create necessary dead time. However, the actual switching output pulse waveform V_s is different from the gate drive waveforms. The V_s waveform has no dead time. Understanding which edges of the switch driver waveforms control the switched output waveform transitions is key to understanding the dead time distortion mechanism. This time relationship was touched on briefly in the discussion of the synchronous buck converter in Chapter 33.

Consider a 500-kHz PWM square wave with 75% duty cycle from the modulator, as shown in Figure 34.4. It will be high for 1500 ns and low for 500 ns. Assume that a fairly large nominal dead time of 50 ns is introduced. After passing through dead time control, the high-side drive pulse will be 1450 ns and the low-side drive pulse will be 450 ns. Assume that $t = 0$ is defined at the turn-off of the low side switch. The high side switch will be activated at $t = 50$ ns and will

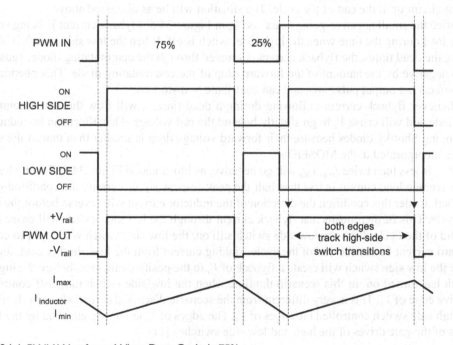

Figure 34.4 PWM Waveforms When Duty Cycle Is 75%

turn off at $t = 1500$ ns. The low side switch will be turned on at $t = 1550$ ns and will be turned off at 2000 ns. We assume no delay in the switches.

The switched output V_s will go high at 50 ns when the high side switch turns on. It will go low at 1500 ns, when the high side switch opens and the flyback process begins. Notice that the output goes low before the low side switch has turned on. The flyback process made it go low, not the turning on of the low side switch. The output will stay low until the high side turns on at 2050 ns. The output will not rise at 2000 ns when the low side switch turns off because the flyback current is still reversing the low side switch. Flyback current is still flowing at the end of the second half-cycle because a large net current is being sourced to the load.

The key observation here is that the switched output waveform V_s follows both rising and falling edges of the high side switch [11]. Detailed timing of the low-side switch has nothing to do with the output edge times under these conditions. This is because the high side is sourcing significant net current to the load and the inductor current remains positive throughout the entire cycle. The positive output pulse width will be 1450 ns, representing a duty cycle of only 72.5%. Some attenuation has thus occurred as a result of dead time.

The situation will be analogous in the case where the PWM duty cycle is small and the low side switch is on most of the time. In this case, both edges of the low side switch will govern the edges of V_s. It is clear that if the dead times for the high side and low side are not identical, the incremental gain of the output stage will be different for positive and negative signals, and distortion will result.

This is not the end of the story. Even if the two dead time intervals are identical, distortion will still result. Recall that there is ripple current in the inductor. When the amplifier is sourcing current to the load the inductor current is at I_{max} at the end of the high side switch time and decreases linearly to I_{min} at the end of the second half-cycle. The difference between I_{max} and I_{min} is the ripple current and the average of I_{max} and I_{min} (I_{avg}) is the net current sourced to the load. As long as I_{max} is at least twice I_{avg}, I_{min} will not go negative and flyback current will persist in the low side switch circuit until the end of the cycle. The situation will be as discussed above.

Notice the small negative-going "ears" on V_s in Figure 34.5. Flyback current is being sourced to the load during the time when the high side switch is off. When the low side switch is also off during the dead times, the flyback current is sourced through the commutating diode, causing V_s to go negative by the amount of the forward drop of the commutating diode. This phenomenon also affects net output pulse area and can contribute to distortion.

Whenever flyback current is flowing during a dead time, it will flow through a commutating diode and will cause V_s to go slightly beyond the rail voltage. This effect can be reduced by employing Shottky diodes because their forward voltage drop is smaller than that of the silicon diodes incorporated in the MOSFETs.

If I_{max} is less than twice I_{avg}, I_{min} will go negative, as illustrated in Figure 34.5. Put another way, if the average load current is less than half the peak-to-peak ripple current, the condition will be satisfied. Under this condition the direction of the inductor current will reverse before the end of the cycle. This further means that flyback current through the low side switch will cease before the end of the cycle. With the low side switch still on, the low side switch will begin to conduct forward current before the end of the cycle, sinking current from the load. In this case, the turn-off of the low side switch will create a flyback of V_s to the positive rail, even before the high side switch has turned on. In this scenario the time when the low side switch turns off controls the positive edge of V_s. This is very different from the scenario discussed earlier where both edges of the high side switch controlled the edges of V_s. The edges of V_s are now controlled by the falling edges of the gate drives of the high and low side switches [11].

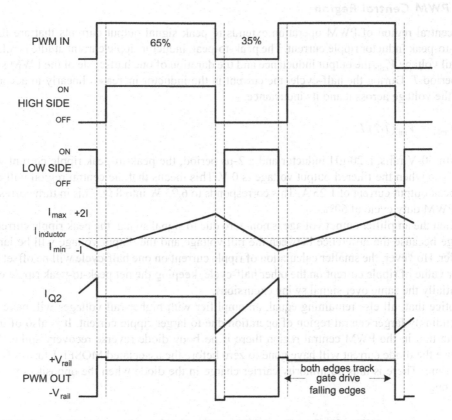

Figure 34.5 Current Waveform of Low Side Transistor Q2 When Duty Cycle Is 65%

PWM Crossover Distortion

If the turn-off edges of both switches control their respective edges of V_s, then the signal attenuation due to dead time pulse narrowing described earlier no longer exists. The duty cycle of V_s is a more faithful replica of the duty cycle created by the PWM modulator. The incremental gain of the output stage is thus higher when the current delivered to the load is smaller than half the peak-to-peak inductor ripple current. We will refer to this as the *central region* of operation.

In the region of 50% modulator duty cycle, the duty cycle of V_s is a faithful replica of the incoming PWM duty cycle, independent of dead time. This region of operation is not unlike the class A region of a linear class AB amplifier where both top and bottom transistors are contributing to the output. When larger signal currents cause the class D output stage behavior to exit this region, crossover distortion occurs because the incremental gain of the output stage decreases. This is akin to *gm* doubling in a linear class AB output stage where output stage gain decreases at currents outside the class A region. If dead time is made no longer than the time required to commutate the output stage cleanly at idle, the transition when exiting the central region will be smoother and PWM crossover distortion will be minimized.

The PWM Central Region

The central region of PWM operation extends to peak signal output currents that are half the peak-to-peak inductor ripple current. The peak-to-peak inductor ripple current at idle is related to the rail voltage V_{rail}, the output inductance and the duration of one half-cycle of the PWM switching period T. During the half-cycle, the current in the inductor increases linearly in accordance with the voltage across it and its inductance.

$$I_{ripple} = V_{rail}(T/2)/L \tag{34.1}$$

With 50-V rails, a 20-µH inductor and a 2-µs period, the peak-to-peak ripple current will be 2.5 A p-p when the filtered output voltage is 0 V. This means that the central region will extend to a peak output current of 1.25 A. This corresponds to 6.25 W into 8 Ω. This in turn corresponds to a PWM duty cycle of 60%.

When the amplifier output voltage is non-zero due to signal swing, the peak ripple current will change because the difference between the rail voltage and the output voltage will be larger or smaller. However, the smaller calculation of ripple current on one half-cycle will be offset by the larger value of ripple current on the other half-cycle, keeping the net peak-to-peak ripple current essentially the same over signal swing excursions.

Notice that, all else remaining equal, an amplifier with higher rail voltages will have a correspondingly larger central region of operation due to larger ripple current. It is also of interest to note that in the PWM central region there is no body diode reverse recovery spike. This is because the diode current will have gone to zero before the associated MOSFET turns off for its dead time. There is thus no minority carrier charge in the diode when the opposite-side switch turns on.

Extending the PWM Central Region

The central region can be extended if the ripple current is increased. This can be accomplished by decreasing the size of the output inductor. If the output inductor value is cut in half, the ripple current will be doubled and the extent of the central region of PWM operation will be doubled (quadrupled in power). In this respect, the amount of ripple current is akin to the amount of quiescent bias current in a linear class AB output stage.

If the inductance is cut in half and the size of the capacitor in the filter is left unchanged, the cutoff frequency of the filter will be increased by 1/2 octave and EMI will be increased. It may be possible to employ a fourth-order filter to restore the high-frequency attenuation, however. Alternatively, the inductance can be cut in half and the capacitance doubled, keeping the cutoff frequency the same. Conduction and core losses will go up as the inductance is reduced, so there is an optimum value of inductance.

Asymmetrical Rise/Fall Times

If a switch is conducting forward current when it turns off, the switched output node will *fly back* to the opposite rail voltage in a *freewheeling* fashion; rise time will be controlled by the inductor and capacitances at the switching node. On the other hand, if a switch is off before the end of its half-cycle period, commutation current will be flowing in the reverse direction and the rise time will be governed by the turn-on current of the opposite switch when it turns on. The resulting rise time at the switching node may be much faster in that case. Asymmetry in switching node rise and fall time will thus result. This phenomenon will generally occur when the signal current is large enough to be outside the central operating region.

Body Diode Conduction Time

Another small error is introduced into the pulse areas by the additional voltage dropped across the body diode during the time it is conducting the flyback current instead of the associated switch. This occurs during the dead time on the switching side whose pulses are narrower when the output stage is operating outside the central region. The effective rail voltage during this time is larger by the junction drop of the conducting body diode. As a result, the area of the smaller pulse is increased slightly. This is evidenced by the "ears" on the low-side pulse in Figure 34.4. This effect is exacerbated when currents are high.

Sliver Pulses

At very high modulation levels the pulses on one switch side become very narrow. Consider a PWM modulator operating at 500 kHz with an output stage that has 20-ns dead times and 20-ns rise and fall times on the switched output. Let the PWM duty cycle be 99%. This means that the idealized *on* time of the low side switch will be 1% of the period of 2000 ns, or 20 ns. It is immediately apparent that the low side switch will never turn on because its dead time equals the PWM modulator off time. Bear in mind that under these conditions high current is being delivered to the load and the timing of V_s will be entirely governed by the leading and trailing edges of the high side switch. The modulator on time is 1980 ns and the high side switch on time is 1960 ns.

When the high side switch opens V_s will transition to the negative rail, taking 20 ns to do so. It will remain there for 20 ns until the high side switch closes. V_s will then rise to the positive rail, taking 20 ns to do so. The effective duration of the low pulse is 40 ns. The effective duration of the high pulse is 2000 − 40 = 1960 ns. This is fortunately the same as the high side switch on time. Even at 99% duty cycle the output stage works largely as expected. This is a bit non-intuitive.

34.6 Bus Pumping

Bus pumping represents the transfer of energy from one power supply to the opposite power supply when a half-bridge output stage is employed [11]. Assume that the low side has been on and sinking current from the load through the inductor. The magnetic field in the inductor represents stored energy. When the low side turns off, the output voltage will quickly transition from the negative rail to the positive rail as the magnetic field collapses. Commutation current will continue to flow in the same direction through the inductor for some time because inductors seek to keep current flowing in the same direction.

The commutation current flowing into the inductor now must come from the positive supply, but it is in a direction that actually seeks to make the positive supply more positive. The commutation current pumps up the positive rail through the high side commutation diode. The process represents an almost lossless transfer of energy from the negative supply to the positive supply. If the duty cycle of the square wave is less than 50% over a long period of time (as with a negative DC output condition or a negative low-frequency half-cycle), more energy will be transferred from the negative supply to the positive supply during this period, representing an average transfer of energy from the negative supply to the positive supply. If the positive supply cannot absorb this energy, its voltage will rise as a result of the pumping.

The pumping occurs even with a resistive load; it does not depend on the reactive nature of a loudspeaker. Bus pumping depends on the reactive behavior of the output inductor. The reactive nature of the loudspeaker can exacerbate pumping, however. The effects of bus pumping are worse at low frequencies because the reservoir capacitors can change their voltage over the signal cycle if the power supply cannot absorb the pump current. If there is another source of heavy

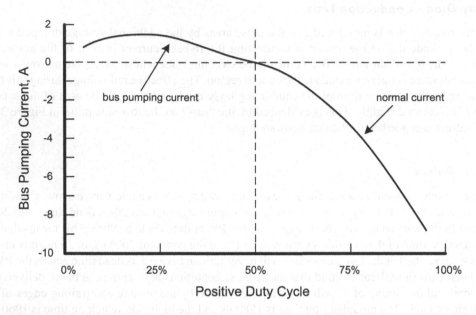

Figure 34.6 Bus Pumping Current as a Function of Modulation Depth. Negative Values Show Normal Bus Sourcing Current

current drawn from the supply in excess of the commutation current that is pumping the supply, then there will be no problem.

Figure 34.6 shows the amount of current flowing back into the positive supply as a function of PWM duty cycle. The pump current is greatest at a duty cycle of about 25%, representing a fairly strong negative output current. The positive supply must be able to sink this current or its voltage will rise. Negative values on the Y axis indicate current that the positive supply must source when the load is being driven positive. These values are shown for context. The values shown are for a half-bridge output stage with ±50-V rails driving a 4-Ω load.

The full bridge is largely immune to the bus pumping effect because the long-term current of the low side from the negative rail is matched by a similar long-term current of the high side from the positive rail that flows on the other side of the bridge. The commutation current is thus returned to the other rail. In effect, the commutation current flows through a closed loop [11].

Power supply variation due to bus pumping must be avoided because any variation will influence the audio signal as a result of the 0 dB open-loop PSRR of an ordinary PWM class D amplifier. Bus pumping is mainly a problem at low frequencies where the rail capacitance may not be large enough to suppress rail voltage increases. Large reservoir capacitors help to mitigate these effects.

Some switching power supply architectures are able to absorb the pump current with very little loss and return it to the opposite supply or to a common supply source. Indeed, the same phenomenon that causes supply pumping can be used in reverse by a switch-mode circuit to return the pump current to the opposite supply. This is not unlike the pump current circulation that occurs naturally in full-bridge architectures.

34.7 Power Supply Rejection

Power supply rejection (PSRR) is one of the first problems a novice designer will have to deal with in designing a class D amplifier; in some rudimentary designs PSRR is literally 0 dB [1, 11].

When either output switch is on, one of the power supply rails is connected directly to the output of the amplifier. This means that there is virtually no power supply rejection inherent to the class D output stage. Indeed, power supply ripple and noise is sampled by the class D process and applied to the output. Recall that the gain of the simple class D amplifier is also defined in terms of the ratio of the power supply rail voltage to the peak voltage of the triangle reference wave. Changes in power supply voltage thus modulate the gain of the output stage.

All of this means that the power supply for the class D output stage must be very quiet and well regulated. Any variation in the power supply voltage caused by the signal will result in gain variations and thus intermodulation distortion. We will see later that several techniques are available to improve power supply rejection. Most of these involve negative feedback in one form or another.

Poor PSRR is the Achilles heel of some class D amplifier designs. PSRR in conventional linear class AB amplifiers is reflective of how much power supply ripple and noise gets added to the program signal. It is very important to understand that PSRR in a class D amplifier is reflective of more than just the addition of noise to the program signal. Because the power supply voltage directly influences the gain of the class D amplifier, lack of PSRR also causes intermodulation distortion in class D amplifiers. This is something that is definitely not mitigated by the use of a full bridge output stage.

Loop Gain Modulation

The fact that power supply fluctuation, ripple and noise can modulate the gain of the PWM amplifier can be problematic for PWM amplifiers that employ negative feedback. If the open-loop gain of the PWM amplifier is modulated, the unity gain frequency of the negative feedback loop will also be modulated. This is not unlike the role that amplitude intermodulation distortion plays in modulating the closed-loop bandwidth in linear feedback amplifiers. When the closed-loop pole frequency is modulated, the in-band phase shift is modulated; this gives rise to *phase intermodulation distortion* (PIM).

Power Supply Feedback to the Triangle Generator

The gain of the PWM amplifier is proportional to the power supply rail voltage and is inversely proportional to the amplitude of the triangle reference voltage. If the same variation that is present on the power supply rails is used to control the amplitude of the reference generator, some of the power supply modulation of the PWM amplifier gain will be canceled out. A key detail here is that the variations on the positive and negative rails may not be the same.

Modulator and Output Stage Gain

The modulator gain in a conventional PWM modulator reflects how much input signal is required to cause a given change in pulse width ratio. It is inversely related to the amplitude of the triangle reference signal. Larger triangle amplitude will result in smaller modulator gain because more input signal amplitude will be required to cause a given change in pulse width. Output stage gain reflects how much average output signal amplitude is created by a given amount of pulse width modulation. It is usually proportional to the power supply voltage.

If the triangle reference signal is generated in such a way that its amplitude is proportional to the same power supply voltage as supplies the output stage, the inverse relationship between modulator gain and output stage gain dependence on power supply voltage (and noise) will tend to cause the overall effect of power supply voltage on signal voltage to cancel out. In essence, this can be thought of as a form of feedforward correction, since this process works even for amplifiers that have no negative feedback. This effect can increase PSRR in most of the audio band by 20 dB or more. There is a caveat, however. Because of transistor voltage drops in the output stage and some other factors, the change in output stage gain with supply voltage may not be exactly the inverse of that of the modulator gain. We will see later that if the amplitude of the reference triangle waveform is somehow made proportional to the peak-to-peak value of the PWM output waveform, this limitation will be mitigated.

34.8 Power Supplies for Class D Amplifiers

The poor power supply rejection of class D amplifiers can place greater performance requirements on the power supply design than in linear amplifiers. Although the use of negative feedback in PWM amplifiers is a powerful tool for lessening the power supply effects, it is always important to strive for the best open-loop linearity before the application of negative feedback. This is no different than the objective in linear amplifier design. Class D amplifiers that do not employ some kind of negative feedback require especially quiet power supplies.

Linear Power Supplies

Conventional class AB power amplifiers employ linear power supplies that are made up of a transformer, rectifier and reservoir capacitor. Such power supplies can also be used for class D amplifiers if they can be made sufficiently quiet and free of voltage variations as a function of signal and bus pumping. Depending on the PSRR characteristics of the amplifier, regulation of the supplies might be needed or desirable. Those regulators must be able to sink current unless they are followed by large reservoir capacitors. Switching regulators would thus likely be a better choice than linear regulators.

Switching Power Supplies

Since class D amplifiers are themselves switching devices, it is only natural that modern switching power supplies be used with class D power amplifiers. They have the advantage of operating at high switching frequencies that reduce ripple and create it at much higher frequencies where it can be more easily filtered. However, if they are not implemented with the usual hefty reservoir capacitors used with linear power supplies, there may not be adequate current reserves to supply the needs of the output stage during high-amplitude bass notes. The need for large reservoir capacitors also derives from the power supply pumping phenomena when half-bridge output stages are employed.

Switching power supplies are inherently regulated with high efficiency because the output power equals the input power to within better than 90%. The regulation can be made soft or stiff. As long as good reservoir capacitors are used with effective high frequency bypassing, the ripple and noise at the output of switching regulators can be made very small and the current reserve can be made as large as that of a linear power supply with the same size reservoir capacitors. The switching supplies can be made stiff and quiet. Unfortunately, sometimes the switching supply invites the use of undersized reservoir capacitors.

If synchronous PWM modulators are used, it is desirable that the switching power supply clock be synchronized with the PWM carrier so that beat frequencies will not be created that might fall into the audio band.

Switching power supplies that employ isolated synchronous buck converters with high side and low side switches also enjoy the property that power can flow in both directions at their output terminals. This means that half-bridge class D amplifiers that create bus pumping may not suffer from power supply rail fluctuations or over-voltage conditions. If isolated synchronous buck converters are employed, two separate converters may have to be used for implementation of the positive and negative power supply rails. Another approach is to employ a linear supply with a mains frequency power transformer to generate raw positive and negative rails. Each of these rails is then passed through a non-isolated buck converter that provides the main operating rails.

34.9 Damping Factor and Load Invariance

Damping factor is the effective output impedance of an amplifier divided into 8 Ω. Damping factor is not just important for accurate bass response. Amplifiers with low damping factor will have their frequency response affected by the variations in loudspeaker load impedance with frequency. They will have poor load invariance. Class D amplifiers that do not employ negative feedback tend to suffer from a poor damping factor because they are operating open loop.

The damping factor of a class D amplifier can be degraded by the impedance in the output filter or by the effective impedance of the power supply; this is an indirect consequence of the lack of power supply rejection. Load invariance is a problem for any class D amplifier that does not incorporate negative feedback around the output filter. However, in-band load dependence will be reduced in amplifiers that incorporate filters with higher cutoff frequencies.

As a point of reference, the impedance of a 20-µH filter inductor at 20 kHz is about 2.4 Ω. This corresponds to a damping factor of only 3.3 at 20 kHz. Define load variance as the change in response at 20 kHz when the load resistance decreases from 8 Ω to 4 Ω. Consider a second-order filter with 20 µH and 0.5 µF. The frequency response at 20 kHz decreases by 1.4 dB when the load resistance decreases from 8 Ω to 4 Ω. The load variance is thus 1.4 dB. This filter also exhibits a +2.8 dB peak in response at 42 kHz when driving the lighter 8-Ω load.

If the filter is designed to be maximally flat when loaded with 8 Ω by employing a 0.16-µF capacitor, then response at 20 kHz decreases by 1.2 dB when the load impedance drops from 8 Ω to 4 Ω. The two filters have very similar load variance at 20 kHz. A respectable commercial class D amplifier that does not incorporate post-filter feedback (PFFB) exhibits output impedance of 0.9 Ω at 20 kHz, corresponding to a damping factor of 8.9 at that frequency [12]. However, with a change in resistive load from no-load to 4 Ω, frequency response changes by only 0.5 dB at 20 kHz. Notably, this amplifier has an outstanding and comprehensive datasheet.

34.10 Summary

The conventional PWM class D amplifier has numerous shortcomings that can be difficult to overcome. For this reason other approaches to class D modulators have been pursued and implemented. Some of these will be discussed in Chapter 35. Some of these are responsible for the significant improvements in performance and sound quality that have been made in class D amplifiers in recent years.

References

1. "High Power Class D Audio Power Amplifier Using IR2011S," International Rectifier Application Note IRAUDAMP1, 2005. Available at www.irf.com.
2. W. Marshall Leach, Jr., *Introduction to Electroacoustics and Audio Amplifier Design*, 2nd ed., Kendall/Hunt, Dubuque, IA, 2001.
3. Eric Gaalaas, "Class D Audio Amplifiers: What, Why, and How," *Analog Dialog*, vol. 40, pp. 1–7, June 2006.

4. "Class D Amplifiers: Fundamentals of Operation and Recent Developments" Application Note 3977, Maxim Integrated Products, 2007.
5. Datasheet, "NCP2820 2.65 W Filterless Class-D Audio Power Amplifier," ON Semiconductor, 2013.
6. Datasheet, "SSM3582 2X31.76 W, Digital Input, Filterless Stereo Class-D Audio Amplifier," *Analog Devices*, 2016.
7. Jun Honda and Jonathan Adams, "How Class D Audio Amplifiers Work," *EE Times*, January 23, 2006.
8. Karsten Neilsen, "A Review and Comparison of Pulse-Width Modulation (PWM) Methods for Analog and Digital Input Switching Power Amplifiers," paper no. 4446, 102d AES Convention, March 1997.
9. AN4789, "Monolithic Shottky Diode In ST F7 LV MOSFET Technology: Improving Application Performance," *STMicroelectronics*, December 2015.
10. "High Power Class D Audio Power Amplifier Using IR2011S," International Rectifier Application Note IRAUDAMP1, 2005. Available at www.irf.com.
11. Jun Honda and Jonathan Adams, "Class D Audio Amplifier Basics," International Rectifier Application Note AN-1071, February 2005.
12. Datasheet, "ICEpower 200AC 200W ICEpower Amplifier," ICEpower a/s, 2016.

Chapter 35

Alternative Class D Modulators

35. INTRODUCTION

There are numerous ways to build a class D power amplifier, but those based on *pulse width modulation* (PWM) are the oldest and still very popular. This chapter will discuss alternatives to traditional PWM modulators. Some are variants of PWM modulators that produce a PWM stream by different means. These include so-called self-oscillating loops that do not require a triangle-wave reference generator and actually rely on oscillation that results from a negative feedback process. These schemes reap the benefits of negative feedback without the struggle to keep it stable. Other approaches to PWM generation depend on high-speed digital logic and/or digital signal processing.

An entirely different group of class D modulators is based on so-called sigma-delta modulators (SDM or ΣΔ). These produce a high-speed bitstream whose bits are of uniform duration, but whose density reflects the amplitude of the audio signal. This is a form of *pulse density modulation* (PDM). Sigma-delta modulators can be implemented in the analog domain or in the digital domain. They depend on a process called *noise shaping* that works in conjunction with oversampling. This technique, used in most audio A/D and D/A converters, shifts most of the inevitable quantization noise out of the audio band to parts of the frequency spectrum above the audio band.

Some class D modulators operate entirely in the digital domain, taking as their input a PCM digital audio signal and converting it to a PWM or ΣΔ output stream. This is attractive in many applications where the signal is already available in digital form.

35.1 Self-Oscillating Loops

Using a triangle reference generator and a comparator is not the only way to generate a PWM stream. There is another type of analog PWM class D amplifier design that comprises a so-called self-oscillating loop [1–4]. In this approach there is no carrier frequency triangle reference source. These designs are asynchronous. While they produce a PWM stream, the frequency of that stream is not fixed. Instead, the loop oscillates on its own as a result of feedback around the loop. Some designs take the feedback from the near side of the filter (pre-filter feedback) while other designs take the feedback from the far side of the filter (post-filter feedback). A number of commercial class D amplifiers use the self-oscillating loop principle.

The self-oscillating loop PWM amplifier often enjoys improved performance over conventional fixed-frequency triangle-based PWM amplifiers because the bandwidth of the feedback loop can be higher than that of a conventional feedback loop that must obey conservative stability criteria.

Self-Oscillation with Pre-Filter Feedback

Consider the PWM amplifier arrangement of Figure 35.1 where pre-filter feedback encloses a class D amplifier whose gain is 20. If additional phase lag is added to the feedback loop, the circuit will become unstable and oscillate. Oscillation will occur even without the triangle reference signal. When the input signal is added to the oscillating loop, the output will be pulse width modulated, as desired.

Figure 35.2 illustrates a conceptual arrangement of a self-oscillating PWM modulator and amplifier. There are numerous ways to arrange the loop and add additional low-pass filtering to the loop to obtain controlled oscillation [1]. The arrangement here illustrates a somewhat general case in which two R-C sections add about 90° to the integrator's 90° at the desired frequency of oscillation, which is in the vicinity of 500 kHz.

Self-Oscillation with the Output Filter

In Section 33.4 the struggle to close the feedback loop around the output filter was discussed. Adequate feedback stability can be difficult to achieve in light of the phase shift introduced by the output filter.

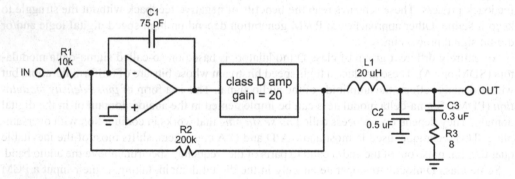

Figure 35.1 PWM Amplifier with Pre-Filter Feedback

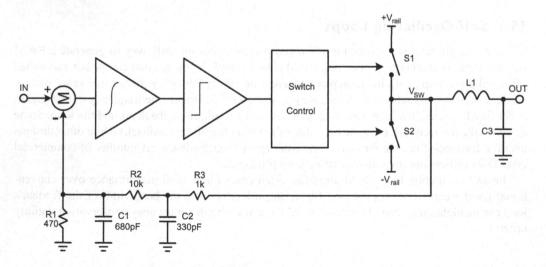

Figure 35.2 Self-Oscillating PWM Class D Amplifier

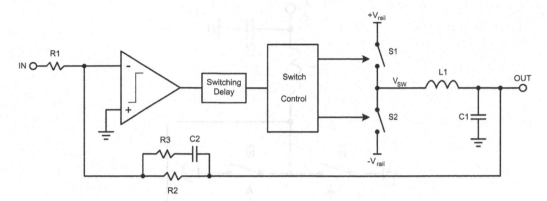

Figure 35.3 Simplified Illustration of a Self-Oscillating PWM Modulator with Post-Filter Feedback

The PWM approach in Reference 1 instead takes advantage of this "problem" by allowing the loop to oscillate in a controlled fashion. It is a self-oscillating design wherein the instability resulting from post-filter feedback (PFFB) is put to good use [1, 2]. The simplified amplifier illustrated in Figure 35.3 oscillates at about 400 kHz, over 10 times the corner frequency of the second-order output filter. At this frequency the phase shift of the filter is nearly 180°, so it does not take much additional phase shift to push the loop into oscillation. This additional phase shift is provided by the switching delay in the forward path. A phase-lead network is placed in the feedback path to control the frequency of oscillation.

The stroke of genius here is that applying an input signal to an otherwise-oscillating class D amplifier will result in the appropriate PWM duty cycle signal corresponding to the audio waveform.

The filter frequency is about 35 kHz and the oscillation frequency is about 400 kHz. The loop gain is almost independent of power supply voltage and depends almost solely on the frequency response of the output filter. Loop gain is about 30 dB, flat across the audio band out to about 35 kHz. The gain crossover frequency is at about 200 kHz, while the closed-loop frequency response is down 3 dB at about 40 kHz. This design obviously profits from post-filter feedback, reducing filter distortion and providing good load invariance.

A discussion of the circuit details can be found in Reference 2. The self-oscillating design is simple and provides high performance, but it is not clear that, with its non-fixed frequency, it can be mated with a direct-digital input.

Self-Oscillation Using a One-Shot

Other self-oscillating loop PWM designs use pre-filter feedback and employ other means to cause the loop to oscillate. One such approach is illustrated in Figure 35.4 [3, 4].

A one-shot multivibrator is central to the operation of this PWM modulator. The cycle begins when the one-shot is triggered and turns on the high side switch for a fixed period. During this period the error between the switched output signal and the input signal is integrated for the duration of the one-shot pulse. The error grows during this period. The high side switch turns off at the end of the one-shot period and the error integrates negatively. The circuit ends a switching cycle when the difference between the audio signal and the PWM waveform is zero. The one-shot will be triggered and the process repeated when the error gets back to zero.

This process seeks to drive the error between the input signal and the integrated output signal to zero (on each cycle). In some respects, this is not radically different from a triangle PWM

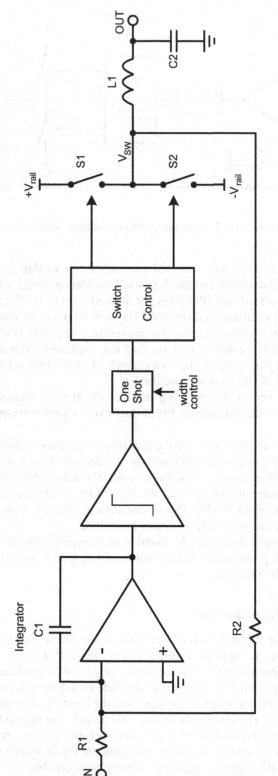

Figure 35.4 A Self-Oscillating PWM Modulator Employing a One-Shot

modulator preceded by an integrated feedback approach where the input signal and feedback are compared by an integrator in the forward path. The technique described here is said to remove switching error faster than any other class D approach [3, 4]. This approach compensates for errors introduced by dead time and non-ideal switching edges because it is responsive to net pulse area. As a result, it also tends to reject power supply deviations and noise.

If a fixed-width one-shot is used, the frequency of the design described will vary widely with signal, since half the period is fixed and the other half of the period must change to accommodate the needs of the input signal amplitude. For this reason, a control circuit is added to constantly adjust the pulse width of the one-shot.

Synchronized Self-Oscillating Loops

One disadvantage of most conventional self-oscillating class D amplifiers is that the PWM carrier frequency is poorly defined and may vary with the signal conditions. This can lead to beat frequency phenomena when multiple channels of class D amplification are in close proximity. For this reason there have been self-oscillating loop designs where some means of frequency synchronization has been incorporated [5, 6]. In some cases synchronization is achieved by employing a form of injection locking.

It is also possible to use more explicit locking techniques such as a *phase-locked loop* (PLL). If the self-oscillating loop has some means to control its frequency, it can be viewed as a VCO and can be locked to a common frequency source if the overall design has sufficient lock range in the presence of the audio signal. The frequency of a self-oscillating loop based on feedback from the near side or far side of the output filter can be controlled by including voltage-dependent phase shift in the loop or by including voltage-dependent feedback coefficients in the loop. The frequency of a multivibrator-based self-oscillating loop can be controlled with a one-shot whose delay is voltage-dependent. A significant design issue in a PLL-based approach is the choice of loop bandwidth and loop order.

35.2 Sigma-Delta Modulators

The PWM class D amplifier is essentially an open-loop device around which negative feedback may be placed. The sigma-delta modulator is an alternative means of generating a bitstream whose average value corresponds to the signal amplitude [7]. Negative feedback is intrinsic to its operation. These modulators are also referred to as delta-sigma modulators. The basis of a commercial implementation of a sigma-delta class D amplifier is described in Reference 8.

Figure 35.5 illustrates a simple first-order sigma-delta modulator (SDM or $\Sigma\Delta$). As in a conventional PWM modulator, it is operated at a fixed frequency. However, instead of producing pulses at a carrier frequency whose width is continuously variable, it produces pulses whose width is in discrete increments of the clock period. This is a form of pulse density modulation. Because the pulses are in increments of the clock period, this signal is quantized in time. As a result, quantization noise is introduced. This is a problem that must be addressed. The simple answer is that the clock for the sigma-delta modulator is at a much higher frequency than the PWM carrier frequency. The $\Sigma\Delta$ modulator fundamentally depends on oversampling where the sampling frequency is larger than twice the maximum frequency of the signal being sampled.

The $\Sigma\Delta$ modulator comprises a summing circuit, an integrator, a comparator and a D-type flip-flop. The output of the flip-flop is the bitstream. As with PWM, the average value of the switched output signal represents the analog output. The switched output signal is fed back to the summer, where the input signal is compared to the average value of the feedback signal. If

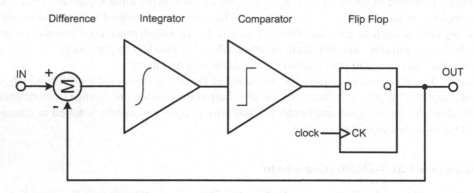

Figure 35.5 A Simple Analog Sigma-Delta Modulator

the input signal is larger than the average value of the feedback signal, the integrator output will move in a positive direction, eventually crossing the threshold of the comparator and causing its output to go high.

On the next positive clock edge the D-type flip-flop will be clocked high and the output will go high for this bit period. On subsequent positive clock edges, the output will continue to be clocked high until the fed-back output catches up with the more positive input and actually passes it. At that point, the integrator output will fall a sufficient amount to go below the comparator threshold and cause the flip-flop to be clocked low on the next positive clock edge. This whole process repeats indefinitely, with the ΣΔ loop always seeking to keep the difference between the input and the reconstructed output small. If the input signal is at zero, the output of the modulator will toggle between high and low levels at the clock rate.

Because the average value of the output bitstream is driven to equal the input signal by feedback, the pulse density in the bitstream will be a faithful representation of the input signal amplitude. In a conventional ΣΔ class D amplifier, the bitstream that is fed back is the bitstream that exists at the output of the amplifier prior to the filter.

It is important to recognize that the feedback signal to the analog summer in Figure 35.5 is not reconstructed; it is either a positive or negative reference voltage in discrete time. Reconstruction of the feedback signal into a continuous-time representation takes place in the integrator. In most class D amplifiers the switched signal fed back is not a fixed reference signal (1-bit A/D), but rather an attenuated version of the actual switched output signal. In this way the pulses fed back are a more faithful representation of the areas of the actual output pulses of the amplifier. If the power supply rail increases, for example, that will be appropriately reflected in the area of the pulse that is fed back. This improves PSRR.

Notice that when a pulse occurs, it will never have a width less than one clock period. This is in contrast to the output of a PWM modulator, where very small pulse slivers can be created when the signal is near its maximum magnitude. Conversely, there is no limit to the width of a pulse in a sigma-delta bitstream. Moreover, there need be no periodic regularity to the time when pulses go from positive to negative or vice versa. There is no carrier frequency.

The negative feedback process that is intrinsic to the operation of the ΣΔ modulator significantly improves PSRR, since the input is being compared to the actual average value of the digital output, regardless of the amplitude (and area) of the pulses that create that average value. In fairness to PWM, this is not unlike the benefit of pre-filter feedback in a PWM amplifier that was illustrated in Figure 35.1.

Oversampling

Because sigma-delta modulators create quantization noise in the time domain, it is important that they run at significantly higher clock frequencies than PWM modulators. This allows for finer granularity of the pulse time intervals. The high rate at which the $\Sigma\Delta$ modulator is clocked is referred to as oversampling. The ratio of the $\Sigma\Delta$ clock rate to the required Nyquist sampling rate is referred to as the *oversampling ratio* (OSR). The Nyquist sampling rate for a 20-kHz analog signal is 40 kHz. A $\Sigma\Delta$ modulator operating at 4 MHz will have an OSR of 100.

Oversampling reduces in-band noise by spreading the total sampling noise power over a much larger range of frequencies [9]. An OSR of 2 will spread the noise over twice the frequency spectrum, cutting the in-band noise power in half and improving SNR by 3 dB.

Increasing the clock rate of a first-order $\Sigma\Delta$ modulator reduces noise by 9 dB per octave of OSR. This larger decrease than 3 dB per octave results from a process called noise shaping that will be discussed later. Increasing the clock rate from 5 MHz to 20 MHz thus reduces the quantization noise by 18 dB. Employing a higher-frequency clock also allows the use of a physically smaller output filter with a higher −3 dB frequency.

There is a limit to the smallest pulse that can be handled by a class D amplifier output stage, especially when necessary dead time margins are considered. This was discussed earlier in connection with sliver pulses produced by PWM modulators at very high or very low duty cycles. The switching rate of the class D output stage also has an upper limit (apart from minimum pulse width) because its power loss increases with the average switching rate. Each time the output state is switched, power is dissipated. The product of the clock frequency and the transition density of the bitstream is what influences heat generation from switching losses.

High-Speed Class D Sigma-Delta Amplifiers

A PWM modulator operating at 500 kHz has pulses that are on average 1 μs in duration with a 2-μs period. At 95% modulation, the minimum pulse width is 100 ns. If this same minimum pulse width is employed for the $\Sigma\Delta$ modulator, then the clock rate will be 10 MHz. Unfortunately, this also means that a zero-output (first-order) $\Sigma\Delta$ "idle pattern" will consist of alternating 100-ns one and zero pulses, corresponding to a 5-MHz square wave. This is 10 times the frequency of the quiescent output frequency of the 500-kHz PWM modulator and may result in excessive switching losses.

Noise Shaping

Noise shaping is a technique whereby the noise power, which is normally flat with frequency, is pushed out of the audio band up to higher frequencies [9]. This is how A/D and D/A converters with a small bit resolution achieve much higher effective resolution in the audio band. The ultimate example of this is the 1-bit converter. Noise shaping is different than the simple reduction of in-band noise achieved by oversampling. With oversampling, the quantization noise power is spread over a wider frequency band that extends beyond the audio band. This dilutes the amount of noise power that lies in the audio band. In general, in-band noise goes down only with the square root of the OSR.

Noise shaping employs the principles of negative feedback to reduce the in-band noise power at the expense of increased out-of-band noise power. In other words, the noise power is spread non-uniformly over the available bandwidth.

The sigma-delta modulator in Figure 35.5 is said to be a first-order modulator because it incorporates only one integrator in its forward path. The gain of the integrator increases at 6 dB per octave as frequency decreases. Modulator quantization noise is injected into the system after the integrator. Recall the input-referred noise analysis approach described for linear amplifiers. There the effective noise contribution of a noise source was decreased by the amount of open-loop gain lying ahead of it. The increased gain of the SDM integrator at low frequencies reduces input-referred quantization noise by 6 dB per octave as frequency decreases. The total noise power at the output of the sampling process is constant, so the noise taken away from the low-frequency end of the spectrum appears as increased noise in the high-frequency portion of the spectrum that lies above the audio band. Oversampling provides a larger amount of out-of-band spectrum in which to dump this noise that has been taken from the in-band portion of the spectrum.

Second-Order Sigma-Delta Modulators

If two integrators are placed in the forward path, the noise-shaping process will have a slope of 12 dB per octave. The input-referred noise will decrease at 12 dB per octave as frequency decreases. Correspondingly, the in-band sampling noise will be decreased by 12 dB for each doubling of the $\Sigma\Delta$ clock frequency. This in-band SNR improvement is on top of the 3 dB per octave reduction in noise that is simply attributable to the higher sampling frequency. The net reduction of in-band noise is then 15 dB for each doubling of the clock frequency. Increasing the clock frequency from 5 MHz to 20 MHz will yield an SNR improvement of 30 dB.

Figure 35.6 illustrates a second-order $\Sigma\Delta$ modulator [8]. It comprises essentially the same architecture as the first-order loop but with a second integrator added in the forward path. The reconstructed output is compared with the input signals of both integrators. Think of the second integrator as the one in Figure 35.5. The second-order $\Sigma\Delta$ modulator has thus had an additional difference circuit and integrator added in front of it. The first-order modulator part of the circuit is now trying to minimize the integrated difference of the input signal and the reconstructed output signal.

Higher-Order Sigma-Delta Modulators

The noise-shaping properties of $\Sigma\Delta$ modulators can be more aggressively exploited by employing high-order loops. These often fall in the range of third to fifth order. The order of the loop is generally the same as the number of integrators in the loop. Higher-order loops provide much more in-band SNR for a given clock frequency. One price paid for these modulators is stability, especially under overload conditions. Higher-order modulators are more often found in DSP implementations and are seen less frequently in the analog implementations used for class D $\Sigma\Delta$ amplifiers.

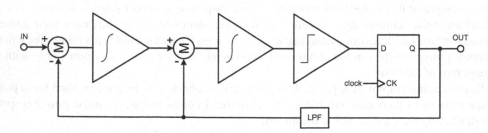

Figure 35.6 A Second-Order Sigma-Delta Modulator

EMI of Sigma-Delta Class D Amplifiers

The pulse pattern generated by the ΣΔ modulator is much more irregular than that from a PWM modulator. This creates a frequency spectrum that is much more spread out and less concentrated at the clock frequency and its harmonics; this results in much less tendency to produce RF tones.

However, idle tone EMI can sometimes be a problem, especially for first-order ΣΔ modulators. When there is no audio signal, a ΣΔ modulator will tend to create an idle tone at half the clock frequency. The more concentrated spectrum of energy under these conditions can create increased EMI. Second-order and higher-order ΣΔ modulators tend to produce more irregular idle tones, and so tend to create a bit less EMI under idle conditions. Idle tones can sometimes extend down into the audio band [11].

The Output Filter

The clock frequency of a ΣΔ modulator is usually much higher than the carrier frequency of a PWM modulator. As a result, the EMI is spread out over a higher frequency band and the output filter can be designed with a higher cutoff frequency. This means that the output filter will be physically smaller and will have a frequency response in the audio band that is less dependent on the load impedance. It may also mean that the output filter will create less distortion as a result of the smaller inductances that can be employed.

Post-Filter Feedback

It is desirable to enclose the output filter within a negative feedback loop to improve the damping factor and make the behavior of the amplifier more load-invariant. As in PWM amplifiers, there are challenges in putting negative feedback around an output filter that contributes a large amount of lagging phase shift.

If the cutoff frequency of the ΣΔ output filter is high, it is possible to enclose the filter in the loop and achieve larger amounts of negative feedback in the audio band with less danger of instability. The governing consideration here is how high the output filter cutoff frequency can be set. Pre-filter feedback will still often be taken to form the ΣΔ modulator itself. This provides a defined closed-loop gain around which additional feedback from the far side of the filter can be taken. Some examples of post-filter feedback with a ΣΔ modulator are described in Reference 8.

35.3 Digital Modulators

In some ways the power amplifier is one of the last things to go digital in many areas of consumer electronics. For this reason, and given the tremendous processing capability made available at low cost by VLSI chips, it is only natural for class D modulators to be implemented in all-digital form.

Class D amplifiers using digital modulators are a natural fit to digital audio sources where the signal often originates in PCM format from an I²S bus. The I²S format (Inter-IC Sound) is a popular standard developed by Philips Semiconductor for transport of digital audio signals. Use of a digital modulator eliminates entirely the analog interface and potentially some other mixed-signal functions. This can be especially attractive in modern home theater receiver applications where most of the audio signals are handled in digital fashion.

Digital Input PWM Amplifiers

Before discussing digital implementations of PWM modulators, it is worth noting that many chips that employ analog PWM modulators include an I²S digital PCM input. These chips provide a convenient middle ground [12]. Examples of these from Texas Instruments include the TAS2770, TAS5825M and TAS3251.

Digital PWM Modulators

The PWM waveform for driving the output stage can be generated digitally. This can be especially attractive when the source material is in digital PCM form. With appropriate processing, PCM can be converted to PWM entirely within the digital domain. PWM outputs can then be generated with great accuracy without reliance on an analog triangle reference generator. Dead time can also be implemented digitally, with separate outputs for the high side and low side switches provided by the digital device.

Unfortunately, the digital generation of PWM brings with it time quantization and associated noise. High clock frequencies must be used to minimize time quantization, but even a 100-MHz clock will result in quantization to 10 ns. This amount of quantization in a 2000 ns PWM carrier period represents a timing precision corresponding to only 6–8 bits. *Digital signal processing* (DSP) must be added to implement noise shaping to drive the resulting quantization noise to frequencies above the audio band. Noise shaping of the PWM signal is usually done with a subsequent $\Sigma\Delta$ modulator implemented with DSP, adding to complexity, but amenable to integration.

Digital Sigma-Delta Modulators

The differencing and integration functions that make up $\Sigma\Delta$ modulators are easily implemented entirely in the digital domain. Moreover, DSP offers great precision and consistency in filter coefficients, allowing the implementation of sophisticated higher-order loops. Examples of digital $\Sigma\Delta$ class D amplifier implementations can be found in References 10 and 13.

Feedback and PSRR

The impairments introduced in a class D amplifier by the output stage are fundamentally analog in nature. For this reason some form of analog feedback must be introduced into the modulator chain if low distortion and good PSRR are to be achieved. This can be challenging in a digital modulator arrangement and usually requires some kind of mixed-signal circuitry, such as an A/D converter, in the feedback signal path. This offsets slightly some of the advantages of the digital modulator approach. Some low-cost all-digital implementations omit the feedback and suffer the consequences.

References

1. Bruno Putzeys, "Power Amplifier," U.S. Patent #7,113,038, September 26, 2006.
2. Bruno Putzeys, "Simple Self-Oscillating Class D Amplifier with Full Output Filter Control," presented at the 118th AES Convention, May 2005.
3. U.S. Patent #6,084,450, "PWM Controller with One Cycle Response," July 4, 2000.
4. "One-Cycle Soundä Audio Amplifiers," PowerPhysics White Paper, 2002. Available at www.power-physics.com.
5. U.S. Patent #7,119,629, "Synchronized Controlled Oscillation Modulator," October 10, 2006.
6. U.S. Patent #6,297,693, "Techniques for Synchronizing a Self-oscillating Variable Frequency Modulator to an External Clock," October 2, 2001.
7. Richard Schreier and Gabor Temes, *Understanding Delta-Sigma Data Converters*, Wiley-IEEE Press, Piscataway, NJ, 2004.
8. U.S. Patent #5,777,512, "Method and Apparatus for Oversampled, Noise-Shaping, Mixed-Signal Processing," July 7, 1998.
9. Malcolm J. Hawksford, "Oversampling and Noise Shaping for Digital to Analogue Conversion," *Reproduced Sound 3*, pp. 151–175, Institute of Acoustics, 1987.

10. U.S. Patent #6,943,717, "Sigma Delta Class D Architecture Which Corrects for Power Supply, Load and H-bridge Errors," September 30, 2005.
11. Enrique Perez Gonzalez and Joshua Reiss, "Idle Tone Behavior in Sigma Delta Modulation," 122nd Convention of the Audio Engineering Society, May, 2007, Vienna, Austria.
12. Bill Schweber, "Trio of Class-D Amplifiers Addresses Today's Home Audio Demands," *Electronic Design*, June 11, 2018.
13. Datasheet, "SSM3582 2X31.76 W, Digital Input, Filterless Stereo Class-D Audio Amplifier," *Analog Devices*, 2016.

Chapter 36

Class D Measurement, Efficiency and Designs

36. INTRODUCTION

This chapter concludes the class D amplifier discussion by considering the amplifier as a whole, ignoring the highly technical implementation details. It is written more from a user perspective, with emphasis on applications that demand high sound quality. For those who demand the highest sound quality and who can compromise on efficiency, there is a middle ground available at the expense of greater complexity. This approach can be referred to as *hybrid class D*. In such a design a class D amplifier provides the lion's share of the power while the actual signal delivered to the loudspeaker comes from a low-power analog class AB amplifier.

Measurement of class D amplifiers requires a different approach in many cases. This is due in part to the fact that class D amplifiers usually have smaller bandwidth than traditional linear amplifiers. Distortion harmonics that lie above the audio band may be seriously attenuated. As a result, the measurement and specification of high-frequency THD (like THD-20) is virtually useless and can be downright misleading. The presence of out-of-band noise and residual PWM carrier at the output of most class D amplifiers further complicates many measurements. Measurement techniques for class D amplifiers are covered in Section 36.2.

36.1 Hybrid Class D

In some cases higher sonic performance can be achieved by combining class D amplifiers with analog power amplification [1]. A simple example of this is to amplify the signal with both class D and class AB amplifiers to the same level. The class AB amplifier is a low-power, high-current amplifier that actually drives the load. Its output stage power supply is floating on the output signal of the class D amplifier. This is somewhat analogous to a linear power amplifier wherein a floating class A amplifier is driven by a class AB amplifier.

Figure 36.1 is a conceptual illustration of a hybrid class D amplifier. The most straightforward approach is to have the class D amplifier drive the entire power supply of the class AB amplifier. That power supply can be either a linear supply or a switcher.

The hybrid class D amplifier has several advantages. It isolates the class D output from the load, taking the output filter out of the signal path and greatly reducing EMI. It also preserves the damping factor that would be attained by a linear amplifier. Finally, it allows the negative feedback to be closed from the output terminals of the amplifier without suffering the consequences of phase shift introduced by the output filter.

Note also that only the output stage of the class AB amplifier needs to be run from the flying rails provided by the class D amplifier. All of the earlier stages can be run from a very clean linear supply because they require very little power. Then, decent PSRR of the linear output stage is all that is needed.

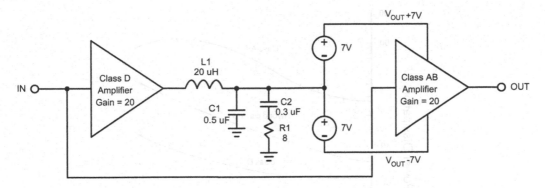

Figure 36.1 Conceptual Diagram of a Hybrid Class D Amplifier

The hybrid class D amplifier is an intelligent trade-off, providing improved sound quality in exchange for a reduction in efficiency. The class AB amplifier can be run at low voltage, but it must still be designed to be able to deliver the full current produced by the amplifier. The use of small local rail voltages in the class AB amplifier section greatly eases safe area requirements for the output transistors.

Figure 36.2 shows estimated power dissipation as a function of output power for a conventional class AB amplifier, a hybrid class D amplifier and a standard class D amplifier, all rated at 200-W/8-Ω. The class AB amplifier within the hybrid class D amplifier is assumed to have ±7-V floating rails. Even with low-voltage rails, the floating class AB amplifier dominates the total power dissipation of the hybrid design. All of its input power is dissipated as heat because it really delivers no added power to the load. Unfortunately, given the need to deliver high current into low-impedance loads and the reality of implementation tolerances, it is very challenging to design a floating class AB amplifier with rail voltages less than about 7 V.

It is interesting to note that the hybrid class D amplifier does not exhibit increased power dissipation at less than full power, even though it includes a class AB amplifier as part of its implementation. The hybrid class D amplifier enjoys its greatest advantage over the class AB amplifier at power output levels between 5 and 50 W. In this region its power dissipation is smaller by a factor of about 4.

Audiophiles usually care most about maximum dissipation as opposed to overall efficiency. These two very different things. Audiophiles don't care as much about power drawn from the outlet. They care about how big they must make their heat sinks in order to achieve a given output power and sound quality. This is why hybrid class D may be attractive for some audiophiles.

36.2 Measuring Class D Amplifiers

Class D amplifiers do not usually have as much bandwidth as linear amplifiers and so they present some challenges for distortion measurements like THD. More importantly, the inevitable high-frequency noise and carrier leak-through at the output can corrupt distortion and SNR readings. At times, high-frequency EMI at the output of a class D amplifier can actually disturb the functionality of sensitive test equipment connected to the amplifier, especially if the slew rate is high. A so-called AES17 low-pass filter is usually required to address this problem. A number of excellent references for measuring class D amplifiers can be found [2, 3, 4, 5].

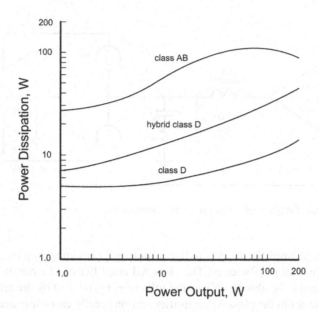

Figure 36.2 Estimated Power Dissipation of Class AB, Hybrid Class D and Standard Class D Amplifiers

Balanced Measurements

The great popularity of *bridge-tied-load* (BTL) class D amplifier arrangements means that the THD analyzer and associated measuring equipment must accept a balanced input. Alternatively, an outboard active circuit must provide a balanced-to-single-ended conversion without signal degradation by the EMI. Many BTL designs are built with two amplifiers and require a balanced input to implement bridging, at least at the input of the class D amplifier IC. Thus, in some cases, the source of the test signal must be balanced. An I²S test signal is often needed for digital-input class D amplifiers. Throughout the remainder of this section, a BTL class D amplifier will be assumed.

The Need for a Measurement Filter

Consider a 200-W/4-Ω class D amplifier whose peak audio voltage output is 40 V differential. Assume that the peak voltage of the unfiltered PWM differential square wave is 45 V, or 90 V p-p. The RMS fundamental of a square wave is 1.25 times the peak value, so here we have a 56 V RMS fundamental at an assumed instantaneous minimum carrier frequency of 200 kHz. Although 200 kHz is quite low for a nominal carrier frequency, bear in mind that some PWM modulators can have their carrier frequencies swoop down to significantly lower frequencies at high modulation (signal) levels. Assume the amplifier has a 50-kHz second-order L-C output filter whose attenuation is about 24 dB at 200 kHz. The carrier component in the output will then be on the order of 3.5 V RMS, with a slew rate of 6.3 V/μs. This amount of 200-kHz carrier residual may be enough to overload or degrade the distortion performance of the active input circuitry of the analyzer. Because it is differential BTL, each side will have 1.75 V RMS with a slew rate of 3.15 V/μs with respect to ground.

It is worth noting that some BTL amplifiers will have the carrier in the common mode, while others will have it in the differential mode. A BTL amplifier that uses an H bridge output stage will usually have it in the differential mode, while a BTL amplifier that is simply made from two identical amplifiers driven out of phase will usually have it in the common mode. Unless otherwise noted, it will be assumed here that the carrier residual is in the differential mode.

The carrier residual may also interfere with measurement accuracy even if it is not causing slew rate limiting or related distortion in the analyzer's active input stages. If the distortion of a 1-V signal fundamental is being measured, there will usually be no attenuation at the front-end of the analyzer. In fact, there might even be 10 dB of front-end gain to bring the signal up to a 3-V analyzer operating level. Bear in mind, even when measuring the distortion of a 1-V fundamental, the raw PWM waveform is still blasting away at its full peak-to-peak amplitude. The auto-ranging level set circuitry in the analyzer will see 3.5 V of differential carrier residual at the input to the analyzer, causing more than 10 dB of auto-ranging error.

The need for some additional filtering between the amplifier and the analyzer, at least some of it passive, is clear. But how much? About 40 dB of additional passive pre-filter loss at 200 kHz should reduce the carrier to 35 mV RMS and 0.063 V/µs differential, enough to avoid slew rate stress on active circuits and auto-ranging errors. If one is to measure the distortion of a 100-mW/4-Ω fundamental, corresponding to 630 mV RMS, the 35-mV carrier residual will have negligible impact on the auto-ranging circuits.

However, what about the distortion measurement itself? What if we want to measure 0.001% distortion on a 100-mV RMS fundamental (1 µV) at 1 kHz? This is not very far from what is achievable at 1 kHz with modern class D amplifiers. Measuring 1 µV in the presence of 56 V RMS of carrier (before the amplifier output filter) implies an extraordinary amount of low-pass filtering to get the carrier residual at least 10 dB below the 1-µV signal.

Assuming that the analyzer has a third-order 80-kHz filter, an additional 24 dB of attenuation will be provided at 200 kHz, a factor of 15.8, bringing the carrier residual down to 222 mV. Taking into account the 40 dB of pre-filtering attenuation assumed above, this brings the carrier residual down to 2.2 mV. This corresponds to a total of 88 dB of attenuation from the unfiltered PWM square wave. This 2.2 mV is only 34 dB down from a fundamental of 100 mV, corresponding to a 2% THD+N measurement floor. This is 66 dB above 0.001%. Total required filter attenuation is now 88 + 66 dB = 154 dB. This requires that the combined attenuation of the pre-filter and AES17 filter must be 106 dB at 200 kHz. If a typical digital audio measurement filter (e.g., AES17) provides 75 dB of attenuation at 200 kHz, then the pre-filter must supply at least 31 dB of attenuation. For measurement accuracy of 0.5 dB, the carrier residual must be down an additional 10 dB, bringing the required pre-filter attenuation to 41 dB. The total of 164 dB from the raw PWM is a lot of filtering! Similar issues pertain to S/N measurements.

Figure 36.3 illustrates a typical filtering chain wherein the pre-filter precedes the THD analyzer and an active AES17 measurement filter precedes the metering circuit in the analyzer. Attenuation at 200 kHz is shown for each filtering block and the amplitude of the carrier residual for this example is shown at each location. The analyzer front-end is assumed to have unity gain.

The AES17 Filter

To deal with the spurious carrier residual and/or quantization noise and other EMI that may be present at the output of digital audio equipment and class D amplifiers, the Audio Engineering Society published a filtering recommendation called AES17 [6]. The 20-kHz brick-wall low-pass filter is placed between the equipment output and measurement instruments like distortion analyzers to prevent overload of their sensitive active input circuits. It may also be placed within the THD analyzer. The filter is very sharp, flat within ±0.1 dB to 20 kHz and then down by 60 dB at 24 kHz. It is much like an anti-alias filter for audio analog-to-digital conversion. Although not required by the standard, a typical AES17 filter will be down at least 75 dB at frequencies above 50 kHz. The 20-kHz brick-wall nature of the filter facilitates reasonably accurate noise

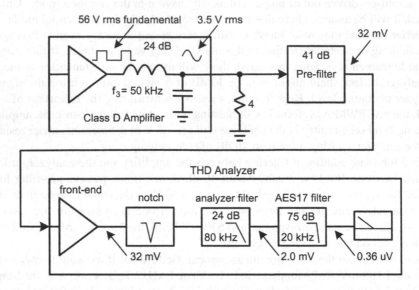

Figure 36.3 Typical Filtering Chain with Pre-Filter and AES17 Filter

measurements in the face of sigma-delta converters using noise shaping wherein noise that has been shaped rises rapidly above 20 kHz. It also establishes a well-defined noise bandwidth of 20 kHz.

The filter usually requires a seventh-order elliptic filter, some or all of which should be implemented with passive components for class D audio amplifier measurements if a pre-filter is not employed. This ensures that neither the filter itself nor sensitive active circuitry in subsequent test instruments is disturbed by the EMI or caused to experience slew rate limiting. In some cases it is sufficient that only the early stages of the filter be passive, but the inductors in at least the early stages of an L-C implementation of the filter must be extremely linear (or in the pre-filter, if used there). This can introduce significant expense. Without the filter, measurements at low signal levels will be especially affected in an adverse way.

In the earlier discussion about making very sensitive THD measurements down to perhaps 0.001% on a 100-mV fundamental, it was estimated that the measurement filter(s) might be required to provide as much as 116 dB of attenuation at 200 kHz. Thus, the AES17 filter's attenuation might fall short by as much as 41 dB under some measurement scenarios if it provides no more than the typical 75 dB of attenuation at higher frequencies like 200 kHz. Additional filtering is thus recommended [2].

Use of the AES17 filter destroys one's ability to assess the true frequency and transient response of the class D amplifier being tested. It also destroys the usefulness and accuracy of THD measurements above 6.7 kHz, even if only second and third harmonics are measured. It also makes it virtually impossible to see what the class D amplifier's output filter is doing. It is usually important to see how that filter's frequency response and transient response changes with different load impedances. In some cases there is a 40-kHz cutoff option available in the AES17 filter, and this helps a little bit for those types of measurements.

Analyzer Pre-Filter

Audio Precision addresses the need for additional filtering beyond the AES17 requirement with an all-passive external pre-analyzer L-C filter called the AUX-0025 [7]. It provides at least 50 dB of attenuation at 250 kHz while maintaining in-band flatness of better than ±0.05 dB up to 20 kHz. Typical attenuation at 60 kHz and 80 kHz is 4.0 and 6.5 dB, respectively. This passive filter provides enough attenuation of the carrier and other EMI to allow the circuitry in the THD analyzer to be active, including the internal brick-wall AES17 filter (which is sometimes implemented digitally). Extremely linear inductors are used in the AUX-0025 filter to allow it to pass large output signals from class D power amplifiers while introducing THD of less than 0.0003% (−110 dB).

Alternatives for the Pre-Filter

It is very desirable to implement the pre-filter without inductors, using a combination of passive R-C filtering and active filters. Figure 36.4 shows such a filter. The signal from the amplifier is first attenuated by 20 dB to bring it down to line level. The filter is a modified sixth-order Chebyshev filter consisting of two third-order Chebyshev filters whose real poles have been grouped together at the front-end to reduce slew rate before active circuitry is encountered. Combined with the quasi-passive filtering of the second-order active section, slew rate of the carrier residual is reduced to a very low level at the output of U1A. In the example above, with a 45-V p-p rectangular PWM signal on each side of the BTL output, the slew rate at the output of U1A is only 0.13 V/μs. This preserves the extremely low distortion capability of the LM4562 op amp. It is important to recognize that even though the op amp is rated at 20 V/μs, achieving extremely low distortion while passing 200 kHz signals requires much, much smaller signal slew rate. The slew rate at the output of U1 would rise to about 0.3 V/μs for a 1000-W/4-Ω amplifier.

The filter is implemented balanced, with a differential-to-single-ended output stage to provide a single-ended output where desired. The filter is flat to better than ±0.1 dB to 20 kHz. It is down about 0.6 dB at 60 kHz and down a little less than 3 dB at 80 kHz. It is down about 60 dB at 200 kHz. When used without an AES17 filter, the 80-kHz bandwidth of this filter enables one to see more of the actual frequency response and transient response of the amplifier being tested.

Alternatives to the AES17 Filter

The AES17 filter is often unnecessarily low in cutoff frequency and steep in its initial cutoff characteristic for some class D amplifiers and measurements. Intended to suppress strong HF signals at the output of class D amplifiers, its stringent brick-wall characteristic is often unnecessary for use with much test equipment, especially if the output of the amplifier is first attenuated to bring it down to line level. The AES17 filter interferes strongly with frequency response, THD, square wave measurements and transient response evaluation. It can also be very expensive to implement, especially in passive form where it provides the most protection to instrumentation input circuits. Its high order requires the use of precision passive components, especially in the active portions of the filter (if present).

There are low-cost and moderate-slope alternatives to the AES17 filter that can be adequate for preservation of the test equipment's performance, even though they do not meet the strict brick-wall cutoff template of the AES17 filter. Active filter approaches, where one or two passive real poles are placed at the front-end of the filter, will often suffice. The sharp-cutoff part of the characteristic can be implemented with much less-expensive active filters later in the chain that will perform very well if implemented with high-performance op amps capable of high slew rate and very low distortion, like the LM4562. If a sharp cutoff characteristic is needed, the use of one or two Bainter active notch filters can be economic and effective in comparison to high-order elliptic filters. Active post-filtering/equalization can be employed to ensure adequate flatness of the in-band response within the AES17 template (or, say, +0 db, −0.2 dB). For example,

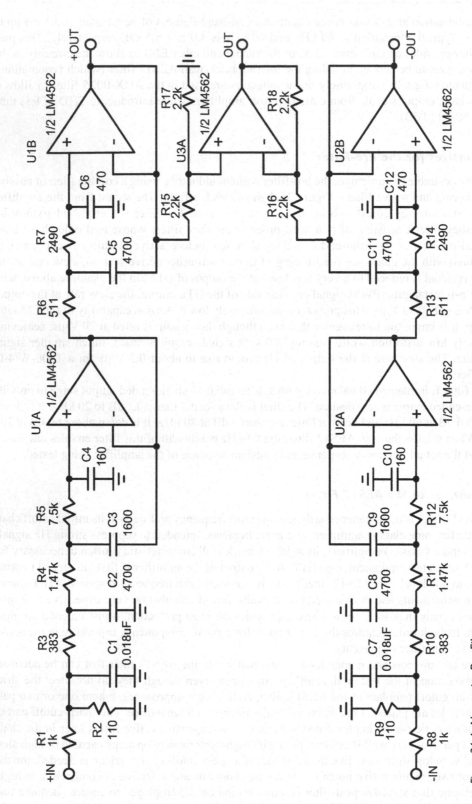

Figure 36.4 Alternative Pre-Filter

a seventh-order Butterworth filter that is down 0.2 dB at 20 kHz will be down 60 dB at 200 kHz. However, such filters may not be adequate for accurate S/N measurements.

The filter in Figure 36.5 is another approach to approximating the AES17 shape adequately, but without such an extreme cutoff slope. It is a fifth-order Chebyshev active filter followed by a Bainter filter that places a high-Q notch at 28 kHz [11, 12]. The notch provides increased attenuation and slope in that region, adequate to counteract the sharply increasing noise in that region when sigma-delta noise-shaping amplifiers are being tested. The Chebyshev filter has been tweaked slightly to improve flatness given the Bainter filter's roll-off characteristic. The filter is down less than 0.05 dB at 20 kHz and falls to −20 dB at 32 kHz. Attenuation at 200 kHz simulates at over 100 dB.

For other tests that may be less sensitive to PWM hash, such as frequency response and transient response, a fifth-, seventh- or ninth-order Bessel filter can be a better choice. In such a case, the real pole of the odd-order Bessel filter can be implemented passively at the front-end of the filter [7, 8, 9, 10, 11, 12]. A seventh-order Bessel filter that is down only 0.25 dB at 20 kHz will be down 30 dB at 200 kHz and start with a real pole at 123 kHz. If it is to be dedicated to power amplifier measurement, a fixed 20 dB input filter of fairly low impedance (e.g., 910 Ω and 100 Ω) can be used. The combination of the attenuator and the passive real pole of the Bessel filter will provide about 25 dB of attenuation at 200 kHz, dropping the carrier level in the earlier example from 1.2 V down to about 66 mV, with a slew rate of only about 0.12 V/μs. This attenuation will be sufficient to allow high performance op amps like the LM4562 to implement the remainder of the filtering task without their distortion performance being compromised. The rated slew rate of the LM4562 is 20 V/μs. The linear phase approximation of the Bessel filter will provide a much better assessment of the transient response of the class D amplifier. It is also relatively easy to make the cutoff frequency of such filters selectable, so as to get the most comprehensive and accurate measurement results from a given class D amplifier. The same kind of approach can be used to implement a balanced-to-single-ended converter, if necessary. Importantly, in this approach no inductors are needed.

Measuring Filterless Class D Amplifiers

The above discussion assumed that the class D amplifier incorporated an L-C output filter to keep the discussion simple and relevant to most audio amplifiers. The filterless class D amplifiers, often used in portable applications and at low power, can with their sharp output edges make things even worse from a measurement filtering perspective [13, 14, 15, 16]. Gone is the 24 dB of carrier attenuation that was assumed in the earlier example. However, there are three mitigating factors that apply in most cases here. The maximum power is usually less than 5 W, with significantly smaller raw PWM voltage swings. Second, it is unlikely that one would need to measure distortion below 0.01%, since these are not audiophile applications and these amplifiers are unlikely to have distortion anywhere near that low. Finally, the carrier residual is virtually always in the common mode. This helps with the needed attenuation for measurement of low THD at low signal levels, but does not relax the slew rate and auto-ranging issues. The first two factors often more than make up for the loss of the output filter attenuation.

However, a major exception to this is the Analog Devices SSM3582 filterless class D amplifier featuring two 32-W channels [16]. It employs a proprietary three-level sigma-delta modulation scheme with a full-bridge power stage to keep EMI down even at these higher power levels. It also boasts typical THD-1 of 0.004% at 5 W into 8 Ω. The chip features I²S digital inputs with sample rates up to 192 kHz. In mono mode it can produce 49 W into a 2-Ω load with a 16-V power supply. The average output switching frequency is typically 300 kHz, with a broad

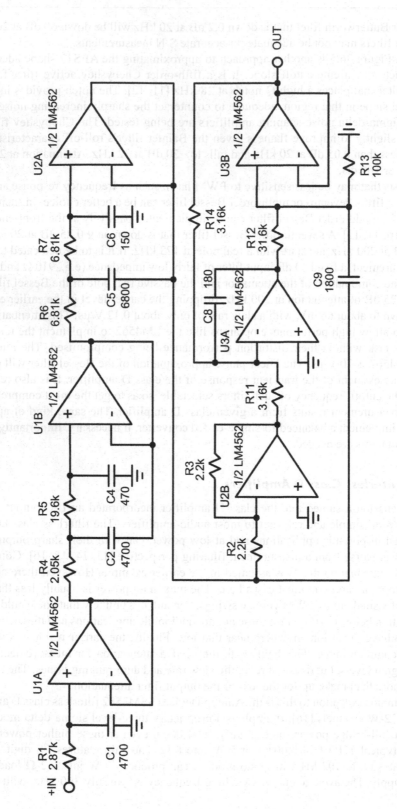

Figure 36.5 Alternative to AES17 Filter

spectrum created by pulses of varying width. This device is well suited to wireless self-powered loudspeakers of high quality and substantial power, such as used in surround systems.

It is possible that higher-power filterless amplifiers will become available in carefully constructed self-powered loudspeakers, and these could add to the filtering burden.

SNR Measurements

Signal-to-noise measurements are also deeply dependent on the above-mentioned filtering. A very good linear amplifier might have input-referred noise of 5 nV/rt Hz and a gain of 30, resulting in 22 µV of noise at its output in a 20-kHz bandwidth, corresponding to 102 dB SNR with respect to 2.83 volts at its output (1 W into 8 Ω). The carrier residual must be attenuated to at least 10 dB below that, or about 7 µV.

Measuring Unwanted High-Frequency Grunge and Its Spectrum

The unwanted high-frequency tones and EMI that the AES17 filter protects instrumentation from can be measured, as it is a quality metric of sorts for the class D amplifier. Too much of it might interfere with optimum performance of other playback equipment. It is even conceivable that ultrasonic hash from a class D amplifier could cause some intermodulation artifacts to be created by the nonlinearities in some tweeters.

Total Harmonic Distortion

THD measurement of class D amplifiers is practical and relevant when conducted at low frequencies like 1 kHz. However, THD measurements are of very limited use when conducted at high frequencies like 20 kHz. This is because the class D amplifier's output filter will block many of the upper harmonics, rendering an optimistic result and low sensitivity to those higher harmonics considered most offensive. If the AES17 filter is in place, all harmonics of a 20-kHz test signal will be blocked. Indeed, only the second harmonic of a 10-kHz test signal will barely manage to get through. THD-1 is a satisfactory basic test, but it provides virtually no information about high-frequency nonlinearities. An inexperienced technician measuring THD-20 + N might actually just be measuring S/N. The output of the analyzer should always be viewed on an oscilloscope to see if the reading is primarily noise and/or carrier residual.

SMPTE IM

The SMPTE IM test employs tones at 60 Hz and 7000 Hz in a 4:1 ratio. It tests intermodulation distortion inflicted on the smaller 7-kHz carrier by the larger 60-Hz aggressor signal. The SMPTE IM test provides a very good measurement of frequency-independent static nonlinearities. An example of such nonlinearity in class D amplifiers is nonlinearity in the triangle reference signal in a PWM modulator.

The test is especially valuable for class D amplifiers because it will show up problems related to power supply rejection (PSRR). Recall that the open-loop gain of many class D amplifiers is proportional to power supply voltage. If the large 60-Hz component of the test signal causes variation in the supply voltage, intermodulation of the 7-kHz carrier will be the direct result. Bus-pumping intermodulation will also be revealed by the SMPTE IM test. Such a test with a 10-Hz or 20-Hz low frequency component can be even more revealing.

CCIF Tests

The 19 + 20 kHz CCIF two-tone test with spectral analysis is an excellent test for class D amplifiers because it stresses the amplifier at high frequencies while producing distortion components that are in-band. This is at least true of the lower IM sidebands. Spectral components at 18 kHz reflect third-order nonlinearities, components at 17 kHz reflect fifth-order nonlinearities and so on. Even-order nonlinearities show up starting at 1 kHz beginning with the second order. It has been pointed out that use of a CCIF test with frequencies of 18.5 kHz and 19.5 kHz can prevent the higher-order even and odd products from overlapping each other in the output spectrum [17]. Even-order products will be at even multiples of 500 Hz and odd-order products will be at odd multiples of 500 Hz.

Transient Intermodulation Distortion

The DIM-30 test that previously was used to measure TIM can prove useful as a test for class D amplifiers, exposing their slew rate limitations and producing in-band distortion products that can be measured with a spectrum analyzer [18, 19, 20]. A 3.15-kHz square wave at f_1 and a 15-kHz sine wave at f_2 are mixed in a 4:1 ratio of peak values. The RMS sum of nine specified IM products, as measured by spectral analysis, is compared to the RMS value of the 15-kHz sine wave to arrive at the DIM distortion value. Of the eight usually significant products, four are even-order and four are odd-order. The odd-order products (like $f_2 - 2f_1$ at 8.7 kHz) are usually of quite similar amplitude, as are the even-order products (like $7f_2 - f_1$ at 7.05 kHz). The odd-order products often dominate substantially the even-order products by 10–20 dB. In the simplest case, taking the RMS sum of the products at 7.05 kHz and 8.7 kHz and multiplying by 2 and comparing it to the 15-kHz amplitude will yield a close answer.

Aliasing

It is very important to sound quality that out-of-band frequency components in the input signal not create aliasing in the class D amplifier, where these frequency components are folded back into the audio band as spectral lines or noise [21, 22, 23, 24, 25]. Modern signal sources often contain energy above 20 kHz, whether it be program material or collateral energy, such as that often present at the output of SACD players.

One way to test for aliasing is to apply a moderately high-level sinusoid that is swept in frequency from 10 kHz to 100 kHz while observing the output of the class D amplifier on a spectrum analyzer. A baseline spectrum analysis should first be done to identify spectral lines and the noise floor present in the absence of the test signal. With the test signal applied, the output of the spectrum analyzer is then evaluated for the presence of new spectral lines or an increase in the noise floor. Input frequencies that cause such results should be noted. This can be a time-consuming test.

A different approach is to apply out-of-band white noise at a fairly high level. This noise should be pre-filtered so that it contains very little energy in the audio band. Once again, the spectrum analyzer results are evaluated before and after application of the test signal.

PSRR

Power supply rejection ratio (PSRR) is not usually measured explicitly in a linear amplifier when the amplifier is measured as a black box. The effects of power supply noise are typically just lumped in with the SNR of the amplifier. The situation is not so simple with class D amplifiers if valid results are to be obtained. This is important because PSRR is often a bigger problem for class D amplifiers than for linear amplifiers.

There are two concerns with PSRR measurement in class D amplifiers. First, ripple and noise on the power supply rails create intermodulation distortion with the audio signal in addition to adding noise. The intermodulation will not be seen in a simple noise measurement when no signal is applied. The second concern is that full-bridge class D amplifiers will not show much of the power supply noise when measured differentially across the speaker terminals because the same rail noise is present in both sides of the bridge [26, 27].

For these reasons PSRR should be measured or inferred from a test that shows up intermodulation distortion. As mentioned above, the SMPTE IM test can reveal PSRR issues if the low-frequency component of the test at 60 Hz causes significant amplitude deviations on the power supply. Measurement of PSRR is preferably carried out with a spectrum analyzer where IM sidebands can be seen. This approach is especially useful in an amplifier development environment where a PSRR test signal can be added to the amplifier's power supply rails for measurement purposes.

Conductive Emissions

As mentioned earlier, EMI can be a problem for class D amplifiers. EMI is categorized into *conductive emissions* and *radiated emissions* [28, 29]. Conductive emissions are those that can be measured electrically at one of the amplifier ports. For a class D amplifier, the most relevant is the output port. Radiated emissions travel through the air and are measured by radio receiver-like instruments. They will not be discussed further here. Conductive emissions are turned into radiated emissions by the antenna formed by speaker cable and the loudspeaker.

On the assumption that conductive emissions below 500 kHz are relatively harmless, a reasonable test can be implemented with a sixth-order high-pass filter connected to the amplifier output and followed by a wideband (10 MHz) true RMS voltmeter like the HP 3400A. The conducted emissions voltage should be measured with the amplifier connected to a 4-W load at no power and 1-kHz full power. The measured voltage should be reported in dBV.

Conductive emissions and radiated emissions for class D amplifiers are subject to the same standards as those for switching power supplies. Those standards are discussed at length in Chapter 20.

36.3 Achievable Performance

The stumbling block to adoption of class D amplifiers in the past has been sound quality. That has changed dramatically in recent years but still has a ways to go for high-end audio.

Efficiency

The typical efficiency of a 100-W class D amplifier employing readily available components is on the order of 90%. An example of putting this to good use is the widespread application of class D amplifiers in subwoofers. Indeed, class D amplifiers are now predominant in both consumer and professional self-powered loudspeaker systems. Here class D amplifiers are so small and inexpensive that it is attractive to implement these loudspeakers as bi-amplified or tri-amplified systems with active crossovers.

Distortion

Getting the distortion down is still a very big challenge for class D amplifier designers, but the ever-higher digital clock speeds available combined with increased DSP sophistication can be expected to yield significant improvements here.

A related issue concerns sound quality that is not addressed by lab measurements. This continues to be a nagging problem in conventional high-end audio and can be expected to be worse with class D if for no other reason than the far smaller amount of design, measurement and listening experience with the class D technology.

Class D amplifiers typically have poor slew rate and rise time, so they should be checked with distortion measurements like DIM and CCIF that produce in-band distortion products.

36.4 Integrated Circuits for Class D Amplifiers

A great many integrated circuits for implementing quality class D amplifiers are available. Most operate on the PWM principle. Below is a list of some good candidates.

Texas Instruments TPA3255	155 wpc Stereo PWM BTL
Texas Instruments TASW5630	240 wpc Stereo PWM BTL
Texas Instruments TAS5176	50 wpc PWM 5.1 CH SE / 3 CH BTL
STMicroelectronics TDA7498E	100 wpc Stereo PWM BTL
NXP Semiconductors TDA8932B	15 wpc Stereo PWM SE/BTL
Maxim MAX13301	4 × 80 W Automotive PWM BTL
Texas Instruments TAS3251*	175 wpc Stereo PWM BTL
Analog Devices SSM3582*	15 wpc Stereo Filterless Sigma Delta
Analog Devices SSM2377	2.5 W Filterless Sigma Delta
Texas Instruments TPA2000D2	1 wpc Stereo Filterless
Maxim MAX9700	1.2 W Filterless PWM
ON Semiconductor NCP2820	2.6 W Filterless PWM

*Digital Input

36.5 Example Class D Amplifiers and Measurements

A number of available class D amplifiers have been measured using some of the techniques discussed above. Space does not permit discussion and presentation of these results, but they are available at the author's website [30].

References

1. Harry Dymond and Phil Mellor, "Switching/Linear Hybrid Audio Power Amplifiers for Domestic Applications, Part 1: The Class-BD Amplifier," preprint no. 8197, 129th Convention of the Audio Engineering Society, November 4–7, 2010.
2. Bruce Hofer, "Measuring Switch-mode Power Amplifiers," Audio Precision, White Paper, October 2003.
3. Claus Neesgaard, "Digital Ausio Measurements," Application Report SLAA114, Texas Instruments, January 2001.
4. R. Palmer, "Guidelines for Measuring Audio Power Amplifier Performance," Application Report SLOA68, Texas Instruments, October 2001.
5. Ken Korzeniowski, "Testing DDX® Digital Amplifiers," Document #13000004–01, Apogee Technology, Inc., 2000.
6. AES17, "AES Standard Method for Digital Audio Engineering—Measurement of Digital Audio Equipment," Audio Engineering Society, 1998.
7. Audio Precision, "Aux-0025 Switching Amplifier Measurement Filter," User's Manual, 2003.
8. Jim Karki, "Active Low-Pass Filter Design," Application Report SLOA049B, 2002.

9. Analog Filter Wizard, Analog Devices. Available at www.analog.com/designtools/en/filterwizard/.

10. Basic Linear Design, Analog Devices, Chapter 8: Analog Filters. Available at www.analog.com.

11. James R. Bainter, "Active Filter has Stable Notch, and Response Can be Regulated," *Electronics*, pp. 115–117, October 2, 1975.

12. "Bainter Notch Filters, MT-203," *Analog Devices*. Available at www.analog.com.

13. Michael D. Score, "Reducing and Eliminating the Class-D Output Filter," Application Report SLOA023, Texas Instruments, 1999.

14. Michael Score and Donald Dapkus, "Filterless Class D Simplifies Audio Amplifier Design," *EE Times*, October 17, 2000.

15. "Filterless Class D Amplifiers," Application Report AN-1497/SNAAO34A, Texas Instruments, 2013.

16. Datasheet, "SSM3582 2X31.76 W, Digital Input, Filterless Stereo Class-D Audio Amplifier," *Analog Devices*, 2016.

17. Bruno Putzeys, private communication.

18. Eero Leinonen, Matti Otala and John Curl, "A Method for Measuring Transient Intermodulation Distortion (TIM)," *Journal of the Audio Engineering Society*, vol. 25, no. 4, pp. 170–177, April 1977.

19. Audio Precision, "DIM 30 and DIM 100 Measurements per IEC 60268–3 with AP2700," 2009.

20. Glen Ballou, *Handbook for Sound Engineers*, 5th ed., Focal Press, Burlington, MA, 2015, pp. 851–852.

21. W. Marshall Leach, Jr., *Introduction to Electroacoustics and Audio Amplifier Design*, 2nd ed., Kendall/Hunt, Dubuque, IA, 2001.

22. Claus Risbo and Lars Neesgaard, "PWM Amplifier Control Loops with Minimum Aliasing Distortion," 120th AES Convention, May 2006.

23. Arnold Knott, "Introduction to Class-D Audio Amplifiers," Harmon/Becker Automotive Systems, 2007.

24. Duncan McDonald, "Class D Audio Power Amplifiers: Interactive Simulations Assess Device and Filter Performance," *EDN*, January 4, 2001.

25. Nuno Pereira and Nuno Paulino, Chapter 2, "Class D Audio Amplifiers and Data Conversion Fundamentals," in *Design and Implementation of Sigma Delta Modulators for Class D Audio Amplifiers Using Differential Pairs*, Springer, Heidelberg, Germany, 2015.

26. Michael Firth and Yang Boon Quek, "The Real Story About Closed-loop, Open-loop Class D Amps," *EE Times-Asia*, no date available.

27. Kim Madsen and Tomas Soerensen, "PSRR for PurePath™ Audio Amplifiers," TI Application Report SLEA049, June 2005.

28. Keith Billings, *Switchmode Power Supply Handbook*, McGraw-Hill, New York, 2010.

29. J. Patrick Donohoe, Professor, Mississippi State University, "ECE4323 EMC Requirements." Available at my.ece.msstate.edu/faculty/donohoe/ece4323EMCreq.pdf.

30. Bob Cordell, "Class D Power Amplifier Measurements." Available at www.cordellaudio.com.

Index

Printed in the United States
by Baker & Taylor Publisher Services